# Handbook of Experimental Pharmacology

Continuation of Handbuch der experimentellen Pharmakologie

# Vol. 66/II

# Glucagon II

Contributors

R. Assan · S. R. Bloom · G. Boden · S. Bonner-Weir
B. Diamant · R.P. Eaton · A.E. Farah · J.B. Field
R.A. Gelfand · J.E. Gerich · J.R. Girard · M. Gormley
J.J. Holst · T. Ishida · J.B. Jaspan · F.W. Kemmer
J. Kolanowsky · C. Kühl · A.M. Lawrence · V. Leclercq-Meyer
P.J. Lefèbvre · H.L.A. Lickley · A.S. Luyckx
W.J. Malaisse · J. Marco · V. Marks · M. Marre · T.J. Merimee
L. Orci · J.P. Palmer · S.B. Pek · J. Picazo · J.M. Polak
K.S. Polonsky · D. Porte, Jr. · A.H. Rubenstein · E. Samols
R.S. Sherwin · R.S. Spangler · M.A. Sperling · R.H. Unger
M. Vranic · D.H. Wasserman · G.C. Weir · S.M. Wood

Editor

P.J. Lefèbvre

Springer-Verlag
Berlin Heidelberg New York Tokyo 1983

Professor PIERRE J. LEFEBVRE, M.D.
Professor of Medicine, University of Liège
Guest Professor, University of Brussels (V.U.B.)
Corresponding Member, Royal Academy of Medicine of Belgium
Chairman, Medical Policlinics, University of Liège
Head, Division of Diabetes, University of Liège
Institut de Médecine, Hôpital de Bavière
Boulevard de la Constitution, 66
4020 Liège, Belgium

With 161 Figures

Library of Congress Cataloging in Publication Data. Main entry under title: Glucagon. (Handbook of experimental pharmacology; v. 66) Bibliography: p. Includes index. 1. Glucagon–Addresses, essays, lectures. I. Lefèbvre. Pierre J. II. Series. [DNLM: 1. Glucagon. W1 HA51L vol. 66 pt. 1–2/WK 801G5656] QP905.H3 vol. 66 [QP572.G5] 615'.1s [612'.34] 83-583

ISBN-13: 978-3-642-69021-1 e-ISBN-13: 978-3-642-69019-8
DOI: 10.1007/978-3-642-69019-8



2122/3130-543210

# List of Contributors

R. Assan, Chef de Service. Hôpital Bichat, Service de Diabétologie – Endocrinologie, 46, Rue Henri Huchard, 75018 Paris, France

S. R. Bloom, Royal Postgraduate Medical School, Hammersmith Hospital, Ducane Road, London W12 OHS, Great Britain

G. Boden, Chief, Division Metabolism/Diabetes, Temple University School of Medicine, 3401 N. Broad Street, Philadelphia, PA 19140, USA

S. Bonner-Weir, Division of Endocrinology, Department of Internal Medicine, Medical College of Virginia, Virginia Commonwealth University, Richmond, VA 23298, USA

B. Diamant, Novo Research Institute, Novo Allé, 2880 Bagsvaerd, Denmark

R. P. Eaton, University of New Mexico, Department of Medicine, School of Medicine, Division of Endocrinology and Metabolism, Albuquerque, NM 87131, USA

A. E. Farah, Vice President for Research, Sterling Drug Inc., Columbia Turnpike, Rensselaer, NY 12144, USA

J. B. Field, Diabetes Research Laboratory, St. Luke's Episcopal Hospital, P.O. Box 20269, Houston, TX 77025, USA

R. A. Gelfand, Division of Endocrinology, Department of Internal Medicine, Yale University School of Medicine, 333 Cedar Street, New Haven, CT 06510, USA

J. E. Gerich, Endocrine Research Unit, Departments of Medicine and Physiology, Mayo Medical School and Mayo Clinic, Rochester, MN 55901, USA

J. R. Girard, Centre de Recherches sur la Nutrition, CNRS, 9, rue Jules Hetzel, 92190 Meudon-Bellevue, France

M. Gormley, Hôpital Bichat, Service de Diabétologie – Endocrinologie, 46, Rue Henri Huchard, 75018 Paris, France

J. J. Holst, Institute of Medical Physiology C, University of Copenhagen, The Panum Institute, Blegdamsvej 3c, 2200 Copenhagen N, Denmark

T. Ishida, Diabetes Research Laboratory, St. Luke's Episcopal Hospital, P.O. Box 20269, Houston, TX 77025, USA

J. B. Jaspan, Department of Medicine, Section of Endocrinology, The University of Chicago, 950 East 59th Street, Box 435, Chicago, IL 60637, USA

F. W. Kemmer, Department of Physiology, Medical Sciences Building, University of Toronto, Toronto, Ontario M5S 1AB, Canada

J. Kolanowski, Endocrine Unit, Department of Physiology and Medicine, University of Louvain, Tour Harvey UCL 5530, Avenue Hippocrate, 55, 1200 Bruxelles, Belgium

C. Kühl, Hvidøre Hospital, Emiliekildevej 1, 2930 Klampenborg, Denmark

A. M. Lawrence, Section of Endocrinology, Loyola University Stritch School of Medicine, and the Hines Veterans Administration Hospital, Hines, IL 60141, USA

V. Leclercq-Meyer, Laboratoire de Médecine Expérimentale, Université Libre de Bruxelles, Boulevard de Waterloo, 115, 1000 Bruxelles, Belgium

P. J. Lefebvre, Head of the Division of Diabetes, Université de Liège, Hôpital Universitaire de Bavière, Institut de Médecine, Boulevard de la Constitution, 66, 4020 Liège, Belgium

H. L. A. Lickley, Department of Physiology, Medical Sciences Building, University of Toronto, Toronto, Ontario M5S 1A8, Canada

A. S. Luyckx, Division of Diabetes, Institute of Medicine, Université de Liège, Hôpital Universitaire de Bavière, Boulevard de la Constitution, 66, 4020 Liège, Belgium

W. J. Malaisse, Laboratoire de Médecine Expérimentale, Université Libre de Bruxelles, Boulevard de Waterloo, 115, 1000 Bruxelles, Belgium

J. Marco, Clinica Puerta de Hierro, Universidad Autonoma de Madrid, San Martin de Porres, 4, Madrid 35, Spain

V. Marks, Department of Biochemistry, University of Surrey, Guilford, Surrey GU2 5XH, Great Britain

M. Marre, Hôpital Bichat, Service de Diabétologie – Endocrinologie, 46, Rue Henri Huchard, 75018 Paris, France

T. J. Merimee, College of Medicine, Division of Endocrinology and Metabolism, University of Florida, Box J-226, JHM Health Center, Gainesville, FL 32610, USA

L. Orci, Institute of Histology and Embryology, University Medical Center, University of Geneva Medical School, 1211 Geneva, Switzerland

J. P. Palmer, Associate Professor of Medicine, Deputy Director, Diabetes Research Center, University of Washington, Seattle, WA 98195, USA

S. B. Pek, Dept. of Internal Medicine, Division of Endocrinology and Metabolism and the Metabolism Research Unit, The University of Michigan, Ann Arbor, MI 48109, USA

J. Picazo, Novo Research Institute, Clinical Research Pharmaceuticals, Sabino de Drana 48-1, Barcelona 28, Spain

J. M. Polak, Royal Postgraduate Medical School, Hammersmith Hospital, Ducane Road, London W12 OHS, Great Britain

K. S. Polonsky, Department of Medicine, Section of Endocrinology, The University of Chicago, 950 East 5th Street, Box 435, Chicago, IL 60637, USA

D. Porte, Jr., Director, Diabetes Research Center, University of Washington, Seattle, WA 98195, USA

A. H. Rubenstein, Department of Medicine, The University of Chicago, 950 East 59th Street, Box 435, Chicago, IL 60637, USA

E. Samols, Chief, Division of Endocrinology, Metabolism and Radionuclide Studies, Department of Medicine, VA Medical Center, and University of Louisville, 800 Zorn Avenue, Louisville, KY 40202, USA

R. S. Sherwin, Division of Endocrinology, Department of Internal Medicine, Yale University School of Medicine, 333 Cedar Street, New Haven, CT 06510, USA

R. S. Spangler, Division of Endocrinology and Metabolism and the Metabolism Research Unit, The University of Michigan, Ann Arbor, MI 48109, USA

M. A. Sperling, University of Cincinnati, Children's Hospital, College of Medicine, Department of Pediatrics, Cincinnati, OH 45229, USA

R. H. Unger, Senior Medical Investigator, Dallas VA Medical Center, 5323 Harry Hines Boulevard, Dallas, TX 75235, USA

M. Vranic, Department of Physiology, Medical Sciences Building, University of Toronto, Toronto, Ontario M5S 1A8, Canada

D. H. Wasserman, Department of Physiology, Medical Sciences Building, University of Toronto, Toronto, Ontario M5S 1A8, Canada

G. C. Weir, Division of Endocrinology, Department of Internal Medicine, Medical College of Virginia, Virginia Commonwealth University, Richmond, VA 23298, USA

S. M. Wood, Royal Postgraduate Medical School, Hammersmith Hospital, Ducane Road, London W12 OHS, Great Britain

# Contents

## Control of Glucagon Secretion

CHAPTER 26

**Ions in the Control of Glucagon Release**

V. LECLERCQ-MEYER and W. J. MALAISSE. With 2 Figures

CHAPTER 27

**Cyclic Nucleotides in the Control of Glucagon Secretion.** G. C. WEIR



CHAPTER 28

**Prostaglandins and Glucagon Secretion.** A. S. LUYCKX and P. J. LEFEBVRE
With 10 Figures



CHAPTER 29

**Hormones in the Control of Glucagon Secretion**
S. B. PEK and R. S. SPANGLER

CHAPTER 30

**Neural Control of Glucagon Secretion.** J. P. Palmer and D. Porte, Jr. With 3 Figures



CHAPTER 31

**Intraislet Insulin-Glucagon-Somatostatin Relationships**
E. Samols, G. C. Weir, and S. Bonner-Weir. With 1 Figure

## Extrapancreatic Glucagon

CHAPTER 33

**Extrapancreatic Glucagon and Its Regulation**
P. J. LEFEBVRE and A. S. LUYCKX. With 5 Figures



## Glucagon in Various Physiological Conditions

CHAPTER 34

**Glucagon and Starvation.** R. A. GELFAND and R. S. SHERWIN. With 7 Figures

CHAPTER 35

**Glucagon and Pregnancy.** C. KÜHL and J. J. HOLST. With 3 Figures



CHAPTER 36

**Glucagon in the Fetus and the Newborn.** J. GIRARD and M. SPERLING
With 14 Figures

CHAPTER 41

**The Renal Handling of Glucagon.** P. J. LEFEBVRE and A. S. LUYCKX
With 1 Figure



## Glucagon in Pathology

CHAPTER 42

**Glucagon Deficiency.** G. BODEN. With 2 Figures

CHAPTER 43

**The Glucagonoma Syndrome.** S. M. Wood, J. M. Polak, and S. R. Bloom
With 10 Figures

CHAPTER 44

**Glucagon in Diabetes Mellitus.** R. H. UNGER and L. ORCI. With 12 Figures



CHAPTER 45

**Glucagon in Human Endocrine and Exocrine Disorders.** A. M. LAWRENCE

CHAPTER 46

**Glucagon and Hyperlipoproteinemias.** R. P. EATON



CHAPTER 47

**Glucagon and Renal Insufficiency**
J. B. JASPAN, K. S. POLONSKY, and A. H. RUBENSTEIN. With 3 Figures



CHAPTER 48

**Glucagon in Cirrhosis of the Liver.** J. MARCO. With 5 Figures

CHAPTER 51

**Glucagon and Catecholamines.** P. J. LEFEBVRE and A. S. LUYCKX
With 1 Figure



CHAPTER 52

**Glucagon and Growth Hormone.** T. J. MERIMEE. With 1 Figure



CHAPTER 53

**Glucagon and the Heart.** A. E. FARAH. With 8 Figures

CHAPTER 55

**Glucagon in the Diagnosis and Treatment of Hypoglycaemia.** V. MARKS



CHAPTER 56

**Miscellaneous Pharmacologic Effects of Glucagon.** P. J. LEFEBVRE

# Contents of Companion Volume 66, Part I

# Control of Glucagon Secretion

CHAPTER 23

# Glucose in the Control of Glucagon Secretion

J. E. GERICH

## A. Introduction

Pancreatic A-cell function is modulated by numerous intracellular and extracellular factors (e. g., nutrients, ions cyclic nucleotides, neurotransmitters, hormones, prostaglandins). Since the primary physiologic role of glucagon is the preservation of normoglycemia, it is not surprising that the A-cell should be exquisitely sensitive to changes in the extracellular concentration of glucose and that glucose should be the major regulator of glucagon secretion. This chapter will attempt to summarize our present knowledge concerning the control of glucagon secretion by glucose; major emphasis will be placed on observations in humans, and this will be supplemented by relevant data from studies in other species and from in vitro experiments.

## B. Effect of Changes in Extracellular Glucose Concentration on Glucagon Secretion

### I. Increases in Extracellular Glucose Concentration

#### 1. In Vivo Studies

Evidence that increases in extracellular glucose concentration suppress glucagon secretion was first provided by studies in dogs (OHNEDA et al. 1969). In humans both ingestion of glucose (HEDING 1971) and infusion of glucose (UNGER et al. 1970), which result in hyperglycemia, normally cause a decrease in plasma glucagon concentrations. For example (Fig. 1), under conditions in which plasma glucose concentrations are increased approximately twofold to 160 mg/dl by either infusion or ingestion of glucose, if basal plasma glucagon concentrations average approximately 100 pg/ml, normally a 40–60 pg/ml decrement in plasma glucagon will be observed. Neither enteric factors nor vagally mediated mechanisms appear to influence suppression of plasma glucagon either by oral or intravenous glucose in normal humans (FINDLAY et al. 1979).

In general, there is no correlation between absolute plasma glucose concentrations achieved and the resultant absolute plasma glucagon concentrations. This may result in part from the fact that the concomitant insulin secretion may also influence suppression of glucagon secretion (ASPLIN et al. 1981). However, a more important reason may be the fact that immunoreactive plasma glucagon is heterogeneous with less than one-half being due to the 3500 dalton, biologically active hormone (see chapter 11); the proportions of the other immunoreactive com-

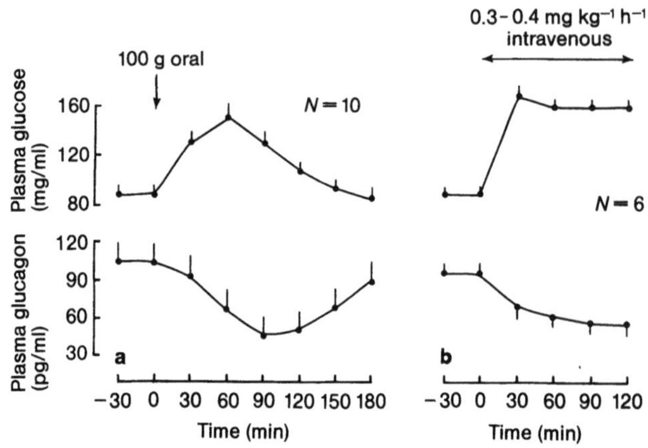


**Fig. 1 a, b.** Effects of oral (**a**) and intravenous (**b**) administration of glucose on plasma glucagon concentrations in normal humans. Mean ± standard error. GERICH (1978); GERICH et al. (1976)

ponents vary from individual to individual and are generally not responsive to changes in plasma glucose concentration (JASPAN and RUBENSTEIN 1977). Consequently, plasma glucagon concentrations as determined in conventional immunoassays usually do not decrease to undetectable levels, even under circumstances when secretion of the hormone from the pancreas is probably totally suppressed (GERICH et al. 1976b; RASKIN et al. 1975). For this reason, estimation of alterations in A-cell function in vivo based on absolute plasma concentrations of glucagon or fractional changes from baseline values may be misleading, and it would appear to be more appropriate to compare increments or decrements in plasma glucagon concentrations for this purpose.

## 2. In Vitro Studies

It is obviously not possible to quantitate the intrinsic suppressive effects of glucose on glucagon secretion in vivo owing to the potential suppressive effects of concomitantly released insulin and because of simultaneous changes in other factors which may potentially influence glucagon secretion (e.g., amino acids and free fatty acids). Similarly, when this question is examined during static incubations of isolated islets in vitro, one must also consider the potential influences of glucagon, insulin, and somatostatin which accumulate in the medium (BUCHANAN and MAWHINNEY 1973; ITOH et al. 1980). Indeed, such accumulation could explain in part why release of glucagon from incubated islets (BUCHANAN and MAWHINNEY 1973; CHESNEY and SCHOFIELD 1969; EDWARDS and TAYLOR 1970; HAHN et al. 1974; GERICH et al. 1979a) is generally less sensitive to suppression by glucose than is release of glucagon in flow-through systems such as perifused rat islets (OLIVER et al. 1976) or perfused rat (GERICH et al. 1974a; PAGLIARA et al. 1974) and canine (CHRISTENSEN and IVERSEN 1973; HERMANSEN 1980) pancreata in which glucagon release can be completely suppressed by glucose (Fig. 2). In these systems, the suppressive effects of glucose are monophasic, are maximal within 5 min, and have been observed with glucose concentrations as low as 1.25 mmol/l.

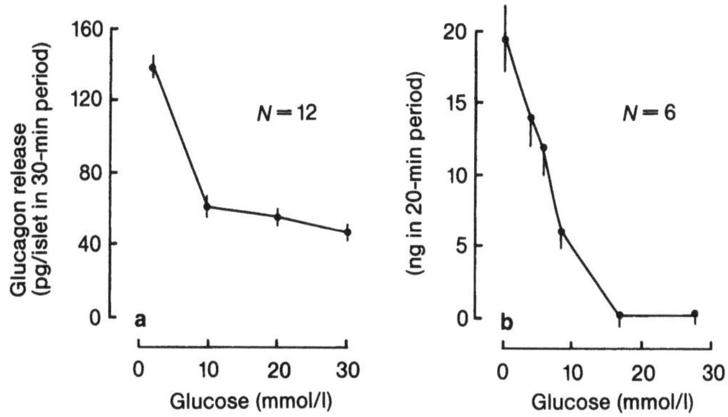


**Fig. 2 a, b.** Comparison of glucose-induced suppression of glucagon release from incubated rat islets (**a**) and from perfused rat pancreata (**b**). Mean ± standard error. GERICH et al. (1974a, 1979a)

Dose-response studies using in vitro flow-through systems indicate that the A-cell is more sensitive to glucose than is the B-cell (CHRISTENSEN and IVERSEN 1973; GERICH et al. 1974a; PAGLIARA et al. 1974; HERMANSEN 1980). Thus, the threshold for suppression of glucagon release (<2.5 mmol/l) is less than the threshold for stimulation of insulin release (>2.5 mmol/l), and half-maximal suppression of glucagon release occurs at 3–6 mmol/l, whereas half-maximal stimulation of insulin release occurs at 8–10 mmol/l. Moreover, glucose concentrations of 6–10 mmol/l generally cause maximal suppression of glucagon release, whereas glucose concentration in excess of 20 mmol/l are usually required for maximal stimulation of insulin release.

## II. Decreases in Extracellular Glucose Concentration

### 1. In Vivo Studies

Evidence that decreases in extracellular glucose concentration increase glucagon secretion was first provided by studies in dogs (OHNEDA et al. 1969). In both dogs (OHNEDA et al. 1969) and in normal humans (OHNEDA et al. 1972; GERICH et al. 1974c; SANTIAGO et al. 1980), decrements from both hyperglycemic and normoglycemic plasma glucose concentrations stimulate glucagon secretion, even in the absence of what is generally held to be bypoglycemia (50 mg/dl). In vivo, insulin-induced hypoglycemia has been most widely used as a model for studying A-cell responses to hypoglycemia (Fig. 3). In normal humans, following injection of conventionally used doses of insulin (0.4–1.5 IU/kg as an intravenous injection), plasma glucose concentrations begin to decrease within 10 min, reach a nadir between 20 and 30 min, and generally return to baseline values between 90 and 120 min. Changes in plasma glucagon concentration parallel those of epinephrine, occur before there are detectable increases in plasma cortisol and growth hormone concentrations, and coincide with changes in glucose production which ultimately restore normoglycemia (GARBER et al. 1976; RIZZA et al. 1979; GERICH et al. 1979b). These observations and studies in which the consequences of inhibition of glucagon secretion have been evaluated (GERICH et al. 1979b; RIZZA et al. 1979) have provided

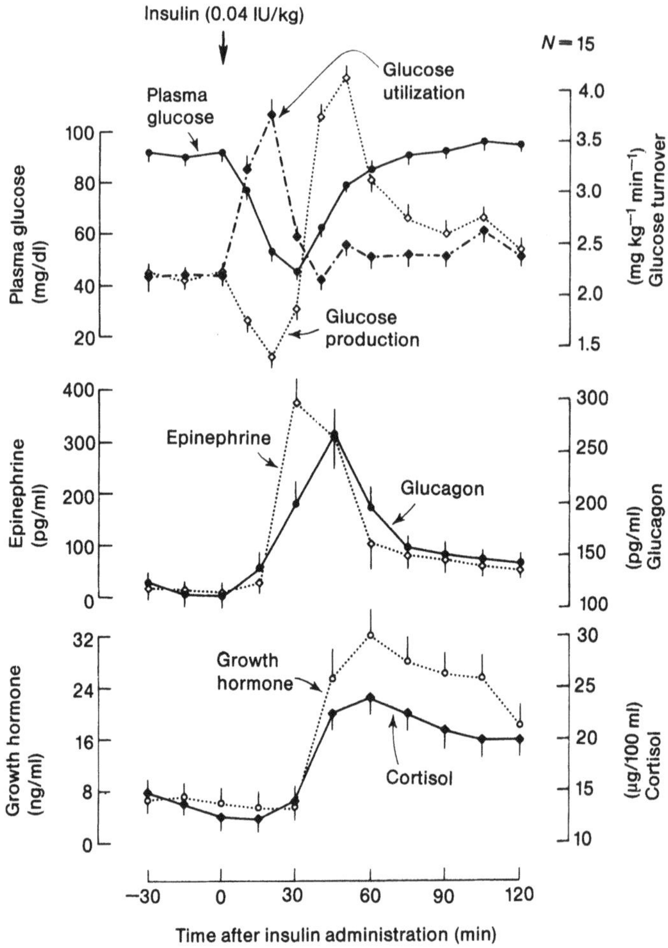


**Fig. 3.** Time course of changes in plasma glucose, glucagon, epinephrine, growth hormone, and cortisol concentrations, and of changes in rates of glucose production and utilization, during insulin-induced hypoglycemia in normal humans. Mean ± standard error. RIZZA et al. (1979)

evidence, discussed in detail in Chapter 37, that glucagon is the major counterregulatory hormone in humans (GERICH et al. 1979b; CRYER 1981). Increments in plasma glucagon concentration are significantly correlated with the magnitude and the rate of the decrease in plasma glucose concentration (Fig. 4) and with its absolute nadir (Fig. 5). Proportionately greater increases in plasma glucagon concentrations seem to occur when the plasma glucose concentration decreases below 40 mg/dl. Responses are monophasic, and the duration of the hypoglycemia generally parallels the duration of the plasma glucagon response.

The extent to which a decrease in plasma glucose concentration per se in the absence of hypoglycemia can act as a stimulus for glucagon release as well as the influence of the rate of fall of plasma glucose concentration on glucagon release are controversial. GERICH et al. (1974b) reported that, after normal subjects had

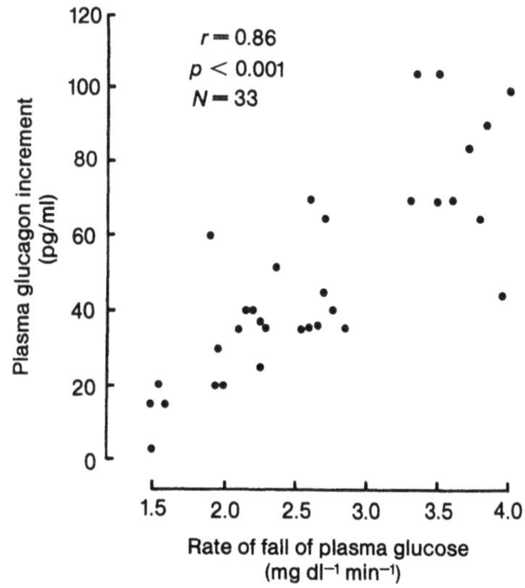


**Fig. 4.** Correlation between rate of fall of plasma glucose concentration following intravenous insulin administration and plasma glucagon response in normal humans. GERICH et al. (1974c), RIZZA et al. (1979)

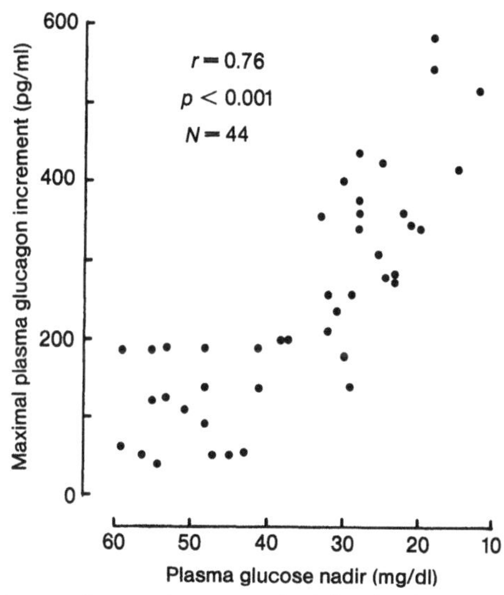


**Fig. 5.** Correlation between plasma glucose nadir following intravenous insulin administration and plasma glucagon response in normal humans. GERICH et al. (1974c, 1979b), RIZZA et al. (1979)

been infused with glucose so as to increase their plasma glucose concentrations to approximately 180 mg/dl for 1 h, plasma glucagon concentrations increased significantly above basal values when plasma glucose concentrations were allowed to decrease to near normoglycemic levels (~115 mg/dl). Moreover, SANTIAGO et al. (1980) observed increases in plasma glucagon concentrations in normal subjects when plasma glucose concentrations were acutely decreased from 200 to 100 mg/dl as well as from 100 to 60 mg/dl. Finally, SACCA et al. (1979) also observed increases in plasma glucagon concentrations when an intravenous infusion of insulin was used to decrease plasma glucose concentrations to approximately 55 mg/dl; however, when these investigators used a lower insulin infusion rate to decrease plasma glucose concentrations less rapidly and to only approximately 70 mg/dl, no increase in plasma glucagon concentration was observed. On the other hand, DE FRONZO et al. (1977) found no increase in plasma glucagon concentration until hypoglycemia had occurred in normal subjects whose plasma glucose concentrations had been previously increased to nearly 300 mg/dl for 1 h. It is possible, however, that this degree of antecedent hyperglycemia, along with concomitant insulin secretion, may have suppressed or delayed A-cell responses (WEIR et al. 1974; GRILL et al. 1979).

## 2. In Vitro Studies

The in vivo studies suggest that there may not be an absolute plasma glucose concentration threshold below which glucagon secretion is stimulated and that, although both the decrement and rate of fall of the plasma glucose concentration may influence glucagon secretion, the absolute plasma glucose concentration appears to be the primary determinant for the magnitude of the increases in plasma glucagon concentrations under hypoglycemic conditions. Nevertheless, the dose-response relationships for the suppressive effects of glucose on glucagon secretion observed in vitro suggest that an absolute threshold should be demonstrable. Surprisingly, this question has not been vigorously examined in in vitro systems where the effects of specific decrements and of different rates of decrease of extracellular concentrations can be precisely determined.

A biphasic increase in glucagon release from the in situ perfused rat pancreas has been observed when perfusate glucose concentrations were decreased from 100 to 25 mg/dl (WEIR et al. 1974); a similar but gradual and monophasic response was observed when perfusate glucose concentrations were decreased from 300 to 25 mg/dl. These observations suggest that the antecedent hyperglycemia may have suppressed the A-cell response to subsequent hypoglycemia and also that the increased rate of fall of the extracellular glucose concentration did not augment glucagon release. With perifused rat islets, an approximately twofold monophasic increase in glucagon release has been observed when the medium glucose concentration was decreased from 100 to approximately 30 mg/dl; and no greater release occurred when medium glucose concentrations were decreased further to zero (OLIVER et al. 1976). These results suggest that a glucose concentration of 30 mg/dl may provide maximal stimulation for glucagon release. The reason for the lack of a biphasic response in this system is unclear. It might be due to damage of islets during their isolation with the collagenese technique, although characteristic biphasic glucagon responses to arginine were observed. Alternatively, with the in situ preparation, neural or other factors may have contributed to the biphasic pattern. Decreases in perfusate glucose concentration are accompanied by release of large amounts of catecholamines from the canine pancreas perfused in vitro, and biphasic glucagon responses to abrupt decreases in media glucose concentration are observed in this preparation (CHRISTENSEN and IVERSEN 1973).

## C. Mechanism of Glucose Action on A-cell Function

The mechanism by which changes in extracellular glucose concentration affect glucagon secretion is poorly understood. This stems at least in part from the inability to assess A-cell glucose metabolism specifically since A-cell normally comprise only 15%–25% of the islet cell mass. Attempts have been made to examine the metabolism of A-cell from islets depleted of B-cells by use of agents which destroy islet B-cells (EDWARDS 1973; MATSCHINSKY et al. 1976a; ÖSTENSON 1980). The results of such studies are subject to reservation since such agents obviously destroy the normal islet architecture (UNGER and ORCI 1981a, b), disturb possible intraislet paracine relationships (HARA et al. 1979), and may in fact alter A-cell metabolism themselves (ÖSTENSON 1980).

At the present time, controversy exists whether glucose alters glucagon secretion as a consequence of its metabolism or whether it may exert its effort through a glucoreceptor without its being metabolized. Obviously both mechanisms may be involved to varying degrees. It is also unclear to what extent the suppressive effects of increases in extracellular glucose concentration are mediated by concomitant changes in insulin and somatostatin release and whether increases in glucagon secretion in response to decreases in extracellular glucose concentrastion represent a form of true stimulation or merely derepression.

### I. A-cell Glucose Metabolism

Using islets from guinea pigs treated with streptozotocin, which contain more than 70% A-cells, EDWARDS (1973) found that A-cells metabolized glucose poorly compared with free fatty acids (octanoate), thus, an increase in the extracellular glucose concentration from 1.7 to 16.7 mmol/l resulted in only a twofold increase in glucose oxidation compared with an almost fivefold increase in the oxidation of octanoate when its concentration was increased from 0.5 to 5.0 mmol/l. However, in another study using islets from streptozotocin-treated guinea pigs (ÖSTENSON 1979), it was found that glucose oxidation increased six fold when extracellular glucose concentrations were increased from 1.7 to 16.7 mmol/l, and was thus accompanied by increases in islet ATP concentrations. The increment in glucose oxidation of islets from the streptozotocin-treated guinea pigs was less than that observed with islets from normal guinea pigs, but was augmented by insulin (3000 μIU/ml).

Somewhat different results have been obtained with islets from streptozotocin-treated rats (MATSCHINSKY et al. 1976a, b): although glucose uptake was less than that from islets from normal rats, no appreciable differences were found in islet glucose, lactate, ATP, or cyclic adenosine monophosphate (cAMP) concentrations, and insulin had no effect on glucose uptake. These overall results are subject to different interpretations: they are consistent with the concepts that A-cells oxidize glucose less well than B-cells, that insulin deficiency caused by streptozotocin impairs A-cell glucose metabolism, or that streptozotocin itself may impair A-cell metabolism. A reversible toxic effect of these agents suggested by the observation that culture for 1 week of islets from guinea pigs treated with either streptozotocin or alloxan restores their glucose metabolism and glucagon secretion to normal (ÖSTENSON 1980).

## II. Effects of Glucose Metabolites and Inhibitors of Glucose Metabolism

Metabolites of glucose such as succinate, α-ketoglutarate, fumarate, and citrate do not suppress glucagon secretion (EDWARDS 1973). However, glyceraldehyde and dihydroxyacetone (HERMANSEN 1980) do inhibit glucagon secretion. Other carbohydrates such as galactose, ribose, and fructose which are poorly metabolized by islets (GRODSKY 1970) do not alter glucagon secretion (HERMANSEN 1980) whereas mannose, which is metabolized by islets, suppresses glucagon secretion (GOTO et al. 1976). The fact that glyceraldehyde is a more potent inhibitor of glucagon secretion than dihydroxyacetone is consistent with concept that metabolism is involved in suppression of glucagon secretion by these substrates since glyceraldehyde is more readily metabolized by islets than is dehydroxyacetone (HELLMAN et al. 1974). However, one should exercise caution in the interpretation of these observations since the metabolism of these substrates has not been specifically examined in A-cells and since concomitant changes in somatostatin and/or insulin secretion may have influenced glucagon release.

Glucose-induced suppression of glucagon release from incubated guinea pig islets can be reversed by metabolic inhibitors such as cyanide, a respiratory poison, 2,4-dinitrophenol, an uncoupler of mitochondrial oxidative phosphorylation, malonate, a citric acid cycle inhibitor, and iodacetate, an inhibitor of glycolysis (EDWARDS and TAYLOR 1970). 2-Deoxyglucose, which inhibits glycolysis, reverses glucose-induced suppression of glucagon release from the perfused canine pancreas (WEIR et al. 1974). These observations also suggest that metabolism of glucose may be necessary for its inhibition of glucagon secretion.

## III. Evidence for a Glucoreceptor Mechanism not Involving Metabolism

Despite the substantial support for the metabolism of glucose being responsible for suppression of glucagon secretion, there appears to be equally compelling evidence to suggest that the effects of glucose in islet A-cell function may also be mediated at least in part via a glucoreceptor mechanism not requiring metabolism of the hexose. Several studies have demonstrated a dissociation between metabolism of glucose and its ability to suppress glucagon secretion. First of all, concentrations of the iodoacetate and iodacetamide which completely block islet glycolysis do not interfere with the suppressive effect of glucose on amino acid-stimulated glucagon release from the perfused rat pancreas (PAGLIARA et al. 1975a). It is thus possible that the inhibitory effects of higher concentrations of these agents on glucose-induced suppression of glucagon release may be explained on the basis of their interaction with sulfhydryl groups related to glucoreceptor sites on A-cell membranes. Secondly, the α anomer of glucose, which apparently is less well metabolized by islets than the β anomer (IDAHL et al. 1976), is a more potent suppressor of glucagon secretion than is the β anomer (GRODSKY et al. 1975). Finally, in patients with insulin-dependent diabetes, both suppression of glucagon by hyperglycemia and stimulation of glucagon by hypoglycemia are impaired (GERICH et al. 1973), a defect compatible with a glucoreceptor abnormality and one difficult to explain solely on the basis of impaired A-cell glucose metabolism.

## IV. Calcium-Potassium and Glucose Action

Other aspects possibly involved in the action of glucose on glucagon secretion such as changes in A-cell membrane potentials CAMP, potassium and calcium fluxes, NADPH/NADP$^+$ ratios, hydrogen ion fluxes, protein phosphorylation, or calmodulin, remain to be examined as systematically as they have been for insulin release (MALAISSE et al. 1979; ZWALICH 1979; ASHCROFT 1980). In most instances, such studies await the development of a suitable experimental model. Recently, however, attention has been focused on the interaction of calcium and glucose in A-cell function.

It is well known that the presence of extracellular calcium is a prerequisite for the stimulation of glucagon and insulin release by various secretagogues (GERICH et al. 1976a). The exact site of action for the calcium effect, however, has not been established. Calcium itself can act as a secretagogue for glucagon, and A-cell responses to calcium are suppressed by glucose (LUNDQUIST et al. 1976; IVERSEN and HERMANSEN 1977). Nevertheless, the presence of extracellular calcium appears to be necessary for optimal suppression of glucagon release by glucose: thus, increases in perfusate glucose concentrations have been found not to suppress glucagon release appropriately from the perfused canine and rat pancreas when such preparations were perfused with calcium-deficient medium; furthermore, normal A-cell responsiveness to the suppressive effects of glucose can be rapidly reinstated when normal medium calcium concentrations are reestablished (LUNDQUIST et al. 1976; IVERSEN and HERMANSEN 1977). Although an increase in glucagon release from the perfused rat pancreas has been observed upon reduction of the perfusate calcium concentration which is paradoxically greater at greater perfusate glucose concentrations (LECLERCQ-MEYER et al. 1975), this phenomenon has not been observed subsequently in studies employing the perfused rat and canine pancreata (LUNDQUIST et al. 1976; IVERSEN and HERMANSEN 1977). Indeed, an abrupt decrease in perfusate calcium concentration has been shown to result in a decrease in glucagon release from the canine pancreas during perfusion with a medium glucose concentration of 200 mg/dl (IVERSEN and HERMANSEN 1977).

The influence of extracellular calcium concentrations on the increase in glucagon secretion occurring when extracellular glucose concentrations are decreased has not been directly examined. Nor has the effect of glucose on A-cell calcium or potassium fluxes been investigated. Glucose has been shown to depolarize islet B-cells (DEAN and MATTHEWS 1968). It has been postulated that glucose-induced changes in B-cell permeability to potassium may cause the opening of voltage-dependent calcium channels with resultant increase in B-cell cytosolic calcium triggering insulin secretion (ATWATER and BEIGELMAN 1976). Whether glucose and potassium also depolarize A-cells has not been examined directly. However, since depolarizing concentrations of potassium stimulate glucagon secretion (EPSTEIN et al. 1978), it seems likely that the inhibitory effects of glucose on glucagon release are due to depolarization of the A-cell (see also Chapter 26).

## V. Mechanism for A-cell Response to Hypoglycemia

### 1. General Considerations

The mechanisms responsible for the stimulation of glucagon secretion by hypoglycemia are unclear, especially in terms of whether the response represents true stimulation of secretion or merely deinhibition. Dose-response studies for the inhibition of glucagon release by hyperglycemia as well as the demonstration of increase of glucagon release in vitro upon decreasing media glucose concentrations are compatible with the response representing mere deinhibition. Theoretically, glucagon secretion could be under tonic inhibition, dependent upon the state of occupancy of a glucoreceptor analogous to the inhibition of lipolysis by insulin. Alternatively, it could be tonically inhibited by some process dependent on energy derived from A-cell metabolism of glucose (EDWARDS 1973) or on some signal ($NADPH/NADP^+$) or substrate generated by A-cell glucose metabolism. A decrease in A-cell metabolism of glucose, which presumably would occur by simple mass action when extracellular glucose concentrations decreased, would then result in deinhibition of glucagon release. Extracellular calcium, potassium, and amino acids and, theoretically, input from the sympathetic and parasympathetic nervous system could act as positive factors promoting glucagon release. Moreover, insulin-induced hypoglycemia is accompanied by decreases in the circulating concentrations of the free fatty acids and ketone bodies (acetoacetic acid and β-hydroxybutyric acid). These substrates inhibit glucagon secretion (see Chap. 25). Such changes would also promote deinhibition of glucagon secretion. It has been demonstrated, for example (LUYCKX et al. 1978), that increasing plasma free fatty acid concentrations by infusion of triglycerides and heparin can completely abolish the plasma glucagon response to insulin-induced hypoglycemia in humans. Although such an effect has not been observed in dogs (MÜLLER et al. 1976), blockade of free fatty acid metabolism of isolated guinea pig islets results in an increase in their release of glucagon at low extracellular glucose concentrations (EDWARDS and TAYLOR 1970).

### 2. Sympathetic and Parasympathetic Modulation

As reviewed in detail in Chap. 30, hypoglycemia increases both sympathetic and parasympathetic nervous system activity. Since both catecholamines (GERICH et al. 1974b; IVERSEN 1973b) and acetylcholine (IVERSEN 1973a) can stimulate glucagon secretion, it is plausible that such neurohumoral factors may participate in the A-cell response to hypoglycemia. Early work in calves demonstrating that maneuvers such as sympathectomy, vagotomy, adrenalectomy, and atropinization decreased plasma glucagon responses to insulin-induced hypoglycemia (BLOOM et al. 1973, 1974a, b), had suggested that both the sympathetic and parasympathetic nervous systems may play an important role in modulating A-cell response to hypoglycemia (see review in Chap. 30). This concept was further supported by observations that vagotomy decreased plasma glucagon responses to insulin-induced hypoglycemia in humans (BLOOM et al. 1974b) and that large amounts of catecholamines were released from the perfused pancreas when perfusate glucose concentrations were decreased (CHRISTENSEN and IVERSEN 1973). The latter finding could provide the basis for an adrenergic influence on A-cell function in an apparently denervated preparation.

Nevertheless, numerous subsequent studies have failed to demonstrate an effect of pharmacologic adrenergic blockade (RIZZA et al. 1979; WALTER et al. 1974; FRIER et al. 1981; LILAVIVAT et al. 1981), adrenalectomy (GERICH et al. 1979b; EN-

SINCK et al. 1976), sympathectomy (PALMER et al. 1976; FRIER et al. 1981), cholinergic blockade (PALMER et al. 1979; FRIER et al. 1981), or vagotomy (PALMER et al. 1979) on plasma glucagon responses to insulin-induced hypoglycemia in humans. Thus, at the present time, the preponderance of evidence suggests that neither the sympathetic nor the parasympathetic nervous systems plays a major role in mediating A-cell responses to hypoglycemia.

## D. Modulatory Effects of Glucose on A-cell Function

### I. Acute Effects

As evidenced by in vitro studies employing flow-through systems (e.g., GERICH et al. 1974a; PAGLIARA et al. 1974), the A-cell responds rapidly to changes in extracellular glucose concentrations. Ambient extracellular glucose concentrations influence A-cell responses to most factors influencing glucagon secretion. Furthermore, the effects of prior ambient extracellular glucose concentrations on glucagon secretion may be longlasting and may influence subsequent A-cell responses. It is well established that glucagon responses to secretagogues such as calcium and amino acids (GERICH et al. 1974; PAGLIARA et al. 1974) are increased if such stimulation occurs at low ambient glucose concentrations and, conversely, that these responses are diminished if A-cells are exposed to these secretagogues under conditions in which ambient glucose concentrations are high. These inhibitory effects of glucose have apparently the same dose-response characteristics as those of glucose alone (GERICH et al. 1974a; PAGLIARA et al. 1974; 1975b).

### II. Prolonged and Chronic Effects

High extracellular glucose concentrations appear to be able to exert a prolonged suppressive effect on islet A-cell function. Thus, decreased release of glucagon from the perfused canine pancreas has been observed when perfusate glucose concentrations were decreased from 300 to 25 mg/dl compared with those when the perfusate glucose concentration was decreased from 100 to 25 mg/dl (WEIR et al. 1974). Moreover, when rat pancreata are exposed in vitro to arginine (ambient glucose concentration 3.9 mmol/l), an intervening infusion of glucose (27.7 mmol/l) decreases glucagon responses to a subsequent exposure to arginine, despite the fact that the high glucose concentrations were terminated 20 min prior to the second arginine exposure (GRILL et al. 1979). As shown in Fig. 6, reciprocal changes in insulin secretion are observed. The fact that this prolonged inhibitory effect of glucose could be mimicked by glyceraldehyde, but not by insulin (GRILL et al. 1979), suggests that it may have been due to metabolism of glucose.

In humans, a high carbohydrate diet decreases basal plasma glucagon concentrations (LEWIS et al. 1977; FUJITA et al. 1975) and plasma glucagon increments following meal ingestion (AHMED et al. 1980; FUJITA et al. 1975) whereas a low carbohydrate diet has the opposite effects (MÜLLER et al. 1971; FUJITA et al. 1975). Total starvation in humans which is accompanied by decreases in plasma glucose concentrations and in glucose utilization by various tissues, increases basal and amino acid-stimulated glucagon secretion (AGUILAR-PARADA et al. 1969; MARLISS

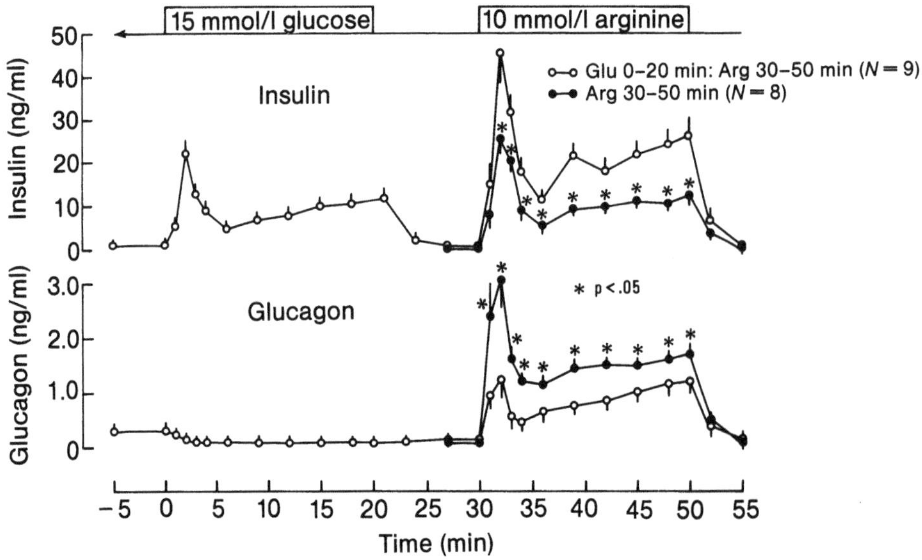


**Fig. 6.** Effects of antecedent glucose infusion on arginine-stimulated glucagon release from the perfused rat pancreas. Mean ± standard error; *open circles* glucose (0–20 min) plus arginine (30–50 min), nine pancreata; *full circles* arginine (30–50 min) alone, eight pancreata. PATTON G and GERICH J (1978, unpublished work)

**Table 1.** Effects of culturing islets at different glucose concentrations on pancreatic A-cell function (SEGERSTRÖM et al. 1976)

| Culture medium glucose concentration (mmol/l) | Glucagon secretion (pg/islet in 60-min-period) | | |
|---|---|---|---|
| | 5.6 mmol/l Glucose | | 27.8 mmol/l Glucose |
| | −5 mmol/l Arginine | +5 mmol/l Arginine | |
| 3.3 | 12±2 | 52± 8 | 16±5 |
| 6.1 | 20±3 | 88±20 | 11±3 |
| 16.7 | 36±6 | 124±25 | 10±2 |

et al. 1970). In contrast, suppression of glucagon release from perfused rat pancreases by glucose is enhanced following deprivation of food for several days (MATSCHINSKY et al. 1980). Whether this discrepancy is due to species difference or factors operative in vivo which are not apparent in vitro and that are unrelated to alterations in A-cell glucose metabolism is unclear.

An intriguing observation has been made (SEGERSTRÖM et al. 1976) that prolonged culture of isolated mouse islets at high extracellular glucose concentrations results in paradoxically increased glucagon secretion (Table 1). Thus, islets cultured in 16 mmol/l glucose subsequently exhibited greater release of glucagon when incubated in the presene of 5 mmol/l glucose and 5 mmol/l glucose plus 5 mmol/l arginine than islets cultured in 3.3 or 6.1 mmol/l glucose. The explanation of this apparent A-cell insensitivity to glucose is unclear, but is reminiscent of phenomena observed in diabetes mellitus. Theoretically, one could postulate that prolonged exposure to high extracellular glucose concentrations could lead to "down-regulation of islet A-cell glucoreceptors". It would, therefore, be of interest to determine whether such culture conditions alter islet A-cell glucose metabolism; a dissociation of glucose metabolism and glucagon release would provide additional support for the glucoreceptor theory.

## References

Aguilar-Parada E, Eisentraut A, Unger RH (1969) Effects of starvation on plasma pancreatic glucagon in normal man. Diabetes 18:717–723

Ahmed M, Nuttall F, Gannon M, Lamusga R (1980) Plasma glucagon and α-amino acid nitrogen response to various dieta in normal humans. Am J Clin Nutr 33:1917–1924

Ashcroft S (1980) Glucoreceptor mechanisms and the control of insulin release and biosynthesis. Diabetologia 18:5–15

Asplin C, Paquette T, Palmer J (1981) In vivo inhibition of glucagon secretion by paracrine beta cell actitivy in man. J Clin Invest 68:314–318

Atwater I, Beigelman P (1976) Dynamic characteristics of electrical activity in pancreatic B-cells. 1. Effect of calcium and magnesium removal. J Physiol (Lond.) 72:769–776

Bloom S, Edwards A, Vaughan N (1973) The role of the sympathetic innervation in the control of plasma glucagon concentration in the calf. J Physiol 233:457–466

Bloom S, Vaughan N, Russell R (1974a) Vagal control of glucagon release in man. Lancet 2:546–549

Bloom S, Edwards A, Vaughan N (1974b) The role of the autonomic innervation in the control of glucagon release during hypoglycaemia in the calf. J Physiol 236:611–623

Buchanan K, Mawhinney W (1973) Glucagon release from isolated pancreas in streptozotocin-treated rats. Diabetes 22:797–800

Chesney T, Schofield J (1969) Studies on the secretion of pancreatic glucagon. Diabetes 18:627–632

Christensen N, Iversen J (1973) Release of large amounts of noradrenaline from the isolated perfused canine pancreas during glucose deprivation. Diabetologia 9:396–399

Cryer P (1981) Glucose counterregulation in man. Diabetes 30:261–264

Dean P, Matthews E (1968) Electrical activity in pancreatic islet cells. Nature 219:389–391

De Fronzo R, Andres R, Bledsoe T, Boden W, Faloona g, Tobin J (1977) A test for the hypothesis that the rate of fall in glucose concentration triggers counterregulatory hormone responses in man. Diabetes 26:445–452

Edwards J (1973) A-cell metabolism and glucagon secretion. Postgrad Med J 49:615–619

Edwards J, Taylor K (1970) Fatty acids and the release of glucagon from isolated guinea-pig islets of Langerhans incubated in vitro. Biochim Biophys Acta 215:310–315

Ensinck J, Walter R, Palmer J, Brodows R, Campbell R (1976) Glucagon responses to hypoglycemia in adrenalectomized man. Metabolism 25:227–232

Epstein G, Fanska R, Grodsky G (1978) The effect of potassium and valinomycin on insulin and glucagon secretion in the perfused rat pancreas. Endocrinology 103:2207–2215

Findlay D, Omond S, Alford F, Chisholm D (1979) Hyperglycemia and glucagon supression: possible importance of the vagus and enteric humoral factors. J Clin Endocrinol 48:13–16

Frier B, Corrall R, Ratcliffe J, Ashby J, McClemont E (1981) Autonomic neural control mechanisms of substrate and hormonal responses to acute hypoglycaemia in man. Clin Endocrinol (Oxf) 14:425–433

Fujita Y, Gotto A, Phil D, Unger RH (1975) Basal and postprotein insulin and glucagon levels during a high and low carbohydrate intake and their relationships to plasma triglycerides. Diabetes 24:552–558

Garber A, Cryer P, Santiago J, Haymond M, Pagliara A, Kipnis D (1976) The role of adrenergic mechanisms in the substrate and hormonal response to insulin-induced hypoglycemia in man. J Clin Invest 58:7–15

Gerich J (1978) On the causes and consequences of abnormal pancreatic alpha cell function in human diabetes mellitus. In: Foà P, Bajaj J, Foà N (eds) Glucagon: its role in physiology and clinical medicine. Springer, Berlin Heidelberg New York, pp 617–641
Gerich J, Langlois M, Noacco C, Karam J, Forsham P (1973) Lack of glucagon response to hypoglycemia in diabetes: evidence for an instrinsic pancreatic alpha cell defect. Science 182:171–173
Gerich J, Charles M, Grodsky G (1974a) Characterization of the effects of arginine and glucose on glucagon and insulin release from the perfused rat pancreas. J Clin Invest 54:833–841
Gerich J, Langlois M, Noacco C, Schneider V, Forsham P (1974b) Adrenergic modulation of pancreatic glucagon secretion in man. J Clin invest 53:1441–1446
Gerich J, Schneider V, Dippe S, Langlois M, Noacco C, Karam J, Forsham P (1974c) Chracterization of the glucagon response to hypoglycemia in man. J Clin Endocrinol Metab 38:77–82
Gerich J, Charles A, Grodsky G (1976a) Regulation of pancreatic insulin and glucagon secretion. Annu Rev Physiol 38:353–388
Gerich J, Langlois M, Noacco C, Lorenzi M, Karam J, Forsham P, Gustafson G (1976b) Comparison of the suppressive effects of elevated plasma glucose and free fatty acid levels on glucagon secretion in normal and insulin-dependent diabetic subjects. J Clin Invest 58:320–325
Gerich J, Lorenzi M, Tsalikian E, Bohannon N, Schneider V, Karam J, Forsham P (1976c) Effects of acute insulin withdrawal and administration on plasma glucagon responses to intravenous arginine in insulin dependent diabetic subjects. Diabetes 25:955–960
Gerich J, Greene K, Hara M, Rizza R, Patton G (1979a) Radioimmunoassay of somatostatin and its application in the study of pancreatic somatostatin secretion in vitro. J Lab Clin Med 93:1009–1017
Gerich J, Davis J, Lorenzi M, Rizza R, Bohannon N, Karam J, Lewis S, Kaplan R, Schultz T, Cryer P (1979b) Hormonal mechanisms of recovery from insulin-induced hypoglycemia in man. Am J Physiol 236(4):E380–E385
Goto Y, Seino Y, Taminato T, Matsukura S, Imura H (1976) Inhibition by mannose of arginine-induced glucagon secretion in the isolated perfused rat pancreas. Horm Metab Res 9:243–244
Grill V, Adamson V, Rundfeldt M, Andersson S, Cerasi E (1979) Glucose memory of pancreatic B- and $A_2$-cells. J Clin Invest 64:700–707
Grodsky G (1970) Insulin and the pancreas. Vitam Horm 28:37–101
Grodsky G, Fanska R, Lundquist I (1975) Interrelationships between $\alpha$ and $\beta$ anomers of glucose affecting both insulin and glucagon secretion in the perfused rat pancreas II. Endocrinology 97:573–580
Hahn H, Ziegler M, Mohr E (1974) Inhibition of glucagon secretion by glucose and glyceraldehyde on isolated islets of wistar rats. FEBS Lett 49:100–102
Hara M, Patton G, Gerich J (1979) Increased somatostatin release from pancreases of alloxan diabetic rats perfused in vitro. Life Sci 24:625–628
Heding L (1971) Radioimmunological determination of pancreatic and gut glucagon in plasma. Diabetologia 7:10–19
Hellman B, Idahl L, Lernmark A, Sehlen J, Taljedal I (1974) The pancreatic B-cell recognition of insulin secretogogues. Comparison of glucose with glyceraldehyde insomers and dihydroxyacetone. Arch Biochem Biophys 162:448–457
Hermansen K (1980) Pancreatic D-cell recognition of D-glucose studies with D-glucose, D-glyceraldehyde, dihydroxyacetone, D-mannopheptulose, D-fructose, D-galactose, and D-ribose. Diabetes 30:203–210
Idahl L, Rahemtulla F, Sehlin J, Taljedal I (1976) Further studies on the metabolism of D-glucose anomers in pancreatic islets. Diabetes 25:450–458
Itoh M, Mandarino Lerich J (1980) Antisomatostatin gamma globulin augments secretion of both insulin and glucagon in vitro: evidence for a physiologic role for endogenous somatostatin in the regulation of pancreatic A- and B-cell function. Diabetes 39:693–696
Iversen J (1973a) Adrenergic receptors and the secretion of glucagon and insulin for its perfused canine pancreas. J Clin Invest 52:2102–2116

Iversen J (1973 b) Effect of acetylcholine on the secretion of glucagon and insulin from the perfused canine pancreas. Diabetes 22:381–387
Iversen J, Hermansen K (1977) Calcium, glucose and glucagon release. Diabetologia 13:297–303
Jaspan J, Rubenstein A (1977) Circulating glucagon plasma profiles and metabolism in health and disease. Diabetes 26:887–902
Leclercq-Meyer V, Rebolledo O, Marchand J, Malaisse W (1975) Glucagon release: paradoxical stimulation by glucose during calcium deprivation. Science 189:897–899
Lewis S, Wallin J, Kane J, Gerich J (1977) Effect of diet composition on metabolic adaptations to hypocaloric nutrition: comparison of high carbohydrate and high fat isocaloric diets. Am J Clin Nutr 30:160–170
Lilavivat U, Brodows R, Campbell R (1981) Adrenergic influence on glucocounterregulation in man. Diabetologia 20:482–488
Lundquist I, Fanska R, Grodsky G (1976) Interaction of calcium and glucose on glucagon secretion. Endocrinology 99:1304–1312
Luyckx A, Gaspard U, Lefèbvre P (1978) Influence of elevated plasma free fatty acids on the glucagon response to hypoglycemia in normal and in pregnant women. Metabolism 27:1033–1040
Malaisse W, Sener A, Herchuelz A, Hutton J (1979) Progress in endocrinology and metabolism. Insulin release: the fuel hypothesis. Metabolism 28:373–386
Marliss E, Aoki T, Unger R, Soeldner J, Cahill G (1970) Glucagon levels and metabolic effects in fasting man. J Clin Invest 49:2256–2270
Matschinsky F, Pagliara A, Hover B, Pace C, Ferrendelli J, Williams A (1976 a) Hormone secretion and glucose metabolism in islets of Langerhans of the isolated perfused pancreas from normal and streptozotocin diabetic rats. J Biol Chem 251:6053–6061
Matschinsky F, Pagliara A, Stillings S, Hover B (1976 b) Glucose and ATP levels in pancreatic islet tissue of normal and diabetic rats. J Clin Invest 58:1193–1200
Matschinsky F, Rujanavech C, Pagliara A, Norfleet W (1980) Adaptations of $\alpha_2$- and $\beta$-cells of rat and mouse pancreatic islets to starvation, to refeeding after starvation, and to obesity. J Clin Invest 65:207–218
Müller W, Faloona G, Unger R (1971) The influence of the antecedent diet upon glucagon and insulin secretion. N Engl J Med 285:1450–1454
Müller W, Aoki T, Flatt J, Blackburn G, Egdahl R, Cahill G (1976) Effects of $\beta$-hydroxybutyrate, glycerol, and free fatty acid infusions on glucagon and epinephrine secretion in dogs during acute hypoglycemia. Metabolism 25:1077–1086
Östenson C (1979) Regulation of glucagon release: effects of insulin on the pancreatic $A_2$-cell of the guinea pig. Diabetologia 17:325–330
Östenson C (1980) Alloxan reversibly impairs glucagon release and glucose oxidation by pancreatic $A_2$-cells. Biochem J 188:201–206
Ohneda A, Aguilar-Parada E, Eisentraut A, Unger RH (1969) Control of pancreatic glucagon secretion by glucose. Diabetes 18:1–10
Ohneda A, Sato M, Matsuda K, Yanbe A, Maruhama Y, Yamagata S (1972) Plasma glucagon response to blood glucose fall, gastrointestinal hormones and arginine in man. Tohoku J Exp Med 107:241–251
Oliver J, Williams V, Wright P (1976) Studies on glucagon secretion using isolated islets of Langerhans of the rat. Diabetologia 12:301–306
Pagliara A, Stillings S, Hover B, Martin D, Matschinsky F (1974) Glucose modulation of amino acid-induced glucagon and insulin release in the isolated perfused rat pancreas. J Clin Invest 54:819–832
Pagliara A, Hover B, Ellerman J, Matschinsky F (1975 a) Iodoacetate and iodoacetamide-induced alterations of pancreatic $\alpha$- and $\beta$-cell responses Endocrinology 97:698–708
Pagliara A, Stillings S, Haymond M, Hover B, Matschinsky F (1975 b) Insulin and glucose als modulators of the amino acid-induced glucagon release in the isolated pancreas of Alloxan and Streptozotocin Diabetic Rats. J Clin Invest 55:244–255
Palmer J, Henry D, Benson J, Johnson D, Ensinck J (1976) Glucagon response to hypoglycemia in sympathectomized man. J Clin Invest 57:522–525
Palmer J, Werner P, Hollander P, Ensinck J (1979) Evaluation of the control of glucagon secretion by the parasympathetic nervous system in man. Metabolism 28:549–552

Raskin P, Fujita Y, Unger RH (1975) Effect of insulin-glucose infusions on plasma glucagon levels in fasting diabetics and nondiabetics. J Clin Invest 56:1132–1138
Rizza R, Cryer P, Gerich J (1979) Role of glucagon, catecholamines, and growth hormone in human glucose counterregulation: effects of somatostatin and combined α- and β-adrenergic blockade on plasma glucose recovery and glucose flux rates after insulin-induced hypoglycemia. J Clin Invest 64:62–71
Sacca L, Sherwin R, Hendler R, Felig P (1979) Influence of continuous physiologic hyperinsulinemia on glucose kinetics and counterregulatory hormones in normal and diabetic humans. J Clin Invest 63:849–857
Santiago J, Clarke W, Shah S, Cryer P (1980) Epinephrine, norepinephrine, glucagon, and growth hormone release in association with physiological decrements in the plasma glucose concentration in normal and diabetic man. J Clin Endocrinol Metab 51:877–883
Segerström K, Andersson A, Lundquist G, Petersson B, Hellerström C (1976) Regulation of the glucagon release from mouse pancreatic islets maintained in tissue culture at widely different glucose concentrations. Diabete 2:45–48
Unger RH, Orci L (1981 a) Glucagon and the A-cell: physiology and pathophysiology (first of two parts). N Engl J Med 304:1518–1524
Unger RH, Orci L (1981 b) Glucagon and the A-cell: physiology and pathophysiology (second of two parts). N Engl J Med 304:1575–1580
Unger RH, Aguilar-Parada E, Müller W, Eisentraut AM (1970) Studies of pancreatic alpha cell function in normal and diabetic subjects. J Clin Invest 49:837–848
Walter R, Dudl R, Palmer J, Ensinck J (1974) The effect of adrenergic blockade on the glucagon responses to starvation and hypoglycemia in man. J Clin Invest 54:1214–1220
Weir G, Knowlton S, Martin D (1974) Glucagon secretion from the perfused rat pancreas. J Clin Invest 54:1403–1412
Zwalich W (1979) Intermediary metabolism and insulin secretion from isolated rat islets of Langerhans. Diabetes 28:252–262

CHAPTER 24

# The Amino Acid-Induced Secretion of Glucagon

R. ASSAN, M. MARRE, and M. GORMLEY

## A. Introduction

An increase in the amino acid concentration in contact with pancreatic A-cells is followed by a stimulation of glucagon release. This phenomenon is observed not only in vitro at some pharmacologic concentrations of amino acid, but also in vivo, during the rise in blood amino acid which follows a protein meal. The magnitude of the A-cell secretory response is modulated by the concomitant physiologic status, i.e., mainly the plasma glucose and insulin concentrations, and by the nutritional status in the preceding days: short-term and long-term influences can modify amino acid-induced glucagon release. As a clinical consequence of this phenomenon, the administration of an amino acid, or a mixture of amino acids has become a standard functional test for exploration of A-cells in vivo as well as in vitro.

In vivo, the amino acid-induced glucagon release has important physiologic consequences on the distribution of substrates: an increase in hepatic glucose output follows the glucagon release. When B-cells are functionally intact and respond concomitantly to the amino acid, the glucose turnover is increased and blood glucose concentration remains normal. In diabetic subjects conversely, the amino acid-induced glucagon release is followed by hyperglycemia and protein wastage.

## B. Phenomenology

The A-cell response to amino acids can be observed in a wide variety of in vitro and in vivo experimental and physiologic situations.

### I. In Vitro

#### 1. Isolated Perfused Pancreas

The infusion of L(+)-arginine hydrochloride into the circuit of an isolated perfused rat pancreas is followed within a few seconds by an acute release of glucagon (ASSAN et al. 1972; PAGLIARA et al. 1974; GERICH et al. 1974a). When the arginine stimulation is in the form of a square wave infusion, the A-cell response is biphasic: an early short response (lasting 3–5 min) is followed by a transient decrease of glucagon output for 5–10 min, then by a secondary sustained phase of release, which lasts as long as the arginine infusion is maintained (Fig. 1). Further stimulations by arginine boluses are able to induce acute A-cell responses. The kinetics of this amino acid-induced glucagon release appear somewhat similar to the biphasic glu-

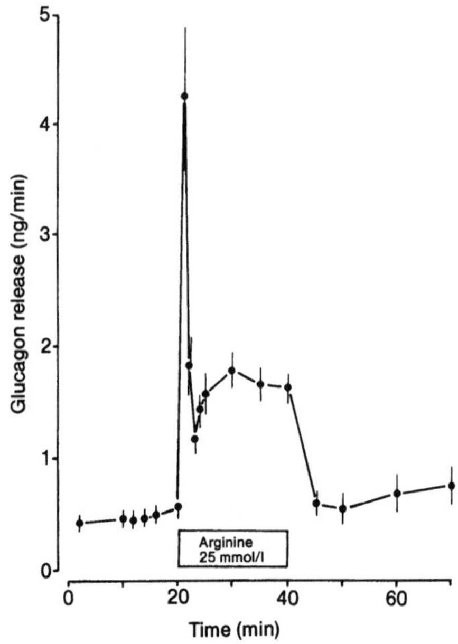


**Fig. 1.** L(+)-Arginine induced glucagon release in isolated perfused rat pancreas. The results presented here are mean values ± standard error of 32 experiments. ASSAN et al. (1977)

cose-induced insulin release, although the slope of the secondary phase of glucagon release is often sharper than that which is observed with insulin release.

The A-cell response is dose related, with a maximal release *V* max for a concentration of approximately 50 mmol/l L(+)-arginine, and a half-maximal response for a concentration of 10 mmol/l (ASSAN et al. 1977). As shown in Table 1, the screening for glucagon-stimulating potencies of a variety of naturally occurring and artifical amino acids suggests that in this perfused rat pancreas system, L(+)-arginine and L(+)-ornithine are among the most potent stimulators. Both amino acids are important intermediates in the urea cycle in the liver. Considering the potent stimulation by glucagon of the enzymes involved in this cycle, this A-cell-stimulating potency may be of some physiologic importance. Citrulline, the third amino acid intermediate in the urea cycle, is strangely devoid of any A-cell-stimulating potency in spite of its close structural relationship with ornithine and arginine. Several gluconeogenic amino acids are also strong A-cell stimulators (Table 1) and this feature can also be interpreted in a teleological manner. However L(+)-leucine, which is an exclusively ketogenetic amino acid, is not devoid of A-cell-stimulatory potency, at least in this rat perfused pancreas system (ASSAN et al. 1977; PEK et al. 1978; TAITY et al. 1978; AKPAN et al. 1981).

Several nonmetabolizable artificical analogs of the naturally occuring *l* amino acids can elicit a glucagon response, such as guanylpiperidine carboxylic acid and homoarginine (the latter being present at low concentrations in normal humans), and also D(+)-arginine and D(+)-leucine (ASSAN et al. 1977; PEK et al. 1978; TAITY et al. 1978). The dose-response relationship, however, suggests that D(+)-arginine is less potent than L(+)-arginine. Similarly, D(+)-leucine, which is transported

**Table 1.** Effect of various amino acids[a] on the first and second phases of glucagon release by the isolated perfused rat pancreas

| Amino acid | Number of experiments | 1st Phase 0–3 min | 2nd Phase 3–20 min | Total 0–20 min |
|---|---|---|---|---|
| L(+)-Arginine | 32 | 7.38±0.94 | 27.76±2.58 | 36.62±4.03[b] |
| D(+)-Arginine | 6 | 6.23±1.97 | 17.32±1.64 | 23.55±2.80[b] |
| Ornithine | 5 | 15.14±1.67 | 35.50±3.40 | 50.54±2.67[b] |
| Homoarginine | 6 | 5.94±1.90 | 17.62±2.96 | 23.47±4.79[b] |
| Guanylpiperidine carboxylic acid | 2 | 15.18 | 12.63 | 31.73 |
| Citrulline | 10 | 0.20±0.22 | 0.16±0.20 | 0.30±0.35 |
| Glycine | 11 | 3.36±0.63 | 13.93±4.31 | 18.55±4.17[b] |
| Homoserine | 4 | 3.63±0.51 | 8.17±1.13 | 13.34±1.70[c] |
| Valine | 11 | 3.22±0.75 | 9.88±3.21 | 13.11±3.39[b] |
| Alanine | 4 | 2.20±0.62 | 12.57±3.77 | 12.43±4.51[b] |
| Glutamic acid | 5 | 2.79±0.34 | 9.27±2.44 | 12.04±2.52[c] |
| Serine | 10 | 2.33±0.22 | 9.21±1.11 | 11.54±1.02[c] |
| Asparagine | 4 | 1.83±0.30 | 9.12±1.26 | 11.02±1.54[c] |
| Tyrosine | 8 | 1.35±0.31 | 10.05±2.28 | 10.93±2.52[b] |
| Leucine | 9 | 1.94±0.25 | 6.91±1.43 | 8.89±1.60[c] |
| Tryptophan | 10 | 1.86±0.22 | 6.91±0.86 | 8.77±0.95[c] |
| Aminoisobutyric acid | 10 | 0.94±0.11 | 5.96±0.94 | 6.94±1.06[c] |
| Cycloleucine | 9 | 0.76±0.10 | 6.02±0.98 | 6.78±1.03[c] |
| Cysteine | 4 | 0.12±0.06 | 0.25±0.05 | 0.40±0.15 |
| Homocystine | 4 | 0.10±0.05 | 0.25±0.08 | 0.35±0.10 |
| Cystine | 4 | 0.10±0.08 | 0.20±0.05 | 0.30±0.10 |
| Phenylalanine | 4 | 0.05±0.02 | 0.20±0.05 | 0.20±0.05 |

[a] All amino acids were tested at a uniform concentration of 25 mmol/l in the presence of glucose (4,4 mmol/l).
Significant differences with basal release for the corresponding 20-min-period are denoted by [b] ($P<0.05$) and [c] ($P<0.005$).
Results are expressed in ng per period and are given as mean ± standard error

through plasma membrane by the same carrier system as L(+)-leucine, is less potent than the latter and the dose dependency is less clear for D(+)-leucine than for its L(+) counterpart (Pek et al. 1978, Taity et al. 1978). Preincubation with D(+)-leucine suppresses the A-cell response to a subsequent stimulation by L(+)-leucine, as if the carrier system were rate limiting and saturated by preexposure to D(+)-leucine.

The same isolated perfused rat pancreas has been used for testing amino acid mixtures in various conditions (Pagliara et al. 1974). The A-cell response to amino acids has been demonstrated in a similar system for the pancreas of birds (Karman and Mialhe 1972), hamsters (Frankel et al. 1974), pigs (Holst et al. 1979a), and dogs (Iversen 1971). The conditions of perfusion can be critical for interpretation of results. For instance, the addition of organic acids and cofactors can profoundly modify the A-cell response in the absence of $Ca^{2+}$ (Gerich et al. 1974b; Leclercq-Meyer et al. 1976). The rat pancreas can be cross-perfused in situ, with conservation of the nerves afferent to the pancreas (Assan et al. 1976) and A-cell responses to arginine can then be characterized in this system.

### 2. Pancreatic Fragments, Islets and Islet Cells

Amino acid-induced glucagon release has also been characterized in pancreatic fragments from adult (ASSAN et al. 1975) or fetal organisms (ASSAN and GIRARD 1975b), islets (EDWARDS et al. 1972; BUCHANAN and MAWHINNEY 1973; CHESNEY and SCHOFIELD 1969), and dispersed islet cells (BOITARD et al. 1981). Tissues and cells can be incubated in static conditions (SAI et al. 1981) or perifused in vitro with a physiologic Krebs-Ringer solution to which the amino acid is added (EDWARDS and HOWELL 1973; AKPAN et al. 1981). The kinetics of glucagon release appear less clearly defined in the case of perifused fragments than in the case of the perfused pancreas (ASSAN et al. 1975). The paracrine influences which presumably prevail inside intact islets (ORCI and UNGER 1975) are suppressed or profoundly disturbed when dispersed islet cells are studied. Finally, important differences can appear when isolated islets are studied, depending upon the topographic origin of islets in one single pancreas (ORCI 1976).

### 3. Gastric A-cells In Vitro

The isolated perfused dog stomach releases glucagon and this release is augmented by the addition of arginine to the perfusate (LEFEBVRE and LUYCKX 1977). Similar results have been obtained with the perfused isolated rat stomach (MARRE et al. 1981). Since this type of preparation does not contain insulin-secreting B-cells, it has proved of particular interest for the study of the role of insulin; thus, the respective roles of glucose and insulin concentrations in the modulation of A-cell responses to amino acids have been clearly analyzed (LEFEBVRE and LUYCKX 1977) as depicted in detail in Chap. 33.

### 4. Other Tissues

Fragments of salivary glands (LAWRENCE et al. 1975) and of hormone-secreting hamster tumors (DUNBAR et al. 1976) have been similarly incubated or perifused for the study of glucagon release in the presence of amino acids.

## II. In Vivo

### 1. Human Studies

The intravenous infusion of L(+)-arginine hydrochloride induces a rise in glucagon concentration in peripheral venous blood (ASSAN et al. 1967; PEK et al. 1968; OHNEDA et al. 1968). This rise is dose related to the rise in blood α-amino nitrogen (ASSAN et al. 1981). The biphasic pattern of release can be detected in peripheral venous blood, in the form of an early peak (around 10–15 min of arginine infusion) followed by a transient pause, then by a further rise lasting as long as the amino acid infusion is performed. A concomitant insulin release occurs in nondiabetics subjects. In these subjects, blood glucose concentration does not vary significantly, but an early slight glucose rise, followed by a secondary slight fall can be detected. In contrast, in diabetic patients, arginine infusion elicits a clear-cut rise in blood glucose (Fig. 2).

Infusion of a mixture of amino acids induces very similar changes (PEK et al. 1969; MÜLLER et al. 1970). Infusion of alanine induces a selective rise in plasma

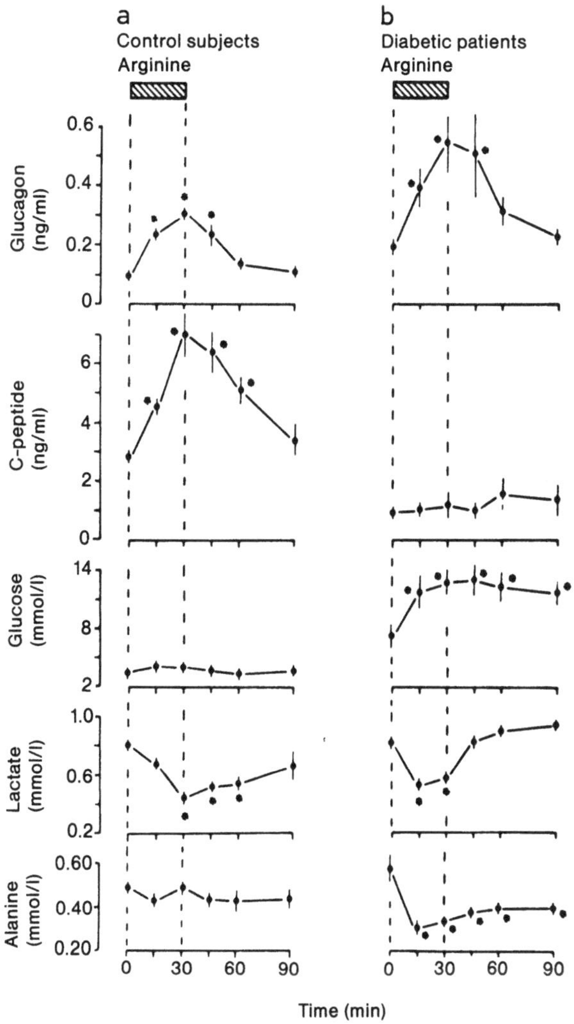


**Fig. 2 a, b.** Effect of arginine on plasma hormone and blood substrate concentrations in 15 control subjects (**a**) and 17 insulin-dependent diabetics. **b** Results are presented as mean values ± standard error. Asterisks indicate statistical significance of differences from basal values in the same group. TIENGO et al. (1982)

glucagon concentration (WISE et al. 1970). Ingestion of a protein meal is also followed by a rise in glucagon concentration in peripheral and portal plasma as illustrated in Fig. 3 (FELIG et al. 1974; BLACKARD et al. 1974; ASSAN 1971).

## 2. Studies in Normal Dogs

In conscious dogs, the intravenous infusion of a mixture of amino acids induces a rise in glucagon in pancreaticoduodenal blood (ROCHA et al. 1972). A similar A-cell response occurs when the amino acid mixture is instilled into the duodenum (OHNEDA et al. 1968). The material secreted displays the immunochemical and bio-

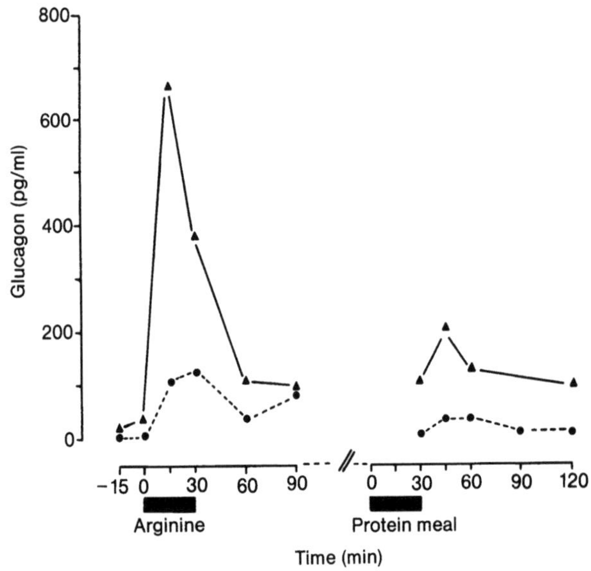


**Fig. 3.** Glucagon concentrations measured concomitantly in peripheral (*circles*) and portal (*triangles*) veins in humans during an arginine intravenous infusion and during a protein meal

logic characteristics of "true" pancreatic glucagon and is different from the "glucagon-like material" released by the gut after administration of glucose (MARCO et al. 1971). The systematic analysis of the respective glucagon-stimulating potencies for various amino acids under these experimental conditions suggests that the main gluconeogenic amino acids are among the most potent A-cell stimulators, a feature which may be physiologically meaningful (UNGER et al. 1969).

### 3. Other Studies

The administration of amino acids similarly stimulates A-cells in vivo in a wide variety of species, such as the rat (MARRE et al. 1979), hedgehog (HOO-PARIS et al. 1982) goose, and duck (KARMAN and MIALHE 1972).

As shown in Fig. 4, the catheterization of portal vein in anesthetized rats allows the kinetic study of the complex endocrine response to the intravenous arginine infusion. The increase in portal glucagon concentration is associated with a biphasic increase in insulin, followed by an increase in somastostatin-like immunoreactivity. Both A- and B-cell responses appear soon after the start of infusion. By contrast, the increase in somatostatin-like immunoreactive material is progressive and reaches its maximal level at the end, or after the end, of the amino acid infusion.

## C. Mechanism of Amino Acid-Induced Glucagon Release

Little is known about the intimate mechanism of the A-cell response to amino acids. By analogy with the mechanism of insulin release (MALAISSE et al. 1981) several points can be delineated:

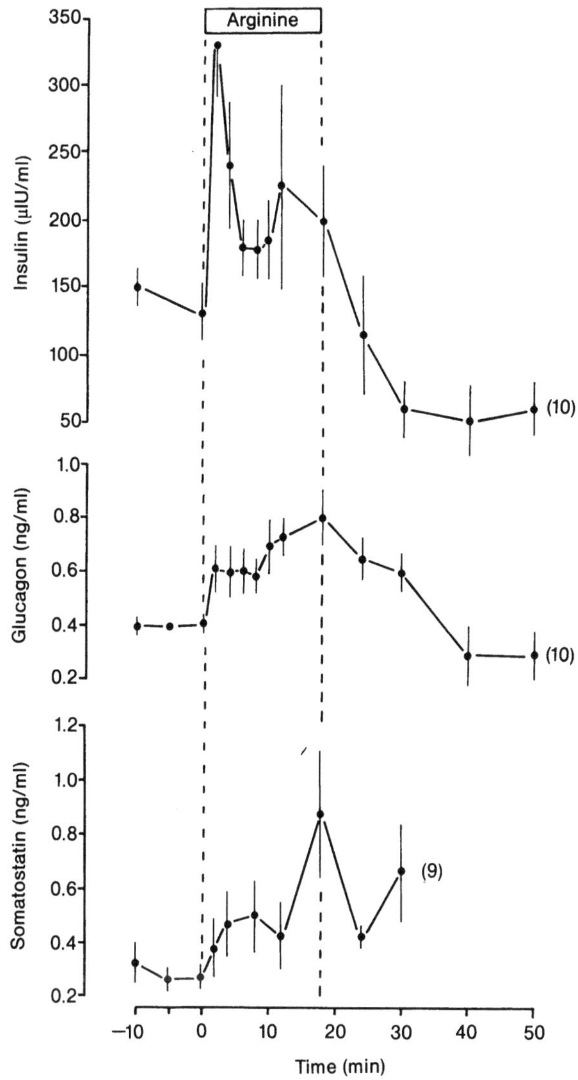


**Fig. 4.** Effects of arginine infusion on insulin, glucagon, and somatostatin immunoreactivities in portal blood from normal 12-h fasted rats. Numbers of experiments are given in parentheses; results are presented as mean ± values standard error

1. Is a microtubular-microfilamentous system involved in A-cell emiocytosis?
2. Do amino acids trigger the glucagon release through their intracellular metabolism or another type of interaction with the A-cell?
3. What are the respective roles of $Ca^{2+}$ movement and the adenylate cyclase system in the stimulus-secretion coupling?
4. A final point, which is more peculiar to the A-cell physiology regards the persistence of glucagon release at relatively low body temperatures, which, in contrast, block the release of insulin.

## I. Contribution of the Microtubular-Microfilamentous System

Microtubules and microfilaments are present in the A-cell (LACY 1962) and their intervention in the emiocytosis of glucagon granules during stimulation by arginine is strongly suggested by studies with agents blocking the microorganelles. Some early studies using incubated islets had suggested a paradoxical enhancement of glucagon release in the presence of colchicine and vinblastine (EDWARDS and HOWELL 1973). These results were apparently discrepant from the well-documented suppression of insulin release by such drugs (MALAISSE et al. 1971; LACY et al. 1972). It has since become apparent, in studies using the isolated perfused rat pancreas, that colchicine and *Vinca* alkaloids do induce a decrease in basal and arginine-stimulated glucagon release; however, when low concentrations and/or short preexposure times are used, the glucagon release is paradoxically augmented (ASSAN et al. 1978; LECLERCQ-MEYER et al. 1974). The paradoxical enhancement of the release is more pronounced in the early phase of response to arginine. These results are in agreement with corresponding studies on B-cells (MALAISSE et al. 1971). They suggest that a superficial treatment of cells by colchicine or the *Vinca* alkaloids modifies selectively the most peripheral tubulin-containing structures of the cells in such a way that a paradoxical enhancement of emiocytosis can occur. Treatment with cytochalasin-B, at concentrations which do not interfere with glucose transport or metabolism enhanced the arginine-stimulated glucagon release (ASSAN et al. 1978) in a way very similar to that which has been documented in the case of insulin release (MALAISSE et al. 1971; LACY et al. 1972).

## II. How Do Amino Acids Trigger Glucagon Release?

Do amino acids trigger the glucagon release through a nonmetabolic interaction with A-cells, or because of their intracellular metabolism? Among the more potent glucagon stimulators are two amino acids which are involved in the urea cycle (in the liver) and some others which are important glucose precursors (also in the liver). Although these observations are attractive for a teleological interpretation, one cannot extrapolate from them to reach conclusions regarding the metabolism of amino acids inside the A-cells.

The glucagon-stimulating potency of some nonmetabolizable amino acids suggests that the metabolism of the stimulatory agent is not a prerequisite for its action on A-cells. The suppression of A-cell response to L-leucine by preexposure to D-leucine has been an argument for thinking that transport of the amino acid is an important factor in its stimulating potency (PEK et al. 1978; AKPAN et al. 1981). But neither D-leucine nor L-leucine appear as major glucagon stimulators. Preincubation with citrulline (which is not a glucagon stimulator) did not modify the A-cell response to a subsequent exposure to L(+)-arginine or ornithine (ASSAN et al. 1977). Some other arguments favor the hypothesis of a necessary metabolic interference. It has been shown in B-cells that a nonmetabolizable amino acid can stimulate, by itself, the oxidation of another metabolizable substrate and then the release of insulin (MALAISSE et al. 1981). The same phenomenon cannot be excluded as regards A-cells.

The decarboxylation of some amino acids into the corresponding amines may be a mechanism by which these substrates stimulate A-cells. Histidine and histamine can stimulate A-cells in vitro and this effect is suppressed by treatment with antihistaminic drugs (JACOBY and BRYCE 1979; PONTIROLI et al. 1979). The gastrin-secreting D-cells of the stomach are stimulated by several amino acids, as are the A-cells. This stimulatory potency is reproduced by the corresponding amines, and is abolished by a treatment with decarboxylase inhibitors (LICHTENBERGER et al. 1982) but we are not aware of similar results as regards A-cells.

Finally, when L(+)-arginine is administered into an isolated perfused rat pancreas, in the absence of glucose or any other oxidizable substrate for 60–90 min, a high glucagon output is maintained throughout the whole stimulation period; the source of energy for this longlasting secretory process would be enigmatic if the amino acid were not concomitanly oxidized (ATTALI et al. 1979).

## III. Contribution of the Adenylate Cyclase System and of $Ca^{2+}$ and Other Ion Fluxes

The administration of theophylline or dibutyryl cyclic AMP has resulted in conflicting results: stimulation (JARROUSSE and ROSSELIN 1975), suppression (WOLLHEIM et al. 1976), or the absence of any effect upon the amino acid-induced glucagon release. Similarly, some apparent contradictions existed about the influence of extracellular $Ca^{2+}$ concentrations. When the perfusate did not include organic acids and other substrates, the omission of $Ca^{2+}$ combined with the addition of ethyleneglycolaminoethyl tetraacetate (EGTA) resulted in a paradoxical increase in glucagon release (LECLERCQ-MEYER et al. 1976, 1978, 1979). Opposite results were obtained when the perfusate included organic acids (LUNDQUIST et al. 1976). These data are reviewed in detail in Chap. 26. Finally, omission of phosphate from the perfusate reduced the second phase of the amino acid-induced glucagon release in vitro (CAMPILLO et al. 1977), a fact which emphasizes the importance of ion fluxes in the A-cell response to arginine.

## IV. Relative Resistance of A-cell Function to Hypothermia

A most remarkable feature of A-cell function is the persistence of the response to amino acid stimulation at the relatively low temperature of 25 °C. This fact has been observed in vivo in hypothermic rats (REACH et al. 1978) and hedgehogs (HOO-PARIS et al. 1982) and in vitro in the perfused rat pancreas (LOUBATIERES-MARIANI et al. 1980) when, in contrast, insulin release was abolished. This arginine-induced glucagon release is different from the A-cell response to cold exposure and the subsequent catecholamine release. The persistence of active glucagon release at relatively low temperature implies the intervention of enzymes with special thermodynamic properties. Such "ecoenzymes" have been documented in hibernating mammals (HOO-PARIS et al. 1982). If present in A-cells, such enzymes may represent an adaptational feature in the physiology of hibernating mammals. In these animals, a high glucagon release is associated with the early stage of arousal and it contributes to the acute release of substrate which is necessary for this energy-consuming process (HOO-PARIS et al. 1982).

**Table 2.** Influence of glucose transport and metabolism inhibitors on arginine-induced glucagon release[a] in the isolated perfused rat pancreas

| Experiments | Glucose concentration (mg/100 ml) | Number of experiments | Basal release (ng/min) | Stimulation by arginine (25 mmol/l (ng) (0–3 min) | (ng) (3–20 min) |
|---|---|---|---|---|---|
| Control experiments | 80 | 32 | 1.78±0.12 | 9.72±0.74 | 42.02± 2.58 |
| | 400 | 14 | 0.26±0.05[b] | 1.67±0.28[c] | 8.64± 0.76[c] |
| Phlorhizin (2.5 mmol/l) | 80 | 4 | 2.56±0.65[d] | 9.90±2.70 | 55.34±16.7 |
| | 400 | 4 | 2.22±0.46[d] | 8.47±2.05[d] | 46.73±11.2[d] |
| 2-Deoxy-glucose (2.4 mmol/l) | 80 | 6 | 2.33±0.37 | 16.73±2.66[d] | 71.87± 6.01[d] |
| | 400 | 3 | 2.19±0.10 | 8.80±1.12 | 45.17± 2.37 |
| Atractyloside (0.1 mmol/l) | 80 | 6 | 2.96±1.09[d] | 10.88±4.76[d] | 58.58± 2.69 |
| | 400 | 6 | 2.90±0.32[d] | 10.45±1.27[d] | 59.46± 5.43[d] |

[a] Results are expressed as mean±standard error
[b] $P<0.01$
[c] $P<0.001$
[d] $P<0.01$

# D. Modulation of Amino Acid-Induced Glucagon Release

Many physiologic and pathophysiologic factors can modify this process. The concomitant concentrations of glucose and insulin in contact with A-cells play by far the predominant role.

## I. Major Role of Glucose and Insulin Concentrations

### 1. Role of Glucose in Nondiabetic Subjects

Concomitant intravenous administration of glucose or oral intake of glucose can blunt or abolish the A-cell response to amino acids in normal subjects (UNGER et al. 1970). In vitro, the A-cell response to a given arginine dose is negatively correlated with the concentration of glucose in the perfusate (PAGLIARA et al. 1974; GERICH et al. 1974a; ATTALI et al. 1979). Some other metabolizable sugars (mannose, D-ribose) and some nonmetabolizable sugars (3-*O*-methylglucose) can suppress, at least partly, the A-cell response to arginine in vitro (ATTALI et al. 1979). However, this suppressive effect is in all instances lower than that of glucose.

The addition of agents that block glucose transport or metabolism alters markedly the suppressive effect of glucose (Table 2). For instance, phlorhizin, which impairs transmembrane glucose transport, and agents such as 2-deoxyglucose or α-hydroxycyanocinnamate, which block glycolysis at various levels, abolish the negative effect of glucose on A-cells (ATTALI et al. 1979). Finally, sodium atractyloside, which blocks the translocation of ATP molecules out of mitochondria, also abolishes the suppressive effect of glucose on A-cells and their response to arginine (ATTALI et al. 1979). These results are consistent with other studies showing that the glucagon release is negatively correlated with ATP production in islets (ÖSTENSON 1979). A "glucose memory" has been described for the effect of glucose on A-cell response to arginine (GRILL et al. 1979).

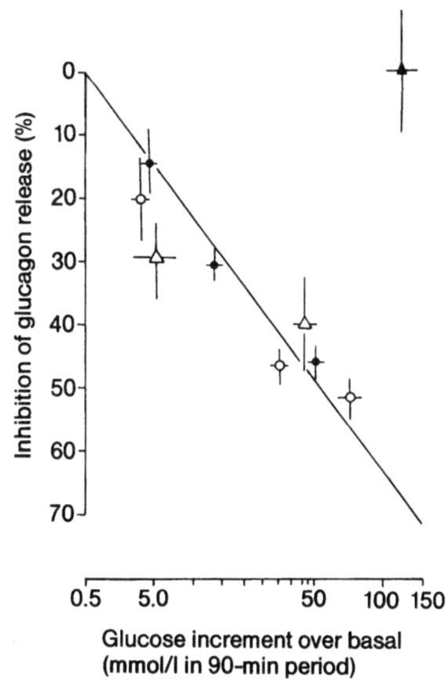


**Fig. 5.** Inhibition of the glucagon response to arginine by glucose in eight high insulin responders (*full circles*), eight low insulin responders (*open circles*), ten uncontrolled diabetics (*full triangles*), and ten tightly controlled diabetics (*open triangles*). ASSAN et al. (1981)

## 2. Role of Insulin

The permissive role of insulin in this suppressive effect of glucose is difficult to demonstrate because of the close interconnection between A- and B-cells in the normal pancreas. The isolated dog stomach represents in this respect a unique experimental model. In the absence of exogenous insulin in the perfusate, a high glucose concentration is unable to block amino acid-induced glucagon release from gastric A-cells (LEFEBVRE and LUYCKX 1977). This result, and the observations made in vivo in diabetic subjects led to the concept that the B-cells might be the "glucose sensors" of A-cells (UNGER 1976).

## 3. Studies in Diabetic Subjects

This major role of insulin in regulating A-cell response to amino acids appears clearly in diabetic subjects. All human diabetic patients, insulin dependent or not, with the notable exception of pancreatectomized patients (BARNES and BLOOM 1976; TIENGO et al. 1982), display an A-cell hyperactivity after intravenous amino acid infusion or ingestion of protein (MÜLLER et al. 1970; GERICH et al. 1975a). The diabetic A-cell hyperreactivity to amino acids is dose dependent (ASSAN et al. 1981). The same hyperreactivity has been observed in vivo in alloxan-treated and streptozotocin-treated dogs (MÜLLER et al. 1971) and rats (MARRE et al. 1979). This hyperreactivity is present in spite of the concomitant hyperglycemia. It is not suppressed by intravenous infusion of glucose (UNGER et al. 1970). Conversely, a strict metabolic control of diabetes, obtained by continous insulin infusion, restores the suppressive effect of glucose (Fig. 5, ASSAN et al. 1981).

The hyperreactivity of diabetic A-cells to amino acids has an important metabolic consequence in diabetic patients: it increases hyperglycemia and nitrogen wastage (RASKIN and UNGER 1978; RIZZA et al. 1979). This observation has led to therapeutic attemps to suppress the hypergluconemia which follows protein meals, by the use of somatostatin (GERICH 1976) and/or optimization of the metabolic control of diabetes (ASSAN et al. 1981). The hyperreactivity of A-cells to amino acids is also found in guinea pigs rendered experimentally diabetic (BUCHANAN et al. 1969), or in the pancreas from a spontaneously diabetic Chinese hamster (FRANKEL et al. 1975), or in diabetic mice (LAUBE et al. 1973).

By contrast, pancreata from rats rendered experimentally diabetic by alloxan or streptozotocin treatment display a lower than normal reactivity to amino acids in vitro (PAGLIARA et al. 1975; WEIR et al. 1976). These discrepancies between in vivo and in vitro behavior of A-cells from the diabetic pancreas may be due to the presence of some stimulating extrapancreatic factors (e. g., catecholamines) which can be present in diabetic animals in vivo, not in the in vitro perfused pancreas.

## II. Other Physiologic Modulations

### 1. Short-Term Modulations

Cholecystokinin-Pancreozymin (CCK-PZ) directly stimulates A-cells (REHFELD et al. 1980). This hormone is released into the blood after a protein meal, and it may then potentiate the direct influence of hyperaminoacidemia in postprandial states (UNGER 1974). In vitro, CCK–PZ potentiates the stimulatory effect of arginine (ASSAN et al. 1975).Conversely, exogenous somatostatin inhibits A-cells and this property has been widely used for the suppression of amino acid-induced glucagon release in vitro (GERICH et al. 1975 b) and in vivo (GERICH et al. 1974 c); the administration of a neutralizing anti-somatostatin immune serum increases glucagon release in vitro (ITOH et al. 1980): this result suggests that endogenous somatostatin, which is released during or after amino acid infusion, might, to some extent, contribute to the prevention of excess glucagon release.

### 2. Long-Term Modulations

Fasting and a protein-rich diet are followed by high basal and arginine-stimulated glucagon levels in plasma (BUCKMAN et al. 1974; EISENSTEIN and STRACK 1978).

## III. Other Pathophysiologic Modulations

### 1. Liver Cirrhosis

As reviewed in Chap. 48, in patients with liver cirrhosis, higher than normal glucagon values are present following arginine infusion (MARCO et al. 1973; SHERWIN et al. 1974; GRECO et al. 1974). One may question, however, if this feature corresponds to an actual oversecretion or to the decreased catabolism of the hormone by the liver, because of the impairment of liver function, or the presence of portocaval shunts.

### 2. Kidney Failure

In uremic patients similarly (see Chap. 47), high glucagon levels are measured (BILBREY et al. 1974) and can be explained by a reduction in catabolism of the hormone, because of renal failure (LEFEBVRE et al. 1974). The metabolic consequences of amino acid-induced hyperglucagonemia are particularly marked in renal-deficient patient: an abnormally high sensitivity of hepatocytes to glucagon is present in this condition (SHERWIN et al. 1976). Thus, in uremic diabetic patients, any rise in glucagon secretion is followed by a marked hyperglycemia because of several contributing factors: a prolonged half-life for the hormone, hypersensitivity to glucagon of the liver, and the accumulation of glucose in extracellular fluid owing to the impairment of glomerular filtration.

### 3. Thyroid Conditions

In hyperthyroidism, the A-cell response to amino acids has been described as near normal, but it is higher than normal in hypothyroidism (SEINO et al. 1974).

### 4. The Somatostatinoma Syndrome

This includes suppression of A-cell response to amino acids. This response is restored by surgical removal of the tumor (KREJS et al. 1979).

### 5. The Glucagonoma Syndrome

In the glucagonoma syndrome (see Chap. 43), the A-cell response to amino acids is sometimes undetectable because of the continous and massive hypersecretion of immunoreactive material (HOLST et al. 1979b).

### 6. Obese Nondiabetic Patients

In obese nondiabetic patients (see Chap. 49), the A-cell response to amino acids has been described as higher (KALKHOFF et al. 1973) or lower (WISE et al. 1972) than normal. The latter situation is consistent with the concept of A-cell suppression by high insulin concentrations in the absence of diabetes. Diabetic obese subjects by contrast display the diabetic hyperreactivity of A-cells to amino acids (TIENGO et al. 1972).

### 7. Pheochromocytoma

Conversely, in patients with pheochromocytoma (see Chap. 45), the insulinopenia is associated with higher than normal A-cell response to amino acids (HAMAJI 1979). Both anomalies disappear after surgical treatment of the endocrine tumor.

### 8. Stress

In a wide variety of stress stimulations: severe burns, trauma, endotoxinic shock, and hypoxia, high glucagon concentrations and higher than normal response to arginine have been described with a concomitant hypoinsulinism (LINSAY et al. 1972; BAUM and PORTE 1980). This normal pattern of stress, and its deleterious consequences on nitrogen metabolism must be taken into account when intravenous amino acid infusions are repeatedly administered to severely ill patients (see also Chap. 30 and 38).

## E. Amino Acid-Induced Glucagon Release and the Regulation of Substrate Distribution

To become physiologically meaningful, a phenomenon such as amino acid-induced glucagon release must fulfil a number of prerequisites. It should: (1) occur in the presence of physiologic (nonpharmacologic) increases in blood amino acid concentration; (2) produce a biologically effective concentration of glucagon in contact with hepatocytes; and (3) be integrated in some understandable teleological scheme, for optimal metabolic and nutritional adaptions.

### I. Physiologic Relevance of the Amino Acid and Glucagon Concentrations

#### 1. Blood Amino Acid Concentration

Although the intravenous infusion of arginine at normal doses produces a pharmacologic rise of this amino acid, the blood amino acid concentration measured after a protein meal does stimulate glucagon release (PEK et al. 1969). The same is true for amino acid mixtures reproducing in vitro the physiologic pattern at concentrations present in normal adults, i. e. 4–6 mmol/l (PAGLIARA et al. 1974; ASSAN et al. 1977). The dose-response studies suggest that A-cell stimulation can occur for minute variations of amino acid concentration in the physiologic range (GERICH et al. 1974a)

#### 2. Plasma Glucagon Concentration and Biologic Efficacy

In experiments where endogenous glucagon release was blocked by the infusion of somatostatin, and the A-cell response to amino acid was mimicked by the infusion of exogenous glucagon, the hepatic glucose output was increased (SACCA et al. 1979). The concentration of glucagon in peripheral venous plasma was in the range of concentrations measured during an arginine infusion test. Assuming that all the immunoreactive material corresponds to biologically active glucagon, these values are at levels which are efficient in stimulating glycogenolysis and gluconeogenesis in the liver (HUTSON et al. 1976). The biologic efficacy of the A-cell response to amino acids also depends upon the precise nature of the material secreted, and upon the concomitant biologic status of target cells. Among the various glucagon-immunoreactive materials circulating in plasma (see Chap. 11), the 3 500 daltons glucagon, reacting with COOH terminal-specific antibodies, rises selectively during arginine infusion in humans. By contrast, the concentration of the high molecular weight material, commonly called big plasma glucagon (BPG), does not vary during the infusion of arginine (Fig. 6).

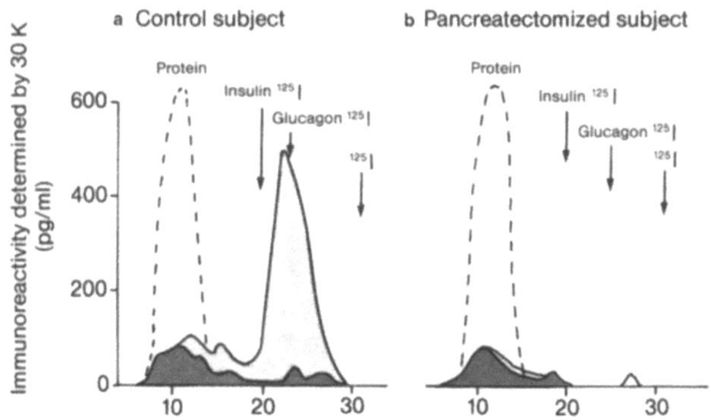


**Fig. 6 a, b.** Elution profiles of glucagon immunoreactivity (as measured with the 3OK antibody on Sephadex G-50 gel chromatography in one control subject (**a**) and one pancreatectomized patient (**b**) before (*cross-hatched*) and during (*plain area*) arginine infusion. *Broken lines* indicate the protein peak

The pancreatic and/or gastric origin of this 3 500 daltons glucagon has been a matter of discussion. In the dog, the stomach contains as much glucagon as does the pancreas, and the dog stomach releases glucagon in response to amino acid infusion (MASHITER et al. 1975). The biologic efficacy of the material is proved by measurements of glucose turnover in dogs. The increase in blood glucose concentration and glucose turnover is even more pronounced in pancreatectomized dogs not treated by insulin (VRANIC et al. 1976; MÜLLER et al. 1978).

In humans, the situation is different and most if not all of the glucagon released during arginine infusion is secreted by the pancreas. Insulin-dependent diabetic patients display, during arginine infusion, a rise in blood glucose and a disappearance of circulating lactate and alanine; both phenomena are correlated in time with the glucagon rise (ASSAN et al. 1981) as illustrated in Fig. 7. In pancreatectomized patients, arginine infusion induces neither the glucagon rise nor the disappearance of circulating lactate and alanine (TIENGO et al. 1982).

These observations are consistent with a scheme in which endogenous glucagon, released under the influence of amino acids, enhances the hepatic glucose output by the stimulation of gluconeogenesis. This scheme is consistent with the lower than normal blood amino acid concentration in patients with glucagonoma (HOLST et al. 1979 b) and the higher than normal lactate and alanine concentration in pancreatectomized patients (TIENGO et al. 1982). The concomitant diversion of amine residues towards ureogenesis is a satellite phenomenon which parallels the enhancement of gluconeogenesis. Indeed, glucagon does stimulate some enzymes of the urea cycle (LIN and SNODGRASS 1979; HUSSON and VAILLANT 1982).

This action accounts for the nitrogen wasting which occurs during higher gluconeogenic states. In normal subjects, the administration of amino acids is followed by certain other hormonal consequences, particularly a biphasic release of insulin (Fig. 8). The overall metabolic consequence of amino acid administration will depend upon the respective concentrations of insulin and glucagon in contact with hepatocytes and of their biologic actions. In normal subjects the concomitant release of glucagon and insulin results in maintenance of normal glucose levels in

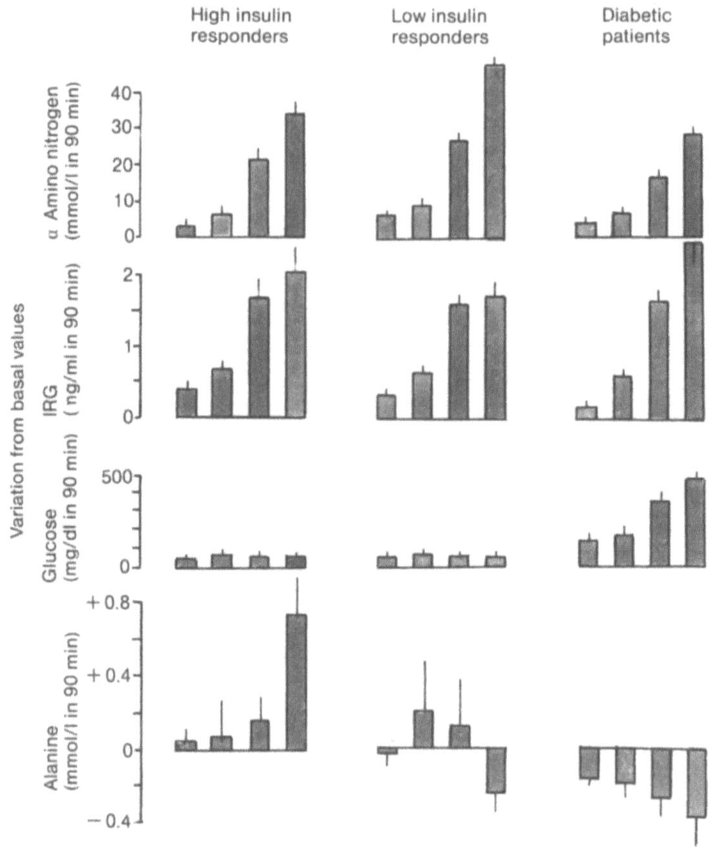


**Fig. 7.** Algebraic sum of variations from basal values of α-amino nitrogen, immunoreactive glucagon, glucose, and alanine during arginine infusion tests performed with increasing doses of arginine. ASSAN et al. (1981)

blood and an increase in glucose turnover. In insulinopenic diabetic subjects, the glucagon release following an amino acid load favors hyperglycemia and nitrogen wasting.

## II. Clinical Correlations

### 1. Normal Subjects

After ingestion of a protein meal, the plasma insulin concentration rises by 60%–100%: this might induce hypoglycemia and a reduction of the hepatic glucose output, two factors which in fact do not occur (PEK et al. 1969; MÜLLER et al. 1970).

UNGER et al. (1969) were the first to suggest that the glucagon secretion which occurs after a protein meal maintains normoglycemia in spite of the amino acid-induced hyperinsulinism. Incidentally, hypoglycemia did occur after a protein meal in the few patients with A-cell deficiency (BLEICHER et al. 1970).

The amino acid-induced hormone rise can be mimicked by infusions of exogenous insulin and glucagon (FELIG et al. 1976). Under such conditions, the infusion

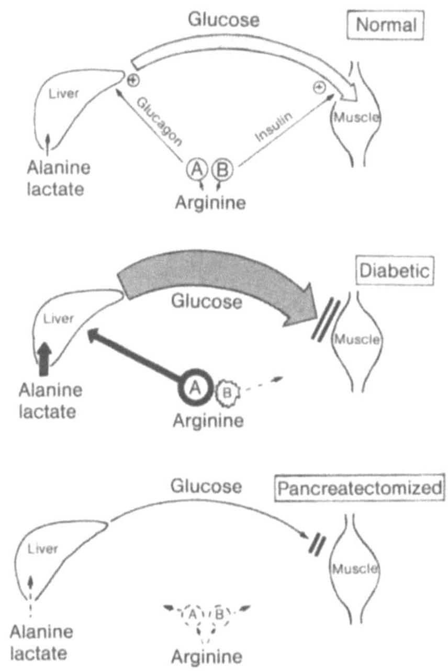


**Fig. 8.** A scheme of the metabolic consequences of A- and B-cell stimulation by arginine in three different situations: normal, diabetic, and pancreatectomized subjects

of glucagon at a rate of 3 ng · kg$^{-1}$ · min$^{-1}$ prevented the reduction in hepatic glucose output and hypoglycemia. Peripheral glucose uptake and hepatic glucose output were in equilibrium. In similar experiments in normal dogs, the glucose turnover was measured during arginine infusion (CHERRINGTON and VRANIC 1973). An increase in hepatic glucose output followed the release of glucagon and an increase in glucose clearance followed the release of insulin. In depancreatized dogs (with a persisting release of glucagon from the stomach), the infusion of arginine was followed by an increase in plasma glucagon concentration and glucose production, but no increase in insulin level and glucose clearance occured, so that the blood glucose concentration rose (CHERRINGTON and VRANIC 1974).

## 2. Insulin-Dependent Patients

The excessive A-cell response to amino acids and the deficient insulin release are associated in such patients with a rise in blood glucose level. Glucose production rises (BOMBOY et al. 1977) owing to the increase in glycogenolysis (CHIASSON et al. 1975) and inhibition of glycogen synthesis by glucagon (see Chaps. 14 and 15). Furthermore, a stimulation of gluconeogenesis is suggested by the disappearance of gluconeogenic substrates from blood.The splanchnic production of glucose which follows arginine infusion in diabetics declines after 60–90 min and stops after 120 min (CHERRINGTON and VRANIC 1974). This is consistent with the so-called evanescent effect of glucagon on the liver. The increase in hepatic glucose output cannot account alone for the persisting hyperglycemia after infusion of arginine in diabetics. A reduction in peripheral glucose clearance, due to insulin deficiency, contributes to the maintenance of high blood glucose values.In noninsulin-dependent diabetic patients, the B-cell sensitivity to amino acids is relatively preserved, while the B-cell response to glucose is reduced and delayed (CERASI and LUFT 1967). For this reason, it was speculated that the addition of arginine might improve the B-cell response to glucose and the glucose tolerance (CERASI 1975): insulin release was actually increased in such conditions, but glucose tolerance was not improved. An excessive A-cell response to arginine was present and this can account for the failure to correct glucose tolerance (CERASI 1975).

In low insulin responders, glucose tolerance is normal in spite of lower than normal B-cell response to glucose (CERASI and LUFT 1967): in these subjects, the A-cell response to arginine was not significantly different from that of normal subjects with a high insulin response. Higher than normal glucose values followed the infusion of arginine alone and the combined infusion of arginine and glucose in nondiabetic low insulin responders. In subjects classified as prediabetic according to other criteria, the A-cell response to amino acids was described as normal (ASSAN et al. 1981; ARONOFF et al. 1976; DAY and TATTERSALL 1975).

## References

Akpan JO, Hurley MC, Lands WE (1981) Insulin and glucagon secretion in essentially fatty acid deficient rats. Acta Diabetol Lat 18:147–156

Aronoff SL, Bennet PH, Rushforth NB, Miller M, Unger RH (1976) Normal glucagon response to arginine infusion in "prediabetic" Pima Indians. J Clin Endocrinol Metab 43:279–286

Assan R (1979) Contribution to the study of glucagon secretion and catabolism in vivo. In: Rodriguez RR, Vallance-Owen J (eds) Excerpta Medica, Amsterdam, pp 610–624

Assan R, Girard JR (1975) Glucagon in the human fetal pancreas. In: Early diabetes in early life, vol 1. Camerini-Davalos RA, Cole HS (eds) Academic Press, New York, pp 115–126

Assan R, Rosselin G, Dolais J (1967) Effets sur la glucagonémie des perfusions et ingestions d'acides aminés. Journ Annu Diabetol Hotel 8:25–41

Assan R, Boillot J, Attali JR, Soufflet E, Ballerio G (1972) Diphasic glucagon release induced by arginine in the perfused rat pancreas. Nature [New Biol] 239:125–126

Assan R, Plouin PF, Girard JR, Slama G, Hautecouverture M, Buneaux JJ (1975) Study of glucagon secretion using *in vitro* perifusion of rat pancreas pieces. Diabete Metab 1:110–111

Assan R, Bourdillat N, Attali JR, Boillot J, Selmi A (1976) Influence of cephalic glucose level on glucagon secretion by the isolated and perfused in situ rat pancreas. Diabetologia 12:377

Assan R, Attali JR, Ballerio G, Boillot J, Girard JR (1977) Glucagon secretion induced by natural and artificial amino-acids in the perfused rat pancreas. Diabetes 26:300–307

Assan R, Soufflet E, Ballerio G, Attali JR, Boillot J, Girard JR (1978) Ambiguous effects of colchicine and vincristine upo A-cell response to arginine. Diabetologia 14:121–127

Assan R, Efendic S, Luft R, Cerasi E (1981) Dose-kinetics of pancreatic glucagon responses to arginine and glucose in subjects with normal and impaired pancreatic B-cell function. Diabetologia 21:452–459

Attali JR, Boillot J, Girard JR, Assan R (1979) Glucagon secretion and glucose metabolism in the pancreas. Journ Annu Diabetol Hotel 20:259–267

Barnes AJ, Bloom SR (1976) Pancreatectomized man: a model for diabetes without glucagon. Lancet 1:219–221

Baum D, Porte D Jr (1980) Stress hyperglycemia and the adrenergic regulation of pancreatic hormones in hypoxia. Metabolism [Suppl 1] 29:1176–1185

Bilbrey GL, Faloona GR, White MG, Knochel JR (1974) Hyperglucagonemia of renal failure. J. Clin Invest 53:841–847
Blackard WG, Nelson NC, Andrews SS (1974) Portal and peripheral vein immunoreactive glucagon concentrations after arginine or glucose infusions. Diabetes 23:199–202
Bleicher SJ, Levy LJ, Zarowitz H, Spergel (1970) Glucagon-deficiency hypoglycaemia: a new syndrome. Clin Res 18:355
Boitard C, Debray-Sachs M, Pouplard A, Assan R, Hamburger J (1981) Lymphocytes from diabetics suppress insulin release in vitro. Diabetologia 21:41–46
Bomboy JE, Lewis SB, Sinclair-Smith BC, Liljenquist JE (1977) Transient stimulatory effect of sustained hyperglucagonemia on splanchnic glucose output in normal and diabetic man. Diabetes 26:177–184
Buchanan KD, Mawhinney W (1973) Glucagon release from isolated pancreas in streptozotocin treated rats. Diabetes 22:797–800
Buchanan KD, Vance JE, Williams RH (1969) Insulin and glucagon release from isolated islets of Langerhans. Diabetes 18:381–386
Buckman MT, Conway MJ, Seibel JA, Eaton PP (1974) Effect of fasting on alanine-stimulated insulin and glucagon secretion. Metabolism 22:1253–1262
Campillo JE, Luyckx AS, Torres MD, Lefebvre PJ (1977) Effect of phosphate omission on arginine-induced insulin and glucagon release by the isolated perfused rat pancreas. FEBS Lett 84:141–143
Cerasi E (1975) Potentiation of insulin release by glucose in man. II. Role of the insulin response and enhancement of stimuli other than glucose. Acta Endocrinol (Copenh) 79:502–510
Cerasi E, Luft R (1967) The plasma insulin response to glucose infusion in healthy subjects and in diabetes mellitus. Acta Endocrinol (Copenh) 55:278–304
Cherrington AD, Vranic M (1973) Effect of arginine on glucose turnover and plasma free fatty acids in normal dogs. Diabetes 22:537–543
Cherrington AD, Vranic M (1974) Effect of interaction between insulin and glucagon on glucose turnover and FFA concentration in normal and depancreatized dogs. Metabolism 23:729–744
Chesney TMcC, Schofield J (1969) Studies on the secretion of pancreatic glucagon. Diabetes 18:627–632
Chiasson JL, Liljenquist JE, Sinclair-Smith BC, Lacy WW (1975) Gluconeogenesis from alanine in normal postabsorptive man. Intra-hepatic stimulatory effect of glucagon. Diabetes 24:574–584
Day JL, Tattersall RB (1975) Glucagon secretion in unaffected monozygotic twins of juvenile diabetics. Metabolism 24:145–151
Dunbar JC, Walsh MF, Foà PP (1976) Secretion of immunoreactive insulin and glucagon in hamsters bearing a transplantable insulinoma. Diabete Metab 2:165–169
Edwards JC, Howell SL (1973) Effect of vinblastine and colchicine on the secretion of glucagon from isolated guinea pig islets of Langerhans. FEBS Lett 30:89–92
Edwards JC, Hellerström C, Petersson B, Taylor KW (1972) Oxydation of glucose and fatty acids in normal and $A_2$-cell-rich pancreatic islets isolated from guinea pigs. Diabetologia 8:93–98
Eisenstein AB, Strack K (1978) Amino-acid stimulation of glucagon secretion by perifused islets of high-protein-fed rats. Diabetes 27:370–376
Felig P, Gusberg R, Hendler R Gump FE, Kinney JM (1974) Concentration of glucagon and insulin: glucagon ratio in the portal and peripheral circulation. Proc Soc Exp Biol Med 147:88–90
Felig P, Wahren J, Sherwin R, Hendler R (1976) Insulin, glucagon and somatostatin in normal physiology and diabetes mellitus. Diabetes 25:1091–1099
Frankel BJ, Gerich JE, Fanska RE, Gerritsen C, Grodsky GM (1974) Abnormal secretion of insulin and glucagon by the in vitro perfused pancreas of the genetically diabetic chinese hamster. J Clin Invest 53:1637–1646
Frankel BJ, Gerich JE, Fanska RE, Gerritsen GC, Grodsky GM (1975) Response to arginine of the perifused pancreas of the genetically diabetic chinese hamster. Diabetes 24:273–279

Gerich JE (1976) Metabolic effects of long-term somatostatin infusion in man. Metabolism [Suppl 1] 25:1505–1507
Gerich J, Charles MA, Grodsky GM (1974a) Characterization of the effects of arginine and glucose on glucagon and insulin release from the perfused rat pancreas. J Clin Invest 54:833–841
Gerich JE, Frankel BJ, Fanska R, West L, Forsham PH, Grodsky GM (1974b) Calcium dependency of glucagon secretion from the in vitro perfused rat pancreas. Endocrinology 94:1381–1385
Gerich JE, Lorenzi M, Schneider V, Kwan C, Karam J, Guillemin R, Forsham P (1974c) Inhibition of pancretic glucagon responses to arginine by somatostatin in normal man and in insulin-dependent diabetics. Diabetes 23:876–880
Gerich JE, Tsalikian E, Lorenzi M, Karam JH, Schneider V, Gustafson G, Bohannon NV (1975a) Normalisation of fasting hyperglucagonemia and excessive glucagon response to IV arginine in human diabetes mellitus by prolonged perfusion of insulin. J Clin Endocrinol Metab 41:1178–1180
Gerich JE, Lovinger R, Grodsky GM (1975b) Inhibition by somatostatin of glucagon and insulin release from the perfused rat pancreas in response to arginine, isoproterenol and theophylline: evidence for a preferential effect on glucagon secretion. Endocrinology 96:749–754
Greco AV, Ghirlanda G, Patrono C, Felidi G, Manna R (1974) Behaviour of pancreatic glucagon, insulin and HGH in liver cirrhosis, after arginine and IV glucose. Acta Diabetol Lat 11:330–339
Grill V, Adamson U, Rundfeldt M, Andersson S, Cerasi E (1979) Glucose memory of pancreatic B and A2 cells: evidence for common time-dependent actions of glucose on insulin and glucagon secretion in the perfused rat pancreas. J Clin Invest 64:700–707
Hamaji M (1979) Pancreatic alpha- and beta-cell function in pheochromocytoma. J. Clin Endocrinol Metab 49:322–325
Holst JJ, von Schenck H, Lindkaer S (1979a) Gel filtration pattern of immunoreactive glucagon secreted by the isolated, perfused, porcine pancreas. Scand J Clin Lab Invest 39:47–52
Holst JJ, Helland S, Ingemannson S, Pedersen NB, von Schenck H (1979b) Functional studies in patients with the glucagonoma synrome. Diabetologia 17:151–156
Hoo-Paris R, Hamsany M, Sutter BCJ, Assan R, Boillot J (1982) Plasma glucose and glucagon concentrations in the hibernating hedgehogs. Gen Comp Endocrinol 46:246–254
Husson A, Vaillant R (1982) Effects of glucocorticosteroids and glucagon on arginosuccinate synthetase, arginino-succinase, and arginase in fetal rat liver. Endocrinology 110:227–232
Hutson NJ, Bromley FT, Assimacopoulos FD, Harper SC, Exton JH (1976) Studies on the adrenergic activation of hepatic glucose output: I-studies on the α-adrenergic activation of phosphorylase and gluconeogenesis and inactivation of glycogen synthase in isolated rat liver parenchymal cells J Biol Chem 251:5200–5208
Itoh M, Mandarino L, Gerich JE (1980) Anti-stomatostatin gamma globulin augments secretion of both insulin and glucagon in vitro: evidence for a physiologic role of endogenous somatostatin in the regulation of pancreatic A- and B-cell function. Diabetes 29:693–698
Iversen J (1971) Secretion of glucagon and insulin from the isolated perfused canine pancreas. J Clin Invest 50:2123–2136
Jacoby JH, Bryce GF (1979) The acute effects of 5 HTP, fluoxetine and quipazine on insulin and glucagon release in the intact rat. Horm Metab Res 11:90–94
Jarrousse C, Rosselin G (1975) Regulation by glucose and cyclic nucleotides of the glucagon and insulin release induced by amino-acids. Diabete Metab 1:135–142
Kalkhoff RK, Gossain VV, Matute ML (1973) Plasma glucagon in obesity: response to arginine, glucose and protein. N Engl J Med 289:465–467
Karman H, Mialhe P (1973) Glucose-glucagon feed-back mechanism in normal and diabetic geese and ducks. Diabetologia 9:74 (abstract)
Krejs GJ, Orci L, Conlon JM, Ravazzola M, Davis GR, Raskin P, Collins SM, McCarthy D, Baetens D, Rubenstein A, Aldor T, Unger RH (1979) Somatostatinoma syndrome: biochemical, morphologic and clinical features. N Engl J Med 301:285–292

Lacy PE (1962) Electron microscopy of the islets of Langerhans. Diabetes 11:509–513
Lacy PE, Walker MM, Finck CJ (1972) Perifusion of isolated rat islets in vitro: participation of the microtubular system in the biphasic release of insulin. Diabetes 20:987–998
Laube H, Füssganger RD, Maier V, Pfeiffer EF (1973) Hyperglucagonemia of the isolated perfused pancreas diabetic mice (db db). Diabetologia 9:400–402
Lawrence AM, Tan S, Hojvat S, Kirsteins L (1975) Salivary gland hyperglycemic factor: an extra-pancreatic source of glucagon-like material. Science 195:70–72
Leclercq-Meyer V, Marchand T, Malaisse WJ (1974) Possible role of microtubular-microfilamentous system in glucagon secretion. Diabetologia 10:215–224
Leclercq-Meyer V, Marchand J, Malaisse WJ (1976) The role of calcium in glucagon release. Interactions between arginine and calcium. Horm Res 76:348–362
Leclercq-Meyer V, Marchand J, Malaisse WJ (1978) The role of calcium in glucagon release. Studies with verapamil. Diabetes 27:996–1004
Leclercq-Meyer V, Marchand J, Malaisse WJ (1979) Calcium-dependency of glucagon-release: its modulation by nutritional factors. Am J Physiol 236:E98–104
Lefèbvre PJ, Luyckx AS (1977) Factors controlling gastric glucagon release. J Clin Invest 59:716–722
Lefèbvre PJ, Luyckx AS, Nizet AH (1974) Renal handling of endogenous glucagon in the dog: comparison with insulin. Metabolism 23:753–761
Lichtenberger LM, Delansorne R, Graziani LA (1982) Importance of amino-acid uptake and decarboxylation in gastrin relase from isolated G cells. Nature 295:698–700
Lin RC, Snodgrass PJ (1979) Induction of urea cycle enzymes of rat liver by glucagon. J Biol Chem 253–2748
Lindsay CA, Wilmore DW, Moylan JA, Faloona GR, Unger RH (1972) Glucagon and the insulin/glucagon (I/G) ratio in burns and trauma. Clin Res 20:802
Loubatieres-Mariani MM, Chapal J, Puech R, Lignon F, Valette G (1980) Different effects of hypothermia on insulin and glucagon secretion from the isolated perfused rat pancreas. Diabetologia 18:329–333
Lundquist K, Fanska F, Grodsky M (1976) Direct calcium-stimulated release of glucagon from the isolated perfused rat pancreas and the effect of chemical sympathectomy. Endocrinology 98:815–818
Malaisse W, Malaisse-Lagae F, Walker MO, Lacy PE (1971) The stimulus-secretion coupling of glucose-induced insulin release. V. The participation of a microtubular-microfilamentous system. Diabetes 20:257–265
Malaisse W, Sener A, Malaisse-Lagae F (1981) Insulin release: reconciliation of the receptor and metabolic hypotheses. Mol Cell Biochem 37:157–165
Marco J, Faloona GR, Unger RH (1971) The glycogenolytic activity of immunoreactive glucagon in plasma. J Clin Invest 50:1650–1655
Marco J, Diego J, Villanuelva ML, Diaz-Fierros M, Valverde I, Segovia JM (1973) Elevated plasma glucagon in cirrhosis of the liver. N Engl J Med 289:1107–1111
Marre M, Bobbioni E, Suarez M, Reach G, Dubois MP, Assan R (1979) Control of gastric glucagon secretion in the acutely pancreatectomized rat. Diabetes 28:213–220
Marre M, Attali JR, Helman A, Miller J, Kronheim S, Assan R (1981) Pancreatic and gastric secretions of somatostatin-like-immunoreactivity (SLI) in rat portal vein: an in vivo and in vitro comparative study. Diabetologia 21:302
Mashiter K, Harding PE, Chou M, Mashiter GD, Stout J, Diamond D, Field JB (1975) Persistent pancreatic glucagon but not insulin response to arginine in pancreatectomized dogs. Endocrinology 96:678–693
Müller WA, Faloona GR, Unger RH (1970) Abnormal alpha cell function in diabetes: response to carbohydrate and protein ingestion. N Eng J Med 283:109–115
Müller WA, Faloona GR, Unger RH (1971) The effect of experimental insulin deficiency on glucagon secretion. J Clin Invest 50:1992–1999
Müller WA, Girardier L, Seydoux J, Berger M, Renold AE, Vranic M (1978) Extrapancreatic glucagon and glucagon-like-immuno-reactivity in depancreatized dogs. J Clin Invest 62:124–132
Östenson CG (1979) Regulation of glucagon release: effects of insulin on the pancreatic A2-cell of the guinea-pig. Diabetologia 17:325–330

Ohneda A, Parada E, Eisentraut M, Unger RH (1968) Characterization of response of circulating glucagon to intra-duodenal und intravenous administration of amino-acids. J Clin Invest 47:2305–2322
Orci L (1976) The microanatomy of the islets of Langerhans. Metabolism [Suppl 1] 11:1303–1313
Orci L, Unger RH (1975) Functional subdivisions of the islets of Langerhans and the possible role of the insular D-cells. Lancet 2:1243–1244
Pagliara AS, Stillings SN, Hover B, Martin DM, Matchinsky FM (1974) Glucose modulation of amino-acid-induced glucagon and insulin release in the isolated perfused rat pancreas. J Clin Invest 54:819–832
Pagliara AS, Stillings SN, Haymond MW, Hover BA, Matchinsky FM (1975) Insulin and glucose as modulators of the amino-acid-induced glucagon release in the isolated pancreas of alloxan and streptozotocin diabetics rats. J Clin Invest 55:244–255
Pek S, Fajans SS, Floyd JC Jr, Knopf RH, Conn JN (1968) Effect of amino-acids on plasma glucagon in man. J Lab Clin Med 72:1003
Pek S, Fajans SS, Floyd JC Jr, Knopf RH, Conn JN (1969) Effects upon plasma glucagon of infused and ingested amino-acids and of protein meals in man. Diabetes 18:523–528
Pek S, Santiago JC, Tai TY (1978) L-Leucine-induced secretion of glucagon and insulin, and the "off-response" to L-Leucine in vitro. I. Characterization of the dynamics of secretion. Endocrinology 103:1208–1218
Pontiroli AE, Micossi P, Foà PP (1979) Effects of histamine, of histidine and of anti-histaminic agents on the release of glucagon and insulin from the rat pancreas. Horm Metab Res 11:100–103
Raskin P, Unger RH (1978) Hyperglucagonemia and its suppression: importance in the metabolic control of diabetes. N Engl J Med 299:433–436
Reach G, Poussier P, Bailey D, Assan R (1978) Acute hypothermia induces insulin lack, glucagon excess and insulin resistance in the rat. Diabetologia 15:265
Rehfeld JF, Larsson LI, Goltermann NR, Schwartz TW, Holst JJ, Jensen SL, Morley JS (1980) Neural regulation of pancreatic hormone secretion by the C-terminal tetrapeptide of CCK. Nature 284:33–38
Rizza R, Verdonk C, Miles J, Service FJ, Gerich JE (1979) Effect of intermittent endogenous hyperglucagonemia on glucose homeostasis in normal and diabetic man. J Clin Invest 63:1119–1123
Rocha DM, Faloona GK, Unger RH (1972) Glucagonstimulating activity of 20 amino-acids in dogs. J Clin Invest 51:2346–2361
Sacca L, Sherwin R, Felig P (1979) Influence of somatostatin on glucagon and epinephrine stimulated hepatic glucose output in the dog. Am J Physiol 236:E113–E117
Sai P, Boitard C, Debray-Sachs M, Pouplard A, Assan R, Hamburger J (1981) Complement-fixing islet cell antibodies from some diabetic patients alter insulin release in vitro Diabetes 30:1051–1057
Seino Y, Goto Y, Taminato T, Ikeda M, Imura H (1974) Plasma insulin and glucagon responses to arginine in patients with thyroid dysfunction. J Clin Endocrinol Metab 38:1136–1140
Sherwin R, Joshi P, Handler R, Felig P, Conn HO (1974) Hyperglucagonemia in Laennec's cirrhosis. The role of portalsystemic shunting. N Engl J Med 290:239–242
Sherwin RS, Bastl C, Finkelstein FO, Fisher M, Black H, Hendler R, Felig P (1976) Influence of uremia and hemodialysis on the turnover and metabolic effects of glucagon. J Clin Invest 57:722–731
Tai TY, Pek S, Santiago JC (1978) L-Leucine-induced secretion of glucagon and insulin, and the "off-response" to L-Leucine in vitro. II. The role of D-glucose. Endocrinology 103:1219–1226
Tiengo A, Assan R, Tchobroutsky G (1972) Metabolic and hormonal patterns after 3 days of total fasting in 27 obese and non obese subjects. Isr J Med Sci 8:821–822
Tiengo A, Bessioud M, Valverde I, Tabbi-Annemi A, Delprato S, Alexandre JH, Assan R (1982) Absence of islet alpha cell function in pancreatectomized patients. Diabetologia 22:25–32

Unger RH (1974) New aspects of glucagon pathology and pathophysiology. In: Malaisse WJ, Pirart J (eds) Proceedings of the eigth congress of the International Diabetes Federation. Excerpta Medica, Amsterdam, pp 137–143

Unger RH (1976) Diabetes and the alpha-cell (Banting Memorial Lecture). Diabetes 25: 136–151

Unger RH, Ohneda E, Aguilar-Parada E, Eisentraut AM (1969) The role of aminogenic glucagon secretion in blood glucose homeostasis. J Clin Invest 48:810–822

Unger RH, Aguilar-Parada E, Müller WA, Eisentraut A (1970) Studies on pancreatic alpha-cell function in normal and diabetic subjects. J Clin Invest 49:837–848

Vranic M, Engerman R, Doi K, Morita S, Yip CC (1976) Extra-pancreatic glucagon in the dog. Metabolism [Suppl 1] 25:1469–1473

Weir GC, Knowlton SD, Atkins RF, McKennan KX, Martin DB (1976) Glucagon secretion from the perfused pancreas of streptozotocin-treated rats. Diabetes 25:275–282

Wise JK, Hendler R, Felig P (1970) Evaluation of alpha-cell function by infusion of alanine in normal diabetic and obese subjects. N Engl J Med 288:484–487

Wise JK, Hendler R, Felig P (1972) Obesity: evidence of decreased glucagon secretion. Science 178:513–514

Wollheim CB, Blondel B, Renold AE, Sharp GW (1976) Stimulatory and inhibitory effects of cyclic AMP on pancreatic glucagon release from monolayer cultures and the controlling role of calcium. Diabetologia 12:269–277

CHAPTER 25

# Free Fatty Acids and Glucagon Secretion

A. S. LUYCKX and P. J. LEFEBVRE

## A. Introduction

The first report dealing with the influence of free fatty acids (FFA) on glucagon secretion was published by SEYFFERT and MADISON (1967). These authors demonstrated that in the dog, an acute elevation in circulating plasma FFA levels resulted in a significant decrease in glucagon secretion. In our laboratory, we investigated the influence of artificial changes in plasma FFA concentrations on plasma glucagon levels in the pancreaticoduodenal vein of anesthetized dogs; we found that the fall in plasma FFA observed during nicotinic acid infusion or after terminating a triglyceride-heparin infusion was accompanied by a significant increase in glucagon secretion (LUYCKX and LEFEBVRE 1970).

During the last 12 years, several investigations have been carried out in order to delineate the possible role of free fatty acids in the regulation of glucagon release. We will review these studies according to the type of experimental protocol used. In vitro studies were performed using isolated guinea pig islets (EDWARDS et al. 1969; EDWARDS and TAYLOR 1970) and the isolated perfused rat pancreas (LUYCKX and LEFEBVRE 1974; CAMPILLO et al. 1979, 1982).

In vivo, the influence of FFA was investigated in dogs (SEYFFERT and MADISON 1967; LUYCKX and LEFEBVRE 1970; MÜLLER et al. 1976), in rats (LUYCKX and LEFEBVRE 1976a), and in ducks (GROSS and MIAHLE 1974; LAURENT and MIAHLE 1978). Finally, a series of reports deal with clinical investigations in normal volunteers (GERICH et al. 1974; ANDREWS et al. 1975; QUABBE et al. 1977; HICKS et al. 1977), in pregnant women (LUYCKX et al. 1975, 1978a), in diabetic subjects (GERICH et al. 1976; TASAKA et al. 1976; LUYCKX and LEFEBVRE 1976b), and in patients with hypertriglyceridemia (TIENGO et al. 1978).

## B. In Vitro Studies

### I. Isolated Islets

Glucagon release is markedly inhibited when isolated guinea pig islets of Langerhans are incubated in the presence of DL-$\beta$-hydroxybutyrate (10 m$M$), octanoate (5 m$M$), or palmitate (0.6 m$M$) (EDWARDS et al. 1969; EDWARDS and TAYLOR 1970). The marked inhibition of glucagon release caused by relatively low levels of fatty acids contrasts with the slight inhibition induced under the same experimental conditions by a high glucose concentration (16.7 m$M$). The inhibition of glucagon release produced by octanoate is not affected by varying the concentration of glu-

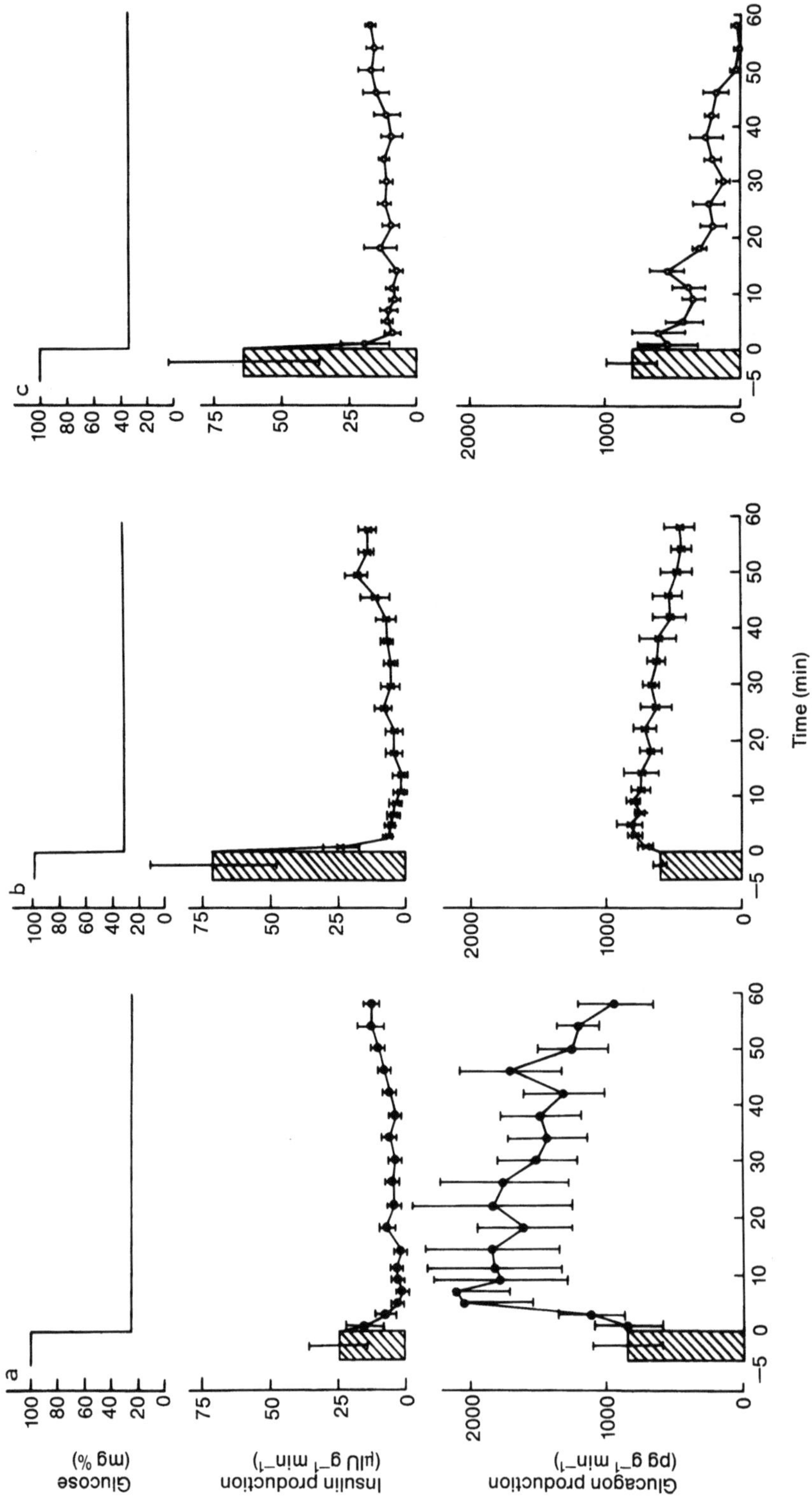

a
b
c
Glucose (mg %)
Insulin production (μIU g⁻¹ min⁻¹)
Glucagon production (pg g⁻¹ min⁻¹)
Time (min)

cose in the incubation medium. This suggests that fatty acid levels may be more important than glucose concentrations in the regulation of glucagon release from guinea pig pancreas. On the same material, the stimulatory effect of arginine (5 m*M*) is completely suppressed when octanoate (5 m*M*) is added to the incubation medium. As EDWARDS and TAYLOR (1970) have stressed, these results again underscore the major role played by fatty acids in the regulation of glucagon secretion, since the inhibition of glucagon release caused by FFA cannot be overcome by metabolites which are known to stimulate glucagon release in the absence of fatty acids. The inhibition of glucagon release caused by octanoate is abolished by the addition of malonate (7 m*M*) or 2,4-dinitrophenol (0.25 m*M*), substances known to prevent energy production from FFA oxidation. Similar results have been obtained when malonate or 2,4-dinitrophenol are added to islets incubated in the presence of $\beta$-hydroxybutyrate.

## II. Isolated Perfused Rat Pancreas

Using the isolated perfused rat pancreas, we have confirmed that there is a marked inhibition of glucagon release when the perfusion medium contains high levels of FFA. Similarly, the unequivocal rise in glucagon release observed when the concentration of glucose in the perfusion medium falls from 100 to 25 mg/100 ml is significantly reduced in the presence of high concentrations of palmitate (2.2 m*M*) or octanoate (2.4 m*M*) in the perfusate (Fig. 1; LUYCKX and LEFEBVRE 1972, 1974). More recent experiments aimed at investigating the effect of oleic acid on the biphasic glucagon secretion evoked in response to 10 m*M* arginine. In the presence of 1.5 m*M* oleic acid, the first phase of glucagon release was unchanged, but the second phase was markedly inhibited. Such an effect was not obtained when oleic acid concentration in the medium was 0.75 m*M* (CAMPILLO et al. 1979).

The isolated perfused rat pancreas was also used to compare the inhibitory effects of oleic and octanoic acid infusions on basal glucagon secretion rate in the absence of glucose in the perfusate. As illustrated by Fig. 2, both fatty acids (1.5 m*M*) inhibited glucagon release; the phenomenon was significant from the first minute of FFA infusion and glucagon release rapidly increased over the basal output when the infusions of fatty acids were interrupted (CAMPILLO et al. 1982). Thus, in two different in vitro systems using the pancreas from two different species, the concentrations of FFA have been found to play a role, not only in regulating basal glucagon release, but also in modulating the glucagon response to hypoglycemia and to arginine infusion.

◄ **Fig. 1 a–c.** Glucagon and insulin secretion rates in response to an acute lowering of the glucose concentration (from 100 to 25 mg/100 ml) in the perfusion medium of the isolated perfused rat pancreas. **a** the mean FFA concentration in the perfusion medium is $680 \pm 136$ µequiv./l ($N=4$); **b** the perfusion medium has been supplemented with palmitate to reach an FFA concentration of 2400 µequiv./l ($N=3$); **c** the perfusion medium has been supplemented with 2400 µequiv./l octanoate. The *hatched columns* correspond to the mean secretion rates during the last 5 min before changing the glucose concentration. After LUYCKX (1975)

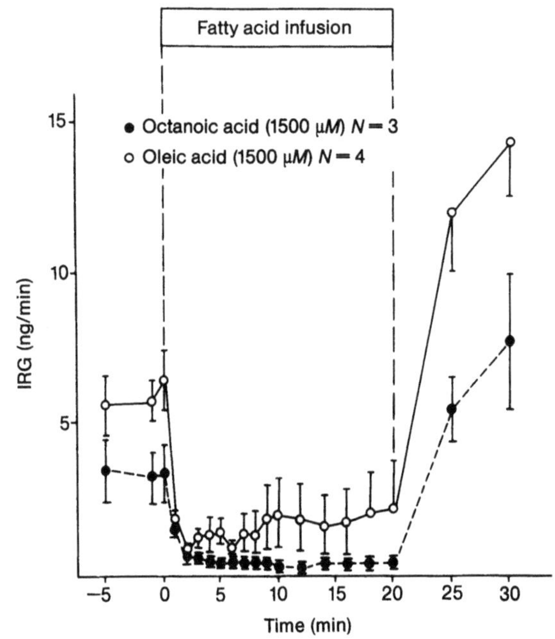


**Fig. 2.** Influence of oleic and octanoic acids on glucagon (IRG) release by the isolated perfused rat pancreas in a glucose-free perfusate. The pancreata were perfused with a standard medium 40 min before the beginning of the fatty acid infusion. The number of perfusions is indicated as *N*. Results are given as mean ± standard error

## C. Experiments in Animals

### I. Dogs

Experiments conducted on the anesthetized dog by SEYFFERT and MADISON (1967) and MADISON et al. (1968) have demonstrated that an increase in plasma FFA, obtained by simultaneous infusion of triglycerides and heparin, depresses peripheral plasma glucagon levels. These results were confirmed by LUYCKX and LEFEBVRE (1970). In addition, the latter group studied the effect of lowering plasma FFA levels with nicotinic acid. In these experiments, blood samples for glucagon determination were obtained from the femoral artery and the pancreaticoduodenal vein. In some animals, the pancreatic blood flow was determined, thus permitting calculation of true "pancreatic glucagon production" per unit time. In all experiments, the fall in plasma FFA observed during nicotinic acid infusion or after termination of triglyceride-heparin infusion was accompanied by an increase in pancreaticoduodenal venous plasma glucagon concentrations (Fig. 3), corresponding to a true increase in pancreatic glucagon production. In these experiments, only negligible and inconstant variations in blood glucose or plasma amino nitrogen were observed, thus ruling out the possibility that the glucagon response to the fall in plasma FFA might have been mediated by a decrease in blood glucose or a rise in plasma amino acid concentration. The rise in pancreaticoduodenal plasma glucagon concentrations was confirmed by decreasing plasma FFA by means of propranolol

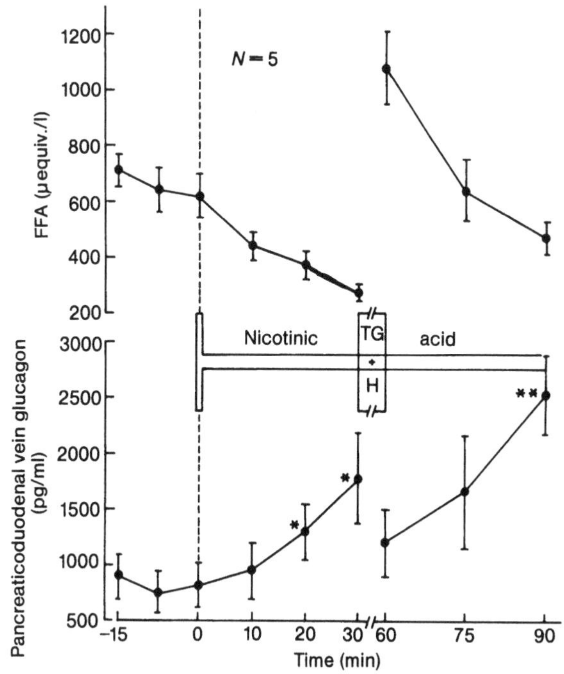


**Fig. 3.** Effect of changes in plasma FFA concentration induced by nicotinic acid infusion and by infusion (TG+H) of a triglyceride emulsion (Lipiphysan) supplemented with heparin (30–60 min) on plasma immunoreactive glucagon measured in the pancreaticoduodenal vein in anesthetized dogs. Statistically significant changes in IRG are indicated by an *asterisk*. After LUYCKX and LEFEBVRE (1970)

or 5-methylpyrazol-3-carboxylic acid administration (LUYCKX and LEFEBVRE 1971).

MÜLLER et al. (1976) investigated the effect of a rise in plasma FFA obtained by Lipomul plus intravenous heparin administration on the glucagon response to insulin-induced hypoglycemia in conscious dogs. In control animals, plasma glucagon rose from about 400 to 1 375 pg/ml after injection of 0.35 IU/kg body weight insulin. When plasma FFA were elevated, the rise in plasma glucagon was only from 200 to 600 pg/ml after injection of 0.8 IU/kg insulin. This important reduction in A-cell response to hypoglycemia was not statistically significant and therefore the authors concluded that raised FFA "failed" to suppress glucagon release. This negative conclusion is to be taken with caution owing to large experimental individual variations and the limited number of animals studied. Incidentally, it is worth mentioning that DL-$\beta$-hydroxybutyrate or glycerol infusions did not influence insulin-induced glucagon rise in the latter study.

## II. Rats

In 1976, we reported the results of a systematic investigation of the influence of nicotinic acid, an antilipolytic agent, on plasma glucagon in overnight fasted rats

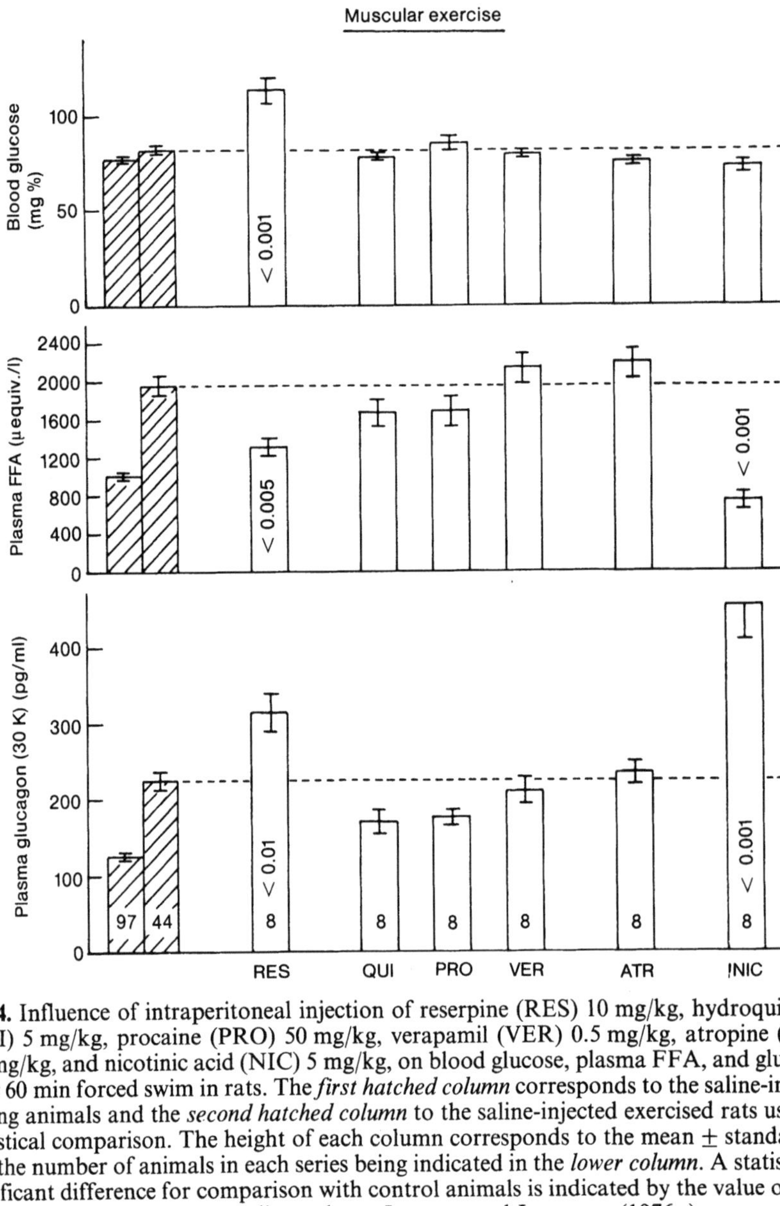


**Fig. 4.** Influence of intraperitoneal injection of reserpine (RES) 10 mg/kg, hydroquinidine (QUI) 5 mg/kg, procaine (PRO) 50 mg/kg, verapamil (VER) 0.5 mg/kg, atropine (ATR) 0.2 mg/kg, and nicotinic acid (NIC) 5 mg/kg, on blood glucose, plasma FFA, and glucagon after 60 min forced swim in rats. The *first hatched column* corresponds to the saline-injected resting animals and the *second hatched column* to the saline-injected exercised rats used for statistical comparison. The height of each column corresponds to the mean ± standard error, the number of animals in each series being indicated in the *lower column*. A statistically significant difference for comparison with control animals is indicated by the value of *P*, in or at the top of the corresponding column LUYCKX and LEFEBVRE (1976a)

submitted either to an insulin-induced hypoglycemia or to a standardized muscular excercise, namely a 60-min forced swim (LUYCKX and LEFEBVRE 1976a). We noticed that, with pretreatment with nicotinic acid, the elevation in plasma FFA normally seen during exercise did not occur, and that exercise-induced hyperglucagonemia was potentiated (Fig. 4). Just as insulin and nicotinic acid produced a fall in plasma FFA which was deeper than that seen following administration of insulin alone, the hyperglucagonemia induced by insulin and nicotinic acid was greater than that produced by insulin alone. These results provide further evidence that the stimulation of glucagon secretion by exercise or hypoglycemia can be modulated by the level of plasma FFA.

### III. Ducks

Glucagon seems to be of particular importance in the regulation of lipid metabolism in birds (see Chap. 19). Conversely, the group of MIAHLE (GROSS and MIAHLE 1974; LAURENT and MIAHLE 1978; FOLTZER and MIAHLE 1980) has demonstrated the important role of FFA in the regulation of glucagon secretion in the duck. Oleate infusion, inducing a rise in plasma FFA up to 2 m*M* was associated with a prompt and marked decline in plasma glucagon in normal ducks. In diabetic ducks, a physiologic increase in plasma FFA still induces a decrease in the A-cell secretion, although the suppressive effect seems slightly reduced when diabetes is present. Augmentations in plasma FFA within a physiologic range were also associated with a decrease in plasma glucagon levels in hypophysectomized animals (FOLTZER and MIAHLE 1980). The effectiveness of the FFA-induced glucagon suppression being maintained in the presence of very small basal concentrations of insulin argues in favor of a direct effect of FFA on the A-cell and supports the concept that "insulin might not be directly involved in the FFA-glucagon feedback in the duck" (LAURENT and MIAHLE 1978).

## D. Studies in Humans

### I. Normal Subjects

The sensitivity of plasma glucagon concentration to relatively small changes in plasma FFA, well within the physiologic range, was clearly demonstrated by GERICH et al. (1974). In that study, elevation of plasma FFA from a mean basal level of 478 ± 36 to 712 ± 55 μ*M* caused basal plasma glucagon to fall approximately 50%. Conversely, lowering of plasma FFA from a basal level of 520 ± 46 to 252 ± 41 μ*M* by nicotinic acid administration raised plasma immunoreactive glucagon (IRG) by 49%. The glucagon response to arginine was diminished following elevation of plasma FFA, but was not altered during nicotinic acid-induced fall in circulating FFA. In subsequent experiments reported by ANDREWS et al. (1975), modest elevations of plasma FFA up to 800 μ*M* caused by oral Lipomul – heparin failed to alter basal and hypoglycemia-induced or arginine-induced glucagon secretion. However, during Intralipid – heparin infusion, raising plasma FFA to approximately 1300 μ*M*, a highly significant suppression of circulating plasma glucagon was observed.

LUYCKX et al. (1978a) have compared, in healthy young women, the influence of insulin-induced hypoglycemia on plasma glucagon either in the basal state, after an overnight fast, or during a period where FFA were raised by infusion of a lipid emulsion supplemented by heparin. In the control test, the increase in plasma glucagon was maximum 30 and 45 min after insulin injection and averaged 130 pg/ml; the infusion of triglycerides and heparin which raised plasma FFA to about

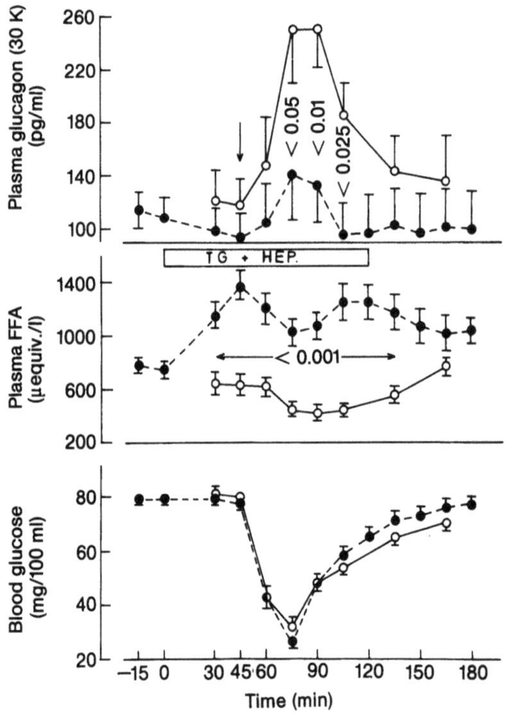


**Fig. 5.** Changes in blood glucose, plasma FFA, and glucagon concentrations following intravenous injection of insulin (0.1 IU/kg at time 45 min) in normal women. Twelve women (*open circles*) were in the basal state after an overnight fast, whereas ten others (*full circles*) were infused with a triglyceride emulsion supplemented with heparin (TG+H) for 45 min before insulin injection, Results are expressed as mean ± standard error

1 300 μ*M* decreased basal plasma glucagon levels and reduced, by about 70%, the glucagon response to hypoglycemia (Fig. 5). The relative importance of the two main metabolic substrates, glucose and free fatty acids, in the regulation of glucagon secretion in normal humans can be discussed on the basis of the data reported by GERICH et al. (1974) and QUABBE et al. (1977). In both studies, infusion of moderate amounts of glucose, increasing blood glucose up to 150-160 mg/dl, resulted in a significant *decrease* in circulating FFA, but nevertheless led to a significant *decline* in plasma IRG, thus suggesting that the influence of glucose predominates over that of FFA. Further studies using nicotinic acid infusion alone or with increasing rates of glucose infusion (QUABBE et al. 1977) reached a similar conclusion. Indeed, the glucagon increase during nicotinic acid-induced FFA depression was completely reversed by glucose, despite the persistance of low FFA plasma concentrations (Fig. 6). All these reports contrast with the negative findings of HICKS et al. (1977); these authors compared the arginine-induced glucagon rise in healthy volunteers infused with saline, nicotinic acid or Intralipid – heparin. After Intralipid – heparin infusion, plasma FFA reached extremely high values, 3.027 ± 0.184 m*M*; the plasma IRG response to intravenous arginine was unaffected by high or low levels of plasma FFA.

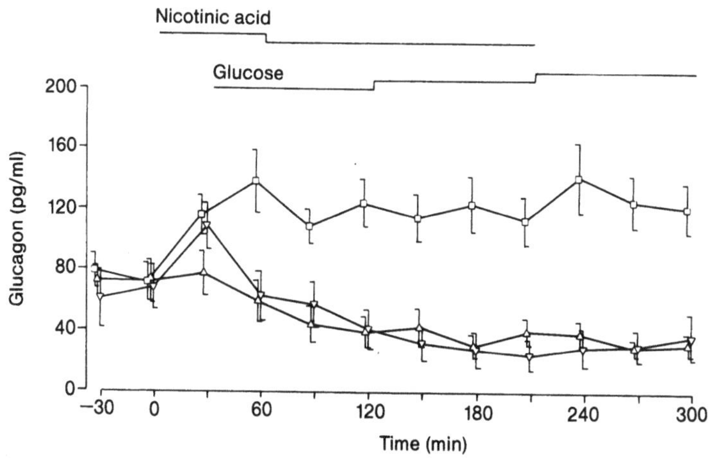


**Fig. 6.** Seven male volunteers were submitted to three tests in a randomized order. After a control period (−30 to 0 min), they received nicotinic acid alone (*squares*), glucose alone (*upward triangles*) or nicotinic acid and glucose (*downward triangles*). Nicotinic acid was infused intravenously at a rate of 1 g/h (0–60 min) and 0.5 g/h (60–210 min) (*squares and downward triangles*). Glucose alone (*upward triangles*) was infused at a rate of 12 g/h (30–120 min), 24 g/h (120–210 min) and 45 g/h (210–300 min). In the last series (*downward triangles*), the subject received the nicotinic acid infusion as well as a glucose infusion at the rates of 24 g/h (30–120 min), 45 g/h (120–210 min), and finally 60 g/h (210–300 min). After QUABBE et al. (1977)

QUABBE et al. (1983a) tested the influence of B-hydroxybutyrate (OHB) infusion either alone or during prevention of the ketone body-induced FFA depression by a concommitant lipid/heparin infusion. Glucagon increased slightly but significantly during the infusion of OHB alone, while it decreased slightly – although non significantly – when the lipid/heparin infusion was added. These changes were small in absolute terms. In this study, the glucagon rise in response to hypoglycemia occured slightly later and the total glucagon response was somewhat less when OHB was infused. QUABBE et al. (1983b) investigated the influence of oral administration of an adenosine derivative (N(6)-allyl-N(6) cyclohexyl-adenosine, BM 11.189), a drug cabable of decreasing plasma FFA. In that study, when FFA were depressed by BM 11.189 alone, no glucagon increased occured. However, the glucagon plasma concentration was decreased when BM 11.189 was ingested during an additional lipid/heparin infusion. These results further confirm the inhibitory effect of FFA elevation on glucagon secretion.

## II. Pregnant Women

Our group has investigated, in healthy pregnant women, the influence of an increase in circulating plasma FFA on glucagon plasma concentrations under basal conditions (LUYCKX et al. 1975). After an overnight fast, the Lipiphysan – heparin-induced increase in plasma FFA up to 2200 μ*M* was associated with a clear-cut decrease in plasma glucagon levels. Plasma glucagon remained suppressed until the FFA levels descended below 1200 μ*M* (Fig. 7). In a different investigation, we

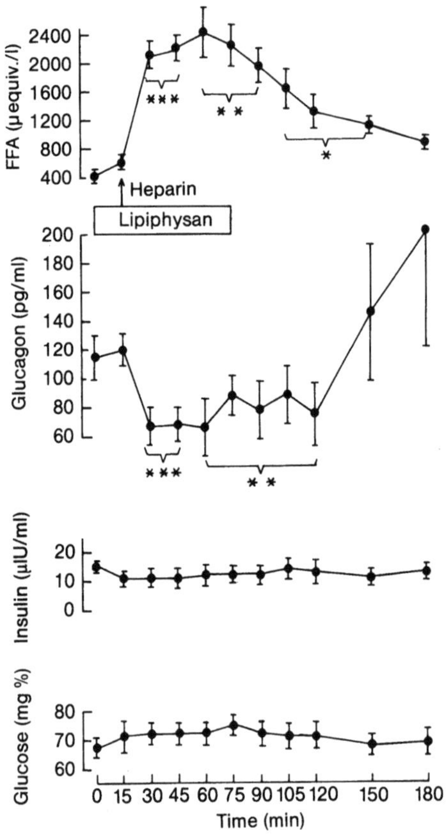


**Fig. 7.** Changes in plasma FFA, glucagon, insulin, and blood glucose induced in six pregnant women by the intravenous infusion of Lipiphysan, 1 ml/min (0–75 min) supplemented with 5 000 IU intravenous heparin 15 min after the perfusion was begun. Results are expressed as mean ± standard error. The Student's *t*-test for paired data obtained from each subject's time 0 measurement was used. *Triple*, *double* and *single asterisks* correspond to $P < 0.001$, 0.01, and 0.02, respectively.

studied the glucagon response to insulin-induced hypoglycemia in women during the last month of gestation; in these conditions, raising plasma FFA to about 1 500 μ*M* completely abolished to hypoglycemia-induced glucagon rise (Fig. 8; LUYCKX et al. 1978 a).

## III. Diabetes

Since insensitivity of the A-cell to glucose is a well-accepted characteristic of diabetes (see Chap. 44), it was of evident interest to investigate the influence of FFA on circulating levels of glucagon in diabetics. In 1976, we reported our results in seven insulin-dependent diabetics after administration of nicotinic acid alone or during a heparin – triglyceride infusion (LUYCKX and LEFEBVRE 1976 b). Infusion of nicotinic acid resulted in a decrease in plasma FFA associated with an increase

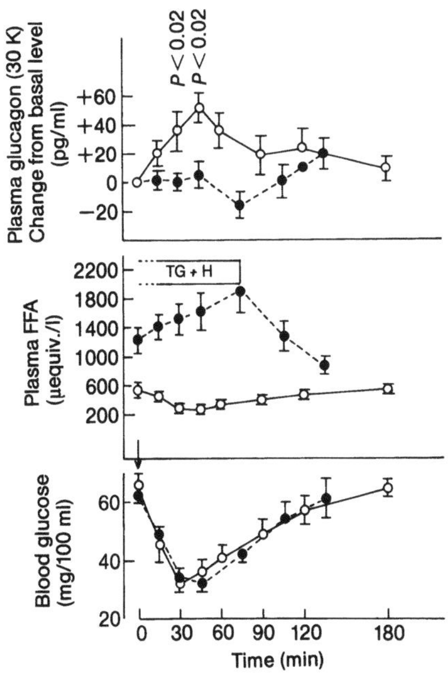


**Fig. 8.** Effect of intravenous insulin-induced hypoglycemia (0.3 IU/kg at time 0) on plasma glucagon in pregnant women. Six women (*open circles*) were in the basal state after an overnight fast whereas eleven others (*full circles*) were infused with a triglyceride emulsion supplemented with heparin (TG+H) for 45 min before insulin injection. Results are expressed as mean ± standard error. LUYCKX et al. (1978a)

in plasma cortisol, glucagon, and growth hormone. Administration of heparin and triglycerides provoked a sharp rise in plasma FFA levels which remained between 2500 and 1800 μ*M* during the period of infusion. During this period, the glucagon and growth hormone rises in response to nicotinic acid were significantly reduced. The cortisol increase was reduced, but the difference was not statistically significant (Fig. 9).

In nine other insulin-dependent diabetics, we confirmed that the nicotinic acid-induced fall in plasma FFA was associated with a significant rise in plasma glucagon, a phenomenon which was completely blocked by somatostatin infusion (Fig. 10; LUYCKX and LEFEBVRE 1976b). In the previously mentioned studies, plasma FFA fell from about 700 – 800 to 400 – 500 μ*M* within 60 min. GERICH et al. (1976) reported that, in insulin-dependent diabetics, glucagon was normally suppressed by a rise in FFA, this observation contrasting with the loss of the glucose-mediated regulation of A-cell function.

## IV. Hypertriglyceridemia

Hyperglucagonemia is seen in various cases of hyperlipoproteinemia (review in EATON 1977; TIENGO et al. 1978; see also Chaps. 20 and 46) and a gross and sustained rise in fasting plasma glucagon was previously reported in hyperlipidemic patients treated with β-piridylcarbinol, a nicotinic acid derivative (MARKS

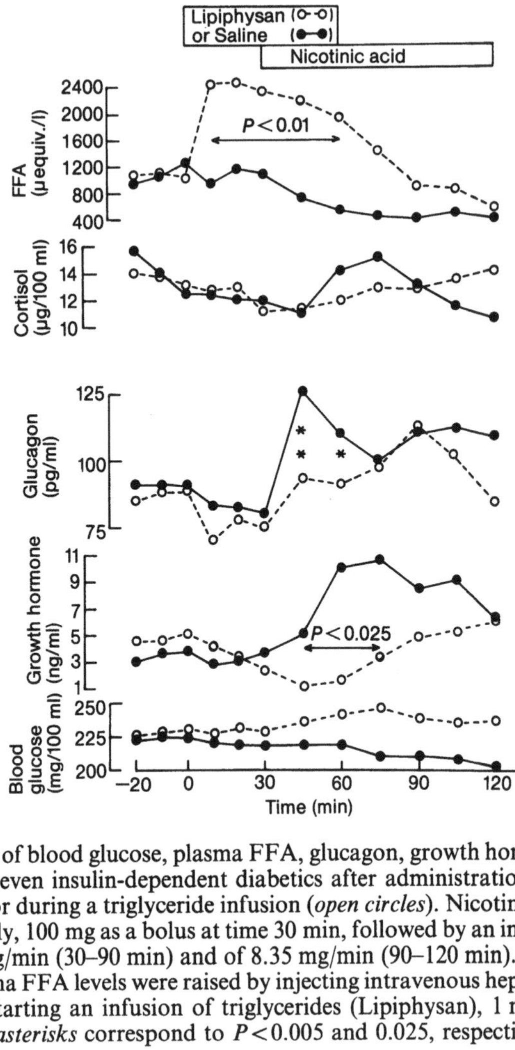


**Fig. 9.** Comparison of blood glucose, plasma FFA, glucagon, growth hormone, and cortisol concentrations in seven insulin-dependent diabetics after administration of nicotinic acid alone (*full circles*) or during a triglyceride infusion (*open circles*). Nicotinic acid was administered intravenously, 100 mg as a bolus at time 30 min, followed by an intravenous infusion at a rate of 16.7 mg/min (30–90 min) and of 8.35 mg/min (90–120 min). In one of the tests (*open circles*), plasma FFA levels were raised by injecting intravenous heparin (5 000 IU) immediately before starting an infusion of triglycerides (Lipiphysan), 1 ml/min (0–60 min). *Double* and *single asterisks* correspond to $P<0.005$ and 0.025, respectively

et al. 1971; MARKS 1976). The effects of clofibrate and halofenate on glucagon secretion are discussed in Chap. 32 and the possible role of glucagon in the pathogenesis of hyperlipoproteinemia is discussed in Chap. 46.

## E. The Modulating Role of Circulating FFA on Glucagon Secretion

The studies summarized here demonstrate that glucagon release from A-cells can be modulated by the availability of FFA in vitro as well as in vivo in animals and in humans. The mechanism (or mechanisms) of this regulatory role and its possible significance in the overall metabolic regulation will be briefly discussed.

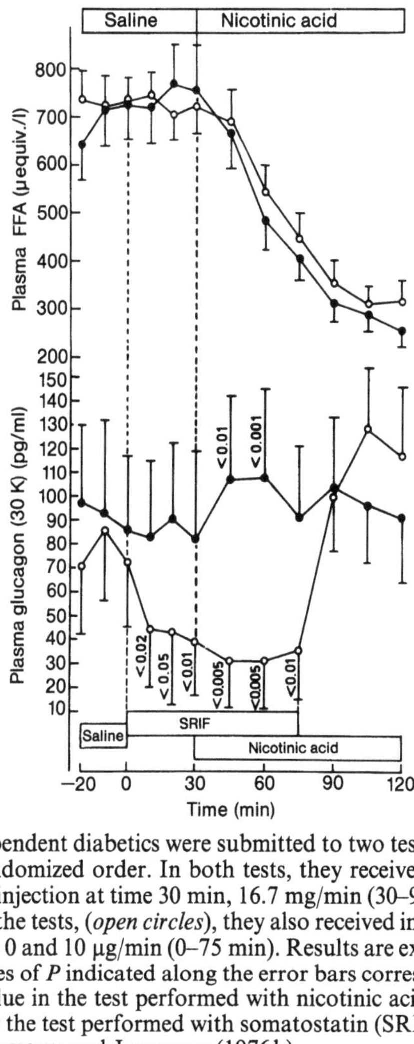


**Fig. 10.** Nine insulin-dependent diabetics were submitted to two tests performed on two different days and in a randomized order. In both tests, they received nicotinic acid, 100 mg as an acute intravenous injection at time 30 min, 16.7 mg/min (30–90 min) and 8.35 mg/min (90–120 min). In one of the tests, (*open circles*), they also received intravenous somatostatin, 250 μg as a bolus at time 0 and 10 μg/min (0–75 min). Results are expressed as mean ± standard error ($N = 9$). Values of $P$ indicated along the error bars correspond to paired comparison with the 30-min value in the test performed with nicotinic acid alone (*full circles*) and with the time 0 value for the test performed with somatostatin (SRIF) *and* nicotinic acid infusion (*open circles*)). LUYCKX and LEFEBVRE (1976b)

## I. Mechanism of Action of FFA on A-cells

EDWARDS et al. (1972) have established that guinea pig A-cell-rich pancreatic islets can oxidize fatty acids and that glucagon release was increased when oxidation of substrates such as glucose, fatty acids, and ketone bodies was impaired (EDWARDS et al. 1969, 1972; EDWARDS and TAYLOR 1970). This finding is compatible with the view that glucagon release increases when the energy level is lower in A-cells, and that, to a certain extent, glucose and FFA can replace each other as energy sources for these cells. Conversely, glucagon secretion is turned off by an increased intracellular provision of energy-yielding fuels, glucose or FFA. The intimate mechanism by which FFA availability suppresses glucagon release remains to be established (LEFEBVRE and LUYCKX 1978, 1979).

## II. Possible Significance of the Role of FFA in the Regulation of A-cell Secretion

Glucose and FFA are the main metabolic fuels on which tissues subsist. It is currently admitted that glucagon plays an important role in fuel homeostasis by mobilizing stored substrates when exogenous sources are lacking or when there is an increased demand of energy sources, as for example, during prolonged exercise (LUYCKX et al. 1978 b). Under these conditions, glucagon can mobilize hepatic glycogen stores and activate both liver gluconeogenesis and adipose tissue lipolysis (see detailed discussion in Chaps. 14, 15, 16, and 19). One could argue that the role of FFA in regulating glucagon release is in fact minor or negligible since, under certain circumstances, like total fast or prolonged muscular exercise, a parallel increase in circulating glucagon *and* FFA is simultaneously observed. A possible interpretation is that, under such circumstances, raised FFA plasma levels prevent an exagerated glucagon rise and thereby restrict liver glycogenolysis and neoglucogenesis. This would explain that, during the fourth hour of prolonged exercise in overnight fasted healthy subjects, lipid oxidation contributes up to 80% of the energy supply (PIRNAY et al. 1977). Our observation that administration of nicotinic acid to exercising rats abolished the rise in plasma FFA and markedly potentiated the glucagon rise (LUYCKX and LEFEBVRE 1976 a) is compatible with this view. Finally, the available results support the *general conclusion* that: (1) the role of glucose predominates over that of FFA in the regulation of glucagon secretion in normal human subjects; and (2) that the regulatory role of FFA (in contrast to that of glucose) is maintained in insulin-dependent diabetics.

## References

Andrews SS, Lopez SA, Blackard WG (1975) Effect of lipids on glucagon secretion in man. Metabolism 24:35–44

Campillo JE, Luyckx AS, Lefèbvre PJ (1979) Effect of oleic acid on arginine-induced glucagon secretion by the isolated perfused rat pancreas. Acta Diabetol Lat 16:287–293

Campillo JE, Luyckx AS, Lefèbvre PJ (1982) Effect of oleic and octanoic acids on glucagon and insulin secretion by the isolated perfused rat pancreas. Horm Metab Res 14:499

Eaton RP (1977) Glucagon and lipoprotein regulation in man. In: Foà PP, Bajaj JS, Foà NL (eds) Glucagon: its role in physiology and clinical medicine. Springer Berlin Heidelberg New York, pp 533–550

Edwards JC, Taylor KW (1970) Fatty acids and the release of glucagon from the guinea-pig islets of Langerhans incubated in vitro. Biochim Biophys Acta 215:310–315

Edwards JC, Howell SL, Taylor KW (1969) Fatty acids as regulators of glucagon secretion. Nature 224:808–809

Edwards JC, Hellerström C, Petersson B, Taylor KW (1972) Oxidation of glucose and fatty acids in normal and in $A_2$-cell rich pancreatic islets isolated from guinea-pigs. Diabetologia 8:93–98

Foltzer C, Miahle P (1980) Pituitary and adrenal control of pancreatic endocrine function in the duck. III. Effects of glucose, oleic, acid, arginine, insulin and glucagon infusions in hypophysectomized or normal ducks. Diabete Metab 6:257–263

Gerich JE, Langlois M, Schneider V, Karam JH, Noacca C (1974) Effect of alterations of plasma free fatty acid levels on pancreatic glucagon secretion in man. J Clin Invest 53:1284–1289

Gerich JE, Langlois M, Noacca C, Lorenzi M, Karam JH, Forsham P (1976) Comparison of the suppressive effects of elevated plasma glucose and free fatty acid levels on glucagon secretion in normal and insulin dependent diabetic subjects. Evidence for selective alpha-cell insensitivity to glucose in diabetes mellitus. J Clin Invest 58:320–325

Gross R, Miahle P (1974) Free fatty acid-glucagon feedback mechanism. Diabetologia 10:277–283
Hicks BH, Taylor CI, Vij SK, Pek S, Knopf RF, Floyd JC, Fajans SS (1977) Effects of changes in plasma levels of free fatty acids on plasma glucagon, insulin and growth hormone in man. Metabolism 26:1011–1023
Laurent F, Miahle P (1978) Effect of free fatty acids and aminoacids on glucagon and insulin secretions in normal and in diabetic ducks. Diabetologia 15:313–321
Lefèbvre PJ, Luyckx AS (1978) Factors controlling glucagon secretion. In: Essman V (ed) Regulatory mechanisms of carbohydrate metabolism. Pergamon, Oxford New York, pp 221–226
Lefèbvre PJ, Luyckx AS (1979) Glucagon: control of secretion and possible role in lipoprotein metabolism. In: Hessel LW, Krans HMJ (eds) Lipoprotein metabolism and endocrine regulations. Elsevier North-Holland Biomedical, Amsterdam Oxford New York, pp 45–52
Luyckx A (1975) Sécrétion de l'insuline et du glucagon. Etude clinique et biologique. Masson, Paris
Luyckx AS, Lefèbvre PJ (1970) Arguments for a regulation of pancreatic glucagon secretion by circulating plasma free fatty acids. Proc Soc Exp Biol 133:524–528
Luyckx AS, Lefèbvre (1971) Facteurs influençant la sécrétion de glucagon pancréatique et extra-pancréatique. In: Austoni M, Scandellari C, Federspil G, Trisotto A (eds) Current topics on glucagon. Cedam, Padova, pp 113–128
Luyckx AS, Lefèbvre PJ (1972) Changes in insulin and glucagon secretion related to concentrations of metabolic substrates in the isolated perfused rat pancreas. INSERM Colloq 99–127
Luyckx AS, Lefèbvre PJ (1974) The role of energy substrates in controlling glucagon secretion. Experimental studies. In: Malaisse W, Pirart J (eds) Diabetes. Proceedings of the VIIIth Congress of the IDF, Brussels, 1973. Excerpta Medica, Amsterdam London New York, pp 190–202
Luyckx AS, Lefèbvre P (1976a) Pharmacological compounds affecting plasma glucagon levels in rats. Biochem Pharmacol 25:2703–2708
Luyckx AS, Lefèbvre PJ (1976b) Effect of somatostatin on metabolic and hormonal changes induced by nicotinic acid in insulin-dependent diabetics. Diabetologia 12:447–453
Luyckx AS, Gérard J, Gaspard U, Lefèbvre PJ (1975) Plasma glucagon levels in normal women during pregnancy. Diabetologia 11:549–554
Luyckx AS, Gaspard U, Lefèbvre PJ (1978a) Influence of elevated plasma free fatty acids on the glucagon response to hypoglycemia in normal and in pregnant women. Metabolism 27:1033–1040
Luyckx AS, Pirnay F, Lefèbvre PJ (1978b) Effect of glucose on plasma glucagon and free fatty acids during prolonged exercise. Eur J Appl Physiol 39:53–61
Madison LL, Seyffert WA, Unger RH, Barker B (1968) Effect of plasma free fatty acids on plasma glucagon and serum insulin concentration. Metabolism 17:301–304
Marks V (1976) Glucagon and lipid metabolism in man. Postgrad Med J 49:615–619
Marks V, Frizel D, Twycross RG, Buchanan KD (1971) Effect of β-pyridylcarbinol on glucose tolerance, plasma glucagon, insulin and growth hormone in man. In: Gey KF, Carlson LA (eds) Metabolic effects of nicotinic acid and its derivatives? Huber, Bern, pp 961–976
Müller WA, Aoki TT, Flatt JP, Blackburn GL, Egdahl RH, Cahill GF JR (1976) Effect of β-hydroxybutyrate, glycerol and free fatty acid infusions on glucagon and epinephrine secretion in dogs during acute hypoglycemia. Metabolism 25:1077–1086
Pirnay F, Lacroix M, Mosora F, Luyckx A, Lefèbvre P (1977) Glucose oxidation during prolonged exercise evaluated with naturally labeled $^{13}C$-glucose. J Appl Physiol 43:258–261
Quabbe HJ, Ramek W, Luyckx AS (1977) Growth hormone, glucagon and insulin response to depression of plasma free fatty acids and the effect of glucose infusion. J Clin Endocrinol 44:383–391
Quabbe HG, Trompke M, Luyckx AS (1983a) Influence of ketone body infusion on plasma growth hormone and glucagon in man. J Clin Endocrinol Metab (in press)

Quabbe HJ, Luyckx AS, L'age M, Schwarz C (1983 b) Growth hormone, cortisol and glucagon concentrations during plasma free fatty acid depression. Different effects of nicotinic acid and an adenosine derivative (BM 11.189). J Clin Endocrinol Metab (in press)
Seyffert WA, Madison LL (1967) Physiologic effect of metabolic fuels on carbohydrate metabolism. I. Acute effect of elevation of plasma free fatty acids on hepatic glucose output, peripheral glucose utilization, serum insulin and plasma glucagon levels. Diabetes 16:765–776
Tasaka Y, Sekine M, Yoshida Y, Hirata Y (1976) Effects of nicotinic acid on the plasma pancreatic glucagon, insulin and free fatty acids in diabetic subjects. Horm Metab Res 8:489
Tiengo A, Nosadini R, Fedele D, Meneghel A, Valerio A, Crepaldi G (1978) The role of glucagon and insulin in endogenous human and rat hyperlipemia (genetic or acquired). In: Crepaldi G, Lefèbvre PJ, Alberti KGMM (eds) Diabetes, obesity and hyperlipidemia. Academic, London New York, pp 21–28

CHAPTER 26

# Ions in the Control of Glucagon Release

V. LECLERCQ-MEYER and W. J. MALAISSE

## A. Introduction

Despite a growing literature, the role of ions in the control of glucagon release remains a controversial subject. Indeed, for most ions so far investigated, both positive and negative influences upon glucagon release have been reported. In our view, one of the reasons which could explain some of these contradictory findings is that ion-induced changes in glucagon release might depend critically on the nutritional environment offered to the A-cell. For instance, it should be remembered that glucose and arginine, i.e., two widely used stimuli for the endocrine pancreas, usually modulate the secretory activity of the A-cell in opposite directions, whereas in the B-cell they may act synergistically. It is thus conceivable that the ionic control of glucagon release differs in relation to the functional response evoked by distinct stimuli. However, and although alternative hypotheses will be duly taken into account, it is with such a scheme in mind that we have undertaken the present chapter.

Among the ions for which experimental data are available, calcium is the one which has been studied the most extensively. Therefore, the major part of this chapter will be devoted to that cation, which apparently plays a versatile role in the regulation of glucagon release: an inhibitory role, a stimulatory role, and possibly also a role in the recognition of externally applied stimuli. Other divalent cations that have been studied include magnesium, manganese, and barium. Although still limited, data are also available for monovalent cations such as potassium and ammonium. As to the anions, to our knowledge, only phosphate and bicarbonate have as yet been studied.

## B. Calcium and Glucagon Release

### I. The Inhibitory Role of Calcium

That an inhibitory role may be ascribed to calcium in the regulation of glucagon release is supported by several findings which, as a rule, have been recorded in vitro, in experimental conditions where glucose was usually the sole exogenous substrate made available to the A-cell.

#### 1. Experimental Data

a) Enhancement of Glucagon Release by Acute Calcium Deprivation

Such a phenomenon has been amply documented using incubated pieces of pancreas obtained from duct-ligated rats (LECLERCQ-MEYER et al. 1973), monolayer cultures of newborn rat pancreas (WOLLHEIM et al. 1976b, 1977; KAWAZU et al. 1981), isolated islets of Langerhans from guinea pigs (EDWARDS 1973; EDWARDS and HOWELL 1973), rats (ASHBY and SPEAKE 1975; CARPENTIER et al. 1977, 1980), or a human islet cell adenoma (BONE et al. 1977), and the isolated perfused rat pancreas (LECLERCQ-MEYER et al. 1975, 1976a, 1977a; OTSUKI et al. 1980). In the latter dynamic model, it was shown that the pattern of glucagon release seen upon calcium deprivation differed in relation to the environmental concentration of glucose. Thus, at low glucose levels ($\leqq 3.3$ m*M*), a sharp and short-lived peak of glucagon secretion was observed (LECLERCQ-MEYER et al. 1975, 1976a, 1977a; LUNDQUIST et al. 1976b; OTSUKI et al. 1980). With increasing glucose concentrations (up to 16.6 m*M*), this early response decreased in amplitude, whereas there was a progressive development of a secondary phase of glucagon release (LECLERCQ-MEYER et al. 1975, 1976a, 1977a). This late phase of glucagon release, which has been consistently seen in our laboratory, was, however, not evident in other studies using a similar perfused rat pancreas preparation (LUNDQUIST et al. 1976b) or the isolated perfused canine pancreas (IVERSEN and HERMANSEN 1977). Whether these negative results reflect differences in experimental design or animal species employed remains to be elucidated. Nevertheless, it should be stressed that, even though calcium deprivation may not always result in a clear stimulation of glucagon release, the glucagon secretory rates are never reduced to zero levels. Thus, a nearly normal glucagon output was recorded in the presence of low glucose concentrations ($\leqq 3.3$ m*M*) with the perfused rat pancreas (GERICH et al. 1974; LECLERCQ-MEYER et al. 1975, 1976a, 1977a; LUNDQUIST et al. 1976a, b; GRODSKY et al. 1977a; OTSUKI et al. 1980) and in the presence of higher (11.1 m*M*) glucose concentrations with the perfused canine pancreas (IVERSEN and HERMANSEN 1977).

### b) Inhibition of Glucagon Release by Increasing Extracellular Calcium

The reintroduction of calcium following a period of calcium deprivation abruptly inhibited the secretion of glucagon in the perfused rat pancreas (LECLERCQ-MEYER et al. 1976a, 1977a). Evidently, this inhibition, which was originally interpreted as being indicative of the reversibility of the secretory response to calcium deprivation, could just as well be taken as evidence for an inhibitory action of calcium in the regulation of glucagon release. Furthermore, upon staircase elevations of the concentration of calcium in the perfusate, an inhibition of glucagon release was also occasionally observed by others using the perfused rat pancreas preparation (LUNDQUIST et al. 1976a, b; GRODSKY et al. 1977a). Incidentally, these authors acknowledged the complex mixed stimulation-inhibition nature of the calcium effects in their experiments, as they were unable to generate a simple glucagon dose-response curve in relation to staircase elevations of the extracellular concentration of calcium. Indeed, such a failure is in marked contrast with the dose-response relationship which had previously been documented by one of these authors in the case of insulin release (GRODSKY 1972).

### c) Inhibition of A-cell Responses by Elevation of Calcium Levels

A small elevation of the extracellular concentration of calcium (from 1 to 2.5 m*M*) was shown to inhibit the stimulation of glucagon release in response to cyclic AMP and the calcium ionophore A23187 (WOLLHEIM et al. 1976a, b).

## 2. Possible Mechanisms of Action

### a) Influx of Calcium Into the A-cell

The possible participation of calcium inflow in the secretory response was studied using the calcium antagonist, verapamil. This drug is thought to impair the transport of calcium in many cells, including the pancreatic B-cell (FLECKENSTEIN 1971; MALAISSE et al. 1977). In the perfused rat pancreas exposed to a fixed concentration of glucose (3.3 or 16.6 m*M*), the infusion of verapamil had little influence on the established steady state of glucagon release, a slight enhancement being seen only at the lowest concentration of glucose (LECLERCQ-MEYER et al. 1978). Moreover, an elevation of the concentration of glucose (from 3.3 to 16.6 m*M*) inhibited the secretion of glucagon in the presence of verapamil as well as in the absence of the drug. Thus, it appears that the A-cell relies but little on inwardly transported calcium for the maintenance of its steady-state function. Such a conclusion would be in agreement with the observation that the secretion of glucagon is never abolished upon calcium deprivation when glucose is the sole exogenous substrate provided to the A-cell. The fact that ethyleneglycolaminoethyl tetraacetate (EGTA) was found to abolish such a residual glucagon release (LUNDQUIST et al. 1976a, b) does not necessarily invalidate the view that only a small influx of calcium is needed in order for the A-cell to maintain a sustained secretion when only glucose is present and that high levels of this cation have an inhibitory action on the secretory process involved in glucagon release.

### b) Cell-to-cell Contact in the Endocrine Pancreas

It might be underlined that much of the evidence for an inhibitory role of calcium in glucagon release has been gained from the indirect finding that an enhancement of glucagon release occurs upon calcium depletion. The withdrawal of calcium is known to cause an alteration in the stability of the cell membranes and uncoupling of cells (FLECKENSTEIN 1971). Certainly, an enlargement of the extracellular space during incubation in calcium-depleted media has been documented in isolated rat islets (CARPENTIER et al. 1977). Furthermore, it was reported that no exocytotic events could be evidenced when the secretion of glucagon was stimulated by low calcium media, at variance to the observation made in the case of arginine (CARPENTIER et al. 1977, 1980). Thus, it was argued that, in contrast to arginine administration, calcium deprivation represented an unphysiologic stimulus for the A-cell (CARPENTIER et al. 1977; WOLLHEIM et al. 1977). However, chromatographic studies performed on pancreatic effluents showed that only glucagon (molecular weight 3500 daltons), and not immunologically related peptides, was liberated during stimulation, whether by low calcium levels or arginine (Fig. 1; LECLERCQ-MEYER et al. 1981). We wish to add that, even if the enhancement of glucagon release upon calcium deprivation results from a destabilizing of the cell membrane, one is still left to find an explanation for the fact that such a procedure results either in a stimulation (glucagon) or an inhibition (insulin) of hormonal output. It is nevertheless conceivable that a disturbance in the normal cell-to-cell contact or coupling in the endocrine pancreas may cause some of the observed effects of calcium deprivation on glucagon release. Gap junctions may connect A- and B-cells (ORCI et al. 1975a).

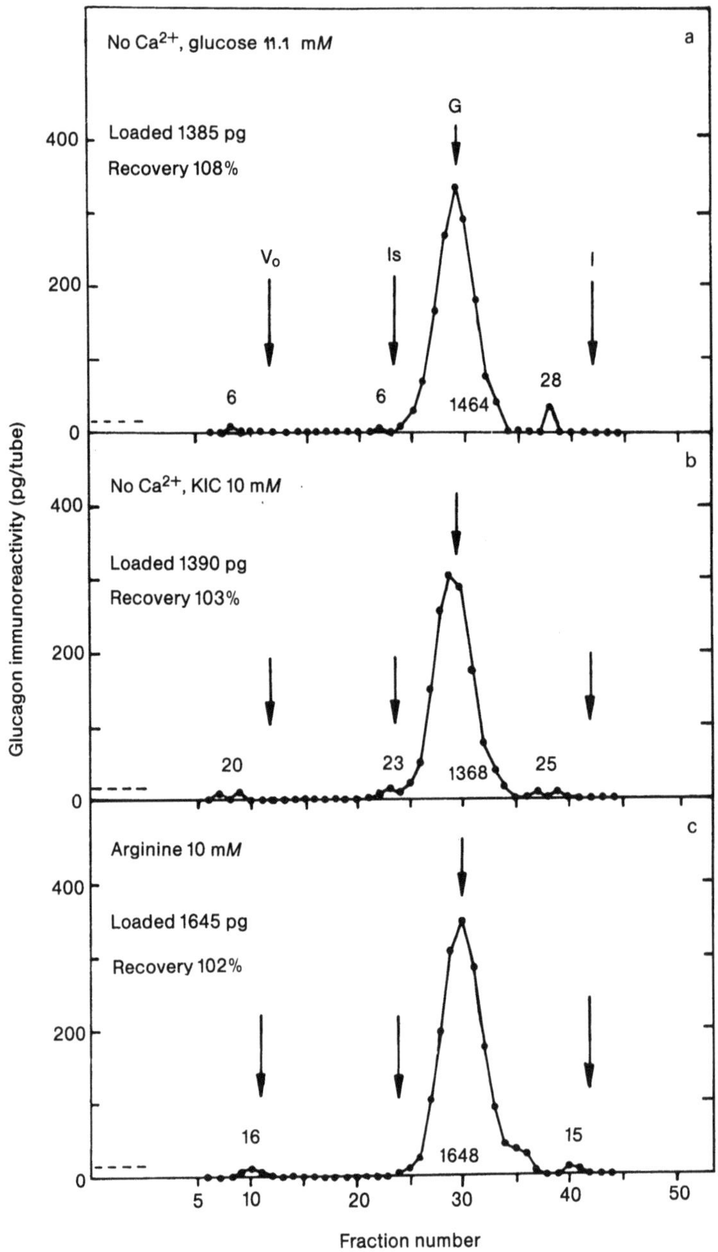


**Fig. 1 a–c.** Elution patterns of glucagon immunoreactivity on Biogel P-30 columns of the effluent collected from the perfused rat pancreas during a period of calcium deprivation in the presence of 11.1 m*M* glucose (**a**) or 10 m*M* 2-ketoisocaproate (KIC, **b**), and during administration of 10 m*M* arginine in the presence of 3.3 m*M* glucose at normal calcium concentration (**c**). Arrows indicate the void volume $V_0$ and the elution of radioactive tracers (*Is*, I-insulin$^{125}$; *G*, I-glucagon$^{125}$; *I*, iodine). *Dotted line* represents the limit of sensivity of the assay (per eluate fraction)

The factors which possibly participate in the intercellular regulation of glucagon release are unknown. They could include metabolic factors, ions, or other molecules sufficiently small to migrate through intercellular junctions (see Chap. 31).

c) Intracellular Metabolism of Glucose

In support of such a possibility is the finding that an enhancement of glucagon release also occurs upon calcium deprivation in the presence of 2-ketoisocaproate (LECLERCQ-MEYER et al. 1981). 2-ketoisocaproate, which is the first metabolic product of leucine, has been shown to be actively metabolized in rat islets and to influence the stimulus-secretion coupling in the B-cell (including the movement of ions) in a manner which paralleled the effects of glucose (HUTTON et al. 1980). Also in the A-cell, 2-ketoisocaproate exerts glucose-like inhibitory effects upon the release of glucagon (LECLERCQ-MEYER et al. 1979a). Moreover, the pattern of the glucagon response to calcium deprivation seen in the presence of 2-ketoisocaproate closely duplicated that observed in the presence of glucose (LECLERCQ-MEYER et al. 1981). The foregoing data, together with the fact that in A-cells, as in B-cells, metabolic fluxes are stimulated in the presence of elevated glucose levels (PETERSSON et al. 1970; EDWARDS et al. 1972; ÖSTENSON et al. 1977) make it conceivable that the inhibitory role of calcium upon glucagon release is somehow linked to the metabolism of exogenous substrates in the A-cells.

d) Adrenergic or Other Mechanisms

Pretreatment of rats with 6-hydroxydopamine abolished the spontaneous release of glucagon seen in the rat pancreas perfused with a calcium-depleted medium (LUNDQUIST et al. 1976a). On the basis of this observation, a participation of adrenergic neurons was suggested. A paracrine effect of insulin or somatostatin might also be postulated (ORCI and UNGER 1975b). It seems, however, premature to speculate extensively on these still hypothetical possibilities.

## II. The Positive Modulating Role of Calcium

That calcium may play a positive modulating role in glucagon release is supported by numerous observations, either based on the use of calcium ionophores (such as A23187) or performed in the presence of such secretagogues as arginine or a mixture of fumarate, glutamate, and pyruvate (FGP). The FGP mixture exerts an arginine-like stimulatory effect upon glucagon release by the perfused pancreas (LECLERCQ-MEYER et al. 1977b).

### 1. Experimental Data

a) Stimulation of Glucagon Release by A23187

A23187 has consistently been found to stimulate the release of glucagon. Such a stimulation has been observed in the presence of arginine or amino acids such as those routinely added to culture media. Thus, an A23187-induced increase in glucagon release was documented using perifused rat islets (ASHBY and SPEAKE 1975) and monolayer cultures of newborn rat pancreas (WOLLHEIM et al. 1976b, 1977;

FUJIMOTO and ENSINCK 1976). The stimulatory effects of A23187 were dependent on the presence of extracellular calcium (ASHBY and SPEAKE 1975; WOLLHEIM et al. 1976b).

b) Stimulation of Glucagon Release by Acute Elevation of Extracellular Calcium

A stimulatory effect of calcium in high concentration was documented both in the absence and presence of arginine or FGP. In the absence of these substrates, a short-lived peak of glucagon output is usually seen upon the reintroduction of calcium to the perfused rat pancreas (LECLERCQ-MEYER et al. 1976a, 1977a, 1979b, 1981; LUNDQUIST et al. 1976a, b; GRODSKY et al. 1977a). This short outburst was originally interpreted as representative of an "off response" to calcium deprivation (LECLERCQ-MEYER et al. 1976a, 1977a), but could just as well represent a direct calcium-induced glucagon release. The latter hypothesis may be substantiated by the observation that a stepwise increase in the extracellular concentration of calcium is able, although not in a staircase pattern, to elicit spikes of glucagon secretion (LUNDQUIST et al. 1976a, b; GRODSKY et al. 1977a). Moreover, supraphysiologic calcium levels (up to 30 m*M*) can evoke a large stimulation of glucagon release in monolayer cultures of newborn rat pancreas (WOLLHEIM et al. 1977).

c) Calcium Dependency of Arginine-Induced and FGP-Induced Glucagon Release

Arginine or FGP fail to induce glucagon release in the absence or extracellular calcium. The calcium dependency of arginine-induced or FGP-induced glucagon release is well documented. Thus, arginine or FGP are unable to induce glucagon release when calcium has not been added to perfusion media (GERICH et al. 1974; LECLERCQ-MEYER et al. 1976b, 1977a). In addition, a stepwise increase in the concentration of extracellular calcium evokes a progressively more prominent response to arginine in the perfused rat pancreas (CAMPILLO et al. 1978) or in culture medium supplemented with amino acids (FUJIMOTO and ENSINCK 1976). Moreover, an excess of calcium potentiates the late phase of arginine-induced or FGP-induced glucagon release (BHATENA et al. 1976; IVERSEN and HERMANSEN 1977; HERMANSEN et al. 1979; HERMANSEN 1980). Finally, an immediate inhibition of glucagon release occurs upon calcium deprivation in the pancreas exposed to arginine or FGP (GERICH et al. 1974; IVERSEN and HERMANSEN 1977; LECLERCQ-MEYER et al. 1977a, 1979b; HERMANSEN 1980).

d) Calcium Modulation of A-cell Secretory Response to Other Stimuli

In the perfused canine pancreas and in a medium supplemented with FGP, acetylcholine-induced glucagon release is amplified by increasing calcium levels (HERMANSEN and SCHWARTZ 1979). Furthermore, the inhibitory effect of somatostatin was reported to be overcome by increased calcium concentrations in the canine pancreas (IVERSEN and HERMANSEN 1980), although this was not the case in the perfused rat pancreas exposed to either glucose or arginine (GRODSKY et al. 1977a).

### 2. Possible Mechanisms of Action

Studies performed with the calcium antagonist, verapamil suggest that inwardly transported calcium is very likely to participate in the positive regulation of glucagon release, in the absence as well as in the presence of arginine or FGP. Thus, in the absence of arginine or FGP, the stimulation of glucagon release normally seen upon the decrease in the concentration of glucose (from 16.6 to 3.3 m*M*) in the perfused rat pancreas was markedly impaired by the drug (LECLERCQ-MEYER et al. 1978). In the case of arginine-induced or FGP-induced glucagon release, the stimulation induced by these secretagogues was markedly impaired by verapamil, both at low and high glucose levels (HERMANSEN and IVERSEN 1977; LECLERCQ-MEYER et al. 1978, 1979b). An increase in the extracellular concentration of calcium was shown to overcome the inhibitory effect of verapamil on FGP-induced glucagon release (HERMANSEN and IVERSEN 1977). Moreover, when verapamil is present throughout the experiments, arginine fails to evoke an enhancement of glucagon release (LECLERCQ-MEYER et al. 1978). The calcium dependency of glucagon release is particularly evident for the late phase of secretion induced by arginine or FGP. In contrast, the first phase seems to rely less than the late phase on inwardly transported calcium (CAMPILLO et al. 1978; LECLERCQ-MEYER et al. 1978). Altogether, the calcium dependency of glucagon release is further supported by the observation that an increase in calcium uptake occurs in A-cells exposed to low glucose concentrations, in contrast to the B-cell, where stimulated uptake is seen at high levels of extracellular glucose (BERGGREN et al. 1979).

## III. The Recognition Role of Calcium

That calcium may be a prerequisite for the recognition of externally applied stimuli is supported by several observations. Thus, during an extended period of calcium deprivation, neither a rise nor a decrease in the concentration of glucose is able to modify the preexisting rates of glucagon release (LUNDQUIST et al. 1976b; LECLERCQ-MEYER et al. 1976a, 1977a; IVERSEN and HERMANSEN 1977). Likewise, the enhancement of glucagon release which occurs upon calcium deprivation in the presence of elevated glucose levels could mean that glucose is no longer recognized by the A-cell (GRODSKY et al. 1977a; LECLERCQ-MEYER et al. 1977a). In addition, several other stimuli have been reported to be dependent on the presence of extracellular calcium for the initiation of secretory response from the A-cell. They include cyclic AMP (WOLLHEIM et al. 1976b), acetylcholine (HERMANSEN and SCHWARTZ 1979), and arginine (GERICH et al. 1974; LECLERCQ-MEYER et al. 1976b). This might, however, not apply for all external stimuli since, for instance, the response of the A-cell to cholecystokinin was found to be greater in the absence than in the presence of calcium in the perfused rat pancreas (OTSUKI et al. 1980). However, a true difference between a recognition and positive regulator role for calcium in the A-cell might be difficult to evaluate.

## IV. The In Vivo Effects of Calcium

Contrasting effects of calcium have been characterized in vivo. Thus, on the one hand, the administration of calcium resulted in a fall in circulating glucagon in dogs (OHNEDA et al. 1974) and humans (STARKE et al. 1981). Prior infusion of calcium also inhibited the subsequent response of the A-cell to arginine and hypoglycemia in humans (STARKE et al. 1980). On the other hand, an increase in glucagon release was documented upon the administration of calcium in dogs (RÖJDMARK et al. 1979) and in a case of human islet cell adenoma (TIENGO et al. 1976). Finally, calcium was reported not to influence glucagon release in dogs (KUZUYA et al. 1974) and normal glucagon responses were observed in the presence of hypercalcemia in hyperthyroid subjects (KALKHOFF et al. 1976). As to the in vivo effects of the calcium antagonist, verapamil, it has been reported either to cause no change in the concentration of plasma glucagon in rats (LUYCKX and LEFEBVRE 1976) or to inhibit arginine-induced glucagon release in humans (GIUGLIANO et al. 1981).

## V. Conclusions

This chapter has demonstrated a complex and versatile role of calcium in glucagon release. It further suggests that metabolic events could be more tightly involved in the inhibitory role than in the positive regulatory role of calcium in the secretory process of the A-cell. The precise mechanisms involved in one or other of these actions of calcium remain to be fully assessed.

As to the in vivo effects of calcium, it seems difficult to predict the outcome of elevated or decreased calcium levels. Possibly, because there is ample provision of circulating nutrients, including amino acids in plasma, a stimulation of glucagon could be the logical response to calcium. However, in the whole organism, calcium may exert its effects in a direct as well as in an indirect manner. For instance, an indirect influence, thought to be mediated through the stimulation of calcitonin release was invoked to explain the inhibitory effect of calcium on glucagon secretion in humans (STARKE et al. 1981).

# C. Other Divalent Cations

## I. Magnesium

In vitro, magnesium seems to exert an overall inhibitory effect upon glucagon release, whether in the absence or presence of arginine or FGP (LECLERCQ-MEYER et al. 1973; HERMANSEN and IVERSEN 1978; HERMANSEN and SCHWARTZ 1979; IVERSEN and HERMANSEN 1980; WOLLHEIM et al. 1976b). Magnesium also inhibited the secretion of glucagon induced by acetylcholine (HERMANSEN and SCHWARTZ 1979). In agreement with these data, the omission of magnesium facilitated the release of glucagon (LECLERCQ-MEYER et al. 1973; IVERSEN and HERMANSEN 1980). Such a facilitation was potentiated by the simultaneous omission of calcium from the medium (LECLERCQ-MEYER et al. 1973). An interference with the calcium movement into the A-cell was postulated for the mechanism of action of this cation upon glucagon release (HERMANSEN and IVERSEN 1978; HERMANSEN and SCHWARTZ 1979).

In vivo, magnesium did not influence the secretion of glucagon in the dog (BÖTTGER et al. 1972).

### II. Manganese

In the canine pancreas perfused in the presence of FGP, manganese influenced the release of glucagon in a dual manner (HERMANSEN and IVERSEN 1978). Thus, the early effect of this cation was to inhibit, whereas the late effect was to stimulate the release of glucagon. Increased extracellular calcium concentrations antagonized the early inhibitory effect of manganese, suggesting that manganese, like magnesium, inhibits glucagon release by competitively blocking calcium influx into the A-cell. As to the late stimulatory effect of this cation, it was suggested to be related to a manganese-induced translocation of organelle-bound calcium.

### III. Miscellaneous Cations

There is little information on the influence of other divalent cations upon glucagon release. In vitro, barium induced a stimulation of glucagon release, which was ascribed to the direct calcium-like stimulatory properties of this cation (WOLLHEIM et al. 1976b, 1977). In contrast, nickel was found to provoke an inhibition of glucagon release, which was ascribed to an interaction of this cation with the plasma membrane (LALLE and TAMARIT 1980). Finally, cobalt chloride causes hyperglucagonemia in the rat (LOCHNER et al. 1964; EATON 1973). The significance of these findings needs further evaluation.

## D. Monovalent Cations

### I. Potassium

Like some of the divalent cations, potassium appears to exert a dual action on the secretory activity of the A-cell. Thus, in the absence of glucose and arginine, potassium stimulated the release of glucagon from the perfused rat pancreas (GRODSKY et al. 1977b; EPSTEIN et al. 1978). This stimulatory effect of potassium was attributed to the depolarizing properties of this monovalent cation, suggesting that depolarization is a positive signal for the A-cell as well as for the B-cell or other secretory cells. A potassium-induced increase in glucagon release was also reported for low-insulin secreting clones derived from a rat islet cell tumor (BHATENA et al. 1980). In contrast, potassium was found to inhibit the secretion of glucagon in the parent cell line derived from that tumor (BHATENA et al. 1980). An inhibitory effect of potassium was also documented in the perfused rat pancreas when the medium was supplemented with arginine (BHATENA et al. 1976; GRODSKY et al. 1977b; EPSTEIN et al. 1978). Moreover, valinomycin, an antibiotic known for its potassium-ionophoretic properties and therefore able to cause hyperpolarization in cells (HENQUIN and MEISSNER 1978), blocked both potassium-induced and arginine-induced glucagon release in the perfused rat pancreas (EPSTEIN et al. 1978). In vivo, increased potassium levels were reported to stimulate the secretion of glucagon in dogs (SANTEUSIANO et al. 1973), although such a finding was not confirmed by others (DE FRONZO et al. 1978).

**Table 1.** Glucagon secretory rates (ng/min) from the perfused rat pancreas in response to $K^+$ depletion or ouabain infusions (Periods 2 and 5) in the absence or presence of arginine. Italicized values shown in parentheses represent secretory rates expressed in percent of the paired control value found prior to $K^+$ depletion or ouabain infusion (Periods 1 and 4, respectively)

| Line No | Experimental | Glucose (m*M*) | No arginine | | | | |
|---|---|---|---|---|---|---|---|
| | | | Period 1 (15–31 min) | Period 2 | | | Period 3 (55–61 min) |
| | | | | Whole (32–46 min) | Early phase (32–37 min) | Late part (37–46 min) | |
| 1 | Controls | 2.8 | 0.33±0.04 *(100)* | 0.28±0.11 | 0.28±0.10 *(83±22)* | 0.28±0.11 *(80±25)* | 0.20±0.09 *(56±20)* |
| 2 | No $K^+$ added | 2.8 | 0.42±0.12 *(100)* | 0.14±0.02 | 0.16±0.03 *(46±5)* | 0.13±0.2 *(39±5)* | 0.18±0.05 *(42±6)* |
| 3 | No $K^+$ added | 8.3 | 0.05±0.08 *(100)* | 0.04±0.01 | 0.04±0.01 *(73±16)* | 0.03±0.01 *(66±24)* | 0.05±0.01 *(91±32)* |
| 4 | Ouabain 0.1 m*M* | 2.8 | 0.30±0.08 *(100)* | 0.32±0.07 | 0.37±0.09 *(124±3)* | 0.29±0.06 *(97±5)* | 0.28±0.11 *(88±12)* |
| 5 | Ouabain 0.5 m*M* | 2.8 | 0.25±0.02 *(100)* | 0.43±0.18 | 0.63±0.30 *(245±95)* | 0.26±0.06 *(103±12)* | 0.15±0.02 *(59±1)* |
| 6 | Ouabain 1 m*M* | 2.8 | 0.28±0.05 *(100)* | 0.29±0.04 | 0.48±0.08 *(193±45)* | 0.19±0.02 *(74±11)* | 0.09±0.01 *(37±5)* |

**Table 1.** (Continued)

| Line No | Experimental | Glucose (m*M*) | Arginine 10 m*M* | | | | |
|---|---|---|---|---|---|---|---|
| | | | Period 4 (67–76 min) | Period 5 | | | Period 6 (100–105 min) |
| | | | | Whole (77–91 min) | Early part (77–80 min) | Late part (84–91 min) | |
| 1 | Controls | 2.8 | 2.34±0.21 (*100*) | 2.12±0.29 | 2.44±0.39 (*103*±*8*) | 1.93±0.27 (*82*±*5*) | 0.97±0.17 (*41*±*4*) |
| 2 | No $K^+$ added | 2.8 | 4.48±0.93 (*100*) | 4.26±0.87 | 4.78±0.86 (*111*±*7*) | 3.70±0.82 (*79*±*10*) | 1.63±0.31 (*38*±*6*) |
| 3 | No $K^+$ added | 8.3 | 0.46±0.06 (*100*) | 0.29±0.05 | 0.32±0.05 (*68*±*2*) | 0.23±0.04 (*49*±*3*) | 0.09±0.01 (*21*±*3*) |
| 4 | Ouabain 0.1 m*M* | 2.8 | 3.31±0.62 (*100*) | 2.22±0.30 | 4.06±0.67 (*125*±*14*) | 1.61±0.19 (*51*±*9*) | 1.28±0.20 (*39*±*3*) |
| 5 | Ouabain 0.5 m*M* | 2.8 | 3.11±0.21 (*100*) | 1.35±0.04 | 3.61±0.37 (*117*±*20*) | 0.52±0.02 (*17*±*1*) | 1.44±0.03 (*47*±*4*) |
| 6 | Ouabain 1 m*M* | 2.8 | 2.88±0.49 (*100*) | 1.59±0.14 | 4.06±0.6[a] (*148*±*15*) | 0.67±0.02 (*25*±*3*) | 1.77±0.43 (*65*±*13*) |

[a] Computed from min 76 to 79

## II. Sodium

A stimulation of glucagon release was observed upon decrease in the extracellular concentration of sodium (RABINOVITCH et al. 1974; BLACKARD et al. 1975). However, such a stimulation was attributable to the concomitant decrease in osmolarity rather than to the sodium ion per se. Using monolayer cultures of newborn rat pancreas, a calcium-dependent stimulatory effect of the sodium ionophore, veratridine was reported (KAWAZU et al. 1981).

## III. The Sodium-Potassium Pump

In the B-cell, the activation of an electrogenic sodium-potassium pump is thought to be associated with repolarization and the silent phases occurring after a burst of electrical activity induced by stimulating glucose levels (ATWATER and MEISSNER 1975; ATWATER et al. 1980; MEISSNER and PREISSLER 1980). The presence of a sodium-potassium pump in the $A_2$-cell is supported by recent findings indicating that both $K^+$ deprivation and ouabain exert marked effects upon glucagon release by the isolated perfused rat pancreas (LECLERCQ-MEYER et al. 1983). As illustrated in Table 1, such effects are not necessarily identical in response to $K^+$ deprivation and ouabain administration, respectively. Moreover, the influence upon glucagon release of a given environmental condition is not identical in the absence or presence or arginine and may differ during the early and late period of exposure to such a condition. Taken as a whole, these data strongly suggest a dual role for monovalent cations in glucagon release.

A stimulatory action of ouabain upon glucagon release has been shown by others (CHESNEY and SCHOFIELD 1969; KAWAZU et al. 1981). The existence of a sodium-potassium pump would be supported by the observation that diphenylhydantoin (an anticonvulsive drug which is thought to act through the activation of the $Na^+$, $K^+$-ATPase) inhibits arginine-induced glucagon release in the perfused rat pancreas (GERICH et al. 1972). In vivo, however, diphenylhydantoin failed to affect plasma glucagon in golden hamsters (TAMBURRANO et al. 1976).

## IV. Ammonium

There seems to be wide disagreement as to the effects of ammonium upon glucagon release. Thus, ammonium was reported to enhance the secretion of glucagon from perifused rat islets (EISENSTEIN and STRACK 1979) and in vivo in dogs (STROMBECK et al. 1978). In contrast, ammonium inhibited arginine-induced glucagon release in the perfused rat pancreas (LECLERCQ-MEYER et al. 1980) and impaired the response of rat islets to arginine (EISENSTEIN and STRACK 1979). Finally, this monovalent cation failed to affect the output of glucagon in the absence of arginine (LECLERCQ-MEYER et al. 1980) and, in vivo, during experimental hyperammonemia in rats and human subjects (SHERWIN et al. 1978; MULLOY and VISEK 1979, 1980).

# E. Anions

To our knowledge, phosphate and bicarbonate are the sole anions so far studied. Omission of phosphate augmented the second phase of arginine-induced glucagon release in the perfused rat pancreas (CAMPILLO et al. 1977). It was thus suggested that phosphate could play an inhibitory role in the secretory activity of the A-cell. At a pH close to physiologic levels, an increase in the concentration of bicarbonate from 25 to 40 m*M* may abolish the inhibitory effect of glucose (5.5 m*M*) upon basal glucagon release, but fails to affect the glucagon response to arginine (CAMPILLO et al. 1981).

## F. Conclusions

The data presently available in the literature suggest that both cations and anions are able to modulate the secretory activity of the A-cell. Many apparent contradictions remain to be solved, perhaps more particularly in vivo. Curiously, however, it appears that, in addition to calcium, other cations and anions may play a versatile role in the control of glucagon release. It is conceivable that such a versatile role of ions is somehow related to the fact that glucose and arginine may act in the A-cell by fundamentally different modalities. Such a tentative hypothesis presents the advantage of remaining open for future testing. Finally, the fact that, in the A-cell, calcium deposits are found on the inner leaflet of the plasma membrane rather than on the outer leaflet, as in the case of the B-cell, might prove to be of fundamental importance in the overall ionic regulation of glucagon release (RAVAZZOLA et al. 1976).

## References

Ashby JP, Speake RN (1975) Insulin and glucagon secretion from isolated islets of Langerhans. The effects of calcium ionophores. Biochem J 150:89–96

Atwater I, Meissner HP (1975) Electrogenic sodium pump in $\beta$-cells of islets of Langerhans. J Physiol (Lond) 247:56P–58P

Atwater I, Dawson CM, Scott A, Eddlestone G, Rojas E (1980) The nature of the oscillatory behaviour in electrical activity from pancreatic $\beta$-cell. Horm Metab Res [Suppl] 10:100–107

Berggren PO, Östenson CG, Petersson B, Hellman B (1979) Evidence for divergent glucose effects on calcium metabolism in pancreatic $\beta$- and $\alpha_2$-cells. Endocrinology 105:1463–1468

Bhatena SJ, Perrino PV, Voyles NR, Smith SS, Wilkins SD, Schally AV, Recant L (1976) Reversal of somatostatin inhibition of insulin and glucagon secretion. Diabetes 25:1031–1040

Bhatena SJ, Voyles NR, Oie HK, Smith SS, Gazdar AF, Recant L (1980) Glucagon secreting clones of rat islet cell tumor. Horm Metab Res 12:632–633

Blackard WG, Kikuchi M, Rabinovitch A, Renold AE (1975) An effect of hyposmolarity on insulin release in vitro. Am J Physiol 228:706–713

Böttger I, Faloona GR, Unger RH (1972) The effect of calcium and other salts upon the release of glucagon- like immunoreactivity (GLI) from the gut. J Clin Invest 51:831–836

Bone AJ, Gumperts RW, Howell SL, Sheldon J, Tellez-Yudilevich M, Tyhurst M, Whittaker PG, Zaheer F (1977) Regulation of insulin and glucagon secretion from a human islet cell adenoma. J Endocrinol 74:273–280

Campillo JE, Luyckx AS, Torres MD, Lefèbvre PJ (1977) Effect of phosphate omission on arginine-induced insulin and glucagon release by the isolated perfused rat pancreas. FEBS Lett 84:141–143

Campillo JE, Luyckx AS, Torres MD, Lefèbvre PJ (1978) Effect of various concentrations of calcium on arginine-induced insulin and glucagon release in vitro. Rev Esp Fisiol 34:191–198

Campillo JE, Luyckx AS, Lefèbvre PJ (1981) Effect of bicarbonate on the arginine-induced insulin and glucagon secretion in vitro. Horm Metab Res 13:552–556

Carpentier JL, Malaisse-Lagae F, Müller WA, Orci L (1977) Glucagon release from rat pancreatic islets. Combined morphological and functional approach. J Clin Invest 60:1174–1182
Carpentier JL, Malaisse-Lagae F, Müller WA, Orci L (1980) Morphology of glucagon release in vitro. An unsettled controversy. In: Andreani D, Lefèbvre PJ, Marks V (eds) Current views on hypoglycemia. Academic Press, London New York, pp 13–20
Chesney TM, Schofield JG (1969) Studies on the secretion of pancreatic glucagon. Diabetes 18:627–632
De Fronzo RA, Sherwin RS, Dillingham M, Hendler R, Tamborlane WV, Felig P (1978) Influence of basal insulin and glucagon secretion on potassium and sodium metabolism. J Clin Invest 61:472–479
Eaton RP (1973) Glucagon secretion and activity in the cobalt chloride-treated rat. Am J Physiol 225:67–72
Edwards JC (1973) A-cell metabolism and glucagon secretion. Postgrad Med J [suppl] 49:611–615
Edwards JC, Howell SL (1973) Effects of vinblastine and colchicine on the secretion of glucagon from isolated guinea-pig islets of Langerhans. FEBS Lett 30:89–92
Edwards JC, Hellerström C, Petersson B, Taylor KW (1972) Oxidation of glucose and fatty acids in normal and $A_2$-cell rich pancreatic islets from guinea-pigs. Diabetologia 8:93–98
Eisenstein AB, Strack IB (1979) Ammonium ion causes glucagon release by perifused rat islets. Fed Proc 38:554
Epstein G, Fanska R, Grodsky GM (1978) The effect of potassium and valinomycin on insulin and glucagon secretion in perfused rat pancreas. Endocrinology 103:2207–2215
Fleckenstein A (1971) Specific inhibitors and promoters of calcium action in the excitation-contraction coupling of heart muscle and their role in the prevention or production of myocardial lesions. In: Harris P, Opie L (eds) Calcium and the heart. Academic Press, London New York, pp 135–138
Fujimoto WY, Ensinck JW (1976) Somatostatin inhibition of insulin and glucagon secretion in rat islet culture: reversal by ionophore A23187. Endocrinology 98:259–262
Gerich JE, Charles MA, Levin SR, Forsham PH, Grodsky GM (1972) In vitro inhibition of pancreatic glucagon secretion by diphenlhydantoin. J Clin Endocrinol 35:823–824
Gerich JE, Frankel BJ, Fanska R, West L, Forsham PH, Grodsky GM (1974) Calcium dependency of glucagon secretion from the in vitro perfused rat pancreas. Endocrinology 94:1381–1385
Giugliano D, Gentile S, Verza M, Passariello N, Giannett G, Varricchio M (1981) Modulation by verapamil of insulin and glucagon secretion in man. Acta Diabetol Lat 18:163–171
Grodsky GM (1972) A threshold distribution hypothesis for packet storage of insulin. II. Effect of calcium. Diabetes [Suppl 2] 21:584–593
Grodsky GM, Lundquist I, Fanska R, Pictet R (1977a) Interrelationship of calcium and somatostatin in the secretion of insulin and glucagon. In: Foà PP, Bajaj JS, Foà NL (eds) Glucagon: its role in physiology and medicine. Springer, Berlin Heidelberg New York, pp 215–230
Grodsky GM, Epstein GH, Fanska R, Karam JH (1977b) Pancreatic action of the sulfonylureas. Fed Proc 36:2714–2719
Henquin JC, Meissner HP (1978) Valinomycin inhibition of insulin release and alteration of the electrical properties of pancreatic islets. Biochim Biophys Acta 543:455–464
Hermansen K (1980) Duodenal contribution to pancreaticoduodenal vein islet hormones during stimulation of the canine pancreas with calcium. Diabetes 29:361–364
Hermansen K, Iversen J (1977) Effect of verapamil on pancreatic glucagon release from the isolated perfused canine pancreas. Scand J Clin Lab Invest 37:139–142
Hermansen K, Iversen J (1978) Dual action of $Mn^{++}$ upon the secretion of insulin and glucagon from the isolated perfused canine pancreas. Possible interactions with $Ca^{++}$. Diabetologia 15:475–479
Hermansen K, Schwartz TW (1979) The influence of calcium on the basal and acetylcholine stimulated secretion of pancreatic polypeptide. Endocrinology 105:1469–1474
Hermansen K, Christensen SE, Ørskov H (1979) Characterisation of somatostatin release from the pancreas. The role of calcium and acetylcholine. Diabetologia 16:261–266

Hutton JC, Sener A, Herchueltz A, Atwater I, Kawazu S, Boschero AC, Somers G, Devis G, Malaisse WJ (1980) Similarities in the stimulus-secretion coupling mechanisms of glucose- and 2-keto acid-induced insulin release. Endocrinology 106:203–219
Iversen J, Hermansen K (1977) Calcium, glucose and glucagon release. Diabetologia 13:297–303
Iversen J, Hermansen K (1980) Characterisation of the inhibitory effect of somatostatin upon insulin and glucagon release in the perfused canine pancreas. Evidence for interaction with calcium. Metabolism 29:151–160
Kalkhoff RK, Gossain VV, Matute ML, Wilson SD (1976) Plasma alpha-cell glucagon in primary hyperthyroidism. Metabolism 25:769–775
Kawazu S, Ikeuchi M, Kikuchi M, Kanazawa Y, Fujimoto WY, Kosaka K (1981) Dual effects of veratridine on glucagon and insulin secretion: dependence on extracellular and intracellular calcium. Diabetes 30:446–450
Kuzuya T, Kajinuma H, Ide T (1974) Effect of intrapancreatic injection of potassium and calcium on insulin and glucagon secretion in dogs. Diabetes 23:55–60
Lalle C, Tamarit J (1980) Glucagon secretion inhibited by nickel in mouse pancreatic islets. Horm Metab Res 12:127–128
Leclercq-Meyer V, Marchand J, Malaisse WJ (1973) The effect of calcium and magnesium on glucagon secretion. Endocrinology 93:1360–1370
Leclercq-Meyer V, Rebolledo O, Marchand J, Malaisse WJ (1975) Glucagon release: paradoxical stimulation during calcium deprivation. Science 189:897–899
Leclercq-Meyer V, Marchand J, Malaisse WJ (1976a) The role of calcium in glucagon release. Interactions between glucose and calcium. Diabetologia 12:531–538
Leclercq-Meyer V, Marchand J, Malaisse WJ (1976b) The role of calcium in glucagon release. Interactions between arginine and calcium. Horm Res 7:348–362
Leclercq-Meyer V, Marchand J, Malaisse WJ (1977a) The versatile role of calcium in glucagon release. In: Foà PP, Bajaj JS, Foà NL (eds) Glucagon: its role in physiology and clinical medicine. Springer Berlin Heidelberg New York, pp 185–213
Leclercq-Meyer V, Marchand J, Malaisse WJ (1977b) An arginine-like effect of the "fumarate+glutamate+pyruvate" mixture on glucagon release. Life Sci 20:1193–1198
Leclercq-Meyer V, Marchand J, Malaisse WJ (1978) The role of calcium in glucagon release. Studies with Verapamil. Diabetes 27:996–1004
Leclercq-Meyer V, Marchand J, Leclercq R, Malaisse WJ (1979a) Interactions of α-ketoisocaproate, glucose and arginine in the secretion of glucagon and insulin from the perfused rat pancreas. Diabetologia 17:121–126
Leclercq-Meyer V, Marchand J, Malaisse WJ (1979b) Calcium dependency of glucagon release: its modulation by nutritional factors. Am J Physiol 236:E98–E104
Leclercq-Meyer V, Marchand J, Malaisse WJ (1980) Evidence for a limited role of NAD(P)H in the nutritional regulation of glucagon release: Studies with menadione and $NH_4Cl$. Acta Diabet Lat 17:23–32
Leclercq-Meyer V, Marchand J, Leclercq R, Malaisse WJ (1981) Calcium deprivation enhances glucagon release in the presence of 2-ketoisocaproate. Endocrinology 108:2093–2097
Leclercq-Meyer V, Marchand J, Malaisse WJ (1983) Effect of $K^+$ deprivation and ouabain upon glucagon release. Endocrinology (in press)
Lochner J, Eisentraut AM, Unger RH (1964) The effect of $CoCl_2$ on glucagon levels in plasma and pancreas of the rat. Metabolism 13:868–873
Lundquist I, Fanska R, Grodsky GM (1976a) Direct calcium-stimulated release of glucagon from the isolated perfused rat pancreas and the effect of chemical sympathectomy. Endocrinology 98:815–818
Lundquist I, Fanska R, Grodsky GM (1976b) Interaction of calcium and glucose on glucagon secretion. Endocrinology 99:1304–1312
Luyckx AS, Lefèbvre PJ (1976) Pharmacological compounds affecting plasma glucagon levels in rats. Biochem Pharmacol 25:2703–2708
Malaisse WJ, Herchuelz A, Levy A, Sener A (1977) Calcium antagonists and islet function. III. The possible site of action of verapamil. Biochem Pharmacol 26:735–740
Meissner HP, Preissler M (1980) Ionic mechanisms of the glucose-induced membrane potential changes in B-cells. Horm Metab Res [Suppl] 10:91–99

Mulloy AL, Visek WJ (1979) Arginine-induced secretion of insulin and glucagon in rats with experimental hyperammonemia. Horm Metab Res 11:527–528
Mulloy AL, Visek WJ (1980) Plasma insulin, glucagon and gut glucagon-like immunoreactivity during experimentally induced hyperammonemia in rats. Proc Soc Exp Biol Med 80:137–140
Östenson CG, Andersson A, Brolin SE, Petersson B, Hellerström C (1977) Effects of insulin on the glucagon release, glucose utilization and ATP content of the pancreatic $A_2$-cells of the guinea-pig. In: Foà PP, Bajaj JS, Foà NL (eds) Glucagon: its role in physiology and clinical medicine. Springer Berlin Heidelberg New York, pp 243–254
Ohneda A, Matsuda A, Horigome K, Ishii S, Yamagata S (1974) Effect of intrapancreatic administration of calcium upon glucagon secretion in dogs. Tohoku J Exp Med 113:301–311
Orci L, Unger RH (1975 b) Functional subdivision of islets of Langerhans and possible role of D-cells. Lancet 2:1243–1244
Orci L, Malaisse-Lagae F, Ravazzola M, Rouiller D, Renold AE, Perrelet A, Unger RH (1975 a) A morphological basis for intercellular communication between $\alpha$- and $\beta$-cells. J Clin Invest 56:1066–1070
Otsuki M, Sakamoto C, Yun H, Maeda M, Morita S, Baba S (1980) Effect of somatostatin and calcium deprivation on cholecystokinin or caerulein-induced insulin and glucagon release from the isolated perfused rat pancreas. Endocrinol Jpn 27:95–102
Petersson B, Hellerström C, Gunnarsson R (1970) Structure and metabolism of the pancreatic islets in streptozotocin treated guinea-pigs. Horm Metab Res 2:313–317
Rabinovitch A, Kikuchi M, Gutzeit AH, Blackard WG, Renold AE (1974) Hyposmolarity as a cause of insulin and glucagon "off-responses" in vitro. Horm Metab Res [Suppl] 5:44–49
Ravazzola M, Malaissa-Lagae F, Amherdt M, Perrelet A, Malaisse WJ, Orci L (1976) Patterns of calcium localization on pancreatic endocrine cells. J Cell Sci 27:107–117
Röjdmark S, Ishida T, Bloom G, Chou CY, Field J (1979) Effect of intraportal calcium infusion on insulin and glucagon secretion and hepatic glucose output in anesthetized dogs. Endocrinology 104:814–821
Santeusiano F, Faloona GR, Knockel JP, Unger RH (1973) Evidence for a role of endogenous insulin and glucagon in the regulation of potassium homeostasis. J Lab Clin Med 81:809–817
Sherwin RS, Fisher M, Besoff J, Snyder N, Hendler R, Conn HO, Felig P (1978) Hyperglucagonemia in cirrhosis: altered secretion and sensivity to glucagon. Gastroenterology 74:1224–1228
Starke A, Keck E, Cüppers HJ, Zaremba A, Zimmermann H, Krüskemper HL (1980) The effect of calcium on glucagon secretion stimulated by arginine and hypoglycemia. In: Andreani D, Lefèbvre PJ, Marks V (eds) Current views on hypoglycemia and glucagon. Academic Press, London New York, pp 487–488
Starke A, Keck E, Berger M, Zimmermann H (1981) Effect of calcium and calcitonin on circulating levels of glucagon and glucose in diabetes mellitus. Diabetologia 20:547–552
Strombeck DR, Rogers Q, Stern JS (1978) Effects of intravenous ammonia infusion on plasma levels of amino acids, glucagon and insulin in dogs. Gastroenterology 74:1165
Tamburrano S, Luyckx AS, Lefèbvre PJ (1976) Studies on the hyperglycemic effect of diphenylhydantoin in normal golden hamsters. Horm Metab Res [Suppl] 6:74–79
Tiengo A, Fedele D, Marchiori E, Nosadini R, Muggeo M (1976) Suppression and stimulation mechanisms controlling glucagon secretion in a case of islet cell tumor producing glucagon, insulin and gastrin. Diabetes 25:408–412
Wollheim CB, Blondel B, Renold AE, Sharp GWG (1976 a) Stimulatory and inhibitory effects of cyclic AMP on pancreatic glucagon release from monolayer cultures and the controlling role of calcium. Diabetologia: 12:269–277
Wollheim CB, Blondel B, Renold AE, Sharp GWG (1976 b) Calcium-induced glucagon release in monolayer culture of the endocrine pancreas. Studies with ionophore A 23187. Diabetologia 12:287–294
Wollheim CB, Blondel B, Renold AE, Sharp GWG (1977) Somatostatin inhibition of pancreatic glucagon release from monolayer cultures and interaction with calcium. Endocrinology 101:911–919

CHAPTER 27

# Cyclic Nucleotides in the Control of Glucagon Secretion

G. C. WEIR

## A. Introduction

One is probably on safe ground in thinking that cyclic AMP is somehow intimately involved in the mechanisms by which glucagon is secreted by the A-cells of the islets of Langerhans. This is largely based upon the assumption that A-cell secretory mechanisms are unlikely to be very different from those of other cells in which a role for cyclic AMP has been more firmly established (CATT and DUFAU 1981; SHARP 1979). It is also based upon an enlarging body of indirect evidence which will be discussed in the following sections, but unfortunately little direct biochemical support is presently available. A-cells along with D-cells and pancreatic polypeptide-containing (PP)-cells represent a minority of the islet cell population (ORCI and PERRELET 1981) and this has made it very difficult to carry out direct biochemical measurements. Propably the best approach has been to use the toxins streptozotocin and alloxan which destroy most of the majority B-cell population and leave small islets containing mostly A-cells (HOWELL et al. 1974; MATSCHINSKY et al. 1976a, b; ÖSTENSON 1979). Even then the proportion of A-cells is still 70% or less of the total number of residual cells, with most of the others probably being D-cells. Howell et al. (1974) used such islets from streptozotocin-treated guinea pigs and were able to measure adenylate cyclase activity, which was stimulated by epinephrine and potassium fluoride. They also found evidence for cyclic AMP-dependent protein kinase activity in these islets. This appears to be the only direct biochemical work published on cyclic AMP and the A-cell and the more indirect approaches are discussed throughout this chapter.

Most information about cyclic AMP's involvement in A-cell secretion comes from what can be termed a "black box" approach. Thus, a compound is exposed to A-cells either in vivo or with in vitro methods such as the isolated perfused pancreas, isolated incubating islets, or neonatal pancreatic monolayers. On the basis of changes in the secretion of glucagon, one can make educated guesses about whether the effect was mediated by cyclic AMP. There are a number of pitfalls in this approach, however. Among other concerns is the likelihood that islets are functioning as intricately coordinated microorgans, and that a change in glucagon secretion is sometimes mediated by various indirect pathways. For instance, there have been a number of suggestions and even some evidence that glucagon secretion can be influenced by the intraislet secretion of insulin and somatostatin, both of which have inhibitory influences upon the A-cell (SAMOLS et al. 1972; WEIR et al. 1976a; ORCI and UNGER 1975; HONEY et al. 1981; ASPLIN et al. 1981). Thus, a given compound could stimulate glucagon secretion and the effect might have been mediated indirectly through suppression of somatostatin secretion. In addition, depending upon the experimental system one is using, indirect effects through neural control of islet function could confuse the situation (PALMER and PORTE (1981).

Some efforts have been made to study A-cells which are not in an intraislet environment to avoid the complicating potential influences of somatostatin and insulin. There is a peculiarity of nature in that the stomachs of some mammals contain A-cells which are not intimately associated with B-cells (SRIKANT et al. 1977; BAETENS et al. 1976; LEFEBVRE and LUYCKX 1981). Results of studies of gastric glucagon secretion provide us with perhaps the best current model of how "pure" A-cells respond to various secretagogues (LEFEBVRE and LUYCKX 1977, 1981; YOSHIDA and KONDO 1980). One must maintain caution, however, because there is no guarantee that pancreatic and gastric A-cells are truly identical. Another model which can be used to learn more about a role for cyclic AMP is glucagon-secreting tumors (LEICHTER 1980). It is unclear, however, whether the process of neoplasia might lead to unique secretory behavior, and in addition these tumors frequently contain other cell types such as B-cells and D-cells (ORCI 1976; BODEN et al. 1977), and this could complicate any interpretation.

## B. Effects of Exogenous Cyclic AMP

Some workers have approached this problem by examining the effects of exogenous cyclic AMP upon glucagon secretion. A background assumption for such an experiment is that there is an intracellular cyclic AMP-dependent protein kinase inside the A-cell which, after being activated by exogenous cyclic AMP, would trigger the secretion mechanism. Indeed, stimulation of glucagon secretion in response to cyclic AMP or dibutyryl cyclic AMP has been found using the perfused dog pancreas (IVERSEN 1970), rat pancreas (WEIR et al. 1975), neonatal pancreatic pieces (JARROUSSE and ROSSELIN 1975a, b), and monolayers (WOLLHEIM et al. 1976), and with islets from streptozotocin-treated guinea pigs (HOWELL et al. 1974). Making the situation less clear, however, another group reported inhibition of glucagon secretion by dibutyryl cyclic AMP in the perfused rat pancreas (TOYOTA et al. 1975). It is disconcerting that we do not know more about how exogenous cyclic AMP acts, but these results at least suggest that activation of an A-cell cyclic AMP-dependent protein kinase leads to glucagon release.

## C. Effects of Phosphodiesterase Inhibitors

Another way to probe for cyclic AMP effects is to use phosphodiesterase inhibitors. Theophylline has been the most commonly used agent and most investigators have found either stimulation of basal glucagon secretion or an enhancement of arginine-stimulated release (CHESNEY and SCHOFIELD 1969; HOWELL et al. 1974; WEIR et al. 1974; FRANKEL et al. 1974; GERICH et al. 1975; GOTO et al. 1976). On the other hand, theophylline has been found to have an inhibitory effect upon early epinephrine-stimulated glucagon secretion (WEIR et al. 1974), and upon long-term secretion by neonatal rat pancreatic monolayer tissue (WOLLHEIM et al. 1976). It is difficult to interpret these inhibitory effects, but the more recent observations that phosphodiesterase inhibitors can stimulate somatostatin secretion (SCHAUDER et al. 1977; GERBER et al. 1981) could provide a clue. Thus, a stimulatory effect upon glucagon secretion could be obscured by somatostatin induced inhibition with perhaps an additional inhibitory influence from insulin (BUCHANAN and MAWHINNEY 1973; PAGLIARA et al. 1975; WEIR et al. 1976a). Aminophylline when given to humans was found to inhibit arginine-stimulated glucagon secretion (MARCO et al. 1972). To complicate matters further, theophylline is thought to be an antagonist for the putative P1 receptors (BURNSTOCK 1978). Other phosphodiesterase inhibitors such as papavarine, theobromine, and 3-isobutyl-1-methylxanthine do not seem to have these effects. It will be of interest to reinvestigate the effects of these agents when their mechanisms of action are more completely understood.

## D. Effects of Agents Thought to Act via Endogenous Cyclic AMP

Probably the most convincing support for cyclic AMP's role in glucagon secretion comes from studies employing agents which almost certainly act via adenylate cyclase activation (CATT and DUFAU 1981; HOWLETT et al. 1979). β-Adrenergic stimulation has been found in a number of studies to release glucagon (IVERSEN 1973; LUYCKX and LEFEBVRE 1973; GERICH et al. 1974; SAMOLS and WEIR 1979). This has been found either with isoproterenol alone or with epinephrine or norepinephrine combined with α-adrenergic blockade. α-Adrenergic stimulation has also been found to enhance glucagon secretion (HARVEY et al. 1974; SAMOLS and WEIR 1979; BAUM et al. 1979), but on the basis of studies in other tissues (EXTON 1981) it is assumed that this effect is not mediated by cyclic AMP, but by an as yet unidentified second messenger. There is recent evidence that cyclic AMP is involved in secretion by gastric A-cells because β-adrenergic stimulation has been shown to stimulate glucagon release in pancreatectomized dogs (YOSHIDA and KONDO 1980). In a patient with an A-cell-containing glucagonoma, it has been reported that epinephrine stimulates glucagon release (LEICHTER et al. 1975), but it is unclear whether this was exerted by α- or β-adrenergic mechanisms.

Other agents which are presumed to stimulate glucagon secretion via adenylate cyclase activation include vasoactive intestinal peptide (VIP) (SCHEBALIN et al. 1977; KANETO et al. 1977) and gastric inhibitory peptide (GIP) (RABINOWITCH and DUPRE 1974; FUJIMOTO et al. 1978), even though stimulatory effects by GIP are not always observed (IPP et al. 1979). These belong to a glucagon-like family of peptides so it is interesting that the A-cell may have receptors which respond to agents which are similar to glucagon. In view of the evidence that insulin seems capable of inhibiting the B-cell (IVERSEN and MILES 1971; LILJENQUIST et al. 1978), and a somatostatin analog can inhibit the D-cell (IPP et al. 1979), one can suspect that glucagon itself may eventually be shown to have an effect upon A-cells.

Another class of agents which may stimulate glucagon secretion via endogenous cyclic AMP includes several purine nucleotides and nucleosides. These may act on the purinergic P1 receptors (BURNSTOCK 1978). It has been shown in the perfused rat pancreas that adenosine, ADP, 2′-AMP, 3′-AMP, and 5′-AMP are all capable of stimulating glucagon secretion (WEIR et al. 1975). In addition, the in vivo administration of a 5′-substituted adenosine analog has been found to stimulate glucagon secretion, even though it had no effect in isolated islets (SCHUTZ et al. 1979). Demonstration that these agents act via adenylate cyclase activation awaits the development of a relatively pure A-cell preparation in which biochemical measurements can be carried out.

Cholera toxin has been shown to stimulate adenylate cyclase in other tissues (CASSEL and PFEUFFER 1978), but its effects upon the A-cell remain unclear. In a study using neonatal pancreatic monolayer cultures, cholera toxin was found to stimulate glucagon secretion during the first 1 h incubation, but to cause inhibition at 3–4 h (WOLLHEIM et al. 1976). These results again raise the interesting possibility that the later inhibitory effects may have been mediated by increases of somatostatin and insulin in the incubation buffer.

There are other agents capable of influencing glucagon secretion, which are thought to act primarily by noncyclic AMP routes of action, even though they can cause changes in cyclic AMP levels in some tissues. Insulin, for instance, can inhibit glucagon secretion and, even though cyclic AMP does not appear to be insulin's second messenger (CZECH 1981), insulin can lower cyclic AMP by activating a low $K_m$ phosphodiesterase (LOTEN et al. 1978). Somatostatin is a potent inhibitor of glucagon secretion (KOERKER et al. 1974) and, although its mechanism of action remains unclear (GERICH 1981), it has been shown to lower cyclic AMP in liver (VINICOR et al. 1977), pituitary (BORGEAT et al. 1974; KANEKO et al. 1974), and islets (EFENDIC et al. 1975; OLIVER and WAGLE 1975; OLIVER et al. 1978). The degree to which these changes in cyclic AMP influence the overall inhibitory effect of somatostatin is not known. A group of agents which may turn out to be important in the regulation of glucagon secretion are the prostaglandins. In some studies reviewed in Chap. 28, they have been shown to stimulate glucagon secretion (PEK et al. 1978; LUYCKX and LEFEBVRE 1978). The degree to which prostaglandins could influence the adenylate cyclase-cyclic AMP system is unclear at the present time and they are presumed to act via other pathways (CATT and DUFAU 1981).

## E. Conclusions

The comments in this chapter are an overview of our very incomplete understanding of the role of cyclic AMP in glucagon secretion. There seemed to be little point in indulging in endless speculation about the numerous other agents and conditions which could provide hints about a cyclic AMP role. There are a numer of obstacles preventing us from being able to determine definitively what biochemical events are occurring within the A-cells. Whereas it has not previously been possible to sort the different islet cells from one another, there is encouraging progress being made with the new techniques of elutriation and density gradients (PIPELEERS and PIPELEERS-MARICHAL 1981) and flow cytometry (NIELSEN et al. 1980; FLETCHER et al. 1981). Short of success with these approaches, one can look forward to developments in histochemistry which may allow us to measure events occurring within individual cells. The approach of using streptozotocin-treated or alloxan-treated islets (HOWELL et al. 1971, 1974; MATSCHINSKY et al. 1976a; ÖSTENSON 1979) remains largely unexplored with regard to cyclic AMP. Furthermore, little attention has been paid to the A-cell-rich dark islets of the avian class (MIKAMI and ONO 1962; WEIR et al. 1976b). We can summarize by stating that cyclic AMP is almost certainly an important mediator of glucagon secretion, but this is based upon assumption and indirect evidence. More direct evidence should be forthcoming.

## References


Asplin CM, Paquette TL, Palmer JP (1981) In vivo inhibition of glucagon secretion by paracrine beta cell activity in man. J Clin Invest 68:314–318

Baetens D, Rufener C, Srikant BC, Dobbs R, Unger R, Orci L (1976) Identification of glucagon-producing cells (A cells) in dog gastric mucosa. J Cell Biol 69:455–464

Baum D, Porte D Jr, Ensinck J (1979) Hyperglucagonemia and alpha-adrenergic receptor in acute hypoxia. Am J Physiol 237:E404–E408

Boden G, Owen OE, Rezvani I, Elfenbein BI, Quickel KE (1977) An islet-cell carcinoma containing glucagon and insulin. Chronik glucagon excess and glucose homeostasis. Diabetes 26:128–137

Borgeat P, Labrie P, Drouin J, Belanger A (1974) Inhibition of adenosine 3′, 5′ monophosphate accumulation in anterior pituitary gland in vitro by growth hormone release inhibiting hormone. Biochem Biophys Res Commun 56:1052–1059

Buchanan KD, Mawhinney WAA (1973) Insulin control of glucagon release from insulin-deficient rat islets. Diabetes 22:801–803

Burnstock G (1978) A basis for distinguishing two types of purinergic receptor. In: Straub R (ed) Cell membrane receptors for drugs and hormones: a multidisciplinary approach. Raven, New York, pp 107–118

Cassel D, Pfeuffer T (1978) Mechanism of cholera toxin action: covalent modification of the guanyl nucleotide binding protein of the adenylate cyclase system. Proc Natl Acad Sci USA 75:2669–2673

Catt KJ, Dufau ML (1981) Hormone action: control of target-cell function by peptide, thyroid, and steroid hormones. In: Felig P, Baxter JD, Broadus AE, Frohman LA (eds) Endocrinology and metabolism. McGraw-Hill, New York Maidenhead Hamburg, pp 61–105

Chesney TM, Schofield JG (1969) Studies on the secretion of pancreatic glucagon. Diabetes 18:627–632

Czech M (1981) Insulin action. Am J Med 70:142–150

Efendic S, Grill V, Luft R (1975) Inhibition by somatostatin of glucose induced 3′,5′-monophosphate (cyclic AMP) accumulation and insulin release in isolated pancreatic islets of the rat. FEBS Lett 55:131–133

Exton JH (1981) The effects of glucagon on hepatic glycogen metabolism and gluconeogenesis. In: Unger RH, Orci L (eds) Glucagon: physiology, pathophysiology and morphology of the pancreatic A-cells. Elsevier, New York

Fletcher DJ, Weir GC, Grogan WM, Maclaren NK (1981) Preparation of islet B cell-enriched fractions by low-angle light scatter flow-cytometry. J Cell Biol 91:393 a

Frankel BJ, Gerich JE, Hagura R, Fanska RE, Gerritsen GC, Grodsky GM (1974) Abnormal secretion of insulin and glucagon by the in vitro perfused pancreas of the genetically diabetic Chinese hamster. J Clin Invest 43:1637–1646

Fujimoto WY, Ensinck JW, Merchant FW, Williams RH, Smith PH, Johnson DG (1978) Stimulation by gastric inhibitory polypeptide of insulin and glucagon secretion by rat islet cultures (39997). Proc Soc Exp Biol Med 157:89–93

Gerber PG, Trimble ER, Wollheim CB, Renold AE, Miller RE (1981) Glucose and cyclic AMP as stimulators of somatostatin and insulin secretion from the isolated perfused rat pancreas: a quantitative study. Diabetes 30:40–44

Gerich JE (1981) Somatostatin. In: Brownlee M (ed) Handbook of diabetes mellitus, vol 1. Garland, New York, pp 297–354

Gerich JE, Langlois M, Noacco C, Schneider V, Forsham PH (1974) Adrenergic modulation of pancreatic glucagon secretion in man. J Clin Invest 53:1441–1446

Gerich JE, Lovinger R, Grodsky GM (1975) Inhibition by somatostatin of glucagon and insulin release from the perfused rat pancreas in response to arginine, isoproterenol, and theophylline: Evidence for a preferential effect on glucagon secretion. Endocrinology 96:749–754

Goto Y, Seino Y, Taminato T, Imura H (1976) Potentiation by theophylline of arginine-induced glucagon secretion in the isolated perfused rat pancreas. Horm Metab Res 8:240–241
Harvey WD, Faloona GR, Unger RH (1974) The effect of adrenergic blockade on exercise-induced hyperglucagonemia. Endocrinology 94:1254–1258
Honey RN, Arimura A, Weir GC (1981) Somatostatin neutralization stimulates glucagon and insulin secretion from the avian pancreas. Endocrinology 109:1971–1974
Howell SL, Edwards JC, Whitfield M (1971) Preparation of B-cell deficient guinea pig islets of langerhans. Horm Metab Res 3:37–43
Howell SL, Edwards JC, Montague W (1974) Regulation of adenylate cyclase and cyclic-AMP dependent protein kinase activities in $A_2$-cell rich guinea pig islets of langerhans. Horm Metab Res 6:48–52
Howlett AC, Sternweis PC, Macik BA, Van Arsdale PM, Gilman AG (1979) Reconstitution of catecholamine-sensitive adenylate cyclase. Association of a regulatory component of the enzyme with membranes containing the catalytic protein and $\beta$-adrenergic receptors. J Biol Chem 254:2287–2295
Ipp E, Rivier J, Dobbs RE, Brown M, Vale W, Unger RH (1979) Somatostatin analogs inhibit somatostatin release. Endocrinology 104:1270–1273
Iversen J (1970) In: Rodrigues RR, Ebling FJG, Henderson I, Assan R (eds) Proceedings of the 7th congress of the international diabetes federation. Excerpta Medica, Amsterdam, p 46
Iversen J (1973) Adrenergic receptors and the secretion of glucagon and insulin from the isolated perfused canine pancreas. J Clin Invest 52:2102–2116
Iversen J, Miles DW (1971) Evidence for a feedback inhibition of insulin on insulin secretion in the isolated perfused canine pancreas. Diabetes 20:1–9
Jarrousse CL, Rosselin G (1975a) Regulation by glucose and cyclic nucleotides of the glucagon and insulin release induced by amino acids. Diabete Metab 1:135–142
Jarrousse CL, Rosselin G (1975b) Interaction of amino acids and cyclic AMP on the release of insulin and glucagon by newborn rat pancreas. Endocrinology 96:168–177
Kaneko T, Oka H, Munemura M, Suzuki S, Yasuda H, Oda T, Yandihara N (1974) Stimulation of guanosine 3′,5′ cyclic monophosphate accumulation in rat anterior pituitary gland in vitro by synthetic somatostatin. Biochem Biophys Res Commun 61:53–57
Kaneto A, Kaneko T, Kajinuma H, Kosaka K (1977) Effect of vasoactive intestinal polypeptide infused inpancreatically on glucagon and insulin secretion. Metabolism 26:781–786
Koerker DJ, Ruch W, Chideckel E, Palmer J, Goodner CJ, Ensinck J, Gale CC (1974) Somatostatin: hypothalamic inhibitor of the endocrine pancreas. Science 184:482–484
Lefèbvre PJ, Luyckx AS (1977) Factors controlling gastric-glucagon release. J Clin Invest 59:716–722
Lefèbvre PJ, Luyckx AS (1981) The physiology of extrapancreatic glucagon. In: Unger RH, Orci L (eds) Glucagon: physiology, pathophysiology and morphology of the pancreatic A cells. Elsevier, New York, pp 335–357
Leichter SB (1980) Clinical and metabolic aspects of glucagonoma. Medicine (Baltimore) 59:100–113
Leichter SB, Pagliara AS, Greider MH, Pohl S, Rosai J, Kipnis DM (1975) Uncontrolled diabetes mellitus and hyperglucagonemia associated with an islet-cell carcinoma. Am J Med 58:285–293
Liljenquist JE, Horwitz DL, Jennings AS, Chiasson J, Keller U, Rubenstein AH (1978) Inhibition of insulin secretion by exogenous insulin in normal man as demonstrated by C-peptide assay. Diabetes 27:563–570
Loten EG, Assimacopoulos-Jeannet FE, Exton JH, Park CR (1978) Stimulation of a low $K_m$ phosphodiesterase from liver by insulin and glucagon. J Biol Chem 253:746–757
Luyckx A, Lefèbvre P (1973) Exercise-induced glucagon secretion. Postgrad Med J 49:620–623
Luyckx A, Lefèbvre P (1973) Exercise-induced glucagon secretion. Postgrad Med J 49:620–623

Luyckx AS, Lefèbvre P (1978) Possible role of endogenous prostaglandins in glucagon secretion by isolated guinea-pig islets. Diabetologia 15:411–416
Marco J, Diaz-Fierros M, Baroja IM, Villanueva ML, Valverde I (1972) Opposite effects of aminophylline on arginine-induced glucagon and insulin secretion in humans. Diabetes 21:289–294
Matschinsky FM, Pagliara AS, Hover BA, Pace CS, Ferrendelli JA, Williams AD (1976a) Hormone secretion and glucose metabolism in islets of Langerhans of the isolated perfused pancreas from normal and streptozotocin diabetic rats. J Biol Chem 251:6053–6061
Matschinsky FM, Pagliara AS, Stillings SN, Hover BA (1976b) Glucose and ATP levels in pancreatic islet tissue of normal and diabetic rats. J Clin Invest 58:1193–1200
Mikami SI, Ono K (1962) Glucagon deficiency induced by extirpation of alpha islets of fowl pancreas. Endocrinology 71:464–473
Nielsen DA, Lernmark A, Steiner DF (1980) Separation of islet cell sub-populations by fluorescence-activated cell sorting. Diabetes [Suppl 2] 29:125
Oliver J, Wagle S (1975) Studies on the inhibition of insulin release, glycogenolysis, and gluconeogenesis by somatostatin in the rat islets of Langerhans and isolated hepatocytes. Biochem Biophys Res Commun 62:772–776
Oliver J, Wright P, Ashmore J (1978) The effect of somatostatin on glucose stimulated adenosine 3′,5′ monophosphate accumulation and glucose oxidation by isolated rat islets of Langerhans. Proc Soc Exp Biol Med 158:458–461
Orci L (1976) Somatostatin-containing cells in the pancreas and the gastrointestinal tract. In: Ciba Foundation Symposium, Polypeptide hormones: molecular and cellular aspects. Little Brown, New York, pp 345–346
Orci L, Perrelet A (1981) The morphology of the A-cell. In: Unger RH, Orci L (eds) Glucaton: physiology, pathophysiology and morphology of the pancreatic A cells. Elsevier, Amsterdam Oxford New York, pp 3–32
Orci L, Unger RH (1975) Functional sub-division of islets of Langerhans an possible role of D-cells. Lancet 2:1243–1244
Östenson CG (1979) Regulation of glucagon release: effects of insulin on the pancreatic $A_2$ cell of the guinea pig. Diabetologia 17:325–330
Pagliara AS, Stillings SN, Haymond MW, Hover BA, Matschinsky FM (1975) Insulin and glucose as modulators of the amino acid-induced glucagon release in the isolated pancreas of alloxan and streptozotocin diabetic rats. J Clin Invest 55:244–255
Palmer JP, Porte D Jr (1981) Control of glucagon secretion: the central nervous system. In: Unger RH, Orci L (eds) Glucagon: physiology, pathophysiology and morphology of the pancreatic A cells. Elsevier, New York, pp 135–160
Pek S, Tai T, Elster A (1978) Stimulatory effects of prostaglandins E-1, E-2, and F-2-alpha on glucagon and insulin release in vitro. Diabetes 27:801–809
Pipeleers DG, Pipeleers-Marichal MA (1981) A method for the purification of single A, B and D cells and for the isolation of coupled cells from isolated rat islets. Diabetologia 20:654–663
Rabinovitch A, Dupré J (1974) Effect of gastric inhibitory polypeptide present in impure pancreozymin-cholecystokinin on plasma insulin and glucagon in the rat. Endocrinology 94:1139–1144
Samols E, Weir GC (1979) Adrenergic modulation of pancreatic A, B and D cells: alpha-adrenergic suppression and beta-adrenergic stimulation of somatostatin secretion, alpha-adrenergic stimulation of glucagon secretion in the perfused dog pancreas. J Clin Invest 63:230–238
Samols E, Tyler JM, Marks V (1972) Glucagon-insulin interrelationships. In: Lefebvre PJ, Unger RH (eds) Glucagon, molecular physiology. Clinical and therapeutic implications. Pergamon, Oxford New York, pp 151–173
Schauder P, McIntosh C, Arends J, Arnold R, Frerichs H, Creutzfeldt W (1977) Somatostatin and insulin release from isolated rat pancreatic islets in response to D-glucose, L-Leucine, α-ketoisocaproic acid or D-glyceraldehyde: Evidence for a regulatory role of adenosine -3′,5′-cyclic monophosphate. Biochem Biophys Res commun 75:630–635

Schebalin M, Said SI, Makhlouf GM (1977) Stimulation of insulin and glucagon secretion by vasoactive intestinal peptide. Am J Physiol 232:E197–E200
Schutz W, Raberger G, Brugger G, Kraupp O (1979) Effect of a 5′-substituted adenosine analogue (744c96) in insulin and glucagon release in isolated rat islets of Langerhans. Arch. Int Pharmacodyn 237:88–97
Sharp GWG (1979) The adenylate cyclase-cyclic AMP system in islets of Langerhans and its role in the control of insulin release. Diabetologia 16:287–296
Srikant CB, McCorkle K, Unger RH (1977) Properties of immunoreactive fractions of canine stomach and pancreas. J Biol Chem 252:1847–1851
Toyota T, Sato S, Kudo M, Abe K, Goto Y (1975) Secretory regulation of endocrine pancreas: cyclic AMP and glucagon secretion. J Clin Endocrinol Metab 41:81–89
Vinicor F, Higdon G, Clark CM Jr (1977) Effects of somatostatin on the hepatic adenylate cyclase system in the rat. Endocrinology 101:1071–1077
Weir GC, Knowlton SD, Martin DB (1974) Glucagon secretion from the perfused rat pancreas. Studies with glucose and catecholamines. J Clin Invest 54:1403–1412
Weir GC, Knowlton SD, Martin DB (1975) Nucleotide and nucleoside stimulation of glucagon secretion. Endocrinology 97:932–936
Weir GC, Knowlton SD, Atkins RF, McKennan KX, Martin DB (1976a) Glucagon secretion from the perfused pancreas of streptozotocin-treated rats. Diabetes 25:275–282
Weir GC, Goltsos PC, Steinberg EP, Patel YC (1976b) High concentration of somatostatin immunoreactivity in chicken pancreas. Diabetologia 12:129–132
Wollheim CB, Blondel B, Renold AE, Sharp WG (1976) Stimulatory and inhibitory effects of cyclic AMP in pancreatic glucagon release from monolayer cultures and the controlling role of calcium. Diabetologia 12:269–277
Yoshida T, Kondo M (1980) Effect of adrenergic agents on the secretion of gastrointestinal immunoreactive glucagon in depancreatized dogs. J Diabetes 29:355–360

# CHAPTER 28

# Prostaglandins and Glucagon Secretion

A. S. LUYCKX and P. J. LEFEBVRE

## A. Introduction

### I. Origin and Metabolism of Prostaglandins

Prostaglandins (PGs) are a group of compounds which have been found to exert numerous and potent physiologic and pharmacologic effects. Their existence was originally reported in seminal fluid about 50 years ago (KURZROK and LIEB 1930; GOLDBLATT 1933; VON EULER 1934), but the structure of the six primary PGs was only described in the early 1960s (review in BERGSTRÖM et al. 1968). PGs are found in all mammalian tissues which also appear able to synthesize them. The precursors of PGs are essential fatty acids with varying degrees of unsaturation, mainly arachidonic acid (C20:4) and $\gamma$-dihomolinolenic acid (C20:3), originating from the plasma membrane phospholipids. In response to various stimuli, these acids are cleaved from the membrane phospholipids by the enzyme phospholipase $A_2$ and are then acted on by a complex group of enzymes given the generic name "prostaglandin synthetase"; the first step involves a cyclooxygenase, which transforms arachidonic acid into cyclic endoperoxides ($PGG_2$ and $PGH_2$); these unstable intermediates, with very short half-lives, are rapidly converted into $PGE_2$, $PGF_{2\alpha}$, thromboxane $A_2$ ($TXA_2$), or prostacyclin ($PGI_2$), depending upon the tissues considered (Fig. 1, review in SAMUELSSON et al. 1980). Prostaglandins of the first series

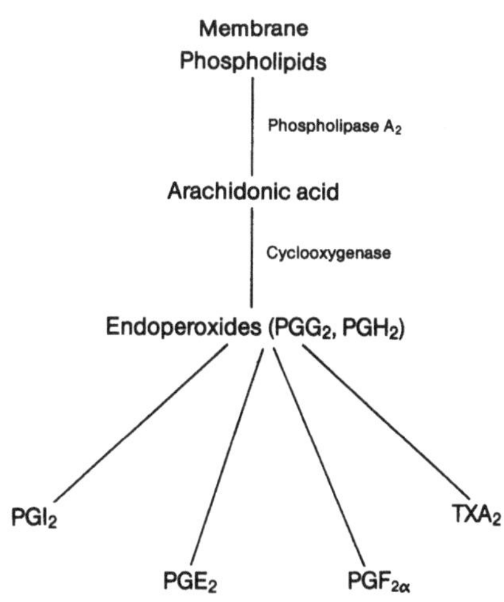


**Fig. 1.** Biosynthesis of various prostaglandins

($PGE_1$, $PGF_{1\alpha}$) are formed by similar steps from $\gamma$-dihomolinolenic acid. Prostaglandins are quickly degraded in vivo and converted mainly into the following metabolites: 6-keto-$PGF_{1\alpha}$ (for $PGI_2$), thromboxane $B_2$ (for $TxA_2$), and 15-keto-13,14-dihydro-$PGE_2$ for $PGE_2$.

Pharmacologic compounds with antiinflammatory action inhibit one or other steps in the biosynthesis of PGs; corticosteroids inhibit the relase of arachidonic acid from membrane phospholipids whereas nonsteroidal antiinflammatory drugs (indomethacin, acetylsalicylic acid, meclofenamate, ibuprofen, etc.) reduce cyclooxygenase activity (VANE 1971; SIMON and MILLS 1980). Sodium salicylate is a weak inhibitor of PG synthesis (HAMBERG 1972).

## II. Prostaglandins as Local or Intracellular Messengers

It is generally admitted that prostaglandins are rapidly synthesized within various cell types and that their physiologic role is to modulate specific biochemical steps within the cell in which they are formed. This role of intracellular messenger is probably similar to that of cyclic AMP (SILVER and SMITH 1975). PGs could also act as local messengers and, by paracrine mechanisms, influence the neighboring cells. Finally, in pathologic circumstances, a massive release of prostaglandins into the circulation could exert "hormone-like" effects. Receptors for PGs have been found in plasma membrane of various cell types and it has been established that PGs activate the adenylate cyclase present in the islets of Langerhans (KUO et al. 1973; JOHNSON et al. 1974; KUEHL 1974).

## III. Methodological Considerations

Before considering the arguments in favor of a role of PGs in the control of glucagon secretion, it is worthwile emphasizing the methodological problems encountered in these studies. Indeed, it is still difficult to ascertain whether the PGs exert a physiologic role in the control of glucagon secretion for the following reasons:

1. Experimental studies measuring glucagon secretion by perfused or incubated pancreatic tissue in the presence of various concentrations of PGs do not necessarily reproduce the physiologic conditions of biosynthesis of these agents.
2. In vivo administration of PGs (most often via the intravenous route) induce general reactions and particularly hemodynamic effects. Thus, the modifications of endocrine secretions could be due to indirect regulatory mechanisms rather than to a direct effect exerted by the PG on the A-cells.
3. Experiments carried out with inhibitors of PG synthesis are not always easy to interpret. Indeed, it is known that indomethacin, which is used to inhibit cyclooxygenase activity, also exerts several other effects, including an inhibitory action on enzymes involved in PG degradation (PACE-ASCIAK and COLE 1975), on phosphodiesterase (WEISS and HAIT 1977; OJEDA et al. 1979), and on cyclic AMP-dependent protein kinase (KANTOR and HAMPTON 1978). It ist thus necessary to carry out experiments not only with indomethacin, but also with other inhibitors of PG synthesis like acetylsalicylic acid (ASA), L8027, ibuprofen, meclofenamate, and other nonsteroidal antiinflammatory agents.

4. On the other hand, in vivo administration of ASA induces significant decreases in several circulating metabolites (free fatty acids, glycerol, $\beta$-hydroxybutyrate) which could modify glucagon secretion (see Chap. 25) by mechanisms independent of any change in PG synthesis.
5. Finally, under certain conditions and in certain animal species, PGs inhibit norepinephrine (NE) release from the sympathetic nerve terminals. As a consequence, inhibition of PG synthesis could result in an increase in the postsynaptic concentration of NE and thus in indirect changes in A-cell secretion (see Chap. 30). Indeed, it has been proposed that PGs of the E series may function as a physiologic brake mechanism for the release of the adrenergic transmitter (HEDQVIST and BRUNDIN 1969; HEDQVIST 1970; HEDQVIST et al. 1971).

These numerous methodological problems probably explain why the results of the various studies dealing with a possible role of PGs in the control of glucagon secretion remain controversial, particularly when comparing results obtained in vitro and in vivo or data obtained with exogenous PGs versus those obtained with inhibitors of PG synthesis. This situation led us to subdivide the present chapter according to the experimental protocols used and to describe separately the results obtained in vitro or in vivo. In each case, we will consider the effects of *exogenous* PGs and later on, the role of *endogenous* PGs, as suspected on the basis of studies using inhibitors of PG synthesis.

## B. Studies In Vitro on the Influence of Prostaglandins on Glucagon Secretion

### I. Effect of Exogenous Prostaglandins

PEK et al. (1976, 1978, 1980) have studied in detail the influence of various PGs on insulin and glucagon release by the isolated, perfused rat pancreas. In the concentration range $10^{-8}$–$10^{-6}$ $M$, $PGE_1$, $PGE_2$, and $PGF_{2\alpha}$ stimulate the release of insulin and glucagon (with a biphasic pattern for both islet hormones) and increase the secretory response to 10 m$M$ L-arginine (PEK et al. 1976, 1978). More recently, PEK et al. (1980) reported that the endoperoxide, $PGH_2$ was also stimulatory, whereas $TXA_2$ and $PGI_2$ had little or no effect. Preliminary observations suggest that prolonged incubations of $PGH_2$ with platelet microsomes could generate a stable metabolite, probably $PGD_2$, which selectively stimulates glucagon-producing cells without affecting insulin release (AKPAN et al. 1979). Using the technique of perfusion of the isolated dog stomach, we observed that $PGE_1$ also stimulates the release of glucagon (Fig. 2); LEFEBVRE and LUYCKX 1978).

Recently, MATSUYAMA et al. (1983) reported that $PGD_2$ induced a glucose dependent stimulation of glucagon and insulin release by the perfused rat pancreas.

### II. Role of Endogenous Prostaglandins

#### 1. Prostaglandin Biosynthesis by Islet Tissue

HAMAMDZIC and MALIK (1977) reported that a rat pancreatic homogenate converted arachidonic acid $^{14}$C-1 into $PGE_2$ and $PGF_{2\alpha}$ in the presence of 6.0 μ$M$ nor-

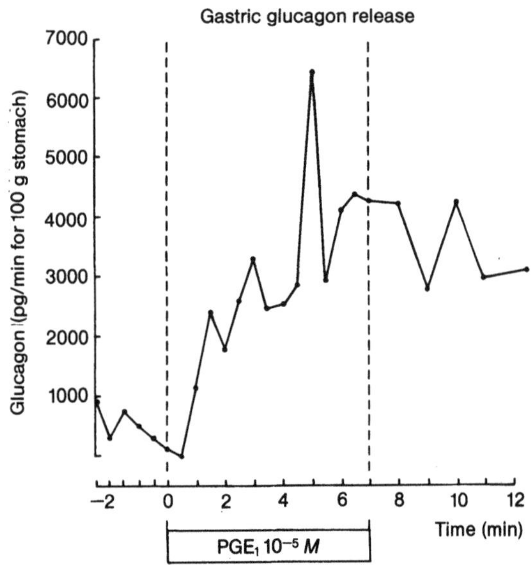


**Fig. 2.** Influence of $PGE_1$ (10 $\mu M$) on glucagon release by the isolated perfused dog stomach. After LEFEBVRE and LUYCKX (1978)

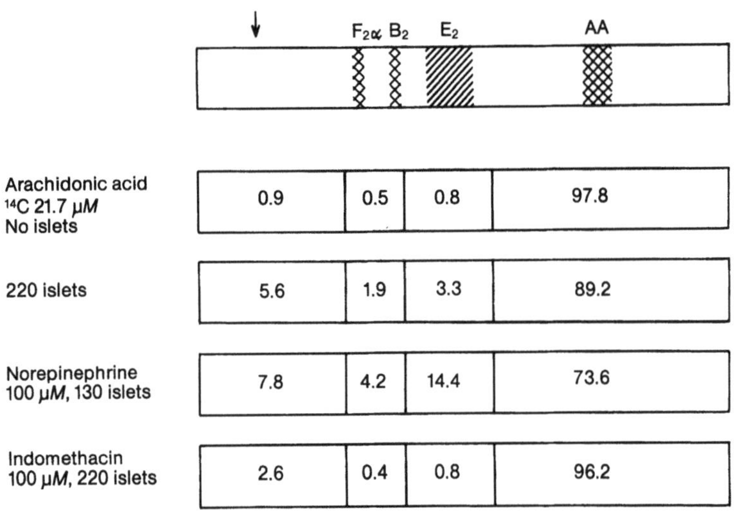


**Fig. 3.** In these experiments, groups of isolated guinea pig islets were incubated in the presence of arachidonic acid $^{14}C$ (21.7 $\mu M$) for 4 h. After incubation, extracts of these islets were submitted to thin layer chromatography on 60 $F_{254}$ silica gel plates (Merck, Darmstadt, Federal Republic of Germany). The *arrow* indicates the origin. The zones of migration of arachidonic acid (AA), prostaglandin $E_2$ ($E_2$), thromboxane $B_2$ ($B_2$), and prostaglandin $F_{2\alpha}$ ($F_{2\alpha}$) were identified on separate plates run in parallel. Control experiments were also performed by incubating arachidonic acid $^{14}C$ under the same conditions, but in the absence of islets

epinephrine, but this biosynthetic activity could not be attributed specifically to the islets of Langerhans. Further observations support the view that islet tissue has the synthetic machinery to synthetize PGs in response to various stimuli. Indeed, we have found that the $PGE_2$ content of isolated guinea pig islets increased under conditions of intensive stimulation of glucagon secretion (LUYCKX and LEFEBVRE 1980b). Moreover, pieces of a human insulinoma incubated in vitro released measurable amounts of $PGE_2$ in the incubation medium, and this release was inhibited by 76% in the presence of indomethacin (Table 1; LUYCKX et al. 1981a). Finally, we observed that isolated guinea pig islets incubated in vitro in the presence of arachidonic acid $^{14}C$-1 synthesize $PGE_2$, since thin layer chromatography of islets extracts revealed the presence of $PGE_2$ $^{14}C$; this biosynthetic activity was stimulated by NE and inhibited by indomethacin (Fig. 3).

METZ et al. (1980) studied the effects of sodium salicylate (20 mg/dl) and indomethacin (2 μg/ml) upon PGE synthesis in monolayer cultures of neonatal rat islets and found that both drugs inhibited PGE synthesis by about 75%. Recently, sonicates of isolated rat islets were found to incorporate radiolabeled arachidonic acid into $PGF_{2\alpha}$ and $PGE_2$ equivalents, in addition to 15-keto-13,14-dihydro metabolites of these primary PGs; moreover, labeled arachidonate was also incorporated into compounds which elute as hydroxy- or hydroperoxyeicosatetraenoic acids on high pressure liquid chromatography (KELLY and LAYCHOCK 1981). From these studies, it can be concluded that the islets of Langerhans synthesize PGs but, until now, it has not been possible to attribute this biosynthetic activity to any specific cell type present in the islets.

## 2. Endogenous Prostaglandins and Glucagon Secretion In Vitro

Using isolated guinea pig islets incubated in vitro we demonstrated that two inhibitors of PG synthesis, L8027 (isopropyl-2 nicotinoyl-3 indole) and indomethacin inhibited arginine- or induced NE-induced glucagon secretion (Fig. 4; LUYCKX and LEFEBVRE 1978). Using similar protocols, we investigated the influence of other inhibitors of cyclooxygenase activity (meclofenamate, ibuprofen) on glucagon release stimulated by arginine, alanine, epinephrine, or isoproterenol. As illustrated by Fig. 5 and 6 and by Table 1, concordant results were found, suggesting that, as originally proposed, prostaglandin synthesis is involved in the stimulus-secretion coupling of the pancreatic A-cells. In a separate series of investigations, we tested several experimental conditions known to stimulate endogenous PG synthesis. For this purpose, isolated guinea pig islets were incubated in a medium supplemented with arachidonic acid. As illustrated by Fig. 7 and Table 2, glucagon release was markedly stimulated by arachidonic acid in the 16.5–66 $\mu M$ concentration range. We also investigated the influence of increasing concentrations of phospholipase $A_2$ added to the incubation medium. As already said, this enzyme cleaves long chain, polyunsaturated fatty acids from phospholipids in the plasma membrane and thereby makes them available for prostaglandin synthesis. Figure 8 shows that phospholipase $A_2$ induces a concentration-dependent release of glucagon.

The fact that the stimulatory effects of arachidonic acid and phospholipase $A_2$ are significantly reduced in the presence of indomethacin or ASA (LUYCKX and LEFEBVRE 1979, 1980a, b) strongly suggests that the stimulatory mechanism really in-

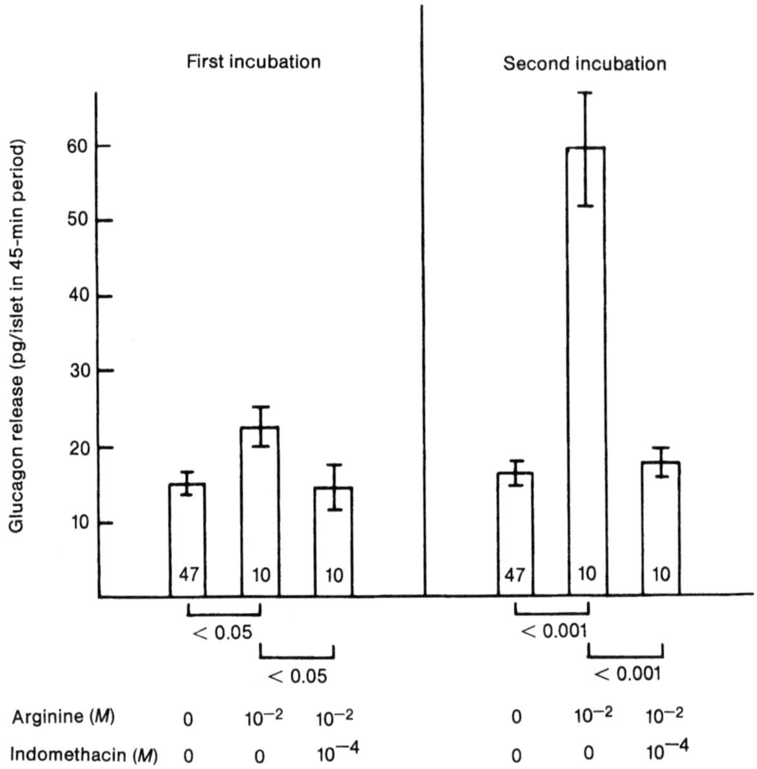


**Fig. 4.** Influence of indomethacin on arginine-stimulated glucagon release by isolated guinea pig islets incubated in vitro (15 islets per vial) during two successive periods of incubation. Incubation medium: Gey and Gey bicarbonate buffer, containing 2 g/l crystallized bovine albumin and no glucose. The number of vials in each series is indicated in the columns. Results are expressed as mean ± standard error. After LUYCKX and LEFEBVRE (1978)

**Table 1.** Influence of ibuprofen on epinephrine- or isoproterenol-induced glucagon release by isolated guinea pig islets. Ibuprofen (100 μ*M*) alone does not alter glucagon release under control conditions. Results are expressed as mean ± standard error; figures in parentheses indicate numbers of vials

| | Glucagon release | |
|---|---|---|
| | (pg/islet in 90-min period) | (% islet glucagon content) |
| Control conditions | 31.1 ± 3.03 (17) | 0.55 ± 0.05 |
| Epinephrine 100 μ*M* | 57.4 ± 6.8 (10) | 1.15 ± 0.14[a] |
| Epinephrine 100 μ*M* + ibuprofen 100 μ*M* | 30.9 ± 2.6 (10) | 0.79 ± 0.06[b] |
| Isoproterenol 100 μ*M* | 60.9 ± 5.8 (9) | 1.05 ± 0.1[a] |
| Isoproterenol 100 μ*M* + ibuprofen 100 μ*M* | 35.5 ± 3.0 (9) | 0.72 ± 0.06[c] |

[a] $P<0.01$ versus control conditions
[b] $P<0.05$ versus epinephrine alone
[c] $P<0.01$ versus isoproterenol alone

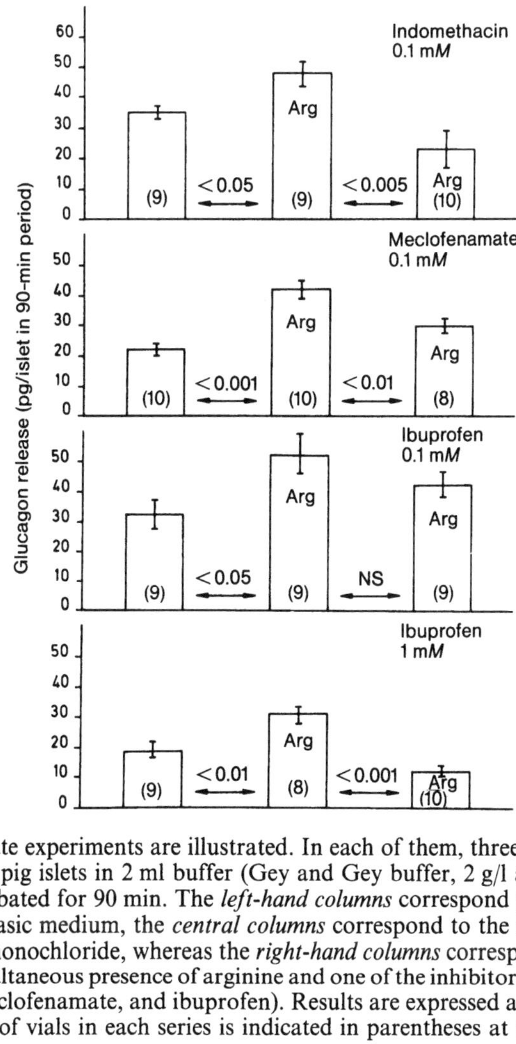


**Fig. 5.** Four separate experiments are illustrated. In each of them, three series of vials containing ten guinea pig islets in 2 ml buffer (Gey and Gey buffer, 2 g/l albumin, 125 mg/dl glucose) were incubated for 90 min. The *left-hand columns* correspond to the control islets incubated in the basic medium, the *central columns* correspond to the islets stimulated by 100 μ*M* arginine monochloride, whereas the *right-hand columns* correspond to the islets incubated in the simultaneous presence of arginine and one of the inhibitors of cyclooxygenase (indomethacin, meclofenamate, and ibuprofen). Results are expressed as mean ± standard error, the number of vials in each series is indicated in parentheses at the bottom of each column

**Table 2.** Glucagon release by isolated islets from normal guinea pigs – influence of arachidonic acid

| Incubation medium | First incubation | Second incubation |
|---|---|---|
| Basic medium[a] $N=5$ | 13.7 ± 1.2 | 14.6 ± 1.7 |
| Basic medium[a] + arachidonic acid 66 μ*M* $N=6$ | 25.4 ± 1.3[b] | 126.2 ± 19.7[b] |

[a] Basic medium: 2.9 μ*M* albumin + 1.7 m*M* glucose
[b] $P < 0.001$ versus basic medium

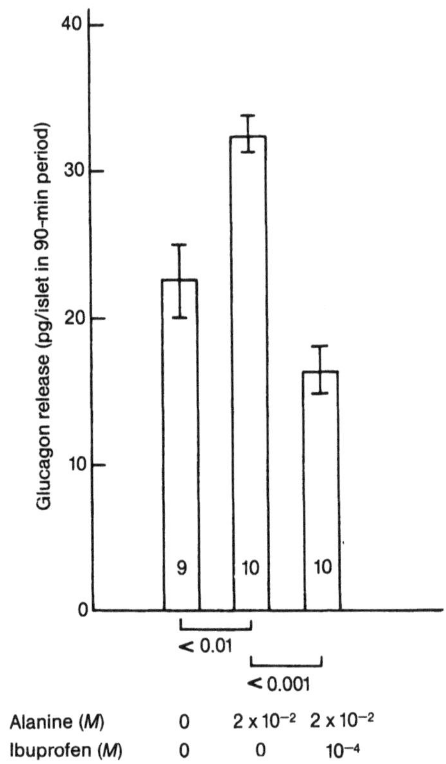


**Fig. 6.** Influence of ibuprofen (100 μ*M*) on the glucagon release induced by alanine (200 μ*M*). The experimental conditions are described in the legend to Fig. 5

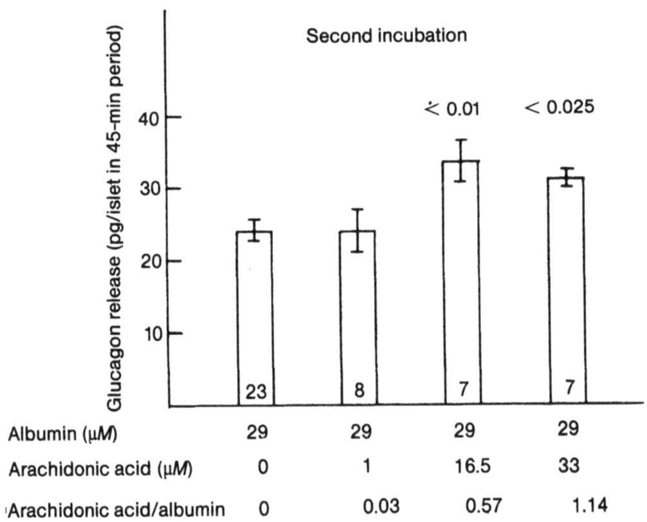


**Fig. 7.** Influence of increasing concentrations of arachidonic acid on glucagon release by isolated guinea pig islets during the second incubation period. Results are expressed as mean ± standard error, the number of incubation vials is indicated at the bottom of each column

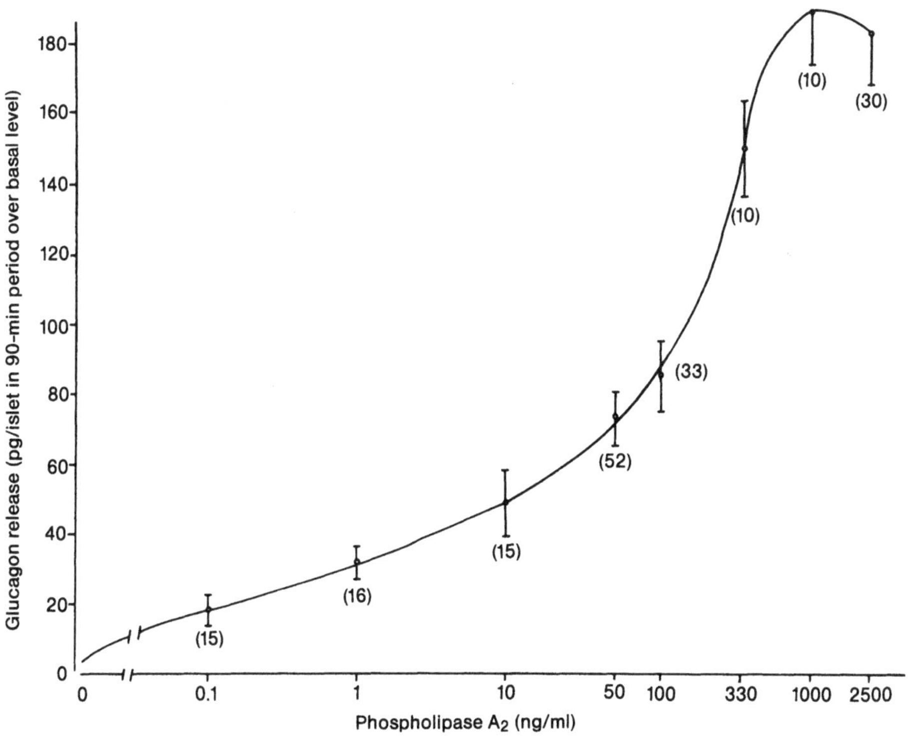


**Fig. 8.** Influence of increasing concentrations of phospholipase $A_2$ (from bee venom, Sigma Chemical Company, St. Louis, Missouri, USA) on glucagon release by isolated guinea pig islets incubated in vitro. Results are expressed as mean ± standard error, the number of vials used for each concentration is indicated in parentheses

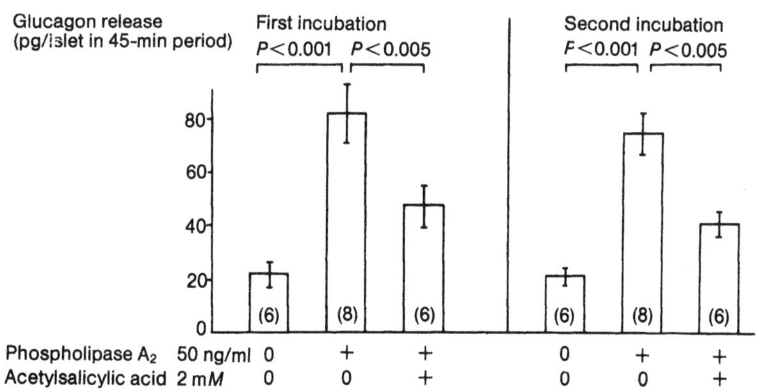


**Fig. 9.** Effect of acetylsalicylic acid on phospholipase $A_2$-induced glucagon release from isolated guinea pig islets incubated in vitro. For experimental conditions, see legends to Figs. 4 and 5

**Table 3.** Influence of indomethacin (100 μ*M*) on the release of $PGE_2$, insulin, and glucagon from pieces of insuloma tissue incubated in vitro. Results are expressed as mean ± standard error; figures in parentheses indicate numbers of vials, (LUYCKX et al. 1981 a)

| | Control medium Gey and Gey buffer + 2 g/l + albumin 5.5 m*M* glucose (5) | Control medium + indomethacin 100 μ*M* (5) | *P* |
|---|---|---|---|
| $PGE_2$ (pmol/2 h) | 3.65 ± 1.3 | 0.89 ± 0.23 | <0.05 |
| Insulin (mIU/2 h) | 10.5 ± 1.2 | 16.8 ± 1.9 | <0.025 |
| Glucagon (pg/2 h) | 708.4 ± 141.8 | 176.0 ± 19.7 | <0.01 |

volves PG biosynthesis rather than any nonspecific or toxic effect on the A-cells (Fig. 9). The simultaneous decrease in $PGE_2$ and glucagon release observed after addition of indomethacin to the incubation medium of pieces of a human insulinoma (Table 3) allows one to extrapolate the results obtained with guinea pig islets to humans (LUYCKX et al. 1981 a).

Recent data of MCCARDEL et al. (1981), however, questioned the mediatory role of endogenous PG synthesis in NE-induced glucagon release by the isolated perfused rat pancreas. Indeed, these authors found that NE ($10^{-6}M$) increased total PGE and glucagon levels in the portal venous effluent, but that flurbiprofen, which inhibited PGE release by 81%, did not prevent the effect of NE on glucagon secretion. In this study, a persistent PG synthesis within the islets could be masked by the reduction in the overall biosynthetic activity of the pancreatic tissue. Thus, similar studies should be carried out using incubated isolated islets before admitting that PG synthesis and NE-induced glucagon release can be dissociated.

# C. Studies In Vico on the Influence of Prostaglandins on Glucagon Secretion

## I. Effect of Exogenous Prostaglandins

### 1. Rats

In rats, intravenous infusion of $PGE_1$ (2 μg min) increased plasma glucagon concentrations. This effect appeared independent of any hypotension-related change in sympathetic activity since it persisted in sympathectomized or in β-blocked animals (SACCA and PEREZ (1976).

### 2. Dogs

In dogs, intravenous infusion of $PGE_2$ (1 μg, $kg^{-1}$, $min^{-1}$) increased glucagon levels in the gastric fundic vein as well as in the inferior vena cava (SCHUSDZIARRA et al. 1980). Interestingly, a concommitant rise in the plasma somatostatin-like immunoreactivity, was observed, thus suggesting that PGs might also be important in the regulation of gastric endocrine function (SCHUSDZIARRA et al. 1981).

### 3. Humans

In humans, GIUGLIANO et al. (1979, 1981) found that intravenous infusion of $PGE_1$ (0.2 μg $kg^{-1}$ $min^{-1}$) in healthy volunteers caused a significant increase in plasma glucagon concentrations. A superimposed infusion of propranolol (0.08 mg/min) suppressed $PGE_1$-induced hyperglucagonemia. Thus, in humans the effect of $PGE_1$ on A-cell secretion seems to be mediated, either by endogenously released catecholamines or by an interaction between the prostaglandin and the A-cell β-adrenoceptor.

## II. Role of Endogenous Prostaglandins

As already discussed in this chapter, it is generally admitted that changes in glucagon secretion secondary to the administration of nonsteroidal antiinflammatory drugs are mediated by a corresponding reduction of PG synthesis and thus argue in favor of a role of endogenous PGs in the regulation of A-cell function. However, several factors must be taken into consideration before admitting such a causal relationship. Most nonsteroidal antiinflammatory compounds are highly bound to plasma albumin, and thus the circulating free fraction (about 10% of the total plasma concentration in the case of indomethacin) could remain below the threshold necessary for modifying glucagon secretion.

Moreover, indomethacin not only inhibits cyclooxygenase activity and prostaglandin synthesis (VANE 1971), but also reduces the activity of some enzymes involved in the degradation of prostaglandins (PACE-ASCIAK and COLE 1975) as well as that of phosphodiesterase (FLORES and SHARP 1972). This latter effect leads to increased intracellular concentrations of cyclic AMP and, as a consequence, would rather stimulate glucagon release as demonstrated by WEIR et al. (1975) (review in WOLLHEIM et al. 1976; see also Chap. 27). On the other hand, indomethacin has alo been shown to inhibit cyclic AMP-dependent protein kinase (KANTOR and HAMPTON 1978), another possible factor of interference in the control of glucagon release. In an attempt to avoid these "nonspecific" actions of indomethacin, other authors have investigated the effects of other inhibitors of PG synthesis like ASA (GIUGLIANO et al. 1978, 1980 a). Here again, it must be kept in mind that the resulting decrease in plasma free fatty acids (FFA) could stimulate glucagon release (see Chap. 25) and thus mask the possible direct action of this compound on A-cells. Most clinical studies failed to demonstrate an effect of inhibitors of prostaglandin synthesis on glucagon secretion: in normal subjects, ASA (3.2 g/day by mouth in four divided doses) did not alter glucagon response to glucose-, arginine- (TORELLA et al. 1979 a), tolbutamide- (TORELLA et al. 1979 b), or insulin-induced hypoglycemia (GIUGLIANO et al. 1981).

In normal subjects and in mild type 2 diabetic subjects, MICOSSI et al. (1978) found a slight stimulatory effect of ASA on basal plasma glucagon levels and a singnificant stimulatory effect on the glucagon responses to oral glucose; they speculated that their observations could be related to some nonspecific irritant effect of ASA on the gastrointestinal mucosa, leading to the release of gastric glucagon. More recently, PRINCE et al. (1981) confirmed that ASA (750 mg four times a day for 5 days and 1 h prior to the test) significantly increased fasting plasma glucagon in normal subjects and in C-peptide-negative diabetics. The suppressed levels following glucose infusion and the stimulated levels following arginine found in normal subjects were not modified by the treatment. No significant changes in either basal on arginine-stimulated glucagon concentrations were observed after treatment with ASA (50 mg $kg^{-1}$ $day^{-1}$ plus 1 g before the second test) in type 1 (insulin-dependent) diabetics (GIUGLIANO et al. 1980a). Thus, none of these results obtained with ASA administration support the view that in vivo PG synthesis is involved in the regulation of glucagon secretion.

Experiments with indomethacin were conducted in rats, dogs, and humans. In rats, FRAME et al. (1977) studied insulin and glucagon responses to intravenous infusions of glucose or arginine after intraperitoneal injection of 10 mg/kg body weight indomethacin 18 or 3 h prior the study and found no difference from the corresponding results obtained in control animals. BYBEE et al. (1978) investigated insulin and glucagon secretion in response to intraduodenal glucose with and without treatment with indomethacin (1.5 mg/kg body weight 45 min before the glucose load) in dogs; both insulin and glucagon secretion increased after treatment. In another study carried out in dogs, SCHUSDZIARRA et al. (1980) found that intravenous infusion of indomethacin (2 mg $kg^{-1}$ $h^{-1}$) decreased PGE levels in the effluent gastric veins, but did not alter plasma glucagon concentrations in either the gastric fundic vein or the inferior vena cava.

We have investigated the influence of a single oral administration of indomethacin (50 mg) on the hormonal and metabolic changes induced by a standardized mixed 500 kcal meal in healthy volunteers. The meal ingestion was followed by a late rise in plasma glucagon levels in controls. Such a rise was not detected when the meal was ingested after administration of indomethacin (LUYCKX et al. 1981b). In contrast, indomethacin did not modify the arginine-induced glucagon rise in a study reported by VIERHAPPER et al. (1980), and also carried out in normal volunteers.

Two studies were devoted to the possible influence of indomethacin on glucagon secretion in insulin-treated diabetics (LUYCKX et al. 1981a). As illustrated by Fig. 10, blood glucose, FFA, and glucagon concentrations were totally unaffected by the two doses of indomethacin tested, both in the basal state and during arginine infusion at two different rates. Similarly, indomethacin did not modify glucagon plasma levels after a standardized meal, preceded by a subcutaneous administration of 15 IU regular insulin (Actrapid Insulin, Novo, Copenhagen) (LUYCKX and LEFEBVRE 1981). Recently GIUGLIANO et al. (1980b, 1981) reported two studies which, in contrast with some of the negative results obtained with ASA or indomethacin already summarized, are in favor of a role of PGs in the regulation of glucagon secretion.

First, they measured glucagon plasma levels during intravenous administration of furosemide, a classical stimulator of PG synthesis, and found that this drug significantly increased the glucagon rise occuring 3–10 min after a 3-g intravenous arginine pulse (GIUGLIANO et al. 1980b). Our own results, however, obtained during a furosemide intravenous infusion in healthy volunteers, failed to detect any significant effect of the same drug on basal plasma IRG levels (LUYCKX et al. 1980).

Second, intravenous infusion of salbutamol (5 μg/min), a $\beta_2$-adrenoceptor agonist, elicited a weak glucagon response in six insulin-dependent diabetics. An in-

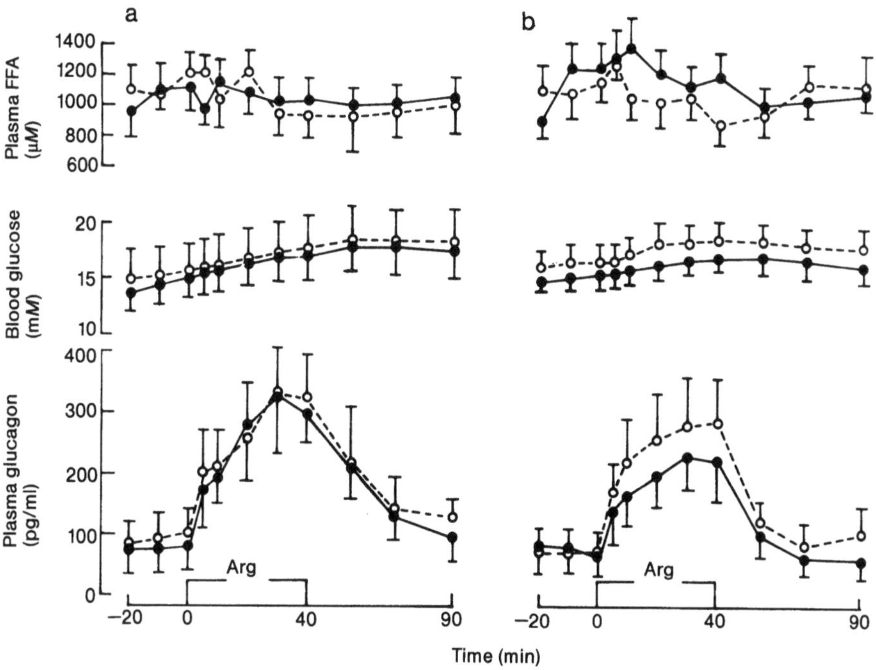


**Fig. 10a, b.** Influence of arginine infusion on blood glucose, plasma free fatty acids (FFA), and glucagon concentrations in two groups of insulin-treated diabetics. Arginine infusion rates (Arg): **a** 11.7 mg $kg^{-1}$ $min^{-1}$; **b** 5.85 mg $kg^{-1}$ $min^{-1}$; (*full circles*) control test; (*open circles*) after indomethacin administration. Low dose (**a**) 25 mg indomethacin orally every 8 h the day before and 50 mg indomethacin 2 h before starting the arginine infusion. High dose (**b**) 25 mg indomethacin orally every 8 h the da before, 2 mg/kg body weight of indomethacin 2 h before and 1 mg/kg immediately before starting the arginine infusion. Results are expressed as mean ± standard error (six subjects in each group)

fusion of lysine acetylsalicylate almost completely abolished this glucagon response to salbutamol (GIUGLIANO et al. 1981) thus suggesting that it was mediated via an increased PG synthesis.

## D. Summary and Conclusions

In vitro, five lines of evidence suggest that a prostaglandin system is involved in the regulation of glucagon secretion:

1. Exogenous PGs stimulate glucagon release and increase glucagon secretion in response to L-arginine (PEK et al. 1976, 1978, 1980).
2. Activation of PG synthesis by phospholipase $A_2$, arachidonic acid, and furosemide increases glucagon release, while the effect of phospholipase $A_2$ or arachidonic acid is reduced by indomethacin or ASA (LUYCKX and LEFEBVRE 1979, 1980a, b).
3. Arginine-, alanine-, epinephrine-, NE- and isoproterenol-stimulated glucagon release are reduced by various inhibitors of prostaglandin synthesis, namely in-

domethacin, ASA, meclofenamate, ibuprofen, and L8027 (LUYCKX and LEFEBVRE 1978; LUYCKX and LEFEBVRE 1981, unpublished work).
4. Indomethacin significantly reduces $PGE_2$ and glucagon release from pieces of human insulinoma incubated in vitro (LUYCKX et al. 1981 a).
5. Intensive stimulation of glucagon release is accompanied by an increased synthesis of $PGE_2$, as evidenced by the incorporation of arachidonic acid $^{14}C$ into $PGE_2$ $^{14}C$ (LUYCKX and LEFEBVRE 1981) and by a rise in the $PGE_2$ content of isolated guinea pig islets of Langerhans incubated in vitro (LUYCKX and LEFEBVRE 1980 b).

In vivo, in rats, dogs, and humans, intravenous infusion of $PGE_1$ or $PGE_2$ increases plasma glucagon levels.

On the basis of these results, it is worthwhile considering pharmacologic agents which inhibit PG synthesis among the compounds to be tested for reducing glucagon secretion, particularly in diabetics. In this respect, most of the results obtained so far with ASA and indomethacin have been negative since treatment with these drugs did not modify the increased plasma glucagon levels observed after stimulation by arginine (LUYCKX et al. 1981 c; GIUGLIANO et al. 1981) or by a mixed meal (LUYCKX and LEFEBVRE 1981). However, the fact that these two compounds exert several effects, apart from reducing cyclooxygenase activity (see Sect. A III), makes it necessary to test other substances in this group before drawing any definite conclusion.

## References

Akpan JO, Hurley MC, Pek S, Lands WEM (1979) The effects of prostaglandins on secretion of glucagon and insulin by the perfused rat pancreas. Can J Biochem 57:540–547
Bergström S, Carlson LA, Weeks JR (1968) The prostaglandins: a family of biologically active lipids. Pharmacol Rev 20:1–48
Bybee DE, Brodie TR, Georges LP, Fletcher JR, Diaz MJ, O'Brian JT (1978) Indomethacin induced augmentation of insulin and glucagon release (Abstr). Diabetes 27:470
Flores AGA, Sharp GWP (1972) Endogenous prostaglandins and osmotic water flow in the toad bladder. Am J Physiol 223:1392–1397
Frame CM, Ganuli S, Christensen R, Voina S, Madding C (1977) The role of endogenous prostaglandins in insulin and glucagon responses in the rat (Abstr). Diabetes 236:397
Giugliano D, Torella R, Siniscalchi N, Improta L, D'Onofrio F (1978) The effect of acetylsalicylic acid on insulin response to glucose and arginine in normal man. Diabetologia 14:359–362
Giugliano D, Torella R, Sgambato S, D'Onofrio F (1979) Effect of α- and β-adrenergic inhibition and somatostatin on plasma glucose, free fatty acids, insulin, glucagon, growth hormone responses to prostaglandin $E_1$ in man. J Clin Endocrinol Metab 48:302–308
Giugliano D, Luyckx AS, Lefèbvre PJ (1980 a) Effect of acetylsalicylic acid on blood glucose, plasma FFA, glycerol, 3-hydroxybutyrate, alanine, C-peptide, glucagon and growth hormone responses to arginine in insulin-dependent diabetics. Diab Metab 6:39–46
Giugliano D, Torella R, Sgambato S, D'Onofrio F (1980 b) Effect of furosemide on insulin and glucagon responses to arginine in normal subjects. Diabetologia 18:293–296
Giugliano D, Torella R, D'Onofrio F (1981) Prostaglandins and the alpha-cell. Prostaglandins Med 6:283–297
Goldblatt MW (1933) A depressor substance in seminal fluid. J Soc Chem India 52:1056
Hamamdzic M, Malik KU (1977) Prostaglandins in adrenergic transmission of isolated perfused rat pancreas. Am J Physiol 232 (2):E201–E209

Hamberg M (1972) Inhibition of prostaglandins synthesis in man. Biochem Biophys Res Commun 49:720–726

Hedqvist P (1970) Studies on the effect of prostaglandins $E_1$ and $E_2$ on the sympathetic neuromuscular transmission in some animal tissues. Acta Physiol Scand 79:1–40

Hedqvist P, Brundin T (1969) Inhibition by prostaglandin $E_1$ of noradrenaline release and of effector responses to nerve stimulation in the cat spleen. Part I. Life Sci 8:389–395

Hedqvist P, Stjärne L, Wennmalm Å (1971) Facilitation of sympathetic neurotransmission in the cat spleen after inhibition of prostaglandin synthesis. Acta Physiol Scand 83:430–432

Johnson DG, Thompson WJ, Williams RH (1974) Regulation of adenylylcyclase from isolated pancreatic islets by prostaglandins and guanosine 5′-triphosphate. Biochemistry 13:1920–1924

Kantor HD, Hampton M (1978) Indomethacin in submicromolar concentrations inhibits cyclic AMP-dependent protein kinase. Nature 276:841–842

Kelly KL, Laychock SG (1981) Prostaglandin synthesis and metabolism in isolated pancreatic islets of the rat. Prostaglandins 21:759–769

Kuehl FA Jr (1974) Prostaglandins: cyclic nucleotides and cell function. Prostaglandins 5:325–340

Kuo WN, Hodgkins DS, Kuo JF (1973) Adenylate cyclase in islets of Langerhans. Isolation of islets and regulation of adenylate cyclase activity by various hormones and agents. J Biol Chem 248:2705–2711

Kurzrok R, Lieb CC (1930) Biochemical studies of human semen. Proc Soc Exp Biol Med 28:268–282

Lefèbvre PJ, Luyckx AS (1978) Stimulation of gastric-glucagon release by prostaglandin $E_1$. Prostaglandins Med 1:419–420

Luyckx AS, Lefèbvre PJ (1978) Possible role of endogenous prostaglandins in glucagon secretion by isolated guinea-pig islets. Diabetologia 15:411–416

Luyckx AS, Lefèbvre PJ (1979) Further studies on the role of prostaglandins in glucagon secretion. In: Abstracts of the fourth international prostaglandin conference, Washington DC, 27–31 May 1979, p 72

Luyckx AS, Lefèbvre PJ (1980a) Further studies on the role of prostaglandins in glucagon secretion. In: Proceedings of the 2nd European symposium on hypoglycemia, Rome, January 1979. Academic Press, London New York, pp 47–56

Luyckx AS, Lefèbvre PJ (1980b) Endogenous prostaglandins modulate glucagon secretion by isolated guinea-pig islets. Adv Prostaglandin Thromboxane Res 8:1299–1302

Luyckx AS, Lefèbvre PJ (1980c) Prostaglandins and glucagon release in vitro: effect of phospholipase $A_2$, indomethacin, somatostatin and theophylline (Abstr). Diabetologia 19:296

Luyckx AS, Lefèbvre PJ (1981) Prostaglandines et hormones pancréatiques. In: Journées annuelles de diabétologie de l'Hôtel Dieu. Flammarion, Paris, pp 27–41

Luyckx AS, Mendoza E, Lefèbvre PJ (1980) Furosemide intravenous infusion in normal man: electrolytic, metabolic and hormonal effects. Lack of changes in basal insulin and glucagon plasma levels. Arch Int Pharmacodyn 248:305–313

Luyckx AS, Deliege M, Jardon-Jeghers CL, Lefèbvre PJ (1981a) Insulin, prostaglandin $E_2$ and glucagon release by human insuloma tissue incubated in vitro, influence of indomethacin. Diabete Metab 7:13–17

Luyckx AS, Gueurten D, Scheen A, Delporte JP, Lefèbvre P, Jaminet F (1981b) Effect of indomethacin on the metabolic and hormonal response to a standardized breakfast in normal subjects. Acta Diabetol Lat 18:259–266

Luyckx AS, Mendoza E, Lefèbvre PJ (1981c) Failure of indomethacin to affect arginine-induced C-peptide and glucagon release in insulin-treated diabetics- Major role of residual B-cell function in conditionning the magnitude of the blood glucose rise after intravenous arginine. Diabetologia 21:376–382

Matsuyama T, Horie H, Namba M, Nonaka K, Tarui S (1983) Glucose dependent stimulation by prostaglandin-$D_2$ of glucagon and insulin in perfused rat pancreas. Life Sci 32:979–982

McCardel B, McAdams MR, Pek SB (1981) Relationship of prostaglandin synthesis to the effects of norepinephrine on islet-hormone secretion (Abstr). Endocrinology 108 (Suppl 1) 144
Metz SA, Fujimoto WY, Robertson RP (1980) A role for prostaglandins as mediators of α-adrenergic inhibition of the acute insulin response to glucose. Prostaglandin Thromboxane Res 8:1291–1294
Micossi P, Pontiroli A, Baron SH, Tamayo RC, Lengel F, Bevilacqua M, Raggi U, Rorbiato G, Foà PP (1978) Aspirin stimulates insulin and glucagon secretion and increases glucose tolerance in normal and diabetic subjects. Diabetes 27:1196–1204
Ojeda SR, Naor Z, Negro-Vilar A (1979) The role of prostaglandins in the control of gonadotropin and prolactin secretion. Prostaglandins Med 5:249–275
Pace-Asciak C, Cole S (1975) Inhibitors of prostaglandin catabolism. I. Differential sensitivity of 9-PGDH, 13-PGR and 15-PGDH to low concentrations of indomethacin. Experientia 31:143–147
Pek S, Tai TY, Crowther R, Fajans SS (1976) Glucagon release precedes insulin release in response to common secretagogues. Diabetes 25: 764–770
Pek S, Tai TY, Elster A (1978) Stimulatory effects of prostaglandins E-1, E-2 and F-2-alpha on glucagon and insulin release in vitro. Diabetes 27:801–809
Pek S, Lands EM, Akpan J, Hurley M (1980) Effect of prostaglandins $H_2$, $D_2$, $I_2$ and thromboxane on in vitro secretion of glucagon and insulin. Adv Prostaglandin Thromboxane Res 8:1295–1298
Prince RL, Larkins RG, Alford FP (1981) The effect of acetylsalicylic acid on plasma glucose and the response of glucose regulatory hormones to intravenous glucose and arginine in insulin treated diabetics and normal subjects. Metabolism 30:293–298
Sacca L, Perez G (1976) Influence of prostaglandins on plasma glucagon levels in the rat. Metabolism 25:127–130
Samuelsson B, Ramwell P, Paoletti R, (eds) (1980) Adv Prostaglandin Thromboxane Res 6
Schusdziarra V, Rouiller D, Jaffe BM, Harris V, Unger RH (1980) Effect of endogenous and endogenous prostaglandin E upon gastric endocrine function in dogs. Endocrinology 106:1620–1627
Schusdziarra V, Rouiller D, Harris V, Wasada T, Unger RH (1981) Effect of prostaglandin $E_2$ upon release of pancreatic somatostatin-like immunoreactivity. Life Sci 28:2099–2102
Silver MJ, Smith JB (1975) Prostaglandins as intracellular messengers. Life Sci 16:1635–1648
Simon LS, Mills JA (1980) Nonsteroidal antiinflammatory drugs. N Engl J Med 302:1179–1185
Torella R, Giugliano D, Siniscalchio N, Sgambato S, D'Onofrio F (1979a) Influence of acetylsalicylic acid on plasma glucose, insulin, glucagon and growth hormone levels following tolbutamide stimulation in man. Metabolism 28:887–889
Torella R, Siniscalchio N, Improta L (1979b) The influence of acetylsalicylic acid on insulin, glucagon and growth hormone plasma levels following glucose and arginine in man. Farmaco 34:131–137
Vane JR (1971) Inhibition of prostaglandin synthesis as a mechanism of action for aspirin-like drugs. Nature New Biol 231:232–235
Vierhapper H, Bratusch-Marrain P, Waldhausl W (1980) Unchanged arginine-induced stimulation of insulin, glucagon, growth hormone and prolactin after pretreatment with indomethacin in normal man. J Clin Endocrinol Metab 50:1131–1134
von Euler US (1934) Zur Kenntnis der pharmakologischen Wirkungen von Nativsekreten und Extrakten männlicher accessorischer Geschlechtsdrüsen. Arch Exp Pathol Pharmakol 175:78–84
Weir GC, Knowlton SC, Martin DB (1975) Nucleotide and nucleoside stimulation of glucagon secretion. Endocrinology 97:932–936
Weiss B, Hait WN (1977) Selective cyclic nucleotide phosphodiesterase inhibitors as potential therapeutic agents. Annu Rev Pharmacol Toxicol 17:441–477
Wollheim CB, Blondel B, Renold AE, Sharp GW (1976) Stimulatory and inhibitory effects of cyclic AMP on pancreatic glucagon release from monolayer cultures and the controlling role of calcium. Diabetologia 12:269–277

CHAPTER 29

# Hormones in the Control of Glucagon Secretion

S. B. PEK and R. S. SPANGLER

## A. Introduction

This is a review of the knowledge on one of the messenger systems involved in the regulation of secretion of glucagon. The messengers in this system are hormones, and, by definition, they are blood borne and reach their target, namely the A-cells of pancreatic islets, through the vascular system. The purpose of this hormonal messenger system is multiple. (1) The availability of hormones that are involved in the regulation of certain metabolic events would be coordinated by each influencing the secretion of the other (Sects. B–E). (2) Hormones which are secreted by cells in the intestinal endothelium could alert the glucagon-producing cell to the approaching work load, namely the assimilation of ingested nutrients (Sect. F). (3) High level coordination of metabolic events would be achieved by the central nervous system via the humoral route (Sects. G, H).

The evidence that the hormonal messenger system is involved in the regulation of glucagon release is circumstantial. The first line of evidence comprises information on the effects of exogenous, administered hormones on glucagon release. The second line of evidence is provided by associations in the changes of circulating levels of glucagon with those of other hormones. The information reviewed in this chapter is not intended to be comprehensive. The cited reports are mostly, if not exclusively, those in which glucagon levels have been quantified by radioimmunoassays using COOH terminal-specific glucagon antisera (see Chap. 10). The actions of other pancreatic islet hormones or of catecholamines on glucagon release have been excluded, since they are covered in other chapters of this Handbook (see Chaps. 22, 30, and 31).

## B. Thyroid Hormones

Thyroid hormones enhance mitochondrial enzymatic activities and the affinity of adrenergic receptors for catecholamines. Thus, thyroid hormones could promote glucagon secretion. Indeed, EDWARDS et al. (1974) observed that, in thyroidectomized calves, replacement with thyroxine resulted in greater increases in plasma levels of glucagon in response to splanchnic nerve stimulation than the increases in thyroidectomized controls not receiving thyroxine. KABADI et al. (1980) reported that, in 13 patients with hyperthyroidism, ingestion of glucose failed to suppress plasma levels of glucagon; 4 of these patients had elevated fasting levels of glucagon as well. When thyroid function returned to normal, plasma glucagon became responsive to glucose. On the other hand, SHIMA et al. (1976) found that, in patients with hyperthyroidism, the increases in plasma glucagon were normal in response to insulin-induced hypoglycemia and subnormal (60% of the increases observed in healthy control subjects) in response to intravenously administered arginine. No information is available on the in vitro direct action of thyroid hormones on the A-cell. Thus, the knowledge on this hormonal interaction is inconclusive.

## C. Calcium-Regulating Hormones

### I. Parathyroid Hormone

In patients with hyperparathyroidism, pancreatitis occurs more often than in the general population. A potential association between parathyroid hormone excess and/or hypercalcemia and glucagon secretion was suspected when patients with acute pancreatitis were found to have elevated plasma levels of glucagon. Indeed, during the era of nonspecific immunoassays for glucagon, hyperglucagonemia was suggested in some patients with parathyroid hyperplasia or adenoma in the absence of pancreatitis (PALOYAN 1967). A systematic study on the in vivo influence of parathyroid hormone on glucagon release has not been conducted with specific glucagon assays. Induction of hypercalcemia in response to excess parathyroid hormone in circulation would complicate the interpretation of the results of such studies. In vitro, REBOLLEDO et al. (1975) administered natural parathyroid hormone (0.15 IU/ml) to the isolated perfused rat pancreas and observed no changes in glucagon secretion at low or high concentrations of glucose or in response to arginine. Likewise, in the isolated rat pancreas, we could not show an effect of the synthetic 1–34 amino acid sequence of bovine parathyroid hormone ($10^{-10}$–$10^{-7}$ $M$ perfused over 10 min) on basal or arginine-stimulated secretion of glucagon (S. B. PEK and G. MAYOR 1975, unpublished work). Thus, any effect of parathyroid hormone on glucagon release is likely to be mediated by the hypercalcemic action of parathyroid hormone.

### II. Calcitonin

STARKE et al. (1981) recently reported that an intravenous infusion of porcine calcitonin (4.5 IU/kg body weight over 2 h) induced a 18%–22% decrease in plasma glucagon circulating levels of both healthy volunteers and hyperglucagonemic insulin-dependent diabetics.

In contrast, GIUGLIANO et al. (1982) recently reported that the suppressive effect of glucose on glucagon release in man was significantly reduced by calcitonin infusion.

### III. Vitamin D

The effects of vitamin D or its biologically active metabolites on glucagon release have not been studied. A direct effect of vitamin D on the pancreas may be anticipated based on the observation that the chick pancreas contains vitamin D-inducible calcium-binding protein (CHRISTAKOS et al. 1979). Preliminary information suggests that any such effects of vitamin D on the pancreatic tissue may be selective and exclude the A-cells: NORMAN et al. (1980) have shown that, in the presence of physiologic concentrations of calcium, the secretion of insulin, but not glucagon, is reduced from pancreata isolated from vitamin D-deficient rats.

## D. Steroid Hormones

### I. Glucocorticoids

The effects of glucocorticoid hormones on glucose homeostasis resemble those of glucagon. Glucocorticoids play a major permissive role in the activation of hepatic gluconeogenesis by glucagon. In order for this synergism between glucocorticoids and glucagon at the target tissue level to operate optimally, a feedback regulation of secretion of glucagon by glucocorticoids could be anticipated. Early studies (VANCE et al. 1968) had suggested that in vitro glucocorticoids stimulate glucagon release from isolated rat islets. In humans, WISE et al. (1973) reported that dexamethasone pretreatment (8 mg/day for 3 days) increased fasting plasma levels of glucagon by approximately 55% and enhanced by 60%–100% the increases in plasma glucagon in response to intravenous infusion of L-alanine (0.15 g/kg) or to protein feeding. On the other hand, in healthy subjects, we observed no effect of a similar pretreatment with dexamethasone on basal levels of glucagon or the increases in plasma glucagon in response to L-arginine (0.41 g/kg intravenously over 30 min; PEK et al. 1973). Furthermore, in another group of healthy subjects, the decreases in plasma glucagon were similar during oral glucose tolerance tests with or without cortisone pretreatment (PEK et al. 1973). MARCO et al. (1973) observed in healthy subjects that, after oral administration of 40 or 60 mg/day prednisolone for 4 days, basal levels as well as arginine-stimulated increases in plasma glucagon were augmented. However, acute administration of prednisolone (100 mg intravenously to healthy subjects) failed to affect plasma levels of glucagon (MARCO et al. 1973). Using the method of isolated, perfused pancreas, LENZEN (1976) noted that glucagon secretion from pancreata obtained from adrenalectomized rats was not adequately suppressed in response to 16.7 m$M$ glucose. When the adrenalectomized animals had been treated with hydrocortisone, in vitro glucose-induced suppression of glucagon release recovered. MARCO et al. (1976) pretreated mice with prednisolone (0.2–0.3 mg/day for 4 days) and observed that the islets isolated from the pancreata of these mice secreted 20% more glucagon than those of control mice in the basal state and in response to arginine. But, in the same study, prednisolone ($5 \times 10^{-5}$ $M$) administered in vitro to isolated mouse islets failed to stimulate glucagon secretion. These observations, though apparently conflicting or contradictory, suggest that under appropriate circumstances, particularly upon prolonged administration, glucocorticoids promote glucagon secretion mainly by indirect mechanisms. (See also Chap. 32.)

### II. Mineralocorticoids

Information is not available on the effect of aldosterone on glucagon release. Recently, BARSEGHIAN and LEVINE (1980) reported that corticosterone, when administered acutely at physiologic concentrations, enhanced arginine-induced secretion of glucagon from the isolated, perfused rat pancreas. Since corticosterone is the major mineralocorticoid hormone in the rat, these results also suggest that the mineralocorticoids promote glucagon secretion.

### III. Sex Steroids

There is no report on any effect of androgenic steroids on glucagon release. The information on female sex steroids is sketchy. BECK et al. (1975) observed that when women were pretreated orally with synthetic estrogens at contraceptive dosages, the increases in plasma glucose which occurred in response to intravenously administered arginine were 25% less than those observed without pretreatment. In women, medroxyprogesterone acetate (10 mg/day orally for 5 days) had no effect on basal or stimulated plasma levels of glucagon (BECK et al. 1977). The same group also reported that after 2 weeks treatment with combined mestranol (80 μg) plus norethindrone (1 mg) daily, the mean peak plasma glucagon response to arginine infusion was suppressed to one-quarter of control levels (BECK et al. 1976).

## E. Placental Hormones

The placenta functions as a major endocrine gland during pregnancy. Most of its hormonal products (polypeptides, glycoproteins, steroids etc.) are analogous to those produced by other endocrine glands, and have been reviewed in other parts of this chapter. A single report exists on the effect on glucagon secretion of an exclusive product of the placenta, chorionic somatomammotropin. When perfused to the isolated rat pancreas over extended periods of time, human chorionic somatomammotropin (10 μg/ml) evoked only a short-lived secretion of glucagon; 11 m*M* glucose suppressed this effect (LAUBE et al. 1972). The physiologic significance of this observation at a huge concentration of the hormone is questionable.

## F. Gastrointestinal Hormones

During the last two decades, a variety of polypeptides have been isolated from the gastrointestinal tissues. Some of these peptides have well-documented biologic actions on various organs. Since the degree of purity of some of the peptides is questionable, biologic actions ascribed to them also remain questionable. Sensitive assays are available to measure the blood levels of some of them in various species, including humans. The role of these polypeptides as the mediators of the signals which reach the pancreatic islets from the gastrointestinal tract during ingestion of nutrients is the subject of intense investigation. Many of the polypeptides isolated from the gastrointestinal tract have structural similarities. Particularly, gastric inhibitory polypeptide (GIP), vasoactive intestinal peptide (VIP), secretin, the porcine heptacosapeptide, bombesin, glicentin, and glucagon form a related family of polypeptides. Another family of polypeptides comprise cholecystokinin, gastrin, cerulein, and motilin.

## I. Gastric Inhibitory Polypeptide

GIP is considered to be a major insulinotropic gastrointestinal peptide. TAMINATO et al. (1977) administered glucose, with or without 1 μg/kg synthetic GIP, intravenously to rats. The suppression of plasma levels of glucagon by glucose was not attenuated significantly by GIP. The same dose of GIP, when injected intravenously without glucose, evoked increases in plasma glucagon. In the isolated rat islets, these workers observed that 1 μg/ml GIP stimulated glucagon release in the presence of 3.3, 8.3, or 16.7 m*M* glucose. Fragments of GIP (amino acid sequences 1–28, 22–43, and 15–43) did not affect glucagon secretion in vitro.

FUJIMOTO et al. (1978) employed 4-day cultures of neonatal rat pancreatic cells and observed that 1 or 10 ng/ml GIP stimulated glucagon release at glucose concentrations of 1.7, 8.3, or 16.5 m*M*. When glucagon release had been stimulated by a mixture of amino acids, GIP augmented the release only at 10 ng/ml. Using an immunoperoxidase technique, these workers demonstrated binding of exogenous GIP to some of the islet cells, but not to fibroblasts, further strengthening the evidence that GIP may have a physiologic role in the regulation of islet hormone secretion. In the same system, FUJIMOTO et al. (1979) also showed that the stimulatory effect of 0.5 ng/ml GIP on glucagon release was not modified by the addition of cholecystokinin or secretin into the incubation media. In dogs, SIRINEK et al. (1980) failed to observe any changes in plasma levels of glucagon in response to intravenous administration of a highly purified preparation of GIP, either as a prolonged infusion (200 and 400 ng $kg^{-1}$ $h^{-1}$, leading to plasma levels of 678 and 1404 pg/ml, respectively), or as a bolus injection (200 ng/kg, maximal plasma level 2150 pg/ml). VERDONK et al. (1980) employed the glucose clamp technique to study the relationship between the changes in plasma levels of GIP and glucagon in response to ingestion of fat (Lipomul) in healthy subjects whose blood levels of glucose were maintained low (43 mg/dl), normal (89 mg/dl), or high (141 mg/dl). No relationship was found between the fat-induced increases in plasma GIP and the levels of glucagon. The results of the in vitro studies with administered GIP assign to GIP a stimulatory role in the regulation of glucagon secretion. However, the latter studies weaken the possibility that the effects of GIP on glucagon release have a physiologic significance.

## II. Vasoactive Intestinal Peptide

VIP, when infused into the pancreatic artery in anesthetized dogs, evoked increases in pancreatic venous plasma levels of glucagon at VIP dosages of 20–400 ng/10 min (OHNEDA et al. 1977). This effect did not appear to be the result of an increase in pancreatic blood flow induced by VIP. Unlike previously observed hyperglycemia at high dosages of VIP, in this study blood levels of glucose did not change. KANETO et al. (1977) infused VIP 50 ng/kg in 10 min into the pancreaticoduodenal artery of anesthetized dogs and observed increases in pancreaticoduodenal venous blood flow and levels of glucagon. Arterial glucagon levels remained unchanged and glucose levels increased gradually. These effects of VIP were not altered by $\beta$-adrenergic blockade induced by propranolol. SCHEBALIN et al. (1977) perfused the cat pancreas in situ with a physiologic buffer solution. At a physiologic concentration of glucose, pulse injections of 0.5 and 5 μg VIP stimulated glucagon release by 158% and 327%, respectively. VIP was ineffective at a high concentration of glucose. In the isolated perfused dog pancreas, JENSEN et al. (1978 a) also observed that 0.03–18.8 n*M* highly purified porcine VIP stimulated glucagon secretion in a dose-dependent manner; VIP was ineffective at glucose concentrations greater than 5.6 m*M*. In a similar system, HERMANSEN (1980) noted that 1 n*M* VIP evoked a 43% rise in portal venous effluent level of glucagon at a glucose concentration of 5.5 m*M*. When the isolated rat pancreas was perfused with 1 n*M* purified porcine VIP, glucagon secretion was stimulated at glucose concentrations of 4.4 m*M* or less; arginine-induced glucagon release was also augmented by VIP (SZECÓWKA et al. 1980 a). Thus, VIP augments glucagon secretion at concentrations which could occur in a physiologic setting, and this effect is significantly modulated by glucose. With the demonstration by immunofluorescence of VIP in pancreatic islets associated with islet cells (BUFFA et al. 1977) and nerve endings (UNGER and ORCI 1981), a local regulatory role may also be assigned to this peptide hormone. A stimulatory effect of VIP infused intraarterially at a dose of 2 μg/min in the isolated perfused dog stomach was also reported by LEFEBVRE and LUYCKX (1980; see also Chap. 33).

## III. Secretin

The physiologic stimulus for the release of secretin from the intestine into the circulation is the lowering of hydrogen ion concentration in the intestinal lumen. Thus, secretin is an ideal candidate for nutrient-induced intestinal signal for islet hormone release. SANTEUSANIO et al. (1972) studied the effect of a highly purified preparation of secretin (Karolinska Institutet, Stockholm, Sweden) on glucagon levels in conscious dogs. Intraportal infusion of 10 IU/min secretin over 20 min resulted in a 50% reduction in glucagon levels in pancreaticoduodenal venous blood. When administered during modest degrees of hyperglycemia induced by intravenous infusion of glucose, secretin was effective in suppressing glucagon secretion at the same dose, as well as at a much lower dose. When secretin was infused intravenously in alloxan-induced diabetic dogs with hyperglucagonemia, peripheral venous levels of glucagon were suppressed. When 14 mequiv. hydrochloric acid was instilled intraduodenally in normal dogs receiving intravenous infusions of glucose, pancreaticoduodenal venous glucagon levels declined, presumably associated with increased secretion of secretin. These observations assigned a physiologic role to secretin as an amplifier of glucagon response to ingested glucose. On the other hand, JENSEN et al. (1978 b) failed to observe any effect of perfusions of pure natural porcine secretin (2.8–278 p*M*) on glucagon release from isolated porcine pancreas in the presence of 3.5–7.5 m*M* glucose. In the latter experiments, secretin was effective in inducing insulin as well as exocrine pancreatic secretion. Thus, the information on the role of secretin in the regulation of glucagon release is inconclusive.

## IV. Porcine Intestinal Heptacosapeptide

A heptacosapeptide with yet to be determined amino acid sequence has been isolated from porcine intestine (MUTT 1978). This polypeptide is likely to have structural similarity to secretin, glucagon, VIP, and GIP. In the isolated, perfused rat pancreas, this polypeptide (3 ng/ml) augmented arginine-induced glucagon release (SZECÓWKA 1980b).

## V. Bombesin

The tetradecapeptide, bombesin was first isolated from frog skin. Subsequently, a similar peptide has been found in mammalian gastrointestinal tract with structural similarity to vasoactive intestinal peptide (MUTT 1978). Bombesin has hyperglycemic action. FALLUCCA et al. (1977, 1978) administered 10 ng $kg^{-1}$ $min^{-1}$ bombesin intravenously to healthy subjects. Modest increases in plasma levels of glucagon were observed. KANETO et al. (1978) administered 20 pmol $kg^{-1}$ $min^{-1}$ bombesin into the pancreatic artery of anesthetized dogs over 10 min. Increases in pancreaticoduodenal venous and arterial levels of glucagon were observed without any arteriovenous step-up across the pancreas. These workers concluded that bombesin must have stimulated the release of glucagon from extrapancreatic sources. On the other hand, in the isolated perfused canine pancreaticoduodenal preparation, 1 n*M* bombesin evoked increases in glucagon secretion by 30% above the basal secretion (HERMANSEN 1980).

## VI. Cholecystokinin

Cholecystokinin (CCK), also referred to as cholecystokinin-pancreozymin, is one of the earliest recognized biologically active products of the intestinal tract. Many of the effects attributed to CCK were based on studies carried out with impure preparations. The confusion created by these early studies, particularly those on the effects of CCK on islet hormone secretion, still lingers. Highly purified natural and synthetic preparations of CCK and its fragments are now available. WILLIAMS and CHAMPAGNE (1979) administered CCK 33 and the carboxyl terminal octapeptide, CCK 8 intravenously to dogs. Only CCK 8 at the highest concentration employed (0.27 μg/kg) evoked modest increases in plasma levels of glucagon. SIRINEK et al. (1980) administered a 99% pure preparation of CCK (Karolinska Institutet, Stockholm, Sweden) intravenously to conscious dogs at a dose assumed to result in physiologic increases in plasma levels of CCK (500 ng $kg^{-1}$ $h^{-1}$); plasma levels of glucagon did not change. On the other hand, in the isolated, perfused dog pancreas, HERMANSEN (1980) demonstrated that 1 n*M* CCK 8 evoked an increment in secretion of glucagon 140% above basal secretion. FUJIMOTO et al. (1979) studied the effect of CCK 8 on glucagon secretion from 4-day cultured neonatal rat pancreatic cells. In the presence of 16.5 m*M* glucose, 1 or 10 ng/ml CCK 8 had no effect on glucagon secretion. When fetal calf serum was added to the media, 10 ng/ml CCK 8 modestly stimulated glucagon release. REHFELD et al. (1980) perfused the isolated porcine pancreas for 5-min periods with $10^{-10}$–$10^{-6}$ *M* COOH terminal tetrapeptide of CCK, namely CCK 4. Glucagon secretion was stimulated at all concentrations tested. More information is needed to assess the role of CCK in regulation of glucagon release. These data suggest that only the short COOH terminal sequences of the polypeptide may have significant activity. The four amino acids at the COOH terminus of CCK are identical to those of gastrin. This tetrapeptide sequence has been identified in nerve terminals in pancreatic islets (REHFELD et al. 1980). Thus, these observations may be interpreted as indicative of neuronal rather than humoral regulation of islet function.

## VII. Gastrin

Gastrin circulates in human blood in several forms: gastrin 34 ("big" gastrin), gastrin 17, gastrin 14 ("minigastrin") in sulfated and nonsulfated forms, and "component-I" (REHFELD 1972; REHFELD et al. 1974). Among the gastrointestinal hormones, gastrin has been in the forefront as a candidate "messenger" of the intestinal signal for islet hormone secretion. Earlier studies on the effect of gastrin on glucagon release have been carried out using relatively crude preparations (BUCHANAN et al. 1969; DUPRÉ et al. 1969; IVERSEN 1971). REHFELD et al. (1978) studied the effect of synthetic human gastrin 17 on plasma glucagon in healthy young adult subjects. Intravenous bolus injections of gastrin 17 in doses from 0.016 to 1 μg/kg resulted in dose-related increases in plasma levels of glucagon, maximal responses occurring 5 min after the injection. When gastrin was injected at the 30th minute of an intravenous infusion of a mixture of fifteen amino acids, amino acid-induced increases in plasma glucagon were not potentiated. JENSEN et al. (1980) perfused the isolated porcine pancreas with various forms of gastrin in the concentration range $10^{-11}$–$10^{-8}$ *M* over 5-min periods. In the presence of 5 m*M* glucose, at the concentration of $10^{-8}$ *M*, gastrin 34, and sulfated forms of gastrin 17 and gastrin 14 stimulated glucagon secretion. None of the gastrins stimulated glucagon release in the presence of 7.5 m*M* glucose. Thus, gastrins seem to affect glucagon secretion only at very high concentrations which may be attained in the circulation of patients with gastrin-producing tumors or with achlorhydria-induced secondary hypergastrinemia.

## VIII. Cerulein

Cerulein is a decapeptide isolated from the skin of frogs. The five amino acids closest to the COOH terminus of cerulein are identical to those of mammalian cholecystokinin and gastrin. When administered into the pancreatic artery of anesthetized dogs, at a dose range of 15–480 ng/min, cerulein increased plasma levels of glucagon in the pancreatic vein in a dose-dependent manner at infusion rates of 120 ng/min or greater (OHNEDA et al. 1978). The high dosages evoked increases in pancreatic blood flow as well. In anesthetized rats, OTSUKI et al. (1979) administered 1–1 000 ng $kg^{-1}$ $min^{-1}$ synthetic cerulein intravenously over 20 min. Glucagon secretion was stimulated at infusion rates greater than 10 ng $kg^{-1}$ $min^{-1}$. These findings resemble the findings with the analogs of gastrin and cholecystokinin.

## IX. Motilin

This intestinal polypeptide has structural similarity to gastrin 34 (MUTT 1978). Very limited information is available on hormonal/metabolic effects of motilin. The only reported study on the effects of motilin on glucagon levels was conducted in healthy subjects by CHRISTOFIDES et al. (1979). Intravenous infusions over 60 min of 0.34 or 0.68 pmol $kg^{-1}$ $min^{-1}$ natural motilin enhanced gastric emptying and meal-induced increases in plasma insulin, but had no effect on plasma levels of glucagon.

## G. Pituitary Hormones

### I. Growth Hormone

Hyperglycemic and diabetogenic effects of growth hormone have been recognized for many years. Prolonged administration of growth hormone to animals results in hyperplasia of pancreatic islets. Thus, stimulatory effects of growth hormone on glucagon secretion had been anticipated. In 1972, GOLDFINE et al. demonstrated that, in patients with active acromegaly, the increases in plasma glucagon in response to intravenously administered arginine were greater than the increases seen in healthy subjects. We observed in healthy subjects that natural human growth hormone when administered over a period of 3 days (10 mg/day intramuscularly) resulted in a doubling of both basal and arginine-stimulated glucagon levels (PEK et al. 1973). In vivo, not only chronic, but also acute increases in growth hormone levels are associated with elevated plasma levels of glucagon. SIREK et al. (1979) administered to anesthetized dogs 10 mg/kg bovine growth hormone intravenously as a pulse and observed increases in portal venous levels of glucagon. Increases in portal levels of glucagon also occurred in conscious dogs in response to a considerably smaller dose of bovine growth hormone (6 μg/kg intravenously). In patients with growth hormone deficiency, no distinct abnormalities in plasma levels of glucagon have been identified (BLACKARD et al. 1973; SIZONENKO et al. 1975). LEVITSKY et al. (1978) studied 32 growth hormone-deficient children before and during therapy with human growth hormone. In these children, before treatment, plasma levels of glucagon were similar to those seen in children with nonendocrine short stature in the fasting state and at the peak of the response to 30-min intravenous infusions of arginine. Treatment with human growth hormone (2 IU twice daily to 2 IU three times weekly) for 5 days, 6 months, or 12–30 months did not modify these parameters. After the infusion of arginine, the return of plasma glucagon to baseline levels was slow before growth hormone treatment and was accelerated during treatment. The latter changes could reflect an enhancement of metabolic clearance of glucagon by growth hormone, since the principal site of clearance of glucagon is the kidney and growth hormone increases renal blood flow.

We have provided conclusive evidence that growth hormone stimulates glucagon secretion directly (TAI and PEK 1976). In the isolated perfused rat pancreas, in the presence of 5.6 m$M$ glucose, bovine growth hormone stimulates glucagon secretion at concentrations $10^{-7}$–$10^{-9}$ $M$, in a dose-related manner. Glucagon response to the perfusion of growth hormone occurs rapidly (within 24–48 s) and dissipates rapidly (within 4 min), even when the perfusion continues over longer periods. Also, a late phase release in glucagon is observed at high concentrations of growth hormone. When the conditions are optimized for energy-requiring secretory process (by the addition of Krebs cycle intermediary substrates), acute phase release of glucagon occurs at the physiologic growth hormone concentrations of

$10^{-10}$ and $10^{-11}$ *M*. We have concluded that endogenous growth hormone, which is secreted in a pulsatile manner, exerts a tonic-stimulatory effect on pancreatic A-cells. Further information on the interrelationship between growth hormone and glucagon can be found in Chaps. 45 and 52.

## II. Adrenocorticotropic Hormone

Information on a direct effect of adrenocorticotropic hormone (ACTH) on glucagon secretion is sketchy. FUSSGÄNGER et al. (1973) perfused the isolated rat pancreas with fractions of porcine pituitary extracts which had lipolytic, but no corticotropic activity. At a concentration of 1 μg/ml, this pituitary preparation evoked biphasic release of insulin and glucagon. On the other hand, a pure preparation of ACTH stimulated the release of insulin, but not glucagon. In the isolated perfused rat pancreas, we observed modest increases in glucagon release in response to a supraphysiologic dose of highly purified native porcine ACTH (1.4 μ*M*) (MACADAMS and PEK 1979). Recently, ACTH-like immunoreactivity has been detected by histochemical techniques in porcine endocrine pancreas (SUNDLER et al. 1979). This observation brings renewed interest in ACTH as a potential regulator of islet hormone secretion, but at the local or "paracrine" level. If genuine ACTH were produced in the pancreatic islets, then its interstitial concentrations could be quite high, approaching the concentration at which we noted an effect of exogenous ACTH on glucagon release.

## III. Endorphins and Enkephalins

Endorphins and enkephalins are endogenous opioid polypeptides which are metabolites of the polypeptide, proopiomelanocortin. The latter is also the precursor of adrenocorticotropic hormone and melanocyte-stimulating hormone. During the last 5 years, large amounts of information have become available on diverse natural and synthetic analogs of these opioid hormones and their receptors in various tissues. Major gaps still exist in the knowledge on the scope of involvement of opioids in various biologic processes.

In monolayer neonatal rat pancreatic cell cultures, in the presence of 16.5 m*M* glucose, leucine-enkephalin, D-alanine-2-leucine-enkephalin, and methionine-enkephalin were found to suppress glucagon secretion in a dose-related manner at concentrations 0.02–2.0 μ*M*; 2 μ*M* enkephalin attenuated glucagon secretory response to arginine and theophylline by 50% and 45%, respectively (KANTER et al. 1980). On the other hand, in healthy subjects, intravenous bolus injections of 0.5, 2.5, and 5 mg synthetic human β-endorphin evoked increases in plasma levels of glucagon (REID and YEN 1981). The latter effect was suggested to be mediated by inhibition of somatostatin release by β-endorphin. These observations provide indirect evidence that islet cells possess receptors for enkephalins and endorphins. The opioid peptides are present in the pancreas as well. FORSSMAN et al. (1977) identified enkephalin immunoreactivity in extracts of human pancreas. BRUNI et al. (1979) found β-endorphin immunoreactivity in extracts of human pancreas and, using immunofluorescence techniques, localized β-endorphin in the pancreatic islets. The significance of these findings in the regulation of islet hormone secretion remains to be determined.

FELDMAN et al. (1983) most recently reported their findings in healthy volunteers and insulin-dependent diabetic patients infused with synthetic human beta-endorphin. In both groups, beta-endorphin increased plasma glucagon concentrations, and this rise was accompanied by a significant increase in blood glucose levels. The threshold for these effects was 0.005 mg. The glucose and glucagon response to beta-endorphin could not be blocked by intravenous naloxone.

## H. Hypothalamic Hormones

The regulation of the function of the endocrine pancreas by the hypothalamus has been the subject of much investigation for many decades. Experiments to induce stereotactic lesions in various parts of the hypothalamus and to monitor the circulating levels of islet hormones were among the earliest forms of investigation. The mechanisms by which such lesions of the hypothalamus could affect hormonal/metabolic events have not been defined. Alterations in neural signals from the hypothalamus to the endocrine glands could be involved. Recent recognition of a variety of polypeptide hormones in the hypothalamus invites the hypothesis that humoral pathways are operative as well. The possibility that these hypothalamic hormones could attain biologically significant concentrations in the systematic circulation is small. On the other hand, the hypothalamic hormones could serve as neurotransmitters and reach the target endocrine glands via peptidergic neurons.

### I. Hypothalamic Lesions

In rats with lesions of the ventromedial hypothalamus (VMH), CHIKAMORI et al. (1980) found that the increases in plasma levels of glucagon induced by arginine were less than those seen in sham-operated control animals. Yet, ROHNER-JEANRENAUD and JEANRENAUD (1980) and KARAKASH et al. (1980) observed in vitro that arginine-induced glucagon release from pancreata isolated from rats with hypothalamic lesions was twofold greater than that from control rats. GOTO et al. (1980) induced VMH lesions in rats. In vitro release of glucagon from pancreata of these rats in response to theophylline or arginine plus theophylline was enhanced, when compared with control rats.

### II. Substance P and Neurotensin

Two polypeptides isolated from the hypothalamus, substance P and neurotensin, were tested for their effects on glucagon secretion. BROWN and VALE (1976) administered 6 μg synthetic substance P and 2 μg neurotensin intravenously to rats and observed hyperglucagonemia as well as hyperinsulinemia. The experiments of PATTON et al. (1976) with the isolated perfused dog pancreas indicated that these effects of substance P and neurotensin (at perfusion rates greater than 100 pmol/min) to stimulate glucagon and insulin are likely to be direct actions on the islet cells. In the isolated canine pancreaticoduodenal preparation, HERMANSEN (1980) found that substance P, at concentrations 0.2–5 n*M*, stimulated glucagon secretion; this effect was attenuated by a high concentration of glucose. DOLAIS-KITABGI et al. (1979) employed neonatal rat islets which had been kept in culture for 48 h. At a glucose concentration of 3 m*M*, 10–100 n*M* neurotensin had a short-lived stimulatory effect on glucagon release. On the other hand, when glucagon secretion had been stimulated with 20 m*M* arginine, neurotensin attenuated this response.

### III. Unidentified Polypeptides

MOLTZ et al. (1977) incubated various regions of the rat brain in vitro. They observed that substances released from segments of the VMH, but not the ventrolateral hypothalamus, stimulated glucagon secretion from isolated rat islets. In these experiments, both substance P and neurotensin (0.1 and 1 μ*M*) also stimulated glucagon release. However, the solutions in which the segments of VMH had been incubated did not contain detectable amounts of immunoreactive substance P or neurotensin, suggesting that the VMH produces substances, other than substance P or neurotensin, which have glucagon-releasing activity. Acid-acetone extracts of the rat VMH or the media in which VMH segments had been incubated were subjected to enzymatic digestion (MOLTZ et al. 1979). Glucagon-releasing activity present in these biologic fluids was abolished by carboxypeptidase or pepsin, unaffected by trypsin, and enhanced by chymotrypsin. Since peptidases altered the glucagon-releasing activity in the VMH products, these products are likely to be polypeptides. Upon Sephadex gel chromatographic fractionation, glucagon-releasing activity was found to be present in late fractions of the columns; the elution position of the activity did not correspond to that of substance P or neurotensin (MOLTZ et al. 1979). These observations clearly indicate that the hypothalamus has the potential of producing several polypeptides which directly stimulate glucagon secretion. The means by which these polypeptides could reach the pancreatic islet and their physiologic role in the regulation of A-cell function should be targets of future investigation.

## J. Conclusions

This chapter exposes one fact: very little is known on the hormonal regulation of secretion of glucagon. (a) Many hormones have not been tested for their possible influences on glucagon release, even though, on the basis of their actions in other tissues, they may be expected to influence the secretory process. (b) In studies with administered hormones, the concentrations employed have been supraphysiologic in many instances, or appropriate dose-response relationships have not been addressed. (c) In some instances, the purity of the exogenous hormone has been inadequate or has not been documented. (d) Even if evidence may have been presented that the administration of a pure hormone at physiologic concentrations affects glucagon release, evidence is frequently missing that the endogenous hormone can do the same thing. (e) In most instances where there is reasonable evidence that the endogenous hormone influences glucagon release, the biologic significance of this interaction has not been defined. (f) In most instances, the mechanism of action of the hormone, or the interaction of the hormone with specific receptors on the A-cell, or at least islet cells, have not been explored. (g) In most if not all instances, no evidence has been presented on the consequence of the neutralization of the active hormone with specific antibodies directed against that hormone or with antagonists of the hormone.

On the positive side, evidence has been provided that the hormones which are involved in the same metabolic processes as glucagon can affect the secretion of glucagon. Many hormones of the gastrointestinal tract, the pituitary, and the hypothalamus influence glucagon release. The structural similarities between some of the gastrointestinal and hypothalamic hormones are intriguing and strengthen the probability that they can act as neurotransmitters as well as hormones. A breakthrough in the understanding of hormone-hormone interactions may be anticipated.

## References


Barseghian G, Levine R (1980) Effect of corticosterone on insulin and glucagon secretion by the isolated perfused rat pancreas. Endocrinology 106:547–551

Beck P, Eaton RP, Arnett DM, Alsever RN (1975) Effect of contraceptive steroids on arginine-stimulated glucagon and insulin secretion in women. I. Lipid physiology. Metabolism 24:1055–1060

Beck P, Arnett DM, Alsever RN, Eaton RP (1976) Effect of contraceptive steroids on arginine-stimulated glucagon, and insulin secretion in women. II. Carbohydrate and lipid physiology in insulin-dependent diabetics. Metabolism 25:23–31

Beck P, Zimmerman DE, Eaton RP (1977) Effect of contraceptive steroids on arginine-stimulated glucagon and insulin secretion in women. III. Medroxyprogesterone acetate. Metabolism 26:1193–1197

Blackard WG, Andrews SS, Lazarus EJ (1973) Effect of growth hormone deficiency on glucagon secretion. Proc Soc Exp Biol Med 143:1042–1044

Brown M, Vale W (1976) Effects of neurotensin and substance-P on plasma insulin, glucagon and glucose levels. Endocrinology 98:819–822

Bruni JF, Watkins WB, Yen SSC (1979) $\beta$-endorphin in the human pancreas. J Clin Endocrinol Metab 49:649–651

Buchanan KD, Vance JE, Williams RH (1969) Insulin and glucagon release from isolated islets of Langerhans. Diabetes 18:381–386

Buffa R, Capella C, Solcia E, Frigerio B, Said SI (1977) Vasoactive intestinal peptide (VIP) cells in the pancreas and gastrointestinal mucosa. Histochemistry 50:217–227

Chikamori K, Nishimura N, Suehiro F, Sato K, Mori H, Saito S (1980) Alterations in glucagon secretion in obese rats with hypothalamic lesions. Horm Metab Res 12:56–59

Christakos S, Friedlander EJ, Frandsen BR, Norman AW (1979) Mode of action of calciferol. XIII. Development and application of a radioimmunoassay for vitamin D-dependent chick intestinal calcium-binding protein and tissue distribution. Endocrinology 104:1495–1503

Christofides ND, Modlin IM, Fitzpatrick ML, Bloom SR (1979) Effect of motilin on the rate of gastric emptying and gut hormone release during breakfast. Gastroenterology 76:903–907

Dolais-Kitabgi J, Kitabgi P, Brazeau P, Freychet P (1979) Effect of neurotensin on insulin, glucagon and somatostatin release from isolated pancreatic islets. Endocrinology 105:256–263

Dupré J, Curtin JD, Unger RH, Waddell RW, Beck J (1969) Effects of secretin, pancreozymin or gastrin on the response of the endocrine pancreas to administration of glucose or arginine in man. J Clin Invest 48:745–757

Edwards AV, Nathanielsz PW, Bloom SR, Vaughan NH (1974) The role of thyroxine in the maintenance of a normal glycogenolytic response to splanchnic nerve stimulation in adrenalectomized calves. Experientia 30:163–164
Fallucca F, DelleFave GF, Gambardella S, Mirabella C, DeMagistris L, Carratù R (1977) Glucagon secretion induced by bombesin in man. Lancet 2:609–610
Fallucca F, DelleFave GF, Gambardella S, Mirabella C, DeMagistris L, Garratù R (1978) Glucagon secretion induced by bombesin in man. Adv Exp Med Biol 106:259–261
Feldman M, Kiser RS, Unger RH, Li CH (1983) Beta-endorphin and the endocrine pancreas. Studies in healthy and diabetic human beings. N Engl J Med 308:349–353
Forssman WG, Helmstaedter V, Feurle G (1977) Relationship of enkephalin and endorphin immunoreactivity with D-cells and G-cells of stomach. Acta Hepatogastroenterol (Stuttg) 24:488
Fujimoto WY, Ensinck JW, Merchant FW, Williams RH, Smith PH, Johnson DG (1978) Stimulation by gastric inhibitory polypeptide of insulin and glucagon secretion by rat islet cultures. Proc Soc Exp Biol Med 157:89–93
Fujimoto WY, Williams RH, Ensinck JW (1979) Gastric inhibitory polypeptide, cholecystokinin, and secretin effects on insulin and glucagon secretion by islet cultures. Proc Soc Exp Biol Med 160:349–353
Fussgänger RD, Schleyer M, Pfeiffer EF (1973) Correlative insulin and glucagon secretion of the perfused rat pancreas following pituitary peptides. Acta Endocrinol [Suppl] (Copenh) 173:105
Giugliano D, Passariello N, Sgambato S, Torella R, D'Onofrio F (1982) Calcitonin modulation of insulin and glucagon secretion in man. Am J Physiol 242 (Endocrinol Metab 5): E206–E213
Goldfine ID, Kirsteins L, Lawrence AM (1972) Excessive glucagon responses to arginine in active acromegaly. Horm Metab Res 4:97–100
Goto Y, Carpenter RG, Berelowitz M, Frohman LA (1980) Effect of ventromedial hypothalamic lesions on the secretion of somatostatin, insulin and glucagon by the perfused rat pancreas. Metabolism 29:986–992
Hermansen K (1980) Effects of substance P and other peptides on the release of somatostatin, insulin and glucagon in vitro. Endocrinology 107:256–263
Iversen J (1971) Secretion of glucagon from the isolated perfused canine pancreas. J Clin Invest 50:2123–2136
Jensen LS, Fahrenkrug J, Holst JJ, Nielsen OV, Schaffalitzky De Muckadell OB (1978a) Secretory effects of VIP on isolated perfused porcine pancreas. Am J Physiol 235:E387–E391
Jensen LS, Fahrenkrug J, Holst JJ, Kuhl C, Nielsen OV, Schaffalitzky De Muckadell OB (1978b) Secretory effects of secretin on isolated perfused porcine pancreas. Am J Physiol 235:E381–E386
Jensen LS, Rehfeld JF, Holst JJ, Fahrenkrug J, Nielsen OV, Schaffalitzky De Muckadell OB (1980) Secretory effects of gastrins on isolated perfused porcine pancreas. Am J Physiol 238:E186–E192
Kabadi UM, Eisenstein AB (1980) Glucose intolerance in hyperthyroidism: role of glucagon. J Clin Endocrinol Metab 50:392–398
Kaneto A, Kaneko T, Kajinuma H, Kosaka K (1977) Effect of vasoactive intestinal polypeptide infused intrapancreatically on glucagon and insulin secretion. Metabolism 26:781–787
Kaneto A, Kaneko T, Nakaya S, Kajinuma H, Kosaka K (1978) Effect of bombesin infused intrapancreatically on glucagon and insulin secretion. Metabolism 27:549–553
Kanter RA, Ensinck JW, Fujimoto WY (1980) Disparate effects of enkephalin and morphine upon insulin and glucagon secretion by islet cell cultures. Diabetes 29:84–86
Karakash C, Rohner-Jeanrenaud F, Hustvedt BE, Jeanrenaud B (1980) Nitrogen handling in adult hypothalamic obese rats. Am J Physiol 238:E32–E37
Laube H, Fussgänger RD, Schröder KE, Pfeiffer EF (1972) Acute effects of human chorionic somatomammotropin on insulin and glucagon release in the isolated perfused pancreas. Diabetes 21:1072–1076

Lefèbvre PJ, Luyckx AS (1980) Neurotransmitters and glucagon release from the isolated, perfused canine stomach. Diabetes 29:697–701
Lenzen S (1976) The effect of hydrocortisone treatment and adrenalectomy on insulin and glucagon secretion from the perfused rat pancreas. Endokrinologie 68:189–197
Levitsky LL, Uehara JA, Marchichow JA, Dumbovic N (1978) Glucagon response to arginine in growth hormone deficient children before and after treatment with growth hormone and in children with non-endocrine short stature. Clin Endocrinol 8:473–481
McAdams MR, Pek S (1979) Inhibitors of prostaglandin synthetic pathway enhance *in vitro* secretion of glucagon and insulin. In: Program, 61st annual meeting of the endocrine society, Anaheim, p 169
Marco J, Calle C, Román D, Diaz-Fierros M, Villanueva ML, Valverde I (1973) Hyperglucagonism induced by glucocorticoid treatment in man. N Engl J Med 288:128–131
Marco J, Calle C, Hedo JA, Villanueva ML (1976) Enhanced glucagon secretion by pancreatic islets from prednisolone-treated mice. Diabetologia 12:307–311
Moltz JH, Dobbs RE, McCann SM, Fawcett CP (1977) Effects of hypothalamic factors on insulin and glucagon release from the islets of Langerhans. Endocrinology 101:196–203
Moltz JH, Dobbs RE, McCann SM, Fawcett CP (1979) Preparation and properties of hypothalamic factors capable of altering pancreatic hormone release in vitro. Endocrinology 105:1262–1267
Mutt V (1978) Progress in intestinal hormone research. Adv Exp Med Biol 106:133–146
Norman AW, Frankel BJ, Heldt AM, Grodsky GM (1980) Vitamin D deficiency inhibits pancreatic secretion of insulin. Science 209:823–825
Ohneda A, Ishii S, Horigome K, Chiba M, Sakai T, Kai Y, Watanabe K, Yamagata S (1977) Effect of intrapancreatic administration of vasoactive intestinal peptide upon the release of insulin and glucagon in dogs. Horm Metab Res 9:447–452
Ohneda A, Horigome K, Ishii S, Kai Y, Chiba M (1978) Effect of caerulein upon insulin and glucagon secretion in dogs. Horm Metab Res 10:7–11
Otsuki M, Sakamoto C, Maeda M, Yuu H, Morita S, Baba S (1979) Effect of caerulein on exocrine and endocrine pancreas in the rat. Endocrinology 105:1396–1399
Paloyan E (1967) Recent developments in the early diagnosis of hyperparathyroidism. Surg Clin North Am 47:61–69
Patton G, Brown M, Dobbs R, Vale W, Unger RH (1976) Effects of neurotensin and substance P on insulin and glucagon release by the perfused dog pancreas. Metabolism 25:1465
Pek S, Fajans SS, Floyd JC Jr, Knopf RF (1973) Clinical conditions associated with elevated plasma levels of glucagon. Excerpta Med Int Congr Ser 312:207–213
Rebolledo O, Leclercq-Meyer V, Marchand J, Leclercq R, Malaisse WJ (1975) Failure of parathormone to affect insulin and glucagon release from the perfused rat pancreas. Horm Metab Res 7:287–290
Rehfeld JF (1972) Three components of gastrin in human serum. Gel filtration studies on the molecular size of immunoreactive serum gastrin. Biochim Biophys Acta 285:364–372
Rehfeld JF, Stadil F, Vikelsøe J (1974) Immunoreactive gastrin components in human serum. Gut 15:102–111
Rehfeld RF, Holst JJ, Kühl C (1978) The effect gastrin on basal and aminoacid-stimulated insulin and glucagon secretion in man. Eur J Clin Invest 8:5–9
Rehfeld JF, Larsson LI, Goltermann NR, Schwartz TW, Holst JJ, Jensen SL, Morley JS (1980) Neural regulation of pancreatic hormone secretion by the C-terminal tetrapeptide of CCK. Nature 284:33–38
Reid RL, Yen SSC (1981) $\beta$-endorphin stimulates the secretion of insulin and glucagon in humans. J Clin Endocrinol Metab 52:592–598
Rohner-Jeanrenaud F, Jeanrenaud B (1980) Consequences of ventromedial hypothalamic lesions upon insulin and glucagon secretion by subsequently isolated perfused pancreases in the rat. J Clin Invest 65:902–910
Santeusanio F, Faloona GR, Unger RH (1972) Suppressive effect of secretin upon pancreatic alpha cell function. J Clin Invest 51:1743–1751

Schebalin M, Said SI, Makhlouf GM (1977) Stimulation of insulin and glucagon secretion by vasoactive intestinal peptide. Am J Physiol 232:E197–E200
Shima K, Sawazaki N, Tanaka R, Morishita S, Tarui S, Nishikawa M (1976) The pancreatic alpha and beta cell responses to l-arginine and insulin-induced hypoglycaemia in hyperthyroidism. Acta Endocrinol (Copenh) 83:114–122
Sirek A, Vranic M, Sirek OV, Vigas M, Policova Z (1979) Effect of growth hormone on acute glucagon and insulin release. Am J Physiol 237:E107–E112
Sirinek KR, Levine BA, Crockett SE, Cataland S (1980) Cholecystokinin and gastric inhibitory polypeptide are not glucagonotropic in dogs. J Surg Res 28:366–372
Sizonenko PC, Rabinovitch A, Schneider P, Paunier L, Wollheim CB, Zahnd G (1975) Plasma growth hormone, insulin and glucagon responses to arginine infusion in children and adolescents with idiopathic short stature, isolated growth hormone deficiency, panhypopituitarism, and anorexia nervosa. Ped Res 9:733–738
Starke A, Keck E, Berger M, Zimmermann H (1981) Effects of calcium and calcitonin on circulating levels of glucagon and glucose in diabetes mellitus. Diabetologia 20:547–552
Sundler F, Alumets J, Håkanson R (1979) Six types of immunohistochemically and ultrastructurally identified endocrine cells in pig pancreas. Scand J Enterology 49:179
Szecówka J, Sandberg E, Efendić S (1980 a) The interaction of vasoactive intestinal polypeptide (VIP), glucose and arginine on the secretion of insulin, glucagon and somatostatin in the perfused rat pancreas. Diabetologia 19:137–142
Szecówka J, Tatemoto K, Mutt V, Efendić S (1980 b) Interaction of a newly isolated intestinal polypeptide (PHI) with glucose and arginine to affect the secretion of insulin and glucagon. Life Sci 26:435–438
Tai T-Y, Pek S (1976) Direct stimulation by growth hormone of glucagon and insulin release from isolated rat pancreas. Endocrinology 99:669–677
Taminato T, Seino Y, Goto Y, Inoue Y, Kadowaki S, Mori K, Nozawa M, Yajima H, Imura H (1977) Synthetic gastric inhibitory polypeptide. Diabetes 26:480–484
Unger RH, Orci L (1981) Glucagon, physiology, pathophysiology, and morphology of the pancreatic A-cells. Elsevier, New York
Vance JE, Kitabchi AE, Buchanan KD, Williams RH (1968) Effect of adrenalectomy on glucagon and insulin release from isolated islets of Langerhans. Clin Res 16:131
Verdonk CA, Rizza RA, Nelson RL, Go VLW, Gerich JE, Service FJ (1980) Interaction of fat-stimulated gastric inhibitory polypeptide on pancreatic alpha and beta cell function. J Clin Invest 65:1119–1125
Williams RH, Champagne J (1979) Effects of cholecystokinin, secretin, and pancreatic polypeptide on secretion of gastric inhibitory polypeptide, insulin and glucagon. Life Sci 25:947–956
Wise JK, Hendler R, Felig P (1973) Influence of glucocorticoids on glucagon secretion and plasma amino acid concentrations in man. J Clin Invest 52:2774–2782

CHAPTER 30

# Neural Control of Glucagon Secretion

J. P. PALMER and D. PORTE, JR.

## A. Introduction

Glucagon secretion by the islets of Langerhans is controlled by a complex interplay of factors, including circulating fuels, hormones, intraislet cell interactions, and neural innervation. The relative importance of each of these factors in normal and pathologic physiology probably varies with different situations and is under active investigation. Opinions regarding the role of the nervous system cover a wide range, including the claim that neural input is the predominant factor regulating the A-cell (BLOOM 1978).

Neural control of glucagon secretion can be mediated via a direct effect of the nervous system on the A-cell, via regulation of islet vasculature, or indirectly via an effect on fuels or orther hormones which secondarily alter A-cell secretion. In this chapter, we will describe the neuroanatomy of the islets and discuss experimental observations in both animals and humans which have contributed to our understanding of the neural A-cell relationship. Also, we will discuss the potential role that the nervous system plays in controlling glucagon secretion in several normal and pathologic states. Finally, since glucagon and glucagon-like peptides are secreted from the gastrointestinal tract in some species, we will briefly discuss neural control of extrapancreatic A-cells.

## B. Anatomic Observations

### I. Neural Pathways

The major neural connection between the central nervous system and the endocrine pancreas is from the ventral hypothalamus via the autonomic nervous system. Several reviews describe this anatomy in detail (Fig. 1; WOODS and PORTE 1974; SMITH and PORTE 1976; SMITH et al. 1979; PORTE et al. to be published; PALMER and PORTE 1981). The efferent output of the autonomic nervous system from the hypothalamus is via the dorsal motor nucleus of the vagus and the sympathetic system in the spinal cord. The retrograde axonal transport of horseradish peroxidase injected into the pancreas of rats has been used to identify cell groups which project to the pancreas. Parasympathetic neurons were found in the medial two-thirds of the dorsal motor nucleus and in the rostral pole of the nucleus ambiguus (LAUGHTON and POWLEY 1979; POWLEY and LAUGHTON 1981). Preganglionic sympathetic fibers in the splanchnic nerves have synapses with postganglionic neurons in the celiac ganglion. Axons from these neurons travel to the pancreas in the mixed

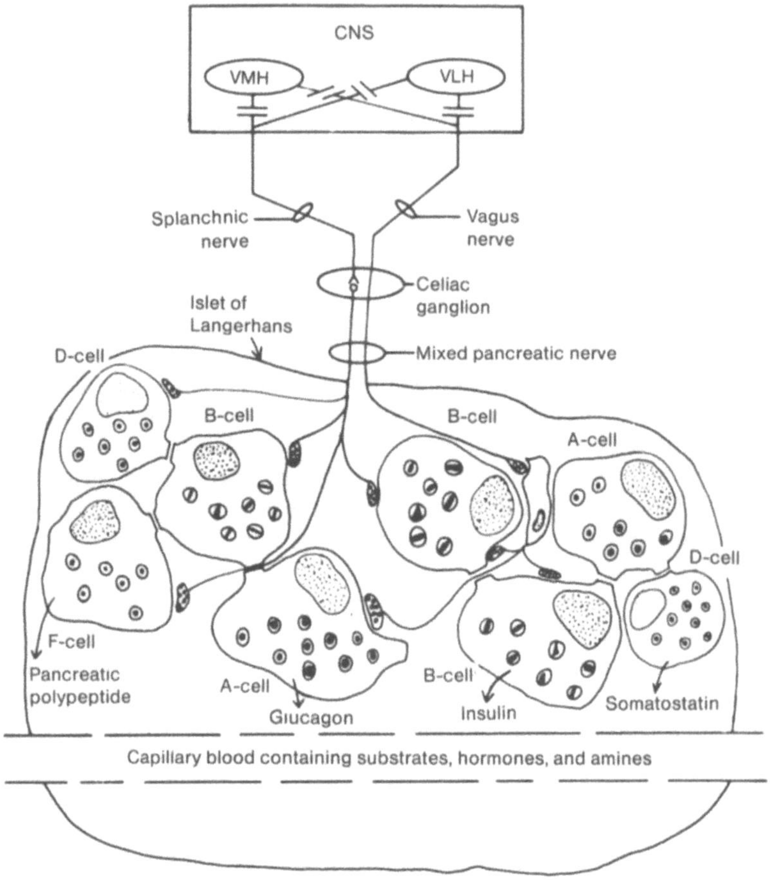


**Fig. 1.** Neural pathways for the central nervous system regulation of islet function. PALMER and PORTE (1981)

pancreatic nerve which follows the arterial system. The parasympathetic preganglionic fibers that innervate the pancreas leave the brain stem in both the right and left branches of the vagus. In the mediastinum, these fibers enter the dorsal vagal trunk which then passes through the celiac ganglion into the mixed pancreatic nerve. Some preganglionic cholinergic fibers may also arrive via the splanchnic nerves and some adrenergic fibers with the vagus (LUNDBERG et al. 1976). The parasympathetic preganglionic fibers establish synapses with short postganglionic fibers near or in the islets themselves. Some nerve fibers from the celiac ganglion do not travel in the mixed pancreatic nerve, but reach the pancreas via pathways that follow minor vascular connections between the pancreas and duodenum. Depending upon the species, the nerve fibers that reach the pancreatic islet are arranged in a peri-insular plexus or may form an intrainsular plexus at the more central portion of the islet. In mammals, typical presynaptic and postsynaptic complexes have not been observed. Nerve terminals end blindly 20–30 nm from islet cells, suggesting that neurotransmitters released from these nerve endings diffuse through the extracellular space of the islet and may affect several types of endocrine cells simultaneously (SMITH and MADSON 1981). This diffusion may in turn be regulated by tight junctions which have been shown to vary, depending upon the functional status of the islet (ORCI 1976; see Chaps. 4 and 31).

Light and/or electron microscopy have demonstrated at least three basic types of axon terminals within the islet. Cholinergic and adrenergic terminals are easily identified by the character of their synaptic vesicles and by stains specific for cholinesterase or catecholamines. The third type of nerve terminal has been found primarily in the islet of teleosts, but has recently also been observed in primates (FORSSMANN and GREENBERG 1978). It has been suggested that these terminals are peptidergic in nature (FORSSMANN and GREENBERG 1978; LARSSON 1980). A variety of peptides which affect glucagon secretion including substance P, somatostatin, enkephalin, vasoactive intestinal peptide (VIP), and cholecystokinin (CCK) have been found in vagal and splanchnic nerves (LUNDBERG et al. 1978) and in nerves wihthin the pancreas (LARSSON 1979). Cholinergic, adrenergic, and peptidergic terminals are all found in relation to A-cells (ESTERHUIZEN et al. 1968; LARSSON 1980; REHFELD et al. 1980). As will be discussed in other parts of this chapter, considerable data exists to support a neurotransmitter role for at least VIP and CCK. While the other peptides are also potential neurotransmitters controlling glucagon secretion, this cannot be stated with as much assurance at this time.

There is evidence that the same axon can innervate more than one islet cell and even more than one islet secretory cell type (PORTE et al. to be published). Also, it appears that, at least in some species, there may be preferential distribution of a given type of nerve terminal to a specific population of islet cells. LUNDQUIST and ERICSON (1978) have reported that, in the mouse, most adrenergic terminals innervate A-cells. Nerves which contain CCK were noted to be primarily associated with A-cells in the cat (REHFELD et al. 1980), but with B-cells in the hamster (J. F. REHFELD 1981, personal communication).

In addition to modulation of effector cells through chemical neurotransmitters, specialized intercellular junctions (gap junctions) with low electrical resistance have been described between nerve endings and B-cells (ORCI et al. 1973). If gap junctions also exist between nerve terminals and A-cells, this would provide a direct ionic mechanism for neural control of glucagon secretion. These junctions have been noted between A- and B-cells and may functionally couple the secretion of many islet cells after electrical and/or chemical stimulation from the nerves (ORCI et al. 1975).

## II. Species Variations

Considerable interspecies variability in the degree of islet innervation and the relative number of adrenergic versus cholinergic fibers has been observed (WOODS and PORTE 1974; SMITH and PORTE 1976). Some investigators have concluded that birds, in contrast to most mammals, have little if any innervation of their islets (KERN and GRUBE 1972; SMITH 1974), but one report claimed to have identified neurons in the islet of the chicken (WATANABE and YASUDA 1977). Histochemical studies to identify adrenergic nerve terminals in islets have revealed the highest number in hamsters, less in dogs and cats, still fewer in mice and rats, and only oc-

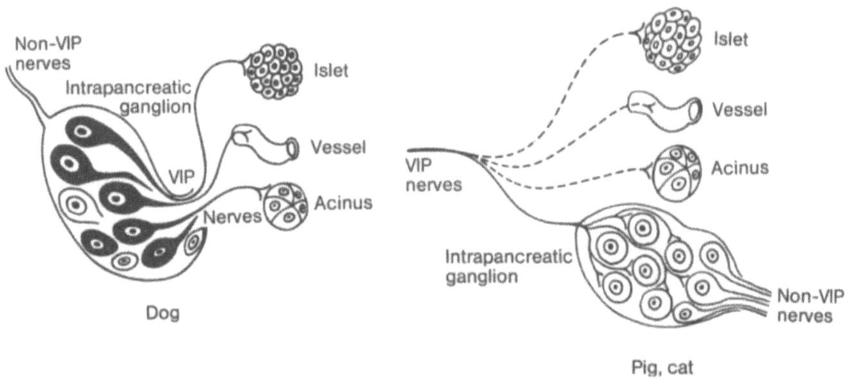


**Fig. 2.** Schematic drawing illustrating VIP nerves in the pancreas of dogs compared with pigs and cats. LARSSON et al. (1978)

casionally positive staining terminals in monkeys, rabbits, pigs, and guine pigs (KERN and GRUBE 1972; CEGRELL 1968). Although rats have relatively few adrenergic fibers, they are relatively rich in cholinergic fibers (MORGAN and LOBL 1966). Dog islets are innervated by both adrenergic and cholinergic terminals (WATARI 1968). The spiny mouse appears to be unique among mammals in that no autonomic fibers have been identified (ORCI et al. 1970).

Teleosts, in marked contrast, have many nerve endings with vesicles ultrastructurally different from either typical adrenergic or cholinergic vesicles. As mentioned previously, these vesicles may be peptidergic or possibly purinergic (SMITH et al. 1979; BRINN 1975). A marked interspecies difference in the location of VIP within nerves and neurons in the pancreas has been observed (Fig. 2). In the pig and cat, VIP-staining nerves are abundant around nerve cell bodies of the intrapancreatic ganglia which themselves do not stain for VIP and are scarce in the islets. In the dog, intrapancreatic ganglia contain strongly VIP-positive nerve cell bodies which give off axons that heavily innervate vessels and the endocrine pancreas. These findings suggest that VIP nerves in the cat and pig are mainly preganglionic and innervate a second type of intrapancreatic neuron, whose transmitter is not VIP, whereas, in the dog, many of the intrapancreatic ganglia are postganglionic VIP nerves which directly contact the effector islet cells (LARSSON et al. 1978). In humans, VIP has been found in cell bodies of the intrapancreatic ganglia and also in fine varicose nerve fibers which form peri-insular networks and appear to contact endocrine cells (BISHOP et al. 1980). Considerable interspecies variability has also been noted for CCK-containing nerves. Cat and mouse islets are most heavily innervated, pig and hamster islets contain less, but still a significant amount of CCK nerve terminals, whereas dog and rat islets contain very few if any CCK-staining nerve fibers (REHFELD 1981).

It is difficult to relate morphological observations in islets to physiologic studies which attempt to evaluate neural influences on A-cell secretion. Also, the apparent absence of certain types of nerve terminals can be due to sampling error and/or weak antigen-antibody reactions due to heterology of the peptides. Nonetheless, anatomic differences probably exist between species and therefore observations in one species may be at variance with those in another.

## C. Experimental Observations

### I. Central Nervous System Studies

A number of experimental protocols have been used to delineate the importance of the hypothalamus in mediating the effects of the central nervous system on glucagon secretion. Electrical stimulation of the ventromedial hypothalamus (VMH) elicits a rise in plasma immunoreactive glucagon (IRG) levels that is not prevented by adrenalectomy or prior treatment with $\alpha$- or $\beta$-adrenergic blockade (FROHMAN and BERNARDIS 1971; FROHMAN et al. 1974; SHIMAZU and ISHIKAWA 1981). Glucagon secretion is also increased by neurochemical (epinephrine, norepinephrine, acetylcholine) stimulation of the VMH. These glucagon responses are not affected by treatment with propranolol, phentolamine, or atropine, but are eliminated by ganglionic blockade with hexamethonium (DE JONG et al. 1977; SHIMAZU and ISHIKAWA 1981; STEFFENS 1981). These results are compatible with the hypothesis that VMH-stimulated glucagon secretion is mediated via the autonomic nervous system, but that the neurotransmitter in the islet is neither cholinergic nor adrenergic. Probably it is peptidergic. Although one study found decreased glucagon levels in animals with destructive VMH lesions (INOUE et al. 1977), several others have found the opposite, namely increased basal and stimulated glucagon levels (GOTO et al. 1980; ROHNER-JEANRENAUD and JEANRENAUD 1980, 1981; BOBBIONI and COSCELLI 1980). Since this hyperglucagonemia is blocked by atropine (ROHNER-JEANRENAUD 1981), it is presumed to be due to cholinergic activity which has been shown to mediate the hyperinsulinemia in these animals (INOUE and BRAY 1977; BERTHOUD and JEANRENAUD 1979).

Data regarding control of glucagon secretion via the ventrolateral hypothalamus (VLH) is much less consistent. Chemical stimulation of the VLH with epinephrine, norepinephrine, and acetylcholine has been reported to have little or no effect on A-cell secretion (SHIMAZU and ISHIKAWA 1981; DEJONG et al. 1977; STEFFENS 1981). Electrical stimulation induces a rise in IRG (HELMAN et al. 1980; PARAMESWARAN et al. 1977) which is abolished by bilateral splanchnic nerve section (HELMAN et al. 1980). Not all investigators have observed a glucagon response to electrical VLH stimulation (SHIMAZU and ISHIKAWA 1981).

Some of the discrepancies in this area may be related to the observation that stimulatory and destructive procedures in the VMH and/or VLH affect not only nerve cell bodies in these locations, but also axons orginating in more distant neurons which are passing near or through the VMH and/or VLH areas. Noradrenergic and/or serotoninergic axons from neurons within the brain stem reticular formation and dopaminergic fibers from the substantia nigra are likely to be important (GOLD 1973; AHLSKOG and HOEBEL 1973; COSCINA and STANCER 1977; FIBIGER et al. 1973) as are peptidergic fibers from the paraventricular nucleus in the anterior hypothalamus (SWANSON and SAWCHENKO 1980).

Another experimental approach to evaluate central nervous system control of glucagon secretion has employed the intraventricular administration of 2-deoxyglucose. In dogs, this results in a severalfold rise in IRG (FROHMAN and NAGAI 1976). No significant effect was observed when similar doses of 2-deoxyglucose were given intravenously, making it very likely that the intraventricular 2-deoxyglucose was affecting glucagon secretion via a central nervous system effect.

Intraventricular 2-deoxyglucose also causes an inhibition of insulin secretion which is probably mediated by the sympathoadrenal system since the inhibition is blocked by adrenalectomy or phentolamine (FROHMAN et al. 1973; MÜLLER et al. 1973). Comparable data are not available for glucagon. Bombesin administered intracisternally to rats also results in hyperglucagonemia. This A-cell stimulation appears to be catecholamine mediated since it is eliminated by adrenalectomy (BROWN et al. 1979).

Although most of the data suggest that the hypothalamus influences A-cell secretion primarily through neural connections, there is some evidence that humoral factors originating within the hypothalamus may also influence the endocrine pancreas via the circulation. One or more factors which stimulate glucagon secretion have been isolated from extracts of both the VMH and the VLH. Although the exact nature of these factors is unknown, it has been shown that they are peptidic in nature, less than 5000 daltons in molecular weight, and are not neurotensin, substance P, or catecholamines (MOLTZ et al. 1977, 1979; HILL et al. 1977).

## II. Nerve Stimulation and Sectioning Studies

Nerve stimulation studies provide strong evidence that neural input can control glucagon secretion. Electrical stimulation of the splanchnic nerves results in a rise in glucagon secretion which is consistently seen in several species including calves, dogs, cats, and sheep. This rise is directly proportional to the rate of stimulation (0.5–4.0 Hz), but is not blocked by propranolol, phentolamine, or atropine (BLOOM 1978; BLOOM and EDWARDS 1975, 1978; BLOOM et al. 1973; MILLER and HORTON 1979; KANETO et al. 1975a; ESTERHUIZEN and HOWELL 1970). Possibly a peptidergic neurotransmitter mediates the glucagon stimulation.

Stimulation of the mixed pancreatic nerve in the dog also results in an increase of glucagon which is not blocked by atropine (MARLISS et al. 1973; GIRARDIER et al. 1976). GIRARDIER et al. (1976) mention in the discussion of their paper that they have observed a blockade of this glucagon response with propranolol, suggesting that the rise is $\beta$-adrenergically mediated. Vagal nerve stimulation in the dog, calf, and pig has been reported to increase glucagon secretion (KANETO et al. 1974; BLOOM et al. 1974a; HOLST et al. 1981a), whereas in the cat, vagal nerve stimulation did not elicit a glucagon response (ESTERHUIZEN and HOWELL 1970). Atropine blocks the glucagon response in the dog, suggesting muscarinic cholinergic transmission whereas in the pig the glucagon rise is not blocked by atropine, but is completely abolished by hexamethonium (KANETO et al. 1974; HOLST et al. 1981b). These observations suggest that a noncholinergic neurotransmitter, possibly VIP or CCK, is responsible for the glucagon response to vagal stimulation in the pig.

The nerve stimulation studies clearly demonstrate that the respectice nerves can influence glucagon secretion, but the relevance of these observations to normal physiology can only be inferred. Measurement of basal glucagon levels after nerve sectioning provides an indication of the tonic action of these nerves on A-cell secretion. No differences in basal levels of glucagon was found between truncal and selective vagotomy patients (RUSSELL et al. 1974) or between normal subjects and truncal vagotomy patients (PALMER et al. 1979). In calves, sectioning the splanchnic nerves and administering atropine did not alter basal glucagon concentrations

(BLOOM et al. 1978). HALTER and PFLUG (1980) assessed the effect of resting sympathetic nervous system activity on the A-cell in humans by measuring glucagon levels before and after high thoracic spinal anesthesia. Basal norepinephrine and epinephrine were suppressed, but basal glucagon was unaffected. Similarly, we have observed normal basal glucagon levels in humans sympathectomized by traumatic cervical cord transection (PALMER et al. 1976). In contrast to these negative results, GIRARDIER et al. (1974) have found that sectioning the mixed pancreatic nerve results in a 60% fall in glucagon level in the fasted conscious dog and that this effect persisted until at least 4 months after sectioning. Also, destruction of peripheral adrenergic neurons in rats by pretreatment with 6-hydroxydopamine reduced basal glucagon secretion by approximately 50% (LUNDQUIST et al. 1976). Taken together, these results suggest that basal glucagon may be more dependent upon neural tone in the dog and rat than in the other species or that only nerve sectioning distal to the celiac ganglion affects basal glucagon secretion.

Studies demonstrating release of CCK and VIP provide additional support for the hypothesis that these peptides are neurotransmitters. CCK is found in synaptic vesicles and is released by calcium in a manner characteristic of neurotransmitters (EMSON et al. 1980). Electrical stimulation of the splanchnic nerve decreases the release of VIP by an α-adrenergic mechanism, whereas vagal nerve stimulation (only high threshold fibers) increases VIP. This latter effect is not blocked by atropine, but is abolished by hexamethonium (SCHAFFALITZKY DE MUCKADEL et al. 1977; FAHRENKRUG et al. 1978a, b). Also, VIP rises during distention of the esophagus and mechanical stimulation of the small bowel, gastrointestinal functions known to be controlled by the autonomic nervous system, but not blocked by cholinergic or adrenergic blocking agents (FAHRENKRUG et al. 1978b).

### III. Infusion Studies

Cholinergic stimulation of IRG which is blocked by atropine has been demonstrated after the administration of acetylcholine in the perfused dog pancreas (IVERSEN 1973a; KANETO and KOSAKA 1974; HOLST et al. 1981c). But, in the isolated perfused porcine pancreas, acetylcholine does not stimulate glucagon release, although it is effective when administered systemically (HOLST et al. 1981c). This suggests that acetylcholine releases glucagon in the pig through muscarinic receptors outside the pancreas. Since acetylcholine releases VIP (FAHRENKRUG et al. 1978a), it ist tempting to speculate that the observed difference between dogs and pigs is related to the described differences in VIP-containing nerves in the two species. Namely, in the dog pancreas there are abundant VIP-staining nerve cell bodies whereas these are not found in the pig and only immunoreactive preganglionic nerve fibers are detected (LARSSON et al. 1978). Acetylcholine has been reported to stimulate IRG in the perfused chicken pancreas (HONEY and WEIR 1980) whereas a preliminary communication reported suppression in ducks (TYLER and KAJINUMA 1972). In humans, the effect of atropine on basal glucagon is controversial with some investigators reporting a small decrement (BLOOM et al. 1974b; HEDO et al. 1978) and others finding no change (PALMER et al. 1979; MCLOUGHLIN et al. 1978). In addition, we were unable to alter basal IRG levels by doses of bethanechol chloride which clearly activated some muscarinic receptors as indicated by dia-

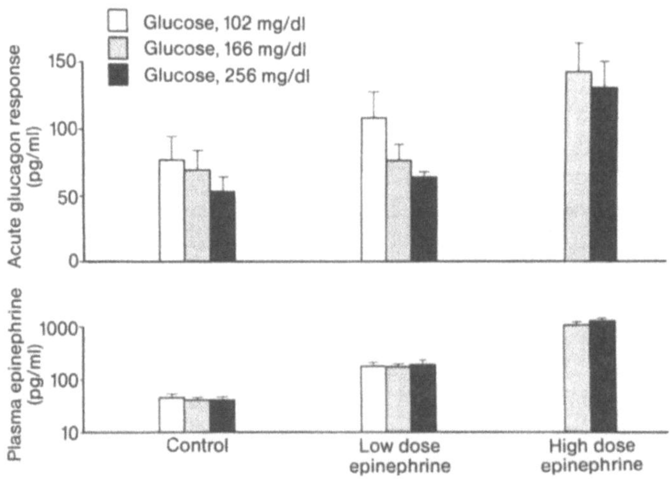


**Fig. 3.** Interaction of epinephrine and glucose concentration on the acute glucagon response to arginine in humans. Epinephrine was infused at 15 ng $kg^{-1}$ $min^{-1}$ (low dose) and 80 ng $kg^{-1}$ $min^{-1}$ (high dose) with glucose concentration maintained at the designated levels. Arginine (5 g) was administered as a bolus intravenously

phoresis, lacrimation, salivation, borborygmi, and an urge to urinate in the volunteers (PALMER et al. 1979).

The administration of adrenergic agonists to perfused rat and dog preparations has been reported by many investigators to stimulate glucagon secretion (SAMOLS and WEIR 1979; KANETO et al. 1975b, 1977a; WEIR et al. 1974; IVERSEN 1973b). Almost all studies agree that β-adrenergic tone is stimulatory. With selective agonists and blockers, this effect was found to be largely mediated via $\beta_2$-adrenoceptors (SAMOLS and WEIR 1979; KANETO et al. 1975b). In birds, the adrenergic system seems to be reversed from that in mammals with β-adrenergic agonism suppressing IRG (TYLER and KAJINUMA 1972). The effect of α-adrenoceptor stimulation is less consistent. IVERSEN (1973b) originally reported no consistent or significant effect of α-adrenergic blockade (phentolamine, dibenzyline) on epinephrine-, norepinephrine-, or isoproterenol-stimulated IRG, although in many of his experiments he only evaluated two or three dogs and the results were quite variable. KANETO et al. (1977a), using the α-adrenergic agonists, methoxamine and phenylephrine, demonstrated inhibition of IRG output that could not be attributed to decreased blood flow. However, SAMOLS and colleagues (SAMOLS and WEIR 1979; SAMOLS et al. 1981) showed α-adrenergic stimulation of IRG using higher doses of epinephrine, norepinephrine, propranolol, and phentolamine.

In humans, infusions of adrenergic agonists frequently elicit a modest rise in IRG (BENSON et al. 1977; GERICH et al. 1972, 1973, 1974, 1976; RAPTIS et al. 1977; SCHADE and EATON 1977, 1978, 1979), but this is not a consistent finding in all normal individuals and indeed has not been observed by all investigators (BENSON et al. 1977; SILVERBERG et al. 1978; CLUTTER et al. 1980). We demonstrated that the hyperglycemia which occurs during epinephrine infusions has an inhibitory influence on the stimulatory effect of catecholamines on A-cell secretion (Fig. 3; BEARD et al. 1981). This may be part of the explanation for these variable results. β-Adrenergic agonists have generally been found to be stimulatory (GERICH et al. 1974, 1976; RAPTIS et al. 1977). GERICH et al. (1974) also observed a fall in IRG levels during infusion of the α-adrenergic agonist, methoxamine.

Experiments employing scorption toxin, dopamine, and L-dopa also provide additional important data regarding adrenergic control of IRG. In the perfused rat pancreas preparation, scorption toxin causes a massive release of norepinephrine, presumably from sympathetic nerve endings, and a 3–4-fold rise in IRG. This IRG response is blocked by both propranolol and phentolamine, suggesting that both α- and β-adrenoceptors are stimulatory (JOHNSON and ENSINCK 1976). Infusions of dopamine stimulate glucagon secretion (LEBLANC et al. 1977; LORENZI et al. 1979). L-Dopa administration also stimulates IRG, probably through conversion to norepinephrine or dopamine in the islet (RAYFIELD et al. 1975; GEORGE and RAYFIELD 1974; GEORGE and BAILEY 1978). Mediation through a central dopaminergic mechanism is unlikely since the more specific dopaminergic agonist, apomorphine does not stimulate the A-cells (LORENZI et al. 1977) and the IRG response to L-dopa is unaffected by ganglionic blockade (GEORGE and BAILEY 1978). Although the IRG response to L-dopa was originally reported to be unaffected by either α- or β-blockade (GEORGE and RAYFIELD 1974), these same authors have shown subsequently that higher doses of either blocker will attenuate the IRG response (GEORGE and BAILEY 1978).

In summary, cholinergic or β-adrenergic stimulation in almost all mammals and under most experimental conditions results in increased glucagon secretion. In certain species, some stimulation may be peptidergically mediated. The role of the α-adrenergic system is less clear, but stimulatory effects have usually been observed. Although there is no completely satisfactory resolution of this inconsistency, it is important to remember that the effects of the adrenergic blocking agents are dose dependent (SAMOLS and WEIR 1979; KANETO et al. 1975b; GEORGE and RAYFIELD 1974; GEORGE and BAILEY 1978), that the blockers themselves may alter the metabolism of infused or endogenous catecholamines (IRVING et al. 1974; LANGER et al. 1977; RIZZA et al. 1980; LILAVIVAT et al. 1981), and that some effects of these agents may be mediated via mechanisms not involving α- or β-adrenoceptors (KOCH-WESER 1975).

The possibility that VIP or CCK function as neurotransmitters in the endocrine pancreas requires that they be able to stimulate islet cell secretion directly. Several groups have demonstrated that VIP is a potent stimulus to glucagon secretion (SCHEBALIN et al. 1977; KANETO et al. 1977b; OHNEDA et al. 1977; ADRIAN et al. 1978). Since the predominant CCK moiety in nerves appears to be the COOH terminal tetrapeptide, the recent demonstration that CCK 4 is a very potent stimulus to the A-cell (REHFELD and JENSEN 1980; REHFELD et al. 1980) further substantiates its potential neurotransmitter role in the islet.

## D. Physiologic and Pathophysiologic Observations

As reviewed in detail in Chap. 38, a variety of stress states have been found to be associated with hyperglucagonemia. Since hyperglycemia is also usually present at these times, neural mechanisms have been suggested as mediating the glucagon elevation because the hyperglycemia would be expected to suppress the A-cell. These states include trauma (LINDSEY et al. 1973, 1974), myocardial infarction (WILLERSON et al. 1974), infection (ROCHA et al. 1973), and burns (WILMORE et al. 1974). In most circumstances, the hypothesis that the elevated glucagon is neurally mediated is based on the concomitant activation of the autonomic nervous system and the absence of other known stimuli. However, a conclusive cause-effect relationship between the nervous system and the glucagon response to stress has been experimentally demonstrated in only a few studies.

Hypoglycemia and exercise are discussed elsewhere in this Handbook (Chaps. 23 and 38). Other conditions include hypoxemia and hypovolemia, both of which are associated with hyperglycemia and elevated glucagon levels. The role of the autonomic nervous system in this state of hyperglucagonemia has been investigated. Hypoxic human infants have elevated glucagon levels and associated slight hyperglycemia (JOHNSTON and BLOOM 1973). In the puppy, we found that hypoxemia results in increased glucagon levels and hyperglycemia which are blocked by the $\alpha$-adrenergic blocker, phenoxybenzamine (BAUM et al. 1979) suggesting mediation via an $\alpha$-adrenergic mechanism. $\beta$-Adrenergic stimulation is unlikely since we have shown that $\beta$-adrenoceptor stimulation is ineffective during hypoxia in puppies (BAUM and PORTE 1976, 1980). In contrast, hyperglucagonemia associated with hypoxia in the calf may not be neurally mediated by adrenergic or cholinergic signals. Splanchnic nerve sectioning abolished the catecholamine response to hypoxia, but did not impair the glucagon rise (BLOOM et al. 1976) and atropine did not block hypoxia-induced hyperglucagonemia in this species (BLOOM et al. 1976).

Hemorrhagic hypotension also results in hyperglucagonemia and hyperglycemia. In cats, adrenalectomy together with bilateral sectioning of the major and minor splanchnic nerves abolished the glucagon rise (JÄRHULT 1975a, b) and, in dogs, infusion of propranolol markedly reduced the hyperglucagonemia (LINDSEY et al. 1973, 1975). Simulated hypotension, accomplished by sectioning the sinus nerves from the carotid baroreceptors, also results in a profound and rapid rise in glucagon (JÄRHULT and HOLST 1977). This reflex is probably mediated via the sympathetic nervous system since the rise in IRG can be blocked by cutting the splanchnic nerves and is markedly reduced by selective sympathetic denervation of the pancreas (HOLST and JÄRHULT 1978). Hypotension, therefore, appears to be a definite example of stress hyperglycemia in which the associated hyperglucagonemia is mediated by the autonomic nervous system. The experimental evidence that autonomic nervous dysfunction may mediate some of the glucagon regulation abnormalities in diabetes mellitus is discussed in Chap. 23.

## E. Neural Control of Gastric Glucagon

Glucagon which is indistinguishable from pancreatic glucagon is secreted from cells of the gastrointestinal tract which are morphologically similar to pancreatic A-cells. The dog fundus is especially rich in these cells and has been used to investigate potential neural control of this extrapancreatic glucagon (see also Chap. 33). Hypoglycemia, acetylcholine, VIP, and high dose $\beta$-adrenergic agonism stimulate stomach glucagon secretion (LEFÈBVRE and LUYCKX 1978, 1980; YOSHIDA and KONDO 1980a, b). However, neither electrical stimulation of the vagus nerves nor local sympathetic nerve terminal release of norephinephrine elicited by scorpion toxin affect gastric glucagon secretion (LEFÈBVRE et al. 1978a, b). This apparent lack of a direct effect of autonomic nerves on gastric glucagon secretion may explain the failure of exercise to increase glucagon levels in pancreatectomized dogs, compared with the threefold increment observed in intact dogs (VRANIC et al. 1976). In contrast, electrical stimulation of the vagal nerves in rats has been reported to stimulate gastric glucagon release (MARRE et al. 1979).

## F. Conclusions

Neural control of glucagon secretion is an important component in the complex interplay of factors which control glucose homeostasis. Substrate levels change markedly during feeding and are critical to islet control. The neuroendocrine modulators change in response to internally perceived signals or as part of an adaptive response to meals and stress from the environment. Alterations of glucagon secretion can be produced by both the traditional amine and the newly described peptide neurotransmitters. Therefore, the appropriate receptor systems are present on the A-cells for neural control. In animals, stimulation of the hypothalamus or the vagus, splanchnic, and pancreatic nerves alters glucagon secretion. Therefore, there are central systems and neural efferents which are capable of regulating A-cell secretion. These effects have been found in all mammalian species studied, but there appears to be considerable variability between species.

During experimental alterations of autonomic activity, parallel changes in glucagon secretion are frequently observed. In the stress response, hyperglycemia is usually observed and may be useful to the organism to improve glucose delivery. Hyperglucagonemia probably plays an important role in maintaining this hyperglycemia and autonomic blockade indicates a major role for the nervous system in stimulating this glucagon secretion. However, insulin-induced hypoglycemia and strenuous exercise in humans are two stress states in which glucose concentration is decreased. In these situations, the A-cell may be stimulated sufficiently by the lowered glucose levels and/or decreased B-cell activity so that amine blockade does not lead to a reduction in glucagon levels. In addition, peptidergic neural stimulation of the A-cells may be involved. During hypoxemia, hypotension, and after the administration of 2-deoxyglucose, hyperglycemia occurs and autonomic stimulation of the A-cells appears to mediate the hyperglucagonemia of these conditions.

It is unlikely that a single factor accounts for the hyperglucagonemia of stress states, rather glucagon secretion should be viewed as controlled by the interplay of input from circulating hormones and fuels, neural signals, interislet cell communications (gap junctions), and paracrine relationships. The extent to which one part of this control system affects IRG release is clearly dependent on the state of the other factors simultaneously affecting the A-cell and often the mechanisms are redundant and/or potentiate one another. As a generalization, the hyperglucagonemia observed during stress states associated with hypoglycemia is probably largely mediated via the decreased glucose concentration directly. However, in stress states associated with hyperglycemia, neural mechanisms probably play a much more important role in mediating the hyperglucagonemia. Therefore, neural control of the A-cell should be viewed as one important component of the control system for the regulation of glucose production and utilization.

*Acknowledgments.* We thank Ms. Sylvia MacMillan for her expert secretarial assistance. The authors' research is supported by funds from the National Institutes of Health (AM 17047, AM 12829), the Diabetes Research Center, The Regional Primate Research Center, the Veterans Administration, and the Juvenile Diabetes Foundation. Dr. Palmer is the recipient of a Clinical Investigator Award (1KO8 AM 00535).

## References

Adrian RE, Bloom SR, Hermansen K, Iversen J (1978) Pancreatic polypeptide, glucagon and insulin secretion from the isolated perfused canine pancreas. Diabetologia 14:413–417
Ahlskog JE, Hoebel BG (1973) Overeating and obesity from damage to a noradrenergic system in the brain. Science 182:166–169
Baum D, Porte D Jr (1976) Beta adrenergic receptor dysfunction in hypoxic inhibition of insulin release. Endocrinology 98:359–366
Baum D, Porte D Jr (1980) Stress hyperglycemia and the adrenergic regulation of pancreatic hormones in hypoxia. Metabolism 29:1176–1185
Baum D, Porte D Jr, Ensinck J (1979) Hyperglucagonemia and the α-adrenergic receptor in acute hypoxia. Am J Physiol 237:E404–E408
Beard JC, Halter JB, Porte D Jr (1981) Does hyperglycemia during stress depend on epinephrine's effects on glucagon and insulin secretion? Diabetes [Suppl 1] 31:11 A
Benson J, Johnson D, Palmer J, Werner P, Ensinck J (1977) Glucagon and catecholamine secretion during hypoglycemia in normal and diabetic man. J CLin Endocrinol Metab 44:459–464
Berthoud HR, Jeanrenaud B (1979) Acute hyperinsulinemia and its reversal by vagotomy after lesions of the ventromedial hypothalamus in anesthetized rats. Endocrinology 105:146–151
Bishop AE, Polak JM, Green IC, Bryant MG, Bloom SR (1980) The location of VIP in the pancreas of man and rat. Diabetologia 18:73–78
Bloom SR (1978) Signals for glucagon secretion. Ciba Found Symp 55:161–179
Bloom SR, Edwards AV (1975) The release of pancreatic glucagon and inhibition of insulin in response to stimulation of the sympathetic innervation. J Physiol (Lond) 253:157–173
Bloom SR, Edwards AV (1978) Certain pharmacological characteristics of the release of pancreatic glucagon in response to stimulation of the splanchnic nerves. J Physiol (Lond) 280:25–35
Bloom SR, Edwards AV, Vaughan NJA (1973) The role of the sympathetic innervation in the control of plasma glucagon concentration in the calf. J Physiol (Lond) 233:457–466
Bloom S, Edwards A, Vaughan N (1974a) The role of the autonomic innervation in the control of glucagon release during hypoglycemia in the calf. J Physiol (Lond) 236:611–623
Bloom S, Vaughan N, Russell R (1974b) Vagal control of glucagon release in man. Lancet 2:546–549
Bloom SR, Edwards AV, Hardy RN, Silver M (1976) Adrenal and pancreatic endocrine responses to hypoxia in the conscious calf. J Physiol (Lond) 261:271–283
Bloom SR, Edwards AV, Hardy RN (1978) The role of the autonomic nervous system in the control of glucagon, insulin and pancreatic polypeptide release from the pancreas. J Physiol (Lond) 280:9–23
Bobbioni E, Coscelli C (1980) Portal levels of glucagon and insulin in VMH-lesioned rats. Horm Metab Res 12:480–481
Brinn JE (1975) Pancreatic islet cytology of ictaluridae (teleostei). Cell Tissue Res 162:357–365

Brown M, Tache Y, Fisher D (1979) Central nervous system action of bombesin: mechanism to induce hyperglycemia. Endocrinology 105:660–665
Cegrell L (1968) The occurrence of biogenic monoamines in the mammalian endocrine pancreas. Acta Physiol Scand [Suppl] 314:1–60
Clutter WE, Bier DM, Shah SD, Cryer PE (1980) Epinephrine plasma metabolic clearance rates and physiologic thresholds for metabolic and hemodynamic actions in man. J Clin Invest 66:94–101
Coscina DV, Stancer HC (1977) Selective blockade of hypothalamic hyperphagia and obesity in rats by serotonin-depleting midbrain lesions. Science 195:416–419
deJong A, Strubbe JH, Steffens AB (1977) Hypothalamic influence on insulin and glucagon release in the rat. Am J Physiol 233:E380–388
Emson PC, Lee CM, Rehfeld JF (1980) Cholecystokinin octapeptide: vesicular localization and calcium dependent release from rat brain in vitro. Life Sci 26:2157–2163
Esterhuizen AC, Howell SL (1970) Ultrastructure of the A-cells of cat islets of Langerhans following sympathetic stimulation of glucagon secretion. J Cell Biol 46:593–631
Esterhuizen AC, Spriggs TLB, Lever JD (1968) Nature of islet-cell innervation in the cat pancreas. Diabetes 17:33–36
Fahrenkrug J, Galbo H, Holst JJ, Schaffalitzky de Muckadell OB (1978a) Influence of the autonomic nervous system on the release of vaso-active intestinal polypeptide from the porcine gastrointestinal tract. J Physiol (Lond) 280:405–422
Fahrenkrug J, Haglund U, Jodal M, Lundgren O, Olbe L, Schaffalitzky de Muckadell OB (1978b) Nervous release of vasoactive intestinal polypeptide in the gastrointestinal tract of cats: possible physiological implications. J Physiol (Lond) 284:291–305
Fibiger HC, Zis AP, McGeer EG (1973) Feeding and drinking deficits after 6-hydroxydopamine administration in the rat: similarities to the lateral hypothalamic syndrome. Brain Res 55:135–148
Forssmann WG, Greenberg H (1978) Innervation of the endocrine pancreas in primates. In: Coupland RE, Forsmann WG (eds) Peripheral neuroendocrine interaction. Springer, Berlin Heidelberg New York, pp 124–133
Frohman LA, Bernardis LL (1971) Effect of hypothalamic stimulation on plasma glucose, insulin, and glucagon levels. Am J Physiol 221:1596–1603
Frohman LA, Nagai K (1976) Central nervous system-mediated stimulation of glucagon secretion in the dog following 2-deoxyglucose. Metabolism [Suppl 1] 25:1449–1452
Frohman LA, Muller EE, Cocchi D (1973) Central nervous system mediated inhibition of insulin secretion due to 2-deoxy-glucose. Horm Metab Res 5:21–26
Frohman LA, Bernardis LL, Strachura ME (1974) Factors modifying plasma insulin and glucose responses to ventromedial hypothalamic stimulation. Metabolism 23:1047–1056
George D, Bailey P (1978) The effect of adrenergic and ganglionic blockers upon the L-dopa-stimulated release of glucagon in the rat. Proc Soc Exp Biol Med 157:1–4
George D, Rayfield E (1974) L-Dopa induced plasma glucagon release. J Clin Endocrinol 39:618–621
Gerich J, Karam J, Forsham P (1972) Reciprocal adrenergic control of pancreatic alpha- and beta-cell function in man. Diabetes 21:332–333
Gerich J, Karam J, Forsham P (1973) Stimulation of glucagon secretion by epinephrine in man. J Clin Endocrinol Metab 37:479–481
Gerich J, Langlois M, Noacco C, Schneider V, Forsham P (1974) Adrenergic modulation of pancreatic glucagon secretion in man. J Clin Invest 53:1441–1446
Gerich J, Lorenzi M, Tsalikian E, Karam J (1976) Studies on the mechanism of epinephrine-induced hyperglycemia in man. Diabetes 25:65–71
Girardier L, Seydoux L, Berger M, Veicsteinas A (1974) Selective pancreatic nerve section. An investigation of neural control of glucagon release in the conscious unrestrained dog. J Physiol (Paris) 74:731–735
Girardier L, Seydoux J, Campfield LA (1976) Control of A and B cells *in vivo* by sympathetic nervous input and selective hyper- or hypoglycemia in dog pancreas. J Physiol (Paris) 72:801–814
Gold RM (1973) Hypothalamic obesity: the myth of the ventromedial nucleus. Science 182:488–490

Goto Y, Carpenter RG, Berelowitz M, Frohmann LA (1980) Effect of ventromedial hypothalamic lesions on the secretion of somatostatin, insulin, and glucagon by the perfused rat pancreas. Metabolism 29:986–990
Halter JB, Pflug AE (1980) Effect of sympathetic blockade by spinal anesthesia on pancreatic islet function in man. Am J Physiol 2:E150–E155
Hedo J, Villanueva M, Marco J (1978) Stimulation of pancreatic polypeptide and glucagon secretion by 2-deoxy-D-glucose in man: evidence for cholinergic mediation. J Clin Endocrinol Metab 47:366–371
Helman AM, Amira R, Nicolaidis S, Assan R (1980) Glucagon release induced by ventrolateral hypothalamic stimulation in the rat. Endocrinology 106:1612–1619
Hill DE, Mayes S, DiBattista D, Lockhard-Eward R, Martin JM (1977) Hypothalamic regulation of insulin release in rhesus monkeys. Diabetes 26:726–731
Holst JJ, Järhult H (1978) Reflex adrenergic control of endocrine pancreas evoked by unloading of carotid baroreceptors in cats. Diabetologia 15:240
Holst JJ, Grøholt R, Schaffalitzky de Muckadell OB, Fahrenkrug J (1981a) Nervous control of pancreatic endocrine secretion in pigs. I. Insulin and glucagon responses to electrical stimulation of the vagus nerves. Acta Physiol Scand 111:1–7
Holst JJ, Grønhold R, Schaffalitzky de Muckadell OB, Fahrenkrug J (1981b) Nervous control of pancreatic endocrine secretion in pigs. II. The effect of pharmacological blocking agents on the response to vagal stimulation. Acta Physiol Scand 111:9–14
Holst JJ, Schaffalitzky de Muckadell OB, Fahrenkrug J, Lindkaer S, Nielsen OV, Schwartz TW (1981c) Nervous control of pancreatic endocrine secretion in pigs. III. The effect of acetylcholine on the pancreatic secretion of insulin and glucagon. Acta Physiol Scand 111:15–22
Honey RN, Weir GC (1980) Acetylcholine stimulates insulin, glucagon and somatostatin release in the perfused chicken pancreas. Endocrinology 107:1065–68
Inoue E, Bray GA (1977) The effects of subdiaphragmatic vagotomy in rats with ventromedial hypothalamic obesity. Endocrinology 95:108–114
Inoue S, Campfield LA, Bray GA (1977) Comparison of metabolic alterations in hypothalamic and high fat diet-induced obesity. Am J Physiol 23:R162–R168
Irving MH, Britton BJ, Wood WG, Padgham C, Carruthers M (1974) Effect of $\beta$-adrenergic blockade on plasma catecholamines in exercise. Nature 248:531–533
Iversen J (1973a) Effect of acetylcholine on the secretion of glucagon and insulin from the isolated, perfused canine pancreas. Diabetes 22:381–387
Iversen J (1973b) Adrenergic receptors and the secretion of glucagon and insulin from the isolated, perfused canine pancreas. J Clin Invest 52:2102–2116
Järhult J (1975a) Osmolar control of the circulation in hemorrhagic hypotension. Acta Physiol Scand [Suppl] 423:29–37
Järhult J (1975b) Role of sympatho-adrenal system in hemorrhagic hyperglycemia. Acta Physiol Scand 93:25–33
Järhult J, Holst JJ (1977) Stimulation of glucagon and inhibition of insulin secretion evoked from carotid baroreceptors. Sep Exp 33:236–237
Johnson D, Ensinck J (1976) Stimulation of glucagon secretion by scorpion toxin in the perfused rat pancreas. Diabetes 25:645–649
Johnston DI, Bloom SR (1973) Plasma glucagon levels in the term human infant and effect of hypoxia. Arch Dis Child 48:451–454
Kaneto A, Kosaka A (1974) Stimulation of glucagon and insulin secretion by acetylcholine infused intrapancreatically. Endocrinology 95:676–681
Kaneto A, Miki E, Kosaka K (1974) Effects of vagal stimulation on glucagon and insulin secretion. Endocrinology 95:1005–1010
Kaneto A, Kajinuma H, Kosaka K (1975a) Effect of splanchnic nerve stimulation on glucagon and insulin output in the dog. Endocrinology 96:143–150
Kaneto A, Miki E, Kosaka K (1975b) Effect of beta and beta$_2$ adrenoreceptor stimulants infused intrapancreatically on glucagon and insulin secretion. Endocrinology 97:1166–1173
Kaneto A, Kajinuma H, Kosaka K (1977a) Effect of alpha adrenoreceptor stimulants infused intrapancreatically on glucagon and insulin secretion. Horm Metab Res 9:267–271

Kaneto A, Kaneko T, Kajinuma H, Kosaka K (1977b) Effect of vasoactive intestinal polypeptide infused intrapancreatically on glucagon and insulin secretion. Metabolism 26:781–786
Kern HF, Grube D (1972) Comparative fine structure and innervation of pancreatic islets. In: Proc 4th Int Congr Endocrinol, Washington DC, pp 224–228
Koch-Weser J (1975) Non-beta-blocking actions of propranolol. N Engl J Med 293:988–990
Langer SZ, Adler-Graschinsky E, Giorgi O (1977) Physiological significance of α-adrenoceptor-mediated negative feedback mechanism regulating noradrenaline release during nerve stimulation. Nature 265:648–650
Larsson LI (1979) Innervation of the pancreas by substance P, enkephalin, vasoactive on intestinal polypeptide and gastrin/CCK immunoreactive nerves. J Histochem Cytochem 27:1283–1284
Larsson LI (1980) New aspects on the neural, paracrine and endocrine regulation of islet function. Front Horm Res 7:14–29
Larsson LI, Fahrenkrug J, Holst JJ, Schaffalitzky de Muckadell OB (1978) Innervation of the pancreas by vasoactive intestinal polypeptide (VIP) immunoreactive nerves. Life Sci 22:773–780
Laughton W, Powley TL (1979) Four central nervous system sites project to the pancreas. Soc Neurosci Abstr 5:46
Leblanc H, Lachelin GCL, Abu-Fadil S, Yen SSC (1977) The effect of dopamine infusion on insulin and glucagon secretion in man. J Clin Endocrinol Metab 44:196–198
Lefèbvre PJ, Luyckx AS (1978) Glucose and insulin in the regulation of glucagon release from the isolated perfused dog stomach. Endocrinology 103:1579–1582
Lefèbvre PJ, Luyckx AS (1980) Neurotransmitters and glucagon release from the isolated, perfused canine stomach. Diabetes 29:697–701
Lefèbvre PJ, Luyckx AS, Moerman E, Bogaert M (1978a) Scorpion venum induced release of noradrenaline does not modify glucagon-output from the isolated perfused dog stomach. Horm Metab Res 10:80–81
Lefèbvre PJ, Luyckx AS, Brassinne AH (1978b) Vagal stimulation and its role in eliciting gastrin but not glucagon release from the isolated perfused dog stomach. Gut 19:185–188
Lilavivat U, Brodows RG, Campbell RG (1981) Adrenergic influence on glucocounterregulation in man. Diabetologia 20:482–88
Lindsey CA, Faloona GR, Unger RH (1973) Glucagon and the insulin: glucagon ratio in severe trauma. Trans Assoc Am Physicians 86:264–271
Lindsey A, Santeusanio F, Braaten J, Faloona G, Unger R (1974) Pancreatic alpha-cell function in trauma. JAMA 227:757–761
Lindsey CA, Faloona GR, Unger RH (1975) Plasma glucagon levels during rapid exsanguination with and without adrenergic blockade. Diabetes 24:313–315
Lorenzi M, Tsalikian E, Bohannon N, Gerich J, Karam J, Forsham P (1977) Differential effects of L-Dopa and apomorphine on glucagon secretion in man: evidence against central dopaminergic stimulation of glucagon. J Clin Endocrinol Metab 45:1154–1158
Lorenzi M, Karam JH, Tsalikian E, Bohannon NV, Gerich JE, Forsham PH (1979) Dopamine during α- or β-adrenergic blockade: hormonal, metabolic, and cardiovascular effects. J Clin Invest 63:310–317
Lundberg J, Ahlman H, Dahlström A, Kewenter J (1976) Catecholamine-containing nerve fibers in the human abdominal vagus. Gastroenterology 70:472–474
Lundberg JM, Höfelt T, Nilsson G, Terenius L, Rehfeld J, Elde R, Said S (1978) Peptide neurons in the vagus, splanchnic and sciatic nerves. Acta Physiol Scand 104:499–501118
Lundquist I, Ericson LE (1978) β-adrenergic insulin release and adrenergic innervation of mouse pancreatic islets. Cell Tissue Res 193:73–85
Lundquist I, Franska R, Grodsky GM (1976) Direct calcium-stimulated release of glucagon from the isolated perfused rat pancreas and the effect of chemical sympathectomy. Endocrinology 98:815–818
Marliss EB, Girardier L, Seydoux J, Wollheim CB, Kanazawa Y, Orci L, Renold AE, Porte D Jr (1973) Glucagon release induced by pancreatic nerve stimulation in the dog. J Clin Invest 52:1246–1259

Marre M, Bobbioni E, Suarez M, Reach G, Dubois MP, Assan R (1979) Control of gastric glucagon secretion in the acutely pancreatectomized rat. Diabetes 28:213–220
McLoughlin J, Hayes J, Buchanan K, Kelly J (1978) Role of neural influences in the release of gastrin, glucagon, and secretin during hypoglycemia in man. Gut 19:632–639
Miller RE, Horton ES (1979) Neural release of glucagon is inhibited by hyperglycemia and enhanced by phentolamine. Diabetes 28:762–768
Moltz JH, Dobbs RE, McCann SM, Fawcett CP (1977) Effects of hypothalamic factors on insulin and glucagon release from the islets of Langerhans. Endocrinology 101:196–202
Moltz JH, Dobbs RE, McCann SM, Fawcett CP (1979) Preparation and properties of hypothalamic factors capable of altering pancreatic hormone release *in vitro*. Endocrinology 105:1262–1268
Morgan CR, Lobl RT (1966) A histochemical study of neuro-insular complexes in the pancreas of the rat. Anat Rec 160:231–238
Müller EE, Frohman LA, Cocchi D (1973) Drug control of hyperglycemia and inhibition of insulin secretion due to centrally administered 2-deoxy-D-glucose. Am J Physiol 224:1210–1217
Ohneda A, Ishii S, Horigome K, Chiba M, Sakai T, Kai Y, Watanabe K, Yamagata S (1977) Effect of intrapancreatic administration of vasoactive intestinal peptide upon the release of insulin and glucagon in dogs. Horm Metab Res 9:447–452
Orci L (1976) Morphofunctional aspects of the islets of Langerhans. The microanatomy of the islets of Langerhans. Metabolism 25 [Suppl] 25:1303–1313
Orci L, Cameron C, Lambert AE, Kanazawa Y, Amherdt M, Stauffacher W (1970) The autonomous nervous system and the B-cell: metabolic and morphological observations made in spiny mice (Acomys cahirinus) and in cultured fetal rat pancreas. Acta Diabet Lat [Suppl 1] 7:184–226
Orci L, Perrelet A, Ravazzola M, Malaisse-Lagae F, Renold AE (1973) A specialized membrane junction between nerve endings and B-cells in islets of Langerhans. Eur J Clin Invest 3:443–445
Orci L, Malaisse-Lagae F, Amherdt M, Ravazzola M, Weisswange A, Dobbs R, Perrelet A, Unger R (1975) Cell contacts in human islets of Langerhans. J Clin Endocrinol Metab 41:841–844
Palmer JP, Porte D Jr (1981) Control of glucagon secretion: the central nervous system. In: Unger RH, Orci C (eds) Glucagon, physiology, pathophysiology and morphology of the pancreatic A-cells. Elsevier, Amsterdam Oxford New York, pp 135–160
Palmer J, Henry D, Benson J, Johnson D, Ensinck J (1976) Glucagon response to hypoglycemia in sympathectomized man. J Clin Invest 57:522–525
Palmer J, Werner P, Hollander P, Ensinck J (1979) Evaluation of the control of glucagon secretion by the parasympathetic nervous system in man. Metabolism 28:549–552
Parameswaran SV, Steffens AB, Hervey GR, deRuiter L (1977) Involvement of a humoral factor in regulation of body weight in parabiotic rats. Am J Physiol 232:R150–R147
Porte D Jr, Woods SC, Smith PH (to be published) Neural control of the pancreatic islet and its relation to stress hyperglycemia. In: Assan R, Girard JR, Marliss EB (eds) Diabetes mellitus: pathophysiologic approach to clinical practice. John Wiley & Sons, New York Chicester
Powley TL, Laughton W (1981) Neural pathways involved in hypothalamic integration of autonomic responses. Diabetologia 20:378–387
Raptis S, Escobar-Jimenez F, Rosenthal J, Ditschuneit H, Pfeiffer E (1977) Somatostatin modulation of pancreatic glucagon, insulin, glucose and free fatty acids following beta adrenergic stimulation. J Clin Endocrinol Metab 44:1088–1093
Rayfield E, George D, Eichern H, Hsu T (1975) L-Dopa stimulation of glucagon secretion in man. N Engl J Med 293:589–591
Rehfeld JF, Jensen SL (1980) The effect of gastrin and cholecystokinin on the endocrine pancreas. Front Horm Res 7:107–118
Rehfeld JF, Larsson LI, Golterman NR, Schwartz TW, Holst JJ, Jensen SL, Morley JS (1980) Neural regulation of pancreatic hormone secretion by the C-terminal tetrapeptide of CCK. Nature 284:33–38

Rizza RA, Cryer PE, Haymond MW, Gerich JE (1980) Adrenergic mechanisms for the effects of epinephrine on glucose production and clearance in man. J Clin Invest 65:682–689

Rocha D, Santeusanio F, Faloona G, Unger RH (1973) Abnormal pancreatic alpha-cell function in bacterial infections. N Engl J Med 288:700–703

Rohner-Jeanrenaud F, Jeanrenaud B (1980) Consequences of ventromedial hypothalamic lesions upon insulin and glucagon secretion by subsequently isolate perfused pancreases in the rat. J Clin Invest 65:902–910

Rohner-Jeanrenaud F, Jeanrenaud B (1981) Possible involvement of the cholinergic system in hormonal secretion by the perfused pancreas from ventromedial-hypothalamic lesioned rats. Diabetologia 20:217–222

Russel R, Thomson J, Bloom S (1974) The effect of truncal and selective vagotomy on the release of pancreatic glucagon, insulin and entero-glucagon. Br J Surg 61:821–824

Samols E, Weir G (1979) Adrenergic modulation of pancreatic A, B, and D cells. J Clin Invest 63:230–238

Samols E, Stagner JI, Weir GC (1981) Autonomic function and control of pancreatic somatostatin. Diabetologia 20:388–392

Schade D, Eaton R (1977) The regulation of plasma ketone body of plasma ketone body concentration by counterregulatory hormones in man. Diabetes 26:989–996

Schade S, Eaton R (1978) The metabolic response to norepinephrine in normal versus diabetic man. Diabetologia 16:433–439

Schade D, Eaton P (1979) The regulation of plasma ketone body concentration by counterregulatory hormones in man. Diabetes 28:5–10

Schaffalitzky de Muckadell OB, Fahrenkrug J, Holst JJ (1977) Release of vasoactive intestinal polypeptide (VIP) by electric stimulation of the vagal nerves. Gastroenterology 72:373–375

Schebalin M, Said SI, Makhlouf GM (1977) Stimulation of insulin and glucagon secretion by vasoactive intestinal peptide. Am J Physiol 232:197–200

Shimazu T, Ishikawa K (1981) Modulation by the hypothalamus of glucagon and insulin secretion in rabbits; studies with electrical and chemical stimulations. Endocrinology 108:605–611

Silverberg A, Shah S, Haymond M, Cryer P (1978) Norepinephrine: hormone and neurotransmitter in man. Am J Physiol 234:E252–E256

Smith P (1974) Pancreatic islets of the coturnix quail. A light and electron microscopic study with special reference to the islet organ of the splenic lobe. Anat Rec 178:567–586

Smith PH, Madson KL (1981) Interactions between autonomic nerves and endocrine cells of the gastroenteropanreatic system. Diabetologia 20:314–324

Smith PH, Porte D Jr (1976) Neuropharmacology of the pancreatic islets. Toxicology 16:269–285

Smith PH, Woods SC, Porte D Jr (1979) Control of the endocrine pancreas by the autonomic nervous system. In: Brooks CM, Koizumi K, Sata A (eds) Integrative functions of the autonomic nervous system. Elsevier/North-Holland Biochemical, Amsterdam Oxford New York, pp 84–97

Steffens AB (1981) The modulatory effect of the hypothalamus on glucagon and insulin secretion in the rat. Diabetologia 20:411–416

Swanson LE, Sawchenko PE (1980) Paraventricular nucleus: a site for the integration of neuroendocrine and autonomic mechanisms. Neuroendocrinology 31:410–417

Tyler J, Kajinuma H (1972) Influence of beta adrenergic and cholinergic agents in vivo on pancreatic glucagon and insulin secretion. Diabetes 21:332

Vranic M, Kawamori R, Pek S, Kovacevic N, Wrenshall GA (1976) The essentiality of insulin and the role of glucagon in regulating glucose utilization and production during strenuous exercise in dogs. J Clin Invest 57:245–255

Watanabe T, Yasuda M (1977) Electron-microscopic study on the innervation of the pancreas of the domestic fowl. Cell Tiss Res 180:454–465

Watari N (1968) Fine structure of nervous elements in the pancreas of some vertebrates. Zeitschr. für Zellforsch. 85:291–314

Weir G, Knowlton S, Martin D (1974) Glucagon secretion from the perfused rat pancreas. J Clin Invest 54:1403–1412
Willerson J, Hutcheson D, Leshin S, Faloona G, Unger RH (1974) Serum glucagon and insulin levels and their relationship to blood glucose values in patients with acute myocardial infarction and acute coronary insufficiency. Am J Med 57:747–753
Wilmore D, Moylan J, Pruitt B, Lindsey G, Faloona G, Unger RH (1974) Hyperglucagonaemia after burns. Lancet 1:73–75
Woods SC, Porte D Jr (1974) Neural control of the endocrine pancreas. Physiol Rev 54:596–619
Yoshida T, Kondo M (1980a) Effect of acetylcholine on the secretion of gut glucagon immunoreactivity and gut glucagon-like immunoreactivity in pancreatectomized dogs. Endocrinol Jpn 27:33–38
Yoshida T, Kondo M (1980b) Effect of adrenergic agents on the secretion of gastrointestinal immunoreactive glucagon in depancreatized dogs. Diabetes 29:355–360

CHAPTER 31

# Intraislet Insulin-Glucagon-Somatostatin Relationships

E. SAMOLS, G.C. WEIR, and S. BONNER-WEIR

## A. Introduction

Anatomic juxtaposition of different islet cells with one another has inspired several observers to propose that these cells might influence one another (HAIST 1965; ORCI et al. 1975). The discovery that glucagon stimulated insulin secretion (SAMOLS et al. 1965, 1966a) and conversely, that insulin suppressed glucagon secretion (SAMOLS et al. 1969, 1970, 1972) led to the hypothesis of an intraislet positive-negative insulin-glucagon feedback (SAMOLS et al. 1970, 1972). The observation that somatostatin (D-cell) is a potent inhibitor of A-cell (KOERKER et al. 1974) and B-cell (ALBERTI et al. 1973) secretion, and that glucagon promotes somatostation release (PATTON et al. 1976, 1977; WEIR et al. 1977, 1978, 1979), appeared to bolster (UNGER and ORCI 1977) the concept that islet cells might interact with one another locally, as secretions from the shooting cell would reach the target cell via the interstitium (dubbed as paracrine effects). This hypothesis, which suggests that we should no longer naively think of islets as being a collection of independently functioning A-, B-, D-, and pancreatic polypeptide-containing (PP)-cells, is being further modified to accomodate new findings in islet microanatomy. It now seems that the islet is an extraordinarily complex microorgan, with a highly sophisticated neural and vascular system, with a multitude of methods for the endocrine cells to communicate or not communicate with one another, and with portal access to the exocrine pancreas.

Although intuition and deduction suggest that the secreting cells of the islet should be coordinated and integrated within the microorgan, and that the microorgans are themselves integrated into a larger functional endocrine organ which discharges its secretions into the portal vein, our understanding of the grand scheme – assuming there is one – is relatively rudimentary. Because of the large amount of data that has emerged in recent years to suggest potential local interactions between islet cells, the evidence that these phenomena actually occur, or are physiologically relevant, needs rigorous critical assessment. Part of the problem is unquestionably technical, in that disturbance of islet microstructure by experimental manipulation may inevitably alter microfunction. The authors have, at times, been daunted by the magnitude of the ratio of speculation: established important physiologic event. At other times, we have been heartened that intraislet concepts do provide an aesthetic resolution of applied problems, as in diabetes mellitus. We have tried to indicate the myriad data which might impinge on intraislet insulin-glucagon-somatostatin interrelationships, and to choose those which currently appear to be most rational.

## B. Overview of Islet Anatomy

It is impossible to understand islet function without appreciating the highly structured nature of this microorgan (ORCI and PERRELET 1981). Most of our knowledge about the anatomy of mammalian islets has been obtained from rodents and humans, and even though there are some differences in other mammalian species, the following overview should be relevant to mammalian islets in general. It has been shown in the rat that there is great variation in islet size with diameters ranging between 40 and 800 μm, and an average sized islet of 250 μm probably contains about 3,000 cells (S. BONNER-WEIR 1981, unpublished work). Islet tissue accounts for about 1.5% of the total volume of the adult pancreas (HOFTIEZER and CARPENTER 1973). Islets contain at least four endocrine cell types: the A-, B-, D-, and PP-cells which contain respectively the peptides glucagon, insulin, somatostatin, and pancreatic polypeptide. B-cells form a central core of the islet which is surrounded by a mantle of A-, D-, and PP-cells, the composition of which depends upon pancreatic location. The head or duodenal portion of the pancreas, which is derived from the ventral pancreatic anlage, and is supplied by the inferior pancreaticoduodenal artery, contains islets whose mantle contains PP-cells comprising about 20% of the total islet cell population, D-cells making up about 5%, and A-cells which are very sparse in number (BENCOSME and LIEPA 1955; ORCI et al. 1976a). The tail and body, or splenic portion, of the pancreas receives its blood supply from the gastroduodenal and splenic arteries, and has its embryonic origin in the dorsal pancreatic anlage. The islets of this large splenic portion are rich in glucagon-containing A-cells, with these cells making up about 20% of the total, and very poor in PP-cells. The D-cell and B-cell frequently is similar in both types of islets. Specialized junctions are present in islets, with tight and gap junctions being found not only between cells of the same type, but also between heterogeneous cells (ORCI et al. 1973, 1975). These may have important influences upon interactions between islet cells, as will be discussed later in this section.

Islets are well known to contain terminals of the autonomic nervous system (see also Chap. 30). On the basis of electron microscopy, characteristic adrenergic and cholinergic nerve endings have been found (SMITH and PORTE 1976). In addition, a third distinct type of nerve fiber, whose function is unclear, has been identified, and this has been called the peptidergic branch of the autonomic nervous system (BLOOM and POLAK 1978). The transmitter of these fibers has not yet been conclusively identified, but islets have been shown to be surrounded by fibers containing vasoactive intestinal peptide (VIP)-like immunoreactivity (LARSSON et al. 1978; BLOOM 1981). (See Sect. D.II for discussion of neural control.)

Our understanding of islet vascularization is still limited (see also Sect. E and Fig. 1), but blood flow may be fundamental to islet functional coordination. Small arterial branches, upon reaching the islet mantle, break up into a glomerular capillary network (WHARTON 1932; FUJITA et al. 1976). Small islets tend to be supplied by one small arteriole, whereas large islets can be supplied by two or more arterioles (BONNER-WEIR and ORCI 1982). In small islets, the capillaries frequently leave the islet, passing through pancreatic exocrine tissue before coalescing to become collecting venules. Contrary to earlier thinking (WHARTON 1932; FUJITA et al. 1976), in the larger islets, which may comprise as much as 60% of the pancreatic islet volume, the capillaries usually coalesce into collecting venules, forming a basket-like network just at the periphery of the islet before forming larger veins. There has been speculation that almost all blood reaching the exocrine pancreas might first pass through islet tissue, but it has recently been possible to demonstrate numerous arteries branching into capillaries which are restricted to exocrine tissue (BONNER-WEIR and ORCI 1982). Pancreatic blood flow has been assessed with trapped microspheres, and it has been estimated that islets receive 13%–20% of the entire pancreatic blood flow, in spite of the fact that islets comprise only 1.5% of pancreatic volume (LIFSON et al. 1980). On the basis of these morphological and physiologic observations, it is assumed that the endocrine pancreas is more richly vascularized than the exocrine pancreas. (For a more detailed description of islet anatomy see Chap. 4).

## C. General Mechanisms of Communication Between Cells

Multicellular organisms have developed a variety of mechanisms which allow cells to communicate with each other, and some of these will be briefly reviewed to help put into perspective the kind of relationships which could occur in islets. We are familiar with such long distance relationships as the *endocrine* system whereby a hormone is secreted by a gland and transported through blood vessels to a distant target organ. A variant of this is called *neuroendocrine*, with the blood-borne hormone released by neural tissue. Communication by the central and peripheral *nervous system* is well known. There are a number of specialized *portal systems* in which hormones exert their major effects only through a restricted vascular compartment. The best described of these are the portal vein, the hypophyseal-pituitary portal system, and the adrenal portal system which carries glucocorticoids from the adrenal cortex to the adrenal medulla (POHORECKY and WURTMAN 1971). There are probably a number of other smaller portal systems which have not yet been adequately defined. It has been proposed for instance that the mammalian gastric antrum contains small vessels which permit locally secreted somatostatin to be carried to nearby gastrin-containing cells (FORSSMAN et al. 1978) and there are some experimental findings which are consistent with such a relationship (SAFFOURI et al. 1979). A similar portal system may functionally connect pancreatic islets with acinar tissue (FUJITA et al. 1976) and it has been thought that there could be a separate intraislet system (FORSSMAN et al. 1978).

*Paracrine* communication is a mechanism through which a mediator is secreted by a cell and moves through an interstitial space to act on a neighboring target cell (FEYRTER 1953). There are a limited number of examples of such a phenomenon, but it may occur in many tissues. This mechanism is thought to be a major form of communication between cells in some primitive organisms such as coelenterates (DOCKRAY 1979). In vertebrates, the local release of mediators from mast cells could be considered to be paracrine in nature. There is good reason to suspect that paracrine effects are very important in the testis and ovary, and there have been numerous recent proposals that paracrine control mechanisms are of major importance in the gut and pancreatic islets (SAMOLS et al. 1966a, 1972; ORCI and UNGER 1975; WEIR et al. 1976a, b; ORCI 1976; FORSSMAN et al. 1978). A related mechanism has been termed *autocrine*, whereby a secreted mediator might influence the cell from which it was secreted so that there might be a feedback relationship. Cell types which could be influenced by an autocrine mechanism include pancreatic B-cells, which can be inhibited by insulin (FRERICHS et al. 1965; IVERSEN and MILES 1971), pancreatic D-cells, which have been shown to be inhibited by somatostatin analogs (IPP et al. 1979), and A-cells, which can be suppressed by glucagon (KAWAI and ROUILLER 1981).

# D. Potential Interactions Within Islets

## I. Potential Interactions Between Cells

### 1. Basic Scheme

If the A-, B-, and D-cells are visualized as the vertices of an equilateral triangle, the potential interactions, whether by short (intraislet) or long (systemic circulation) loops are: (1) glucagon may stimulate somatostatin and/or insulin secretion; (2) somatostatin may inhibit glucagon and/or insulin secretion; (3) insulin may inhibit glucagon secretion – its effect on somatostatin has been variable, including stimulation, inhibition, and no effect. Thus, there is a potential positive-negative glucagon-insulin feedback, a potential positive-negative glucagon-somatostatin feedback, and a potential negative somatostatin-insulin feedback.

### 2. Effect of Glucagon Upon B- and D-cells

Exogenous glucagon stimulates B-cell (SAMOLS et al. 1965, 1966a) and D-cell (PATTON et al. 1976, 1977; WEIR et al. 1977, 1978, 1979) secretion. In the isolated canine pancreas, the insulin and somatostatin stimulatory responses are dose dependent, occur as a biphasic pattern, are seen between 15 and 45 s after the glucagon stimulus (WEIR et al. 1978, 1979), and begin at infusate glucagon concentration of approximately 100 pg/ml (KAWAI and ROUILLER 1981).

Factors modifying the effect of glucagon on the B-cell have been noted in detail in Chap. 22. A raised glucose concentration clearly potentiates glucagon's stimulation of the B-cell, but not of the D-cell (WEIR et al. 1979). The absence of enhancement of glucagon-induced D-cell secretion by glucose is perhaps surprising, because glucose does promote D-cell secretion, albeit modestly (SCHAUER et al. 1976; WEIR et al. 1977; IPP et al. 1977). Although it is argued that this may reflect a relatively minor influence of glucose on the D-cell as compared with the B-cell (HONEY and WEIR 1980), it can also be rationalized that the D-cell response to glucose is extremely important for suppression of the A-cell (see Sect. F).

### 3. Effect of Somatostatin Upon A- and B-cells

The major portion of pancreatic islet somatostatin appears to be identical to hypothalamic somatostatin 14 (SS 14). Somatostatin very rapidly and potently inhibits A- and B-cell secretion in vivo (ALBERTI et al. 1973; KOERKER et al. 1974) and in vitro (IVERSEN 1974; EFENDIC et al. 1974; JOHNSON et al. 1975; GERICH et al. 1975). In the isolated, perfused canine pancreas, the inhibitory effect may be observed within 60 s, and the potency is remarkable for being demonstrable at a concentration of $\geqq 10$ pg/ml (SAMOLS and HARRISON 1976a; KAWAI and ROUILLER 1981). In humans, intravenous infusion of synthetic somatostatin, 2 μg/h (raising somatostatin levels by 20–30 pg/ml) lowers circulating glucagon and insulin levels (ZYZNAR et al. 1981).

Somatostatin is probably equipotent as an inhibitor of A-cell compared with B-cell secretion, with weaker effects on the PP-cell (HERMANSEN and SCHWARTZ 1979). Accordingly, some studies have suggested that somatostatin selectively inhibits insulin (EFENDIC et al. 1975; RAPTIS et al. 1977) while others found a suppressive preference for glucagon (GERICH et al. 1975; BHATHENA et al. 1976; SAMOLS and HARRISON 1976a). Secretion of insulin and glucagon by virtually all known stimulants (including glucose, glucagon, arginine, β-adrenergic or cholinergic agonism, phosphodiesterase inhibitors or cAMP, tolbutamide, and divalent cations) have been inhibited by appropriate concentrations of somatostatin in mammals (GERICH 1981). There is normally an offsetting dose-response relationship between the inhibitory effects of somatostatin on the one hand, and the secretagogue potency on the other. Strangely, somatostatin apparently stimulates glucagon secretion in the duck (STROSSER et al. 1980), but suppresses the chicken A-cell (HONEY et al. 1981).

Although there is evidence that the pancreas contains several molecular species of somatostatin (PATEL and REICHLIN 1978), most of these have been inadequately characterized. However, the presence of natural SS 28 as a minor component (BENOIT et al. 1980) is interesting, because it is about 10–100 times more potent than SS 14 as inhibitor of the arginine-stimulated B-cell (MANDARINO et al. 1981; BROWN et al. 1981), but equipotent with SS 14 with respect to the arginine-stimulated A-cell; i.e., SS 28 should selectively inhibit the B-cell. The dissociation of biologic activity of synthetic somatostatin analogs on the A- and B-cells suggests that the somatostatin receptors on the A-cell are different from those on the B-cell (EFENDIC et al. 1975; SARANTAKIS et al. 1978; WEIR et al. 1980). However, of scores of synthetic analogs tested, the selectivity of suppression for either the B- or the A-cell has, at best, been relative (ADRIAN et al. 1981). These selective analogs may appear to have promise as agents to test for potential paracrine effects of A- and B-cells on one another. However, for paracrine "proof", absolute specificity of suppression, which has not been demonstrable to date, is probably necessary. Relative specificity of suppression has been sufficient to show that somatostation analogs (and by inference, somatostatin itself) inhibit D-cell secretion (IPP et al. 1979).

### 4. Effect of Insulin Upon A- and D-cells

Insulin can suppress glucagon secretion in vivo and in vitro (SAMOLS et al. 1970, 1972; SAMOLS and HARRISON 1976b; WEIR et al. 1976b) while glucose itself does not suppress the A-cell in the absence of insulin in vivo and in vitro (see Sect. F and Chap. 22). Suppression of circulating glucagon by 20%–30% is readily achieved with exogenous insulin, raising insulin levels by 30–200 μIU/ml (SERVICE et al. 1978; ASPLIN et al. 1981) during normoglycemia maintained with a glucose clamp in humans. These, and other observations discussed in part in Sect. F and in detail in Chap. 22, suggest that glucose does not suppress glucagon in the absence of an appropriate insulin concentration.

Assuming for the moment that this argument is correct, there is also an important corollary implicit in the concept of intraislet feedback. As exogenous insulin also inhibits endogenous insulin secretion it induces a counterforce, by intraislet feedback, augmenting glucagon secretion, so that the glucagon releasing activity of glucagon secretagogues (e.g., arginine) is augmented (ASPLIN et al. 1981). Therefore, the "resultant" of several secretory "vector" forces may have been misinterpreted by investigators, particularly with respect to "primary A-cell secretory defects" in diabetes (see Sect. L).

This exegesis is perhaps partly supported by in vitro observations. The concentrations of insulin (~25 mIU/ml) necessary in vitro for A-cell suppression are very high, perhaps because the A-cells are normally accustomed to being surrounded by such high insulin levels (SAMOLS and HARRISON 1976b; WEIR et al. 1976b). Conditions which reduce endogenous insulin secretion facilitate A-cell suppression by exogenous insulin (WEIR et al. 1976), and the gastric A-cell is "glucose blind" but exquisitely insulin sensitive (LEFEBVRE and LUYCKX 1978). Nevertheless, an overview of published work (see Chap. 22) suggests that suppression of glucagon by exogenous insulin is relatively easy with low insulin concentrations in vivo, but relatively difficult, requiring high insulin concentrations, in vitro, without an entirely satisfactory explanation for this discrepancy. It may be that the 30% suppression of glucagon by "physiologic" levels of exogenous insulin during normoglycemia in humans is close to the maximum achievable during normoglycemia *in the presence of intact islets and intraislet B-cells.* For a larger suppression of glucagon in vivo (or in vitro), perhaps at least one of the following requirements is necessary: (1) endogenous insulin secretion plus hyperglycemia; (2) ablation of intraislet B-cells (plus the presence of functioning D-cells – see next paragraph); or (3) enormous levels of insulin.

Studies in mammalian pancreas have either failed to show any significant effect of exogenous insulin on pancreatic somatostatin secretion (PATTON et al. 1977; WEIR et al. 1979), implied suppression (SCHAUER et al. 1978), or demonstrated direct suppression (GERBER et al. 1981; ROUILLER et al. 1981). However, in the isolated chicken pancreas (HONEY and WEIR 1979), exogenous insulin induces a prompt and clear-cut stimulation of D-cell secretion. It is not clear whether these different responses represent species variability, or the complexity of intraislet adjustments.

An indirect suppressive effect of insulin on D-cell secretion has also been inferred (see Sect. L) because pancreatic somatostatin content, and secretion, are increased in various diabetic states. Correction of this hypersomatostatinism and of the loss of D-cell sensitivity to glucose in streptozotocin-induced diabetic islets (TRIMBLE et al. 1981a) by exogenous insulin treatment in vivo, but not by in vitro insulin administration (HERMANSEN et al. 1979) suggests that insulin's effect is accomplished by normalizing the metabolic state of the animal, perhaps because of correction of high circulating levels of somatostatinotropic agents such as fatty acids (WASADA et al. 1981) or glucagon (PATTON et al. 1977; WEIR et al. 1978, 1979).

### 5. Pancreatic Polypeptide as an Ignored Entity and Why

Pancreatic polypeptide originates from the PP-cell (LARSSON et al. 1974), which is localized, in humans, almost entirely in the pancreas (ADRIAN et al. 1978a). In addition to being scattered among the exocrine cells, the PP-cell is a major cell in the islet mantle, particularly in the duodenal lobe. Why then is pancreatic polypeptide usually ignored in a discussion of intraislet interactions?

First, infused pancreatic polypeptide has had no unequivocal influence on insulin and glucagon secretion in the isolated perfused pancreas (SAMOLS and HARRISON 1976b; IPP et al. 1978c) or in humans (ADRIAN et al. 1978b). Second, those effects which were seen in a high dose infusion, were attributable to somatostatin contamination of the bovine pancreatic polypeptide (BPP) preparation (SAMOLS and HARRISON 1976a), making one wonder whether somatostatin may have contributed to the reported (LIN et al. 1977; LIN and CHANCE 1978) suppression of pancreatic exocrine secretion by BPP. The effect of BPP on pancreatic exocrine secretion was "modest" in humans (ADRIAN et al. 1978b). However, if the exocrine suppression is validated as a true biologic action of pancreatic polypeptide, the latter may well have local intrapancreatic effects on acinar cells, and could also reach the exocrine pancreas through the insular-acinar portal system (see Sect. K).

Fortunately, BPP has been synthesized (ARIMURA et al. 1979) and its high dose intravenous injection into rats acutely suppressed portal plasma somatostatin levels, presumed to represent gut and pancreatic secretory inhibition. Although BPP did not influence somatostatin release from the isolated canine pancreas (IPP et al. 1978a), the studies with synthetic pancreatic polypeptide raise, for the first time, a possible intraislet paracrine role for pancreatic polypeptide.

### 6. Other Potential Islet Mediators

The cells of the islets are part of a larger "gastroenteropancreatic" (GEP) system, and in addition to the major four cell types, rarer $D_1$- and P-cells (SOLCIA et al. 1978), whose secretions are not known, have been detected in the human pancreas. However, there are reports of peptide immunoreactivity in A-cells, or unidentified cells, for gastric inhibitory peptide (SMITH et al. 1977), cholecystokinin (CCK) (GRUBE et al. 1978a), $\beta$-endorphin (GRUBE et al. 1978b), secretin (RUFENER et al. 1976), corticotrophin (LARSSON 1977), and glicentin, a peptide purified (MOODY et al. 1978) from gut glucagon immunoreactivity (SAMOLS et al. 1966b). Vasoactive intestinal peptide (VIP), but not somatostatin (LARSSON 1980), is present in islet nerves. The pancreatic islet cells of many species contain dopamine and/or serotonin (LUNDQUIST 1971; FELDMAN 1979) and kallikrein has been demonstrated in human B-cells (OLE-MOIYOI et al. 1979). All of these putative intraislet chemical messengers have been shown to influence the secretion of insulin and most have been described as stimulating or inhibiting glucagon and somatostatin secretion. Their role in physiology has not been characterized.

In addition to the hormones mentioned: neurotransmitters, neurohormones, and cybernins (local regulatory factors), it should be remembered that intraislet cell products may not necessarily be in their characteristic circulating monomeric final form. Intraislet insulin could theoretically form dimers and hexamers (with receptor effects which differ from monomers). C-peptide has been shown to inhibit B-cell secretion (TOYOTA et al. 1975), so it is presumably possible that prohormones, prohormone remnants, or polymers of insulin, glucagon, and somatostatin could influence intraislet interrelationships.

## II. Neural Control

It is clear that autonomic signals modulate the secretory activity of various islet endocrine cells, thus providing a mechanism by which the central nervous system can integrate digestive and metabolic functions. There is now evidence, that, in addition to central integration, there is local (peripheral) regulation of autonomic neurotransmission in the islet microorgan by presynaptic mechanisms (SAMOLS and STAGNER 1981). As nerves and endocrine cells frequently share certain common peptides and/or amines, the distinction between hormones, neurotransmitters, paracrine and neurocrine transmission becomes a little semantic. Nevertheless, a number of basic neural control mechanisms have been defined, and are relevant to the consideration of insulin-glucagon-somatostatin interrelationships. For a comprehensive review of neural control see Chap. 30).

### 1. Cholinergic Agonism

Acetylcholine (with blockade of its effects by atropine) profoundly stimulates pancreatic polypeptide secretion (SAMOLS et al. 1977, 1978), strongly stimulates insulin and glucagon secretion, and either inhibits (SAMOLS et al. 1977, 1981) or has no effect (HERMANSEN 1980) on D-cell secretion in the isolated perfused canine pancreas. In the isolated porcine pancreas, electrical stimulation of the vagus (with blockade of its effects by atropine) inhibits somatostatin output (HOLST et al. 1981). Whether the inhibition of somatostatin contributes to the stimulatory effect of parasympathetic agonism on A-, B-, and PP-cells is one of the teasing neural-paracrine speculations.

### 2. Adrenergic Agonism

$\beta$-Adrenergic agonism stimulates A-, B-, D-, and PP-cell secretion (SAMOLS et al. 1977, 1978), $\alpha$-adrenergic agonism inhibits B- and D-cell secretion and, in the isolated canine pancreas, mildly stimulates A-cell secretion (SAMOLS and WEIR 1979). As the $\alpha$-adrenergic effects on the A-cell have been controversial (SMITH and MADSON 1981), it is remotely possible that the variable results could be caused by different degrees of intraislet interaction, i.e., of lessening of B- and D-cell tonic inhibition of A-cell to release glucagon. Electrical stimulation of the splanchnic nerve suppresses B- and D-cell secretion and stimulates A-cell secretion (MILLER and HORTON 1979; ROY et al. 1981).

### 3. Local Presynaptic Autonomic Regulation

If new concepts of local autonomic regulation (LANGER et al. 1977; BERTHELSEN and PETTINGER 1977) apply to the pancreatic islets and vessels, theoretically the isolated pancreas preparation should retain some autoregulation of adrenergic neurotransmission, and this is true for the canine pancreas, in which pancreatic arterioles are constricted by a presynaptic $\alpha_2$ postsynaptic $\alpha_1$ system and B-cells are inhibited by a presynaptic $\alpha_2$ postsynaptic $\alpha_2$ system (SAMOLS and STAGNER 1981). It has been suggested that the suppressive effect of somatostatin on A- and B-cell secretion might be mediated by $\alpha$-adrenergic agonism (SMITH et al. 1976), because phentolamine "blocked" somatostatin's inhibitory effects. Phentolamine has a higher affinity for $\alpha_2$ receptors than for $\alpha_1$ receptors, and the effects of blockade of presynaptic receptors, causing increased norepinephrine output, are different from blockade of postsynaptic receptors. The resultant effect is stimulation of insulin secretion by phentolamine, rather than the proposed blocking action. One may speculate whether $\alpha_2$ receptors, which are autoinhibitory on noradrenergic nerve terminals, are also autoinhibitory for intra-B-cell catecholamines.

As is evident, distinguishing local autonomic regulation of A-, B-, and D-cells from paracrine events has not been easy because of nonspecific investigative tests. The challenge becomes even more formidable when it is recalled that local autoregulatory presynaptic modulation by prostaglandins, serotonin, dopamine, neuropeptides, and ATP have been described in other organs (WESTFALL 1977), and if present in the islets, could account for a multitude of contradictory results deduced from agonist and antagonist infusions. However, in studies applying current concepts, specific dopaminergic receptors were found to be absent from A-, B-, and D-cells (SAMOLS et al. 1981).

#### 4. Peptidergic and Purinergic Regulation

The function of the islet peptidergic nervous system (LARSSON 1980; SMITH and MADSON 1981) is uncharted, with great potential intraislet modulatory influence, and with VIP and CCK known to stimulate B-, A-, and D-cell secretion, while endorphins may inhibit the D-cell (IPP et al. 1977; 1978b, c). The presence of a purinergic nervous system (BURNSTOCK 1979) in the islets remains unproven.

### III. Potential Role of Gap Junctions as Determinants of Coordination Between Islet Cells

Sections C and E were concerned with the ways in which islet cells might communicate with one another through interstitial and vascular spaces, but there may be important communication which occurs by direct contact of cells via gap junctions (MEDA et al., 1982). These specialized junctions are known to occur not only between homogeneous cell types, but also between heterogeneous cells (ORCI et al. 1973, 1975). They occupy a relatively small area of the surface of islet cells as compared with other secretory tissues (MEDA et al. 1979, 1980a), and are not randomly distributed among B-cells in that they are less frequent in the center of the islet B-cell core and more frequent in B-cells adjacent to the non-B-cell mantle (MEDA et al. 1980b). Gap junctions are thought to allow the passage between cells of electrical impulses, ions, and small molecules of molecular weight 1,200 daltons or less (SIMPSON et al. 1977; STAEHELIN 1974). Studies with islet cells in monolayer have shown that 6-carboxyfluorescein can rapidly pass from one cell to as many as eight others (KOHEN et al. 1979). Furthermore, the passage of glucose-6-phosphate and uridine $^3H$ nucleotides has been demonstrated (KOHEN et al. 1979; MEDA et al. 1981), making it seem possible that numerous other small mediators could be transferred. Because these transfers are presumably made through gap junctions and because gap junctions are found between both homogeneous and heterogeneous islet cell types, one can speculate that these junctions have an important role in coordinating the function of an entire islet. It is attractive to think in terms of gap junctions coordinating islet secretion, but these junctions could just as easily have an important role in islet growth and development or other intraislet metabolic interrelationships.

Data has accumulated indicating that gap junctions may coordinate the electrical activity of B-cells. It is well known that the electrical activity of B-cells changes after islets are exposed to insulin secretagogues (DEAN and MATTHEWS 1970). Furthermore, when membrane potential is measured by two microelectrodes, each of which is monitoring a separate B-cell within the same islet, it has been shown that the oscillations of electrical activity which occur during glucose stimulation are synchronous in the two cells, indicating that these two B-cells are coupled (MEISSNER 1976). The concept of coupling has been strengthened by recent elegant studies in which current flowing into one microelectrode-pierced cell can cause a change of membrane potential in a neighboring cell within a mouse islet (EDDLESTONE and ROJAS 1980). There is evidence that B-cells are heterogeneous with regard to granularity (ORCI 1974) and also to the different thresholds which are required for glucose to stimulate electrical activity (BEIGELMAN et al. 1977). Therefore, one can postulate that there are pacemaker cells which respond to particular glucose concentrations and, in turn, recruit other B-cells which otherwise might not have responded to glucose (MEDA et al. 1980c). Some experimental evidence suggests that communication between B-cells may enhance their secretory response to glucose. Acutely dispersed islet cells respond poorly to glucose, but when they are allowed to reaggregate, their responsiveness improves (HALBAN et al. 1979). A potentially important observation is that the number and size of gap junctions on B-cells increases when islet tissue is stimulated with either glucose or glibenclamide (MEDA et al. 1980c). Thus, secretagogues may enhance the ability of B-cells to become electrically coupled. In summary, these observations and considerations permit the development of attractive hypotheses about the role of gap junctions in B-cell secretion. The potential contributions of gap junctions to the function and interrelationships of A-, D-, and PP-cells remain even less clear. New insights into these questions should emerge when the secretory behavior of truly separated islet cells is better understood.

## E. Anatomic Determinants of Islet Regulation

The manner in which islet secretory products could influence islet function can not be understood without appreciating the extraordinarily complex ways in which islet structure could be influential. As a starting consideration there is persuasive physiologic data which indicates that there is spatial restriction and compartmentalization of islet secretory products. As best shown by perfused pancreas systems, it is known that very low concentrations of somatostatin and glucagon introduced into the arterial supply of islets can profoundly alter the secretion of other islet hormones (SAMOLS and HARRISON 1976a; WEIR et al. 1979; KAWAI and ROUILLER 1981). Because these concentrations are lower than those leaving pancreatic draining veins they must be far lower than concentrations found in certain parts of the islet interstitial space. To be specific, at a time when the somatostatin concentration in the venous effluent is over 100 pg/ml, the arterial infusion of less than

100 pg/ml can inhibit insulin and glucagon secretion (SAMOLS and HARRISON 1976a; KAWAI and ROUILLER 1981). It is obvious that the concentration of somatostatin in that portion of interstitial space into which D-cells release somatostatin must be extremely high, almost certainly more than 1,000-fold higher than those quantities which can influence A- and B-cell function. Because the islet contains so few D-cells it seems almost certain that, without functional barriers, these high concentrations must at times be in contact with portions of the plasma membrane of A- and B-cells. If there were a single interstitial space for each islet there could be diffusion of high concentrations of somatostatin to practically the entire plasma membrane surface of most A- and B-cells, but if this were the case it seems inconceivable that exogenous concentrations of 100 pg/ml could exert any effect. Therefore, there must be areas of the A- and B-cell plasma membrane which cannot be reached by the diffusion of locally secreted somatostatin, and thus there must be some sort of physiologic or anatomic compartmentalization.

There are a number of ways in which this compartmentalization could take place. First, the cell types within a given islet are not homogeneously distributed, and B-cells form a central core with a peripheral mantle made up of either A- and D-cells or PP- and D-cells, depending upon the location of the islet within the pancreas (ORCI and PERRELET 1981). Thus, it could be difficult for glucagon or somatostatin to diffuse through an interstitial space to reach the central B-cells. It is of interest that stimulated B-cells, which are in the center of an islet, have a different ultrastructure, with regard to density of various intracellular organelles, from stimulated B-cells which are adjacent to the peripheral non-B-cell mantle (KOLOD et al. 1981). There is not yet a good explanation for these differences, but they raise the question as to whether the peripheral B-cells are somehow influenced by the adjacent mantle cells. As a separate issue one could argue that because the A-, D- and PP-cells of the mantle are all either adjacent or close to B-cells, they are likely to be influenced by the local secretion of insulin.

A second kind of compartmentalization could be created by tight junctions, which are narrow areas of fusion between homogeneous and heterogeneous islet cell types. These tight junctions could create barriers within the interstitial spaces such that secretory products could be shunted through specific channels, either directly to capillaries or even to a target cell. It has been stated that the tight junctions in islets are relatively short and discontinuous compared with those seen in other tissues. Although this might reduce the likelihood of their exerting an important compartmentalization role, it certainly does not rule it out (ORCI 1976). It is noteworthy that tight junctions exhibit plasticity and become more numerous when islets are exposed to high levels of glucose, but the physiologic significance of this is unclear (ORCI 1976).

A concept which has received little attention has been that of possible polarity of islet cells. In most epithelial tissues, cells have clear polarity in their orientation, with obvious basolateral and apical surfaces. Thus, it is not surprising that one can sometimes see a columnar-like arrangement of cells in a sectioned mammalian islet (GOLDSTEIN and DAVIS 1968; HELLERSTRÖM 1977; LIKE 1977). Mammalian D-cells have been frequently found to have dendritic-like cytoplasmic extensions (FUJITA 1967; LEITER et al. 1979; LARSSON et al. 1979). In the glucagon-rich dark islets of chickens, A-cells are frequently arranged in a columnar pattern (S. BONNER-WEIR

1979, unpublished work), and in glucagonomas in humans a similar columnar arrangement has also been seen (LEICHTER et al. 1975). Thus, there are precedents for islet cells having polarity and it will not be surprising if this is eventually demonstrated to be true for the majority of islet cells. Such a polarity could facilitate compartmentalization, in that peptides might be secreted from an apical surface almost directly into a capillary and have very little contact with neighboring cells. Not only secretion, but also receptors might be polar. Thus, there may be a domain of the cell surface which has a high receptor concentration and which is exposed to afferent capillaries and arterioles. Other domains of that cell's plasma membrane could be deficient in receptors and thus unresponsive to stimulatory peptides in the interstitial space. This could help explain why the infusion of low concentrations of somatostatin and glucagon into the vascular space can have such a marked effect upon islet secretion (SAMOLS and HARRISON 1976a; KAWAI and ROUILLER 1981). Yet another process which could be occurring is that of receptor downregulation (GAVIN et al. 1974). Thus, cells which are bathed by a high concentration of peptide could autoregulate their receptor number and become resistant to the action of that peptide. A more speculative but tantalizing consideration is that there could be downregulation of receptor number on a restricted domain of a given cell's surface, but preservation of receptor concentration on a different domain. This could be a mechanism which could cause cells to be resistant to the secretory products of their neighbors, but sensitive to the same peptide infused arterially.

A fourth consideration is that of interstitial fluid flow. Such a flow might exist in islets, but at this point it is impossible to speculate upon its direction or speed. Presumably, this flow is controlled by islet blood flow and perhaps is influenced by lymphatic drainage as well. If we understood the movement of interstitial fluid in the islet, we would probably have a much clearer idea as to what local secretion phenomena were likely to occur.

The last major consideration is that of the islet vasculature. There could be a microportal circulation within islets whereby a given peptide could be secreted into a capillary, transported for a short distance, and then leave the capillary and act on a target cell. Some recent information on the anatomy of the islet vasculature in the rat permits some speculation on this (BONNER-WEIR and ORCI 1982). In rat islets, the afferent vessel, an arteriole, enters the islet at a gap of the non-B-cell mantle, thus directly entering the B-cell core, whereupon it branches into numerous capillaries which then traverse the core (Fig. 1a, b). As the capillaries leave the core, they pass through the mantle of non-B-cells and either coalesce at the edge of the mantle forming an overlying network of collecting venules (Fig. 1b), or pass through exocrine tissue before coalescing (Fig. 1a). With this pattern as a basis, one can make educated guesses about what kinds of portal communication are possible in the rat islet. It does not appear that glucagon or somatostatin are likely to be carried by vessels to the B-cell core. It is likely, however, that insulin can be carried vascularly to the mantle and it is also possible that glucagon and somatostatin can be carried via vessels to other parts of the mantle, even though this latter possibility is more speculative. Whereas it is very difficult to speculate on the flow of interstitial fluid within the islet, if Starling's hypothesis holds, a premise which is far from certain, it seems unlikely, in view of the vascular pattern, that there is much flow from non-B-cell mantle to B-cell core. However, this does not preclude a local dif-

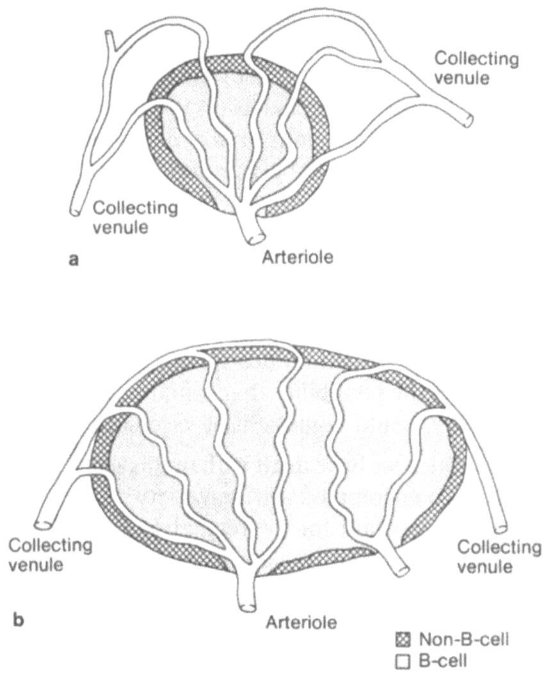


**Fig. 1 a, b.** Diagrammatic summary of combined data from corrosion casts and serial reconstructions of rat islets. The vascular pattern differs somewhat between small islets (**a**) those less than 160 μm in diameter, and large islets (**b**) those 260 μm or greater in diameter. Intermediate-sized islets (not shown) resemble the pattern of large islets. In all islets, short arterioles enter the islet at discontinuities of the non-B-cell mantle and branch into capillaries that form the glomerular structure occupying the area of the B-cell core. (Fewer capillaries are drawn for diagrammatic purposes.) After traversing the B-cell mass, capillaries penetrate the mantle as the blood leaves the islet. In small islets (**a**), efferent capillaries often pass through exocrine tissue before coalescing into collecting venules. However, in large islets (**b**) capillaries coalesce at the edge of the islet and run along the mantle as collecting venules before reaching a vein

fusion whereby secreted glucagon or somatostatin could bathe a peripheral layer of B-cells via the interstitial fluid. Although the rat is often used as the paradigm of islet anatomy, it remains to be seen if this vascular pattern is found in other species.

Another aspect of the islet vasculature which remains unexplained is that of alteration of flow. Small vessels are well known to be able to constrict and dilate and this capability might exert some control over islet secretion. Because islets are innervated by the autonomic nervous system (see Sect. D. II), it is reasonable to assume that vessel tone could be influenced by the local release of norepinephrine, VIP, or a number of other vasoactive mediators. One could, however, argue that changes in flow might not change secretion very much because the major determinants of secretion are presumably the arterial concentrations of glucose, amino acids, catecholamines, and small peptides which enter an interstitial fluid space. It can be predicted that there is rapid equilibration between plasma and interstitial fluid and it is difficult to see how alteration in regional islet flow would create lasting gradients in the concentrations of these secretagogues. Likewise, unless there is unexpected reuptake of secreted peptide by islet cells, it could be expected that whatever is secreted would rapidly leave an islet even if flow is altered. On the other hand, it seems that physiologic variations in flow could alter islet hormone output if paracrine or portal communication between cells was disturbed. For instance, if there was a particular interstitial space in which a certain concentration of peptide exerted a paracrine, or even autocrine effect, a change in blood flow might cause this concentration to either increase or decrease. Likewise, an alteration in islet blood flow could markedly alter the concentrations of secreted material within the islet vasculature and thus influence a local portal control system. A provocative observation is that somatostatin can reduce splanchnic blood flow (WAHREN and FELIG 1976) and this raises the possibility that somatostatin, either secreted locally or circulating systemically, could regulate islet vasculature.

The considerations which we have dealt with in this section are admittedly highly speculative, but they do encompass a wide variety of factors which could ultimately prove to be very important for the regulation of islet secretion. The value of such speculation is that it permits the construction of hypotheses which can be tested as the sophistication of our tools increases.

## F. Indirect Experimental Evidence for Interaction Between Islet Cells

While hormones of the islets have the capacity to influence the secretion of other islet hormones, and while it seems that the microanatomic geography virtually invites certain local interactions, it has been extraordinarily difficult to design, or interpret, experiments to test the hypothesis convincingly. Any methodological intervention to measure changes may well alter either normal anatomic avenues and/or physiologic events, so that the uncertainty (see Chap. 22) has not been circumvented. In theory, the most suitable model for experimental testing is the isolated perfused pancreas, which preserves islet microanatomy better than preparations of isolated islets or cells, and permits one to localize the origin of changes to the pancreas better than when using in vivo systems. Broadly speaking, it is the effects of sulfonylureas, and especially the effects of glucose on the isolated perfused pancreas (without recirculation), which provide evidence that B- and/or D-cell secretion suppresses A-cell secretion by a local mechanism.

In studies utilizing the isolated perfused canine pancreas, it has been shown that tolbutamide mildly stimulates glucagon secretion when perfusate glucose concentration is normal, but when the perfusate glucose concentration is low, particularly when amino acids are present, there is suppression of glucagon secretion (SAMOLS and HARRISON 1976b, 1977) sometimes following a brief stimulation. Because the suppression could be related to absolute enhancement of tolbutamide-stimulated insulin and somatostatin secretion (in the presence of amino acids) or to relative enhancement of tolbutamide-induced insulin and somatostatin secretion (in the presence of low glucose concentrations), one can suggest that the rapid increase in somatostatin and/or insulin caused intraislet suppression of glucagon (SAMOLS and HARRISON 1976b, 1977; SAMOLS et al. 1978). Thus, it has been proposed that tolbutamide normally primarily stimulates A-, B-, D-, and PP-cells and that a secondary intraislet event (somatostatin and/or insulin) converts the A-cell response to suppression (SAMOLS et al. 1978). Similarly, in the rat pancreas perfused with a normal glucose concentration, glibenclamide markedly enhances the effect of arginine on somatostatin release and inhibits the glucagon released by arginine, suggesting paracrine suppression by the enhanced somatostatin release (EFENDIC et al. 1980).

The effect of glucose on the isolated pancreas is probably the paramount argument in favor of direct local effect of B-cells (see Chap. 23) and D-cells upon pancreatic A-cells. Because glucose suppresses glucagon secretion very poorly in insulin-dependent diabetes mellitus (UNGER et al. 1970; SPERLING et al. 1977) and in streptozotocin- or alloxan-induced diabetes in vivo or in vitro (LAUBE et al. 1973; BUCHANAN and MAWHINNEY 1973a, b; PAGLIARA et al. 1975; HERMANSEN et al. 1979), it can be argued that the pancreatic A-cell is not suppressed by glucose in the absence of insulin. Also, because basal diabetic hyperglucagonism can be inhibited by the administration of insulin in vivo (BRAATEN et al. 1974; SPERLING et al. 1977) or in vitro (BUCHANAN and MAWHINNEY 1973a, b; WEIR et al. 1976b; ÖSTENSON 1979), and because exogenous insulin suppresses glucagon secretion when the glucose concentration is kept constant in vivo and in vitro (SAMOLS et al. 1972; SAMOLS and HARRISON 1976b; WEIR et al. 1976b; SERVICE et al. 1978), it can be inferred that the pancreatic A-cell is "insulin sensitive" (LEFEBVRE and LUYCKX 1979). Moreover, the gastric A-cell is not suppressed by glucose alone, but is exquisitely sensitive to insulin (LEFEBVRE and LUYCKX 1978). SAMOLS et al. (1972) deduced that insulin acts directly on the A-cell (see Chap. 22 for further discussion) and it has been proposed that glucose normally suppresses the A-cell via glucose-induced insulin secretion. However, this suggestion is disputed (MATSCHINSKY et al. 1980) and another possible intraislet suppressor needs to be considered. It is clear that the A-cell, whether pancreatic, gastric, or tumorous, will invariably be suppressed by somatostatin (GERICH 1981; ADRIAN et al. 1981). Therefore, because an infusion of glucose suppresses A-cell secretion in the isolated pancreas, and it is assumed that glucose cannot act directly on the A-cell, it is possible that glucagon suppression is mediated by insulin and/or somatostatin, as glucose markedly stimulates the B-cell and modestly stimulates D-cell (SCHAUER et al. 1976; PATTON et al. 1977; WEIR et al. 1978) secretion.

Assuming this argument to be correct, can one assess the relative importance of insulin versus somatostatin as local suppressors of the A-cell as agents for glucose? In terms of in vivo studies, exogenous insulin, if given so that normoglycemia is maintained in dogs (SAMOLS et al. 1972) or humans (SERVICE et al. 1978), suppresses glucagon. Under similar conditions, exogenous insulin also appears to inhibit pancreatic somatostatin secretion in vivo in the dog (ROUILLER et al. 1981), suggesting that insulin alone (without the help of somatostatin) may be sufficient to act as the local mediator of glucose-induced A-cell suppression. On the other hand, it is certainly known from infusion of exogenous somatostatin that somatostatin alone may be sufficient to inhibit the A-cell, as it simultaneously inhibits B-cell secretion. Analysis of animal diabetes models suggests that glucose-induced insulin and somatostatin responses are both defective. Glucose perfusion of the isolated pancreas from streptozotocin-treated dogs shows that absent A-cell suppression is associated with a deficient insulin response and poor somatostatin stimulation (HERMANSEN et al. 1979).

Before accepting this thesis, one more seeming paradox needs consideration. In streptozotocin-induced and alloxan-induced diabetes, there is now good evidence that pancreatic somatostatin is hypersecreted (HARA et al. 1979; SCHUSDZIARRA et al. 1981; TRIMBLE et al. 1981 a), possibly because the D-cell is stimulated by a high free fatty acid level (WASADA et al. 1981) and/or by hyperglucagonemia. Therefore, it could be argued that glucose-induced A-cell suppression in insulin-dependent diabetes mellitus (IDDM) is defective purely because of insulin deficiency in the face of excess somatostatin secretion. Favoring this proposition are the observations that somatostatin, if given in sufficient concentration, will invariably inhibit A-cell secretion (GERICH 1981; ADRIAN et al. 1981) and one would imagine that paracrine concentrations of somatostatin should be extraordinarily high in the islets of IDDM. Mitigating this proposition are the observations that the inhibitory effects of somatostatin may be temporary (POLONSKY et al. 1981) and that streptozotocin-induced diabetic pancreatic D-cells are unresponsive to changes in glucose concentration (HERMANSEN et al. 1979; TRIMBLE et al. 1981 a).

There are other observations on streptozotocin-induced diabetic A-cell suppression, notably the tolbutamide studies of MATSCHINSKY et al. (1980) which are better explained – if they are to be interpreted in terms of intraislet local interactions – as tolbutamide-induced A-cell suppression by somatostatin, despite insulin deficiency. Therefore it currently seems reasonable to suggest that both B- and D-cells contribute to the rapid suppression of the A-cell by glucose and, under various experimental conditions, one or the other may assume greater importance.

Very high concentrations of insulin (20–25 mIU/ml, are necessary to suppress glucagon secretion in the perfused rat pancreas or canine pancreas. Also, the endogenous insulin secretion needs to be low, such as during glucopenia (SAMOLS and HARRISON 1976 b), alloxan-induced diabetes (PAGLIARA et al. 1975), streptozotocin-induced diabetes, and during the infusion of epinephrine (WEIR et al. 1976). The high exogenous insulin dose is thought to be required so as to mimic concentrations which may theroretically be found in the interstitial fluid bathing A-cells, an argument which can be used to support the hypothesis of intraislet insulin-glucagon feedback. When endogenous insulin secretion is high, with high levels of glucose or arginine in the medium, exogenous insulin does not suppress glucagon (WEIR et al. 1976 b), suggesting that maximal suppression by endogenous insulin has already occurred. Alternatively, it may be, when arginine is used, that inhibition of endogenous insulin by exogenous insulin potentiates the glucagon released by arginine (ASPLIN et al. 1981) by intraislet feedback, so that exogenous insulin-induced suppression of the A-cell is overwhelmed by the endogenous counterforce induced by concomitant B-cell suppression.

The latter concept seems to be an extremely important facet in the understanding of the negative insulin-glucagon feedback, and of its relevance to human disease. ASPLIN et al. (1981) infused exogenous insulin to normal human subjects, in whom the blood glucose concentration was kept normoglycemic with a glucose clamp. This procedure caused: (1) a fall in plasma C-peptide, suggesting suppression of the B-cell by exogenous insulin; (2) potentiation of the glucagon response to arginine, suggesting that local suppression of B-cell secretion, by intraislet feedback, enhances stimulation of A-cell secretion because, in the absence of available intraislet B-cells, in IDDM, the identical procedure (infusion of exogenous insulin) caused an inhibition of the glucagon response to arginine. This experiment not only strongly supports an intraislet effect of B-cell secretion on the A-cell, but provides the necessary link for the concept of an intraislet hypothesis which explains the effects of exogenous insulin on A-cell behavior in various forms of diabetes mellitus (see Sect. L). Could the study of ASPLIN et al. (1981) also be explained by changes induced in the D-cell? This seems less likely, although it is conceivable that the insulin in this study suppressed islet D-cell secretion (ROUILLER et al. 1981; GERBER et al. 1981), an event which could explain enhanced glucagon secretion in the normal subjects, but which would not account for the exogenous insulin-induced inhibition of the glucagon response to arginine in IDDM. One other study using isolated perfused pancreases merits mention. In the canine pancreas, the infusion of morphine produces rapid suppression of somatostatin secretion which is followed 2 min later by a rise in insulin and glucagon secretion (IPP et al. 1978b). This result suggests that there may be a tonic intraislet suppressive effect of D-cell secretion upon the A- and B-cells.

Having summarized the "best" indirect evidence we could muster in support of the premise that B- and D-cell secretion locally modulate A-cell secretion, we now consider evidence circumstantiating the concept that local A-cell secretion modulates B- or D-cell secretion. The best evidence of a positive intraislet A-cell effect on the B-cell (see Chap. 22 for a full discussion of the evidence) exploits the difference between the glucagon-rich splenic lobe islets and the glucagon-poor duodenal lobe islets (TRIMBLE et al. 1981b; TRIMBLE and RENOLD 1981). When these islets are isolated and either incubated or perfused, more glucose-stimulated insulin is secreted by the glucagon-rich islets than by their duodenal equivalents, despite identical islet size and D-cell content. Splenic islets secreted more insulin and more glucagon in the presence of 16.7 m*M* glucose. Although the difference in glucagon secretion between the splenic and duodenal islets is probably the explanation for the difference in the rate of insulin secretions, it seems on the surface to be inappropriate, occurring under a condition (high glucose) which usually suppresses glucagon secretion. Therefore, it needs to be shown that there is not a technical artifact caused in the process of islet isolation by disruption of specialized anatomy, which normally prevents local glucagon secretion from reaching adjacent B-cells. Such a disruption could allow B-cells to be exposed to glucagon in an unphysiologic fashion. However, the clever conceptual approach of TRIMBLE and RENOLD (1981) may be validated if a similar variance in the amplification of insulin secretion is obtained when the isolated perfused splenic lobe is compared with that of the duodenal lobe.

The indirect experimental evidence that A-cell secretion locally modulates D-cell is scanty. SORENSON et al. (1979) proposed a positive paracrine effect of glucagon on B- and D-cells to explain the effects of norepinephrine in isolated perfused rat islets. Although norepinephrine normally inhibited B- and D-cell secretion, it did not do so when the dose was sufficiently high to induce maximal stimulation of the A-cell.

## G. Efforts to Demonstrate Local Interactions Directly with Immune Neutralization

An attractive approach to the problem of the local secretion hypothesis is to use specific antibodies to neutralize the effects of a given peptide rapidly within the islet extracellular space. Such a neutralization could, if effective, lead to perturbations of islet secretion which could then be attributed to a deficiency of the neutralized peptide. There are however, some fundamental difficulties with this approach, particularly if the local interaction occurs in the interstitium rather than via vascular pathways. If islet local interactions are exerted by paracrine secretion via an interstitial space, the concentrations of a peptide may be very high within this space and it may be next to impossible to get enough antibodies into this space to obtain effective neutralization. Studies carried out with labeled IgG antibodies suggest that vascularly infused antibodies enter interstitial spaces poorly in the pancreas as a whole, but these studies did not address what happens in the endocrine pancreas (SCHUSDZIARRA et al. 1980). If, on the other hand, the local interaction is mediated through blood vessels such that the modulatory peptide must enter and then leave a vessel before reaching its target, it can be predicted that neutralization would be more likely to succeed. In a vessel, as compared with an interstitial space, the peptide should be at a lower concentration and the antibodies at a higher concentration, which should facilitate neutralization.

There have been several studies in which large amounts of somatostatin antiserum have been injected into rats and dogs intravenously (SCHUSDZIARRA et al. 1978a; TANNENBAUM et al. 1978). In several of these, no changes in the concentration of circulating insulin and glucagon were observed, but in another there was an increase in plasma insulin (SCHUSDZIARRA et al. 1980). These in vivo studies provide little evidence either for or against a local regulatory role for somatostatin. In the positive experiment showing an increase of insulin, it is not possible to know whether the neutralization of somatostatin occurred within the islet or in the extraislet vascular space. Based upon current assessment of circulating somatostatin concentrations (ZYZNAR et al. 1981), there may well be a high enough level to exert a continuous tonic suppression upon the islet, such that neutralization would be predicted to lead to enhanced insulin or glucagon secretion. Thus, the positive studies do not demonstrate an intraislet effect of somatostatin, and conversely the negative studies do not exclude a local role. There is no assurance that enough antibodies reached either the appropriate interstitial space or islet vascular space to neutralize intraislet (i.e., "short-loop") somatostatin effects adequately.

Another approach has been to add somatostatin antisera to incubating isolated islets (TANIGUCHI et al. 1977; BARDEN et al. 1977; ITOH et al. 1980; SCHATZ and KULLEK 1980; TANIGAWA et al. 1981). Even though positive results have not been obtained in all studies, this method can cause increases in both insulin and glucagon secretion. These results are, however, subject to the same criticisms which were leveled at in vitro islet studies (see Sect. F). There is good reason to believe that incubated isolated islets are very different from intact in vivo islets. Certainly, there is no blood flow through isolated islets, a defect which would presumably alter the normal flow of interstitial fluid. In addition, it seems likely that specialized junctional complexes may be distorted by the harsh enzymatic and mechanical process used to isolate islets. Finally, within the medium in the incubation vial, there might well be a concentration of secreted somatostatin surrounding the islets which is high enough to exert a continual suppressive influence upon abnormally exposed A- and B-cells. Thus, it would not be surprising if a high concentration of antibodies could not only neutralize the somatostatin surrounding the islets, but also penetrate into spaces which would be inaccessible in in vivo islets. The positive results obtained with incubating isolated islets, while consistent with the local secretion hypothesis, could thus be explained by the unphysiologic peculiarities of these islets.

Infusion of antiserum into isolated perfused pancreata should be a preferable approach in that islet anatomy would be expected to be preserved. There are nevertheless at least two caveats: (1) the neural supply to the islets in these isolated systems is disrupted and this could conceivably create anatomic and functional artifacts; (2) the perfusate used is only a semiphysiologic substitute and this could also perhaps lead to an artifactual result. In any event, several laboratories have infused somatostatin antisera into perfused rat and canine pancreases and, to our knowledge, the results have been uniformly negative. Very few of these negative results have appeared in the literature (Arimura et al. 1978; Sorenson et al. 1980), but unpublished negative results have been obtained in our laboratories, and we have received several reports, by personal communication, of negative findings from other laboratories. As indicated earlier in this section, these negative findings do not rule out local effects of somatostatin, as it is not possible to know whether enough antibodies could have reached a critical interstitial or vascular space to cause efficient neutralization.

Positive results have, however, been obtained by infusing somatostatin antisera into the isolated perfused chicken pancreas (Honey et al. 1981). In separate experiments, three different somatostatin antisera elicited rapid marked increases in glucagon secretion. Insulin secretion was also stimulated, but the increases were less marked and of shorter duration than those of glucagon. These data indicate that intraislet somatostatin in the chicken exerts some tonic suppressive influence upon A-cells. Interpretation of the insulin results is more complex because somatostatin could be acting directly upon B-cells or indirectly through locally released glucagon, which is a potent stimulator of insulin secretion. Insight into the mechanism of the suppression of glucagon secretion by glucose can be gleaned from these experiments because neutralization of somatostatin caused a marked stimulation of glucagon secretion, even when the perfusate glucose concentration was at a concentration known to cause effective suppression of secretion. This finding indicates that, in the chicken, local somatostatin secretion may be a powerful contributing factor to the way in which glucose suppresses the A-cell, and it remains to be determined whether glucose has any direct suppressive effect at all upon the A-cell. It is not clear whether the antiserum used in these experiments was exerting its effect in an interstitial space or a vascular space, even though one might favor the latter for the reasons given in the first paragraph of this section. Therefore, even though local secretory effects appear to be present, it is not possible to be sure that true paracrine mechanisms are operative and it is attractive to postulate that avian islets contain a local portal circulation through which somatostatin is carried to A-cells. It is important to emphasize that these results are obtained in the chicken and, even though they are likely to reflect the physiology of other avian species, they may provide only limited insight into the function of mammalian islets.

There are marked variations in the anatomy of avian and mammalian islets. While most of the islet secretory controls are similar, there are a few striking differences. In particular, insulin and acetylcholine have stimulatory effects upon somatostatin secretion in the chicken (HONEY and WEIR 1979, 1980), but inhibitory effects in mammals (GERBER et al. 1981; ROUILLER et al. 1981; SAMOLS et al. 1981), even though the inhibition by insulin has not been consistently demonstrated (WEIR et al. 1979). Nonetheless, despite these dissimilarities, it would seem peculiar for there to be marked differences in the mechanisms through which glucose stimulates insulin secretion and inhibits glucagon release in these two vertebrate classes. Thus, nonmammalian models might provide important clues to mammalian puzzles, as long as the limitations of such a comparative approach are recognized.

In summary, it has been suggested that a paracrine system of secretory control within the islets is favored if neutralization of one islet hormone alters the release of other islet hormones. This inference is probably not correct for most of the published studies, in which it seems more likely that the changes in release of the other islet hormones were caused by specific neutralization of one hormone, either in a nonphysiologic manner (in vitro islet studies) or in the intravascular space (in vivo studies, in vitro chicken pancreas).

## H. Arguments Against Islet Interactions

Despite the many attractive aspects of the concept that islet secretory products influence neighboring cells, much has been written about why interactions between A- and B-cells, in particular, are unlikely to occur. The most complete commentary on this point of view has recently been made by MATSCHINSKY et al. (1980). The crux of their argument is that, under many experimental conditions, they have found a poor correlation between stimulation of insulin secretion and inhibition of glucagon secretion (PAGLIARA et al. 1975, 1977; MATSCHINSKY et al. 1976). An example of this approach is provided by their fasting studies, in which rats were fasted for 3 days and then refed for an additional 3 days. Insulin secretion, as assessed by both glucose and amino acid stimulation, fell markedly with fasting, but neither glucagon stimulation by amino acids nor glucagon suppression by glucose was altered significantly during this period. Thus, the rapidity and degree of glucagon suppression by glucose was essentially normal in the face of markedly reduced insulin secretion. The difficulty in interpreting these kinds of data is that it is very hard to know how much local insulin might be required for glucagon suppression by glucose. Even though very little insulin is measured in the venous effluent of the perfused pancreas, the intraislet concentration might be considerably higher and sufficient to contribute to the suppression. One could also speculate that A-cell sensitivity to insulin could change during fasting such that glucagon might be suppressed by much smaller concentrations of insulin than usual.

Additional arguments used by MATSCHINSKY et al. (1980) include their finding that 3-hydroxybutyrate is a powerful stimulator of insulin secretion, but does not enhance A-cell suppression by glucose. In addition they have found "that tolbutamide inhibits stimulated glucagon release in the isolated perfused pancreas taken from starved animals, again in the absence of concomitant changes of insulin release" (MATSCHINSKY et al. 1980). In both of these situations, the point remains that we do not know how much intraislet insulin might be required for suppression of the A-cell. Nonetheless, these results and arguments do raise important issues which should make us cautious about assigning too much importance to local insulin secretion. Their hypothesis is that the A-cell requires "an optimal insulin level for maximal glucose sensitivity" (MATSCHINSKY et al. 1980), but even though this hypothesis remains viable, there is very little evidence that glucose alone in the presence of unchanging insulin or somatostatin concentrations can inhibit glucagon secretion. MATSCHINSKY et al. (1980) were careful to point out that the possible local effects of somatostatin were not evaluated in their studies. We, therefore, reiterate our premise that somatostatin could be playing an important role in the glucagon suppression (see Sects. F and L).

## J. Oscillation of Secretion Suggesting Coordination Between Islets

Oscillations in insular secretion have been noted in monkeys, humans, and dogs in vivo, and originally such oscillations were thought to be reactive to release of hepatic glucose in bursts (ANDERSON et al. 1967; GOODNER et al. 1977; LONG et al. 1979). When the perturbations of blood glucose were stabilized in vitro in the canine pancreas (STAGNER et al. 1980), it was shown that there is an intrinsic rhythm of hormone secretion by the pancreas (STAGNER et al. 1980). Insulin (10-min period), somatostatin (10-min period), and glucagon (8.6-min period) demonstrated regular sustained cyclic secretion, which was not influenced by cholinergic or adrenergic blockade (STAGNER and SAMOLS 1980). Because of the large size of the canine pancreas it seems that the production of in vitro cycles requires not only the presence of a pacemaker within the pancreas, but also the coordination of islets by pacemaker-islet communication, presumably by a neural system. The possible role of the peptidergic nerves supplying the endocrine pancreas, and ramifying from pancreatic ganglia (LARSSON and REHFELD 1979; KAMEL et al. 1979), which might function as pacemaker, is difficult to test without specific antagonists. That B- and D-cell secretion, which share other similarities, should be in phase might perhaps be explained by B- and D-cells experiencing similar oscillations in intracellular metabolism, or by a paracrine link. The latter explanation could also apply to the glucagon cycles 90° out of phase with insulin reported in monkeys (GOODNER et al. 1977), but would not explain the in vitro glucagon cycles.

It has recently become apparent that when the canine pancreas is perfused in vitro so that the splenic (dorsal) and uncinate (ventral) lobes are separately perfused, they each have the same cycle period, and are apparently in phase when perfused with a normal glucose concentration (STAGNER and SAMOLS 1982). However, a difference in behavior of cycles from each lobe (whether detected by split drainage or by separate isolated lobe perfusion) emerges during infusion of a glucose concentration of 200 mg/dl, when the ventral cycles shorten and the splenic cycles lengthen by a small, but significant time period (STAGNER and SAMOLS 1982). These changes would militate against proposed statistical theories (DE HAËN et al. 1981) of oscillations, and favor a coordinating intraislet system for each lobe.

## K. Consideration of an Islet-Acinar Portal System

The distribution of endocrine pancreatic tissue as small islets scattered throughout pancreatic exocrine tissue strongly suggests that the two types of tissue are somehow functionally related (HENDERSON 1969; FUJITA et al. 1976). In some islets, there is a unique vascular arrangement (Sect. E) whereby capillaries leave islet tissue and traverse the exocrine pancreas before coalescing to form veins (WHARTON 1932; FUJITA et al. 1976; BONNER-WEIR and ORCI 1982). This specialized vascular pattern and the experimental data suggesting that islets receive approximately 13%–20% of total pancreatic blood flow (LIFSON et al. 1980) make it seem very likely that exocrine pancreatic tissue is exposed to high concentrations of locally released islet hormones. Somatostatin (inhibition), insulin (stimulation), and pancreatic polypeptide (inhibition) are each able to influence pancreatic exocrine secretion, even though there is little evidence that glucagon has any significant effect (BODEN et al. 1975; CREUTZFELDT et al. 1975; KANNO et al. 1976; LIN et al. 1977; SAITO et al. 1980). Thus, there is good reason to think that what has been called the islet-acinar portal system does exist anatomically and probably functionally. Unfortunately, we have only vague ideas as to why such influences would be beneficial to pancreatic acinar function, but redundancy of control mechanisms in nature is frequently observed. There is good reason to think that insulin exerts some physiologic effects in that it has been found that pancreatic amylase output is reduced in insulin-deficient human or experimental diabetes, but increased after insulin treatment (CHEY et al. 1963; OHARA 1976). These experiments do not prove that the insulin released locally is responsible for these effects because they could equally well be explained by the effects of peripherally circulating insulin.

Recently, experiments have been carried out in the perfused rat pancreas which provide indirect evidence that endogenous insulin released by glucose enhances cholecystokin-induced pancreatic exocrine secretion (SAITO et al. 1980). There is also morphological evidence suggesting the presence of such a functional portal system, in that so-called peri-insular halos have fequently been described (JAROTZKY 1899; SERGEYEVA 1938). These halos consist of a rim of exocrine tissue surrounding islets with a thickness which varies (but is roughly 50 μm) and which is composed of large acinar cells with increased numbers of zymogen granules. It is of interest that the relationship between the concentrations of amylase, lipase, and chymotrypsin in this peri-insular area of the exocrine pancreas differs considerably from determinations made on tissue further away from islets, which is termed teleinsular (MALAISSE-LAGAE et al. 1975). These observations indicate that the exocrine pancreas is heterogeneous with regard to structure and suggest heterogeneity in function as well, with these diversifications perhaps being determined by the islet-acinar portal system.

## L. Islet Interrelationships in Diabetes

In considering the pathophysiology of islets in diabetes, the traditional focus has been upon insulin secretion. However, there has been growing appreciation of complex alterations in other aspects of islet function, some of which may be caused by disruptions of coordination between islet cells. It is now well accepted that there is marked insulin deficiency in IDDM associated with reduction of B-cell number

(National Diabetes Data Group 1979). In noninsulin-dependent diabetes mellitus (NIDDM) there is much current interest in the role of insulin resistance (Olefsky and Kolterman 1981), but in many patients with this form of diabetes there is evidence for a modest reduction in B-cell number and low insulin secretion (Gepts 1957; Westermark and Grimelius 1973; Kosaka et al. 1977; Gepts and Lecompte 1981). Within the past decade and as reviewed in Chap. 44, it has been found that plasma glucagon is elevated, either in an absolute or relative sense, in both types of diabetes (Unger et al. 1970; Aronoff et al. 1976; Unger and Orci 1981 a, b). At first glance, this was perceived as paradoxical, in that hyperglycemia might have been expected to inhibit glucagon secretion. This apparent incongruity led to the concept of a primary A-cell defect in diabetes (Unger et al. 1972; Unger 1976). Thus, it was thought that both B-, and A-cells shared a similar defect characterized as "blindness" to the effects of glucose. This argument has fallen out of favor as evidence has accumulated which suggests that the hyperglucagonemia can all be explained on the basis of defective insulin secretion.

Because a fundamental abnormality of the A-cell in diabetes is its inappropriate responsivity to glucose, it is worth reiterating some current concepts of how glucose normally suppresses glucagon secretion. These concepts are also discussed in Sect. E, and extensively in Chap. 22. It seems likely that glucose suppression of the A-cell is at least partially mediated by the intraislet secretion of insulin. Whereas circulating plasma insulin concentrations could be contributing, it is clear that glucose alone can effectively suppress glucagon secretion in isolated perfused pancreas preparations in which no exogenous insulin is delivered arterially (Iversen 1974; Weir et al. 1974; Pagliara et al. 1974). Intraislet somatostatin could have an important role in this process because its secretion is stimulated by glucose (Schauer et al. 1976; Patton et al. 1977; Weir et al. 1978), but even though some work in the avian species supports this concept (Honey et al. 1981), little experimental data are available to help us understand the situation in mammals. Therefore, although there has been a tendency to try to explain suppression in terms of intraislet insulin, the potential role of intraislet somatostatin must not be ignored.

As further introduction to our consideration of the abnormal glucagon secretion of diabetes, a variety of A-cell responses in normal humans will be considered within the context of a new hypothesis. Initial premises are that insulin has a direct inhibitory effect upon the A-cell (Samols et al. 1972) and that glucose suppression of glucagon secretion is largely mediated by intraislet insulin. Recent experiments by Asplin et al. (1981) have suggested that exogenous insulin may also exert a stimulatory effect upon the A-cell through the indirect route of inhibiting intraislet insulin. Infusions of insulin, which raised plasma concentrations to either 40 or 80 μIU/ml, enhanced the glucagon response to arginine in normal humans and this coincided with suppression of intraislet insulin secretion as measured with plasma C-peptide determinations. The hypothesis was further supported by the finding that similar insulin infusions in subjects with IDDM, who had minimal intraislet insulin secretion, led to an inhibition of arginine-stimulated insulin secretion.

Thus, one can expect that infusions of exogenous insulin can either stimulate or inhibit glucagon secretion, with this response being dependent upon the activity of intraislet insulin secretion. The glucagon response to arginine in normal humans should therefore be viewed as a result of counterbalancing forces, a direct stimu-

**Table 1.** Glucagon secretion in diabetics and nondiabetics

| | Response to oral or intravenous glucose | Response to oral or intravenous glucose+insulin | Response to intravenous arginine or oral protein | Response to intravenous arginine or oral protein+insulin |
|---|---|---|---|---|
| Nondiabetic | Suppression | ? Augmented[a] | "Normal" | Augmented[a] |
| IDDM | No change or increase | "Normal suppression" | Increased[b] | "Normalized", i.e., suppressed[c] |
| NIDDM | | | | |
| Lean | No change or increase | "Normal suppression" | Increased[b] | "Normalized"[c] |
| Obese | No change or increase | "Normal suppression" | Increased[b] | Not "normalized"[c] |

[a] In comparison with the response obtained without simultaneous infusion of exogenous insulin
[b] In comparison with the nondiabetic response to arginine
[c] In comparison with the response obtained without simultaneous infusion of exogenous insulin

latory effect of arginine upon the A-cell and an inhibitory influence of intraislet insulin. Arginine-stimulated intraislet somatostatin might also contribute to inhibition of the A-cell. In another experimental situation, it has been shown that when insulin is infused into fasting normal humans with enough glucose to prevent hypoglycemia, plasma glucagon concentrations fall (RASKIN et al. 1975). One way to explain this is to argue that intraislet insulin secretion has been kept low by the relatively low plasma glucose concentration and that infused insulin is therefore able to suppress the A-cell. One could make predictions about a similar hypothetical experiment in which enough glucose was infused to raise intraislet insulin secretion, and expect that the response of the A-cell to insulin infusions might shift from suppression to enhancement. Thus, the direct inhibitory effect of the exogenous insulin would be overwhelmed by its indirect stimulatory influence. The prediction for this experiment is designated by a question mark in Table 1.

The abnormalities (see Table 1) of glucagon secretion in IDDM (UNGER et al. 1970; MÜLLER et al. 1970; AYDIN et al. 1977) can be at least partially explained by these concepts and the assumption that extensive B-cell loss results in minimal intraislet insulin. An intravenous or oral glucose challenge is unable to suppress glucagon secretion (AYDIN et al. 1977; SPERLING et al. 1977) because of the lack of inhibition of the A-cell by intraislet insulin. When exogenous insulin is injected or infused along with glucose (RASKIN et al. 1975; AYDIN et al. 1977; SPERLING et al. 1977; UNGER and ORCI 1981 a, b), this insulin can suppress glucagon secretion, and whether glucose helps in this situation is unclear. The glucagon response to arginine or a protein meal in IDDM is excessive as compared with that seen in normal subjects (UNGER et al. 1970; RASKIN et al. 1976, 1978; PALMER et al. 1976a) and this is presumably accounted for by the deficiency of intraislet insulin. Thus, the normal counterbalancing inhibitory effect of intraislet insulin is missing, and the stimulatory effect of arginine or protein becomes more apparent. It is said that infusions of insulin "normalize" this excessive response in IDDM (RASKIN et al. 1976, 1978; GERICH et al. 1976), but this "normalization" is an abnormal response (i.e., suppression) which is produced by abnormal mechanisms, as the normal result is potentiation of the A-cell response. The normal intraislet suppression mechanism is now replaced by the inhibitory effect of exogenous insulin, which would not be inhibitory to the glucagon response to arginine in the presence of normal islets.

The situation in NIDDM is more difficult to explain, but the same basic premises can be applied. The glucagon responses to glucose and arginine, with or without insulin infusions, in lean subjects with NIDDM are similar to those seen in IDDM (KAWAMORI et al. 1980). Insulin secretion has frequently been found to be markedly deficient in NIDDM and was low in those studies in which insulin "normalized" the glucagon responses to arginine (KAWAMORI et al. 1980). All of the arguments which were used to explain the abnormalities of glucagon secretion in IDDM can be applied to the lean subjects with NIDDM. These arguments would break down if one were to consider lean subjects with mild NIDDM and some preservation of endogenous insulin secretion, but they seem reasonable to apply to the more insulin-deficient cases.

A notable difference in the responses of obese subjects with NIDDM is that insulin infusions do not "normalize" the glucagon responses to arginine or a protein meal (RASKIN et al. 1976; BENNETT et al. 1976; AYDIN et al. 1977). Perhaps this critical difference can be explained by the preservation of insulin secretion which can be found in these subjects. First, it should be recalled that these subjects may have deficient insulin responses to a glucose challenge, but preservation of the insulin responses to arginine (DECKERT et al. 1972; PALMER et al. 1976a). Thus, when glucose and exogenous insulin are infused together, there is minimal change in the endogenous insulin response, allowing exogenous insulin to suppress the A-cell (RASKIN et al. 1975; AYDIN et al. 1977). In response to arginine or a protein meal, however, although an intraislet insulin secretory response may be present, it is sufficiently ineffective to allow hyperresponsiveness of the A-cell. Once there is an insulin response to arginine, changes during the infusion of exogenous insulin depend on two counterbalancing forces. On the one hand, exogenous insulin may be (partly) effective in suppressing A-cells. On the other hand, suppression of B-cells by exogenous insulin may enhance A-cell secretion. The latter "force" usually dominates in normal subjects (ASPLIN et al. 1981). In obese diabetic subjects, this "force" may be subnormal, because suppression of B-cells by exogenous insulin is impaired in obesity (ELAHI et al. 1981). The net mean effect of exogenous insulin on arginine-stimulated glucagon release may be zero because of the forces canceling one another.

This hypothesis, which can be tested experimentally, may be summarized. In obese NIDDM, intraislet insulin secretion is low enough to allow hyperresponsiveness of the A-cell to arginine, but high enough to prevent the typical effects of exogenous insulin on the glucagon response to arginine seen in either IDDM ("normalization") or in normal subjects (potentiation). Although perhaps too facile an explanation, this hypothesis seems to account for most of the published data.

There are several other abnormalities of glucagon secretion in diabetes, including the very high plasma glucagon concentrations found in diabetic ketoacidosis (ASSAN et al. 1969; MÜLLER et al. 1973). These high concentrations are presumably accounted for by increased glucagon secretion produced by deficient intraislet insulin secretion and stress-induced autonomic nervous system activation (SAMOLS and WEIR 1979), and probably, in addition, by decreased glucagon clearance caused by a dehydration-related fall of renal blood flow (LEFEBVRE et al. 1976; see also Chap. 41). Another alteration of glucagon secretion is the increase in plasma glucagon concentration which can be seen after an oral glucose challenge (BUCHANAN and MCCARROLL 1972; DAY and ANDERSON 1973; AYDIN et al. 1977). This is probably best explained by a stimulatory effect of gut hormones and parasympathetic activation which is unmasked by deficient intraislet insulin. A more difficult phenomenon to explain is the increase of plasma glucagon which is sometimes seen after intravenous glucose is given to subjects with IDDM (BRAATEN et al. 1974). We are unaware of any in vitro precedent for a stimulatory effect of glucose upon the A-cell and one might speculate upon possible influences of the autonomic nervous system, but at present we can not provide any attractive explanation. A third abnormality of glucagon secretion found in diabetics is a failure of glucagon secretion to increase during insulin-induced hypoglycemia (GERICH et al. 1973, 1974; MAHER et al. 1977). It has been suggested that the normal increase is produced by activation of the autonomic nervous system and that autonomic neuropathy prevents this from occurring in diabetics (MAHER et al. 1977) and there is evidence for (MADSBAD et al. 1981) and against (ENSINCK et al. 1976; PALMER et al. 1976b) this hypothesis. An interesting alternative explanation, proposed by ASPLIN et al. (1981) speculates that the increase in glucagon secretion during insulin-induced hypoglycemia is normally secondary to a suppression of local intraislet B-cell release, so that the normal A-cell response cannot occur in the absence of intraislet B-cells, as occurs in IDDM.

Studies with experimental diabetes in animals provide further evidence that hyperglucagonemia is caused by insulin deficiency. Elevated plasma glucagon concentrations have been found in mammals made diabetic with the B-cell toxins, streptozotocin, and alloxan, (MÜLLER et al. 1971; PAGLIARA et al. 1975), and also with the administration of anti-insulin antiserum (MÜLLER et al. 1971). A disquieting aspect of this is that, despite the hyperglucagonemia of chronic streptozotocin-induced diabetic rats, the perfused isolated pancreata of these animals have been found in two studies to release less glucagon than normal controls (PAGLIARA et al. 1975; WEIR et al. 1976b). This raises the question as to whether the increases of plasma glucagon in diabetes might be at least partially accounted for by decreased glucagon clearance or by an extrapancreatic source of glucagon (BODEN 1981). It is surprising that there have been so few studies of glucagon clearance in diabetes (ALFORD et al. 1976; see also Chap. 39).

In discussing intraislet relationships in diabetes we would be remiss to give somatostatin only passing attention, but surprisingly little is known about pancreatic somatostatin secretion in diabetes. Part of this difficulty is explained by the many problems encountered in developing a reliable radioimmunoassay for somatostatin in human plasma. Even when an accepted method is established, it will be difficult to determine whether changes in concentration are accounted for by pancreatic or extrapancreatic somatostatin (GERICH 1981; TABORSKY 1981). We also know very little about the morphology or morphometry of D-cells in human diabetes. It has been reported that the ratio of the number of B- to D-cells in islets falls dramatically in IDDM (ORCI et al. 1976b, c), but this is presumably a reflection of B-cell deficiency and thus we have no information yet about the absolute number of pancreatic D-cells in human diabetes.

Information about somatostatin is, however, being obtained from animal models of diabetes. A consistent finding has been increased pancreatic somatostatin content in streptozotocin-induced diabetic rats (PATEL and WEIR 1976; PATEL et al. 1978; MATSUSHIMA et al. 1978; BERELOWITZ et al. 1979), and a study using quantitative morphometry indicates that this is accompanied by an absolute increase in pancreatic D-cell number (MCEVOY and HEGRE 1977; BAETENS et al. 1978). Furthermore, the secretion of pancreatic somatostatin is elevated when assessed with the perfused rat pancreas (HARA et al. 1979; KADOWAKI et al. 1980; WEIR et al. 1981). Consistent with these findings is the observation that plasma somatostatin concentrations are increased in alloxan-induced diabetic dogs (SCHUSDZIARRA et al. 1977), but it is unclear whether this increment is a reflection of increased pancreatic secretion.

It has been difficult to find a good explanation for the somatostatin changes found in streptozotocin-induced diabetes. An initially attractive hypothesis was that insulin deficiency was responsible for the increased D-cell activity. It was found, for instance, that insulin treatment or pancreatic transplantation could cause a reduction in the increased pancreatic somatostatin concentration (KADOWAKI et al. 1980; BERELOWITZ et al. 1979) and, furthermore, that insulin treatment could lower the elevated plasma or portal vein somatostatin concentrations found in experimental diabetic models (SCHUSDZIARRA et al. 1978b; PATEL et al. 1980; KAZUMI et al. 1980). In addition, an increase in gastric antral somatostatin was found in some hypoinsulinemic diabetic states (PATEL et al. 1976, 1978; TRIMBLE et al. 1980; CHIBA et al. 1981) and this has been found to be accompanied by increased secretion (CHIBA et al. 1981). There are a number of problems with this hypothesis, however. It has been difficult to demonstrate a direct suppressive effect of insulin upon the D-cell in mammals (PATTON et al. 1977; SCHAUDER et al. 1978; WEIR et al. 1979; KADOWAKI et al. 1979), even though a recent report shows such suppression in the perfused rat pancreas (GERBER et al. 1981). Furthermore, increased pancreatic somatostatin has been found in hyperinsulinemic ob/ob and db/db mice (MAKINO et al. 1979; BERELOWITZ et al. 1980) and decreased pancreatic somatostatin has been found in the hypoinsulinemic diabetic Chinese hamster (PETERSSON et al. 1977) and Baltimore Biological Laboratory Wistar rat (PATEL et al. 1980). In addition, increased islet somatostatin content has been found in the hyperinsulinemic obese Zucker rat (TRIMBLE et al. 1980) and increases of pancreatic and gastric somatostatin are found in hyperinsulinemic dexamethasone-treated rats (BONNER-WEIR et al. 1981). One might suggest that increased paracrine glucagon secretion causes these changes, but glucagon secretion by the perfused pancreata of streptozotocin-induced diabetic rats has been found to be either reduced (PAGLIARA et al. 1975; WEIR et al. 1976) or normal (KADOWAKI et al. 1980). The possibility that hyperglycemia leads to these changes does not fit well with the available data (PATEL et al. 1976, 1977), and the suggestion that circulating free fatty acids (WASADA et al. 1981) or 3-hydroxybutyrate and lactate (HERMANSEN 1980) could be playing a role needs to be explored in more detail. Thus, at the present time, the factors responsible for the derangements of D-cell function in experimental diabetes remain a mystery.

We also have little idea about what effect increased somatostatin secretion or content might have upon the A- and B-cells. It has been reported that glucose does not stimulate somatostatin secretion from the perfused pancreas of streptozotocin-induced diabetic dogs (HERMANSEN et al. 1979). This could at least partially explain why glucagon secretion is poorly suppressed by glucose in diabetes. One can wonder whether increased islet somatostatin is further suppressing the already defective insulin secretion in diabetes. The marked decrease in B-cell mass which is particularly evident in IDDM could permit the residual B-cells to be anatomically closer to D-cells and therefore more likely to be influenced by their paracrine secretion. Unfortunately, there are no experimental data available to support this hypothesis. If there are increased paracrine effects from somatostatin secretion, they do not seem to be powerful enough to reduce the increased glucagon secretion which is thought to occur in human diabetes. In summary, we know that somatostatin has very potent inhibitory effects upon A- and B-cells, and that D-cells function abnormally in experimental diabetes, but it is frustrating that we have such a poor understanding as to what causes these derangements or what effects these derangements have on the diabetic state.

## M. Overview and Conclusions

At this point the reader may feel slightly cheated, having been confronted by a bewildering array of potential intraislet interactions, with the constant admonition that none of these has been conclusively proven. Therefore, we shall reexamine the major theoretical concept. The preceding arguments tend to sustain the thesis that intraislet interactions are not implausible. This means that we have two main theories about islet cell secretion, and that, in certain respects, they are radically different. The first – the orthodox theory – states that the individual islet cells function independently of one another. The second – the intraislet interaction theory – postulates that the individual islet cells influence one another's secretions. While we have pointed to the many potential intraislet mediators, and potential routes of communication (Sects. B, C, D, E, K) we shall concentrate in this overview on the paracrine aspect – the thesis that insulin, glucagon, and somatostatin influence the secretion of A-, D-, and B-cells, via the interstitium, without preceding systemic circulation.

Which is more likely to be correct – independent islet cell function or intraislet A-, D-, and B-cell interrelationships? One criticism of the intraislet (paracrine) theory might be that it is not a real theory, but an imaginary concept. Our rebuttal would be to grant that the whole idea was initially extraordinarily unorthodox, but it did not – as occurs with imaginary concepts – grow from improbable scientific foundations. With the knowledge that glucagon promoted insulin secretion (SAMOLS et al. 1965), coupled with observations on the anatomic juxtaposition of A- and B-cells (HAIST 1965), did it require a great leap of the imagination to suggest (SAMOLS et al. 1966a) "that glucagon secretion adjacent to the B-cells may, in conjunction with changes in glucose levels, determine insulin secretion"? The hallmark of a successful theory is that it predicts correctly facts that were not known when the theory was presented, or better still, which were then known incorrectly (CRICK 1981). In broad terms it is true that the theory predicted the suppressive effect of insulin on glucagon secretion, and the possibility that insulin deficiency might cause hyperglucagonism (SAMOLS et al. 1972). It could be argued that the concept was a harbinger of interrelationships between somatostatin and the B- and A-cells (ALBERTI et al. 1973; KOERKER et al. 1974; PATTON et al. 1977; WEIR et al. 1978; GERBER et al. 1981). According to the alternative, or orthodox theory of independent islet cell function: (1) the insulinotropic effect of glucagon would not have presaged other examples of interaction between islet cells and the secretory products of their neighbors; (2) diabetic hyperglucagonism would be ascribed to a primary A-cell defect.

A good theory should make predictions which are testable, stage by stage. Once the broad premises – the potential for B-, A-, and D-cell interactions were established (Sect. D) by "long-loop" effects – the theory could perhaps be criticized for the extreme paucity of relevant evidence that intraislet interactions actually occur, i.e., that insufficient progress has been made despite our present sophisticated technology. There is some merit to this argument, but several rebuttals are in order. First, the essential difficulty resides within the nature of the theory, postulating six interactions occurring within a domain (the islet) whose integrity is essential for the performance of a decisive experiment. If islets are removed from the pancreas, the experiments invariably can be criticized as possibly creating nonphysiologic communications between the incubation medium and the islet cells (Sects. E, F, G). If the pancreas remains intact, the complexity of interpretation has so far precluded unequivocal experiments (Sects. F, G, H). In mammals, it seems that any attempt to demonstrate local interactions directly with immune neutralization is essentially a flawed approach (Sect. G) because of the anatomy of the "paracrine" compartment.

Second, the lack of knowledge of intraislet anatomy has clouded the testing of predictions. If the paracrine compartment were a simple "static" space, and nonpolar A-, B- and D-cells had unlimited access and exposure to this space, one would predict that the islet cells would respond only to very high concentrations of islet hormones, exceeding those high levels normally to be expected in that hypothetical paracrine space. This concept was perhaps supported by the big doses of exogenous insulin required to suppress glucagon secretion in the in vitro isolated perfused pancreas (Sects. D and F), but was undermined by the efficacy of low concentrations of somatostatin and glucagon modulating islet cell secretion (SAMOLS and HARRISON 1976a; WEIR et al. 1979; KAWAI and ROUILLER 1981). The primary question then becomes – how are intraislet interactions prevented (Sects. B, C, E)? According to current concepts of anatomy we can predict that paracrine interactions between A- and D-cells are more likely than other interactions because of the preferential, perhaps obligatory, association between these two cell types (ORCI et al. 1981). Against a background of new observations on the islet vasculature (BONNER-WEIR and ORCI 1982) and postulated interstitial flow, we have been able to make "educated guesses" about the likelihood of other interactions, (Sect. E), favoring the probability that local insulin suppresses glucagon over the possibility that local glucagon or somatostatin influence B-cell secretion.

A third rebuttal to the anti-paracrine critic is that intraislet interactions can explain: (1) why glucose itself has not been clearly shown to suppress A-cell secretion in the absence of an appropriate insulin and/or somatostatin response (Sect. D, F,

L; Chap. 22); (2) why sulfonylureas can both stimulate and suppress glucagon (Sect. D); and (3) why exogenous insulin administration has different effects on glucagon responses in normal subjects, IDDM, and NIDDM (Sect. L). The latter hypothesis has been particularly dependent on the elegant studies of ASPLIN et al. (1981), whose work represents the best in vivo support to date of an intraislet interaction, the suppression of glucagon by endogenous B-cells. The alternative theory of independent islet cell function does not satisfactorily explain these phenomena. With respect to the latter theory, it has not yet been shown that glucose can modulate A- or B-cell secretion in the absence of either glucagon or somatostatin.

The final rebuttal is that new knowledge affords the opportunity for new testable predictions. The anatomic differences between the duodenal lobe (polypeptide-rich) and splenic lobe (glucagon-rich) were used by TRIMBLE and RENOLD (1981) to suggest that glucagon does indeed promote glucose-induced insulin secretion in the islets. If these observations are confirmed in the isolated perfused pancreatic lobes, another step in deciding the intraislet question one way or the other will have been taken.

Therefore, we must concede that the hypothesis of intraislet interaction theory of A-, D-, and B-cells on one another is indeed a valid scientific theory. On the other hand, our periodic exasperation with the subject is understandable, with too much speculation chasing after too few facts. Looking back over the past 15 years, it almost seems as though the onus of proof has been transferred – in view of the anatomy and the known potential interactions – from the positive to the negative; i.e., the burden is to prove that paracrine interactions are not occurring, and if not, why not? The task is formidable, but the problem is soluble. Although it is conceivable that the various potential intraislet interactions, paracrine, and nonparacrine, described in this chapter play only a small part in physiology, our thesis suggests that some major roles are not unlikely.

*Acknowledgments.* G. C. W. and S. B. W. wish to extend particular appreciation to L. ORCI and P. MEDA, pioneers in this area, who generously shared ideas and concepts in long hours of discussion. E. S. wishes to extend particular appreciation to GABRIELE H. SAMOLS and CONNIE S. MEYERS for long hours of research and secretarial assistance.

## References

Adrian TE, Bloom SR, Besterman HS, Bryant MG (1978a) PP-physiology and pathology. In: Bloom SR (ed) Gut hormones. Churchill Livingstone, Edinburgh London, pp 254–260

Adrian TE, Greenberg GR, Besterman HS, McCloy RF, Chadwick VS, Barnes AJ, Mallinson CN, Baron JH, Alberti KGMM, Bloom SR (1978b) PP infusion in man – summary of initial investigations. In: Bloom SR (ed) Gut hormones. Churchill Livingstone, Edinburgh London, pp 265–267

Adrian TE, Barnes AJ, Long RG, O'Shaughnessy DJ, Brown MR, Rivier J, Vale W, Blackburn AM, Bloom SR (1981) The effect of somatostatin analogs on secretion of growth, pancreatic, and gastrointestinal hormones in man. J Clin Endocrinol Metab 53:675–681

Alberti KGMM, Christensen NJ, Christensen SE, Hansen AAP, Iversen J, Lundbaek K, Seyer-Hansen K, Orskov H (1973) Inhibition of insulin secretion by somatostatin. Lancet 2:1299–1301

Alford FP, Bloom SR, Nabarro JDN (1976) Glucagon metabolism in man. Studies on the metabolic clearance rate and the plasma acute disappearance time of glucagon in normal and diabetic subjects. J Clin Endocrinol Metab 42:830–838

Anderson GE, Kologlin Y, Papadopoulos C (1967) Fluctuation in postabsorptive blood glucose in relation to insulin release. Metab Clin Exp 16:586–596
Arimura A, Coy DH, Chihara M, Fernandez-Durango R, Samols E, Chihara K, Meyers C, Schally A (1978) Somatostatin. In: Bloom SR (ed) Gut hormones. Churchill Livingstone, Edinburgh London, pp 437–445
Arimura A, Meyers CA, Case WL, Murphy WA, Schally AV (1979) Suppression of somatostatin levels in the hepatic portal and systemic plasma of the rat by synthetic human pancreatic polypeptide. Biochem Biophys Res Commun 89:913–918
Aronoff SL, Bennett PH, Rushforth NB, Miller M, Unger RH (1976) Arginine-stimulated hyperglucagonemia in diabetic Pima Indians. Diabetes 25:404–407
Asplin CM, Paquette TL, Palmer JP (1981) In vivo inhibition of glucagon secretion by paracrine beta cell activity in man. J Clin Invest 68:314–318
Assan R, Hautecouverture G, Guillemant S, Dauchy F, Protin P, Derot M (1969) Evolution de paramètres hormonaux (glucagon, cortisol, hormone somatotrope) et énergétiques (glucose, acids gras libres, glycerol) dans dix acido-cétoses diabétiques graves traitées. Pathol Biol (Paris) 17:1095–1105
Aydin I, Raskin P, Unger RH (1977) The effect of short-term intravenous insulin administration on the glucagon response to a carbohydrate meal in adult-onset and juvenile type diabetes. Diabetologia 13:629–636
Baetens D, Stefan Y, Ravazzola M, Malaisse-Lagae F, Coleman DL, Orci L (1978) Alteration of islet cell populations in spontaneously diabetic mice. Diabetes 27:1–7
Barden N, Lavoie M, Dupont A, Côté J, Côté J-P (1977) Stimulation of glucagon release by addition of anti-somatostatin serum to islets of Langerhans in vitro. Endocrinology 101:635–638
Beigelman PM, Ribolet R, Atwater I (1977) Electrical activity of mouse pancreatic beta cells. J Physiol (Paris) 73:201–217
Bencosme SA, Liepa E (1955) Regional differences of the pancreatic islets. Endocrinology 57:588–593
Bennett PH, Aronoff SL, Unger RH (1976) Evidence for an insulin-independent alpha-cell abnormality in human diabetes. Metabolism 25 [Suppl 1]:1527–1529
Benoit R, Böhlen P, Brazeau P, Ling N, Guillemin R (1980) Isolation and characterization of rat pancreatic somatostatin. Endocrinology 107(6):2127–2129
Berelowitz MB, Shapiro B, Pimstone B, Kronheim S (1979) Growth hormone-release inhibitory hormone-like immunoreactivity in pancreas and gut in streptozotocin diabetes in the rat and response to insulin administration. Clin Endocrinol 10:195–198
Berelowitz M, Coleman DL, Frohman LA (1980) Temporal relationship of tissue somatostatin-like immunoreactivity to metabolic changes in genetically obese and diabetic mice. Diabetes 29:717–723
Berthelsen S, Pettinger WA (1977) A functional basis for classification of α-adrenergic receptors. Life Sci 21:595–606
Bhathena SJ, Perrino PV, Voyles NR, Smith SS, Wilkins SD, Coy DH, Schally AV, Recant L (1976) Reversal of somatostatin inhibition of insulin and glucagon secretion. Diabetes 25(11):1031–1040
Bloom SR (1981) Control of glucagon secretion. In: Unger RH, Orci L (eds) Glucagon. Physiology, pathophysiology, and morphology of the pancreatic A-cells. Elsevier, New York, pp 99–109
Bloom SR, Polak JM (1978) Peptidergic versus purinergic. Lancet 1:93
Boden G (1981) Extrapancreatic glucagon in human subjects. In: Unger RH, Orci L (eds) Glucagon. Physiology, pathophysiology and morphology of the pancreatic A-cells. Elsevier, New York, pp 349–357
Boden G, Sivitz M, Owen OE, Essa-Kouwar N, Landor JH (1975) Somatostatin suppresses secretin and pancreatic exocrine secretion. Science 190:163–165
Bonner-Weir S, Trent DF, Zmachinski CJ, Clore ET, Weir GC (1981) Limited B cell regeneration in a B cell deficient rat model: studies with dexamethasone. Metabolism 30:914–918
Bonner-Weir S, Orci L (1982) New perspectives on the microvasculature of the islets of Langerhans in the rat. Diabetes 31:883–889

Braaten JT, Faloona GR, Unger RH (1974) The effect of insulin on the alpha-cell response to hyperglycemia in long-standing alloxan diabetes. J Clin Invest 53:1017–1021
Brown M, Reubi J-C, Perrin M, Vale W, Rivier J (1981) Somatostatin-28 (SS-28): selective actions on pancreatic B-cells and brain. Diabetes [Suppl 1]30:28 A
Buchanan KD, Mawhinney WAA (1973a) Insulin control of glucagon release from insulin-deficient rat islets. Diabetes 22:801–803
Buchanan KD, Mawhinney WAA (1973b) Glucagon release from isolated pancreas in streptozotocin-treated rats. Diabetes 22:797–800
Buchanan KD, McCarroll AM (1972) Abnormalities of glucagon metabolism in untreated diabetes mellitus. Lancet 2:1394–1395
Burnstock G (1979) Interactions of cholinergic, adrenergic, purinergic and peptidergic neurons in the gut. In: Brooks CMcC, Koizumi K, Sato A (eds) Integrative functions of the autonomic nervous system. University of Tokyo Press, Tokyo; Elsevier/North-Holland Biomedical, Amsterdam Oxford New York, pp 145–158
Chey WY, Shay H, Shuman CR (1963) External pancreatic secretion in diabetes mellitus. Ann Intern Med 59:812–821
Chiba T, Kadowaki S, Taminato T, Abe H, Chihara K, Matsukura S, Goto Y, Seino Y, Fujita T (1981) Concentration and secretion of gastric somatostatin in streptozotocin-diabetic rats. Diabetes 30:188–191
Creutzfeldt W, Lankisch PG, Fölsch UR (1975) Hemmung der Sekretin und Cholezystokinin-Pankreozymin-induzierten Saft- und Enzymsekretion des Pankreas und der Gallenblasenkontraktion beim Menschen durch Somatostatin. Dtsch Med Wochenschr 100(20):1135–1138
Crick F (1981) Life itself: its origin and nature. Simon and Schuster, New York
Day JL, Anderson JA (1973) Abnormalities of glucagon metabolism in diabetes mellitus. Clin Endocrinol 2:211–217
Dean PM, Matthews EK (1970) Glucose-induced electrical activity in pancreatic islet cells. J Physiol (Lond) 210:255–264
Deckert T, Lauridsen UB, Madsen SN (1972) Serum insulin following isoprenaline in normal and diabetic persons. Horm Metab Res 4:229–232
De Haën C, Gorray KC, Howland HC (1981) A model for oscillatory and biphasic insulin secretion. Diabetes [Suppl 1]30:114 A
Dockray GJ (1979) Evolutionary relationships of the gut hormones. Fed Proc 38:2295–2301
Eddlestone GT, Rojas E (1980) Evidence of electrical coupling between mouse pancreatic B-cells. J Physiol (Lond) 303:76 P–77 P
Efendić S, Luft R, Grill V (1974) Effect of somatostatin on glucose induced insulin release in isolated perfused rat pancreas and isolated rat pancreatic islets. FEBS Lett 42:169–172
Efendić S, Luft R, Sievertsson H (1975) Relative effects of somatostatin and two somatostatin analogues on the release of insulin, glucagon and growth hormone. FEBS Lett 58:302–305
Efendić S, Enzmann F, Nylén A, Uvnäs-Wallensten K, Luft R (1980) Sulphonylurea (Glibenclamide) enhances somatostatin and inhibits glucagon release induced by arginine. Acta Physiol Scand 108:231–233
Elahi D, Nagulesparan M, Hershcopf R, Muller D, Hidaka H, Abraham M, Tobin J, Gingerich R, Unger R, Andres R (1981) Insulin resistance of the B-cells of obese humans. Diabetes [Suppl 1]30:14 A
Ensinck J, Walter R, Palmer J, Brodows R, Campbell R (1976) Glucagon responses to hypoglycemia in adrenalectomized man. Metabolism 25:227–232
Feldman JM (1979) Species variation in the islets of Langerhans (editorial). Diabetologia 16:1–4
Feyrter F (1953) Über die peripheren endokrinen (parakrinen) Drüsen des Menschen, 2nd edn. Maudrich, Vienna
Forssman WG, Helmstaedter V, Mühlmann G, Feurle GE (1978) Immunohistochemistry and ultrastructure of somatostatin cells with special reference to the gastroenteropancreatic (GEP) system. Metabolism 27 [Suppl 1]:1179–1191
Frerichs H, Reich V, Creutzfeldt W (1965) Insulin secretion in vitro. Klin Wochenschr 43:136–141

Fujita T (1967) D cell, the third endocrine element of the pancreatic islet. Arch Histol Jpn 29:1–40
Fujita T, Yanatori Y, Murakami T (1976) Insulo-acinar axis, its vascular basis and its functional and morphological changes caused by CCK-PZ and caerulein. In: Fujita T (ed) Endocrine gut and pancreas. Elsevier, Amsterdam, pp 347–357
Gavin JR, Roth J, Neville DM, De Meyts P, Buell DN (1974) Insulin-dependent regulation of insulin receptor concentrations: a direct demonstration in cell culture. Proc Natl Acad Sci USA 71:84–88
Gepts W (1957) Contribution à l'étude morphologique des îlots de Langerhans au cours du diabète. Ann Soc Sci Med Nat Brux 10:5–108
Gepts W, Lecompte PM (1981) The pancreatic islets in diabetes. Am J Med 70:105–115
Gerber PPG, Trimble ER, Wollheim CB, Renold AE (1981) Effect of insulin on glucose- and arginine-stimulated somatostatin secretion from the isolated perfused rat pancreas. Endocrinology 109:279–283
Gerich JE (1981) Somatostatin. In: Brownlee M (ed) Handbook of diabetes mellitus, vol 1. Garland STPM Press, New York, pp 297–354
Gerich J, Langlois M, Noacco C, Karam J, Forsham P (1973) Lack of glucagon response to hypoglycemia in diabetes: evidence for an intrinsic pancreatic alpha-cell defect. Science 182:171–173
Gerich J, Schneider V, Dippe S, Langlois M, Noacco C, Karam J, Forsham P (1974) Characterization of the glucagon response to hypoglycemia in man. J Clin Endocrinol Metab 38:77–82
Gerich J, Lovinger R, Grodsky G (1975) Inhibition by somatostatin of glucagon and insulin release from the perfused rat pancreas in response to arginine, isoproterenol and theophylline: evidence for a preferential effect on glucagon secretion. Endocrinology 96:749–754
Gerich JE, Lorenzi M, Tsalikian E, Bohannon NV, Schneider V, Karam JH, Forsham PH (1976) Effects of acute insulin withdrawal and administration on plasma glucagon responses to intravenous arginine in insulin-dependent diabetic subjects. Diabetes 25:955–960
Goldstein MB, Davis EA Jr (1968) The three-dimensional architecture of the islets of Langerhans. Acta Anat (Basel) 71:161–171
Goodner CJ, Walike BC, Koerker DJ, Ensinck JW, Brown AC, Chideckel WE, Palmer J, Kalnosy L (1977) Insulin, glucagon and glucose exhibit synchronous, sustained oscillations in fasting monkeys. Science 195:177–179
Grube D, Maier V, Raptis S, Schlegel W (1978a) Immunoreactivity of the endocrine pancreas. Evidence for the presence of cholecystokinin-pancreozymin within the A-cell. Histochemistry 56:13–35
Grube D, Voigt DH, Weber E (1978b) Pancreatic glucagon cells contain endorphin-like immunoreacitivity. Histochemistry 59:75–79
Haist RE (1965) Effects of changes in stimulation on the structure and function of islet cells. In: Leibel BS, Wrenshall GA (eds) On the nature and treatment of diabetes. Excerpta Medica, Amsterdam, pp 12–30
Halban P, Wollheim CB, Blondel B, Renold AE (1979) Evidence for the importance of contact between pancreatic islet cells in glucose-induced insulin release. Diabetes [Suppl] 28:394
Hara M, Patton G, Gerich J (1979) Increased somatostatin release from pancreases of alloxan diabetic rats perfused in vitro. Life Sci 24:625–628
Hellerström C (1977) Growth pattern of pancreatic islets in animals. In: Volk BW, Wellman KF (eds) The diabetic pancreas. Plenum, New York, pp 61–98
Henderson JR (1969) Why are the islets of Langerhans? Lancet 2:469–470
Hermansen K (1980) Secretion of somatostatin from the normal and diabetic pancreas, studies in vitro. Diabetologia 19:492–504
Hermansen K, Schwartz TW (1979) Differential sensitivity to somatostatin of pancreatic polypeptide, glucagon and insulin secretion from the isolated perfused canine pancreas. Metabolism 28:1229–1233
Hermansen K, Orskov H, Christensen SE (1979) Streptozotocin diabetes: a glucoreceptor dysfunction affecting D cells as well as B and A cells. Diabetologia 17:385–390

Hoftiezer V, Carpenter AM (1973) Comparison of streptozotocin and alloxan-induced diabetes in the rat including volumetric quantitation of the pancreatic islets. Diabetologia 9:178–184
Holst JJ, Jensen SL, Nielsen OV, Schwartz TW (1981) Nervous control of pancreatic somatostatin secretion in pigs. Diabetologia 21(3):283–284
Honey RN, Weir GC (1979) Insulin stimulates somatostatin and inhibits glucagon secretion from the perfused chicken pancreas-duodenum. Life Sci 24:1747–1750
Honey RN, Weir GC (1980) Acetylcholine stimulates insulin, glucagon and somatostatin release in the perfused chicken pancreas. Endocrinology 107:1065–1068
Honey RN, Arimura A, Weir GC (1981) Somatostatin neutralization stimulates glucagon and insulin secretion from the avian pancreas. Endocrinology 109:1971–1974
Ipp E, Dobbs RE, Arimura A, Vale W, Harris V, Unger RH (1977) Release of immunoreactive somatostatin from the pancreas in response to glucose, amino acids, pancreozymin-cholecystokinin and tolbutamide. J Clin Invest 60:760–765
Ipp E, Dobbs RE, McCorkle K, Harris V, Unger RH (1978a) Further evidence for similar secretory responses of pancreatic somatostatin and insulin. Clin Res 26:33 A
Ipp E, Dobbs RE, Unger RH (1978b) Morphine and beta-endorphin influence the secretion of the endocrine pancreas. Nature 276:190–191
Ipp E, Dobbs R, Unger RH (1978c) Vasoactive intestinal peptide stimulates pancreatic somatostatin release. FEBS Lett 90:76–78
Ipp E, Rivier J, Dobbs RI, Brown M, Vale W, Unger RH (1979) Somatostatin analogs inhibit somatostatin release. Endocrinology 104:1270–1273
Itoh M, Mandarino L, Gerich JE (1980) Antisomatostatin gamma globulin augments secretion of both insulin and glucagon in vitro: evidence for a physiological role for endogenous somatostatin in the regulation of pancreatic A and B cell function. Diabetes 29:693–696
Iversen J (1974) Inhibition of pancreatic glucagon release by somatostatin: in vitro. Scand J Clin Lab Invest 33:125–129
Iversen J, Miles DW (1971) Evidence for a feedback inhibition of insulin secretion in the isolated perfused canine pancreas. Diabetes 20:1–9
Jarotzky AJ (1899) Ueber die Veraenderungen in der Groesse und im Bau der Pankreaszellen bei einigen Arten der Inanition. Virchows Arch Pathol Anat 156:409–450
Johnson D, Ensinck J, Koerker D, Palmer J, Goodner CJ (1975) Inhibition of glucagon and insulin secretion by somatostatin in the rat pancreas perfused in situ. Endocrinology 96:370–374
Kadowaki S, Taminato T, Chiba T, Mori K, Abe H, Goto Y, Seino Y, Matsukura S, Nozawa M, Fujita T (1979) Somatostatin release from isolated perfused rat pancreas: possible role of endogenous somatostatin on insulin release. Diabetes 28:600–603
Kadowaki S, Taminato T, Chiba T, Goto Y, Nozawa M, Seino Y, Matsukura S, Fujita T (1980) Reversal of the enhanced somatostatin release from the isolated, perfused diabetic rat pancreas after the amelioration of diabetes by whole pancreas transplantation. Diabetes 29:742–746
Kamel J, Mikhail Y, Beshir S (1979) Study on the innervation of the pancreas of the rat. Acta Anat (Basel) 104:237–241
Kanno T, Veda N, Saito A (1976) Insulo-acinar axis: a possible role of insulin potentiating the effects of pancreozymin in the pancreatic acinar cell. In: Fujita T (ed) Endocrine gut and pancreas. Elsevier, Amsterdam New York, pp 335–346
Kawai K, Rouiller D (1981) Evidence that the islet interstitium contains functionally separate "arterial" and "venous" compartments. Diabetes [Suppl 1) 30:14A
Kawamori R, Shichiri M, Kikuchi M, Yamasaki Y, Abe H (1980) Perfect normalization of excessive glucagon responses to intravenous arginine in human diabetes mellitus with the artificial beta-cell. Diabetes 29:762–765
Kazumi T, Utsumi M, Yoshino G, Ishihara G, Hirose K, Makimura H, Baba S (1980) Somatostatin concentration responds to arginine in portal plasma: effects of fasting, streptozotocin diabetes, and insulin administration in diabetic rats. Diabetes 29:71–73
Koerker DJ, Ruch W, Chideckel E, Palmer J, Goodner CJ, Ensinck J, Gale CC (1974) Somatostatin: hypothalamic inhibitor of the endocrine pancreas. Science 184:482–484

Kohen E, Kohen C, Thovell B, Mintz DA, Rabinovitch A (1979) Intercellular communication in pancreatic islet monolayer cultures: a microfluorometric study. Science 204:862–865
Kolod E, Meda P, Perrelet A, Orci L (1981) Influence of intra-islet environment on B cell function. Experientia 37:650
Kosaka K, Hagura R, Kuzuya T (1977) Insulin responses in equivocal and definite diabetes, with special reference to subjects who had mild glucose intolerance but later developed definite diabetes. Diabetes 26:944–952
Langer SZ, Adler-Graschinsky E, Giorgi O (1977) Physiological significance of α-adrenoceptor-mediated negative feedback mechanism regulating noradrenaline release during nerve stimulation. Nature 265:648–650
Larsson L-I (1977) Corticotropin-like peptides in central nerves and in endocrine cells of gut and pancreas. Lancet 2:1321–1322
Larsson L-I (1980) New aspects on the neural, paracrine and endocrine regulation of islet function. Front Horm Res 7:14–29
Larsson L-I, Rehfeld JR (1979) Peptidergic and adrenergic innervation of pancreatic ganglion. Scand J Gastroenterol 14:433–437
Larsson L-I, Sundler F, Håkanson R, Pollock HG, Kimmel JR (1974) Localisation of APP, a postulated new hormone, to a pancreatic endocrine cell type. Histochemistry 42:377–382
Larsson L-I, Fahrenkrug J, Schaffalitzky de Muckadell OB (1978) Innervation of the pancreas by vasoactive intestinal polypeptide (VIP) immunoreactive nerves. Life Sci 22:773–780
Larsson L-I, Goltermann N, de Magistris L, Rehfeld JF, Schwartz TW (1979) Somatostatin cell processes as pathways for paracrine secretion. Science 205:1393–1395
Laube H, Fussgänger RD, Maier V, Pfeifer EF (1973) Hyperglucagonemia of the isolated perfused pancreas of diabetic mice. Diabetologia 9:400–402
Lefèbvre PJ, Luyckx AS (1978) Glucose and insulin in the regulation of glucagon release from the isolated perfused dog stomach. Endocrinology 103:1579–1582
Lefèbvre PJ, Luyckx AS (1979) Glucagon and diabetes: a reappraisal. Diabetologia 16:347–354
Lefèbvre PJ, Luyckx AS, Nizet AH (1976) Independence of glucagon and insulin handling by the isolated perfused dog kidney. Diabetologia 12:359–365
Leichter SB, Pagliara AS, Greiden MH, Pohl S, Rosai J, Kipnis DM (1975) Uncontrolled diabetes mellitus and hyperglucagonemia associated with an islet cell carcinoma. Am J Med 58:285–293
Leiter EH, Gapp DA, Eppig JJ, Coleman DL (1979) Ultrastructure and morphometric studies of delta cells in pancreatic islets from C 57 BL/KS diabetic mice. Diabetologia 17:297–310
Lifson N, Kraminger KG, Mayrand RR, Lander EJ (1980) Blood flow to the rabbit pancreas with special reference to the islets of Langerhans. Gastroenterology 79:466–473
Like AA (1977) Spontaneous diabetes in animals. In: Volk BW, Wellman KF (eds) The diabetic pancreas. Plenum, New York, pp 381–423
Lin T-M, Chance RE (1978) Spectrum of gastrointestinal actions of bovine PP. In: Bloom SR (ed) Gut hormones. Churchill Livingstone, Edinburgh, pp 242–246
Lin TM, Evans DC, Chance RE, Spray GF (1977) Bovine pancreatic peptide: action on gastric and pancreatic secretion in dogs. Am J Physiol 232(3):E311–E315
Long DA, Matthews DR, Phil D, Petro J, Turner RC (1979) Cyclic oscillations of basal plasma glucose and insulin concentrations in human beings. N Engl J Med 301:1023–1027
Lundquist I (1971) Insulin secretion. Its regulation by monoamines and acid amyloglucosidase. Acta Physiol Scand [Suppl] 372:3–47
Madsbad S, Hilsted J, Krarup T, Tronier B, Schwartz T (1981) The importance of autonomic neuropathy and of residual β-cell function for the pancreatic endocrine response to hypoglycemia in insulin dependent diabetics. Diabetes [Suppl 1] 30:44A
Maher TD, Tanenberg RJ, Greenberg BZ, Hoffman JE, Doe RP, Goetz FC (1977) Lack of glucagon response to hypoglycemia in diabetic autonomic neuropathy. Diabetes 26:196–200

Makino H, Matsushima Y, Kanatsuka A, Yamamoto M, Kumagai A, Nishimura M (1979) Changes in pancreatic somatostatin content in spontaneously diabetic mice, as determined by radioimmunoassay and immunohistochemical methods. Endocrinology 104:243–247

Malaisse-Lagae F, Ravazzola M, Robberecht P, Vandermeers A, Malaisse WJ, Orci L (1975) Exocrine pancreas: evidence for topographic partition of secretory function. Science 190:795–797

Mandarino L, Stenner D, Blanchard W, Nissen S, Gerich J, Ling N, Brazeau P, Bohlen P, Esch F, Guillemin R (1981) Selective effects of somatostatin-14, -25 and -28 on in vitro insulin and glucagon secretion. Nature 291:76–77

Matschinsky FM, Pagliara AS, Hover BA, Pace CS, Ferrendelli JA, Williams A (1976) Hormone secretion and glucose metabolism in islets of Langerhans of the isolated perfused pancreas from normal and streptozotocin diabetic rats. J Biol Chem 254(19):6053–6064

Matschinsky FM, Rujanavech C, Pagliara AS, Norfleet WT (1980) Adaptations of α- and β-cells of rat and mouse pancreatic islets to starvation, to refeeding after starvation, and to obesity. J Clin Invest 65:207–218

Matsushima Y, Makino H, Kanatsuka A, Yamamoto M, Kumagai A (1978) Immunohistochemical changes of somatostatin cells in the pancreatic islets of rats after streptozotocin administration. Endocrinol Jpn 25:111–115

McEvoy RC, Hegre OD (1977) Morphometric quantitation of the pancreatic insulin-, glucagon-, and somatostatin-positive cell populations in normal and alloxan-diabetic rats. Diabetes 26:1140–1146

Meda P, Perrelet A, Orci L (1979) Increase of gap junction between pancreatic B-cells during stimulation of insulin secretion. J Cell Biol 82:441–448

Meda P, Halban P, Perrelet A, Renold AE, Orci L (1980a) Gap junction development is correlated with insulin content in the pancreatic B cell. Science 209:1026–1028

Meda P, Denef JF, Perrelet A, Orci L (1980b) Non-random distribution of gap junctions between pancreatic B-cells. Am J Physiol 238:C114–C119

Meda P, Perrelet A, Orci L (1980c) Gap junctions and B-cell funtion. In: Malaisse WJ, Taljedal IB (eds) Biochemistry and biophysics of the pancreatic B-cell. Thieme, Stuttgart, pp 157–162

Meda P, Amherdt M, Perrelet A, Orci L (1981) Metabolic coupling between cultured pancreatic B cells. Exp Cell Res 133(2):421–430

Meda P, Perrelet A, Orci L (1982) Endocrine cell interactions within the islets of Langerhans. In: Pitts JD (ed) Functional integration of cells in animal tissues. Cambridge University Press, Cambridge, pp 113–131

Meissner HP (1976) Electrophysiological evidence for coupling between beta cells of pancreatic islets. Nature 262:502–504

Miller RE, Horton ES (1979) Neural release of glucagon is inhibited by hyperglycemia and enhanced by phentolamine. Diabetes 28:762–768

Moody AJ, Jacobsen H, Sundby F (1978) Gastric glucagon and gut glucagon-like immunoreactants. In: Bloom SR (ed) Gut hormones. Churchill Livingstone, Edinburgh, pp 369–378

Müller WA, Faloona JR, Aguilar-Parada E, Unger RH (1970) Abnormal alpha-cell function in diabetes: responses to carbohydrate and protein ingestion. N Engl J Med 282:109–115

Müller W, Faloona G, Unger RH (1971) The effects of experimental insulin deficiency on glucagon secretion. J Clin Invest 50:1992–1999

Müller WA, Faloona JR, Unger RH (1973) Hyperglucagonemia in diabetic ketoacidosis. Am J Med 54:52–57

National Diabetes Data Group (1979) Classification and diagnosis of diabetes mellitus and other categories of glucose intolerance. Diabetes 28:1039–1057

Östenson C-G (1979) Regulation of glucagon release: effects of insulin on the pancreatic $A_2$-cell of the guinea pig. Diabetologia 17:325–330

Ohara H (1976) Clinical and experimental studies on the entero-insular and insulo-acinar correlations. In: Fujita T (ed) Endocrine gut and pancreas. Elsevier, Amsterdam, pp 321–333

ole-Moiyoi O, Pinkus GS, Spragg J, Austen KF (1979) Identification of human glandular kallikrein in the beta cell of the pancreas. N Engl J Med 300(23):1289–1294
Olefsky JM, Kolterman OG (1981) Mechanisms of insulin resistance in obesity and non insulin-dependent (Type II) diabetes. Am J Med 70:151–168
Orci L (1974) A portrait of the pancreatic B-cell. Diabetologia 10:163–187
Orci L (1976) The microanatomy of the islets of Langerhans. Metabolism 25:1303–1313
Orci L, Perrelet A (1981) The morphology of the A cell. In: Unger RH, Orci L (eds) Glucagon. Physiology, pathophysiology, and morphology of the pancreatic A-cells. Elsevier, New York, pp 3–32
Orci L, Unger RH (1975) Functional subdivision of islets of Langerhans and possible role of D cells. Lancet 2:1243
Orci L, Unger RH, Renold AE (1973) Structural coupling between pancreatic islet cells. Experientia 29:1015–1018
Orci L, Malaisse-Lagae F, Rouiller D, Renold AE, Perrelet A, Unger RH (1975) A morphological basis for intercellular communication between alpha and beta cells in the endocrine pancreas. J Clin Invest 56:1066–1070
Orci L, Baetens D, Ravazzola M, Stefan Y, Malaisse-Lagae F (1976a) Pancreatic polypeptide and glucagon: non-random distribution in pancreatic islets. Life Sci 19:1811–1816
Orci L, Baetens D, Rufener C, Amherdt N, Ravazzola M, Studer P, Malaisse-Lagae F, Unger RH (1976b) Hypertrophy and hyperplasia of somatostatin-containing D-cells in diabetes. Proc Natl Acad Sci USA 73:1338–1342
Orci L, Baetens D, Ravazzola M, Malaisse-Lagae F, Amherdt M, Rufener C (1976c) Somatostatin in the pancreas and the gastrointestinal tract. In: Fujita T (ed) Endocrine gut and pancreas. Elsevier, Amsterdam New York, pp 73–88
Orci L, Stefan Y, Bonner-Weir S, Perrelet A, Unger RH (1981) "Obligatory" association between A and D cells demonstrated by bipolar islets in neonatal pancreas. Diabetologia 21:73–74
Pagliara AS, Stillings SN, Hover B, Martin DM, Matschinsky F (1974) Glucose modulation of amino acid-induced glucagon and insulin release in the isolated perfused rat pancreas. J Clin Invest 54:819–832
Pagliara AS, Stillings SN, Haymond MW, Hover BA, Matschinsky FM (1975) Insulin and glucose as modulators of the amino acid-induced glucagon release in the isolated pancreas of alloxan and streptozotocin diabetic rats. J Clin Invest 55:244–255
Pagliara AS, Stillings SN, Zawalich WS, Williams AD, Matschinsky FM (1977) Glucose and 3-O-methylglucose protection against alloxan poisoning of pancreatic alpha and beta cells. Diabetes 26:973–979
Palmer JP, Benson JW, Walter RM, Ensinck JW (1976a) Arginine-stimulated acute phase of insulin and glucagon secretion in diabetic subjects. J Clin Invest 58:565–570
Palmer J, Henry D, Bensen J, Johnson D, Ensinck J (1976b) Glucagon response to hypoglycemia in sympathectomized man. J Clin Invest 57:522–525
Patel Y, Reichlin S (1978) Somatostatin in hypothalamus, extrahypothalamic brain and peripheral tissues of the rat. Endocrinology 102:523–530
Patel YC, Weir CG (1976) Increased somatostatin content of islets from streptozotocin-diabetic rats. Clin Endocrinol 5:191–194
Patel YC, Orci L, Bankier A, Cameron DP (1976) Decreased pancreatic somatostatin concentration in spontaneously diabetic mice. Endocrinology 99:1415–1418
Patel YC, Cameron DP, Stefan Y, Malaisse-Lagae F, Orci L (1977) Somatostatin: widespread abnormality in tissues of spontaneously diabetic mice. Science 198:930–931
Patel YC, Cameron DP, Bankier A, Malaisse-Lagae F, Ravazzola M, Studer P, Orci L (1978) Changes in somatostatin concentration in pancreas and other tissues of streptozotocin diabetic rats. Endocrinology 103:917–923
Patel YC, Wheatley T, Malaisse-Lagae F, Orci L (1980) Elevated portal and peripheral blood concentration of immunoreactive somatostatin in spontaneously diabetic (BBL) Wistar rats: suppression with insulin. Diabetes 29:757–761
Patton GS, Dobbs RE, Orci L, Vale W, Unger RH (1976) Stimulation of pancreatic immunoreactive somatostatin (IRS) release by glucagon. Metabolism [Suppl 1] 25:1499
Patton GS, Ipp E, Dobbs RE, Orci L, Vale W, Unger RH (1977) Pancreatic immunoreactive somatostatin release. Proc Natl Acad Sci USA 74:2140–2143

Petersson B, Elde R, Efendić S, Hökfelt T, Johansson O, Luft R, Cerasi E, Hellerström C (1977) Somatostatin in the pancreas, stomach and hypothalamus of the diabetic Chinese hamster. Diabetologia 13:463–466
Pohorecky LA, Wurtman RJ (1971) Adrenocortical control of epinephrine synthesis. Pharmacol Rev 23:1–35
Polonsky K, Jaspan J, Pugh W, Dhorajiwald J, Abraham M, Blix P, Moossa AR (1981) Insulin and glucagon breakthrough of somatostatin suppression. Importance of portal vein hormone measurements. Diabetes 30:664–669
Raptis S, Escobar-Jimenez F, Rosenthal J, Ditschuneit HH, Pfeiffer EF (1977) Somatostatin modulation of pancreatic glucagon, insulin, glucose and free fatty acids following beta-adrenergic stimulation. J Clin Endocrinol Metab 44:1088–1093
Raskin P, Fujita Y, Unger RH (1975) Effect of insulin-glucose infusions on plasma glucagon levels in fasting diabetics and non-diabetics. J Clin Invest 56:1132–1138
Raskin P, Aydin I, Unger RH (1976) Effect of insulin on the exaggerated glucagon response to arginine stimulation in diabetes mellitus. Diabetes 25:227–229
Raskin P, Aydin I, Yamamoto T, Unger RH (1978) Abnormal alpha cell function in human diabetes: the response to oral protein. Am J Med 64:988–997
Rouiller D, Schusdziarra V, Unger RH (1981) Insulin inhibits somatostatin-like immunoreactivity release stimulated by intragastric HCl. Diabetes 30:735–738
Roy MW, Jones MS, Lee KC, Miller RE (1981) Splanchnic neural stimulation suppresses secretion of pancreatic somatostatin by $alpha_2$, insulin by $alpha_2$, post synaptic receptors. Endocrinology [Suppl] 108:105
Rufener C, Amherdt M, Baetens D, Yanaihara N, Orci L (1976) Immunofluorescent localization of secretion in pancreatic monolayer culture. Histochemistry 47:171–173
Saffouri B, Weir GC, Bitar K, Makhlouf G (1979) Stimulation of gastrin secretion from the perfused rat stomach by somatostatin antiserum. Life Sci 25:1749–1753
Saito A, Williams JA, Kanno T (1980) Potentiation of cholecystokin-induced exocrine secretion by both exogenous and endogenous insulin in isolated and perfused rat pancreata. J Clin Invest 65:777–782
Samols E, Harrison J (1976a) Remarkable potency of somatostatin as a glucagon suppressant. Metabolism [Suppl 1] 25:1495–1497
Samols E, Harrison J (1976b) Intraislet negative insulin-glucagon feedback. Metabolism 25:1443–1447
Samols E, Harrison J (1977) Tolbutamide: stimulator and suppressor of glucagon secretion. In: Foà PP, Bajaj JS, Foà NL (eds) Glucagon. Its role in physiology and clinical medicine. Springer, Berlin Heidelberg New York, pp 699–710
Samols E, Stagner J (1981) Modulation of insulin: local presynaptic $\alpha_2$ but postsynaptic $\alpha_2$ adrenoceptors. Diabetes [Suppl 1] 30:44A
Samols E, Weir GC (1979) Adrenergic modulation of pancreatic A, B and D cells: alpha-adrenergic suppression and beta-adrenergic stimulation of somatostatin secretion, alpha-adrenergic stimulation of glucagon secretion in the perfused dog pancreas. J Clin Invest 63:230–238
Samols E, Marri G, Marks V (1965) Promotion of insulin secretion by glucagon. Lancet 2:415–416
Samols E, Marri G, Marks V (1966a) The interrelationship of glucagon, insulin and glucose. Diabetes 15:855–866
Samols E, Tyler J, Megyesi C, Marks V (1966b) Immunochemical glucagon in human pancreas, gut and plasma. Lancet 2:727–729
Samols E, Tyler J, Marks V, Mialhe P (1969) The physiologic role of glucagon in different species. In: Gual C, Ebling FJG (eds) Progress in endocrinology. Excerpta Medica, Amsterdam, pp 206–219
Samols E, Tyler J, Kajinuma H (1970) Influence of the sulfonamides on pancreatic humoral secretion and evidence of an insulin-glucagon feedback system. In: Rodriguez RR, Vallance-Owen I (eds) Proceedings Seventh International Diabetes Federation Congress. Excerpta Medica, Amsterdam, pp 636–655
Samols E, Tyler J, Marks V (1972) Glucagon-insulin interrelationships. In: Lefèbvre PJ, Unger RH (eds) Glucagon. Molecular physiology, clinical and therapeutic implications. Pergamon, Oxford, pp 151–173

Samols E, Weir GC, Patel YC, Loo SW, Gabbay KH (1977) Autonomic control for somatostatin and pancreatic peptide secretion by the isolated perfused canine pancreas. Clin Res 25:499A

Samols E, Weir GC, Ramseur R, Day JA, Patel YC (1978) Modulation of pancreatic somatostatin by adrenergic and cholinergic agonism and by hyper- and hypoglycemic sulfonamides. Metabolism [Suppl 1] 27:1219–1221

Samols E, Stagner JI, Weir GC (1981) Autonomic function and control of pancreatic somatostatin. Diabetologia [Suppl] 20:388–392

Sarantakis D, Teichman J, Fenichel R, Lien E (1978) [des-ALA$^{1}$, GLY$^{2}$]-HIS$^{4,5}$ D-TRP$^{8}$-somatostatin. A glucagon-specific and long acting somatostatin analog. FEBS Lett 92(2):153–155

Schatz H, Kullek U (1980) Studies on the local (paracrine) actions of glucagon, somatostatin and insulin in isolated islets of rat pancreas. FEBS Lett 122:207–310

Schauder P, McIntosh C, Arends J, Arnold R, Frerichs H, Creutzfeldt W (1976) Somatostatin and insulin release from isolated rat pancreatic islets stimulated by glucose. FEBS Lett 68:225–227

Schauder P, McIntosh C, Panten V, Arends J, Arnold R, Frerichs H, Creutzfeldt W (1978) Dynamics of somatostatin and insulin release from isolated rat pancreatic islets: evidence for intraislet interactions between B cells and D cells. Metabolism 27:1211–1214

Schusdziarra V, Dobbs RE, Harris V, Unger RH (1977) Immunoreactive somatostatin levels in plasma of normal and alloxan diabetic dogs. FEBS Lett 81:69–72

Schusdziarra V, Rouiller D, Arimura A, Unger RH (1978a) Antisomatostatin serum increases levels of hormones from the pituitary and the gut, but not the pancreas. Endocrinology 103:1956–1959

Schusdziarra V, Rouiller D, Harris V, Conlon JM, Unger RH (1978b) The response of plasma somatostatin-like immunoreactivity to nutrients in normal and alloxan diabetic dogs. Endocrinology 103:2264–2273

Schusdziarra V, Zyznar E, Rouiller D, Boden G, Brown JC, Arimura A, Unger RH (1980) Splanchnic somatostatin: a hormonal regulation of nutrient homeostasis. Science 207:530–532

Schusdziarra V, Rouiller D, Harris V, Unger RH (1981) Origin of peripheral venous hypersomatostatinemia in alloxan-diabetic dogs. Endocrinology 109:1107–1110

Sergeyeva MA (1938) Microscopic changes in the pancreatic gland of the cat produced by sympathetic and parasympathetic stimulation. Anat Rec 71:319–395

Service FJ, Nelson RL, Rubenstein AH, Go VLW (1978) Direct effect of insulin on secretion of insulin, glucagon, gastric inhibitory polypeptide, and gastrin during maintenance of normoglycemia. J Clin Endocrinol Metab 47(3):488–493

Simpson I, Rose B, Lowenstein WR (1977) Size limit of molecules permeating the junctional membrane channels. Science 195:294–296

Smith PH, Madson KL (1981) Interactions between autonomic nerves and the endocrine cells of the gastroenteropancreatic system. Diabetologia [Suppl] 20:314–324

Smith PH, Porte D (1976) Neuropharmacology of the pancreatic islets. Annu Rev Pharmacol Toxicol 16:269–285

Smith PH, Woods SC, Porte D Jr (1976) Phentolamine blocks the somatostatin-mediated inhibition of insulin secretion. Endocrinology 98:1073–1076

Smith PH, Mechant FW, Johnson DG, Fujimoto WL, Williams RH (1977) Immunocytochemical localization of a gastric inhibitory polypeptide-like material within the A-cells of the endocrine pancreas. Am J Anat 149:585–590

Solcia E, Polak JM, Pearse AGE, Forssmann WG, Larsson L-I, Sundler F, Lechago J, Grimelius L, Fujita T, Creutzfeldt W, Gepts W, Falkmer S, Lefranc G, Heitz Ph, Hage A, Buchan AMJ, Bloom SR, Grossman MI (1978) Lausanne 1977 classification of gastroenteropancreatic endocrine cells. In: Bloom SR (ed) Gut hormones. Churchill Livingstone, Edinburgh, pp 40–48

Sorenson RL, Elde RP, Seybold V (1979) Effect of norepinephrine on insulin, glucagon and somatostatin secretion in isolated perifused rat islets. Diabetes 28:899–904

Sorenson RL, Lindell DV, Elde RP (1980) Glucose stimulation of somatostatin and insulin release from the isolated perfused rat pancreas. Diabetes 29:747–751

Sperling MA, Aleck K, Voina S (1977) Suppressibility of glucagon secretion by glucose in juvenile diabetes. J Pediatr 90:543–547
Staehelin LA (1974) Structure and function of intercellular junctions. Int Rev Cytol 39:191–283
Stagner J, Samols E (1980) In vitro insulin cycles from the isolated canine pancreas – possible control by an intrinsic non-adrenergic nervous system. Diabetes [Suppl 2] 29:101
Stagner J, Samols E (1982) Disparate cyclic hormone release between the dorsal and ventral lobes of the in vitro canine pancreas. Sixty-fourth annual meeting of the endocrine society, San Francisco, June 1982, p 331
Stagner J, Samols E, Weir G (1980) Sustained oscillations of insulin, glucagon and somatostatin from the isolated canine pancreas during exposure to a constant glucose concentration. J Clin Invest 65(4):939–942
Strosser MT, Cohen L, Harvey S, Mialhe P (1980) Somatostatin stimulates glucagon secretion in ducks. Diabetologia 18:319–322
Taborsky GJ Jr (1981) Contribution of pancreatic secretion and extraction to plasma somatostatin-like immunoreactivity. Diabetes [Suppl 1] 30:1 A
Tanigawa K, Kuzuya H, Sakurai H, Seino Y, Seino S, Tsuda K, Imura H (1981) Effects of antisomatostatin and antiglucagon sera on insulin release from isolated Langerhan's islets of rats. Horm Metab Res 13:78–80
Taniguchi H, Utsumi M, Hasegawa M, Kobayashi T, Watanabe Y, Murakami K, Seki M, Tsu Tsu A, Makimura H, Sakoda M, Baba S (1977) Insulin release from rat islets treated by somatostatin antiserum. Diabetes 26:700–702
Tannenbaum GS, Epelbaum J, Colle E, Brazeau P, Martin J (1978) Antiserum to somatostatin reverses starvation-induced inhibition of growth hormone but not insulin secretion. Endocrinology 102:1909–1914
Toyota T, Abe K, Kudo M, Kimura K, Goto Y (1975) Inhibitory effects of synthetic rat C-peptide 1 on insulin secretion in the isolated perfused rat pancreas. Tohuku J Exp Med 117:79–83
Trimble ER, Renold AE (1981) Ventral and dorsal areas of rat pancreas: islet hormone content and secretion. Am J Physiol 240:E422–E427
Trimble ER, Herberg L, Renold AE (1980) Hypersecretion of pancreatic somatostatin in the obese Zucker rat: effects of food restriction and age. Diabetes 29:889–894
Trimble ER, Gerber PPG, Renold AE (1981 a) Abnormalities of pancreatic somatostatin secretion corrected by in vivo insulin treatment of streptozotocin-diabetic rats. Diabetes 30:865–867
Trimble ER, Halban PA, Wollheim CB, Renold AE (1981 b) Functional differences between rat islets of ventral and dorsal pancreatic origin: a reflection of endogenous glucagon content? Diabetes [Suppl 1] 30:14 A
Unger RH (1976) Diabetes and the alpha cell (Banting memorial lecture). Diabetes 25:136–151
Unger RH, Orci L (1977) Hypothesis: the possible role of the pancreatic D-cell in the normal and diabetic states. Diabetes 26:241–244
Unger RH, Orci L (1981 a) Diabetes: morphofunctional relationships of the A-cells. In: Unger RH, Orci L (eds) Glucagon. Elsevier, New York, pp 369–376
Unger RH, Orci L (1981 b) Glucagon and the A cell: physiology and pathophysiology. N Engl J Med 304:1518–1524, 1575–1580
Unger RH, Aguilar-Parada E, Müller WA, Eisentraut A (1970) Studies of pancreatic alpha cell function in normal and diabetic subjects. J Clin Invest 49:837–848
Unger RH, Madison LL, Müller WA (1972) Abnormal alpha cell function in diabetes: response to insulin. Diabetes 21:301–307
Wahren J, Felig P (1976) Influence of somatostatin on carbohydrate disposal and absorption in diabetes mellitus. Lancet 2:1213–1216
Wasada T, Howard B, McCorkle K, Harris V, Unger RH (1981) High plasma free fatty acid levels contribute to the hypersomatostatinemia of insulin deficiency. Diabetes 30:358–361
Weir GC, Knowlton SD, Martin DB (1974) Glucagon secretion from the perfused rat pancreas: studies with glucose and catecholamines. J Clin Invest 54:1403–1412

Weir GC, Atkins RF, Martin DB (1976a) Glucagon secretion from the perfused rat pancreas following acute and chronic streptozotocin. Metabolism 25:1519–1521
Weir GC, Knowlton SD, Atkins RF, McKennan KX, Martin DB (1976b) Glucagon secretion from the perfused pancreas of streptozotocin-treated rats. Diabetes 25:275–282
Weir GC, Samols E, Ramseur R, Day JA, Patel YC (1977) Influence of glucose and glucagon upon somatostatin secretion from the isolated perfused canine pancreas. Clin Res 25(3):403A
Weir GC, Samols E, Day JA, Patel YC (1978) Glucose and glucagon stimulate the secretion of somatostatin from the perfused canine pancreas. Metabolism [Suppl 1] 27:1223–1226
Weir GC, Samols E, Loo S, Patel YC, Gabbay KH (1979) Somatostatin and pancreatic polypeptide secretion: effects of glucagon, insulin and arginine. Diabetes 28:35–40
Weir GC, Schwarz JA, Mathe CJ (1980) Inhibition of glucagon and insulin secretion from the perfused rat pancreas by a B-cell-selective somatostatin analog. Metabolism 29(1):68–70
Weir GC, Clore ET, Zmachinski CJ, Bonner-Weir S (1981) Islet secretion in a new experimental model for non-insulin-dependent diabetes. Diabetes 30:590–595
Westermark P, Grimelius L (1973) The pancreatic islet cells in insular amyloidosis in human diabetic and non-diabetic adults. Acta Pathol Microbiol Scand [A] 81:291–300
Westfall T (1977) Local regulation of adrenergic neurotransmission. Physiol Rev 57(4):659–728
Wharton GK (1932) The blood supply of the pancreas with special reference to that of the islands of Langerhans. Anat Rec 53:55–81
Zyznar ES, Pietri AO, Harris V, Unger RH (1981) Evidence for the hormonal status of somatostatin in man. Diabetes 30:883–886

CHAPTER 32

# Pharmacologic Compounds Affecting Glucagon Secretion

A. S. LUYCKX

## A. Introduction

The data accumulated in recent years leave little doubt that excessive plasma glucagon levels contribute to the metabolic abnormalities of diabetes (reviews in UNGER 1978; LEFEBVRE and LUYCKX 1979; see also Chap. 44). On the other hand, patients with glucagonoma often display an impressive clinical and biologic picture whose origin is best confirmed by its marked and rapid improvement by inhibitors of glucagon secretion (review in LUYCKX and LEFEBVRE 1981; see also Chap. 43). These considerations imply that a search for drugs capable of reducing glucagon secretion is necessary. On the other hand, it would be of interest to know if a given compound can increase glucagon release since its administration would be potentially harmful to glucose metabolism.

The rest of this chapter is divided into ten sections which are based on the type of compounds considered and/or on their mode of action. Although it sometimes looks arbitrary, this "classification" was found useful to review a very heterogeneous series of drugs. We will essentially focus our attention on compounds which were indeed found capable of modifying glucagon secretion either in vitro or in vivo in animals or humans. However, we will occasionally mention substances which gave "negative" results if their pharmacologic actions suggest some effect that experimental investigations failed to detect. Analogs of glucagon with agonist or antagonist effects have been considered in Chap. 1 and 13.

## B. Drugs Used in the Treatment of Diabetes

Insulin, sulfonylureas, and biguanides will be considered in this section.

### I. Insulin

Insulin is probably the most physiologic and the most potent glucagon suppressor (see also Chaps. 31 and 44). Numerous observations support the concept that the A-cell belongs to the family of insulin-sensitive cells and that insulin is necessary for permitting glucose entry in the A-cell and subsequent inhibition of glucagon release. The normal pancreatic A-cell, in close contact inside the islets with B-cells, is in all probability usually exposed to high concentrations of insulin; therefore, its sensitivity to additional exogenous insulin may be difficult to demonstrate. In that respect, the A-cell of the canine gastric fundus represents an interesting model of an A-cell isolated in an organ devoid of B-cells; we have demonstrated that insulin concentrations as low as 5–10 μIU/ml exert some inhibition of glucagon release from the canine gastric A-cell (see Chap. 33). If one thus accepts the idea that insulin, when present, may permit glucose to inhibit glucagon secretion, many of the data reported on glucagon levels in diabetes can be interpreted. Among these are:

1. The hypersuppressibility of plasma glucagon after oral glucose administration reported by HATFIELD et al. (1977) in obese, hyperinsulinemic patients, contrasting with the lack of suppressibility in lean, insulin-deficient diabetic subjects.
2. The excessively high circulating levels of plasma glucagon despite extreme hyperglycemia in diabetic ketoacidosis and the prompt decrease in plasma IRG, as soon as insulin is given to these patients (ASSAN et al. 1969; MÜLLER et al. 1973).
3. The hypersuppressibility of plasma glucagon by oral or intravenous glucose in late pregnancy where hyperinsulinism is usually found (DANIEL et al. 1974; LUYCKX et al. 1975 a).

As reviewed in detail elsewhere (see Chap. 44) plasma glucagon levels are usually elevated (absolute or relative hyperglucagonaemia) in *insulin-dependent diabetics*. It is likely that, in these patients, glucagon excess may result from insulin lack. Indeed, insulin therapy usually normalizes the exaggerated glucagon response to amino acids (GERICH et al. 1975 b; RASKIN et al. 1976), reduces basal glucagon levels (GERICH et al. 1975 b; RASKIN and UNGER 1978 a, b), but fails fully to correct the glucagon suppressive effect of glucose or does it only at supraphysiologic concentrations (AYDIN et al. 1977; SEINO et al. 1978). The problem is that insulin is injected or infused peripherally so that the insular A-cells "see" insulin levels which are probably far below those observed within the islets after glucose-induced insulin secretion. Nevertheless, using the artificial B-cell system, SHICHIRI et al. (1979) demonstrated that the response of plasma glucagon to oral glucose was normalized in both nonobese, noninsulin-dependent diabetics and insulin-dependent diabetics; KAWAMORI et al. (1980) established that the exaggerated glucagon response to arginine was also perfectly normalized in both types of diabetics, thus demonstrating that insulin lack was responsible for the abnormal A-cell secretion.

## II. Sulfonylureas

The effect of sulfonylureas on A-cell function has led to conflicting results which we will briefly review. Various investigators using isolated systems have reported that sulfonylureas suppress glucagon secretion (SAMOLS et al. 1969, 1971; LAUBE et al. 1971) while others did not observe any effect (BUCHANAN et al. 1969; CHESNEY and SCHOFIELD 1969; LOUBATIERES et al. 1974; KAJINUMA et al. 1974). LOUBATIERES (et al. (1975) reported that glibenclamide had no effect on glucagon secretion when an isolated rat pancreas was perfused with a medium containing 150 mg% glucose, but that the same drug *stimulated* glucagon release in the absence of glucose. Interestingly enough, SAMOLS and HARRISON (1976 a, b) reported the opposite effect on the isolated perfused canine pancreas, namely an *inhibition* of glucagon release at low or zero glucose concentration and, on the contrary, a stimulation of glucagon secretion when the glucose concentration in the perfusate was at physiologic levels. EFENDIC et al. (1981) investigated the effect of HB 699 on glucagon release by the isolated perfused rat pancreas. This compound, 4- [2- (5- chloro-2-methoxybenzamido)-ethyl]benzoic acid, which is the left residue of glibenclamide and is devoid of the sulfonylurea moiety, suppressed basal and arginine-stimulated glucagon secretion, whereas it stimulated insulin and somatostatin release.

In anesthetized dogs, a stimulation of glucagon release was reported after sulfonylurea administration, and attributed to the hypoglycemia induced by these drugs (AGUILAR-PARADE et al. 1969). In normal humans, PEK et al. (1972) were unable to demonstrate any significant modification of basal glucagon levels and amino acid-induced glucagon secretion after intravenous tolbutamide or oral chlorpropamide administration. Similarly, MARCO and VALVERDE (1973) showed that neither glipizide nor glibenclamide have any influence on basal glucagon levels and A-cell responsiveness to arginine in normal subjects. The same results were obtained for glibenclamide by BAUMEISTER et al. (1975) who additionally reported that sulfonylurea-induced hypoglycemia was followed by a reactive hyperglucagonemia. In diabetics, LORETI et al. (1974) reported a chlorpropamide-induced lowering of the circadian variations of "total plasma glucagon-like immunoreactivity" and speculated that this change may partly explain the blood glucose lowering effect of this type of antidiabetic drug. In contrast, BOHANNON et al. (1982) recently reported that a sustained tolbutamide infusion induced a significant rise in circulating glucagon in insulin-deficient diabetic subjects. After prolonged treatment with glibenclamide (ANDREANI et al. 1973) or various sulfonylureas (SEINO et al. 1977), a reduction of the glucagon response to arginine has been reported in maturity-onset diabetics. A similar finding was reported by FALLUCCA et al. (1976, 1978) when glibenclamide was added to insulin for 1 or 6 months in insulin-dependent diabetics, TSALIKIAN et al. (1977) reported that sulfonylureas, alone or in conjunction with biguanides, reduced basal and postprandial plasma glucagon circulating levels in maturity-onset diabetics. Finally, LECOMTE et al. (1977) observed that a satisfactory control of diabetes obtained with the combination of diet and glipizide was *not* accompanied by any reduction in the circulating levels of plasma glucagon. On the contrary, significantly higher plasma glucagon concentrations were observed when control was achieved, with highest values accompanying, or following, the lowest blood glucose levels (Table 1). All these data clearly show that the glucagon response to sulfonylureas varies greatly according to the species involved, the route of administration of the drug, the duration of treatment, and the type of subject (normal or diabetic) investigated.

## III. Biguanides

Morphological studies have been done on the effects of guanidine derivatives on the A-cells of the islets of Langerhans. DAVID (1952) and CREUTZFELDT and MOENSCH (1958) observed degranulation and hydropic degeneration of A-cells in animals following administration of high doses of synthaline A and phenformin. Similarly, MUNGER (1962) noted degranulation and hydropic degeneration of A-cells in rabbits after administration of synthaline A. In animal studies, LANGSLOW

1 PODOLSKY and LAWRENCE (1978) found that, in maturity-onset diabetics, chlorpropamide did not significantly modify the paradoxical glucagon rise after an oral glucose load

**Table 1.** Plasma glucagon concentrations (pg/ml) in nine patients treated by diet alone (I), diet plus placebo (II), and diet plus glipizide (III) (LECOMTE et al. 1977)

| Time (h) | 8 | 10 | 12 | 14 | 18 | 20 | 22 | $\sum$ Glucagon |
|---|---|---|---|---|---|---|---|---|
| I. Diet alone | $64 \pm 14$ | $83 \pm 18$ | $69 \pm 2$ | $114 \pm 31$ | $93 \pm 22$ | $90 \pm 19$ | $78 \pm 21$ | $579 \pm 125$ |
| II. Diet plus placebo | $71 \pm 22$ | $76 \pm 16$ | $59 \pm 18$ | $87 \pm 21$ | $83 \pm 19$ | $75 \pm 18$ | $80 \pm 15$ | $523 \pm 131$ |
| III. Diet plus glipizide | $95 \pm 28$ | $108 \pm 24$ | $109 \pm 28$ | $147 \pm 29^{b}$ | $143 \pm 35^{b}$ | $120 \pm 27^{a}$ | $114 \pm 25^{b}$ | $836 \pm 213$ |
| Paired comparison III versus II | N.S. | N.S. | $P < 0.02$ | $P < 0.05$ | $P < 0.05$ | $P < 0.02$ | N.S. | $P < 0.01$ |

Results are expressed as mean ± standard error
[a] $P < 0.02$ and [b] $P < 0.01$ by paired comparison versus 8 h values
N.S. = not significant

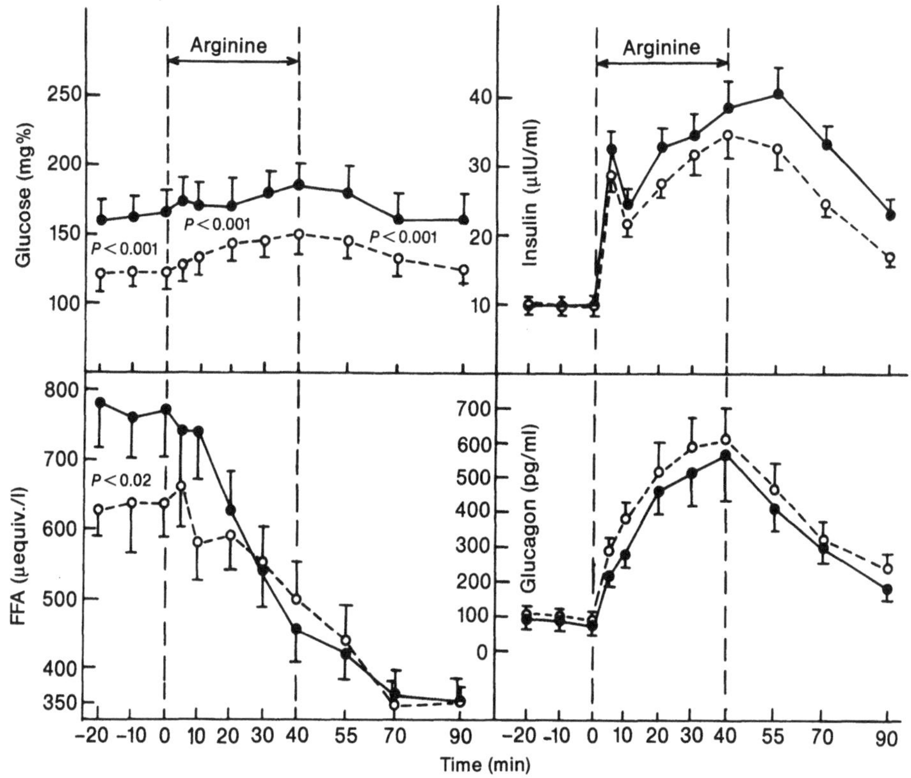


**Fig. 1.** Blood glucose, plasma free fatty acids (FFA), insulin, and glucagon (antiserum 30K) response to a 40-min arginine infusion (11.7 mg $kg^{-1}$ $min^{-1}$ of L-arginine monochlorhydrate) in eleven type II diabetics. The subjects were tested after placebo treatment (*full circles*) and also after treatment with metformin (*three times* 500 mg/day) for 15 days (*open circles*). CARPENTIER et al. (1975)

et al. (1973) demonstrated a rise in glucagon levels in fed chickens after synthalin A administration; this effect was not observed in fasted chickens. In 1974, BERGER et al. found that the administration of metformin enhanced arginine-stimulated glucagon secretion in nondiabetic subjects. We investigated the possible influence of metformin (500 mg, three times a day) on the arginine-induced glucagon response in 11 noninsulin-dependent diabetics. As illustrated in Fig. 1, the reduction in blood glucose and plasma free fatty acid levels was not accompanied by any significant change in insulin or glucagon plasma concentrations during the test (CARPENTIER et al. 1975). In a group of ten obese patients with mild to moderate glucose intolerance, a 14-day administration of butylbiguanide did not modify the plasma glucagon concentrations after an overnight fast and during an oral glucose load (LEFEBVRE et al. 1978). It seems therefore reasonable to conclude that biguanides have little or no effect on pancreatic A-cell function.

## C. Drugs Related to the Cholinergic System

Several studies have established the importance of the cholinergic system in the control of glucagon secretion (ALRIC et al. 1972; IVERSEN 1973 b). In calves, BLOOM

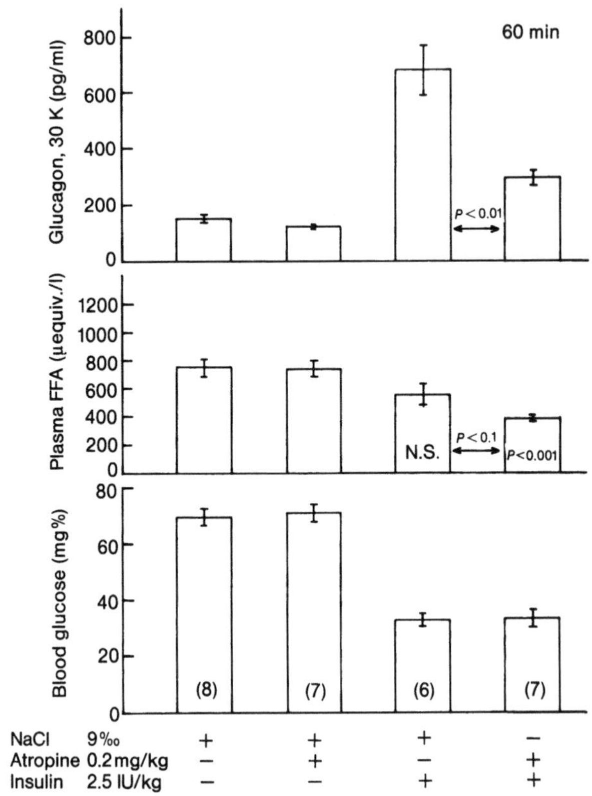


**Fig. 2.** Effect of the intraperitoneal injection of atropine, insulin, and atropine plus insulin on blood glucose, plasma FFA, and glucagon in overnight fasted rats. The control group received saline. All animals were killed 60 min after the injection. The height of each column corresponds to the mean ± standard error, the number of animals is indicated inside the lower column. LUYCKX and LEFEBVRE (1976a)

et al. (1974a) demonstrated that atropine lowered fasting plasma glucagon levels, substantially delayed the glucagon response to insulin-induced hypoglycemia, and almost completely blocked the glucagon rise evoked by the infusion of 2-deoxy-D-glucose (BLOOM and POLAK 1980). In rats, we found that administration of atropine sulfate (0.2 mg/kg body weight) decreased basal plasma glucagon concentrations and reduced by about 50% the stimulatory action of insulin-induced hypoglycemia on glucagon secretion (Fig. 2), but did not alter exercise-induced glucagon release (LUYCKX and LEFEBVRE 1976a). In humans, the effects of atropine as well as those of truncal versus selective vagotomy were studied in details by BLOOM et al. (1974b): in normal volunteers, intravenous administration of atropine (15 μg/kg) produced a significant fall in plasma glucagon levels and reduced by 33% the rise in plasma glucagon induced by arginine infusion. Thus, in various species and experimental conditions, atropine appears as an unequivocal inhibitor of glucagon release (see also Chap. 30).

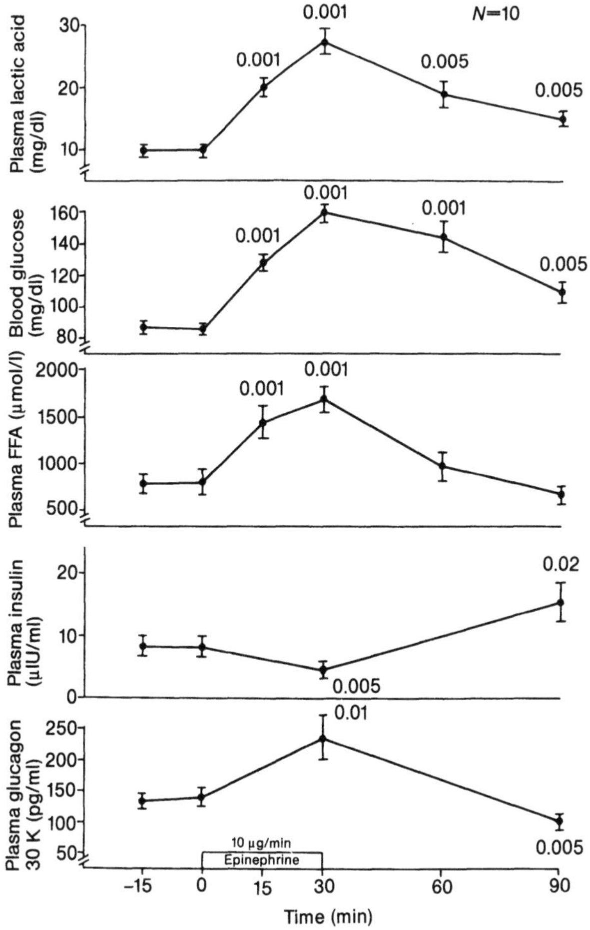


**Fig. 3.** Influence of an epinephrine intravenous infusion (10 μg/min) on plasma lactic acid, blood glucose, plasma FFA, insulin, and glucagon concentrations in ten overnight fasted normal subjects. Results are expressed are mean ± standard error. The statistical comparison (P value) corresponds to the paired comparison with the corresponding value at time zero. SCHEEN et al. (1982)

## D. Drugs Related to the Sympathetic Nervous System

Morphological investigations have demonstrated the extensive ramifications of adrenergic nerve endings in pancreatic islets of subprimate species and numerous physiologic experiments support the concept of a modulation of glucagon release by the sympathetic nervous system (see Chap. 30). Infusion of catecholamines and their agonists have been shown to increase glucagon secretion in vitro (LECLERCQ-MEYER et al. 1971; IVERSEN 1973a) and in vivo in various species including humans (GERICH et al. 1973; SCHADE and EATON 1978; see also Fig. 3). Thus, any compound affecting the sympathetic nervous system could, at least from a theoretical point of view, modify glucagon secretion. The purpose of this section is to review the data collected from in vivo studies in animals and humans.

**Table 2.** Effect of (+)- and (−)-propranolol on forced-swim-induced rise in plasma FFA and glucagon (LUYCKX et al. 1975)

| | Resting controls | | 60-min swim | | |
|---|---|---|---|---|---|
| | Saline ($N=8$) | | (+)-Propranolol 5 mg/kg ($N=8$) | | (−)-Propranolol 5 mg/kg ($N=6$) |
| Blood glucose (mg/100 ml) | 82.5 ± 2.3 | | 73.9 ± 2.5 | | 87.4 ± 4.6 |
| Free fatty acids (μequiv./l) | 590 ± 45 | $P<0.001$ | 1.338 ± 133 | $P<0.05$ | 981 ± 47 |
| Insulin (μIU/ml) | 22.5 ± 2.2 | | 24.8 ± 1.9 | | 22.4 ± 2.1 |
| Glucagon (pg/ml) | 131.8 ± 8.3 | $P<0.001$ | 291.6 ± 25.6 | $P<0.01$ | 170.0 ± 15.2 |

Results are expressed as mean ± standard error ($N$ = number of animals in the series)
$P$ = level of significance (Student's $t$-test)

## I. Reserpine

Glucagon secretion was studied in rats 24 h after intraperitoneal administration of reserpine at a dose of 10 mg/kg. Pretreatment with reserpine significantly increased basal plasma glucagon, but did not modify the glucagon rise induced either by insulin-induced hypoglycemia or by muscular exercise (LUYCKX and LEFEBVRE 1976a). Using a similar protocol, but a lower dose of reserpine (5 mg/kg), JACOBY and BRYCE (1978) did not observe any change in basal plasma glucagon.

## II. Beta-Adrenergic Blocking Agents

In 1974, we reported that, in rats, the exercise-induced rise in glucagon secretion was completely abolished by previous intraperitoneal administration of propranolol at a dose of 5 mg/kg whereas practolol (12.5 mg/kg) was devoid of any effect. We concluded that $\beta_2$-adrenoceptors of the A-cells were involved in glucagon secretion under these conditions (LUYCKX and LEFEBVRE 1974). Later on, we extended these studies and demonstrated that (+)-propranolol, in contrast with (−)-propranolol, did not alter exercise-induced glucagon secretion (Table 2). Since both isomers of the drug are equipotent as membrane-stabilizing agents whereas (−)-propranolol is about 100 times more effective than (+)-propranolol as a $\beta$-adrenergic blocker, we concluded that the effect of propranolol was indeed due to its $\beta$-blocking effect. Pindolol, another $\beta$-blocker was also found to inhibit exercise-induced glucagon release when injected at doses of 0.1 or 0.05 mg/kg (Tables 3 and 4; LUYCKX et al. 1975b).

In agreement with our results, LINDSEY and FALOONA (1973) reported that shock-induced hyperglucagonemia was also blocked by propranolol in dogs. On the contrary, HARVEY et al. (1974) reported that propranolol potentiated glucagon rise in rats submitted to very intensive exercise, which was indeed associated with a pronounced hypoglycemia. Under these conditions, it seems likely that glucagon rise is controlled not only by sympathetic activity, but also by blood glucose

**Table 3.** Effect of pindolol (0.1 mg/kg) on resting and exercised rats (LUYCKX et al. 1975)

| | Saline | | | | Pindolol (0.1 mg/kg) |
|---|---|---|---|---|---|
| | Resting controls | | | | 60-min swim |
| | ($N$=8) | | ($N$=8) | | ($N$=8) |
| Blood glucose (mg/100 ml) | 72.4 ± 2.2 | $P<0.001$ | 101.8 ± 2.5 | | 93.7 ± 2.9 |
| Free fatty acids (μequiv./l) | 771 ± 74 | $P<0.02$ | 1.093 ± 82 | $P<0.01$ | 1.565 ± 94 |
| Insulin (μIU/ml) | 7.1 ± 1.4 | $P<0.001$ | 23.5 ± 2.9 | | 20.7 ± 3.8 |
| Glucagon (pg/ml) | 90.0 ± 5.8 | | 78.8 ± 5.6 | | 81.2 ± 12.3 |

Results are expressed as mean ± standard error ($N$ = number of animals in each series)
$P$ = level of significance (Student's $t$-test)

**Table 4.** Effect of pindolol (0.05 mg/kg) on forced-swim-induced rise in plasma free fatty acids and glucagon (LUYCKX et al. 1975)

| | Saline | | Pindolol (0.05 mg/kg) | |
|---|---|---|---|---|
| | Resting controls ($N$=7) | | 60-min swim | |
| | | | ($N$=8) | ($N$=7) |
| Blood glucose (mg/100 ml) | 85.5 ± 3.4 | | 92.1 ± 4.7 | 98.5 ± 6.0 |
| Free fatty acids (μequiv./l) | 763 ± 74 | $P<0.05$ | 1,055 ± 94 | 1,180 ± 78 |
| Insulin (μIU/ml) | 16.5 ± 1.2 | | 17.7 ± 1.4 | 17.0 ± 2.3 |
| Glucagon (pg/ml) | 100.2 ± 16.6 | $P<0.02$ | 194.4 ± 24.3 | 147.8 ± 14.2 |

Results are expressed as mean ± standard error ($N$ = number of animals in each series)
$P$ = level of significance (Student's $t$-test)

changes. We have previously demonstrated that, in rats, propranolol did not prevent hypoglycemia-induced glucagon secretion which was, on the contrary, exaggerated, probably because of the more marked decrease in blood glucose (LUYCKX and LEFEBVRE 1974).

The inhibitory effect of propranolol on exercise-induced glucagon release, which is well demonstrated in rats and dogs, is not observed in humans. Indeed, GALBO et al. (1976) demonstrated that pretreatment with intravenous propranolol (0.15 mg/kg) did not diminish the glucagon response in humans exercised at 60% of their individual maximum oxygen uptake and we have reported that, in humans, the exercise-induced glucagon secretion was probably essentially mediated through the decrease in glucose availibility to the A-cell (LUYCKX et al. 1978a). Similarly, in healthy volunteers, infusion of propranolol failed to alter the pattern of glucagon release in response to 84 h of fast or to insulin-induced hypoglycemia (WALTER et al. 1974). Finally, ATKINSON et al. (1981) investigated the influence of propranolol and of phentolamine on epinephrine-induced glucagon rise in massively obese subjects. This rise was reduced by phentolamine and completely blocked by propranolol.

In summary, many investigations support the concept that, in animals, glucagon secretion is stimulated by β-adrenoceptor stimulants (IVERSEN 1973a; KANETO et al. 1975) and inhibited by β-adrenergic blockers (LUYCKX et al. 1975b), but the importance of these control mechanisms in humans seems doubtful (WALTER et al. 1974; GALBO et al. 1976). For further discussion see Chap. 30.

## III. Alpha-Adrenergic Blocking Agents

Several reports indicated that phentolamine (an α-adrenergic blocking agent) increased glucagon secretion in various experimental conditions (IVERSEN 1973b). For instance, GERICH et al. (1974a) demonstrated a highly significant increase in plasma glucagon concentrations in human subjects during phentolamine infusion and EATON et al. (1972) found that phentolamine infusion enhanced the glucagon response to infused alanine in conscious dogs. Similarly, in rats, we found that intraperitoneal injection of 25 mg/kg phentolamine led to a slight increase in plasma glucagon (LUYCKX and LEFEBVRE 1974). More recently, using an innervated, cross-perfused, canine pancreas-stomach-duodenum preparation, MILLER and HORTON (1979) reported that the immunoreactive glucagon secretion rate evoked by splanchnic nerve stimulation was enhanced by phentolamine. This series of data suggest that the glucagon-secreting cells may be tonically inhibited by α-adrenergic influences. Other studies do not fit with this concept. IVERSEN (1973a) did not find any change in the glucagon rise subsequent to catecholamine infusion when he added phentolamine to the perfusate and HARVEY et al. (1974) found that the increased glucagon concentrations subsequent to exercise in rats were blocked by phentolamine. The discrepancies between the results obtained with phentolamine administration could be due to the fact that this compound blocks not only the postsynaptic, $\alpha_1$-adrenoceptors, but also the presynaptic, $\alpha_2$-adrenoceptors, the latter effect increasing norepinephrine release by the sympathetic nerve endings.

## IV. Clonidine

Clonidine is an antihypertensive drug which mainly activates central $\alpha_2$-adrenoceptors and thereby reduces sympathetic nervous system activity and suppresses plasma catecholamine levels. Under certain conditions, however, clonidine behaves as a peripheral $\alpha_1$-adrenergic stimulator. This latter mechanism probably explains the fall in plasma insulin reported after a single dose of 0.5 mg clonidine given to normal subjects. *No effect* on plasma glucagon was detected in this study (METZ et al. 1978).

## V. L-Dopa, Dopamine, and Bromocriptine

Studies in rhesus monkeys (GEORGE and RAYFIELD 1974) and in humans (RAYFIELD et al. 1975) have demonstrated that the administration of a catecholamine precursor, L-dopa, increases plasma glucose, insulin, and glucagon levels. The authors speculated that these effects might be mediated through the conversion of L-dopa to dopamine and subsequent stimulation of dopaminergic receptors of the islet cells. The possibility of a central dopaminergic pathway for glucagon stimulation was tested by LORENZI et al. (1977): in six normal subjects, L-dopa (500 mg orally) induced a rise in plasma glucagon whereas apomorphine (0.6 mg subcutaneously) did not affect glucagon secretion in control subjects nor in juvenile-onset, insulin-dependent diabetics. Since apomorphine is considered to interact specifically with postsynaptic dopamine receptors, its inability to stimulate glucagon release argues in favor of the absence of dopamine-sensitive steps in the pathway of glucagon stimulation (LORENZI et al. 1977). The lack of stimulation of glucagon after administration of piribedil, another specific dopamine receptor stimulant (CORRODI et al. 1971; JENNER et al. 1973; THORNER et al. 1976) confirms this view.

Since both α- and β-adrenergic blockers are ineffective in modifying the glucagon response to L-dopa, it is unlikely that norepinephrine mediates the glucagon rise induced by this compound (GEORGE and RAYFIELD 1974). In fact, intravenous infusion of dopamine (4 μg $kg^{-1}$ $min^{-1}$), in normal subjects, elicited a simultaneous rise in plasma insulin and glucagon. These results suggest that glucagon release in response to L-dopa in humans is mediated through the stimulation of dopaminergic receptors in the A-cells of the pancreas via conversion of L-dopa to dopamine (LEBLANC et al. 1977).

Bromocriptine, a long-acting dopamine agonist, was tested for its ability to alter basal and arginine-stimulated glucagon release in acromegalic subjects (FEDELE et al. 1980). In agreement with previous results in a smaller series of subjects (THORNER et al. 1975), bromocriptine had no effect on glucagon plasma levels.

## VI. Diazoxide and Tolmesoxide

Conflicting results have been published regarding the influence of diazoxide on glucagon secretion. SAMOLS et al. (1972) reported that intravenous diazoxide (2.1–2.5 mg $kg^{-1}$ $min^{-1}$) causes hyperglucagonemia in ducks. Further studies, however showed a *suppressive* effect of diazoxide on glucagon secretion in vitro (SAMOLS and HARRISON 1975) and in vivo (ALTSZULER et al. 1977). The results of SAMOLS and HARRISON were obtained with the isolated perfused dog pancreas and those of ALTSZULER et al. with intravenous administration of the compound (16.5 mg/kg given in 10 min), also in the dog.

A detailed investigation of the effect of diazoxide (325 μmol/l) on glucagon release by the isolated perfused rat pancreas was published by URDANIVIA et al. (1979). Basal glucagon secretion was not modified by the drug while the glucagon secretion in response to L-arginine, L-leucine, and prostaglandin $F_{2\alpha}$ ($PGF_{2\alpha}$) was reduced. The results obtained in vitro are in sharp contrast with those reported in vivo. Indeed, EATON and SCHADE (1980), who administered diazoxide by stomach tube in the rat (30 mg/kg), reported a highly significant rise in plasma catecholamine and glucagon concentrations. The increase in glucagon reached 115% of the basal level and was statistically significant 30 and 60 min after the administration of diazoxide. We did not observe systematic changes in plasma glucagon after oral administration of 300 mg diazoxide in normal or in obese subjects (LUYCKX and LEFEBVRE 1972).

We investigated the influence of another vasodilating compound, tolmesoxide, on glucagon plasma levels in hypertensive subjects. After intravenous infusion of tolmesoxide at doses ranging between 0.5 and 3.5 mg/kg, we found a systematic rise in circulating plasma glucagon. It is not known whether this represents a direct effect of tolmesoxide on A-cells or merely an indirect effect mediated through the activation of the sympathetic system resulting from drug-induced hemodynamic changes (SCHEEN et al. 1981).

## E. Serotonin and Serotonin Antagonists

Several studies support the concept of a serotoninergic mechanism in the control of A-cell function. Whereas tryptophan increased glucagon secretion from isolated mouse pancreatic islets, serotonin itself (4 mmol/l) consistantly inhibited glucagon output, both in the basal state and in the presence of arginine (MARCO et al. 1977); 5-hydroxy-tryptophan had no effect on basal glucagon release, but reduced the effect of arginine. MARCO et al. (1977) suggested that the inhibition of glucagon secretion by 5-hydroxy-tryptophan could be attributed to its conversion to serotonin, a phenomenon which was indeed shown to take place in that species (EKHOLM et al. 1971).

These in vitro results offer a valuable interpretation for the results previously reported in humans by the same authors (MARCO et al. 1976b), who demonstrated in healthy volunteers, that treatment with two known antagonists of serotonin (cyproheptatine and methysergide) and one agent which depletes tissue serotonin by blocking its synthesis (*p*-chlorophenylalanine) potentiated the glucagon responses to arginine infusion and to insulin-induced hypoglycemia. The doses utilized were 16 mg daily for 2 days for cyproheptadine (Periactin), 9 mg daily for 2 days for methysergide (Deseril) and 2 g daily for 4 days for *p*-chlorophenylalanine (Fenclomine). In contrast with these results, JACOBY and BRYCE (1978), using a protocol based on intraperitoneal injection of exogenous serotonin (5–25 mg/kg) in overnight fasted rats, reported a stimulatory effect on glucagon release, which was blocked by the serotonin antagonist, methysergide, and partially reduced following α-adrenergic blockade with phentolamine. Recent studies using isolated rat islets incubated in vitro concluded that glucagon secretion was stimulated by 5-hydroxytryptophan but *unaffected* by serotonin (FURMAN et al. 1981).

Thus, it is likely that A-cells, like other members of the APUD system, are subject to a serotoninergic control mechanism. However, its precise function remains unclear because differences between the various species studies, between in vitro and in vivo results, and between direct and indirect effects preclude any overall conclusion which would encompass the data so far published.

## F. Drugs Affecting Ionic Concentrations and/or Fluxes

It is beyound the scope of this section to review in detail the numerous studies devoted to the role of calcium, sodium, phosphate, and bicarbonate on glucagon secretion (see Chap. 26). Nevertheless, a certain number of these data will be briefly summarized insofar as they contribute to the understanding of the effect of various pharmacologic agents considered here.

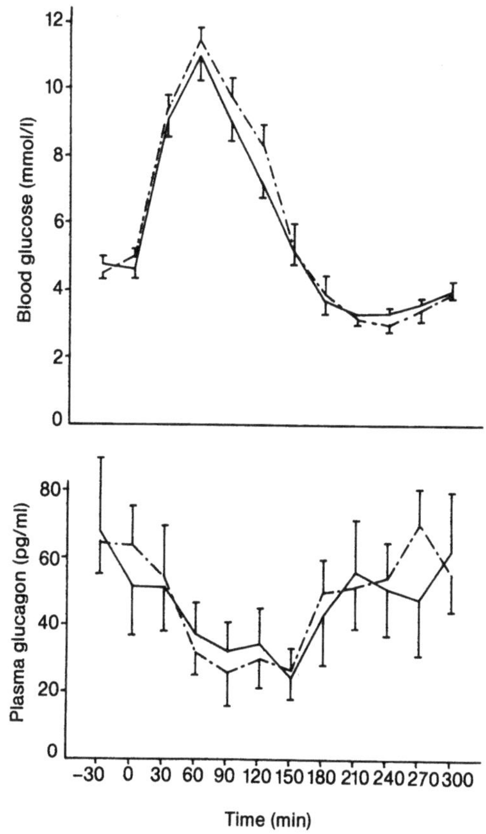


**Fig. 4.** Lack of effect of a calcium gluconate infusion (0.22 mmol/ml; infusion rates: 0.20 mmol/min from −30 to 0 min and 0.04 mmol/min from 0 to 300 min) on plasma glucagon levels during an oral glucose tolerance test (75 g glucose given at time zero). The patients were selected on the basis of a blood glucose level above 8.7, 7.7, and 6.6 mmol/l, respectively at 60, 90, and 120 min of the test. Ten subjects were infused with calcium gluconate (*full line*) and with physiologic saline (*broken line*) in a randomized order. Results are expressed as mean ± standard error. VEXIAU et al. (1982)

## I. Calcium, Calcitonin, and Vitamin D

As has been summarized by LECLERCQ-MEYER et al. (1977) and in Chap. 26, calcium has at least three different roles in the secretory activity of the pancreatic A-cell:

a. A cellular accumulation of calcium may mediate the glucagonotropic action of arginine (CAMPILLO et al. 1978)
b. A sufficient amount of extracellular calcium seems a prerequisite for the identification of a high concentration of glucose by the A-cell (LECLERCQ-MEYER et al. 1975)
c. During prolonged exposure to a low glucose level, the accumulation of intracellular calcium may exert a feedback inhibitory effect upon the secretory process.

It is therefore not surprising that pharmacologic compounds affecting either calcium concentration or fluxes across the plasma membrane influence glucagon secretion. In healthy subjects, calcium infusion (15 mg/kg in 120 min) was found ineffective in altering basal circulating glucagon levels, but it lowered plasma glucagon levels stimulated by arginine or insulin-induced hypoglycemia (STARKE et al. 1980). In a group of patients with glucose intolerance and reactive hypoglycemia, we found that calcium infusion did not modify plasma glucagon under basal conditions and after oral glucose administration (Fig. 4). Similar studies were conducted in insulin-dependent diabetic subjects. STARKE et al. (1981) reported that infusion of calcium gluconate reduced basal plasma glucagon by about 25% in these subjects. Since the same effect was observed during porcine or salmon calcitonin infusion (4.5 IU/kg in 120 min) in the absence of significant change in circulating calcium concentration, the authors concluded that calcitonin (endogenous or exogenous) depressed elevated circulating glucagon levels in insulin-dependent diabetics; similar results were obtained in healthy volunteers.

These in vivo results contrast with the lack of effect of calcitonin when added to the perfusate of the isolated rat pancreas (TAMARIT-RODRIGUEZ et al. 1978). On the contrary, the suppressive effect of calcium administration on glucagon secretion in vivo is in agreement with previous studies in dogs (OHNEDA et al. 1974; TAMARIT and MAYOR 1978).

The effect of vitamin D deficiency on insulin and glucagon release was determined using the isolated perfused rat pancreas originating from vitamin D-deficient or repleted animals. While vitamin D-deficient rats exhibited a marked decrease in insulin secretion, their pancreatic glucagon secretion remained normal (NORMAN et al. 1980).

## II. Verapamil and Procaine

The fact that calcium is involved in the process of glucagon secretion prompted us to test the influence of verapamil, which has calcium antagonistic properties (DEVIS et al. 1975), and of procaine hydrochloride, which inhibits the binding and facilitates the release of calcium by phospholipid membranes (HOPE-GILL et al. 1974).

When injected at a dose of 0.5 mg/kg in overnight fasted rats, verapamil hydrochloride failed to influence glucagon plasma concentrations either in the basal state or during stimulation by muscular exercise or insulin-induced hypoglycemia (LUYCKX and LEFEBVRE 1976a). Contrasting with these in vivo negative results, verapamil (10–20 μmol/l) inhibited the arginine-induced glucagon release as well as the glucagon output increase in response to a sudden lowering of the glucose concentration. These results obtained with the isolated perfused rat pancreas by LECLERCQ-MEYER et al. (1978) demonstrate that the verapamil-induced blockade of inward calcium transport resulted, at least in certain conditions, in a reduced glucagon secretion.

In intact rats, administration of procaine hydrochloride (50 mg/kg) reduced basal glucagon plasma concentrations, but did not prevent exercise-induced or hypoglycemia-induced glucagon release (LUYCKX and LEFEBVRE 1976a).

### III. Veratridine, Ouabain, and Hydroquinidine

KAWAZU et al. (1981) investigated the influence of the sodium ionophore, veratridine (10, 25, or 50 μmol/l) on glucagon secretion using monolayer cultures of newborn rat pancreas. In the presence of a "normal" calcium concentration (1.3 mmol/l), veratridine caused a dose-dependent enhancement of glucagon secretion. This effect was attenuated or lost when extracellular calcium was omitted and was also dependent on intracellular calcium, thus suggesting that elevation of intracellular sodium by veratridine could act via mobilization of calcium from intracellular storage sites, as has been suggested for insulin-secreting cells (review in KAWAZU et al. 1981). The $NA^+$, $K^+$-ATPase inhibitor, ouabain was also tested by the same authors. Ouabain stimulated glucagon release at high glucose concentration (20 mmol/l), but was devoid of effect in the absence of glucose (KAWAZU et al. 1981). These in vitro results support the existence of veratridine-sensitive sodium channels in A-cells.

Quinidine and its derivatives are supposed to reduce outward transport of sodium across the cell membrane (VAUGHAN-WILLIAMS 1970). In vivo in rats, administration of dihydroquinidine gluconate (5 mg/kg) did not modify the secretion of glucagon in response to muscular exercise or insulin-induced hypoglycemia (LUYCKX and LEFEBVRE 1976a).

## G. Drugs Affecting Lipid Metabolism

EATON et al. (1974, 1976) have suggested that some forms of acquired and genetic endogenous hyperlipemia (type IV) "may be maintained by a reduction in the lipid-lowering actions of glucagon relative to the lipid-elevating actions of insulin" (see Chaps. 17 and 46). It is thus interesting to mention that several drugs, including nicotinic acid, clofibrate, and halofenate, used in the treatment of these disorders, can alter either the glucagon levels or the ratio of circulating insulin and glucagon concentrations.

Nicotinic acid, which lowers plasma free fatty acids, increases plasma glucagon in dogs, rats, and humans (see review in Chap. 25). On the other hand, a gross and sustained rise in fasting plasma glucagon has been reported in hyperlipidemic patients treated with β-piridylcarbinol, a nicotinic acid derivative (MARKS et al. 1971; MARKS 1973).

Glucagon secretion after arginine infusion was studied by TIENGO et al. (1975, 1978) in a group of patients with primary endogenous hypertriglyceridemia before and after a 2–3-months period of clofibrate therapy. The arginine-induced glucagon rise, which was already elevated in hypertriglyceridemic patients when compared with normal controls, was further increased after administration of clofibrate. This observation was in agreement with the previously reported data of EATON and SCHADE (1974) showing a potentiation of arginine-stimulated glucagon secretion together with a reduction in insulin secretion in humans. Finally, FERY and BALASSE (1978) found that the glucagon response to a protein meal was enhanced, but not to a significant degree after 2 day clofibrate for 8 days in normal or mildly hypertriglyceridemic subjects. In this case, as well as in the previous studies (EATON and SCHADE 1974; TIENGO et al. 1975), the most striking feature was the reduction of insulin secretion associated with clofibrate administration, resulting in a fall in the ratio of insulin to glucagon circulating concentrations. Experimental studies were conducted in Zucker rats with genetic endogenous hyperlipemia before and after treatment with another hypolipemic drug, halofenate; before treatment, arginine injection was associated with a reduction in plasma glucagon concentrations, while a striking rise was observed in the halofenate-treated animals (EATON et al. 1976). The participation of changes in the insulin:glucagon ratio in the lipid-lowering effect of drugs is discussed in Chap. 46.

## H. Drugs Acting on the Central Nervous System

### I. Diphenylhydantoin, Diazepam, and Haloperidol

Diphenylhydantoin and diazepam, two potent anticonvulsivant agents, have been shown to inhibit arginine-induced glucagon secretion from the isolated perfused pancreas. Confirming earlier results by GERICH et al. (1972), PETRACK et al. (1976a) reported that diphenylhydantoin (0.01–0.05 mmol/l) markedly suppresses the second phase of glucagon release by the isolated perfused rat pancreas in response to 20 mmol/l arginine, leaving the first phase release relatively unaffected. On the contrary, when injected intraperitoneally at a dose of 100 mg/kg into normal hamsters, diphenylhydantoin induced a marked increase in blood glucose accompanied by a significant decrease in plasma insulin and a significant 40% increase in plasma glucagon (TAMBURRANO et al. 1976).

Although the chemical structures of diphenylhydantoin and diazepam appear unrelated, their three-dimensional conformations are closely related and, moreover, according to PETRACK et al. (1976b), might be similar to the apparent active site of somatostatin. Accordingly, these authors reported that diazepam, like diphenylhydantoin, inhibited arginine-induced glucagon release by the perfused rat pancreas in a dose-dependent manner, with a 50% inhibitory concentration ($IC_{50}$) of approximately 65 μmol/l. However, the concentrations of the drug which are active in vitro are more than ten fold greater than the blood levels reached after chronic administration of massive doses of the drug. HERMANSEN (1978) investigated the effect of haloperidol, a dopaminergic antagonist, on insulin and glucagon secretion, using the isolated perfused canine pancreas and reported that this compound inhibited basal and stimulated glucagon secretion. However, it was later reported by the same author (HERMANSEN 1980) that the 1% ethanol used for dissolving haloperidol might be responsible for the observed inhibition of glucagon secretion. This was indeed demonstrated by SAMOLS and STAGNER (1980).

### II. Morphine, Endorphins, and Enkephalins

Endorphins and enkephalins are widely distributed in the nervous system and gastrointestinal tract of animals and endorphin-like substances have been demonstrated, by immunocytochemical techniques, in the A-cells of the islet of Langerhans of the rat (GRUBE et al. 1978). IPP et al. (1978) have shown that morphine and β-endorphin enhance glucagon release from the isolated perfused dog pancreas and REID and YEN (1981) reported that β-endorphin stimulated the secretion of insulin and glucagon in humans. Using pancreatic islet cell cultures, KANTER et al. (1980) found that morphine (20 µmol/l) significantly enhanced glucagon secretion while enkephalin (1.2 µmol/l) inhibited arginine-induced glucagon secretion. In this system, the effects of both enkephalins and morphine were blocked by naloxone (20 µmol/l) which by itself had no effect. KANTER et al. (1980) concluded that enkephalin and morphin had "disparate" effects upon glucagon secretion, but seemed capable of directly influencing secretion from islet cells. It is nevertheless conceivable that opiates also influence somatostatin secretion from D-cells and thereby, by paracrine mechanisms, other islet cells (see Chap. 31). In humans, intravenous infusion of a long-acting enkephalin did not alter plasma glucagon levels (STUBBS et al. 1978). Work is still in progress to elucidate the role of morphine and related compounds on glucagon release: IPP et al. (1980) recently reported that an intravenous bolus injection of 0.5 mg/kg morphine increased plasma glucagon in normal as well as in alloxan-induced diabetic dogs. TOYOTA et al. (1980) demonstrated with the isolated perfused rat pancreas that Met-Enkephalin suppressed the acetylcholine-induced glucagon secretion and IPP et al. (1981) reported that subcutaneous injection of morphine stimulated glucagon secretion in the dog, an effect which was not blocked by pretreatment with propranolol.

## J. Hormonal Steroids

This section gives a brief summary of the effects of glucocorticoids and contraceptive steroids on glucagon secretion. A more detailed discusion is given in Chap. 29.

### I. Glucocorticoids

The influence of glucocorticoids on basal and arginine-stimulated glucagon secretion in normal humans was studied by MARCO et al. (1973). Acute intravenous injection of prednisolone (100 mg) failed to alter plasma glucagon values; on the contrary, oral administration of prednisolone (40 mg daily for 4 days) increased fasting glucagon levels as well as arginine-stimulated glucagon secretion. The same authors reported an enhanced glucagon secretion from prednisolone-treated mice (MARCO et al. 1976a).

### II. Contraceptive Steroids

In normal women, BECK et al. (1975) have observed that glucagon secretion in response to intravenous arginine was blunted during treatment with mestranol plus norethindrone. A comparable suppression of aminogenic glucagon secretion was observed in normal women during treatment with ethinylestradiol alone, but not with norethindrone alone. In insulin-dependent diabetic women, after 2 weeks treatment with combined mestranol (80 µg) plus norethindrone (1 mg) daily, the mean peak glucagon response to arginine infusion was suppressed to one-quarter of control levels (BECK et al. 1976).

The reduction in circulating glucagon after an oral glucose load was studied in three groups of normal women before and after a 6 month's treatment with three recently developped oral contraceptives. The triphasic administration of ethynyl oestradiol and levonorgestrel (Logynon, Trigynon, Trinordiol) had no influence on plasma glucagon. On the contrary, the two associations containing desogestrel as progestagen (Ovidol and Marvelon) moderately increased plasma glucagon in the basal state as well as after glucose administration the latter change being statistically significant at the 90th and 120th min of the OGTT. These alterations were not associated with a significant alteration of glucose tolerance (LUYCKX A, GASPARD U, ROMUS M, LEFEBVRE P 1983, unpublished results).

## K. Drugs Affecting the Mitotic Spindle

Microtubules and microfilaments are involved in the process of emiocytosis of the α-granules by the pancreatic A-cells and several studies have been devoted to the influence of drugs known to interfere with tubulin. The anti-tubulin agents studied were essentially vinblastine, vincristine, and colchicine.

Using isolated guinea pig islets incubated in vitro. EDWARDS and HOWELL (1973) reported that vinblastine and colchicine increased basal and arginine-stimulated glucagon secretion and LECLERCQ-MEYER et al. (1974), using incubated pieces of pancreas, also demonstrated that these two drugs potentiated arginine-induced glucagon release. Further studies by ASSAN et al. (1978) demonstrated that the situation was probably more complex since the effect of these agents was dependent upon their concentration and also upon the time of exposure of the isolated perfused rat pancreas, a 40-min stimulation with arginine (25 mmol/l) being selected as a standardized stimulus. Colchicine, $10^{-4}$ mol/l, enhanced the early phase of glucagon release after a short exposure time (35 min), but with longer exposure, the potentiation vanished and was replaced by an inhibition of glucagon release affecting both phases of secretion, but more the second phase than the first. For a constant exposure time of 45 min, a potentiating effect was noticed in the concentration range $10^{-7}$–$10^{-6}$ mol/l, whereas values in the range $10^{-5}$–$10^{-3}$ mol/l were inhibitory. These authors obtained similar results with vincristine and concluded that a limited disturbance of the microtubule system was associated with facilitated glucagon release, but that an extended disruption of the microtubular apparatus reduced both phases of glucagon secretion. Cytochalasin-B (usually 10 μg/ml), which exerts its effect on the cell web, facilitates basal and arginine-stimulated glucagon release (WOLLHEIM et al. 1973; LECLERCQ-MEYER et al. 1974; ASSAN et al. 1978). We investigated the influence of vinblastine on blood glucose, plasma insulin, and glucagon concentrations during an arginine infusion test in patients treated for Hodgkin's disease. The tests were performed 120 min and 15 days after intravenous injection of 10 mg vinblastine; we found that vinblastine administration was followed by a 35% decrease in the insulin response and by a very slight increase in the glucagon response (J.L. CARPENTIER and A.S. LUYCKX 1973, unpublished work).

## L. Somatostatin and Somatostatin Analogs

As reviewed by UNGER (1978) somatostatin is the only glucagon suppressant thus far tested in human diabetics and remarkable improvement in the control of hyperglycemia has been observed when a constant intravenous infusion of somatostatin was added to insulin therapy in juvenile diabetics (GERICH et al. 1974b, 1975a, 1976; GERICH 1978; UNGER and RASKIN 1978; RASKIN and UNGER 1978).

The usefulness of somatostatin itself is limited by the large number of effects exerted by this peptide, including inhibition of insulin and growth hormone secretion, inhibitory effects on the gastrointestinal system, influence on platelets (MIELKE et al. 1975; KOERKER et al. 1975), and possibly on hemodynamic parameters (ROSENTHAL et al. 1976, 1977, 1978) as well as by the short duration of action of somatostatin (SS-14).

Considerable effort has been devoted to the design and synthesis of analogs of somatostatin possessing increased selectivity towards glucagon secretion and improved pharmacokinetic characteristics. This was worthwhile since DOBBS et al. (1977) had shown that somatostatin analogs administered subcutaneously to severely alloxan-induced diabetic dogs as a supplement to suboptimal insulin therapy reduced both their hyperglycemia and hypertriglyceridemia. Owing to the very rapid developments in this field, the data given here under are inevitably incomplete.

GRANT et al. (1976) replaced the cystine bridges of somatostatin and prepared two peptides (I and II) with the sequences 1–10 (Wy-19, 840) and 1–11 (Wy-40, 056), respectively. Neither altered the release of glucagon (tests in rats, in vivo, arginine stimulation) at dose levels 30–60 times higher than that needed for somatostatin. Three other analogs [D-Cys$^{14}$]-, [Ala$^{2}$, D-Cys$^{14}$]-, and [D-Trp$^{8}$, D-Cys$^{14}$]-somatostatin were studied by MEYERS et al. (1977). These peptides sharply suppressed the release of growth hormone in vitro and glucagon in vivo, but had lesser effects on insulin secretion in vivo. BROWN et al. (1977) compared the effects of [D-Cys$^{14}$]-, [D-Trp$^{8}$]- and [D-Trp$^{8}$, D-Cys$^{14}$]somatostatin analogs. [D-Cys$^{14}$]somatostatin was found more potent in inhibiting glucagon and growth hormone secretion than it was in inhibiting insulin secretion. [D-Trp$^{8}$]somatostatin, was 8–10 times more potent than SS-14 in inhibiting all three hormones secretion and [D-Trp$^{8}$, D-Cys$^{14}$]somatostatin was more potent that [D-Cys$^{14}$]somatostatin and retained its relative selectivity for inhibiting the secretion of glucagon and growth hormone.

SCHUSDZIARRA et al. (1978) compared the effects of these three analogs in fasted, insulin deprived, alloxan-induced diabetic dogs. When injected subcutaneously, [D-Trp$^{8}$]somatostatin and [D-Trp$^{8}$, D-Cys$^{14}$]somatostatin had a more prolonged action in inhibiting glucagon secretion than unmodified somatostatin. More recently, MÄRKI et al. (1979) reexamined in unanesthetized rats, the "glucagon selectivity" of the [D-Cys$^{14}$]- and [D-Trp$^{8}$, D-Cys$^{14}$]somatostatin analog and concluded that, although they show partial *kinetic* dissociation of effects on glucagon and insulin, they were not truly selective inhibitors of glucagon release.

The clinical evaluation of the efficacy of the [D-Trp$^{8}$, D-Cys$^{14}$]somatostatin analog (also called GHRIH-A supplied by Serono SPA, Rome, Italy) as a glucagon suppressant was rather disappointing: in normal humans, RAPTIS et al. (1978a) reported, during an arginine test, a marked decrease in the growth hormone response, a modest although significant reduction of glucagon values, and no changes in insulin secretion. The lack of glucagon-suppressive effect in insulin-dependent diabetics before and after meals (RAPTIS et al. 1978b; LENTI et al. 1979) favors the view that the human A-cell is less suppressible by this somatostatin analog than the rat's or dog's A-cell. Indeed, in acromegalics as well as in mild diabetics, no significant changes in plasma pancreatic glucagon were noted by Trovati et al. (1980), after bolus intravenous injection of 25 μg of the compound followed by 25 μg/h over 60–120 min. Furthermore, the same dosage did not enhance the glucose-induced suppression of glucagon release in healthy humans (Bratusch-Marrain et al. 1981). Considerable interest has been focused on naturally occurring somatostatin peptides and particularly on SS-28 which was found six times more active in blocking insulin than glucagon release (Coy et al. 1981). In healthy volunteers and in acromegalics, synthetic SS-28 was found five times more potent than SS-14 regarding inhibition of the three hormones, growth hormone, insulin, and glucagon. This was not confirmed by Rodriguez-Arnao et al. (1981) who compared, in normal subjects, the relative potency of SS-14 and SS-28 (called here Pro-SS) on the growth hormone, insulin, and glucagon responses to an intravenous arginine infusion. Infused at equimolar doses (2 μg/min SS-14 and 4 μg/min SS-28) these authors concluded that SS-28 appeared more potent than SS-14 in inhibiting growth hormone and insulin responses and seemed to have a more prolonged effect. Both peptides were found equipotent in inhibiting the glucagon response to arginine.

Recently, Sarantakis et al. (1981) reported that the dipeptide Ala-Gly at positions 1 and 2 was not involved in the biologic activity and that the specificity towards glucagon inhibition was connected with a basic amino acid at position 4 (Lys, Arg, His) and diverse amino acid substitutions at position 5 (His, Glu, Tyr); all these analogs had D-Trp at position 8. Very promising results were presented in abstract from by Hirschman (1981) about a low molecular weight, orally active, long-acting peptide capable of reducing glucagon release in alloxan-induced diabetic dogs.

Experience with somatostatin analogs in the treatment of the glucagonoma syndrome is still limited. A remarkable decrease in plasma levels of glucagon was found during infusion or [D-Trp$^8$, D-Cys$^{14}$]somatostatin at doses of 0.5–8.0 μg/min (intravenous infusion). This decrease was associated with severe hypoglycemia. [desAA$^{1,2,4,5,12,13}$, D-Trp$^8$, D–Cys$^{14}$] somatostatin at doses of 5 and 10 mg subcutaneously greatly suppressed plasma glucagon for more than 12 h with concomitant reduction of the blood glucose level (Bloom et al. 1980).

## References


Aguilar-Parada E, Eisentraut AM, Unger RH (1969) Effect of HB419 (glibenclamide)-induced hypoglycemia on pancreatic glucagon secretion. Horm Metab Res [Suppl] 1:48–50

Alric R, Loubatières-Mariani MM, Loubatières AL, Puech R (1972) Récepteurs cholinergiques et sécrétion de glucagon. Etude sur le pancréas isolé et perfusé du rat. C R Soc Biol (Paris) 166:1030–1033

Altszuler N, Hampshire J, Moraru E (1977) On the mechanism of diazoxide-induced hyperglycemia. Diabetes 26:931–935

Andreani D, Fallucca F, Iavicoli M, De Pirro R, Gambardella S, Menzinger G (1973) Comportamento dell'insulina (IRI), dell'ormone della crescita (GH) e del glucagone (IRG) plasmatici in un gruppo di soggetti diabetici dopo trattamento con glibenclamide e glibenclamide associata a fenformina. In: Atti giornate di diabetologia del Mediterraneo, Rodi, 1–3 Nov. 1973, p 252

Assan R, Hautecouverture G, Guillemant S, Dauchy F, Protin P, Derot M (1969) Evolution de paramètres hormonaux (glucagon, cortisol et hormone somatotrope) et énergétiques (glucose, acides gras libres, glycérol) dans dix acidocétoses diabétiques graves traitées. Pathol Biol (Paris) 17:1095–1105

Assan R, Soufflet E, Ballerio G, Attali JR, Boillot J, Girard JR (1978) Ambiguous effects of colchicine and vincristine upon $A_2$-cell response to arginine. Diabetologia 14:121–127
Atkinson RL, Dahms WT, Bray GA, Sperling MA (1981) Adrenergic modulation of glucagon and insulin secretion in obese and lean humans. Horm Metab Res 13:249–253
Aydin I, Raskin P, Unger RH (1977) The effect of short-term intravenous insulin administration on the glucagon response to a carbohydrate meal in adult-onset and juvenile-type diabetes. Diabetologia 13:629–636
Baumeister G, Wagner H, Stahl M (1975) Klinisch experimentelle Untersuchungen zur Beeinflussung der pankreatischen Glukagon Inkretion durch Tolbutamid und Glibenclamid. Klin Wochenschr 53:571–574
Beck P, Eaton RP, Arnett DM, Alsever RN (1975) Effect of contraceptive steroids on arginine-stimulated glucagon and insulin secretion in women. I. Lipid physiology. Metabolism 24:1055–1056
Beck P, Arnett DM, Alsever RN, Eaton RP (1976) Effect of contraceptive steroids on arginine-stimulated glucagon and insulin secretion in women. II. Carbohydrate and lipid physiology in insulin-dependent diabetics. Metabolism 25:23–31
Berger W, Stahl M, Ohnhaus E, Goschke H (1974) Pancreatic glucagon, plasma insulin and blood glucose responses to arginine infusion in non-diabetic subjects following biguanide pretreatment. Horm Metab Res 6:165
Bloom SR, Polak HM (1980) Control of pancreatic hormone release by islet innervation. In: Andreani D, Lefèbvre PJ, Marks V (eds) Current views on hypoglycemia and glucagon. Academic Press, London New York, pp 37–46
Bloom SR, Edwards AV, Vaughan NJA (1974a) The role of the autonomic innervation in the control of glucagon release during hypoglycemia in the calf. J Physiol (Lond) 236:611–623
Bloom SR, Vaughan NJA, Russell RCG (1974b) Vagal control of glucagon release in man. Lancet 2:546–549
Bloom SR, Adrian TE, Mallinson CN, Polak JM, Brown MR, Vale W, Rivier JE (1980) The glucagonoma syndrome and the effect of a new long acting subcutaneous somatostatin. In: Andreani D, Lefèbvre PJ, Marks V (eds) Current views on hypoglycemia and glucagon. Academic Press, London New York, pp 127–135
Bohannon NV, Lorenzi M, Grodsky GM, Karam JH (1982) Stimulatory effects of tolbutamide infusion on plasma glucagon in insulin-dependent diabetic subjects. J Clin Endocrinol Metab 54:459–462
Bratusch-Marrain P, Waldhaüsl W, Nowotny P (1981) Impairment by D-Trp$^8$-Cys$^{14}$-somatostatin of intravenous glucose tolerance in healthy man. In: Abstracts of the 2nd international symposium on somatostatin, Athens, June 1–3. 1981. Abstract book, Serono symposia, p 17
Brown M, Rivier J, Vale W (1977) Somatostatin: analogs with selected biological activities. Science 196:1467–1468
Buchanan KD, Vance JE, Williams RH (1969) Insulin and glucagon release from isolated islets of Langerhans. Effect of enteric factors. Diabetes 18:381–386
Campillo JE, Luyckx AS, Torres MD, Lefèbvre PJ (1978) Effect of various concentrations of calcium on arginine-induced insulin and glucagon release by the isolated perfused rat pancreas. Rev Esp Fisiol 34:191–198
Carpentier JL, Luyckx AS, Lefèbvre PJ (1975) Influence of metformin on arginine-induced glucagon secretion in human diabetes. Diabete Metab 1:23–28
Chesney TM, Schofield JG (1969) Studies on the secretion of pancreatic glucagon. Diabetes 18:627–632
Corrodi H, Fuxe K, Ungerstedt U (1971) Evidence for a new type of dopamine receptor stimulating agent. J Pharm Pharmacol 23:989–991
Coy DH, Murphy WA, Meyers CA, Fries JL (1981) Recent developments in the search for selective and more active somatostatin peptides. In: Abstracts of the 2nd international symposium on somatostatin, Athens, June 1–3, 1981. Abstract book, Serono symposia, p 26
Creutzfeldt W, Moensch A (1958) Vergleichende Untersuchungen mit den blutzuckersenkenden Quinidinderivaten Synthalin B und Phenyläthylbiguanid (DBI); ein Beitrag zur Frage der sog. A-Zellgifte. Endokrinologie 36:167

Daniel RR, Metzger BE, Freinkel N, Faloona GR, Unger RH, Nitzan M (1974) Carbohydrate metabolism in pregnancy. XI. Response of plasma glucagon to overnight fast and oral glucose during normal pregnancy and in gestational diabetes. Diabetes 23:771–776
David JC (1952) Hydropic degeneration of the α cell of the pancreatic islets produced by synthalin A. J Pathol 64:575–584
Devis G, Somers G, Van Obberghen E, Malaisse WJ (1975) Calcium antagonists and islet function. I. Inhibition of insulin release by verapamil. Diabetes 24:547–551
Dobbs R, Schusdziarra V, Rivier J, Brown N (1977) Somatostatin analogs as glucagon suppressants in diabetes (Abstr). Diabetes [Suppl 1] 26:360
Eaton RP, Schade DS (1974) Effect of clofibrate on arginine stimulated glucagon and insulin secretion in man. Metabolism 23:445–454
Eaton RP, Schade DS (1980) The effect of diazoxide-induced hormonal secretion on plasma triglyceride concentration in the rat. Diabetologia 18:301–306
Eaton RP, Conway M, Buckman M (1972) Role of α-adrenergic blockade on alanine-induced hyperglucagonemia. Metabolism 21:371–373
Eaton RP, Schade DS, Conway M (1974) Hypothesis: decreased glucagon activity: a mechanism for both genetic and acquired hyperlipemia. Lancet 2:1545–1548
Eaton RP, Oase R, Schade DS (1976) Altered insulin and glucagon secretion in treated genetic hyperlipemia: a mechanism of therapy? Metabolism 25:245–249
Edwards JC, Howell SL (1973) Effects of vinblastine and colchicine on the secretion of glucagon from isolated guinea-pig islets of Langerhans. FEBS Lett 30:89–92
Efendic S, Enzmann F, Gutniak M, Nylen A, Zoltobrocki M (1981) The effect of HB 699 (acyl-amino-alcyl benzoic acid) on the release of islet hormones. In: Abstracts of the 2nd international symposium on somatostatin, Athens, June 1–3, 1981. Abstract book, Serono symposia, p 37
Ekholm R, Ericson LE, Lundquist I (1971) Monoamines in the pancreatic islets of the mouse. Diabetologia 7:339–348
Fallucca F, Iavicoli M, Menzinger G, Gambardella S, Andreani D (1976) Arginine stimulated glucagon and growth hormone secretion in insulin-treated diabetic patients: effect of added sulphonylureas (Abstr). Diabetologia 12:389
Fallucca F, Iavicoli M, Menzinger G, Mirabella C, Andreani D (1978) Glucagon and growth hormone secretion in insulin-treated diabetics: effects of added sulfonylureas. Metabolism 27:5–11
Fedele D, Molinari M, Meneghel A, Valerio A, Muggeo M, Tiengo A (1980) Bromocriptine acute effect on insulin, glucagon and growth hormone levels in acromegalic patients. J Endocrinol Invest 3:149–153
Fery F, Balasse EO (1978) Influence of short-term administration of clofibrate on insulin and glucagon response to protein ingestion in man. Arch Int Pharmacodyn 235:341–350
Furman BL, Itoh M, Gerich JE (1981) Effect of 5-hydroxytryptamine (5HT) and its precursor 5HTP on insulin and glucagon secretion in vitro. (Abstr). Diabetologia 20:669
Galbo H, Holst JJ, Christensen NJ, Hilsted J (1976) Glucagon and plasma catecholamines during beta-receptor blockade in exercising man. J Appl Physiol 40(6):855–863
George DT, Rayfield EJ (1974) L-Dopa induced plasma glucagon release. J Clin Endocrinol Metab 39:618–621
Gerich JE (1978) On the causes and consequences of abnormal glucagon secretion in human diabetes mellitus. In: Foà PP, Bajaj JS, Foà NL (eds) Glucagon: its role in physiology and clinical medicine. Springer, Berlin Heidelberg New York, pp 617–641
Gerich JE, Charles MA, Levin SR, Forsham PH, Grodsky GM (1972) In vivo inhibition of pancreatic glucagon secretion by diphenylhydantoin. J Clin Endocrinol Metab 35:823–824
Gerich JE, Karam JH, Forsham PH (1973) Stimulation of glucagon secretion by epinephrine in man. J Clin Endocrinol Metab 37:479–481
Gerich JE, Langlois M, Noacco C, Schneider V, Forsham PH (1974a) Adrenergic modulation of pancreatic glucagon secretion in man. J Clin Invest 53:1441–1446
Gerich JE, Lorenzi M, Schneider V, Karam JH, Rivier J, Guillemin R, Forsham PH (1974b) Effects of somatostatin on plasma glucose and glucagon levels in human diabetes mellitus. N Engl J Med 291:544–547

Gerich JE, Lorenzi M, Bier D, Schneider V, Tsalikian E, Karam J, Forsham P (1975a) Prevention of human diabetic ketoacidosis by somatostatin: evidence for an essential role of glucagon. N Engl J Med 292:985–989
Gerich JE, Tsalikian E, Lorenzi M, Schneider V, Bohannon N, Gustafson G, Karam J (1975b) Normalization of fasting hyperglucagonemia and excessive glucagon responses to intravenous arginine in human diabetes mellitus by prolonged infusion of insulin. J Clin Endocrinol Metab 41:1178–1180
Gerich JE, Lorenzi M, Bier DM, Tsalikian E, Schneider V, Karam JH, Forsham PH (1976) Effects of physiologic levels of glucagon and growth hormone on human carbohydrate and lipid metabolism. Studies involving administration of exogenous hormone during suppression of endogenous hormone secretion with somatostatin. J Clin Invest 57:875–884
Grant N, Clark D, Garsky V, Jaunakais I, Mc Gregor W, Sarantakis D (1976) Dissociation of somatostatin effects. Peptides inhibiting the release of growth hormone but not glucagon or insulin in rats. Life Sci 19:629–632
Grube D, Voight KH, Weber E (1978) Pancreatic glucagon cells contain endorphin-like immunoreactivity. Histochemistry 59:75–79
Harvey WD, Faloona GR, Unger RH (1974) The effect of adrenergic blockade on exercise-induced hyperglucagonemia. Endocrinology 94:1254–1258
Hatfield HH, Banasiak MF, Driscoll T, Kim HJ, Kalkhoff RK (1977) Glucose suppression of glucagon: relationship to pancreatic beta cell function? J Clin Endocrinol Metab 44:1080–1087
Hirschman R (1981) Design, synthesis and biological evaluation of an orally active, long acting analog of somatostatin. In: Abstracts of the 2nd international symposium on somatostatin, Athens, June 1–3, 1981. Abstract book, Serono symposia, p 68
Hermansen K (1978) Haloperidol, a dopaminergic antagonist: somatostatin-like inhibition of glucagon and insulin release from the isolated, perfused canine pancreas. Diabetologia 15: 343–347
Hermansen K (1980) Haloperidol, and insulin and glucagon secretion. Diabetologia 19:84
Hope-Gill H, Vydelingum N, Kissebah AH, Tulloch BR, Fraser TR (1974) Stimulation and enhancement of the adipose tissue insulin response by procaine hydrochloride: evidence for a role of calcium in insulin action. Horm Metab Res 6:457–463
Ipp E, Dobbs R, Unger RH (1978) Morphine and beta-endorphine influence the secretion of the endocrine pancreas. Nature 276:190–191
Ipp E, Schusdziarra V, Harris V, Unger RH (1980) Morphine-induced hyperglycemia: role of insulin and glucagon. Endocrinology 107:461–463
Ipp E, Moosa AR, Masabay A, Pugh W (1981) Morphine-induced hyperglycemia: role of insulin and glucagon (Abstr). Diabetes [Suppl 1] 30:325
Iversen J (1973a) Adrenergic receptors and the secretion of glucagon and insulin from isolated perfused canine pancreas. J Clin Invest 52:2102–2116
Iversen J (1973b) Effect of acetylcholine on the secretion of glucagon and insulin from the isolated perfused canine pancreas. Diabetes 22:381
Jacoby JH, Bryce GF (1978) The acute pharmacologic effects of serotonin on the release of insulin and glucagon in the intact rat. Arch Int Pharmacodyn 235:254–270
Jenner P, Taylor AR, Campbell DB (1973) Preliminary investigation of the metabolism of piribedil (ET 495); a new central dopaminergic agonist and potential antiparkinson agent. J Pharm Pharmacol 25:749–750
Kajinuma H, Kuyuta T, Ide T (1974) Effects of hypoglycemic sulfonamides on glucagon and insulin secretion in ducks and dogs. Diabetes 23:412–417
Kaneto A, Miki E, Kosaka K (1975) Effect of $beta_1$ and $beta_2$ adrenoreceptor stimulants infused intrapancreatically on glucagon and insulin secretion. Endocrinology 97:1166–1173
Kanter RA, Ensinck JW, Fujimoto WY (1980) Disparate effects of enkephalin and morphin upon insulin and glucagon secretion by islet cell cultures. Diabetes 29:84–86
Kawamori R, Shichiri M, Kikuchi M, Yamasaki Y, Abe H (1980) Perfect normalization of excessive glucagon responses to intravenous arginine in human diabetes with the artificial beta-cell. Diabetes 29:762–765

Kawazu S, Ikeuchi M, Kikuchi M, Kanazawa Y, Fujimoto WY, Kosaka K (1981) Dual effects of veratridine on glucagon and insulin secretion. Dependence upon extracellular and intracellular calcium. Diabetes 30:446–450

Koeker JD, Harker LA, Goodner CJ (1975) Effects of somatostatin on hemostasis in baboons. N Engl J Med 293:476–479

Langslow DR, Freeman BM, Buchanan KD (1973) The contrasting response of fed and starved chickens to synthalin A (Abstr). In: Abstracts of the meeting of the British Diabetic Association

Laube JR, Fussgänger R, Goberna R, Schröder K, Straub K, Sussman K, Pfeiffer EF (1971) Effects of tolbutamide on insulin and glucagon secretion of the isolated perfused rat pancreas. Horm Metab Res 3:238–242

Leblanc H, Lachelin GCL, Abu-Fadil S, Yen SSC (1977) The effect of dopamine infusion on insulin and glucagon secretion in man. J Clin Endocrinol Metab 44:196–198

Leclercq-Meyer V, Brisson R, Malaisse WJ (1971) Effect of adrenaline and glucose on release of glucagon and insulin in vitro. Nature New Biol 231:248–249

Leclercq-Meyer V, Marchand J, Malaisse WJ (1974) Possible role of microtubular-microfilamentous system in glucagon secretion. Diabetologia 10:215–224

Leclercq-Meyer V, Rebolledo O, Marchand J, Malaisse WJ (1975) Glucagon release: paradoxical stimulation by glucose during calcium deprivation. Science 189:897–899

Leclercq-Meyer V, Marchand J, Malaisse WJ (1977) The versatile role of calcium in glucagon release. In: Foà PP, Bajaj JS, Foà NL (eds) Glucagon: its role in physiology and clinical medicine. Springer, Berlin Heidelberg New York, pp 185–213

Leclercq-Meyer V, Marchand J, Malaisse WJ (1978) The role of calcium in glucagon release. Studies with verapamil. Diabetes 27:996–1004

Lecomte MJ, Luyckx AS, Lefèbvre PJ (1977) Plasma glucagon and clinical control of maturity-onset type diabetes. Effects of diet, placebo and glipizide. Diab Metab 3:239–243

Lefèbvre PJ, Luyckx AS (1979) Glucagon and diabetes: a reappraisal. Diabetologia 16:347–354

Lefèbvre PJ, Luyckx AS, Mosora F, Lacroix M, Pirnay F (1978) Oxidation of an exogenous glucose load using naturally labelled $^{13}$C-glucose. Effect of butylbiguanide therapy in obese mildly diabetic subject. Diabetologia 14:39–45

Lenti G, Pagano G, Trovati M, Lorenzati R (1979) The use of a somatostatin analogue in insulin-dependent and insulin-independent diabetics controlled by artificial pancreas. Excerpta Med Int Congr Ser 148:138

Lindsey CA, Faloona GR (1973) Adrenergic blockade in shock-induced hyperglucagonemia (Abstr). Diabetes [Suppl 1] 22:51

Lorenzi M, Tsalikian E, Bohannon NV, Gerich JE, Karam JH, Forsham PH (1977) Differential effects of L-Dopa and apomorphine on glucagon secretion in man: evidence against central dopaminergic stimulation of glucagon. J Clin Endocrinol Metab 45:1154–1158

Loreti L, Sugase T, Foà PP (1974) Diurnal variations of serum insulin, total glucagon, cortisol, glucose and free fatty acids in normal and diabetic subjects before and after treatment with chlorpropamide. Horm Res 5:278–292

Loubatières AL, Loubatières-Mariani MM, Alric R, Ribes G (1974) Tolbutamide and glucagon secretion. Diabetologia 10:271–276

Loubatières AL, Loubatières-Mariani MM, Chapal J, Blayac JP (1975) Action du glibenclamide sur les sécrétions de glucagon et d'insuline étudiées sur le pancréas isolé et perfusé du rat. C R Soc Biol (Paris) 169:1568–1571

Luyckx AS, Lefèbvre PJ (1972) Effets métaboliques et hormonaux du diazoxide chez le sujet normal, chez quatre patients porteurs d'une tumeur pancréatique insulino-sécrétante et dans un cas d'hypoglycémie réactionnelle. In: Austoni M, Scandellari C, Trisotto A, Federspil G (eds) Hypoglycaemia and diazoxide. Proceedings of the international symposium on hypoglycaemia and diazoxide, Venice, 15–16 October 1971. Cedam, Padova, pp 207–224

Luyckx AS, Lefèbvre PJ (1974) Mechanisms involved in the exercise-induced increase in glucagon secretion in rats. Diabetes 23:81–93

Luyckx AS, Lefèbvre PJ (1976a) Pharmacological compounds affecting plasma glucagon levels in rats. Biochem Pharmacol 25:2703–2708

Luyckx AS, Lefèbvre PJ (1981) Les glucagonomes. Diab Metab (Paris) 7:289–300
Luyckx AS, Gérard J, Gaspard U, Lefèbvre P (1975a) Plasma glucagon levels in normal women during pregnancy. Diabetologia 11:549–554
Luyckx AS, Dresse A, Cession-Fossion A, Lefèbvre PJ (1975b) Catecholamines and exercise-induced glucagon and fatty acid mobilization in the rat. Am J Physiol 229(2):376–383
Luyckx AS, Pirnay F, Lefèbvre P (1978a) Effect of glucose on plasma glucagon and free fatty acids during prolonged exercise. Eur J Appl Physiol 39:53–61
Märki F, Kamber B, Rink H, Sieber P (1979) Non-selective inhibition of basal glucagon release by [D-$Cys^{14}$]-analogues of somatostatin in the rat. J Endocrinol 81:315–323
Marco J, Valverde I (1973) Unaltered glucagon secretion after seven days of sulphonylurea administration in normal subjects Diabetologia Suppl. 9:317–319
Marco J, Calle C, Roman D, Diaz-Fierros M, Villanueva ML, Valverde I (1973) Hyperglucagonism induced by glucocorticoid treatment in man. N Engl J Med 288:128–131
Marco J, Calle C, Hedo JA, Villanueva ML (1976a) Enhanced glucagon secretion by pancreatic islets from prednisolone-treated mice. Diabetologia 12:307–311
Marco J, Hedo JA, Martinell J, Calle C, Villanueva ML (1976b) Potentiation of glucagon secretion by serotonin antagonists in man. J Clin Endocrinol Metab 42:215–211
Marco J, Hedo JA, Villanueva ML (1977) Inhibition of glucagon release by serotonin in mouse pancreatic islets. Diabetologia 13:585–588
Marks V (1973) Glucagon and lipid metabolism in man. Postgrad Med J 49:615–619
Marks V, Frizel D, Twycross RG, Buchanan KD (1971) Effect of pyridylcarbinol on glucose tolerance, plasma glucagon, insulin and growth hormone in man. In: Gey KF, Carlson LA (eds) Metabolic effects of nicotinic acid and its derivatives. Huber, Bern, pp 961–976
Metz S, Halter JB, Roberston RP (1978) Induction of defective insulin secretion and impaired glucose tolerance by clonidine. Diabetes 27:554–562
Meyers C, Arimura A, Gordin A, Fernandez-Durango R, Coy DH, Schally AJ, Drouin J, Ferland L, Beaulieu M, Labrie F (1977) Somatostatin analogs which inhibit glucagon and growth hormone more than insulin release. Biochem Biophys Res Comm 74:630–636
Mielke CH, Gerich JE, Lorenzi M, Tsalikian E, Rodvien R, Forsham PH (1975) The effect of somatostatin on coagulation and platelet function in man. N Engl J Med 293:480–483
Miller RE, Horton ES (1979) Neural release of glucagon is inhibited by hyperglycemia and enhanced by phentolamine. Diabetes 28:762–768
Müller W, Faloona G, Unger RH (1973) Hyperglucagonemia in diabetic ketoacidosis: its prevalence and significance. Am J Med 54:52–57
Munger BL (1962) The secretory cycle of the pancreatic islet alpha cell. Lab Invest 11:885–901
Norman AW, Frankel BJ, Heldt AM, Grodsky GM (1980) Vitamin D deficiency inhibits pancreatic secretion of insulin. Science 209:823–825
Ohneda A, Matsuda K, Horigome K, Ishii S, Yamagata S (1974) Effect of intrapancreatic administration of calcium upon glucagon secretion in dogs. Tohoku J Exp Med 113:301–311
Pek S, Fajans JS, Floyd JC Jr, Knopf RF, Conn JW (1972) Failure of sulfonylureas to suppress plasma glucagon in man. Diabetes 21:216–223
Petrack B, Czernick AJ, Itterly W, Ansell J, Chertock H (1976a) On the suppression of insulin and glucagon release by diphenylhydantoin (Abstr). Diabetes [Suppl 1] 25:380
Petrack B, Czernik AJ, Itterly W, Ansell J, Chertock H (1976b) Diazepam: in vitro effects on glucagon and insulin release. Biochem Biophys Res Commun 73:934–939
Podolsky S, Lawrence AM (1978) Effect of long term sulfonylurea therapy on glucagon levels in maturity onset diabetes mellitus. In: Foà PP, Bajaj JS, Foà NL (eds) Glucagon: its role in physiology and clinical medicine. Springer, Berlin Heidelberg New York, pp 711–721
Raptis S, Zoupas CH, Maier V, Beischer W, Rosenthal J (1978a) Effect of a somatostatin analog on glucoregulatory hormones in man. Acta Endocrinol [Suppl] (Copenh) 87:60
Raptis S, Zoupas CH, Mallas E, Moulopoulos S (1978b) Effect of a new somatostatin analog in insulin dependent diabetics (Abstr). Diabetologia 15:264

Raskin P, Unger RH (1978a) Effect of insulin therapy on the profiles of plasma immunoreactive glucagon in juvenile-type and adulttype diabetics. Diabetes 27:411–419
Raskin P, Unger RH (1978b) Hyperglucagonemia and its suppression: importance in the metabolic control of diabetes. N Engl J Med 299:433–436
Raskin P, Aydin I, Unger RH (1976) The effect of insulin on the exaggerated glucagon response to arginine stimulation in diabetes mellitus. Diabetes 25:227–229
Rayfield EJ, George DT, Eichner HM, Hsu HT (1975) L-dopa stimulation of glucagon secretion in man. N Engl J Med 293:589–591
Reid RL, Yen SSC (1981) Beta-endorphin stimulates the secretion of insulin and glucagon in humans. J Clin Endocrinol Metab 52:592–594
Rodriguez-Arnao MD, Gomez-Pan A, Rainbow SJ, Woodhead S, Owens DB, Wass JAH, Besser GM, Coy DH, Comaru-Schally AM, Schally AV, Hall R (1981) Prosomatostatin (Pro-S-S) effects on anterior pituitary and pancreatic hormones in man. Comparison with somatostatin (S-S). In: Abstracts of the 2nd international symposium on somatostatin, Athens, June 1–3, 1981. Abstract book, Serono symposia, p 146
Rosenthal J, Raptis S, Escobar-Jimenez F, Pfeiffer EF (1976) Inhibition of furosemide-induced hyperreninemia by growth hormone release-inhibiting hormone in man. Lancet 1:772–774
Rosenthal J, Escobar-Jimenez F, Raptis S (1977) Prevention by somatostatin of rise in blood pressure and plasma renin mediated by beta receptor stimulation. Clin Endocrinol (Oxf) 6:455–461
Rosenthal J, Raptis S, Zoupas C, Escobar-Jimenez F (1978) Hemodynamic and renin responses to somatostatin in essential hypertension. Metabolism 27:1361–1363
Samols E, Harrison J (1975) Inhibition of glucagon secretion by diazoxide. Clin Res 23:14A
Samols E, Harrison J (1976a) Intraislet negative insulin-glucagon feedback. Metabolism [Suppl 1] 25:1443–1447
Samols E, Harrison J (1976b) Tolbutamide stimulates glucagon secretion. In: Abstracts of the Vth international congress of endocrinology, Hamburg, June 18–24, 1976, p 343
Samols E, Stagner JI (1980) Reinterpretation of the effect of Haloperidol and ethanol on insulin secretion. Diabetologia 19:81–83
Samols E, Tyler JM, Miahle P (1969) Suppression of pancreatic glucagon release by the hypoglycemic sulphonylureas. Lancet 1:174–176
Samols E, Tyler JM, Kajinuma H (1971) Influence of the sulfonamides on pancreatic humoral secretion and evidence for an insulin-glucagon feedback system. Excerpta Med Int Congr Ser 23:636–655
Samols E, Tyler JM, Marks V (1972) Glucagon-insulin interrelationships. In: Lefèbvre PJ, Unger RH (eds) Glucagon: molecular physiology, clinical and therapeutic implications. Pergamon Oxford New York, pp 151–173
Sarantakis D, Teichman J, Chai SY, Lien E (1981) Structure-activity studies of glucagon selective analogs of somatostatin. In: Abstracts of the 2nd international symposium on somatostatin, Athens, June 1–3, 1981. Abstract book, Serono symposia, p 150
Schade DS, Eaton RP (1978) The metabolic response to norpinephrine in normal versus diabetic man. Diabetologia 15:433–439
Scheen A, Luyckx AS, De Graeve J (1981) Cardiovascular effects of intravenous tolmesoxide in hypertensive patients. Arch Int Pharmacodyn 251:322–334
Scheen A, Cession-Fossion A, Scheen-Lavigne M, Luyckx AS (1982) Effect of protein supplemented fasting on metabolic and hormonal responses to epinephrine infusion in obese subjects. Horm Metab Res 14:240–245
Schusdziarra V, Rivier J, Dobbs R, Brown M, Vale W, Unger RH (1978) Somatostatin analogs as glucagon suppressants in diabetes. Horm Metab Res 10:563–565
Seino Y, Goto Y, Kurahachi H, Sakurai H, Ikeda M, Kadowaki S, Inoue Y, Mori K, Taminoto T, Imura H (1977) Alteration of plasma glucagon response to arginine after treatment in patients with diabetes mellitus, Cushing's syndrome and hypothyroidism. Horm Metab Res 9:28–32
Seino Y, Ikeda M, Kurahachi H, Taminato T, Sakurai H, Goto Y, Inoue Y, Kadowaki S, Kozaburo M, Imura H (1978) Failure to suppress plasma glucagon concentrations by orally administered glucose in diabetic patients after treatment. Diabetes 27:1145–1150
Shichiri Y, Kawamori R, Abe H (1979) Normalization of paradoxical secretion of glucagon in diabetics who were controlled by the artificial beta cell. Diabetes 28:272–275

Starke A, Keck E, Cüppers HJ, Zaremba A, Zimmermann H, Krüskemper HL (1980) The effect of calcium on glucagon secretion stimulated by arginine and hypoglycemia. In: Andreani D, Lefèbvre P, Marks V (eds) Current views on hypoglycemia and glucagon. Academic Press, London, pp 487–488

Starke A, Keck E, Berger M, Zimmermann H (1981) Effects of calcium and calcitonin on circulating levels of glucagon and glucose in diabetes mellitus. Diabetologia 20:547–552

Stubbs WA, Jones A, Edwards CR, Delitala G, Jeffcoate WJ, Ratter SJ (1978) Hormonal and metabolic responses to an enkephalin analogue in normal man. Lancet 2:1225

Tamarit J, Mayor P (1978) Effect of $Ca^{2+}$ intravenous load on insulin and glucagon secretion (Abstr). Diabetologia 15:275

Tamarit-Rodriguez J, Cebeira M, Tamarit J, Garcia C, Roncero J (1978) Calcitonin modulation of insulin and glucagon release by the isolated and perfused rat pancreas (Abstr). Diabetologia 15:275

Tamburrano S, Luyckx AS, Lefèbvre PJ (1976) Studies on the hyperglycemic effect of diphenylhydantoin in normal golden hamsters. In: Andreani D, Lefèbvre P, Marks V (eds) Hypoglycemia. Proceedings of the European Symposium, Rome. Thieme, Stuttgart, pp 74–79

Thorner MO, Chait A, Aitken M, Benker G, Bloom SR, Mortimer CH, Sanders P, Stuart Mason A, Besser GM (1975) Bromocriptine treatment of acromegaly. Br Med J 1:299–303

Thorner MO, Wass JAH, Jones A, Bloom SR, McLeod RM (1976) Effect of piribedil infusions on circulating growth hormone (GH), insulin, glucagon, cortisol and blood sugar in man (Abstr). In: Abstracts o the Vth international congress of endocrinology, Hamburg, July 1976, p 24

Tiengo A, Muggeo M, Assan R, Crepaldi G (1975) Glucagon secretion in primary endogenous hypertriglyceridemia before and after clofibrate treatment. Metabolism 24:901–914

Tiengo A, Nosadini R, Fedele D, Meneghel A, Valerio A, Crepaldi G (1978) The role of glucagon and insulin in endogenous human and rat hyperlipemia (genetic or acquired). In: Crepaldi G, Lefèbvre P, Alberti KGMM (eds) Diabetes, obesity and hyperlipidemias. Academic, London, pp 21–28

Toyota T, Ikeda M, Kay T, Miura M, Goto Y (1980) Effects of metenkephalin on the release of insulin and glucagon (Abstr). In: Programme and abstracts of the sixth international congress of endocrinology, Melbourne, p 595

Trovati M, Massara F, Camanni F, Molinatti GM, Lorenzati R, Pagano GF (1980) Effect of the somatostatin analog D-$TRP^8$, D-$Cys^{14}$ on glucose insulin, pancreatic glucagon and growth hormone plasma levels in acromegalics and mild diabetics. J Endocrinol Invest 3:189–192

Tsalikian E, Dunphy T, Bohannon N, Lorenzi M, Gerich J, Forsham P, Kane JP, Karam JH (1977) The effect of chronic oral antidiabetic therapy on insulin and glucagon responses to a meal. Diabetes 26:314–321

Unger RH (1978) Role of glucagon in the pathogenesis of diabetes: the status of the controversy. Metabolism 27:1691–1709

Unger RH, Raskin P (1978) The effect of glucagon and of its suppression upon diabetic hyperglycemia. Diabetes 27 [Suppl 1] 27:455

Urdanivia E, Pek S, Santiago JC (1979) Inhibition of glucagon secretion by diazoxide in vitro. Diabetes 28:26–31

Vaughan-Williams EM (1970) Classification of anti-arrhytmic drugs. In: Sandøe E, Flensted-Jensen E, Olesen KH (eds) Symposium on cardiac arrhytmias. Södertälje, Sweden Astra, pp 449–472

Vexiau P, Cathelineau G, Luyckx A, Lefèbvre P, Canivet J (1982) Glucagon is not involved in intravenous calcium infusion induced improvement in glucose tolerance or in correction of reactive hypoglycemia (Abstr). Diabetologia 23:208

Walter RM, Dudl RJ, Palmer JP, Ensinck JW (1974) The effect of adrenergic blockade on the glucagon responses to starvation and hypoglycemia in man. J Clin Invest 54:1214–1220

Wollheim CB, Marliss EB, Blondel B, Orci L, Like A, Renold AE (1973) A-cell function in monolayer cultures of newborn rat pancreas (Abstr). Diabetologia 9:96

# Extrapancreatic Glucagon

CHAPTER 33

# Extrapancreatic Glucagon and Its Regulation

P. J. LEFEBVRE and A. S. LUYCKX

## A. Introduction

The A-cell of the islets of Langerhans of the pancreas has long been considered to be the only site of origin of glucagon. This dogma was challenged, in the mid-1970s, when three groups, working independently, almost simultaneously reported that the plasma of totally depancreatized dogs contained normal, or even increased, quantities of a material immunometrically indistinguishable from pancreatic glucagon by radioimmunoassays regarded as highly specific for this hormone (VRANIC et al. 1974; MATSUYAMA and FOÀ 1974; MASHITER et al. 1975). These observations prompted a series of experiments which soon firmly established that the fundus of the dog stomach possesses A-cells identical to the pancreatic A-cells and that it contains and releases a material identical, by all criteria, to pancreatic glucagon. In fact, the presence of glucagon (then called the "hyperglycemic glycogenolytic factor") in the canine stomach had been mentionned 35 years ago by SUTHERLAND and DE DUVE (1948) who had written:

> One might therefore conclude that the α-cells [of the islets of Langerhans] are the site of formation of the glycogenolytic factor. However, the presence of an apparently identical factor in the upper two-thirds of the gastric mucosa of the dog raises the question of specificity, unless it were shown that the stomach mucosa contains cell types related to the α-cells of the pancreas.

Most of the studies on extrapancreatic glucagon have been performed in dogs, and more precisely on this canine gastric glucagon; they will be analyzed in some detail in the present chapter. Available data on extrapancreatic glucagon in other animal species will be reviewed, a section will be devoted to the presence, reported in several species, of glucagon in the salivary glands and another to the presence of glucagon in the central nervous system. Finally, the question, still open, of the existence of an extrapancreatic source of glucagon in humans will be briefly considered (see also Chap. 42).

## B. Extrapancreatic Glucagon in the Canine Stomach

### I. Presence of A-cells

Since 1968, various groups (ORCI et al. 1968; CAVALLERO et al. 1969, 1970; SOLCIA et al. 1970; KUBES et al. 1974) have reported that the oxyntic glandular mucosa of the canine stomach contains cells resembling pancreatic A-cells. The existence of "true" A-cells in the dog gastric fundus has been established by cytologic, cytochemical, and immunochemical procedures (LARSSON et al. 1975; BAETENS et al.

1976); electron microscope examination has demonstrated that these cells contain secretory granules with a round dense core surrounded by a clear halo, indistinguishable from secretory granules of pancreatic A-cells (BAETENS et al. 1976). The number of these fundic A-cells was particularly high in the stomach of a dog studied 5 years after pancreatectomy (RAVAZZOLA et al. 1977). More recently, RAVAZZOLA and ORCI (1978) and RAVAZZOLA et al. (1979) reported that the same A-cell population in the islets of Langerhans and in the stomach oxyntic mucosa of the dog was, by immunofluorescence studies, positive for both glucagon and glicentin (a component of gut glucagon-like immunoreactivity GLI) and suggested that glicentin might be a common biosynthetic precursor of both glucagon and gut GLIs (see Chap. 7). Such a view is confirmed by the report by RAVAZZOLA and ORCI (1980 a) that enzymatic digestion with trypsin and carboxypeptidase B can transform ileal glicentin-containing L-cells into glucagon-containing cells. Moreover, RAVAZZOLA et al. (1981 a) have provided convincing evidence that, within the pancreatic A-cell, biotransformation of glicentin into glucagon indeed occurs.

## II. Presence of Glucagon

SUTHERLAND and DE DUVE (1948) were the first to suggest the presence of glucagon in the dog gastric fundus. The first detailed analysis of canine gastric glucagon was reported by MORITA et al. (1976) who showed that the extracted material was similar to pancreatic glucagon in various analytic systems. Similar studies were conducted by SRIKANT et al. (1977), who, in addition, showed that fundic immunoreactive glucagon of molecular weight 3 500 daltons ($IRG^{3500}$) was more active than pancreatic glucagon in stimulating liver adenylate cyclase activity. In contrast, DOI et al. (1979) found that pancreatic porcine glucagon and purified glucagon from dog stomach fundus have the same biologic activities (glycogenolysis, ureogenesis, gluconeogenesis) in isolated rat hepatocytes. From 105 canine stomachs they isolated 1.5 μg of a material apparently identical to pancreatic glucagon because: (a) their molecular weights, elution properties in ion exchange chromatography, and their electrophoretic mobility are indistinguishable; and (b) both hormones elicited identical biologic effects in isolated rat hepatocytes.

## III. Control of Gastric Glucagon Release In Vitro

The isolated dog stomach, perfused with whole canine blood, provides a unique tool for studying the factors controlling gastric glucagon release (LEFEBVRE and LUYCKX 1977, 1978 a, b, c, d, 1980 a, b; LEFEBVRE et al. 1976, 1978 a, b, 1979, 1981).

### 1. Role of Glucose and Insulin

At basal concentrations of glucose ($101 \pm 5$ mg/100 ml) and insulin ($8 \pm 1$ μIU/ml, the release of glucagon from the stomach is indeed very modest ($279 \pm 84$ pg/100 g stomach in 1-min period) and even in some experiments replaced by a discrete, but significant glucagon uptake or degradation (LEFEBVRE and LUYCKX 1978 d). In this isolated system, hyperglycemia alone does not further reduce this small basal gastric glucagon release. In contrast, hypoglycemia ($32 \pm 1$ mg/100 ml) slightly, but

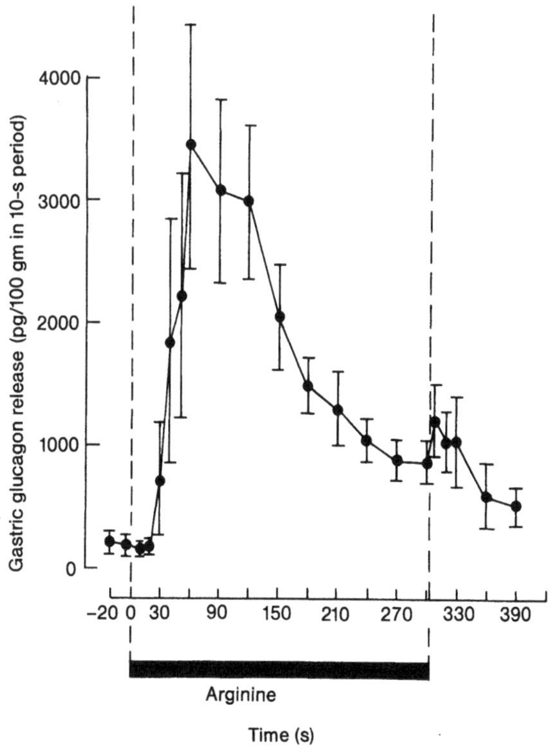


**Fig. 1.** Gastric glucagon release in response to intraarterial arginine infusion in the isolated perfused dog stomach. Results are expressed as mean ± standard error for five experiments. LEFEBVRE and LUYCKX (1977)

significantly increased gastric glucagon release (610 ± 79 pg/100 g in 1-min period). The effect of hypoglycemia was doubled when the low concentration of insulin present in the perfusing blood (collected from overnight fasted normal blood donor dogs) was neutralized by an excess of anti-insulin serum. This finding demonstrates that even the presence of very small quantities of insulin (5–10 μIU/ml) is sufficient to suppress gastric glucagon release. This exquisite sensitivity to insulin is probably one of the major characteristics of the extrapancreatic A-cell. Cytoglycopenia, induced by 2-deoxy-D-glucose, a nonmetabolizable glucose analog which blocks intracellular glucose metabolism, is also a potent stimulator of gastric glucagon release as it is for pancreatic glucagon (LEFEBVRE and LUYCKX 1978d).

## 2. Stimulation by Arginine

Arginine hydrochloride infused intraarterially, at a concentration calculated to reach an arterial plasma concentration of about 10 m*M*, induced a rapid rise in gastric glucagon output (Fig. 1; LEFEBVRE et al. 1976). This effect was completely inhibited by somatostatin (100 ng/ml). Arginine-induced gastric glucagon release was not modified by either massive hyperinsulinemia (10000 μIU/ml at a "normal" blood glucose concentration) or hyperglycemia alone (312 ± 19 mg/100 ml at

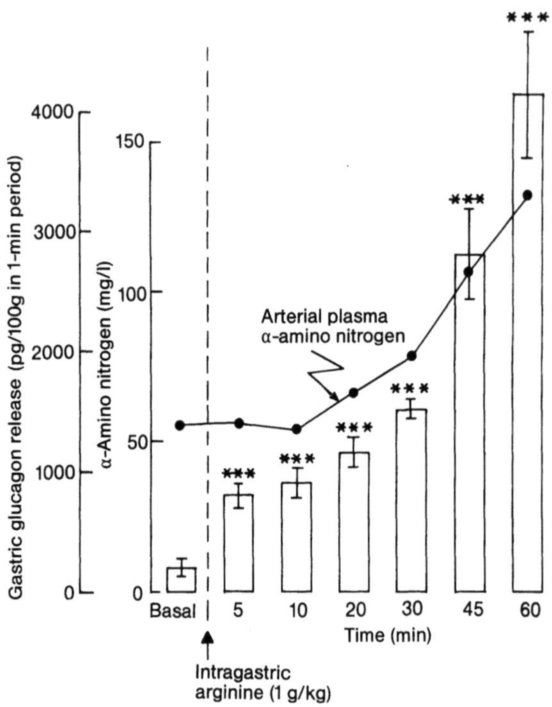


**Fig. 2.** Response of gastric glucagon to the instillation in the lumen of the stomach of arginine (1.0 g/kg body weight) in a representative experiment. Each histogramm represents the mean of three determinations. The rise in gastric glucagon release precedes the rise in the plasma concentration of α-amino nitrogen. *Triple asterisks* correspond to $P<0.01$ versus basal release. LEFEBVRE and LUYCKX (1980a)

a low plasma insulin level: $7 \pm 2$ μIU/ml). In contrast, arginine-induced gastric glucagon release was reduced by about 40% when hyperglycemia ($319 \pm 13$ mg/100 ml) was concomitant with a moderate ($115 \pm 20$ μIU/ml) hyperinsulinemia (LEFEBVRE and LUYCKX 1977). Arginine, when introduced intraluminally, also stimulated gastric glucagon release, an effect which was not observed when saline alone was present in the stomach lumen (LEFEBVRE and LUYCKX 1980a). As shown in Fig. 2, depicting a typical experiment, gastric glucagon release was significantly increased before any rise in arterial plasma α-amino nitrogen (resulting from gastric resorption of arginine in these experiments where recirculation of the blood was permitted). This finding suggests a possible direct stimulation of the gastric A-cell by arginine from the stomach lumen (LEFEBVRE and LUYCKX 1980a).

## 3. Role of the Autonomic Nervous System

Contrary to the pancreatic A-cell (BLOOM 1976), the canine gastric A-cell does not seem to be directly under the control of the autonomic nervous system. The electrical stimulation of both dorsal and ventral vagal trunks of the isolated perfused canine stomach failed to elicit gastric glucagon release either at low or normal blood glucose concentrations, when, in contrast, gastrin release was markedly stimulated (LEFEBVRE et al. 1978a). Interestingly, the intraarterial infusion of carbamylcholine or acetylcholine ($5 \times 10^{-6} M$) stimulated the release of both gastrin and glucagon (LEFEBVRE et al. 1979; LEFEBVRE and LUYCKX 1980b). At this time, there is no clear explanation for this discrepancy, i.e., the lack of effect of the electrical stimulation of the vagus nerves contrasting with the effectiveness of the carbamylcholine infusion. A possibility is that vagal nerve endings do not enter into close contact with the gastric A-cells; this would strikingly differentiate the gastric A-cell from the well-innervated pancreatic A-cell. If this were true, one would have to admit that the dog gastric A-cell, receiving no or little innervation from the vagus, would nevertheless possess the receptor for the neurotransmitter (or its analogs) since it responds to carbamylcholine [1]. Another possibility is that somatostatin, also released (LEFEBVRE and LUYCKX 1980a) in our system by the electrical stimulation of the vagus nerves simultaneously inhibits gastric glucagon release. The close interrelationships between gastric A- and D-cells in the canine stomach (J. M. POLAK 1979, personal communication) and the complete inhibition of carbamylcholine-induced gastric glucagon release by exogenous somatostatin (LEFEBVRE et al. 1981) support this interpretation. Similar relationships of these two types of endocrine cells have been observed in the stomach of the monkey by HELMSTAEDTER et al. (1977).

JOHNSON and ENSINCK (1976) have demonstrated that the toxin (or the venom) of the scorpion Leiurus quinquestriatus releases norepinephrine from the sympathetic nerve endings in the perfused rat pancreas and also releases glucagon, thus suggesting that pharmacologic stimulation of the adrenergic nerve endings in this system can elicit a rapid release of glucagon. In similar experiments, it has been reported (LEFEBVRE et al. 1978b) that intraarterial scorpion venom infusion (50 μg/min) elicits a massive release of norepinephrine from the isolated perfused dog stomach without any significant release of gastric glucagon. Similarly, exogenous norepinephrine does not significantly modify glucagon release from the isolated perfused dog stomach (LEFEBVRE and LUYCKX 1980b) while vasoactive intestinal peptide (VIP) unequivocally stimulates it (Fig. 3; LEFEBVRE and LUYCKX 1980b).

#### 4. Possible Role of Prostaglandins

Intraarterial infusion of prostaglandins $E_1$ ($PGE_1$) induced a clear-cut, dose-related increase in gastric glucagon release (LEFEBVRE and LUYCKX 1978c). This finding is interesting in view of the role that endogenous prostaglandins seem to play in the control of pancreatic glucagon release (see Chap. 28).

## IV. Secretion of Gastric Glucagon In Vivo

In the normal conscious dog, the gastric fundus is not a major source of glucagon: gastric vein glucagon levels were found similar to vena cava glucagon levels and always well below pancreaticoduodenal vein plasma glucagon concentrations

1 An argument in favor of this concept has been found in the demonstration that electrical vagal stimulation elicited a modest and transient release of gastric glucagon when the blood perfusing the stomach contained a cholinesterase inhibitor, like physostigmine (P. J. LEFEBVRE and A. S. LUYCKX 1981, unpublished work)

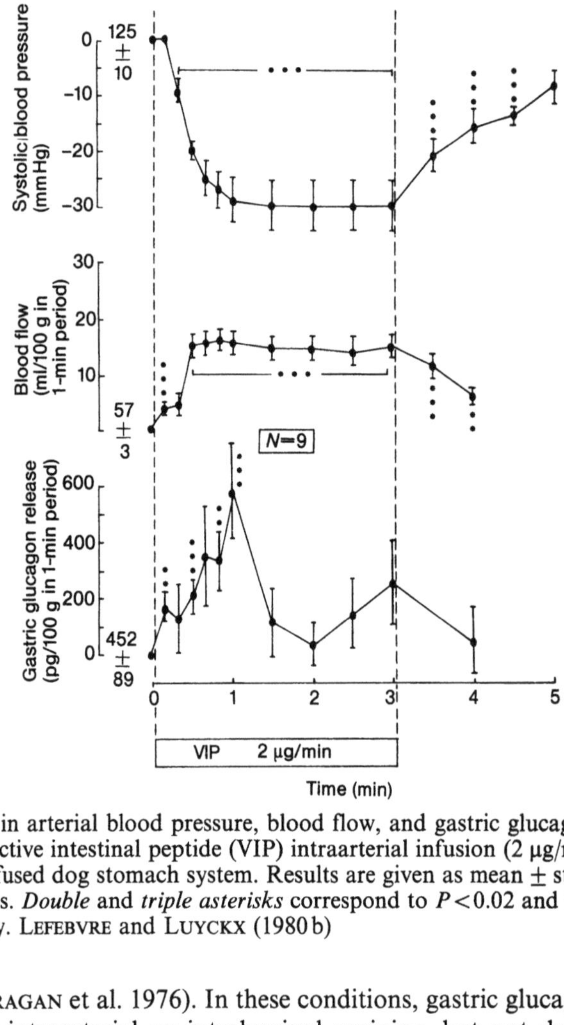


**Fig. 3.** Changes in arterial blood pressure, blood flow, and gastric glucagon release in response to vasoactive intestinal peptide (VIP) intraarterial infusion (2 µg/min for 3 min) in the isolated perfused dog stomach system. Results are given as mean ± standard error for nine experiments. *Double* and *triple asterisks* correspond to $P < 0.02$ and 0.01 versus baseline, respectively. LEFEBVRE and LUYCKX (1980b)

(MUÑOZ-BARRAGAN et al. 1976). In these conditions, gastric glucagon release was stimulated by intraarterial or intraluminal arginine, but not during insulin or phlorhizin-induced hypoglycemia.

In depancreatized dogs, the stomach fundus appeared as a major source of glucagon (BLAZQUEZ et al. 1976; OHNEDA et al. 1979), being able to release approximately one-quarter of the amount released by a normal pancreas (MÜLLER et al. 1978); gastric glucagon secretion was stimulated by intravenous or intragastric arginine administration (BLAZQUEZ et al. 1976); it was not influenced by intravenous or intragastric glucose administration (BLAZQUEZ et al. 1976), but was markedly suppressed by extremely low doses (0.0015 IU $kg^{-1}$ $min^{-1}$) of insulin. Similar results have been reported in alloxan-induced diabetic dogs (BLAZQUEZ et al. 1977). In the depancreatized dog, insulin-induced *hypoglycemia* did not increase circulating plasma glucagon levels (MATSUYAMA et al. 1978) as was expected from the potent inhibitory action exerted by insulin on the gastric A-cell (LEFEBVRE and LUYCKX 1978a). Extrapancreatic glucagon in the dog has a diabetogenic effect during acute insulin deficiency (ROSS et al. 1978). Available data suggest that pancreatic and extrapancreatic glucagon have similar clearance rates (MÜLLER et al. 1978). As in the isolated perfused dog stomach, acetylcholine stimulates extrapancreatic glucagon release in the pancreatectomized dog, an effect reduced by atropine (YOSHIDA and KONDO 1980a). In conscious, depancreatized dogs, untreated with insulin, large amounts (0.6 μg $kg^{-1}$ $min^{-1}$) of epinephrine, norepinephrine, and isoproterenol stimulated the release of extrapancreatic glucagon in the portal vein. This effect was completely abolished by simultaneous infusion of the specific $\beta$-adrenoceptor blocking compound, propranolol, but not by $\alpha$-adrenergic blockade with phentolamine (YOSHIDA and KONDO 1980b). Hypophysectomy reduced plasma circulating levels of glucagon in depancreatized dogs, an effect reversed by hypophysis transplantation (YOSHIDA and KONDO 1980c). LICKLEY et al. (1981) reported that the intravenous infusion of epinephrine (0.1 μg $kg^{-1}$ $min^{-1}$) in unanesthetized, pancreatectomized dogs induced the release of two glucagon moieties of molecular weight 3500 and 9000.

## V. Physiologic and Pathophysiologic Relevance of Extrapancreatic Glucagon

Abdominal evisceration in the dog results in zero values of circulating glucagon (LEFEBVRE and LUYCKX 1976), suggesting that all significant sources of glucagon in this animal are located in the abdominal organs. The A-cells of the islets of Langerhans of the pancreas and of the oxyntic mucosa of the gastric fundus are most likely to be the only significant sources of glucagon. Scarce cells resembling the A-cells have been described in the dog duodenum and jejunoileum (LARSSON et al. 1975); their secretion is minimal and they do not respond to arginine, thus the stomach seems to be the major source of extrapancreatic glucagon in the dog (MÜLLER et al. 1978). In the normal animal, gastric glucagon release is small compared with pancreatic glucagon release (MUÑOZ-BARRAGAN et al. 1976); the extreme sensitivity of the gastric A-cell to insulin or, more likely, to glucose in the presence of insulin (LEFEBVRE and LUYCKX 1978a), probably explains why gastric glucagon secretion is usually suppressed and can only be stimulated by a potent stimulus such as arginine (MASHITER et al. 1975; MÜLLER et al. 1978). In contrast, in the diabetic dog, deprived of insulin, gastric glucagon release occurs (BLAZQUEZ et al. 1976, 1977) and contributes to the diabetic state (ROSS et al. 1978). Its suppression by somatostatin (DOBBS et al. 1975) is accompanied by a lowering of blood glucose, although this has not been confirmed (MATSUYAMA et al. 1979). The hyperglucagonemia of the insulin-deprived diabetic dog is rapidly reversed by the administration of low doses of insulin (BLAZQUEZ et al. 1976, 1977), a phenomenon due again to the extreme sensitivity of the extrapancreatic A-cell to insulin.

A characteristic of the fundic A-cell of the dog is that it does appear to be less under the control of the autonomic nervous system (LEFEBVRE et al. 1978a, b; LEFEBVRE and LUYCKX 1980a, b), than the pancreatic A-cell (BLOOM 1976). This probably explains the observations of VRANIC et al. (1976) who reported that strenuous exercise, which is known to be a potent stimulus of pancreatic glucagon secretion, probably mainly through stimulation of the adrenergic system (LUYCKX and LEFEBVRE 1974; LUYCKX et al. 1975), does not induce any rise in the plasma glucagon levels of depancreatized dogs.

## C. Gastrointestinal Glucagon in Other Animal Species

Morphological studies have demonstrated the presence of A-cells in the oxyntic glands of the stomach of the pig (LARSSON et al. 1975), the cat (LARSSON et al. 1975) and other felines (KITAMURA et al. 1982), the rat (LARSSON et al. 1975), and the monkey (HELMSTAEDTER et al. 1977). Only occasionally (1 cell in 5–10 sections) were glucagon-immunoreactive cells seen in the ileum and proximal colon of the pig (LARSSON et al. 1975). A material, identical by all criteria to pancreatic glucagon, has been extracted from a 50-fold purified porcine duodenal extract by SASAKI et al. (1975). Pancreatectomy in the pig results in high circulating levels of glucagon (HOLST 1978). Chromatographic studies, however, have indicated that, after pancreatectomy, only one of two circulating components with glucagon immunoreactivitiy was found in the plasma; the higher molecular weight of this component probably explains the fact that this half-life is different from the one of exogenous glucagon (HOLST 1978). In the cat, extrapancreatic glucagon secretion is observed after pancreatectomy and is reversed by prolonged insulin therapy (CHISHOLM et al. 1978); the role of this extrapancreatic glucagon in the metabolic disturbances resulting from pancreatectomy remains to be established; in that species, the circulating immunoreactivity after removal of the pancreas seems to correspond in molecular weight to both glucagon and a larger molecular weight glucagon precursor. In the rat, extrapancreatic glucagon originates from the stomach (MARRE et al. 1979), and maybe also (see Sect. D) from the salivary glands. The work of O'CONNOR et al. (1979) suggests that polypeptides with a molecular weight about 3 000–4 000 reacting with glucagon-specific antisera can be mobilized from the perfused rat intestine following stimulation by glucose. The relationship of these compounds to glucagon remains to be established.

A puzzling observation is the persistence of circulating immunoreactive glucagon (and insulin) in the eviscerated rat, even with a functional liver (PENHOS et al. 1975); it has been suggested that persistent IRG circulating levels in eviscerated rats represent immunoreactive materials with slow rates of degradation, although an unresponsive extravisceral source of glucagon can not be ruled out (SMITH et al. 1978).

## D. Glucagon and the Salivary Glands

DUNBAR et al. (1976) were the first to suggest the presence of a glucagon-related peptide in the submaxillary glands. Subsequently, LAWRENCE et al. (1976) showed that extracts of homogenates of rat, mouse, rabbit, and human submaxillary salivary glands contain a significant quantity of a material reacting with an antibody considered specific for glucagon. The highest concentrations were found in the submaxillary glands of rats and mice (5 000–7 000 ng/g), with lower concentrations in similar glands from humans and rabbits. Trace amounts were detected in the sublingual and parotid glands, with the exception of the rat parotid gland where 20–80 ng/g was measured. The molecular weight of this material appeared to be approximately 70 000; its release in vitro was stimulated by low glucose concentrations and by arginine, it was inhibited by high concentrations of glucose. The presence of a glucagon-related peptide in the submaxillary glands has been confirmed by BATHENA et al. (1977). In contrast with the previous study, no change in the release of this compound occured in vitro in response to arginine or glucose. Moreover, the molecular weight of the compound isolated by BHATHENA and his co-workers was 29 000; urea dissociated it into smaller molecular weight fragments including that of 3 500, thus suggesting that it represents an aggregate of smaller glucagon molecules. This material was not extractable by the classical acid-alcohol procedure, but only with an acid saline medium. The physiologic significance and pathophysiologic relevance of rat salivary "glucagon" remain to be established (SMITH et al. 1979). Recently, PEREZ-CASTILLO and BLAZQUEZ (1980 a) confirmed the presence of high amounts of immunoreactive glucagon in acid-saline extracts of human submaxillary (18.5 ± 2.5 ng/g wet weight) and parotid (3.5 ± 0.3 ng/g wet weight) glands. After fractionation of the acid-ethanol extracts on Biogel P-30 columns or gel elctrophoresis, an immunoreactive peak of 3 500 daltons was always obtained. Arginine, epinephrine, and low glucose concentrations stimulated glucagon release from both salivary glands. Active glucagon biosynthesis by these glands was established by the incorporation of L-tryptophan-$^3$H into a 3 500 daltons polypeptide with specific immune reaction with the "glucagon-specific" antiserum 30K. These findings indicate that the salivary glands represent a source of extrapancreatic glucagon in humans and may therefore contribute to the circulating levels of the hormone (see Sect. G).

## E. Other Extrapancreatic Localizations of Glucagon

Significant amounts of glucagon have been found by PEREZ-CASTILLO and BLAZQUEZ (1980 b) in acid-alcohol extracts of rat thymus (0.75 ± 0.09 ng/100 mg wet tissue), adrenals (0.86 ± 0.13 ng/100 mg), thyroid (2.43 ± 0.9 ng/100 mg), and hypophysis (4.01 ± 0.60 ng/100 mg). Active glucagon biosynthesis by these organs was established by the incorporation of labeled tryptophan into a 3 500 molecular weight polypeptide with specific immune reaction with 30K antiserum.

High concentrations of immunoreactive glucagon, measured with antibodies directed against the COOH terminal region of the molecule have been found in the canine hypothalamus: 1.88 ± 0.25 and 0.40 ± 0.09 ng/mg protein for the anterior and the mediobasal hypothalamus respectively (CONLON et al. 1979). GLI has also been found in the dog hypothalamus, amygdala, and mesencephalon (CONLON et al. 1979). TAGER et al. (1980) also reported the presence of glucagon-related peptides in rat hypothalamus, cortex, thalamus, cerebellum, and brain stem. These peptides showed immunoreactivity with anti-glucagon sera directed towards the central portion of the hormone, but not with antisera specific for the free COOH terminus of glucagon. However, digestion of these peptides with trypsin plus carboxypeptidase B released the immunoreactive COOH terminal fragment of pancreatic glucagon from these larger forms. These findings, together with various immunocytochemical studies, suggested that glucagon-containing peptides that have undergone the intestinal type of post translational modification are present in some neuronal cells of the rat brain. SASAKI et al. (1980) confirmed the presence of glucagon and glucagon-like immunoreactivity in the canine hypothalamus and brain stem and TOMINAGA et al. (1981) reported on the species differences of glucagon-like materials in the brain.

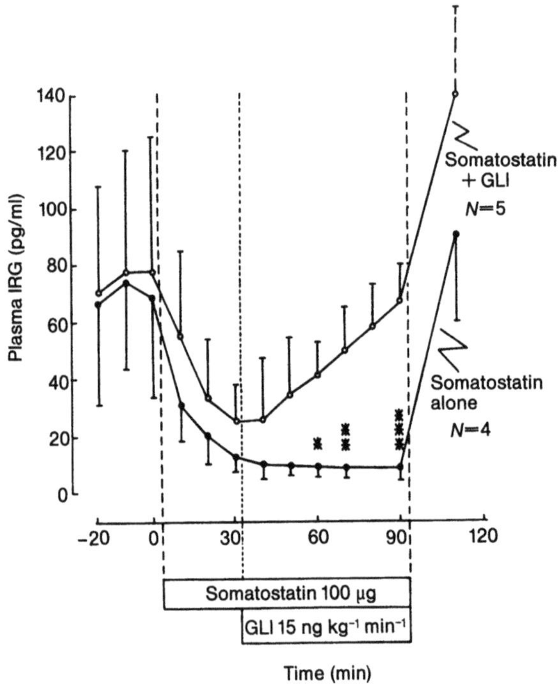


**Fig. 4.** Generation of glucagon (IRG) from intravenously infused glucagon-like immunoreactive material (GLI) in piglets. The endogenous secretion of insulin and glucagon was inhibited by somatostatin. *Single*, *double*, and *triple asterisks* correspond to $P<0.05$, $P<0.01$, and $P<0.005$ versus baseline, respectively. KORANYI et al. (1981)

## F. In Vivo Generation of Glucagon from Glucagon-Like Immunoreactive Peptides

Recent morphological investigations have shown that glucagon and glicentin immunoreactivity are topologically segregated in the granule of the pancreatic A-cell (RAVAZZOLA and ORCI 1980b) and that ileal glicentin-containing L-cells can be transformed into glucagon-containing cells by in vitro enzymatic digestion with trypsin and carboxypeptidase B (RAVAZZOLA and ORCI 1980a). These findings support the hypothesis presented elsewhere (see Chap. 6) of glicentin as a glucagon precursor.

Recently, KORANYI et al. (1981) have shown that, in anesthetized pigs, in which endogenous glucagon secretion was blocked by somatostatin, intravenous infusion of purified porcine GLI induced, in vivo a progressive generation of glucagon, as measured by an antiserum directed towards the COOH terminus of glucagon (Fig. 4). This IRG generated in vivo from GLI was biologically active since it prevented the hypoglycemia otherwise induced by somatostatin (Fig. 5).

## G. Extrapancreatic Glucagon in Humans

As reviewed by BODEN (see Chap. 42), the presence of extrapancreatic A-cells in human tissues has been difficult to identify (ORCI 1976; SASAGAWA et al. 1974;

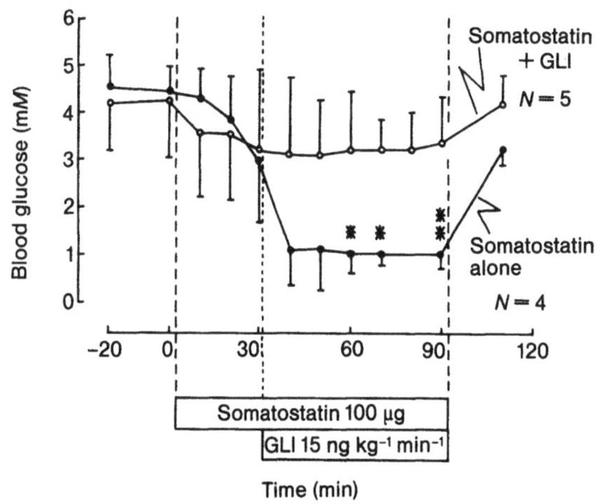


**Fig. 5.** The transformation of GLI into IRG (see Fig. 4) prevents somatostatin-induced hypoglycemia in normal piglets. *Single* and *double asterisks* indicate $P<0.05$ and $P<0.01$, versus baseline, respectively. KORANYI et al. (1981)

MUÑOZ-BARRAGAN et al. 1977). More recently, the presence of typical A-cells has been reported in the human fetal gastric fundus (RAVAZZOLA et al. 1981 b; ITO et al. 1981; BUCHAN et al. 1982). The amounts of glucagon recovered from human fundic mucosa are 10–100 times lower than those found in canine fundi (DOI et al. 1979). Most of the patients having undergone duodenopancreatectomy and/or hemigastrectomy still have very small, but clearly detectable, levels of circulating immunoreactive glucagon ($IRG^{3500}$), as detailed in Chap. 42. BRINGER et al. (1981) reported a series of ten pancreatectomized patients in whom basal IRG levels, using 30K antiserum, averaged $69 \pm 8$ pg/ml. In two of these patients as well as in two patients reported previously by WERNER and PALMER (1978), a clear-cut rise in circulating plasma glucagon was observed during intravenous arginine infusion. Such findings, however, appear quite unusual since, in most of these patients, arginine infusion is usually ineffective (see review in Chap. 42; TIENGO et al. 1980; TRONIER et al. 1981).

## H. Conclusions

The existence of significant extrapancreatic sources of glucagon is accepted in dogs and some other animal species, but remains to be firmly established in humans (for human studies see Chap. 42). In the dog, the major source of extrapancreatic glucagon is the gastric fundus. In that species, gastric glucagon secretion appears to be more dependent on metabolic or hormonal control than on neural factors. The main characteristic of the gastric A-cell is its great sensitivity to glucose in the presence of insulin. This probably explains why: (1) gastric glucagon release is minimal in normal animals; and (2) complete insulin deprivation is needed to permit gastric glucagon release in depancreatized animals. The isolated perfused dog stomach probably represents, at this time, a unique model for studying A-cell function in the absence of insulin. Such studies are difficult with pancreatic A-cells which are in close contact (in the islets of Langerhans) with the B-cells and, therefore, probably exposed to high basal concentrations of insulin.

## References


Baetens D, Rufener C, Srikant CB, Dobbs R, Unger RH, Orci L (1976) Identification of glucagon-producing cells (A-cells) in dog gastric mucosa. J Cell Biol 69:455–464

Bhathena SJ, Smith SS, Voyles NR, Penhos JC, Recant L (1977) Studies on submaxillary gland immunoreactive glucagon. Biochem Biophys Res Commun 74:1574–1581

Blazquez E, Muñoz-Barragan L, Patton GS, Orci L, Dobbs RE, Unger RH (1976) Gastric A-cell function in insulin-deprived depancreatized dogs. Endocrinology 99:1182–1188

Blazquez E, Muñoz-Barragan L, Patton GS, Dobbs RE, Unger RH (1977) Demonstration of gastric glucagon hypersecretion in insulin-deprived alloxan diabetic dogs. J Lab Clin Med 89:971–977

Bloom SR (1976) Blood glucose control by direct islet innervation. In: Andreani D, Lefèbvre P, Marks V (eds) Hypoglycemia. Thieme, Stuttgart, pp 85–90

Bringer J, Mirouze J, Marchal G, Pham TC, Luyckx AS, Lefèbvre P, Orsetti P (1981) Glucagon immunoreactivity and antidiabetic action of somatostatin in the totally duodenopancreatectomized and gastrectomized man. Diabetes 30:851–856

Buchan AMJ, Bryant MG, Stein BA, Gregor M, Ghatei MA, Morris JF, Bloom SR, Polak JM (1982) Pancreatic glucagon in human-fetal stomach. Histochemistry 74:515–520

Cavallero C, Solcia E, Vassallo G, Capella C (1969) Cellule endocrine della mucosa gastroenterica ed ormoni gastro-intestinali. Rend RR Gastroenterol 1:51–61

Cavallero C, Capella C, Solcia E, Vassallo G, Bussolati G (1970) Cytology, cytochemistry and ultrastructure of glucagon-secreting cells. Acta Diabetol Lat 7:542–556

Chisholm DJ, Alford FP, Harewood MS, Findlay DM, Gray BN (1978) Nature and biologic acitivity of "extrapancreatic glucagon": studies in pancreatectomized cats. Metabolism 27:261–273

Conlon JM, Samson WK, Dobbs RE, Orci L, Unger RH (1979) Glucagon-like polypeptides in canine brain. Diabetes 28:700–702

Dobbs R, Sakurai H, Sasaki H, Faloona G, Valverde I, Baetens D, Orci L, Unger RH (1975) Glucagon: role in the hyperglycemia of diabetes mellitus. Science 187:544–546

Doi K, Prentki M, Yip C, Müller WA, Jeanrenaud B, Vranic M (1979) Identical biological effects of pancreatic glucagon and a purified moiety of canine gastric immunoreactive glucagon. J Clin Invest 63:525–531

Dunbar JC, Silverman H, Kirman E, Foà PP (1976) Salivary gland and kidney glucagon in the rat (Abstr). Fed Proc 35:218

Helmstaedter V, Feurle GE, Forssmann WG (1977) Relationship of glucagon-somatostatin and gastrin-somatostatin cells in the stomach of the monkey. Cell Tissue Res 177:29–46

Holst JJ (1978) Extrapancreatic glucagon. Digestion 17:168–190

Ito S, Iwanaga T, Kusumoto Y, Sudo N, Sano M, Suzuki T, Shibata A (1981) Is glucagon in the human gastric fundus? Horm Metab Res 13:419–422

Johnson DG, Ensinck JW (1976) Stimulation of glucagon secretion by scorpion toxin in the perfused rat pancreas. Diabetes 25:645–649

Kitamura N, Yamada J, Yamashita T (1982) Immunocytochemical study on the glucagon cells in the feline gastric glands. Jpn J Vet S 44:849–851

Koranyi L, Peterfy F, Szabo J, Török A, Guoth M, Tamas G Jr (1981) Evidence for transformation of glucagon-like immunoreactivity of gut into pancreatic glucagon in vivo. Diabetes 30:792–794

Kubeś L, Jirasek K, Lomsky R (1974) Endocrine cells of the dog gastrointestinal mucosa. Cytologia (Tokyo) 39:179–194

Larsson L-I, Holst JJ, Håkanson R, Sundler F (1975) Distribution and properties of glucagon immunoreactivity in the digestive tract of various mammals: an immunohistochemical and immunochemical study. Histochemistry 44:281–290

Lawrence AM, Tan S, Hojvat S, Kirsteins L (1976) Salivary gland hyperglycemic factor: an extra-pancreatic source of glucagon-like material. Science 195:70–72
Lefèbvre PJ, Luyckx AS (1976) Plasma glucagon after kidney exclusion: experiments in somatostatin-infused and in eviscerated dogs. Metabolism 25:761–768
Lefèbvre PJ, Luyckx AS (1977) Factors controlling gastric-glucagon release. J Clin Invest 59:716–722
Lefèbvre PJ, Luyckx AS (1978a) Le glucagon extrapancréatique: données expérimentales et importance clinique. Journ Annu Diabetol Hôtel Dieu: 195–207
Lefèbvre PJ, Luyckx AS (1978b) Gastric-glucagon: physiology and pathology. In: Grossmann M, Speranza V, Basso N, Lezoche E (eds) Gastrointestinal hormones and pathology of the digestive system. Plenum, New York London, pp 173–181
Lefèbvre PJ, Luyckx AS (1978c) Stimulation of gastric-glucagon release by prostaglandin $E_1$. Prostaglandins Med 1:419–420
Lefèbvre PJ, Luyckx AS (1978d) Glucose and insulin in the regulation of glucagon release from the isolated perfused dog stomach. Endocrinology 103:1579–1582
Lefèbvre PJ, Luyckx AS (1980a) Extrapancreatic glucagon: experimental studies using the isolated perfused dog stomach. In: Andreani D, Lefèbvre PJ, Marks V (eds) Current views on hypoglycemia and glucagon. Academic Press, London, pp 27–35
Lefèbvre PJ, Luyckx AS (1980b) Neurotransmitters and glucagon release from the isolated perfused canine stomach. Diabetes 29:697–701
Lefèbvre PJ, Luyckx AS, Brassinne AH, Nizet AH (1976) Glucagon and gastrin release by the isolated perfused dog stomach in response to arginine. Metabolism 25 [Suppl 1]: 25:1477–1479
Lefèbvre PJ, Luyckx AS, Brassinne AH (1978a) Vagal stimulation and its role in eliciting gastrin but not glucagon release from the isolated perfused dog stomach. Gut 19:185–188
Lefèbvre PJ, Luyckx AS, Moerman E, Bogaert M (1978b) Scorpion venom-induced release of noradrenaline does not modify glucagon output from the isolated perfused dog stomach. Horm Metab Res 10:80–81
Lefèbvre PJ, Luyckx AS, Brassinne AH (1979) Stimulation de la libération de glucagon gastrique par la carbamylcholine. Ann Endocrinol (Paris) 40:249–250
Lefèbvre PJ, Luyckx AS, Brassinne AH (1981) Inhibition by somatostatin of carbamylcholine-induced gastrin and glucagon release from the isolated perfused canine stomach. Gut 22:793–797
Lickley HLA, Kemmer FW, Gray DE, Kovacevic N, Hatton TW, Perez G, Vranic M (1981) Chromatographic pattern of extrapancreatic glucagon and glucagon-like immunoreactivity before and during stimulation by epinephrine and participation of glucagon in epinephrine-induced hepatic glucose overproduction. Surgery 90:186–194
Luyckx AS, Lefèbvre PJ (1974) Mechanisms involved in the exercise-induced increase in glucagon secretion in rats. Diabetes 23:81–93
Luyckx AS, Cession-Fossion A, Dresse A, Lefèbvre PJ (1975) Role of catecholamines in exercise-induced glucagon and free fatty acid mobilization in rat. Am J Physiol 229:376–383
Marre M, Bobbioni E, Suarez M, Reach G, Dubois MP, Assan R (1979) Control of gastric glucagon secretion in the acutely pancreatectomized rat. Diabetes 28:213–220
Mashiter K, Harding PE, Chou M, Mashiter GD, Stout J, Diamond D, Field JB (1975) Persistent pancreatic glucagon but not insulin response to arginine in pancreatectomized dogs. Endocrinology 96:678–693
Matsuyama T, Foà PP (1974) Plasma glucose, insulin, pancreatic and enteroglucagon levels in normal and depancreatized dogs. Proc Soc Exp Biol Med 147:97–102
Matsuyama T, Tanaka R, Shima K, Nonaka K, Tarui S (1978) Lack of gastrointestinal glucagon response to hypoglycemia in depancreatized dogs. Diabetologia 15:471–474
Matsuyama T, Tanaka R, Shima K, Tarui S (1979) Failure of somatostatin to decrease blood glucose by suppression of extrapancreatic glucagon in severely diabetic depancreatized dogs. Endocrinol Jpn 25:529–532
Morita S, Doi K, Yip CC, Vranic M (1976) Measurement and partial characterization of immunoreactive glucagon in gastrointestinal tissues of dogs. Diabetes 25:1018–1025

Müller WA, Girardier L, Seydoux J, Berger M, Renold AE, Vranic M (1978) Extrapancreatic glucagon and glucagon-like immunoreactivity in depancreatized dogs. J Clin Invest 62:124–132
Muñoz-Barragan L, Blazquez E, Patton GS, Dobbs RE, Unger RH (1976) Gastric A-cell function in normal dogs. Am J Physiol 231:1057–1061
Muñoz-Barragan L, Rufener C, Srikant CB, Dobbs RE, Shannon WA Jr, Baetens D, Unger RH (1977) Immunocytochemical evidence for glucagon-containing cells in the human stomach. Horm Metab Res 9:37–39
O'Connor FA, Conlon JM, Buchanan KD, Murphy RF (1979) The use of perfused rat intestine to characterise the glucagon-like immunoreactivity released into serosal secretions following stimulation by glucose. Horm Metab Res 11:19–23
Ohneda A, Kobayashi T, Nihei J, Umezu M, Sakai T, Sanoyama K (1979) Response of extrapancreatic glucagon to arginine in dogs. Tohoku J Exp Med 129:45–54
Orci L (1976) Polypeptide hormones: molecular and cellular aspects. Ciba Found Symp 41:344
Orci L, Forssmann WG, Forssmann W, Roullier C (1968) Electron microscopy of the intestinal endocrine cells. Comparative study. In: Bocchiarelli SD (ed) Electron microscopy. 4th European Regional Conference. Tipografia Poliglotto Vaticana, Rome, pp 369–370
Penhos JC, Ezequiel M, Lepp A, Ramey ER (1975) Plasma immunoreactive insulin (IRI) and immunoreactive glucagon (IRG) after evisceration with and without a functional liver. Diabetes 24:637–640
Perez-Castillo A, Blazquez E (1980a) Synthesis and release of glucagon by human salivary glands. Diabetologia 19:123–129
Perez-Castillo A, Blazquez E (1980b) Tissue distribution of glucagon, glucagon-like immunoreactivity and insulin in the rat. Am J Physiol 238:E258–266
Ravazzola M, Orci L (1978) Immunofluorescence and immunochemistry of glicentin (gut GLI-1) immunoreactive cells: their relationship with glucagon producing cells. Diabetologia 15:264–265
Ravazzola M, Orci L (1980a) Transformation of glicentin-containing L-cells into glucagon-containing cells by enzymatic digestion. Diabetes 29:156–158
Ravazzola M, Orci L (1980b) Glucagon and glicentin immunoreactivity are topologically segregated in the α granule of the human pancreatic A-cell. Nature 284:66–67
Ravazzola M, Baetens D, Engerman E, Kovacevic N, Vranic M, Orci L (1977) Endocrine cells in oxyntic mucosa of a dog 5 years after pancreatectomy. Horm Metab Res 9:480–483
Ravazzola M, Siperstein A, Moody AJ, Sundby F, Jacobsen H, Orci L (1979) Glicentin immunoreactive cells: their relationship to glucagon-producing cells. Endocrinology 105:499–508
Ravazzola M, Orci L, Perrelet A, Unger RH (1981a) Immunocytochemical quantitation of glicentin and glucagon during maturation of A-cell secretory granules. Diabetologia 21:319
Ravazzola M, Unger RH, Orci L (1981b) Demonstration of glucagon in the stomach of human fetuses. Diabetes 30:879–882
Ross G, Lickley L, Vranic M (1978) Extrapancreatic glucagon in control of glucose turnover in depancreatized dogs. Am J Physiol 234:E213–E219
Sasagawa T, Kobayashi S, Fujita T (1974) Electron microscope studies on the endocrine cells of the human gut and pancreas. In: Fujita T (ed) Gastro-enteropancreatic endocrine system. A cell biological approach. Igahu Shoin, Tokyo, pp 17–38
Sasaki H, Rubalcava B, Baetens D, Blazquez E, Srikant CB, Orci L, Unger RH (1975) Identification of glucagon in the gastrointestinal tract. J CLin Invest 56:135–145
Sasaki H, Ebitani I, Tominaga M, Yamatani K, Yawata Y, Hara M (1980) Glucagon-like substance in the canine brain. Endocrinol Jpn 27:135–140
Smith SS, Bhathena SJ, Nompleggi D, Penhos JC, Recant L (1978) Studies on persistent circulating immunoreactive glucagon (IRG) and immunoreactive insulin (IRI) found in eviscerated rats with a functional liver. Diabetologia 14:177–184
Smith S, Mazur A, Voyles N(1979) Is submaxillary gland immunoreactive glucagon important in carbohydrate metabolism? Metabolism 28:343–347

Solcia E, Vassallo G, Capella C (1970) Cytology and cytochemistry of hormone-producing cells of the upper gastro-intestinal tract. In: Creutzfeldt W (ed) Origin, physiology and pathophysiology of the gastrointestinal hormones. Schattauer, Stuttgart, pp 3–29
Srikant CB, McCorkle K, Unger RH (1977) Properties of immunoreactive fractions of canine stomach and pancreas. J Biol Chem 252:1847–1851
Sutherland EW, de Duve C (1948) Origin and distribution of the hyperglycemic-glycogenolytic factor of the pancreas. J Biol Chem 175:663–674
Tager H, Hohenboken M, Markese J, Dinerstein RJ (1980) Identification and localization of glucagon-related peptides in rat brain. Proc Natl Acad Sci USA 77:6229–6233
Tiengo A, Bessioud M, Valverde I, Tabbi A, Alexandre J, Assan R (1980) Absence de glucagon chez les humains pancréatectomisés. Diab Metab 6:172
Tominaga M, Ebitani I, Marubashi S, Kamimura T, Katagiri T, Sasaki H (1981) Species difference of glucagon-like materials in the brain. Life Sci 29:1577–1581
Tronier B, Kåresen R, Aune S (1981) Glucagon immunoreactivity in man after pancreas resection and total pancreatectomy measured by different antisera. Horm Metab Res 13:56–57
Vranic M, Pek S, Kawamori R (1974) Increased "glucagon immunoreactivity" in plasma of totally depancreatized dogs. Diabetes 23:905–912
Vranic M, Kawamori R, Pek S, Kovacevic N, Wrenshall GA (1976) The essentiality of insulin and the role of glucagon in regulating glucose utilization and production during strenuous exercise in dogs. J Clin Invest 57:245–255
Werner PL, Palmer JP (1978) Immunoreactive glucagon responses to oral glucose, insulin infusion and deprivation, and somatostatin in pancreatectomized man. Diabetes 27:1005–1012
Yoshida T, Kondo M (1980a) Effect of acetylcholine on the secretion of gut glucagon immunoreactivity and gut glucagon-like immunoreactivity in pancreatectomized dogs. Endocrinol Jpn 27:33–38
Yoshida T, Kondo M (1980b) Effect of adrenergic agents on the secretion of gastrointestinal immunoreactive glucagon in depancreatized dogs. Diabetes 29:355–360
Yoshida T, Kondo M (1980c) The effect of hypophysectomy and hypophysis-transplantation on the secretion of gut glucagon immunoreactivity and gut glucagon-like immunoreactivity in depancreatized dogs. Endocrinol Jpn 27:77–81

# Glucagon in Various Physiological Conditions

CHAPTER 34

# Glucagon and Starvation

R. A. GELFAND and R. S. SHERWIN

## A. Introduction

When food supply is interrupted, survival is dependent upon a highly integrated metabolic response directed at maintaining glucose homeostasis and conserving body protein. Initially, the predominant metabolic requirement is the maintenance of a continuing supply of blood glucose for utilization by obligate glucose-consuming tissues, especially brain. Because liver glycogen stores are rapidly exhausted, continuing hepatic glucose production to maintain normoglycemia depends on an early acceleration of hepatic gluconeogenesis. However, since amino acids represent the sole precursors for de novo glucose synthesis, the demand for continuing glucose production imposes a steady drain on the body's protein stores. Inasmuch as significant depletion of body protein (beyond 30%–50%) is generally fatal, survival during prolonged starvation ultimately depends upon a reduction of glucose consumption and a shift toward maximal utilization of fat, the body's major storage fuel. The overall response to starvation may thus be characterized as biphasic, with the early, gluconeogenic phase most pronounced during the first 3–5 days, and the later, protein-conserving phase dominating after 2–4 weeks (SAUDEK and FELIG 1976).

For the primitive human, few metabolic adaptations could have rivaled in survival value the capacity to endure prolonged periods without food. Today, problems related to starvation are encountered not only in underdeveloped parts of the world, but in highly developed countries as well, in patients suffering from a variety of debilitating catabolic illnesses. Management of such patients would be facilitated by an understanding of the hormonal mechanisms mediating the normal adaptive response to starvation. The crucial role of the decline in plasma insulin as the primary regulatory signal governing the transition from the fed to fasted state has long been emphasized (CAHILL et al. 1966; CAHILL 1970, 1971). In the last decade, however, the potential role of glucagon in the adaptation to fasting has attracted increasing attention. This chapter reviews the metabolic events which characterize starvation in humans, with emphasis on glucagon's potential regulatory function.

## B. The Postabsorptive State

After an overnight fast, the concentrations of hormones and substrates which were altered by meal ingestion during the preceding day have returned to baseline, and the rate of fuel consumption is closely matched by endogenous fuel production.

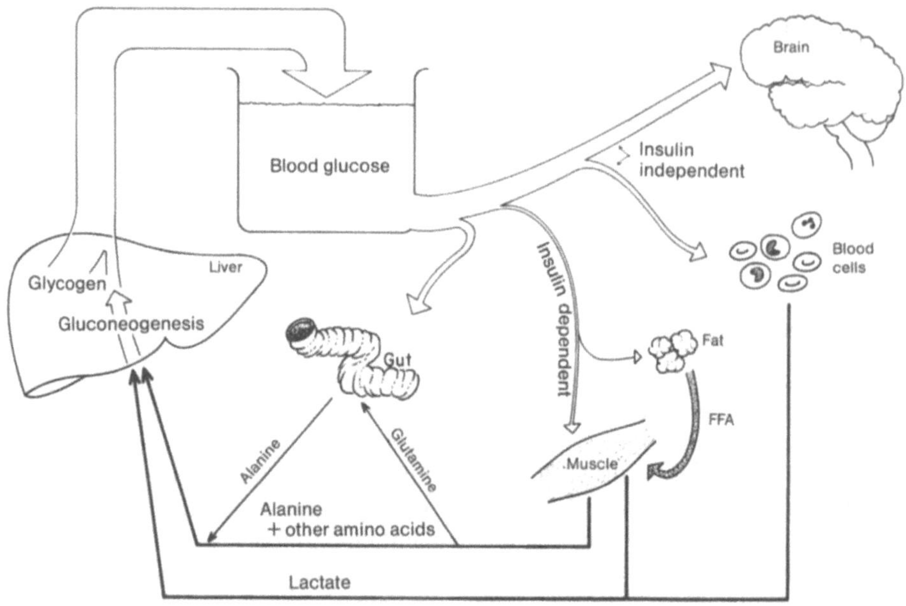


**Fig. 1.** Glucose homeostasis in the postabsorptive state in normal humans

The postabsorptive state thus serves as a useful reference point, as it represents the period of metabolic transition from the fed to the fasted condition.

In normal humans, the decline in circulating insulin to postabsorptive levels results in a marked diminution of glucose uptake by peripheral insulin-sensitive tissues (muscle and adipose tissues) and a shift toward utilization of fatty acids as energy-yielding fuels. Glucose consumption, nevertheless, continues by noninsulin-dependent tissues (brain, renal medulla, and formed elements of the blood) and the splanchnic bed (SACCA et al. 1982), so that total glucose uptake occurs at a rate of 2.0–2.5 mg $kg^{-1}$ $min^{-1}$ (SACCA et al. 1979), or 200–250 g/day. The major site of glucose uptake is the brain, which in the postabsorptive state is critically dependent on an ongoing supply of glucose for oxidative metabolism (REINMUTH et al. 1965). Maintenance of blood glucose homeostasis in this circumstance is achieved by the hepatic production of glucose at rates equal to those of tissue utilization.

The hepatic processes involved in the release of glucose into the bloodstream consist of glycogenolysis and gluconeogenesis. On the basis of studies employing splanchnic balances of gluconeogenic substrates (FELIG 1973) and the rate of disappearance of liver glycogen in biopsy samples (HULTMAN and NILSSON 1971), it has been estimated that approximately 70%–80% of hepatic glucose production is derived from glycogenolysis with gluconeogenesis contributing the remainder. The resynthesis of glucose from the glycolytic intermediate, lactate accounts for at least one-half of the gluconeogenic component, and the conversion of glycogenic amino acids comprises most of the remainder. Alanine, whose release from muscle (FELIG et al. 1970) and uptake by liver (FELIG et al. 1969a) predominates over that of other amino acids, is the major amino acid contributing to glucose synthesis. Conversion of fat-derived glycerol and recycled pyruvate contribute less than 2%

**Table 1.** Effects of glucagon, insulin, and somatostatin on total glucose production and gluconeogenesis

| | Total glucose production | Gluconeogenesis |
|---|---|---|
| Somatostatin (SRIF) | Transient ↓ | ↑ |
| SRIF + Glucagon | Transient ↑ | ↑↑ |
| SRIF + Insulin | ↓↓ | Transient ↓ |

and 1%, respectively, to total glucose production (FELIG 1973). Fuel homeostasis in the postabsorptive state is summarized in Fig. 1.

Regarding the hormonal factors regulating glucose metabolism in postabsorptive humans, it has long been suspected that both glucagon and insulin modulate glucose release by the liver. The discovery of somatostatin and its suppressive effects on glucagon and insulin secretion has provided, for the first time, a means of assessing the contribution of both islet hormones to the maintenance of hepatic glucose production and gluconeogenesis. Interestingly, studies using somatostatin suggest that the manner in which glucagon and insulin interact to regulate total glucose production and gluconeogenesis may differ (Table 1). This may be explained by the fact that total glucose production in the postabsorptive state is, for the most part, derived from glycogenolysis, a process which is regulated independently of gluconeogenesis (CHERRINGTON et al. 1981). Data relating to total glucose production in the postabsorptive state thus cannot be directly extrapolated to more prolonged fasting, when endogenous glucose production is totally accounted for by gluconeogenesis.

Perhaps the most compelling evidence that basal glucagon secretion is important in maintaining glucose production in the postabsorptive state derives from studies in which somatostatin is infused to suppress plasma glucagon, and circulating insulin is prevented from falling by exogenous hormone infusion. In this circumstance, a sustained 70%–75% reduction in hepatic glucose production occurs both in dogs (CHERRINGTON et al. 1979) and humans (LILJENQUIST et al. 1977), indicating a continuing effect of basal concentrations of glucagon in opposing the inhibitory actions of insulin on the liver. However, when the rate of hepatic gluconeogenesis was estimated in the dog by infusing labeled alanine and determining its conversion of glucose, CHERRINGTON et al. (1979) observed only a small, transient (< 15%) decrease in hepatic gluconeogenesis. Taken together, these findings suggest that the presence of glucagon is critical for the maintenance of hepatic glycogenolysis in the postabsorptive state, but does not appreciably affect the rate of hepatic gluconeogenesis when basal levels of insulin are present.

The restraining influence of basal concentrations of insulin on postabsorptive glucose production is also evident when plasma insulin is suppressed by somatostatin, and circulating glucagon is maintained at the basal level by exogenous glucagon infusion (LILJENQUIST et al. 1977; CHERRINGTON et al. 1978; SHERWIN et al. 1977 b). Under these conditions hepatic glucose production promptly increases 2–3-fold. In contrast, when somatostatin alone is infused, hepatic glucose production declines (CHERRINGTON et al. 1978; SHERWIN et al. 1977 a; ALTSZULER et al. 1976).

Although these studies suggest that the stimulatory effect of insulin deficiency on glucose production is dependent on the presence of glucagon, the influence of glucagon appears to be short-lived. When the infusions are extended beyond 2–3 h, the rate of hepatic glucose production in each instance returns toward baseline and ultimately is no different whether or not glucagon is replaced (CHERRINGTON et al. 1978; SHERWIN et al. 1977b). Thus, when insulin is deficient, a sustained stimulatory effect of glucagon on total hepatic glucose production is difficult to demonstrate.

It should be emphasized, however, that these observations are of limited relevance when fasting is extended and hepatic glucose production is exclusively dependent on gluconeogenesis. When the influence of basal glucagon on gluconeogenesis is directly quantitated in insulin-deficient postabsorptive dogs by means of isotopic techniques, a different pattern of response emerges. Combined suppression of glucagon and insulin with somatostatin results in a progressive twofold increase in the conversion of labeled alanine to glucose (CHERRINGTON et al. 1978). When basal glucagon concentrations are restored by intraportal glucagon infusion, a further twofold increase in gluconeogenesis is observed (CHERRINGTON et al. 1978). Most importantly, these effects on gluconeogenesis are sustained, in contrast to glucagon's transient influence on total glucose production in the insulin-deficient state (Table 1). These data indicate that while insulin deficiency per se enhances gluconeogenesis, this effect is exaggerated by the presence of basal concentrations of glucagon. Glucagon's most pronounced effects on gluconeogenesis are thus observed when insulin is deficient (Table 1). This point deserves emphasis when glucagon's potential regulatory function during starvation is considered, since glucose homeostasis at this time is critically dependent on gluconeogenesis, and hypoinsulinemia characterizes the hormonal milieu.

## C. Glucoregulatory Hormones in Starvation

The metabolic events which characterize starvation can best be appreciated in the context of the hormonal alterations which occur in this setting. Starvation induces characteristic changes in circulating concentrations of insulin and glucagon which are illustrated in Fig. 2. Insulin levels exhibit a rapid decline to concentrations about 50% below baseline within 3–5 days (CAHILL et al. 1966). Thereafter, plasma insulin levels remain relatively stable as fasting continues (OWEN et al. 1969; MARLISS et al. 1970). The hypoinsulinemia results exclusively from suppression of insulin secretion, since the metabolic clearance rate of insulin is unchanged (DEFRONZO et al. 1978). The decline in circulating insulin closely parallels the blood glucose concentration, which drops progressively over the first 3–5 days, and then stabilizes at levels about 15–20 mg/dl below baseline for the subsequent duration of the fast (CAHILL et al. 1966; OWEN et al. 1969; MARLISS et al. 1970). Whether the fall in insulin secretion is solely due to the decline in glucose levels has recently been questioned (LILAVIVATHANA et al. 1978).

Glucagon, in contrast, rises early in starvation to peak concentrations 50%–100% above basal on about the third day, after which there is a gradual decline over the ensuing weeks toward postabsorptive levels (MARLISS et al. 1970; FISHER et al. 1976). Interestingly, as shown in Fig. 3, the early rise in glucagon results not

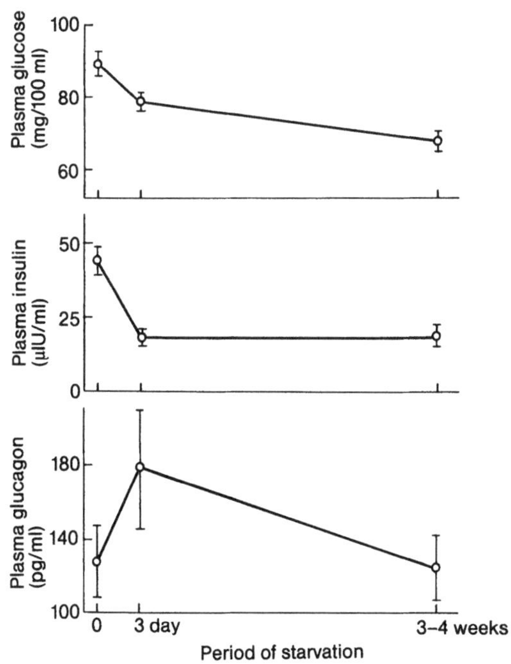


**Fig. 2.** Changes in plasma glucose, insulin, and glucagon during brief and prolonged starvation in humans

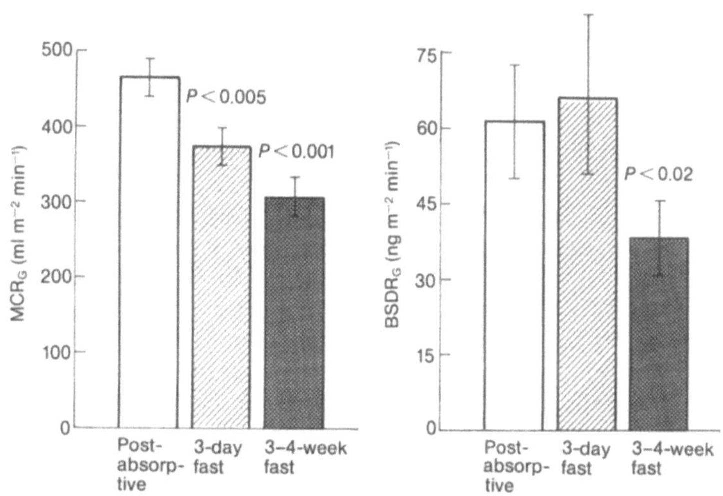


**Fig. 3.** Influence of short-term (3 day) and prolonged starvation (3–4 weeks) on the metabolic clearance rate ($MCR_G$) and basal systemic delivery rate ($BSDR_G$) of glucagon in obese subjects. *P* values represent the significance of differences from the postabsorptive state. After FISHER et al. (1976)

from augmented secretion, but rather reflects a 20% reduction in its metabolic clearance rate (FISHER et al. 1976). As starvation continues for 3–4 weeks, decreased glucagon secretion accounts for the return of plasma glucagon toward baseline, despite a further reduction in glucagon disposal (FISHER et al. 1976). The recent suggestion (BURMAN et al. 1980) that the fall in triiodothyronine levels with fasting may mediate this decrease in glucagon catabolism requires confirmation. The mechanism underlying the ultimate fall in glucagon secretion has not been established.

# D. Metabolic Alterations in Starvation

## I. The Early Phase

Since liver glycogen is limited to about 70 g after an overnight fast, while glucose consumption occurs at a rate of approximately 200–250 g/day, hepatic glycogen stores are rapidly dissipated very early in the course of fasting. By 72 h, glycogen is essentially absent from liver biopsy samples (HULTMAN and NILSSON 1971). Thus, the initial phase of starvation is characterized by an acceleration of hepatic gluconeogenesis to meet ongoing tissue demands (GARBER et al. 1974). The overall dependence of glucose production on gluconeogenesis from protein is reflected in the active rate of urinary nitrogen excretion early in starvation (CAHILL et al. 1966; OWEN et al. 1969).

The augmentation of gluconeogenesis and the maintenance of glucose homeostasis are mediated by both hepatic and extrahepatic events. The release of alanine and other glycogenic amino acids from muscle increases (POZEFSKY et al. 1976a) and the rate of hepatic conversion of labeled alanine to glucose rises threefold (CHIASSON et al. 1977). The enhancement of glucose synthesis from amino acids is not, however, solely a function of increased availability of these precursors, since plasma levels of alanine and other glycogenic amino acids actually fall (FELIG et al. 1969; POZEFSKY et al. 1976a), despite their increased release from muscle. These observations indicate that intrahepatic gluconeogenic mechanisms are specifically stimulated during short-term fasting. An additional factor contributing to glucose homeostasis at this time is the increased release of free fatty acids (FFA) from adipose tissue. Peripheral oxidation of fatty acids spares glucose for use by brain, while their oxidation by liver activates key gluconeogenic enzymes and furnishes the energy and reducing power necessary for glucose synthesis (WILLIAMSON et al. 1966).

These metabolic alterations coincide with the development of hypoinsulinemia and the attainment of peak increments in plasma glucagon at 72 h of starvation. The manner in which these hormonal changes regulate fuel homeostasis in the early fasted state is depicted in Fig. 4. Insulin deficiency promotes all aspects of this metabolic response, by virtue of its pronounced effects on both liver and peripheral tissues. The effects of glucagon, in contrast, are confined to the liver.

That the early fall in plasma insulin represents a sine qua non of successful metabolic adaptation to fasting is a well-established clinical observation. Certain patients with insulinoma who do not exhibit absolute hyperinsulinemia, nonetheless experience life-threatening hypoglycemia within 72 h of fasting, because of failure of plasma insulin concentrations to fall below postabsorptive levels (SHERWIN and FELIG 1981). The decline in plasma insulin with fasting directly stimulates hepatic gluconeogenesis, results in virtual cessation of glucose uptake by insulin-sensitive tissues, and promotes enhanced release of FFA from adipose tissue and amino acids from muscle (Fig. 4). These effects of hypoinsulinemia may be exag-

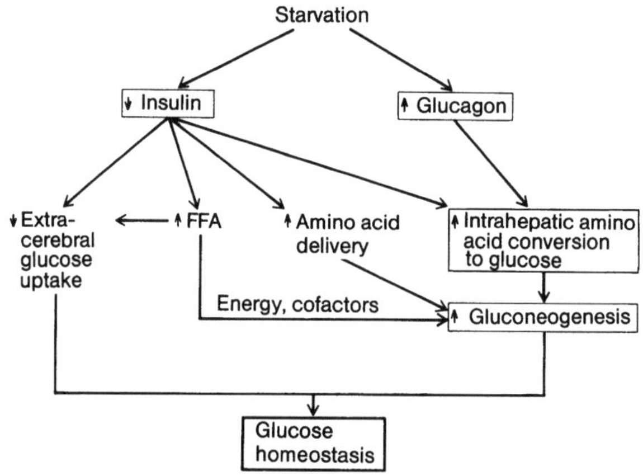


**Fig. 4.** Hormonally mediated metabolic alterations maintaining glucose homeostasis during fasting

gerated by tissue resistance to the action of insulin in fasting humans (DeFronzo et al. 1978). The modest hyperglucagonemia of early starvation, while stimulating hepatic gluconeogenesis, probably exerts no major influence on peripheral fuel metabolism. Although higher concentrations of glucagon have recently been reported to stimulate lipolysis directly (Schneider et al. 1981), other similar studies have demonstrated no effect of glucagon on lipolysis, glucose utilization, or amino acid exchange across the human forearm (Pozefsky et al. 1976b).

As discussed previously, stimulation of gluconeogenesis by hypoinsulinemia was clearly demonstrated by studies conducted in the postabsorptive state using somatostatin (Table 1). Selective suppression of plasma insulin levels by infusion of somatostatin and replacement doses of exogenous glucagon in postabsorptive dogs enhanced the conversion of alanine $C^{14}$ to glucose $C^{14}$ 3–4-fold above basal (Cherrington et al. 1978). Hypoinsulinemia stimulated gluconeogenesis in these experiments whether or not basal glucagon was replaced; however, the effect was significantly more pronounced in the presence of basal glucagon (Table 1). It will be recalled that this influence of basal glucagon was not observed when basal concentrations of insulin were maintained (Cherrington et al. 1979). Thus, insulin deficiency not only itself stimulates hepatic gluconeogenesis, but apparently also renders the liver more sensitive to the stimulatory influence of glucagon on gluconeogenesis.

A specific role for glucagon in stimulating gluconeogenesis had long been suspected on the basis of earlier in vitro data demonstrating marked enhancement by glucagon of alanine uptake and conversion to glucose in perfused rat liver (Mallette et al. 1969). While the somatostatin data indicate that gluconeogenesis is stimulated by basal levels of glucagon when insulin is deficient, other studies suggest that an elevation of plasma glucagon above basal may stimulate hepatic gluconeogenesis, even without coexisting insulin deficiency. Chiasson et al. (1974, 1975) showed that pharmacologic infusion of glucagon (15–50 ng $kg^{-1}$ $min^{-1}$) causes a doubling in hepatic glucose synthesis from alanine in both humans and animals. Lower dose infusions (5 ng $kg^{-1}$ $min^{-1}$), achieving more physiologic

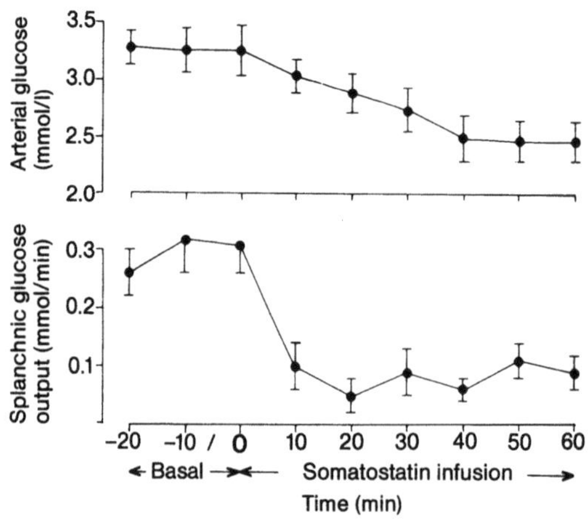


**Fig. 5.** Changes in arterial glucose and splanchnic glucose output produced by somatostatin infusion in 60-h fasted humans. After WAHREN et al. (1977)

elevations in glucagon, however, exerted considerably less effect (CHIASSON et al. 1974). More recently, studies by CHERRINGTON et al. (1981) have suggested that the influence of glucagon may have been underestimated in previous experiments, because of the inhibitory effects of the hyperinsulinemia induced by glucagon administration. Using somatostatin to prevent the counterregulatory rise in plasma insulin, these workers demonstrated a pronounced stimulation of gluconeogenesis in the dog when plasma glucagon was increased 150%–200% by physiologic glucagon infusion (2.6 ng $kg^{-1}$ $min^{-1}$) (CHERRINGTON et al. 1981). In addition, this enhancement of gluconeogenesis was associated with an increase in the rate of disappearance of labeled alanine from plasma, an effect not seen in earlier infusion studies (CHIASSON et al. 1975), in which gluconeogenesis was nonetheless stimulated. Hyperglucagonemia may thus stimulate alanine uptake, although its major effect appears to be on intrahepatic gluconeogenic pathways.

These observations suggest that, in early starvation, the rise in plasma glucagon may contribute to the acceleration of gluconeogenesis which occurs as glycogen stores are exhausted. Although the increment in plasma glucagon induced by starvation (50%–100%) is less than that during glucagon infusion (150%–300%), the concomitant fall in plasma insulin with fasting would be expected to exaggerate the influence of even modest hyperglucagonemia. Direct demonstration of the importance of endogenous glucagon secretion in maintaining hepatic glucose production in the early fasted state has been provided by the studies of KELLER et al. (1977) in 72 h fasted dogs, and WAHREN et al. (1977) in 60 h fasted humans (Fig. 5). These investigators showed that inhibition of insulin and glucagon secretion by somatostatin infusion produced a prompt fall in hepatic glucose production within 1 h, by 35% and 70%, respectively. These findings suggest that, as insulin levels fall in early starvation, the rate of hepatic gluconeogenesis becomes critically dependent upon circulating glucagon levels. In contrast to the postabsorptive state, where the dominant factor controlling gluconeogenesis is the restraining influence of basal insulin levels, with the development of fasting hypoinsulinemia, the stimulatory effect of glucagon assumes major importance. These data in the fasted state thus complement the somatostatin data obtained postabsorptively. Together, they imply a crucial role for glucagon in the maintenance of hepatic glucose production during fasting.

In summary, early in starvation, hypoinsulinemia and hyperglucagonemia act in concert to maximize hepatic gluconeogenesis. Hypoinsulinemia enhances the delivery of substrate amino acids from muscle to liver, where increased quantities of glucose are synthesized under the influence of both the elevation in plasma glucagon and the depression in plasma insulin. Hypoinsulinemia further contributes to glucose homeostasis by reducing extracerebral glucose consumption and by increasing the availability of FFA for oxidative metabolism by muscle and liver. One final factor which may contribute to the enhancement of gluconeogenesis during starvation is the decline in the plasma levels of glucose itself. Since hyperglycemia per se has been shown to inhibit hepatic alanine extraction and its conversion to glucose (SHULMAN et al. 1980), it is conceivable that a reduction in glucose concentration may enhance these processes.

## II. Prolonged Starvation

Because de novo glucose synthesis depends solely on protein-derived amino acids, the conservation of body protein necessary for survival during prolonged starvation depends on a gradual reduction in the demand for gluconeogenesis. This is mediated largely by a shift toward the use of fat-derived substrates rather than glucose as oxidative fuels. In addition, basal metabolic rate and oxygen consumption decline (KEYS et al. 1950), which may be related to the fall in circulating levels of triiodothyronine (VAGENAKIS et al. 1975). The increasing reliance on fat as fuel and decreased breakdown of body protein were both demonstrated almost 70 years ago by the pioneering studies of BENEDICT (1915). The alterations in circulating substrate concentrations and interorgan fuel fluxes which underlie this adaptation have since been elucidated.

Plasma levels of FFA approximately double during the first 3–7 days of fasting, after which they remain reatively stable (OWEN et al. 1969; MARLISS et al. 1970). The increased delivery of FFA from adipose tissue to liver, where they are oxidatively catabolized, serves to promote the marked acceleration of hepatic ketogenesis that occurs during the first few days of fasting. Peak rates of hepatic ketone production (80–115 gm/24 h) are achieved by the third day, and are maintained as long as fasting continues (OWEN et al. 1969; GARBER et al. 1974). Despite relatively constant or even slightly declining rates of ketone production after the third day, blood ketone levels continue to rise throughout the first 3–4 weeks of fasting (OWEN et al. 1969), as their extraction and oxidation by muscle progressively decline (OWEN and REICHARD 1971). As a consequence, the central nervous system is supplied in increasing abundance with these fat-derived substrates, which after prolonged starvation in humans supply over one-half of the brain's energy requirements (OWEN et al. 1967). This reduction in cerebral glucose consumption is crucial in reducing the need for hepatic gluconeogenesis, which in turn is essential for the conservation of body protein. The diminution in gluconeogenesis with prolonged

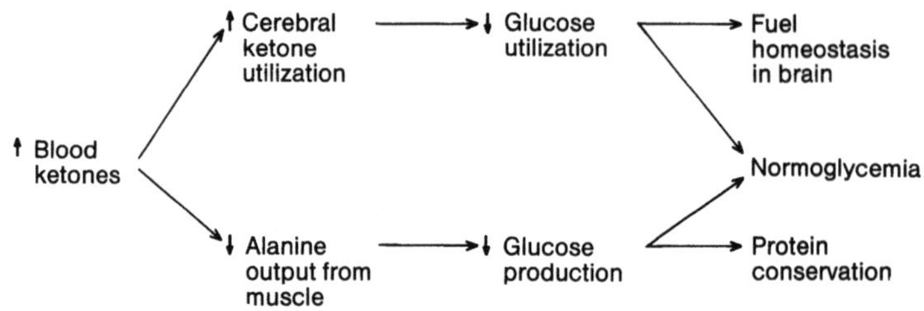


**Fig. 6.** Role of ketone bodies in maintaining normoglycemia and promoting protein conservation during starvation

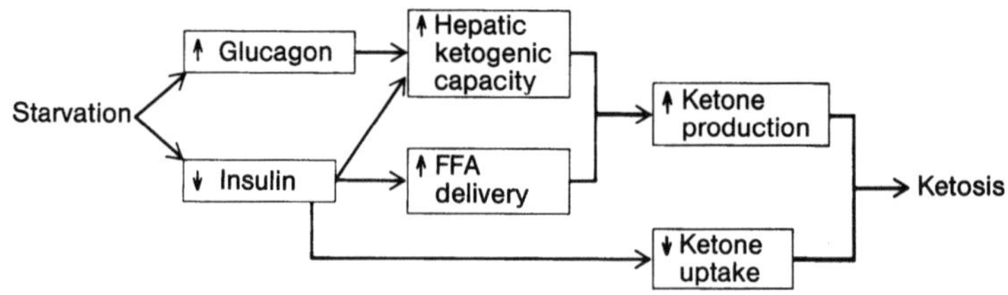


**Fig. 7.** Interaction of insulin and glucagon in promoting starvation-induced ketosis

fasting is associated with a progressive decline in plasma alanine levels (FELIG et al. 1969 a), as a consequence of a marked fall in muscle alanine release to rates considerably below those in the postabsorptive state (FELIG et al. 1970). It is this reduction in alanine availability, rather than an inhibition of hepatic gluconeogenic mechanisms, that ultimately accounts for the decrease in hepatic glucose synthesis (FELIG et al. 1969 b). It has been suggested that ketone bodies themselves may mediate this response by acting directly on muscle to retard protein breakdown and reduce alanine release (SHERWIN et al. 1975; PALAIOLOGOS and FELIG 1976). Figure 6 illustrates the central role of ketosis in fuel homeostasis and protein conservation during prolonged starvation.

The formation and accumulation of ketone acids involve three distinct metabolic events (Fig. 7): (1) delivery of FFA from adipose tissue; (2) hepatic oxidation of FFA to acetyl-CoA, from which ketones are formed (hepatic ketogenic capacity); and (3) a reduction in ketone utilization by peripheral tissues. Insulin deficiency, which activates each of these steps, is critical not only for the development of fasting hyperketonemia, but for all ketotic states. Hypoinsulinemia activates the enhanced release from adipose tissue of FFA (step 1), an absolute prerequisite for the development of ketosis. Insulin deficiency also has a direct ketogenic effect on the liver itself (step 2), as evidenced by the demonstration that in vivo administration of anti-insulin serum enhances the ability of rat liver to synthesize ketones when perfused with oleic acid (MCGARRY et al. 1975). Finally, by virtue of its influence on ketone turnover (step 3) (SHERWIN et al. 1976), hypoinsulinemia may enhance the magnitude of the hyperketonemic response in any state of increased ketone production.

The potential contribution of glucagon to fasting ketosis derives largely from the hormone's effect on the ketogenic capacity of the liver[1]. The mechanisms of hepatic ketogenesis and its regulation have been extensively studied by McGARRY and FOSTER (McGARRY et al. 1975; McGARRY and FOSTER 1976, 1977, 1981; McGARRY 1979) who have also reviewed them in Chap. 17. Maximal ketogenesis requires, in addition to a sufficient rate of delivery of FFA, a change in the metabolic "set" of the liver, such that an increased fraction of incoming FFA enters the intramitochondrial β-oxidative pathway by which β-hydroxybutyric and acetoacetic acids are formed. The enhancement of this pathway in ketotic states appears to be mediated by an activation of the carnitine acyltransferase system, which acts to shuttle FFA into mitochondria. An important role for glucagon in enhancing hepatic ketogenic capacity was suggested by these workers' demonstration that livers obtained from rats given glucagon exhibit augmented ketone production from oleic acid in vitro, analagous to the results obtained with anti-insulin serum (McGARRY et al. 1975). These results suggested that either insulin deficiency or glucagon excess were capable of changing the metabolic set of the liver toward ketogenesis.

Studies in humans, however, have demonstrated that only in circumstances of insulin deficiency does glucagon exert an important influence on hepatic ketogenic capacity. Normal subjects receiving physiologic glucagon infusion (3 ng $kg^{-1}$ $min^{-1}$) followed by administration of heparin to raise plasma FFA, exhibit no greater increment in plasma ketones than observed when similar FFA elevations are induced during saline control infusion (SCHADE and EATON 1976). This finding is not attributable to compensatory hyperinsulinemia, as identical results are obtained when basal insulin is fixed by combined infusion of somatostatin and replacement insulin (MILES et al. 1981). These observations indicate that hyperglucagonemia, even in the presence of elevated FFA, cannot promote ketosis in normal humans when basal concentrations of insulin are present. In contrast, when physiologic hyperglucagonemia is induced in either insulin-withdrawn diabetics (SCHADE and EATON 1975), or in normal subjects receiving somatostatin without insulin replacement (MILES et al. 1981), an augmentation in the hyperketonemic response to FFA elevation is observed. Further evidence for an important ketogenic role of glucagon in the setting of insulin deficiency derives from the observation that suppression of glucagon by somatostatin in insulin-withdrawn type I diabetics markedly delays and reduces the magnitude of hyperketonemia (GERICH et al. 1975).

These results suggest that during fasting, when insulin levels are low, hyperglucagonemia may contribute to the development of ketosis by enhancing the stimulatory effect of hypoinsulinemia on hepatic ketogenic capacity. However, studies by KELLER et al. (1977) examining the hormonal control of fasting ketogenesis in dogs, failed to demonstrate such an effect of glucagon in the maintenance of hepatic ketone production. These investigators measured hepatic ketone production

---

1 The possibility that glucagon might be involved in the mobilization of FFA by stimulation of adipose tissue lipolysis has been critically analyzed in Chap. 19. While there is little doubt that glucagon is concerned in starvation-induced adipose tissue lipolysis in some species (rodents, birds), there is no definite proof that such a mechanism is operating in humans

in 3-day fasted dogs during infusion of somatostatin, alone and in combination with intraportal replacement of either insulin or glucagon. Hepatic effects were assessed independent of substrate supply by infusion of Intralipid and heparin which produced similar 2–3-fold increments in plasma FFA in all groups. When both insulin and glucagon were suppressed by somatostatin infusion, a slight but insignificant increase in ketone production was observed at both low and high FFA levels. In contrast, maintenance of glucagon at preinfusion levels during somatostatin-induced insulinopenia was associated with a significant enhancement in ketone output at both FFA levels. These results confirm an important ketogenic influence of glucagon, at least when insulin is suppressed to levels below those observed in 3-day fasted dogs. However, when insulin levels were maintained at 3-day fasted levels by exogenous hormone infusion and plasma glucagon was suppressed by somatostatin, hepatic ketone production was unaltered at both low and high FFA levels as compared with 3-day fasted saline control dogs. These experiments suggest that while the decline in plasma insulin during short-term fasting is sufficient to initiate ketosis, insulin still exerts a significant restraining influence which predominates over the ketogenic potential of glucagon. Only with more profound insulin deficiency did the ketogenic action of glucagon become manifest.

Although these studies in dogs do not exclude a contributory role of glucagon in fasting ketogenesis in humans, they do suggest that insulin plays the predominant regulatory role. This conclusion is supported by studies in humans as well. When plasma insulin is suppressed with somatostatin in postabsorptive subjects to levels similar to those observed during fasting, ketosis is promoted whether or not glucagon is replaced (SHERWIN et al. 1977b). These observations are complemented by experiments in fasting humans that more directly address this issue. After a 60 h period of starvation, suppression of both insulin and glucagon with somatostatin results in a significant rise in the net splanchnic output and arterial concentrations of $\beta$-hydroxybutyrate (WAHREN et al. 1977). Although increased supply of substrate (FFA levels more than doubled during somatostatin infusion) may have largely accounted for this augmentation of ketogenesis, it is noteworthy that this occurred even in the virtual absence of basal glucagon secretion. These results again underscore the primacy of insulin lack, rather than glucagon excess, in the initiation and maintenance of ketosis during starvation.

## E. Summary and Conclusions

Successful adaptation to starvation involves an elegantly coordinated metabolic response allowing continued delivery of fuels to tissues and conservation of body protein. The data reviewed in this chapter indicate an important role for glucagon, as well as insulin, in the hormonal orchestration of this response. Glucagon's influence is essentially confined to the liver, where its effects are antagonized by those of insulin. Insulin, in contrast, exerts marked peripheral as well as hepatic effects. After an overnight fast, maintenance of hepatic glucose production, which is largely dependent on glycogenolysis, requires the action of basal glucagon to oppose the tonic restraining influence exerted by basal insulin. As fasting continues for 24–72 h, plasma insulin levels decline and plasma glucagon rises, as hepatic glucose production becomes solely dependent on gluconeogenesis. The fall in plasma insulin subserves glucose hometostasis by reducing extracerebral consumption of glucose, mobilizing glucogenic amino acids from muscle, and directly promoting intrahepatic gluconeogenic processes. This stimulatory effect of hypoinsulinemia on hepatic glucose synthesis is accentuated by glucagon. The important contribution of glucagon in promoting hepatic gluconeogenesis in starvation is indicated by the sharp reduction in hepatic glucose production observed when glucagon secretion is suppressed by somatostatin in 60-h fasted humans.

During prolonged starvation, protein dissolution is minimized by an increasing reliance on fat-derived fuels, rather than glucose, for oxidative metabolism. As blood ketone bodies rise during 3–4 weeks of fasting, they progressively replace glucose as fuel for brain, the major obligate glucose-consuming organ in the postabsorptive state. The development of hyperketonemia depends on enhanced release of substrate FFA from adipose tissue, increased hepatic conversion of fatty acids to ketone bodies via the $\beta$-oxidative pathway, and reduced uptake of ketones by extracerebral tissues, principally muscle. Each of these processes is promoted by the decline in plasma insulin, which appears to be the major factor regulating fasting ketosis. Glucagon, which is known to stimulate intrahepatic ketogenesis in circumstances of insulin deficiency, may contribute to ketone production during fasting. However, this has proven difficult to demonstrate experimentally.

Appreciation of glucagon's importance in postabsorptive and fasting glucose homeostasis has advanced at a remarkable pace. Just over a decade ago, glucagon received little more than passing mention in an authoritative review of human starvation (Cahill 1970). Today, glucagon has become a topic of intense interest and, at times, heated controversy. Only further research can clarify its precise role in starvation as well as a variety of other physiologic and pathophysiologic states.

## References


Altszuler N, Gottlieb B, Hampshire J (1976) Interaction of somatostatin, glucagon, and insulin on hepatic glucose output in the normal dog. Diabetes 25:116–121

Benedict FG (1915) A study of prolonged fasting. Publication No 203. Carnegie Institute, Washington, DC

Burman KD, Smalleridge RC, Jones L, Ramos EA, O'Brian JT, Wright FD, Wartofsky L (1980) Glucagon kinetics in fasting: physiological elevations in serum 3,5,3′-triiodothyronine increase the metabolic clearance rate of glucagon. J Clin Endocrinol Metab 51:1158–1165

Cahill GF Jr (1970) Starvation in man. N Engl J Med 282:668–675

Cahill GF Jr (1971) Physiology of insulin in man. Diabetes 20:785–799

Cahill GF Jr, Herrera MG, Morgan AP, Soeldner JS, Steinke J, Levy PL, Reich GA Jr, Kipnis DM (1966) Hormone-fuel interrelationships during fasting. J Clin Invest 45:1751–1769

Cherrington AD, Chiasson JL, Liljenquist JE, Jennings AS, Keller U, Lacy WW (1976) The role of insulin and glucagon in the regulation of basal glucose production in the postabsorptive dog. J Clin Invest 58:1407–1418

Cherrington AD, Lacy WW, Chiasson JL (1978) Effect of glucagon on glucose production during insulin deficiency in the dog. J Clin Invest 62:664–677

Cherrington AD, Liljenquist JE, Shulman GI, Williams P, Lacy WW (1979) Importance of hypoglycemia-induced glucose production during isolated glucagon deficiency. Am J Physiol 236:E263–E271

Cherrington AD, Williams PE, Shulman GI, Lacy WW (1981) Differential time course of glucagon's effect on glycogenolysis and gluconeogenesis in the conscious dog. Diabetes 30:1980–1987

Chiasson JL, Cook J, Liljenquist JE, Lacy WW (1974) Glucagon stimulation of gluconeogenesis from alanine in the intact dog. Am J Physiol 227:19–23
Chiasson JL, Liljenquist JE, Sinclair-Smith BC, Lacy WW (1975) Gluconeogenesis from alanine in normal postabsorptive man: intrahepatic stimulatory effect of glucagon. Diabetes 24:574–584
Chiasson JL, Liljenquist JE, Lacy WW, Jennings AS, Cherrington AD (1977) Gluconeogenesis: methodological approaches in vivo. Fed Proc 36:229–235
DeFronzo RA, Soman V, Sherwin RS, Hendler R, Felig P (1978) Insulin binding to monocytes and insulin action in human obesity, starvation, and refeeding. J Clin Invest 62:204–213
Felig P (1973) The glucose-alanine cycle. Metabolism 22:179–207
Felig P, Owen OE, Wahren J, Cahill GF Jr (1969a) Amino acid metabolism during prolonged starvation. J Clin Invest 48:584–594
Felig P, Marliss E, Owen OE, Cahill GF Jr (1969b) Role of substrate in the regulation of hepatic gluconeogenesis. Adv Enzyme Regul 7:41–46
Felig P, Pozefsky T, Marliss E, Cahill GF Jr (1970) Alanine: key role in gluconeogenesis. Science 167:1003–1004
Fisher M, Sherwin RS, Hendler R, Felig P (1976) Kinetics of glucagon in man: effects of starvation. Proc Natl Acad Sci USA 73:1734–1739
Garber AJ, Menzel PH, Boden G, Owen OE (1974) Hepatic ketogenesis and gluconeogenesis in humans. J Clin Invest 54:981–989
Gerich JE, Lorenzi M, Bier DM, Schneider V, Tsalikian E, Karam JH, Forsham PH (1975) Prevention of human diabetic ketoacidosis by somatostatin. N Engl J Med 292:985–989
Hultman E, Nilsson LH (1971) Liver glycogen in man: effect of different diets and muscular exercise. Adv Exp Med Biol 11:143–151
Keller U, Chiasson JL, Liljenquist JE, Cherrington AD, Jennings AS, Crofford OB (1977) The roles of insulin, glucagon, and free fatty acids in the regulation of ketogenesis in dogs. Diabetes 26:1040–1051
Keys A, Brozck J, Henschel A, Mickelson O, Taylor HL (1950) The biology of human starvation. University of Minnesota Press, Minneapolis
Lilavivathana V, Campbell RG, Brodows RG (1978) Control of insulin secretion during fasting in man. Metabolism 27:815–821
Liljenquist JE, Mueller GL, Cherrington AD, Keller U, Chiasson JL, Perry JM, Lacy WW, Rabinowitz D (1977) Evidence for an important role of glucagon in the regulation of hepatic glucose production in normal man. J Clin Invest 59:369–374
Mallette LE, Exton JH, Park CR (1969) Control of gluconeogenesis from amino acids in the perfused rat liver. J Biol Chem 244:5713–5723
Marliss EB, Aoki TT, Unger RH, Soldner S, Cahill GF Jr (1970) Glucagon levels and metabolic effects in fasting man. J Clin Invest 49:2256–2271
McGarry JD (1979) New perspectives in the regulation of ketogenesis. Diabetes 28:517–523
McGarry JD, Foster DW (1976) Ketogenesis and its regulation. Am J Med 61:9–13
McGarry JD, Foster DW (1977) Hormonal control of ketogenesis: biochemical considerations. Arch Intern Med 137:495–501
McGarry JD, Foster DW (1981) Ketogenesis. In: Unger RH, Orci L (eds) Glucagon: physiology, pathophysiology, and morphology of the pancreatic A-cells. Elsevier, New York, p 273
McGarry JD, Wright P, Foster D (1975) Hormonal control of ketogenesis: rapid activation of hepatic ketogenic capacity in fed rats by antiinsulin serum and glucagon. J Clin Invest 55:1202–1209
Miles J, Nissen S, Haymond D (1981) Interaction of insulin, glucagon, and free fatty acids on ketone body production in man. Diabetes [Suppl 1] 30:63A
Owen OE, Reichard GA Jr (1971) Human forearm metabolism during progressive starvation. J Clin Invest 50:1536–1545
Owen OE, Morgan AP, Kemp HG, Sullivan JM, Herrera MG, Cahill GF Jr (1967) Brain metabolism during fasting. J Clin Invest 46:1589–1595
Owen OE, Felig P, Morgan AP, Wahren J, Cahill GF Jr (1969) Liver and kidney metabolism during prolonged starvation. J Clin Invest 48:574–583

Palaiologos G, Felig P (1976) Effects of ketone bodies on amino acid metabolism in isolated rat diaphragm. Biochem J 154:709–716

Pozefsky T, Tancredi RG, Moxley RT, Dupre J, Tobin JD (1976a) Effects of brief starvation on muscle amino acid metabolism in nonobese man. J Clin Invest 57:444–449

Pozefsky T, Tancredi RG, Moxley RT, Dupre J, Tobin JD (1976b) Metabolism of forearm tissues in man: studies with glucagon. Diabetes 25:128–135

Reinmuth OM, Scheinberg P, Bourne B (1965) Total cerebral blood flow and metabolism. Arch Neurol 12:49–66

Sacca L, Sherwin R, Hendler R, Felig P (1979) Influence of continuous physiologic hyperinsulinemia on glucose kinetics and counterregulatory hormones in normal and diabetic humans. J Clin Invest 63:849–857

Sacca L, Vigorito C, Cicala M, Ungaro B, Sherwin RS (1982) Mechanisms of epinephrine-induced glucose intolerance in normal humans: role of the splanchnic bed. J Clin Invest 69:284–293

Saudek CD, Felig P (1976) The metabolic events of starvation. Am J Med 60:117–126

Schade DS, Eaton RP (1975) Glucagon regulation of plasma ketone body concentration in human diabetes. J Clin Invest 56:1340–1344, 1975

Schade DS, Eaton RP (1976) Modulation of fatty acid metabolism by glucagon in man. IV. Effects of a physiologic hormone infusion in normal man. Diabetes 25:978–983

Schneider SH, Fineberg SE, Blackburn GL (1981) The acute metabolic effects of glucagon and its interactions with insulin in forearm tissue. Diabetologia 20:616–624

Sherwin RS, Felig P (1981) Hypoglycemia. In: Felig P, Baxter JD, Broadus AE, Frohmann LA (eds) Endocrinology and metabolism. McGraw-Hill, New York, p 869

Sherwin RS, Hendler R, Felig P (1975) Effect of ketone infusions on amino acid and nitrogen metabolism in man. J Clin Invest 55:1382–1390

Sherwin RS, Hendler RG, Felig P (1976) Effect of diabetes mellitus and insulin on the turnover and metabolic response to ketones in man. Diabetes 25:776–784

Sherwin RS, Hendler R, DeFronzo R, Wahren J, Felig P (1977a) Glucose homeostatis during prolonged suppression of glucagon and insulin secretion by somatostatin. Proc Natl Acad Sci USA 74:348–352

Sherwin RS, Tamborlane W, Hendler R, Sacca L, DeFronzo RA, Felig P (1977b) Influence of glucagon replacement on the hyperglycemic and hyperketonemic response to prolonged somatostatin infusion in normal man. J Clin Endocrinol Metab 45:1104–1107

Shulman GI, Lacy WW, Liljenquist JE, Keller U, Williams PE, Cherrington AD (1980) Effect of glucose, independent of changes in insulin and glucagon secretion, on alanine metabolism in the conscious dog. J Clin Invest 65:496–505, 1980

Vagenakis AG, Burger A, Portnay GI, Rudolph M, O'Brian JT, Azizi F, Arky RA, Nicod P, Ingbar SH, Braverman LE (1975) Diversion of peripheral thyroxine metabolism from activating to inactivating pathways during complete fasting. J Clin Endocrinol Metab 44:1002–1005

Wahren J, Efendic S, Luft R, Hagenfeldt L, Bjorkman O, Felig P (1977) Influence of somatostatin on splanchnic glucose metabolism in postabsorptive and 60-hour fasted humans. J Clin Invest 59:299–307

Williamson JR, Kreisberg RA, Felts PW (1966) Mechanism for the stimulation of gluconeogenesis by fatty acids in perfused rat liver. Proc Natl Acad Sci USA 56:247–254

CHAPTER 35

# Glucagon and Pregnancy

C. KÜHL and J.J. HOLST

## A. Introduction

The interest in obtaining knowledge of glucagon secretion in pregnancy emanates from two characteristic features of this state. First, pregnancy is known to exert a diabetogenic stress on carbohydrate metabolism and second, the metabolic responses to starvation and food intake are characteristically modified by gestation. Evaluation of the possible contributory role of glucagon in bringing about these metabolic changes in pregnancy is obviously of interest.

### I. Impaired Glucose Tolerance in Pregnancy

In normal pregnancy, several physiologic changes take place, the sum of which tends to reset glucose homeostasis in the direction of diabetes. Thus, glucose tolerance gradually deteriorates (LIND et al. 1973; KÜHL 1975) for which reason pregnancy is often called "diabetogenic". The reason for the deterioration of glucose tolerance in pregnancy is still not fully understood (KÜHL 1977). Theoretically, any one of the mechanisms outlined in Table 1, or any combination of these mechanisms, might be involved. This chapter will mainly focus on the role of glucagon in the first of these possibilities, i.e., A-cell function in human and animal pregnancy.

### II. Metabolic Adaptations to Pregnancy

During intrauterine life, the human fetus is completely dependent upon maternal supply of glucose, free fatty acids, and amino acids. The increasing demands of the growing fetus are met by certain changes in the metabolic responses to caloric deprivation in the pregnant mother in order to ensure a continous supply of fuels for energy metabolism for the mother and the fetus.

Pregnancy is associated with a conversion from a glucose-utilizing fuel economy to catabolism of mainly lipids and protein, and with a drain of maternal en-

**Table 1.** Possible pathophysiologic mechanisms for the diabetogenicity of pregnancy

| |
|---|
| a) Secretory function of endocrine pancreas altered by pregnancy |
| b) Changed blood levels of hormones and other constituents with anti-insulin effect |
| c) Changes in insulin metabolism |
| d) Changes at the receptor level in target organs of insulin |
| e) Diminished potentiation of insulin secretion by insulinotropic gastrointestinal hormones |

ergy products, nutrients being mobilized for the fetus at the expense of the mother. Maternal adaptations to dietary deprivation result in a metabolic profile (e.g., increase in fasting plasma free fatty acid levels and in production of ketone bodies, but a decrease in fasting plasma glucose and amino acid levels) which has been designated "accelerated starvation" (FREINKEL et al. 1972). Contrariwise, when the mother is fed, mechanisms are evoked which favor maternal anabolism (e.g., the conversion of a larger proportion of ingested glucose into circulating triglyceride, a mechanism which, in view of the relative impermeability of the placenta to esterified lipids, would assure retention of some of the carbohydrate excess in the mother for later mobilization). Therefore, "accelerated starvation" in the fasting state and a compensating "facilitated anabolism" in the fed state would seem to be present at the same time in late pregnancy (FREINKEL et al. 1973). The possible contributory role of glucagon in the induction of the metabolic adaptation to pregnancy is the subject of this chapter.

## B. Plasma Glucagon in the Fasted State

### I. Changes After Overnight Fasting

There is some uncertainty about the influence of pregnancy on overnight fasting plasma glucagon levels. In normal human pregnancy, the plasma glucagon concentration in the fasting state has been reported to be decreased (KÜHL and HOLST 1976) or increased (LUYCKX et al. 1975) in midpregnancy, whereas both unchanged (LUYCKX et al. 1975; LEBLANC et al. 1976; LORRAIN et al. 1977) and elevated (DANIEL et al. 1974; KÜHL and HOLST 1976; KÜHL et al. 1977; METZGER et al. 1977; HORNNES and KÜHL 1980; KITZMILLER et al. 1980; HORNNES et al. 1981a, b) glucagon levels have been reported in late pregnancy.

In late gestational diabetic pregnancy, fasting plasma glucagon is reported to be either unchanged (KÜHL and HOLST 1976; LORRAIN et al. 1977; KITZMILLER et al. 1980) or enhanced (DANIEL et al. 1974; HORNNES et al. 1981a). In insulin-dependent diabetic women in the last trimester of pregnancy, data on fasting plasma glucagon are sparse and contradictory; both unchanged (KITZMILLER et al. 1980) and increased (FALLUCCA et al. 1979) levels have been reported. Data on fasting plasma glucagon concentrations in animal pregnancy are no less confusing. Thus, decreased (METZGER et al. 1974) or unchanged (SAUDEK et al. 1975) levels have been found in rats, whereas, in the rhesus monkey, fasting glucagon levels were unaffected by pregnancy (CHEZ et al. 1974).

The discrepant results concerning fasting plasma glucagon levels in gestation are probably due to the fact that, compared with postpartum, changes in fasting plasma glucagon concentrations in pregnancy, if any, are small. Detection of small differences in plasma glucagon levels makes heavy demands on the specificity and accuracy of the glucagon assay and, furthermore, it is mandatory that the pregnant women serve as their own nonpregnant controls. With these reservations in mind, it is probably permissible to conclude that the available evidence is for an enhancing effect of late human pregnancy on fasting plasma glucagon concentrations. As regards other stages of pregnancy and pregnancy in gestational diabetics and insulin-dependent diabetic women, data are still too sparse to allow final conclusions on fasting plasma glucagon to be drawn. Changes do, however, seem to be small.

It is generally accepted that the fasting plasma insulin concentration gradually increases in human pregnancy (for review, see KÜHL 1977) and, on a molar basis, the increase in plasma insulin is larger than that of glucagon. The molar insulin : glucagon ratio in peripheral plasma (UNGER 1971) is therefore increased in late human pregnancy (DANIEL et al. 1974; KÜHL and HOLST 1976). A similar finding has been reported in pregnant rats (METZGER et al. 1974; SAUDEK et al. 1975). Insulin and glucagon exert antagonistic effects on hepatic glycogenolysis and gluconeogenesis (MACKRELL and SOKAL 1969; PARK and EXTON 1972) and both hormones are profoundly involved in the maintenance of constant plasma glucose levels during fasting (ALFORD et al. 1974; GERICH et al. 1975; CHERRINGTON et al. 1976). The decrease in fasting plasma glucose concentration during pregnancy (KÜHL 1977) might therefore be explained by the increased molar insulin : glucagon ratio found in this state. However, presumably owing to a decrease in the number of insulin receptors (BECK-NIELSEN et al. 1979), pregnancy is a state of insulin resistance (BURT and DAVIDSON 1974; LIND et al. 1977). An increase in the molar insulin : glucagon ratio, therefore, does not by itself lead to a decreasing fasting plasma glucose concentration in gestation. It was recently demonstrated that the systemic glucose production rate is increased by about one-fifth in normal women in late pregnancy (KALHAN et al. 1979). Therefore, the decline in fasting glucose concentrations in pregnancy could be the result of an increase in the apparent volume of distribution of glucose rather than a reflection of the enhanced molar insulin : glucagon ratio.

## II. Changes After Prolonged Fasting and Insulin-Induced Hypoglycemia

In human pregnancy, there are no metabolic signs of accelerated starvation after a 12-h fast, but even in the absence of physical activity plasma ketone bodies and free fatty acids begin to rise when food is withheld for a further 4–6 h (RAVNIKAR et al. 1978). Since ketonuria during pregnancy has been associated with impaired intellectual development in the offspring, studies on changes in plasma glucagon during prolonged fasting have only been carried out in pregnant animals.

After a 48-h fast, plasma insulin falls to comparably low levels in pregnant and virgin rats. A small rise in plasma glucagon concentrations is seen in virgin, but not in pregnant rats, despite well-developed hypoglycemia in the latter. Thus, glucagon fails to increase in the fasted pregnant state, despite the stimulus of a lower plasma glucose (SAUDEK et al. 1975).

METZGER et al. (1974) fasted pregnant and virgin rats for 120 h. Plasma glucose fell to sustained hypoglycemic levels in the pregnant rats whereas glucose declined, but not to hypoglycemia in the virgin rats. Despite the hypoglycemia, higher levels of plasma insulin persisted in the pregnant rats and the molar insulin : glucagon ratio was not significantly different from that of the normoglycemic virgin rats. These studies indicate that accelerated starvation in pregnancy cannot be ascribed to relative glucagon excess. Rather, the preservation of a normal or enhanced molar insulin : glucagon ratio, despite prevailing hypoglycemia (METZGER et al.

1974), may provide a mechanism during fasting in pregnancy for restraining maternal catabolism.

Normal pregnant women subjected to insulin-induced hypoglycemia also exhibit a reduced glucagon response as compared with the nonpregnant state (LUYCKX et al. 1978; LIND et al. 1979). Thus, both in human and animal pregnancy, a reduced glucagon response to hypoglycemia is a characteristic feature.

### III. Placental Transfer of Glucagon and Morphology of A-cells in Pregnancy

Glucagon does not cross the placenta in either direction (ADAM et al. 1972; JOHNSTON et al. 1972; ALEXANDER et al. 1973; CHEZ et al. 1974; MOORE et al. 1974). Plasma glucagon determined in samples from pregnant women and animals is therefore solely of maternal origin.

Studies of the islets of Langerhans in human and rat pregnancy have revealed a highly significant increase in the ratio of B-cells to A-cells which is probably due to an absolute increase in the number of B-cells (HELLMAN 1960; VAN ASSCHE et al. 1978). The ultramicroscopic appearence of B-cells from pregnant rats suggests hyperinsulinsm of the individual B-cells. In contrast, no ultrastructural changes of the A-cells of the pregnant rats were observed (AERTS and VAN ASSCHE 1975). These morphological changes are in accordance with the enhanced molar insulin: glucagon ratio in plasma from pregnant women and animals (see Sect. B. I).

## C. Plasma Glucagon in the Fed State

### I. Response to Glucose Administration

In pregnancy, glucose tolerance gradually deteriorates, in spite of steadily increasing levels of plasma insulin, both in the fasting state and after oral intake of glucose or meals (KÜHL 1977). The high plasma levels of immunoreactive insulin are not due to an inappropriately high share of the biologically almost inactive proinsulin (PHELPS et al. 1975; KÜHL 1976). Likewise, there is no evidence for a pathogenic role of changes in insulin degradation in gestation (BURT and DAVIDSON 1974; BELLMANN and HARTMANN 1975; KÜHL et al. 1981). Hence, the diabetogenicity of pregnancy is not caused by inappropriate changes in insulin secretion or degradation.

Abnormalities of glucagon secretion in diabetes are well documented and an essential role of glucagon in the pathogenesis of diabetes has been proposed (UNGER 1971; UNGER and ORCI 1975; LEFEBVRE and LUYCKX 1979). Since, as previously mentioned, the decrease of glucose tolerance in pregnancy is not related to a lack of insulin, abnormalities of glucagon secretion might instead be involved. Against this background, KÜHL and HOLST (1976) subjected a group of normal pregnant women to an oral glucose tolerance test in mid- and late pregnancy and again about 6 weeks postpartum (Fig. 1). Glucose tolerance decreased gradually in gestation, in spite of an exaggerated insulin response. Furthermore, during the test, the suppression of plasma glucagon below basal levels was more pronounced and

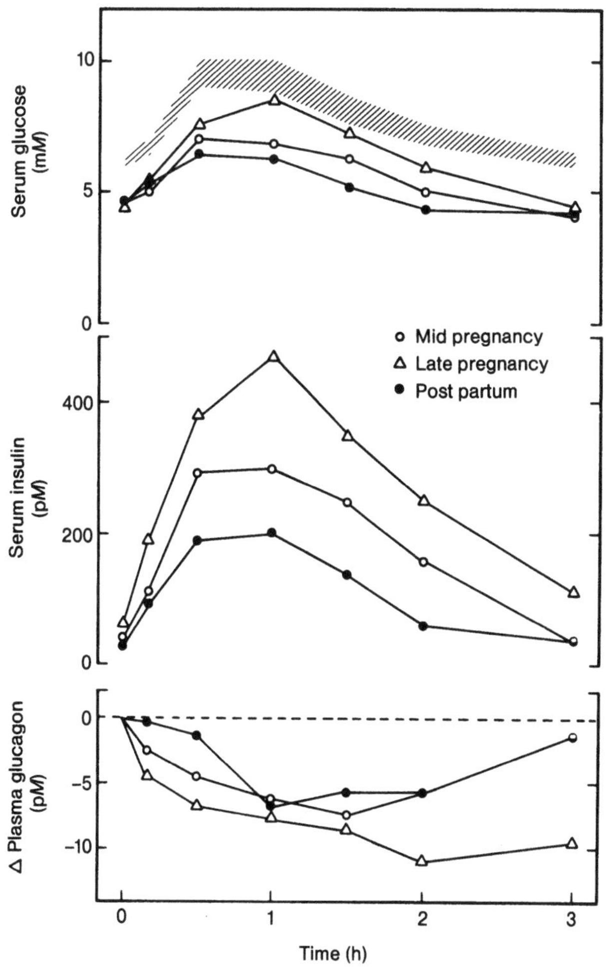


**Fig. 1.** Mean changes in serum glucose, serum insulin, and plasma glucagon in 8 normal women subjected to oral glucose tolerance tests in midpregnancy (*open circles*), late pregnancy (*triangles*), and postpartum (*full circles*). The lower demarcation of the *hatched area* represents the mean +2 standard deviations glucose concentration curve and the upper the mean +3 standard deviations curve, both calculated from oral glucose tolerance tests carried out in 46 normal non-pregnant women (KÜHL 1975). Adapted from KÜHL and HOLST (1976)

sustained in late pregnancy as compared with postpartum. Similar findings have been reported by others (DANIEL et al. 1974; HORNNES et al. 1981 a). In agreement with these findings, LUYCKX et al. (1975), using the intravenous glucose tolerance test, found a more rapid and more pronounced decrease in plasma glucagon during hyperglycemia in normal women late in pregnancy as compared with a group of nonpregnant controls. These findings contrast with the well-known decreased suppressibility of glucagon by glucose seen in diabetes (UNGER et al. 1970; BUCHANAN and MCCARROLL 1972).

The exaggerated insulin response and the enhanced and prolonged decline of plasma glucagon following oral glucose administration in late pregnancy are both

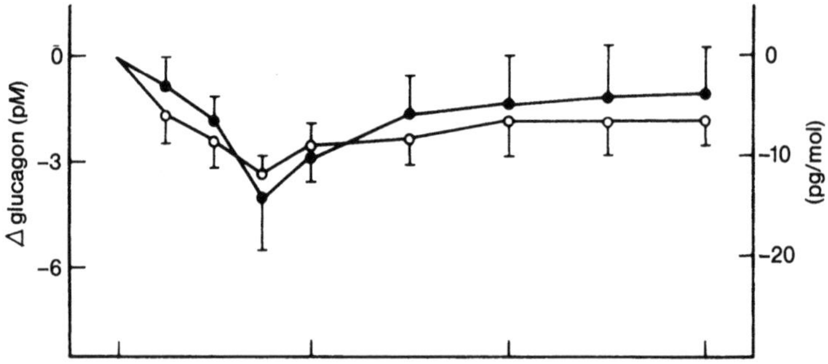


**Fig. 2.** Changes in plasma glucagon concentrations during isoglycemic stimulation in seven normal women in late pregnancy (*open circles*) and postpartum (*full circles*). Values are mean ± standard error. Adapted from HORNNES and KÜHL (1980)

consistent with the concept of "facilitated anabolism" (FREINKEL et al. 1973). The increased suppressibility of glucagon during hyperglycemia in gestation is probably due to the higher plasma glucose levels reached after administration of exogenous glucose in this state. Thus, if plasma glucose levels are similarly enhanced by graded intravenous glucose infusions in the same women in late pregnancy and postpartum, identical suppressions of plasma glucagon below fasting levels are seen (Fig. 2; HORNNES and KÜHL 1980). Contrariwise, the insulin response to this "isoglycemic" stimulus was found to be almost fourfold greater in pregnancy than postpartum. Again, these results are in accordance with the concept of pregnancy as a state of facilitated anabolism.

Using perifused rat pancreatic islets, KALKHOFF and KIM (1978) found almost identical suppression of glucagon secretion by 16.7 m*M* glucose in islets from pregnant and virgin rats. This in vitro finding supports the in vivo finding by HORNNES and KÜHL (1980) that, in normal pregnancy, the enhanced suppression of glucagon secretion during hyperglycemia is due to an increased glycemic stress on the A-cell.

Also, gestational diabetics seem to display significant reductions in plasma glucagon after oral glucose ingestion in late pregnancy (DANIEL et al. 1974; KÜHL and HOLST 1976; HORNNES et al. 1981a) whereas in postpartum, normal suppression (KÜHL and HOLST 1976) and lack of suppression have both been reported (DANIEL et al. 1974; HORNNES et al. 1981a). In a group of pregnant women with suspected or mild diabetes mellitus, a "paradoxical" increase in plasma glucagon after oral glucose ingestion in late pregnancy was observed (LORRAIN et al. 1977). In contrast, the same individuals exhibited a significant suppression of plasma glucagon during an intravenous glucose tolerance test. The discrepant results concerning glucagon secretion in gestational diabetics may be explained by the heterogeneity of this group (PEDERSEN 1977). Furthermore, there is no general agreement as to the definition of gestational diabetes.

## II. Response to Amino Acids

Information on the effect of amino acids on glucagon secretion in human pregnancy is scanty. In one study, oral intake of 10 g L-alanine dissolved in 100 ml water elicited a greater rise in plasma glucagon in a group of normal pregnant women investigated in late pregnancy as compared with 6 weeks postpartum (KITZMILLER et al. 1980). Despite the rise in plasma glucagon, blood glucose remained unchanged antepartum and postpartum, probably because of a concomitant rise in plasma insulin levels (KITZMILLER et al. 1980). Since a group of insulin-dependent pregnant diabetics exhibited a mean increment in plasma glucagon after administration of alanine which was comparable to that of the normal subjects, it is unlikely that the hyperglucagonemia observed could contribute to the well-known increased requirements for insulin during pregnancy in diabetics.

Using an intravenous infusion of arginine (30 g over 30 min), FALLUCCA et al. (1979) found identical glucagon responses in pregnant insulin-dependent diabetics and healthy controls. Furthermore, both groups exhibited greater glucagon responses in late gestation than postpartum. These results agree well with those of KITZMILLER et al. (1980). The glucagon response to intravenous alanine was enhanced in pregnant rats (SAUDEK et al. 1975) whereas a 6 m$M$ solution of a mixture of 20 amino acids elicited similar glucagon releases from perifused rat islets of pregnant and nonpregnant control rats (KALKHOFF and KIM 1978).

## III. Response to Mixed Meals

Amino acids are well-known stimulators of glucagon secretion (OHNEDA et al. 1968; UNGER et al. 1969). Mixed meals contain, besides protein, glucose and fat which may modulate the glucagon secretion following stimulation by the amino acid components of the ingested meal (PAGLIARA et al. 1975; ANDREWS et al. 1975). For this reason, data on glucagon secretion following the ingestion of mixed meals are more difficult to interpret than those obtained after stimulation with pure solutions of amino acids or glucose.

Normal women in late pregnancy subjected to a mixed meal consisting of 9 g carbohydrate, 13 g lipid, and 35 g protein, showed a 36% lower plasma glucagon response when compared with the response obtained in the same subjects about 2 months postpartum (Fig. 3) (KÜHL et al. 1977). Even though the mean insulin response to the meal was unaffected by pregnancy, plasma glucose increased slightly but significantly in pregnancy. These findings bring further support to the conclusion that changes in glucagon secretion per se are not involved in the pathogenesis of the diabetogenicity of pregnancy. Anyhow, the reduced glucagon response to protein meals in pregnancy fits well into the concept of facilitated anabolism during this state (FREINKEL et al. 1973).

Studies employing mixed meals of different compositions have yielded results which diverge slightly from those already cited. METZGER et al. (1977) administered a liquid formula meal consisting of 50 g carbohydrate, 25 g protein, and 10 g lipid to normal women in late pregnancy and postpartum. The glucagon response to the meal remained unaffected by pregnancy but, in accordance with the findings of KÜHL et al. (1977), plasma glucose increased more in pregnancy than postpartum. Similar findings have recently been reported by HORNNES et al. (1981 a) who used a mixed meal consisting of 20 g carbohydrate, 49 g protein, and 30 g lipid. It therefore seems that, in normal late pregnancy, the glucagon response to a mixed meal

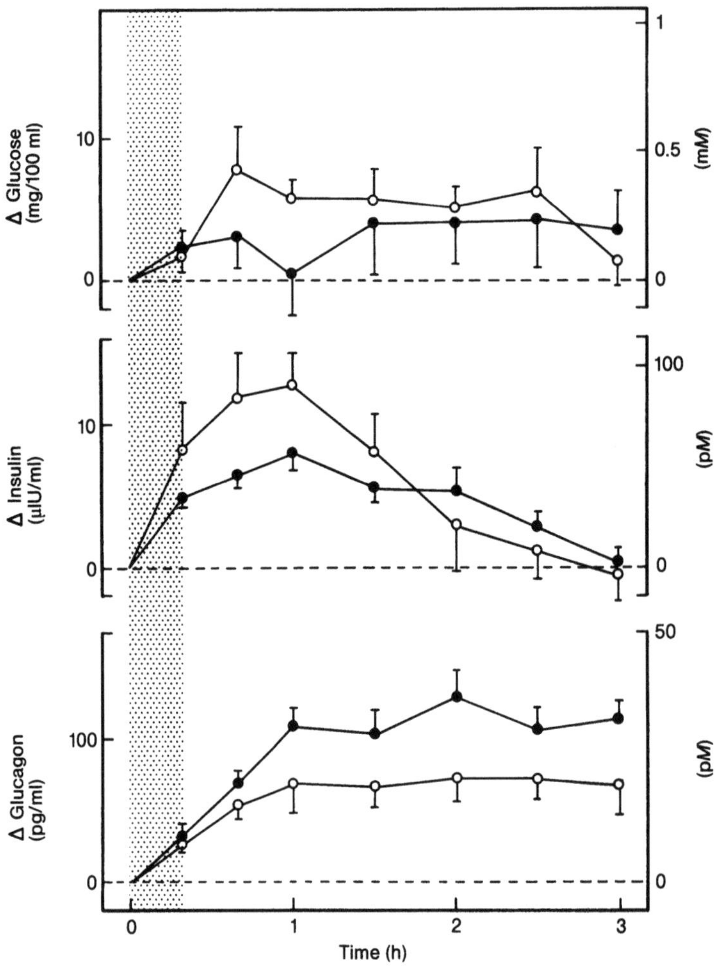


**Fig. 3.** Changes in plasma glucose, insulin, and glucagon concentrations following the ingestion of a protein-rich meal in normal late pregnancy (*open circles*) and postpartum (*closed circles*). The *stippled area* represents duration of the meal. Values are mean ± standard error. Adapted from KÜHL et al. (1977)

is, probably dependent upon the composition of the meal, either decreased or unchanged, but not enhanced. These findings also contrast to the exaggerated glucagon response to a protein-rich meal in nonpregnant diabetic patients (MÜLLER et al. 1970).

## D. Summary and Conclusions

The data available concerning glucagon secretion in pregnancy do no support the hypothesis that changes in glucagon secretion are involved in the pathogenesis of the diabetogenicity of pregnancy. Thus, in the fasting state, the molar insulin : glucagon ratio is increased whereas the opposite finding is a characteristic feature of diabetes (UNGER 1971). Following glucose administration, glucagon suppression below fasting levels is either unchanged or exaggerated in late normal and gestational diabetic pregnancy. This contrasts to the lack of suppressibility found in nonpregnant diabetics (UNGER et al. 1970; BUCHANAN and MCCARROL 1972). Ingestion of alanine is associated with a larger increment in plasma glucagon in normal women in the last trimester of pregnancy than postpartum. However, owing to a concomitant rise in plasma insulin, the hyperglucagonemia seen after oral administration of alanine is not followed by a rise in plasma glucose concentrations. Finally, after a mixed meal, the glucagon response is either reduced or unaffected in normal women in late pregnancy whereas an exaggerated response is found in nonpregnant diabetics (MÜLLER et al. 1970).

Whereas it is evident that the changes observed in glucagon secretion during pregnancy are incompatible with a pathogenic role of glucagon in the diabetogenicity of pregnancy, most of the changes reported fit well into the concept of facilitated anabolism in late pregnancy (FREINKEL et al. 1973). This means that glucagon – together with insulin – apparently plays a role in metabolic adaptations to pregnancy.

The evidence outlined in this chapter seems to exclude the possibility that glucose tolerance deteriorates in pregnancy because of diabetes-like changes in the secretory function of the endocrine pancreas (see Table 1). Likewise, neither insulin degradation nor proinsulin secretion seem to change in pregnancy (KÜHL 1976; KÜHL et al. 1981). An impaired gastrointestinal stimulation of insulin secretion might play a minor contributory role (HORNNES et al. 1978, 1981 a), but the major pathogenic factor is unquestionably the severe insulin resistance that develops in gestation. The insulin resistance in pregnancy is connected with a decrease in the number of insulin receptors on its target cells (BECK-NIELSEN et al. 1979) and it is also reflected by a decreased decline in plasma glucose following intravenously administered insulin (BURT and DAVIDSON 1974; LIND et al. 1977). Most pregnant women are able to counteract the insulin resistance by increasing their insulin secretion. Thus, about 99% of all pregnant women retain a normal glucose tolerance in pregnancy (KÜHL 1977). However, when the capacity of insulin secretion is not sufficiently large to meet the gradually developing insulin resistance in pregnancy, glucose intolerance develops and the woman becomes a gestational diabetic.

## References

Adam PA, King KC, Schwartz R, Teramo K (1972) Human placental barrier to 125-I-glucagon early in gestation. J Clin Endocrinol 34:772–782

Aerts L, Van Assche FA (1975) Ultrastructural changes of the endocrine pancreas in pregnant rats. Diabetologia 11:285–289

Alexander DP, Assan R, Britton HG, Nixon DA (1973) Impermeability of the sheep placenta to glucagon. Biol Neonate 23:391–402

Alford FP, Bloom SR, Nabarro JDN, Hall R, Besser GM, Coy DH, Kastin AJ, Schally AV (1974) Glucagon control of fasting glucose in man. Lancet 2:974–977

Andrews SS, Lopez A, Blackard WG (1975) Effects of lipids on glucagon secretion in man. Metabolism 24:35–44

Beck-Nielsen H, Kühl C, Pedersen O, Bjerre-Christensen C, Toftegaard Nielsen T, Klebe JG (1979) Decreased insulin binding to monocytes from normal pregnant women. J Clin Endocrinol Metab 49:810–814

Bellmann O, Hartmann E (1975) Influence of pregnancy on the kinetics of insulin. Am J Obstet Gynecol 122:829–833
Buchanan KD, McCarroll AM (1972) Abnormalities of glucagon metabolism in untreated diabetes mellitus. Lancet 2:1394–1395
Burt RL, Davidson IWF (1974) Insulin half-life and utilization in normal pregnancy. Obstet Gynecol 43:161–170
Cherrington AD, Chiasson JL, Liljenquist JE, Jennings AS, Keller U, Lacy WW (1976) The role of insulin and glucagon in the regulation of basal glucose production in the postabsorptive dog. J Clin Invest 58:1407–1418
Chez RA, Mintz DH, Epstein MF, Fleichman AR, Oakes GK, Hutchinson DL (1974) Glucagon metabolism in non-human primate pregnancy. Am J Obstet Gynecol 120:690–696
Daniel RR, Metzger BE, Freinkel N, Faloona GR, Unger RH, Nitzan M (1974) Carbohydrate metabolism in pregnancy. XI. Response of plasma glucagon to overnight fast and oral glucose during normal pregnancy and in gestational diabetes. Diabetes 23:771–776
Fallucca F, Russo A, Tinelli FP, Giangrande L, Pedulla C, De Gado F, Pachi A (1979) Influence of pregnancy on glucagon levels in insulin-dependent diabetic women. Acta Diabetol Lat 16:359–364
Freinkel N, Metzger BE, Nitzan M, Hare JW, Shambaugh GE, Marshall RT, Surmaczynska BZ, Nagel TC (1972) "Accelerated starvation" and mechanisms for the conservation of maternal nitrogen during pregnancy. Isr J Med Sci 8:826–839
Freinkel N, Metzger BE, Nitzan M, Daniel R, Surmaczynska BZ, Nagel TC (1973) Facilitated anabolism in late pregnancy: Some novel maternal compensations for accelerated starvation. In: Malaisse WJ, Pirart J, Vallance-Owen J (eds) Diabetes. Proceedings of the Eighth Congress of the International Diabetes Federation. Excerpta Medica, Amsterdam London New York, p 474
Gerich JE, Lorenzi M, Hane S, Gustafson G, Guillemin R, Forsham PH (1975) Evidence for a physiologic role of pancreatic glucagon in human glucose homeostasis: studies with somatostatin. Metabolism 24:175–182
Hellman B (1960) The islets of Langerhans in the rat during pregnancy and lactation, with special reference to the changes in the B/A cell ratio. Acta Obstet Gynaecol Scand 39:331–342
Hornnes PJ, Kühl C (1980) Plasma insulin and glucagon responses to isoglycemic stimulation in normal pregnancy and post partum. Obstet Gynecol 55:425–427
Hornnes P, Kühl C, Klebe JG (1978) Diminished gastrointestinal potentiation of insulin secretion in human pregnancy. Diabetologia 15:165–168
Hornnes PJ, Kühl C, Lauritsen KB (1981 a) Gastrointestinal insulinotropic hormones in normal and gestational diabetic pregnancy: response to oral glucose. Diabetes 30:504–509
Hornnes PJ, Kühl C, Lauritsen KB (1981 b) Gastro-enteropancreatic hormones in normal pregnancy: response to a protein rich meal. Eur J Clin Invest 11:345–349
Johnston DI, Bloom SR, Greene KR, Beard RW (1972) Failure of the human placenta to transfer pancreatic glucagon. Biol Neonate 21:375–380
Kalhan SC, D'Angelo LJ, Savin SM, Adam PAJ (1979) Glucose production in pregnant women at term gestation. Sources of glucose for human fetus. J Clin Invest 63:388–394
Kalkhoff RK, Kim H-J (1978) Effects of pregnancy on insulin and glucagon secretion by perifused rat pancreatic islets. Endocrinology 102:623–631
Kitzmiller JL, Tanenberg RJ, Aoki TT, Tabatabaii A, Gleason R, Jewett JF, Hare JW, Soeldner JS (1980) Pancreatic alpha cell response to alanine during and after normal and diabetic pregnancies. Obstet Gynecol 56:440–445
Kühl C (1975) Glucose metabolism during and after pregnancy in normal and gestational diabetic women. I. Influence of normal pregnancy on serum glucose and insulin concentration during basal fasting conditions and after a challenge with glucose. Acta Endocrinol (Copenh) 79:709–719
Kühl C (1976) Serum proinsulin in normal and gestational diabetic pregnancy. Diabetologia 12:295–300
Kühl C (1977) Serum insulin and plasma glucagon in human pregnancy – on the pathogenesis of gestational diabetes. A review. Acta Diabetol Lat 14:1–8

Kühl C, Holst JJ (1976) Plasma glucagon and the insulin: glucagon ratio in gestational diabetes. Diabetes 25:16–23

Kühl C, Hornnes P, Klebe JG (1977) Effect of pregnancy on the glucagon response to protein ingestion. Horm Metab Res 9:206–209

Kühl C, Hornnes PJ, Faber OK (1981) Hepatic insulin extraction in human pregnancy. Horm Metab Res 13:71–72

Leblanc H, Anderson JR, Yen SSC (1976) Glucagon secretion in late pregnancy and the puerperium. Am J Obstet Gynecol 125:708–710

Lefèbvre PJ, Luyckx AS (1979) Glucagon and diabetes: a reappraisal. Diabetologia 16:347–354

Lind T, Billewicz WZ, Brown G (1973) A serial study of changes occurring in the oral glucose tolerance test during pregnancy. J Obstet Gynaecol Br Commonw 80:1033–1039

Lind T, Bell S, Gilmore E, Huisjes HJ, Schally AV (1977) Insulin disappearance rate in pregnant and non-pregnant women, and in non-pregnant women given GHRIH. Eur J Clin Invest 7:47–51

Lind T, Burne JM, Kühl C (1979) Metabolic changes in pregnancy relevant to diabetes. In: Sutherland HW, Stowers JM (eds) Carbohydrate metabolism in pregnancy and the newborn. Springer, Berlin Heidelberg New York, p 32

Lorrain J, Robillard R, Cartier A, Ross SA, Dupre J, Boskovic A, Hinse MC (1977) Glucagon secretion in pregnant women with suspected or mild diabetes mellitus. Can Med Assa J 116:1289–1290

Luyckx AS, Gérard J, Gaspard U, Lefèbvre PJ (1975) Plasma glucagon levels in normal women during pregnancy. Diabetologia 11:549–554

Luyckx AS, Gaspard U, Lefèbvre PJ (1978) Influence of elevated plasma free fatty acids on the glucagon response to hypoglycemia in normal and in pregnant women. Metabolism 27:1033–1040

Mackrell DJ, Sokal JE (1969) Antagonism between the effects of insulin and glucagon on the isolated liver. Diabetes 18:724–732

Metzger B, Pek S, Hare J, Freinkel N (1974) Relationships between glucose, insulin and glucagon during fasting in late gestation in the rat. Life Sci 15:301–308

Metzger BE, Unger RH, Freinkel N (1977) Carbohydrate metabolism in pregnancy. XIV. Relationships between circulating glucagon, insulin, glucose and amino acids in response to a "mixed meal" in late pregnancy. Metabolism 26:151–156

Moore WMO, Ward BS, Gordon C (1974) Human placental transfer of glucagon. Clin Sci Mol Med 46:125–129

Müller WA, Faloona GR, Aguilar-Parada E, Unger RH (1970) Abnormal alpha-cell function in diabetes. Response to carbohydrate and protein ingestion. N Engl J Med 283:109–115

Ohneda A, Aguilar-Parada E, Eisentraut AM, Unger RH (1968) Characterization of response of circulating glucagon to intraduodenal and intravenous administration of amino acids. J Clin Invest 47:2305–2322

Pagliara AS, Stillings SN, Haymond MW, Hover BA, Matschinsky FM (1975) Insulin and glucose as modulators of the amino acid-induced glucagon release in the isolated pancreas of alloxan and streptozotocin diabetic rats. J Clin Invest 55:244–255

Park CR, Exton JH (1972) Glucagon and the metabolism of glucose. In: Lefèbvre PJ, Unger RH (eds) Glucagon. Pergamon, Oxford New York, p 77

Pedersen J (1977) The pregnant diabetic and her newborn, 2nd edn. Williams & Wilkins, Baltimore

Phelps RL, Bergenstal R, Freinkel N, Rubenstein AH, Metzger BE, Mako M (1975) Carbohydrate metabolism in pregnancy. XIII. Relationships between plasma insulin and proinsulin during late pregnancy in normal and diabetic subjects. J Clin Endocrinol 41:1085–1091

Ravnikar V, Metzger BE, Freinkel N (1978) Is there a risk of "accelerated starvation" in normal human pregnancy? Diabetes 27:Abstr 132

Saudek CD, Finkowski M, Knopp RH (1975) Plasma glucagon and insulin in rat pregnancy. Roles in glucose homeostasis. J Clin Invest 55:180–187

Unger RH (1971) Glucagon and the insulin: glucagon ratio in diabetes and other catabolic illnesses. Diabetes 29:834–838
Unger RH, Orci L (1975) The essential role of glucagon in the pathogenesis of diabetes mellitus. Lancet 1:14–16
Unger RH, Ohneda A, Aguilar-Parada E, Eisentraut AM (1969) The role of aminogenic glucagon secretion in blood glucose homeostasis. J Clin Invest 48:810–822
Unger RH, Aguilar-Parade E, Müller WA, Eisentraut AM (1970) Studies of pancreatic alpha cell function in normal and diabetic subject. J Clin Invest 49:837–848
Van Assche FA, Aerts L, De Prins F (1978) A morphological study of the endocrine pancreas in human pregnancy. Br J Obstet Gynaecol 85:818–820

CHAPTER 36

# Glucagon in the Fetus and the Newborn

J. GIRARD and M. SPERLING

## A. Introduction

In adult mammals, insulin and glucagon exert diametrically opposing actions upon hepatic metabolism (UNGER and ORCI 1981). Glucagon, through its powerful glycogenolytic and gluconeogenic effects on the liver, promotes hepatic glucose production, while insulin opposes both these actions and can cause a storage of glucose in hepatic glycogen under appropriate conditions (see Chaps. 14, 15, 16, and 37). Glucagon inhibits hepatic lipogenesis and triglyceride production by the liver and it stimulates ketone body production (see Chapt. 17).

It has been proposed that the islets of Langerhans, by varying the relative concentrations of these two hormones in the portal blood, possess the biologic capacity of controlling the disposition of key nutrients and endogenous substrates in a manner appropriate to the prevailing exogenous fuel supply and energy requirements (UNGER 1974; UNGER and ORCI 1981). The idea that A- and B-cells of the islets of Langerhans act as a functional unit has been strengthened by the finding of a structural coupling between these different endocrine cells (ORCI et al. 1975). The validity of this concept has now been clarified in a variety of physiologic and pathologic conditions, including fasting, exercise, stress, and diabetes (see Chaps. 34, 38, and 44).

There are two periods during the development of the mammal when pancreatic hormones could be expected to play an important role in the regulation of liver metabolism: birth and weaning. Indeed both of these periods are associated with marked changes in the source and composition of nutrients. While the "diet" of the fetus is high in carbohydrate and low in fat (BATTAGLIA and MESCHIA 1978), that of the newborn is rich in fat and low in carbohydrate (JENNESS 1974). A comparison of the amount of glucose supplied from the milk with the rate of glucose utilization indicates that an active endogenous glucose production is essential to cover the glucose needs of the newborn of several species including the human (GIRARD 1981 a). Immediately after birth and during the suckling period, the newborn is dependent on its capacity for efficient hepatic gluconeogenesis and ketogenesis (GIRARD 1981 b) and glucagon could be expected to play an important role in the triggering and regulation of these metabolic pathways. In contrast, the weaning period is characterized by a progressive change from a high fat and low carbohydrate diet to a high carbohydrate and low fat diet, and is accompanied by a decrease in hepatic gluconeogenesis and ketogenesis and an increase in hepatic lipogenesis and glycogenesis. A fall in plasma glucagon could be expected to play a role in this adaptation. The aim of the present chapter is to review the factors controlling glucagon secretion during development, and to discuss the role that this hormone plays in the metabolic adaptations occurring at birth and during weaning period.

## B. Ontogenesis of Glucagon in Pancreas and Plasma

In most species including humans, the pancreas appears at approximately 20–25 somites of embryonic development. In humans, this stage corresponds to the fourth week of gestation. The morphological aspects of the ontogenic differentiation of the pancreas are reviewed in Chap. 5 and therefore will not be discussed here.

### I. Rat

The presence of glucagon in fetal rat pancreas was first determined by bioassay (OKUNO et al. 1964) and later by radioimmunoassay (ORCI et al. 1969; RALL et al. 1973; PICTET and RUTTER 1972; BLAZQUEZ et al. 1972; JAROUSSE et al. 1973). Immunoreactive glucagon was present as early as the 11th day of gestation (the gestation length in the rat is 22 days) and the total content rose 500-fold at term. The concentration of glucagon in fetal pancreas showed a maximal value at 14 days gestation (PICTET and RUTTER 1972). During the last 4 days gestation, the glucagon concentration in fetal pancreas varied between 2 and 4 μg/g wet weight (GIRARD et al. 1973b).

### II. Rabbit

Glucagon has been found in fetal rabbit pancreas at 13 days gestation (the gestation length in the rabbit is 32 days). The glucagon content of the pancreas increased 1,500-fold at term (ASSAN et al. 1974). The concentration of glucagon in fetal pancreas was maximal at 26 days (30 ng/g) and decreased in late gestation (20 ng/g).

### III. Sheep

Glucagon is detected in the fetal pancreas of the sheep at a relatively early stage of gestation (60 days; term is approximately 145 days). Higher glucagon concentrations are found in the pancreas of fetuses aged 100–117 days (ALEXANDER et al. 1971; BASSET 1977) than in fetuses near term, i.e., 130–143 days (ALEXANDER et al. 1971).

### IV. Human

Glucagon was detectable in fetal pancreas at 6 weeks and its content sharply increased until 26 weeks (ASSAN and BOILLOT 1973; SCHAEFFER et al. 1973; ASSAN and GIRARD 1975). At this stage of gestation the concentration of glucagon in fetal pancreas (10 μg/g) was higher than in adult pancreas (2–5 μg/g). Thus, the early appearance of glucagon in fetal pancreas is well documented. The changes in glucagon concentration in fetal pancreas are similar in the several species studied, i.e., the highest concentrations being reached at midgestation and then decreasing until term.

## C. Glucagon Secretion in the Fetus

The presence of glucagon at concentrations in the same range as those in maternal plasma has been found in the plasma of fetuses from several species: rat (GIRARD et al. 1973 b, 1974; BLAZQUEZ et al. 1974); rabbit (ASSAN et al. 1974); sheep (ALEXANDER et al. 1971; FISER et al. 1974a; BASSETT 1977; SHELLEY and GIRARD 1981); monkey (CHEZ et al. 1974); and in the umbilical cord of the human at delivery (BLAZQUEZ et al. 1974; BLOOM and JOHNSTON 1972; JOHNSTON and BLOOM 1973; MILNER et al. 1973; SPERLING et al. 1974; WISE et al. 1973; WILLIAMS et al. 1979). The earliest stage where the presence of glucagon has been reported in fetal plasma is 18 days in the rat (GIRARD et al. 1974; BLAZQUEZ et al. 1974), 24 days in the rabbit (ASSAN et al. 1974), 59 days in the sheep (ALEXANDER et al. 1971), and 15 weeks in the human (ASSAN and BOILLOT 1973).

### I. Impermeability of the Placenta to Glucagon

It is widely accepted that the placenta is impermeable to glucagon. Infusion of large amounts of glucagon or of radioiodine-labeled glucagon to the mother was not associated with any increase in fetal plasma glucagon or in transfer of radioactive glucagon in the fetal rat (GIRARD et al. 1973 b; BLAZQUEZ et al. 1972), sheep (ALEXANDER et al. 1973; SPERLING et al. 1973; ASSAN et al. 1974), monkey (CHEZ et al. 1974), and human (ADAM et al. 1972; JOHNSTON et al. 1972; MOORE et al. 1974; SPELLACY and BUHI 1976). This implies that glucagon found in fetal plasma is secreted by fetal pancreas and that observed changes in fetal plasma glucagon will reflect the activity of fetal pancreas without interference of maternal transfer across the placenta.

### II. Control of Glucagon Secretion in the Fetus

#### 1. Changes in Glucose Concentration

a) Acute Changes

In the adult free of stress or diabetes, hyperglycemia is the most potent glucagon suppressor and hypoglycemia is one of the most effective glucagon stimuli (see Chap. 23). So, it is of interest whether glucose–glucagon relationships occur to the same extent during fetal life. As fetal blood glucose concentration varies in parallel with maternal blood glucose concentration, it is relatively easy to increase fetal glycemia by infusing glucose to the mother and to decrease fetal glycemia by infusing insulin to the mother. This technique has been used to study the effects of acute changes in fetal glycemia on the secretion of glucagon by the rat fetus in utero (GIRARD et al. 1974). As expected, hyperglycemia suppressed glucagon secretion in the mother and hypoglycemia stimulated maternal glucagon secretion (Figs. 1

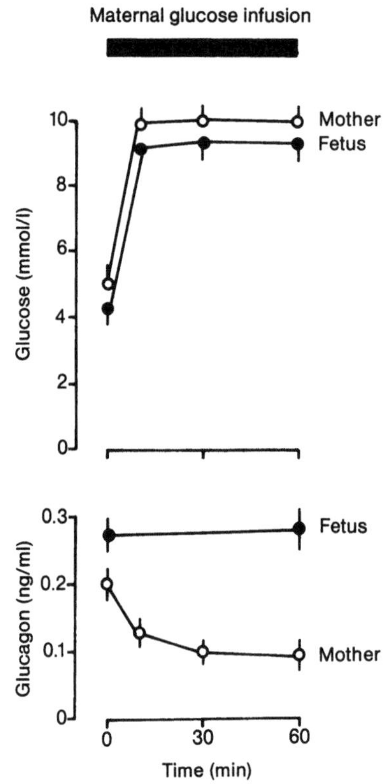


**Fig. 1.** Effects of a 1-h hyperglycemia on plasma glucagon concentrations in term pregnant rats and their fetuses. Under pentobarbital anesthesia, the pregnant rats were infused intravenously with D-glucose (a prime injection of 300 mg over 2 min, followed by a perfusion at a rate of 6 mg/min). GIRARD et al. (1974); GIRARD and ASSAN (1981)

and 2). In contrast, fetal plasma glucagon remained unchanged during fetal hyperglycemia or hypoglycemia (Figs. 1 and 2). Similar results have been obtained in exteriorized fetal lambs and monkeys under anesthesia during acute studies (CHEZ et al. 1974; ALEXANDER et al. 1976). However, anesthesia of the mother, exposition and handling of the uterus, and exteriorization of the fetus before blood sampling are associated with some degree of stress. Thus, it was of crucial interest to know whether glucagon secretion by the fetus was also insensitive to changes in fetal glycemia in a species such as the sheep, in which indwelling carotid artery and jugular vein catheters can be inserted in the fetus and studies can be performed on conscious animals. In fetal sheep, hyperglycemia or hypoglycemia induced by glucose or insulin infusion to the fetus did not influence glucagon secretion (FISER et al. 1974a; SHELLEY and GIRARD 1981).

These findings have been corroborated by in vitro studies in which glucagon released from isolated islets of Langerhans or pieces of pancreas of rat (BAJAJ and BUCHANAN 1977; SODOYEZ-GOFFAUX et al. 1979), mouse (LERNMARK and WENNGREN 1972), and human fetuses (SCHAEFFER et al. 1973) was not influenced by large variations in medium glucose concentration. Thus, the unresponsiveness of fetal

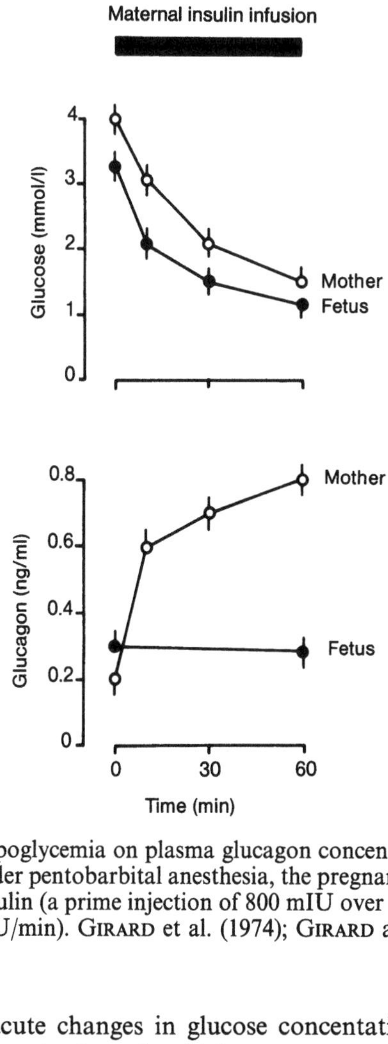


**Fig. 2.** Effects of a 1-h hypoglycemia on plasma glucagon concentrations in term pregnant rats and their fetuses. Under pentobarbital anesthesia, the pregnant rats were infused intravenously with porcine insulin (a prime injection of 800 mIU over 2 min, followed by a perfusion at a rate of 13 mIU/min). GIRARD et al. (1974); GIRARD and ASSAN (1981)

pancreatic A-cells to acute changes in glucose concentation is now well documented. The acute changes in glycemia which may spontaneously occur in the fetus in utero seem relatively unimportant in influencing glucagon secretion by the fetal pancreas.

### b) Long-Term Changes

Several lines of evidence suggest that long-term changes in fetal blood glucose concentration could influence the secretion of glucagon by fetal pancreas. When chronically catheterized pregnant rats are continuously infused with glucose during the last 3 days gestation, thus chronically raising fetal blood glucose concentration from 4 to 10 mmol/l, there is a twofold decrease in fetal plasma glucagon at term (Fig. 3; KTORZA et al. 1981). Moreover, fasting of pregnant rats during the last 4 days gestation, thus chronically decreasing fetal blood glucose concentration from 3 to 2 mmol/l, is accompanied by a twofold increase in fetal plasma glucagon (Fig. 4; GIRARD et al. 1977b). Studies performed in the sheep have shown that fetal

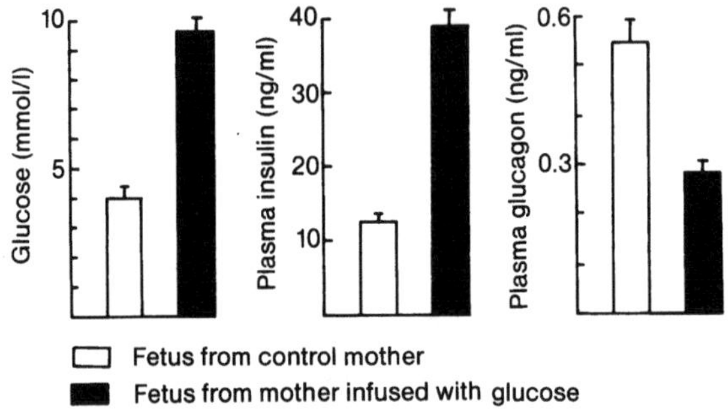


**Fig. 3.** Effects of chronic hyperglycemia during the last 3 days gestation on plasma glucagon and insulin concentrations in term rat fetuses. Unrestrained pregnant rats were infused with glucose from day 18 of gestation until term at a rate of 8 mg/min. KTORZA et al. (1981)

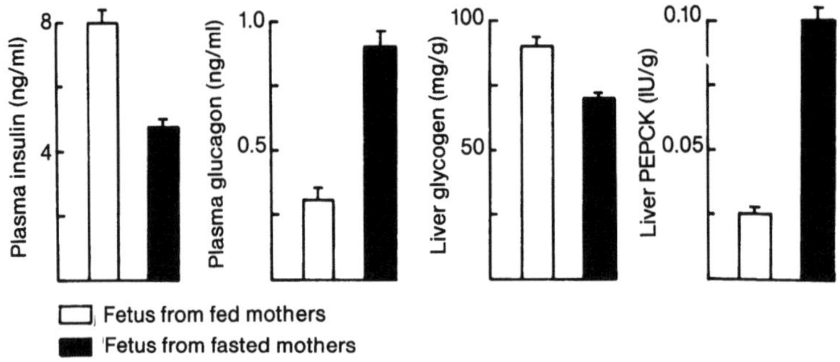


**Fig. 4.** Effects of chronic hypoglycemia, induced by fasting pregnant rats during the last 4 days gestation, on fetal plasma glucagon and insulin concentrations and on fetal liver glycogen concentrations and PEPCK activity. GIRARD et al. (1977b)

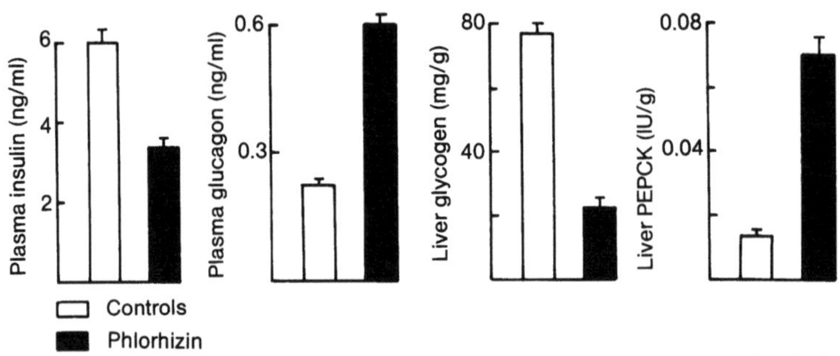


**Fig. 5.** Effects of phlorhizin-induced hypoglycemia in term pregnant rats on fetal plasma glucagon and insulin concentrations and on fetal liver glycogen concentration and PEPCK activity. FREUND et al. (1980)

plasma glucagon is not affected after 7 days maternal starvation (SCHREINER et al. 1980a, b), but is increased by 9 days and remains elevated after 11 days maternal starvation (SCHREINER et al. 1981). More recently it has been shown that severe hypoglycemia induced in term rat fetuses for 24 h by injecting pregnant rats with phlorhizin is associated with a twofold increase in fetal plasma glucagon (Fig. 5;

FREUND et al. 1980). All these data suggest that long-term exposure to high or low blood glucose levels can affect the secretion of glucagon by the fetal pancreas. These findings may have relevance to human fetal glucagon secretion in infants of hyperglycemic diabetic mothers and in nutritionally deprived small-for-gestational-age infants, especially immediately after birth (see Sect. D.II).

### 2. Effects of Amino Acids

Abundant evidence has accumulated in the last 10 years to indicate that hyperaminoacidemia does exert a powerful stimulatory action upon the A-cell of the adult pancreas (see Chap. 24). There are some discrepancies about the sensitivity in vivo of fetal A-cells to amino acids. Infusion of alanine or glycine in chronically catheterized fetal lambs failed to cause significant stimulation of glucagon secretion (FISER et al. 1974a; BASSETT 1977). Similarly, alanine was ineffective in stimulating glucagon release in the fetal monkey (CHEZ et al. 1974). In contrast, arginine injection to exteriorized fetal lambs (ALEXANDER et al. 1976) or to rat fetuses in utero (GIRARD et al. 1974) induced a significant increase in fetal plasma glucagon. Similarly, alanine infusion to women during labor caused a 50%–70% increase in umbilical cord plasma glucagon as well as maternal glucagon (WISE et al. 1973), suggesting that the human fetal pancreatic A-cells are responsive to aminogenic stimulation, since amino acids readily cross the placenta, whereas glucagon does not.

In vitro studies have shown that glycine, alanine, and arginine markedly stimulate glucagon release from isolated islets of Langerhans or pieces of pancreas of rat (ASSAN et al. 1974; SODOYEZ-GOFFAUX et al. 1979), mouse (LERNMARK and WENNGREN 1972), sheep (BASSETT et al. 1977), and human (ASSAN et al. 1974; SCHAEFFER et al. 1973). The reason why fetal pancreatic A-cells are sensitive to amino acids in vitro and much less, or not at all in vivo could be due to the already very high level of circulating amino acids in fetal blood (see GIRARD et al. 1979), thus blunting the effect of additionally infused amino acids.

### 3. Effects of Neurotransmitters

In adults, plasma glucagon increases markedly in response to adrenergic stimulation (see Chap. 30). Injection of epinephrine or norepinephrine to the term rat fetus in utero (GIRARD et al. 1974) or infusion of epinephrine to the chronically catheterized fetal lamb (SPERLING et al. 1980) increase fetal plasma glucagon (Fig. 6). As propranolol markedly attenuated the rise in fetal plasma glucagon during epinephrine infusion in the fetal lamb (SPERLING et al. 1980), this suggests that $\beta$-adrenoceptors are involved in the effects of epinephrine (Fig. 6). L-Dopa injected intravenously to the monkey fetus increased fetal plasma glucagon (EPSTEIN et al. 1977). Injection of serotonin or dopamine to the rat fetus in utero did not change fetal plasma glucagon (GIRARD et al. 1974, 1975). In contrast, acetylcholine injection to the term rat fetus increased fetal plasma glucagon, although the rise was of smaller magnitude than that observed after injection of catecholamines (GIRARD et al. 1974).

In vitro studies have also shown that epinephrine and norepinephrine stimulate glucagon secretion from fetal pancreas of rat, human (ASSAN et al. 1974), and sheep

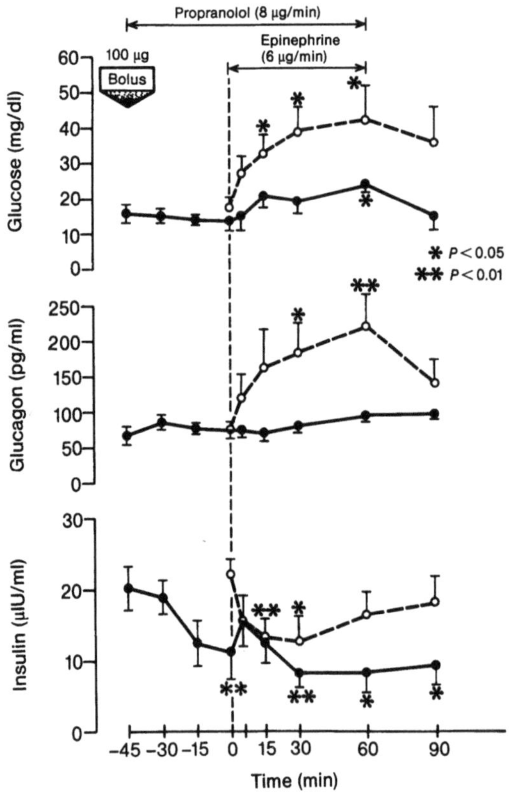


**Fig. 6.** Effects of epinephrine infusion without *(open circles)* and with *(full circles)* concomitant prime infusion of the $\beta$-blocker, propranolol to the sheep fetus in utero. Epinephrine alone raises glucose and glucagon concentration, while insulin is suppressed. Propanolol alone does not affect glucose or glucagon, but suppresses insulin; propanolol prevents the effects of epinephrine on glucose and glucagon and exaggerates the fall in insulin. SPERLING et al. (1980)

(BASSETT 1977). These data indicate that the sympathetic and parasympathetic control of glucagon secretion is already present in term fetuses of several species.

## 4. Effects of Hypoxia

Hypoxia induced by ligation of uterine arteries results in a marked rise in fetal plasma glucagon (GIRARD and ASSAN 1981). In the chronically catheterized fetal lamb, plasma glucagon was increased in hypoxic fetuses when the pH fell below 7.30 (SHELLEY and GIRARD 1981). The glucagon levels in umbilical blood of infants with fetal distress (pH < 7.2 and hypoxic) are twofold higher than in normal infants (JOHNSTON and BLOOM 1973). The mechanisms by which hypoxia results in increased glucagon release have not been clarified, but could be due to activation of the sympathetic nervous system. Indeed, hyperglucagonemia during acute hypoxia in the dog is largely mediated by adrenergic mechanisms involving $\alpha$-adrenoceptor activation (BAUM et al. 1979).

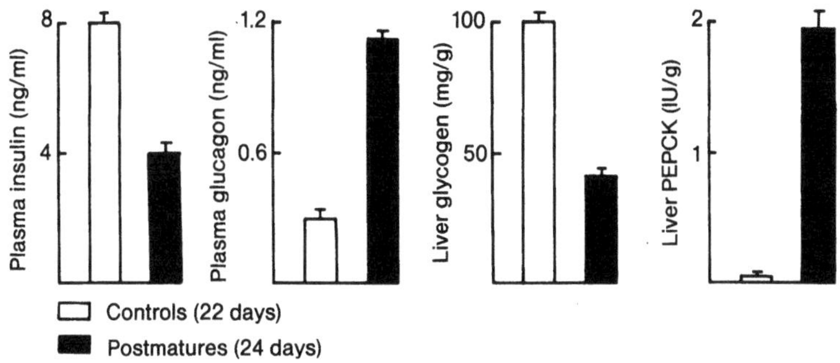


**Fig. 7.** Effects of prolongation of pregnancy by 2 days on fetal plasma glucagon and insulin concentrations and on liver glycogen concentration and PEPCK activity. Gestation was prolonged by daily injecting pregnant rats with 2.5 mg progesterone from day 20 of gestation until day 23. GIRARD et al. (1977a)

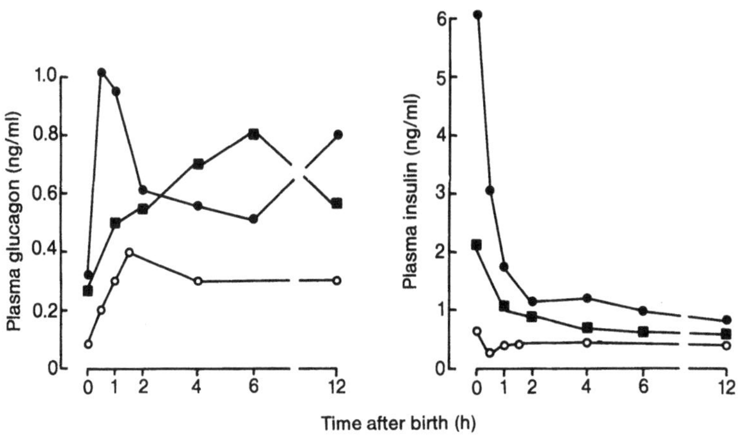


**Fig. 8.** Changes in plasma insulin and glucagon concentration in fasting newborn of several species: *open circles* human; *squares* rabbit; and *full circles* rat. SPERLING et al. (1974); CALLIKAN et al. (1979); GIRARD et al. (1973c)

### 5. Prolonged Pregnancy

In the rat, pregnancy can be prolonged by 2 days by injecting progesterone daily from day 19 of pregnancy. A marked rise in plasma glucagon occured in postmature fetuses (Fig. 7; GIRARD et al. 1977a; PORTHA et al. 1978a). When pregnant rats were infused with glucose during prolonged pregnancy, fetal hyperglucagonemia was partly prevented (KTORZA et al. 1981). The factors which could be responsible for the stimulation of glucagon secretion by the fetal pancreas during prolonged gestation have not been clarified, but could involve hypoxia or increased catecholamine levels (see Sects. C.III.3 and 4).

## D. Glucagon Secretion in the Newborn

At birth, the maternal supply of nutrients to the fetus ceases abruptly and the circulating substrates change rapidly. There is a marked fall in plasma amino acid

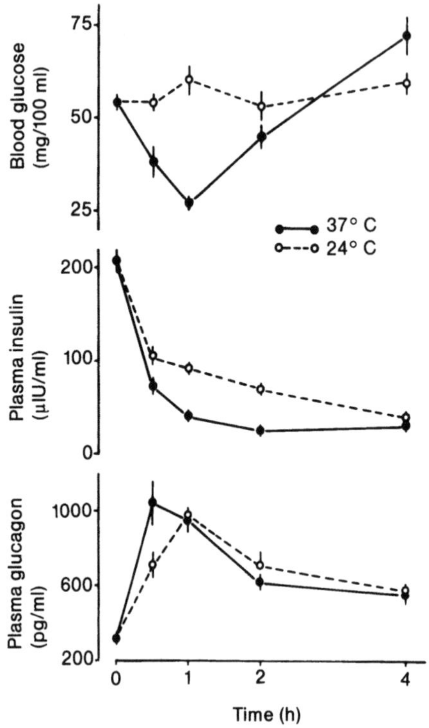


**Fig. 9.** Effect of environmental temperature on blood glucose and plasma insulin and glucagon concentrations in newborn rats during the first hours of extrauterine life. KERVRAN et al. (1976)

concentrations and a transient decline of blood glucose concentration in rat (GIRARD et al. 1973c) and human (SPERLING et al. 1974) newborns. In contrast, in the newborn sheep, blood glucose concentration, which was already very low in utero, shows very little change (GRAJWER et al. 1977). During the immediate postnatal period a significant increase in plasma glucagon occurs in all the species studied (Fig. 8). This was first described in the rat (GIRARD et al. 1972) and then confirmed in other species: human (LUYCKX et al. 1972; SPERLING et al. 1974; BLAZQUEZ et al. 1974); sheep (GRAJWER et al. 1977); and rabbit (CALLIKAN et al. 1979). In the rat, the rise in plasma glucagon was observed both after cesarean section (GIRARD et al. 1972, 1973c; PORTHA et al. 1978a; DIMARCO et al. 1978) or vaginal delivery (BLAZQUEZ et al. 1974; GIRARD et al. 1977a) and it occured similarly, both in fasted and suckled neonates (GIRARD et al. 1980).

## I. Evidence for a Role of the Sympathetic Nervous System in the Neonatal Surge of Glucagon

It has been postulated by several authors that the transient postnatal hypoglycemia which occurs in the rat maintained at thermoneutrality (37 °C), might trigger the release of glucagon by the pancreas. In fact, this is not the case since similar changes occur in the newborn maintained at 24 °C where blood glucose levels do not fall (Fig. 9; KERVRAN et al. 1976). Furthermore, the surge of glucagon is also observed in species (sheep, rabbit) in which there is no fall of blood glucose at birth (GRAJWER et al. 1977; CALLIKAN et al. 1979). Finally, the insensitivity of newborn A-cells to changes in glucose concentration is well documented, both in vivo (GIRARD et al. 1976; FISER et al. 1974b; LUYCKX et al. 1972) and in vitro (EDWARDS et al. 1972; LERNMARK and WENNGREN 1972; BLAZQUEZ et al. 1974; BAJAJ et al. 1977; SODOYEZ-GOFFAUX et al. 1979).

Although amino acids have been reported to be effective stimulators of glucagon release by the newborn pancreas (EDWARDS et al. 1972; MARLISS et al. 1973; JAROUSSE and ROSSELIN 1975; SPERLING et al. 1974), it is unlikely that they play a role in the surge of glucagon at birth, since plasma amino acids fall dramatically in the immediate postnatal period (LINDBLAD 1970; CALLIKAN et al. 1979; GIRARD et al. 1973c). A more likely mechanism to explain the acute rise in plasma glucagon at birth might relate to an adrenergic stimulation in response to the stress of birth: transient hypoxia, cold exposure, or cord cutting (GRAJWER et al. 1977). Epinephrine and norepinephrine levels are increased severalfold in umbilical blood of human newborns (LAGERCRANTZ and BISTOLETTI 1977; ELIOT et al. 1980) and in sheep (ELIOT et al. 1981). And both of these hormones are capable of stimulating glucagon release by the fetal pancreas (GIRARD et al. 1974; MARLISS et al. 1973; SPERLING et al. 1980). A direct stimulation of neonatal pancreas through sympathetic nerves is also possible since it has been shown that nervous structures are present adjacent to endocrine islets in the rat fetus (PERRIER 1970). As an $\alpha$-adrenergic blocking drug (phentolamine) partially inhibited the rise in plasma glucagon in the newborn rat, while the $\beta$-adrenergic blocking drug (propanolol) did not, it has been suggested that an $\alpha$-adrenergic mechanism is involved in the neonatal surge of glucagon (GIRARD et al. 1975).

## II. Glucagon Secretion in Newborn Infants of Diabetic Mothers

Infants born to diabetic mothers (IDM) have an increased incidence of hypoglycemia in the immediate postnatal period. For a long time, this has been considered as the result of their hyperinsulinemia and of their increased glucose utilization. More recently, it has been reported that IDM also have a decreased glucose production during the perinatal period (KALHAN et al. 1977). The spontaneous increase in plasma glucagon which occurs in normal human newborn infants (BLOOM and JOHNSTON 1972; SPERLING et al. 1974) is blunted in IDM (BLOOM and JOHNSTON 1972; WILLIAMS et al. 1979). The postnatal surge in plasma glucagon is reduced with increasing severity of the maternal diabetes (KÜHL et al. 1980). This failure of glucagon release could contribute in association with hyperinsulinemia to the defect of hepatic glucose production in IDM. At present, we do not know the factors responsible for the failure; they may be related to deficient catecholamine release at birth (LIGHT et al. 1967; STERN et al. 1968) although a recent report did not confirm previous observations (YOUNG et al. 1979). In addition, chronic hyperglycemia and hyperinsulinemia in utero may be factors, since glucose infusion alone supresses glucagon in IDM, while glucose with insulin in necessary to suppress glucagon in normal infants (LUYCKX et al. 1972; MASSI-BENEDETTI et al. 1974).

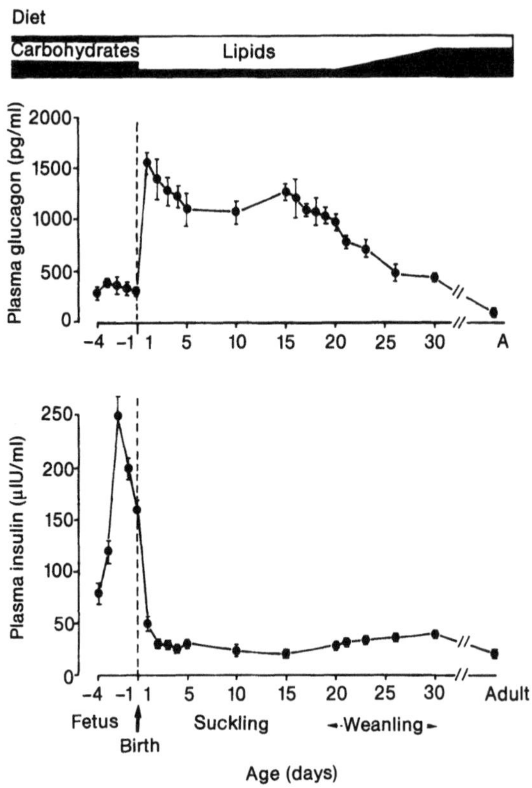


**Fig. 10.** Plasma insulin and glucagon concentrations in suckling and weanling rats. GIRARD et al. (1977a)

## III. Glucagon Secretion During the Suckling Period

Most mammals are fed a diet rich in fat and low in carbohydrate (JENNESS 1974). An increase in plasma glucagon is associated with the first feed in the rat (GIRARD et al. 1975, 1980). A high concentration of plasma glucagon has been observed in suckling rats (Fig. 10; BLAZQUEZ et al. 1972; GIRARD et al. 1977a; BEAUDRY et al. 1977), but also in breast-fed babies (SPERLING et al. 1974), suckling lambs (FISER et al. 1974b), dogs (HETENYI et al. 1976), and rabbits (CSER et al. 1977). The factors which stimulate the release of glucagon during the suckling period have not been clarified. The release of a gut hormone in response to fat and protein ingestion could be involved. Gastric inhibitory polypeptide (GIP) is released during a fat meal in adult humans and dogs and GIP stimulates the secretion of glucagon by the newborn rat pancreas (BATAILLE et al. 1977). Further research in this area is needed to undestand what factor is responsible for the hyperglucagonemia of suckling neonates.

## IV. Glucagon Secretion During the Weaning Period

When rat pups remain with their mother, weaning occurs progressively between 20 and 30 days. It consists primarily in a decrease in fat and an increase in carbohy-

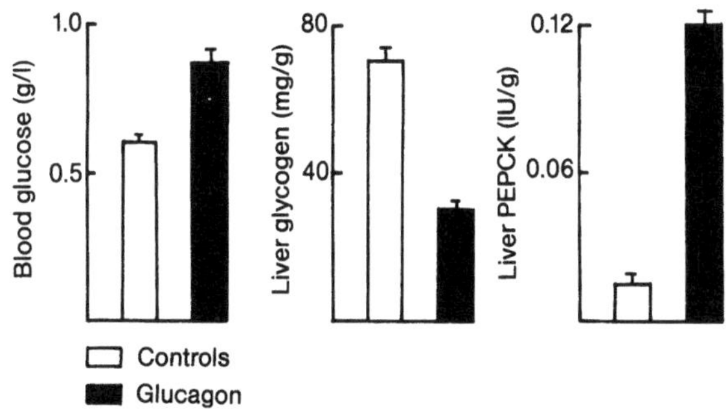


**Fig. 11.** Effect of glucagon injection into term rat fetuses. Under pentobarbital anesthesia of the mother, 3–5 fetuses in one uterine horn were injected substaneously through the uterus with 10 μg zinc glucagon. The same number of fetuses in the opposite horn received the same volume of vehicle and served as ideal controls since glucagon does not cross the placenta. Blood glucose, liver glycogen, and liver PEPCK activity were measured 4 h after glucagon or vehicle injection into the fetuses. GIRARD et al. (1973 a)

drate content of the diet when the rat pups begin progressively to nibble on standard laboratory chow (caloric percentage: carbohydrate 55%, protein 24%, and fat 21%), and still continue to suck the mother, but less frequently. Weaning is complete at 30 days of age. A progressive decrease in plasma glucagon occurs at weaning (Fig. 10; BLAZQUEZ et al. 1974; GIRARD et al. 1977 a). Premature weaning in the rat can be performed 18 days after birth. Weaning on to a high carbohydrate diet is followed by a marked fall in plasma glucagon whereas weaning on to a high fat diet led to a maintenance of plasma glucagon as high as during suckling (GIRARD et al. 1977 a). These data suggest that the fall in plasma glucagon, in association with the increased carbohydrate and decreased fat content of the diet, could be due to the disappearance of a gut hormone secretion in response to fat ingestion.

## E. Metabolic Effects of Glucagon

### I. In the Fetus

During late fetal life, exogenous glucagon produces several metabolic effects in the fetus. Infusion of glucagon in fetal lambs during the last third of gestation results in significant increases in their plasma glucose concentration (BASSETT and THORNBURN 1971; WARNES et al. 1977). The effect is probably due entirely to stimulation of fetal hepatic glycogenolysis since no significant gluconeogenesis can be demonstrated in these conditions (WARNES et al. 1977). Glucagon injection in term rat (GIRARD et al. 1973 a) and monkey (CHEZ et al. 1974) fetuses also results in an increase of blood glucose. In the rat, injection of glucagon produces a marked decrease in fetal liver glycogen concentration (GIRARD et al. 1973 a) and the premature appearance of phosphoenolpyruvate carboxykinase (PEPCK) (Fig. 11), the rate-limiting enzyme of liver gluconeogenesis (YEUNG and OLIVER 1968; HANSON et al. 1973; GIRARD et al. 1973 a). Furthermore, injection of glucagon to the rat fetus, induces a decrease in fetal plasma amino acids and stimulates amino acid uptake by fetal liver (GIRARD et al. 1976). Glucagon has also been reported to stimulate glycogenolysis (PLAS and NUNEZ 1975) and to induce PEPCK (BULANYI et al. 1979) in cultured fetal rat hepatocytes. Glucagon also stimulates glycogenolysis and gluconeogenesis in isolated perfused liver or isolated hepatocytes from sheep, guinea pig, rabbit (JONES et al. 1981; JONES and ROLPH 1981), and human (12–20-week) fetuses (ADAM et al. 1978). It also induces glycogenolysis, gluconeogenesis, and amino acid accumulation in human fetal liver explants in organ culture (SCHWARTZ 1974; SCHWARTZ and RALL 1975a, b; SCHWARTZ et al. 1975). Most of these effects of exogenous glucagon were achieved with doses now known to be pharmacologic rather than physiologic. Recent evidence suggests that the fetal liver is relatively resistant to physiologic increases of plasma glucagon in part due to a diminished number of fetal hepatic glucagon receptors (see Sect. F).

## II. In the Neonate

As exogenous glucagon was capable of reproducing in utero the metabolic changes in glucose and amino acid metabolism which normally occur immediately after birth, i.e., liver glycogenolysis, induction of hepatic PEPCK, and appearance of gluconeogenesis (BALLARD 1971), it has been postulated that this hormone might play an important role during this critical period of development (GIRARD et al. 1977a; SPERLING and GANGULI 1980). Several lines of evidence suggest that it is the increase in plasma glucagon, coupled with the drop plasma insulin that occur immediately after birth in the rat (GIRARD et al. 1973c) that trigger the induction of liver PEPCK. First, the induction of liver PEPCK after glucagon injection in the rat fetus markedly potentiated by simultaneous administration of anti-insulin serum, designed to neutralize the endogenous insulin released in response to the glucagon injection (GIRARD et al. 1977a). Second, in several situations in which a rise in plasma glucagon and a decrease in plasma insulin are induced experimentally in the rat fetus (see Figs. 4, 5, and 6), namely prolonged maternal starvation (GIRARD et al. 1977b), phlorhizin administration to the pregnant rat (FREUND et al. 1980), and prolongation of pregnancy (GIRARD et al. 1977a; PORTHA et al. 1978a) there is a marked increase in liver PEPCK in the fetus (PEARCE et al. 1974; GIRARD et al. 1977a; PORTHA et al. 1978b; FREUND et al. 1980). Third, when pregnant rats are infused chronically with glucose, in order to increase fetal plasma insulin and to lower fetal plasma glucagon, the rise in liver PEPCK which normally occurs in postmature fetuses is markedly impaired (Fig. 12; KTORZA et al. 1981). The essential role of glucagon in glucose homeostasis in the newborn sheep has been demonstrated (SPERLING et al. 1977). Infusion of somatostatin in newborn lambs aged 24–72 h suppressed both insulin and glucagon and induced hypoglycemia (SPERLING et al. 1977). When glucagon was infused with somatostatin, producing an insulin deficiency concomitantly with high plasma glucagon, the blood glucose level was restored to normal (SPERLING et al. 1977). When insulin was infused with somatostatin, producing a glucagon deficiency with unopposed high insulin, profound hypoglycemia resulted (Fig. 13). Moreover, it has been reported that human newborns suffering from glucagon deficiency develop a severe hypoglycemia, which can be corrected by exogenous glucagon administration (SHERWOOD et al. 1974; VIDNES and OYASAETER 1977; KOLLEE et al. 1978). All these data support the view that glucagon plays a crucial role in neonatal glucose homeostasis.

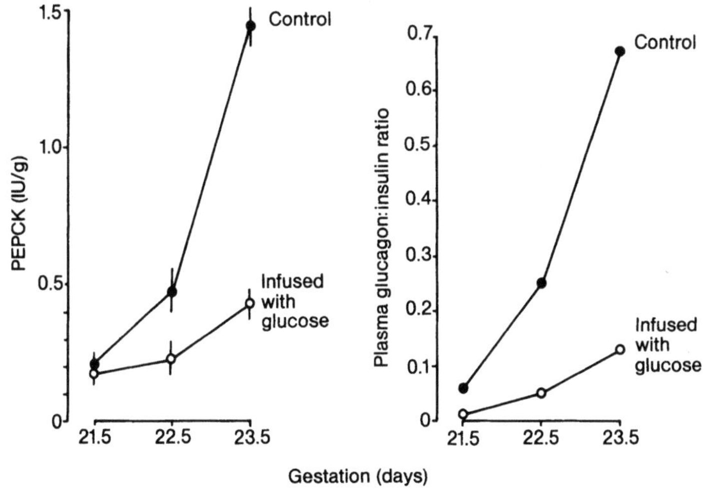


**Fig. 12.** Effects of chronic glucose infusion in pregnant rats during the prolongation of pregnancy on fetal plasma glucagon: insulin molar ratio and fetal liver PEPCK activity. KTORZA et al. (1981)

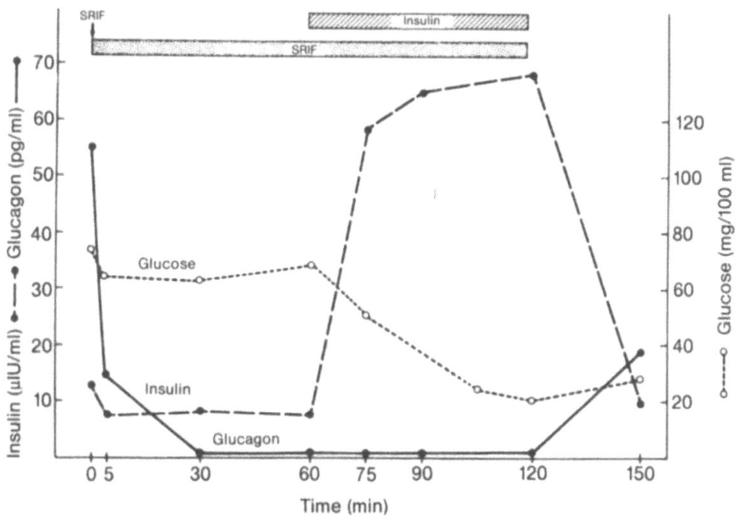


**Fig. 13.** Effects of unopposed insulin action on blood glucose in the newborn lamb. Somatostatin (SRIF) alone suppresses insulin and glucagon and causes a small fall in blood glucose. Reinfusion of insulin while glucagon remains suppressed results in profound hypoglycemia, demonstrating the important role of glucagon. SPERLING et al. (1977)

## III. During the Suckling Period

During the suckling period, the activity of gluconeogenesis and ketogenesis is high in the liver of the rat, while activity of lipogenesis is very low (see reviews by SNELL and WALKER 1973; BAILEY and LOCKWOOD 1973). The high glucagon and the low insulin levels in the blood of suckling rats (BLAZQUEZ et al. 1972; GIRARD et al.

1977a; BEAUDRY et al. 1977) are appropriate for the high gluconeogenic and ketogenic activities and the low lipogenic activity in the liver. However, a low insulin:glucagon molar ratio (see Fig. 10) in an organism whose growth rate is very rapid does not fit with the concept of UNGER (1974) in which maximal anabolism must occur with a high insulin:glucagon ratio. This apparent discrepancy can perhaps be explained by recent observations performed in patients receiving total parenteral nutrition. Normocaloric lipid infusion has been demonstrated to promote positive nitrogen balance with a low plasma insulin:glucagon ratio; glucose infusion produced the same effect, but with a high plasma insulin:glucagon ratio (JEEJEEBHOY et al. 1976). Similarly, when newborn infants are infused for total parenteral nutrition with either 20% glucose and a nitrogen source, or 12% glucose, a nitrogen source and soybean fat emulsion (Intralipid), weight gain and positive nitrogen balance are of equal magnitude (ASCH et al. 1975). However, plasma insulin and glucagon levels and the insulin:glucagon ratio are remarkably different. With 20% glucose in the absence of Intralipid, insulin and glucagon are significantly lower, resulting in a higher insulin:glucagon ratio than with 12% glucose plus Intralipid, when the molar ratio of insulin:glucagon is reversed (ASCH et al. 1975). Thus, studies in adult and newborn humans suggest that high plasma insulin levels seem required for maximal anabolism only when carbohydrates provide the calories to meet the energy needs of the body. By contrast, anabolism can occur normally at normal or low plasma insulin levels when calories are supplied in the form of lipids. If the observations in adult and newborn humans apply to suckling rats, one could explain how rapid growth can occur with low plasma insulin levels, since they are fed a high fat diet, and how they could simultaneously perform active gluconeogenesis which is supported by the high plasma glucagon levels.

## IV. During the Weaning Period

As the milk diet is replaced by laboratory chow, a progressive rise in lipogenesis and a progressive decrease in gluconeogenesis and ketogenesis occur in the liver of the rat (VERNON and WALKER 1968; BAILEY and LOCKWOOD 1973). During this period, a decrease in plasma glucagon and a significant increase in plasma insulin is observed. The rise in insulin:glucagon ratio during weaning can explain the shift in hepatic metabolism occurring at this time. Similarly, when the rats are weaned prematurely to a high carbohydrate diet after 18 days of age there is a rapid fall in hepatic gluconeogenesis (HAHN and KIRBY 1973, 1974; VERNON and WALKER 1968) which is accompanied by a fall in plasma glucagon and a rise in plasma insulin (Fig. 14). In contrast, when the rats are weaned prematurely to a high fat diet after 18 days of age, the rates of hepatic gluconeogenesis and ketogenesis remain as high as they were during suckling (HAHN and KIRBY 1973, 1974; VERNON and WALKER 1968) whereas the rise in liver lipogenesis is prevented (HAHN and KIRBY 1973, 1974; BAILEY and LOCKWOOD 1973). In this situation, plasma glucagon remains very high and plasma insulin very low (Fig. 14). When newborn rats are weaned on to a high carbohydrate diet, but are injected with exogenous glucagon, the rate of liver gluconeogenesis remains at a high level whereas the rise in hepatic lipogenesis is suppressed (HAHN and KIRBY 1974). This suggests that changes in

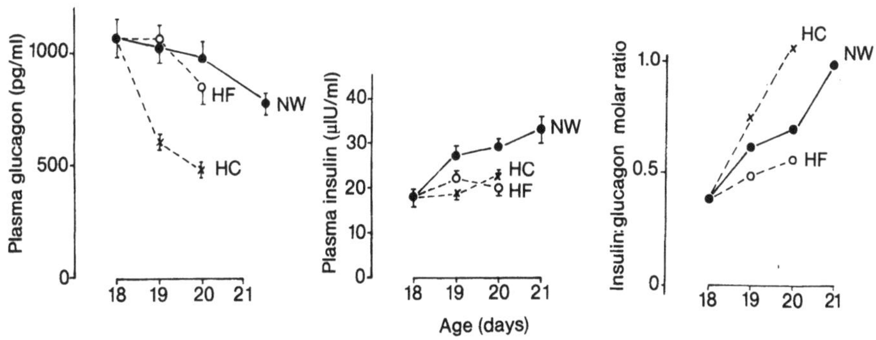


**Fig. 14.** Plasma insulin and glucagon concentrations in suckling rats which remain with their mother (NW) and in pups weaned prematurely at 18 days on a high fat (HF) or a high carbohydrate (HC) diet. GIRARD et al. (1977a)

plasma insulin and glucagon during the weaning period are attended by important modification of hepatic metabolism.

The effects on glucagon and insulin secretion of breast milk feeding, formula feeding, and the transition to solid foods during weaning in human neonates have not been carefully delineated. Caution is therefore necessary for extrapolating results from studies in rats to the situation in human newborns. Nevertheless, consistent similarity in the hormonal adaptation at birth in several mammalian species (SPERLING and GANGULI 1980) and the insulin or glucagon responses to various nutrients suggest that the findings in rats will be applicable to humans. That infusion of lipids to human newborns elevates plasma glucagon has been demonstrated (ASCH et al. 1975), as has a marked glycemic response to lipid infusion (VILEISIS et al. 1982). Oral feeding of alanine at a dose of 500 mg/kg produces significant increases in plasma glucagon, glucose, and insulin in healthy newborns, but the glucagon response could be prevented by a constant glucose infusion (FISER et al. 1975). In small-for-gestational-age infants, alanine feeding increases plasma glucagon, but no glycemic response occurs, suggesting diminished glycogen stores in such infants (WILLIAMS et al. 1975). The remarkable homology of spontaneous and nutrient-induced changes in glucagon secretion and its significance for normal energy homeostasis during the perinatal period among various species now also extends to the glucagon receptor.

## F. Glucagon Receptors in the Perinatal Period

For glucagon to exert its biologic effect on liver glucose output, it must first bind to specific membrance receptors and stimulate cAMP production by activating adenylate cyclase (LEVEY 1975; FREYCHET 1976: see also Chap. 13). Studies on liver membranes from fetal and newborn rat (BLAZQUEZ et al. 1976; VINICOR et al. 1976; SPERLING et al. 1979), guinea pig (GANGULI et al. 1980), and rabbit (GANGULI et al. 1981) all demonstrate significant reduction of glucagon receptor number in fetal and newborn liver when compared with adult. In addition, cAMP production by liver plasma membrane in response to graded doses of glucagon is virtually absent in the fetus and markedly impaired in the newborn animal (BLAZQUEZ et al. 1976;

VINICOR et al. 1976; SPERLING et al. 1979; GANGULI et al. 1980, 1981). There is a high degree of correlation between receptor number and cAMP production in response to any dose of glucagon used in these in vitro studies which also demonstrate gradual progressive increase of glucagon receptor number and cAMP response to glucagon with increasing postnatal maturation. In the rat, glucagon receptor number and cAMP responsiveness of liver plasma membrane obtained at 21 days postnatal life are still only 40% of the values obtained from adult livers. Also there is rapid change from fetal to neonatal life: virtually no cAMP production by fetal liver membranes and some responsiveness within hours of delivery (GANGULI et al. 1981). In contrast, insulin receptor number and affinity of liver plasma membranes, monocytes, and erythrocytes are markedly increased when compared with adult, and there is a gradual reduction to adult levels after birth (NEUFELD et al. 1980; SINHA et al. 1981).

Taken in conjunction, these findings of opposite characteristics in glucagon and insulin receptors between fetus and adult permit a hypothesis for the preponderance of anabolic processes while catabolic processes are limited in utero, and the rapid activation of catabolism after birth. In utero, the high insulin and low glucagon receptor number facilitate anabolism and limit catabolism. After birth, the increase in glucagon receptors which rapidly become linked to cAMP production, and the gradual decrease in insulin receptors permit activation of catabolism, but do not preclude future anabolism and growth. The hormonal changes at birth: a rapid increase in glucagon and a decrease in insulin secretion, are in accord with this concept.

Finally, these findings suggest that activation of glycogenolysis by physiologic elevations of glucagon would be limited in the fetus compared with the adult, by virtue of the reduction of fetal liver glucagon receptors. Indeed, in fetal sheep late in gestation, infusion of glucagon at 5 ng $kg^{-1}$ $min^{-1}$ which raised fetal plasma glucagon from basal levels of approximately 100 pg/ml to levels of approximately 400 pg/ml did not increase blood glucose or glucose production rates, whereas the same dose of glucagon markedly increased blood glucose and glucose production rates in adult sheep. Only when glucagon was infused at pharmacologic doses of 50 ng $kg^{-1}$ $min^{-1}$ to the fetus (plasma glucagon 4,000–5,000 pg/ml) were increases in blood glucose and glucose production observed (M. A. SPERLING et al. 1981, unpublished work). These findings support the hypothesis concerning the role of glucagon receptors in the fetus and newborn. They also suggest that previous observations concerning the effects of pharmacologic doses of glucagon on fetal metabolism should be interpreted with caution before endowing them with physiologic significance. Further study is necessary to elucidate the exact role of glucagon in perinatal metabolism.

## References

Adam PAJ, King KC, Schwartz R, Teramo K (1972) Human placental barrier to $^{125}$I-glucagon early in gestation. J Clin Endocrinol 34:772–785

Adam PAJ, Schwartz AL, Rahiala EL, Kekomaki M (1978) Glucose production in midterm human fetus. 1. Autoregulation of glucose uptake. Am J Physiol 234:E560–E567

Alexander DP, Assan R, Britton HG, Nixon DA (1971) Glucagon in the fetal sheep. J Endocrinol 51:597–598

Alexander DP, Assan R, Britton HG, Nixon DA (1973) Impermeability of the sheep placenta to glucagon. Biol Neonate 23:391–402
Alexander DP, Assan R, Britton HG, Fenton E, Redstone D (1976) Glucagon release in the sheep fetus. 1. Effect of hypo and hyperglycemia and arginine. Biol Neonate 30:1–10
Asch MJ, Sperling M, Fiser RH, Leake R, Moore TC, Oh W (1975) Metabolic and hormonal studies comparing three parenteral nutrition regimens in infants. Ann Surg 182:62–65
Assan R, Boillot J (1973) Pancreatic glucagon and glucagon-like material in tissues and plasma from human fetuses 6–26 weeks old. Pathol Biol (Paris) 21:149–155
Assan R, Girard J (1975) Glucagon in the human fetal pancreas. In: Camerini-Davalos RA, Cole HS (ed) Early Diabetes in Early Life, Academic Press, London New York, pp 115–126
Assan R, Attali JR, Ballerio G, Girard JR, Hautecouverture M, Kervran A, Plouin PF, Slama G, Soufflet E, Tchobroutky, Tiengo A (1974) Some aspects of the physiology of glucagon. In: Malaisse WJ, Pirart J (eds) Diabetes. Excerpta Medica, Amsterdam London New York, pp 144–179
Bailey E, Lockwood EA (1973) Some aspects of fatty acid oxidation and ketone body formation and utilization during development of the rat. Enzyme 15:239–253
Bajaj JS, Buchanan KD (1977) Glucose homeostasis in the newborn, In: Foà PP, Bajaj JS, Foà NL (eds) Glucagon: its role in physiology and clinical medicine. Springer, Berlin Heidelberg New York, pp 583–593
Ballard FJ (1971) Gluconeogenesis and the regulation of blood glucose in the neonate. In: Rodriguez RR, Vallance-Owen J (eds) Diabetes. Excerpta Medica, Amsterdam London New York, pp 592–600
Bassett, JM (1977) Glucagon, insulin and glucose homeostasis in the fetal lamb. Ann Rech Vet 8:362–373
Bassett JM, Thornburn GD (1971) The regulation of insulin secretion by the ovine fetus in utero. J Endocrinol 62:59–74
Bassett JM, Hunziker V, Madill D (1977) Glycine and alanine regulation of glucagon secretion in foetal and post-natal lambs. J. Physiol (Lond) 275:51–52P
Bataille D, Jarousse C, Vauclin N, Gespach C, Rosselin G (1977) Effect of vasoactive intestinal peptide (V.I.P.) and gastric inhibitory peptide (G.I.P.) on insulin and glucagon release by perifused newborn rat pancreas. In: Foà PP, Bajaj JS, Foà NL (eds) Glucagon: its role in physiology and clinical medicine. Springer, Berlin Heidelberg New York, pp 255–269
Battaglia FC, Meschia G (1978) Principal substrates of fetal metabolism. Physiol Rev 58:499–527
Baum D, Porte D Jr, Ensinck J (1979) Hyperglucagonemia and α-adrenergic receptors in acute hypoxia. Am J Physiol 237:E404–E408
Beaudry MA, Chiasson JL, Exton JH (1977) Gluconeogenesis in the suckling rat. Am J Physiol 233:E175–E180
Blazquez E, Sugase T, Blazquez M, Foà PP (1972) The ontogeny of metabolic regulation in the rat, with special reference of the development of insular function. Acta Diabetol Lat [Suppl 1] 9:13–34
Blazquez ET, Sugase M, Blazquez M, Foà PP (1974) Neonatal changes in the concentration of rat liver cyclic AMP and of serum glucose, FFA, insulin, pancreatic and total glucagon in man and in the rat. J Lab Clin Med 83:957–967
Blazquez E, Rubalcava B, Montesano R, Orci L, Unger RH (1976) Development of insulin and glucagon binding and the adenylate cyclase response in liver membranes of the prenatal, postnatal, and adult rat: evidence of glucagon "resistance". Endocrinology 98:1014–1023
Bloom SR, Johnston DI (1972) Failure of glucagon release in infants of diabetic mothers. Br Med J 4:453–454
Bulanyi GS, Steele JG, McGrath MC, Yeoh GCT, Oliver IT (1979) Hormonal regulation of phosphoenolpyruvate carboxykinase in cultured fetal hepatocytes from rat. Eur J Biochem 102:93–100
Callikan S, Ferre P, Pegorier JP, Marliss EB, Assan R, Girard JR (1979) Fuel metabolism in fasted newborn rabbits. J Dev Physiol 1:2670–281

Chez RA, Mintz DH, Epstein MF, Fleischman AR, Oakes GK, Hutchinson DL (1974) Glucagon metabolism in nonhuman primate pregnancy. Am J Obstet Gynecol 120:690–694
Cser A, Girard JR, Goode M, Heim T, Assan R, Milner RDG (1977) Effect of racemic, dextro-, levo-propranolol and isoxuprine on the metabolic and endocriner response to cold in the newborn rabbit. Eur J Clin Invest 7:491–496
DiMarco PN, Ghisalberti AV, Martin CE, Oliver IT (1978) Perinatal changes in liver corticosterone, serum insulin and plasma glucagon and corticosterone in the rat. Eur J Biochem 87:243–247
Edwards JC, Asplund K, Lundquist G (1972) Glucagon release from the pancreas of the newborn rat. J Endocrinol 54:493–504
Eliot RJ, Lam R, Leake RD, Hobel CJ, Fisher DA (1980) Plasma catecholamine concentration in infants at birth and during the first 48 h of life. J Pediatr 96:311–315
Eliot RJ, Klein AH, Glatz TH, Nathanielsz PW, Fisher DA (1981) Plasma norepinephrine, epinephrine and dopamine concentrations in maternal and fetal sheep spontaneous parturition and in premature sheep during cortisol-induced parturition. Endocrinology 108:1678–1682
Epstein M, Chez RA, Oakes GK, Mintz DH (1977) Fetal pancreatic glucagon responses in glucose intolerant nonhuman primate pregnancy. Am J Obstet Gynecol 127:268–272
Fiser RH Jr, Erenberg A, Sperling MA, Oh WH, Fisher DA (1974 a) Insulin-glucagon substrate interrelations in the fetal sheep. Pediatr Res 8:951–955
Fiser RH Jr, Phelps DL, Williams PR, Sperling MA, Fisher DA, Oh WH (1974 b) Insulin-glucagon substrate interrelationships in the neonatal sheep. Am J Obstet Gynecol 120:944–950
Fiser RH, Williams PR, Fisher DA, Delamater PV, Sperling MA, Oh WH (1975) The effect of oral alanine on blood glucose and glucagon in the human newborn infant. Pediatrics 56:78–81
Freund N, Kervran A, Assan R, Geloso JP, Girard JR (1980) Fetal metabolic response to phloridzin-induced hypoglycemia in pregnant rats. Biol Neonate 38:321–327
Freychet P (1976) Interaction of polypeptide hormones with cell membrane specific receptors: Studies with insulin and glucagon. Diabetologia, 12:83–100
Ganguli S, Whitsett J, Voina S, Velayo N, Sperling MA (1980) A difference between fetal and adult hepatic glucagon receptors coupled to adenylate cyclase: evidence for the postnatal appearance of a new population of glucagon receptors. Endocrine Society Program. Williams and Wilkins, Baltimore
Ganguli S, Sinha M, Harris P, Sperling MA (1981) Differential maturation of insulin and glucagon receptors and the adenylate cyclase system in rabbit liver plasma membrane. Diabetes [Suppl 1] 30:8A
Girard JR (1981 a) Glucose homeostasis in the perinatal period: the critical role of pancreatic hormones and exogenous substrates in the rat. Ciba Found Symp 86:234–246
Girard JR (1981 b) Fuel homeostasis during the perinatal period. In: Ritzen M, Larsson A (eds) Biology of normal human growth. Raven, New York, pp 193–202
Girard JR, Assan R (1981) Glucagon secretion during the perinatal period. In: DeMeyer R (ed) Metabolic adaptation to extrauterine life. Nijhoff, The Hague, pp 241–257
Girard JR, Bal D, Assan R (1972) Glucagon secretion during the early postnatal period in the rat. Horm Metab Res 4:168–170
Girard JR, Caquet D, Bal D, Guillet I (1973 a) Control of rat liver phosphorylase, glucose-6-phosphatase and phosphoenolpyruvate carboxykinase activities by insulin and glucagon during the perinatal period. Enzyme 15:272–285
Girard J, Assan R, Jost A (1973 b) Glucagon in the rat foetus. In: Comline KS, Cross KW, Dawes GS, Nathanielsz PW (eds) Foetal and neonatal physiology. Cambridge University Press, Cambridge pp 456–461
Girard JR, Cuendet GS, Marliss EB, Kervran A, Rieutort M, Assan R (1973 c) Fuels hormones and liver metabolism at term and during the early postnatal period in the rat. J Clin Invest 52:3190–3200
Girard JR, Kervran A, Soufflet E, Assan R (1974) Factors affecting the secretion of insulin and glucagon by the rat fetus. Diabetes 23:310–317

Girard JR, Kervran A, Assan R (1975) Functional maturation of the A cell in the rat. In: Camerini-Davalos RA, Cole HS (eds) Early diabetes in early life. Academic Press, London New York, pp 57–71
Girard JR, Guillet I, Marty J, Assan R, Marliss EB (1976) Effects of exogenous hormones and glucose on plasma levels and hepatic metabolism of amino acids in the fetus and in the newborn rat. Diabetologia 12:327–337
Girard JR, Ferre P, Kervran A, Pegorier JP, Assan R (1977a) Role of the insulin-glucagon ratio in the changes of hepatic metabolism during development of the rat. In: Foà PP, Bajaj JS, Foà NL (eds) Glucagon: its role in physiology and clinical medicine. Springer, Berlin Heidelberg New York, pp 563–581
Girard JR, Ferre P, Gilbert M, Kervran A, Assan R, Marliss EB (1977b) Fetal metabolic response to maternal fasting in the rat. Am J Physiol 232:E456–E463
Girard JR, Ferre P, Pintado EI (1979) Fuel metabolism in the mammalian fetus. Ann Biol Anim Biochim Biophys 19:181–197
Girard JR, Ferre P, Pegorier JP, Leturque A, Callikan S (1980) Factors involved in the development of hypoglycemia in fasting newborn rats. In: Andreani D, Lefèbvre P, Marks V (eds) Current views on hypoglycemia and glucagon. Academic Press, London New York, pp 343–352
Grajwer LA, Sperling MA, Sack J, Fisher DA (1977) Possible mechanisms and significance of the neonatal surge in glucagon secretion: studies in newborn lambs. Pediatr Res 11:833–836
Hahn P, Kirby L (1973) Immediate and late effects of premature weaning and of feeding a high fat or high carbohydrate diet to weaning rats. J Nutr 103:690–696
Hahn P, Kirby LT (1974) The effects of catecholamines, glucagon and diet on enzyme activities in brown fat and liver of the rat. Can J Biochem 52:739–743
Hanson RW, Fisher L, Ballard FJ, Reshef L (1973) The regulation of phosphoenolpyruvate carboxykinase in fetal rat liver. Enzyme 15:97–110
Hetenyi G Jr, Kovacevic N, Hall SEH, Vranic M (1976) Plasma glucagon in pups, decreased by fasting, unaffected by somatostatin or hypoglycemia. Am J Physiol 231:1377–1382
Jarousse C, Rosselin G (1975) Interaction of amino acids and cyclic AMP on the release of insulin and glucagon by newborn rat pancreas. Endocrinology 96:168–177
Jarousse C, Rancon F, Rosselin G (1973) Hormonogenèse perinatale de l'insuline et du glucagon chez le rat. C R Acad Sci [D] (Paris) 276:585–588
Jeejeebhoy KN, Anderson GH, Nakhooda AF, Greenberg GR, Sanderson I, Marliss EB (1976) Metabolic studies in total parenteral nutrition with lipid in man: comparison with glucose. J Clin Invest 57:125–136
Jenness R (1974) Biosynthesis and composition of milk. J Invest Dermatol 63:109–188
Johnston DI, Bloom SR (1973) Plasma glucagon levels in the term human infant and effect of hypoxia. Arch Dis Child 48:451–454
Johnston DI, Bloom SR, Green KR, Beard RW (1972) Failure of the human placenta to transfer pancreatic glucagon. Biol Neonate 21:375–380
Jones CT, Rolph TP (1981) Metabolic events associated with the preparation for the fetus for independent life. Ciba Found Symp 86:214–228
Jones CT, Rolph T, Baud G, Michael E (1981) Factors regulating blood glucose in the fetus and in the newborn. In: DeMeyer R (ed) Metabolic adaptation to extra-uterine life. Nijhoff, The Hague, pp 55–78
Kalhan SC, Savin SM, Adam PAJ (1977) Attenuated glucose production rate in newborn infants of insulin dependent diabetic mothers. N Eng J Med 296:375–376
Kervran A, Gilbert M, Girard JR, Assan R, Jost A (1976) Effect of environmental temperature on glucose-induced insulin response in the newborn rat. Diabetes 25:1026–1030
Kollee LA, Monnens LA, Cejka V, Wilms RH (1978) Persistent neonatal hypoglycemia due to glucagon deficiency. Arch Dis Child 53:422–424
Ktorza A, Girard J, Kinebanyan MF, Picon L (1981) Hyperglycemia induced by glucose infusion in unrestrained pregnant rat during the last 3 days of gestation: metabolic and hormonal changes in the mother and the fetus. Diabetologia 21:569–574

Kühl C, Mølsted-Pedersen L, Pedersen J, Skouby SO, Winkel S (1980) Plasma insulin, glucagon and the molar insulin; glucagon ratio in newborn infants of diabetic mothers. In: Andreani D, Lefèbvre PJ, Marks V (eds) Current views on hypoglycemia and glucagon. Academic Press, London New York, pp 397–407

Lagercrantz H, Bistoletti P (1977) Catecholamine release in the newborn infant at birth. Pediatr Res 11:889–893

Lernmark A, Wenngren BI (1972) Insulin and glucagon release from the isolated pancreas of foetal and newborn mice. J Embryol Exp Morphol 28:607–614

Levey GS (1975) The glucagon receptor and adenylate cyclase. Metabolism 24:301–309

Light IJ, Sutherland JM, Loggie JM, Gaffney TE (1967) Impaired epinephrine release in hypoglycemic infants of diabetic mothers. N Enbl J Med 277:394–398

Lindblad BS (1970) The venous plasma free amino acid levels during the first hours of life. Acta Paediatr Scand 59:13–20

Luyckx A, Massi-Benedetti F, Falorni A, Lefèbvre PJ (1972) Presence of pancreatic glucagon in the portal plasma of human neonates. Differences in the insulin and glucagon responses to glucose between normal infants and infants from diabetic mothers. Diabetologia 8:296–300

Marliss EB, Wollheim CB, Blondel B, Ori L, Lambert AE, Stauffacher W, Like AA, Renold AE (1973) Insulin and glucagon release from monolayer cell cultures of pancreas from newborn rat. Eur J Clin Invest 3:16–26

Massi-Benedetti F, Falorni A, Luyckx A, Lefèbvre P (1974) Inhibition of glucagon secretion in the human newborn by the simultaneous administration of glucose and insulin. Horm Metab Res 6:392–396

Milner RDG, Chauksey SK, Mickleson KNP, Assan R (1973) Plasma pancreatic glucagon and insulin: glucagon ratio at birth. Arch Dis Child 48:241–242

Moore WMO, Ward BS, Gordon C (1974) Human placental transfer of glucagon. Clin Sci 46:125–129

Neufeld ND, Scott M, Kaplan SA (1980) Ontogeny of the mammalian insulin receptor: studies of human and rat liver plasma membranes. Dev Biol 78:151–160

Okuno G, Price S, Grillo TAI, Foà PP (1964) Development of phosphorylase and phosphorylase activating (glucagon-like) substances in the rat embryo. Gen Comp Endocrinol 4:446–451

Orci L, Lambert AE, Rouiller C, Renold AE, Samols E (1969) Evidence for the presence of A-cells in the endocrine fetal pancreas of the rat. Horm Metab Res 1:108–110

Orci L, Malaisse-Lagae F, Ravazzola M, Rouiller C, Renold AE, Perrelet A, Unger RH (1975) A morphological basis for intercellular communication between A- and B-cells in the endocrine pancreas. J Clin Invest 46:1066

Pearce PH, Buirchell BJ, Weaver PK, Oliver IT (1974) The development of phosphopyruvate carboxylase and gluconeogenesis in neonatal rats. Biol Neonate 24:320–329

Perrier H (1970) Evolution de l'ultrastructure du pancréas chez le foetus de rat. Diabetologia 6:605–615

Pictet R, Rutter WJ (1972) Development of the embryonic endocrine pancreas. In: Freinkel N, Steiner DF (eds) Endocrine pancreas. Williams and Wilkins, Baltimore, pp 25–66

Plas C, Nuñez J (1975) Glycogenolytic response to glucagon of cultured fetal hepatocytes. Refractoriness following prior exposure to glucagon. J Biol Chem 250:5304–5311

Portha B, Picon L, Rosselin G (1978a) Postmaturity in the rat: high levels of glucagon in the plasma of the foetus and the neonate. J Endocrinol 77:153–154

Portha B, LePrevost E, Picon L, Rosselin G (1978b) Postmaturity in the rat: phosphorylase, glucose-6-phosphatase and phosphoenolpyruvate carboxykinase activities in the fetal liver. Horm Metab Res 10:141–144

Rall LB, Pictet RL, Williams RH, Rutter WJ (1973) Early differentiation of glucagon-producing cells in embryonic pancreas: a possible developmental role of glucagon. Proc Nat Acad Sci USA 70:3478–3482

Schaeffer LD, Wilder ML, Williams RH (1973) Secretion and content of insulin and glucagon in human fetal pancreas slices in vitro. Proc Soc Exp Biol Med 143:314–319

Schreiner RL, Nolen PA, Bonderman PW, Moorehead HL, Gresham EL, Lemons JA, Escobedo MD (1980a) Fetal and maternal hormonal response to starvation in the ewe. Pediatr Res 14:103–108
Schreiner RL, Lemons JA, Gresham EL (1980b) Effect of maternal malnutrition on singleton and twin pregnancies in the sheep. Nutr Rep Int 21:525–530
Schreiner RL, Lemons JA, Gresham EL (1981) Metabolic and hormonal response to chronic maternal fasting in the ewe. Ann Nutr Metab 25:38–47
Schwartz A (1974) Hormonal regulation of amino acid accumulation in human fetal liver explants. Effects of dibutyryl cyclic AMP, glucagon and insulin. Biochim Biophys Acta 362:276–289
Schwartz AL, Rall TW (1975a) Hormonal regulation of glycogen metabolism in human fetal liver. 2. Regulation of glycogen synthase activity. Diabetes 24:1113–1122
Schwartz AL, Rall TW (1975b) Hormonal regulation of incorporation of alanine-U-$^{14}$C into glucose in human fetal liver explants. Effect of dibutyryl cyclic AMP, glucagon, insulin and triamcinolone. Diabetes 24:650–657
Schwartz AL, Raiha NCR, Rall TW (1975) Hormonal regulation of glycogen metabolism in human fetal liver. 1. Normal development and effects of dibutyryl cyclic AMP, glucagon and insulin in liver explants. Diabetes 24:1101–112
Shelley HJ, Girard JR (1981) Plasma insulin and glucagon in well oxygenated and hypoxic fetal lambs. In: DeMeyer R (ed) Metabolic adaptation to extrauterine life. Nijhoff, The Hague, pp 261–279
Sherwood WG, Chance GW, Toews CJ, Martin JM, Marliss EB (1974) A new syndrome of familial pancreatic agenesis: essential role of glucagon in neonatal gluconeogenesis (abst). Pediatr Res 8:438
Sinha MK, Ganguli S, Sperling MA (1981) Disappearance of erythrocyte insulin receptors during maturation in sheep. Diabetes 30:411–415
Snell K, Walker DG (1973) Gluconeogenesis in the newborn rat: the substrates and their quantitative significance. Enzyme 15:40–81
Sodoyez-Goffaux F, Sodoyez JC, Devos CJ, Foà PP (1979) Insulin and glucagon secretion by islets isolated from fetal and neonatal rats. Diabetologia 16:121–123
Spellacy WN, Buhi WC (1976) Glucagon, insulin and glucose levels in maternal and umbilical cord plasma with studies of placental transfer. Obstet Gynecol 47:291–294
Sperling MA, Ganguli S (1980) Neonatal glucose homeostasis. In: Waldhausl WK (ed) Diabetes 1979. Proceedings of the 10th Congress of the International Diabetes Federation. Excerpta Medica, Amsterdam London New York, pp 752–757
Sperling MA, Erenberg A, Fiser RH, Oh W, Fisher DA (1973) Placental transfer of glucagon in sheep. Endocrinology 93:1435–1438
Sperling MA, DeLamater PV, Phelps D, Fiser RH, Oh W, Fisher DA (1974) Spontaneous and amino acid stimulated glucagon secretion in the immediate postnatal period. Relation to glucose and insulin. J Clin Invest 53:1159–1166
Sperling MA, Grajwer L, Leake RD, Fisher DA (1977) Effects of somatostatin (SRIF) infusion on glucose homeostasis in newborn lambs: evidence for a significant role of glucagon. Pediatr Res 11:962–967
Sperling MA, Christensen RA, Ganguli S, Anand R (1980) Adrenergic modulation of pancreatic hormone secretion in utero: studies in fetal sheep. Pediatr Res 14:203–208
Sperling MA, Ganguli S, Christensen R, Voina S (1983) Maternal diabetes does not alter postnatal development of the hepatic glucagon receptor-adenylate cyclase system in the rat. Pediatr Res 17:101–106
Stern L, Ramos A, Leduc J (1968) Urinary catecholamines excretion in infants of diabetic mothers. Pediatrics 42:498–605
Unger RH (1974) Alpha and beta-cell interrelationships in health and disease. Metabolism 23:518
Unger RH, Orci L (1981) Glucagon and the A-cell: physiology and pathophysiology. N Engl J Med 304:1518–1524, 1575–1580
Vernon RG, Walker DG (1968) Adaptative behaviour of some enzymes involved in glucose utilization and formation in rat liver during the weaning period. Biochem J 196:331

Vidnes J, Oyasaeter S (1977) Glucagon deficiency causing severe neonatal hypoglycemia in a patient with normal insulin secretion. Pediatr Res 11:943–949
Vileisis RA, Cowett RM, Oh W (1982) Glycemic response to lipid infusion in the premature neonatr. J Pediat 100:108–112
Vinicor F, Higdon JF, Clark JF, Clark CM (1976) Development of glucagon sensitivity in the neonatal rat liver. J Clin Invest 58:571–578
Warnes DM, Sedmark RF, Ballard FJ (1977) The appearance of gluconeogenesis at birth in sheep. Activation of the pathway associated with blood oxygenation. Biochem J 162:627–634
Williams PR, Fiser RH, Sperling MA, Oh W (1975) Effects of oral alanine feeding on blood glucose, plasma glucagon and insulin concentrations in small-for-gestational age infants. N Engl J Med 292:612–614
Williams PR, Sperling MA, Racasa Z (1979) Blunting of spontaneous and alanine-stimulated glucagon secretion in newborn infants of diabetic mothers. Am J Obstet Gynecol 133:51–56
Wise JK, Lyall SS, Hendler R, Felig P (1973) Evidence of stimulation of glucagon secretion by alanine in the human fetus at term. J Clin Endocrinol Metab 37:345–348
Yeung D, Oliver IT (1968) Factors affecting the premature induction of phosphopyruvate carboxylase in neonatal rat liver. Biochem J 108:325–331
Young JB, Cohen WR, Rappaport EB, Lansberg L (1979) High plasma norepinephrine concentrations at birth in infants of diabetic mothers. Diabetes 28:697–699

CHAPTER 37

# Glucagon as a Counterregulatory Hormone

J. E. GERICH

## A. Glucose Counterregulation, an Overview

The term "glucose counterregulation" refers to the physiologic processes which defend the organism against hypoglycemia. These processes must not only prevent the occurrence of hypoglycemia, but, should it occur, must then restore normoglycemia. The significance of these processes for homeostasis stems from the obligatory consumption of glucose by various tissues, the most important of which is the brain. In the postabsorptive state, utilization of glucose by brain averages approximately 1.0 mg $kg^{-1}$ $min^{-1}$ (HUANG et al. 1980). This is about one-half of the total amount of glucose produced by the liver. Although brain can increase its extraction of glucose in the face of a decreasing plasma glucose concentration, its capacity to do so is limited (LUND-ANDERSEN 1979). Once the plasma glucose concentration decreases below 40 mg/dl, brain glucose uptake decreases so that cerebral and other important neural functions become impaired (EISENBERG and SELTZER 1962). Should this situation persist or worsen, irreversible brain damage and ultimately death may occur.

To prevent hypoglycemia in the postabsorptive state, when exogenous sources of glucose and other nutrients are not available, glucose counterregulatory processes must ensure that there is appropriate mobilization of endogenous substrates to satisfy the fuel needs of the organism – needs which may be increased by factors such as exercise, infection, or trauma. These processes must also be operative in intraprandial periods to prevent hypoglycemia and ensure efficient storage of exogenous substrates; for example, following meal ingestion plasma insulin concentrations increase to levels which can suppress hepatic glucose production (FELIG et al. 1975). If the liver did not resume its output of glucose following absorption of ingested carbohydrate, hypoglycemia would occur. The resultant mobilization of recently ingested substrates would thwart the anabolic effects of prandially secreted insulin and thus disturb the body's nutrient economy.

The plasma glucose concentration is determined by the balance between rates of glucose production and rates of glucose utilization. If demands for glucose as a substrate are increased and if this increase in glucose utilization is matched by an appropriate increase in glucose production, the plasma glucose concentration will remain constant; if, however, glucose utilization exceeds glucose production, the plasma glucose concentration will decrease. Thus, glucose counterregulation may include those processes which directly and indirectly affect both the production and utilization of glucose. For example, epinephrine may directly stimulate glucose production and may also affect it indirectly by stimulation of lipolysis. The latter would provide free fatty acids and ultimately ketone bodies which can substitute for glucose as sources of energy. This sparing action will decrease demands for glucose, and glucose utilization will decrease. Such an adaptation, which occurs during fasting (HAVEL 1972), permits the steady state to be maintained by a lower rate of glucose production.

Theoretically, hormonal, neural, and antoregulatory factors may all participate in glucose counterregulation (CRYER 1981; GERICH et al. 1981). In this chapter, the role of glucagon in the prevention of hypoglycemia and the restoration of normoglycemia will be placed in perspective to these other potential factors. The specific effects of glucagon on glycogen metabolism and gluconeogenesis, its role in the regulation of hepatic glucose output, and the influence of glucose on glucagon secretion are discussed in detail in other Chapters of this book (see Chaps. 14–16).

## B. Glucagon in the Prevention of Hypoglycemia

### I. The Postabsorptive State

In the postabsorptive state, rates of glucose production and glucose utilization are closely matched so that over the course of a 10–20 h fast the plasma glucose concentration remains relatively stable (RIZZA et al. 1979c). Such tight coupling of the moment-to-moment control of glucose production and utilization requires processes with rapid onsets of action and short half-fives. Thus, it is unlikely that hormones such as cortisol and growth hormone, which have long biologic half-lives and which generally require several hours to initiate their action (BAXTER and FUNDER 1979), would play an appreciable role. Moreover, the lack of effects of adrenergic and cholinergic blockade on glucose flux and plasma glucose concentrations in postabsorptive humans (CLARK et al. 1979; WALTER et al. 1974: PALMER et al. 1979; FRIER et al. 1981) and the preservation of fasting normoglycemia in sympathectomized and vagotomized individuals (PALMER et al. 1976; FRIER et al. 1981) and in glucocorticoid-treated, adremalectomized patients (ENSINCK et al. 1976; GERICH et al. 1979) suggests that, at least under basal conditions, the sympathetic and parasympathetic nervous systems also play no appreciable role. Although the liver is capable of autoregulating its output of glucose (SOSKIN et al. 1938; BERGMAN and BUCOLO 1974; LILJENQUIST et al. 1979b; SACCA et al. 1978, 1979a), it is unlikely that such a process in itself or a mere decrease in insulin secretion are capable of maintaining normoglycemia in the absence of other hormonal and neural modulation (RIZZA et al. 1979a; DERI et al. 1981).

In contrast, there is considerable evidence that glucagon, via its action on glucose production, is a major factor responsible for the prevention of hypoglycemia in the postabsorptive state. First of all, as discussed elsewhere in this text (see Chaps. 14–16), glucagon is a potent stimulator of hepatic glucose production; for example, on a molar basis, glucagon is ten times more potent than epinephrine (SOKAL et al. 1964). Second, its effects are rapid in onset and termination, characteristics which are a prerequisite for involvement in the moment-to-moment control of glucose homeostasis. Third, physiologic concentrations of glucagon can antagonize and overcome the effects of physiologic concentrations of insulin (PARRILLA et al. 1974).

That these actions of glucagon on the liver are important for maintenance of normoglycemia in the postabsorptive state is evident from consideration of the pivotal role of this organ in the regulation of glucose homeostasis. In the postabsorptive state, most of glucose utilization is obligatory and occurs in noninsulin-sensitive tissues (CAHILL 1970). Thus, it is likely that glucose production by the liver rather than glucose utilization be extrahepatic tissues is the major site subject to acute modulation for ensuring an appropriate balance between these two processes to maintain normoglycemia and to prevent hypoglycemia.

Further evidence for an important role for glucagon in the maintenance of normoglycemia has come from studies which have examined the effects on plasma glucose concentrations of glucagon deficiency produced by the administration of antiglucagon serum in the rat (GREY et al. 1970; FROHMAN et al. 1970; BARLING and BELOFF-CHAIN 1973; EPAND and DOUGLAS 1973; HOLST et al. 1978). In all studies, fasting hypoglycemia or normoglycemia in the face of decreased plasma insulin concentrations was observed. The strongest and most clear-cut evidence has been obtained in studies in which somatostatin, an inhibitor of insulin and glucagon secretion (GERICH et al. 1975c) which itself does not affect glucose metabolism (BYRNE et al. 1977; CHERRINGTON et al. 1977) has been employed to produce deficiencies of these hormones in humans (ALFORD et al. 1974; GERICH et al. 1975; LILJENQUIST et al. 1977; WAHREN et al. 1976), and in experimental animals (ALTSZULER et al. 1976; CHERRINGTON et al. 1976, 1978, 1979; SHERWIN et al. 1977; BLAUTH et al. 1977; DERI et al. 1981; LICKLEY et al. 1979).

Infusion of somatostatin alone, which produces a combined deficiency of insulin and glucagon, causes a decrease in plasma glucose concentrations in normal humans (ALFORD et al. 1974; GERICH et al. 1975, 1981; WAHREN et al. 1977; LILJENQUIST et al. 1977) and in experimental animals (ALTSZULER et al. 1976; SHERWIN et al. 1977; CHERRINGTON et al. 1976, 1978; LICKLEY et al. 1978), which is due to a decrease in glucose production. As shown in Fig. 1, plasma glucose concentrations and glucose production remain suppressed for almost 2 h. These observations indicate that, in the absence of sustained secretion of glucagon, neither decreases in insulin secretion nor decreases in plasma glucose concentration nor the actions of other remaining counterregulatory processes are sufficient to initiate an appropriate increase in glucose production to restore normoglycemia promptly although ultimately this does occur. Studies employing isotopic estimation of gluconeogenesis in dogs (CHERRINGTON et al. 1977) and those using the hepatic venous catheterization technique in glycogen-depleted humans (WAHREN et al. 1977) indicate that suppression of glucose production under these circumstances involves inhibition of both glycogenolysis and gluconeogenesis. Based on these observations, it has been estimated that basal glucagon secretion is responsible for sustaining approximately 60%–70% of basal hepatic glucose production (CHERRINGTON et al. 1977).

The importance of basal glucagon secretion in counteracting the effects of basal insulin secretion on hepatic glucose production has been further examined in studies in which somatostatin has been infused along with sufficient insulin to produce an isolated deficiency of glucagon (ALFORD et al. 1974; LILJENQUIST et al. 1977; ALTSZULER et al. 1976; CHERRINGTON et al. 1976, 1977; JENNINGS et al. 1977; LICKLEY et al. 1979; GERICH et al. 1981). Under such conditions (Fig. 2a), decreases in plas-

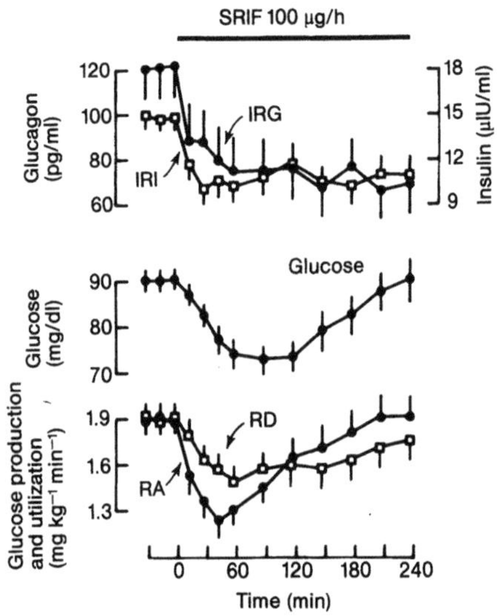


**Fig. 1.** Effect of infusion of somatostatin (SRIF) (combined deficiency of glucagon and insulin) on plasma glucose concentrations and rates of glucose production in 8 normal humans. Mean ± standard error. *IRG*, immunoreactive glucagon; *IRI*, immunoreactive insulin; *RD*, glucose utilization; *RA*, glucose production. Adapted from GERICH et al. (1981)

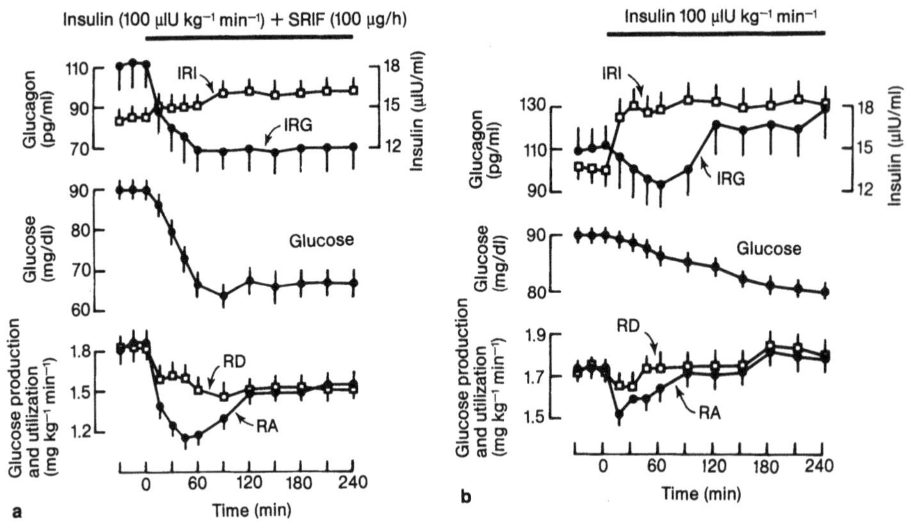


**Fig. 2 a, b.** Effect of infusion of somatostatin (SRIF) plus insulin (isolated glucagon deficiency) (**a**) and infusion of insulin alone (**b**) on plasma glucose concentrations and rates of glucose production in 8 normal humans. Mean ± standard error. Adapted from GERICH et al. (1981)

ma glucose concentrations are more profound that those observed when both insulin and glucagon secretion was inhibited (Fig. 1) or when insulin alone is infused (Fig. 2b). This occurs as a result of a more marked and sustained decrease in hepatic glucose production. These observations demonstrate the severe hypoglycemia and near total paralysis of glucose counterregulation would occur if insulin se-

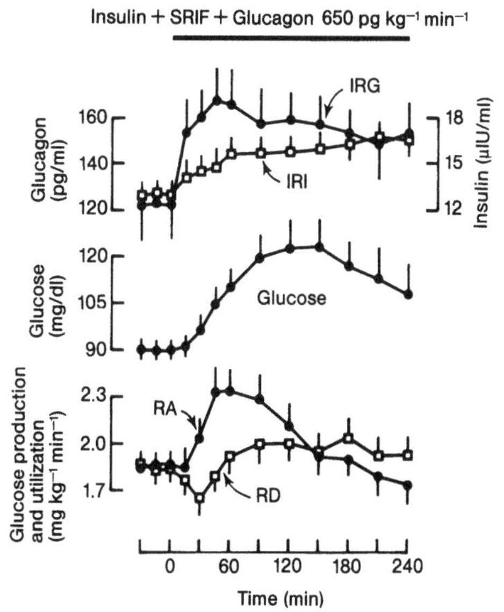


**Fig. 3.** Effect of infusion of somatostatin (SRIF), insulin, and glucagon (relative glucagon excess) on plasma glucose concentrations and rates of glucose production in 8 normal humans. Mean ± standard error. Adapted from GERICH et al. (1981)

cretion did not decrease during hypoglycemia and if glucagon were not available to augment hepatic glucose production. These observations also indicate that, under such conditions, other counterregulatory processes (e.g., autoregulatory, neural, hormonal) are relatively ineffectual, at least on a short-term basis.

Another aspect of the effect of glucagon as a counterregulatory hormone is the effect of an increase in glucagon secretion relative to that of insulin. This has been studied in experiments in which isolated deficiency of insulin or relative glucagon excess has been produced by infusion of somatostatin and glucagon (ALTSZULER et al. 1976; ALFORD et al. 1974; JENNINGS et al. 1977; CHERRINGTON et al. 1977, 1978; LICKLEY et al. 1979) or somatostatin, insulin, and glucagon (LILJENQUIST et al. 1979; SHULMAN et al. 1978; GERICH et al. 1981) and in experiments in which glucagon secretion was stimulated under conditions in which insulin secretion was not possible (RIZZA et al. 1979b). Under such conditions (Fig. 3) hyperglycemia is observed which can be explained primarily on the basis of an increase in glucose production that has not been accompanied by an appropriate increase in glucose utilization. Normally, such hyperglycemia would not occur because, as is illustrated by the experiments shown in Fig. 4 comparing the effects of arginine-induced glucagon secretion in normal subjects and in insulin-dependent diabetic subjects, there would be a compensatory increase in insulin secretion which would suppress glucose production and augment glucose utilization. Nevertheless, these results demonstrate how the potent counterregulatory actions of glucagon might adversely affect glucose homeostasis if pancreatic B-cell function were impaired so that appropriate compensatory changes in insulin secretion were not possible, as is the case in diabetes mellitus. Thus, a minor imbalance in the pancreatic A–B-cell interrelationship which is programmed to maintain normoglycemia and prevent hypoglycemia can quite readily lead to hyperglycemia.

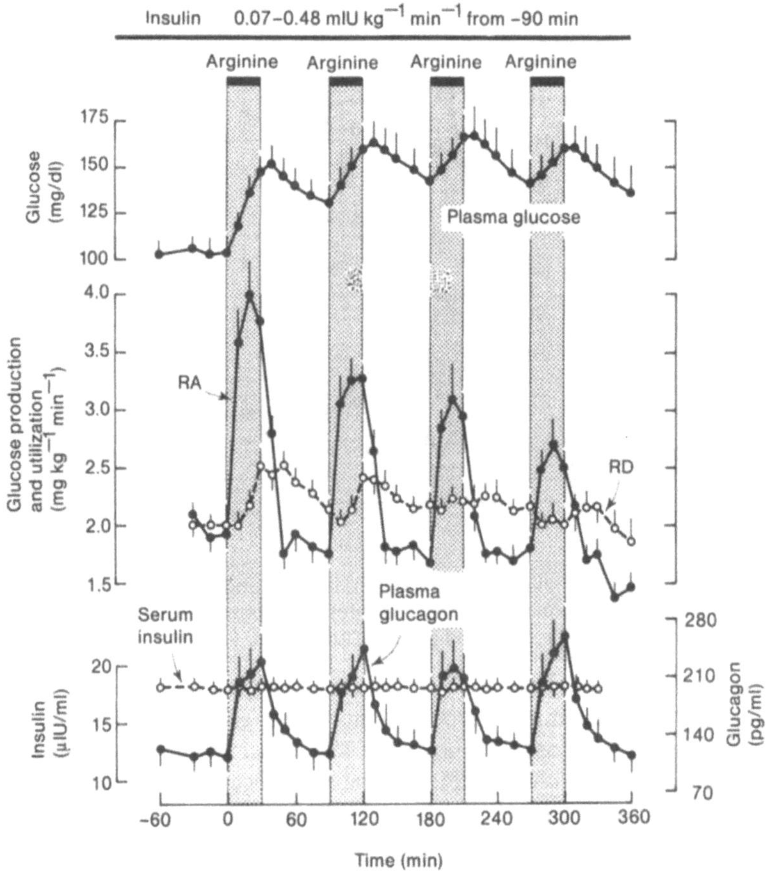


**Fig. 4.** Comparison of the effects of arginine-induced hyperglucagonemia in 9 normal and 9 diabetic humans. Mean ± standard error. Adapted from RIZZA et al. (1979a)

It should be pointed out that, although current evidence indicates that glucagon is the primary hormone responsible for counteracting the effects of insulin on a moment-to-moment basis in the maintenance of normoglycemia in the postabsorptive state, it is likely that epinephrine and other counterregulatory processes assume more importance when glucagon secretion is impaired (see Sect. D) and thus serve as a secondary line of defense against hypoglycemia. Growth hormone and cortisol, while probably not involved in moment-to-moment actions on glucose homeostasis, appear to play an important role in its long-term control by altering sensitivity of tissues to insulin, glucagon, and epinephrine although, at the present time, it is controversial whether these actions are the results of alterations in receptor or postreceptor events (GERICH et al. 1981).

## II. The Intraprandial State

Meal ingestion provides for the replenishment of endogenous fuel stores that have been depleted during fasting. Although the gastrointestinal absorption is probably

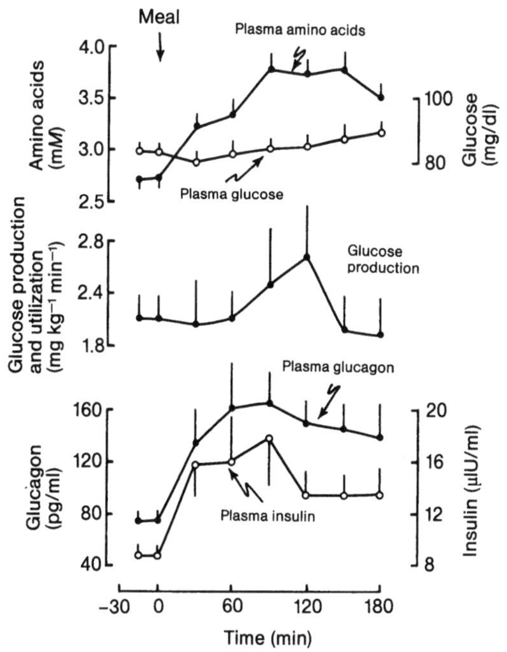


**Fig. 5.** Changes in plasma amino acid and glucose concentrations, rates of glucose production, and circulating concentrations of glucagon and insulin following ingestion of a predominantly protein-containing meal in 7 normal humans. Mean ± standard error. Adapted from WAHREN et al. (1976)

complete within 4–5 h, assimilation of absorbed nutrients probably continues beyond this period. This anabolic flux of nutrients is not only influenced by the circulating concentration of substrates, but also by insulin whose actions persist long after its plasma concentration has returned to basal values. This point is exemplified by the prolonged increases in glucose clearance observed postprandially (RIZZA et al. 1981 b; SACCA et al. 1981). During ingestion of mixed meals or those predominantly containing protein, postprandial plasma insulin concentrations increase to levels that ordinarily would be sufficient to suppress hepatic glucose production totally (RIZZA et al. 1981 a; SACCA et al. 1981). It has been suggested (UNGER et al. 1969) that the increase in plasma glucagon observed under such conditions (GERICH et al. 1975 b; WAHREN et al. 1976), which presumably results from the accompanying hyperaminoacidemia, functions to prevent hypoglycemia that might occur if the actions of insulin were unopposed. The importance of such a role for glucagon lies in the fact that if hypoglycemia were to occur during the interprandial period of anabolism, the stimulation of epinephrine, growth hormone, and cortisol secretion as part of a counterregulatory effort to restore normoglycemia would result in mobilization of substrates from endogenous stores and thus lead to a catabolic state.

The original evidence for such a role for glucagon (UNGER et al. 1969) can be summarized as follows. In dogs, there is a negative correlation between increases in plasma glucagon and maximal decreases in plasma glucose following infusion of amino acids; furthermore, suppression of glucagon secretion by infusion of glucose or a triglyceride emulsion plus heparin, or selective diminution of glucagon secretion due to partial pancreatectomy, augments this decrease in plasma glucose which can be prevented either by infusion of glucagon or by augmentation of glucagon secretion. Subsequent support for this concept in humans has come from studies demonstrating that splanchnic glucose output following ingestion of a predominantly protein meal increases, despite increases in plasma insulin concentrations, and that it parallels increases in plasma glucagon concentrations (Fig. 5; WAHREN et al. 1976). Additional evidence for an important role for glucagon has come from studies in patients with insulin-dependent diabetes in whom glucagon secretion has been inhibited by somatostatin; when glucagon secretion was inhibited by somatostatin, plasma glucose concentrations decreased following ingestion of mixed meals and insulin administration. In contrast, without suppression of glucagon secretion by somatostatin, plasma glucose concentrations increased following ingestion of mixed meals and administration of the same dose of insulin (GERICH et al. 1975b; RASKIN and UNGER 1978).

## C. Glucagon in the Restoration of Normoglycemia

Of the possible mechanisms involved in the other aspect of glucose counterregulation, namely restoration of normoglycemia following development of hypoglycemia, both the time course of the homeostatic changes and the results of studies employing pharmacologic blockade of the actions or secretion of counterregulatory hormones provide evidence supporting the view that glucagon is probably the most important factor.

Insulin-induced hypoglycemia has generally been used as a model for examining counterregulatory mechanisms involved in the restoration of normoglycemia. Nevertheless, one should consider the possibility that the conclusions drawn from such a model may not necessarily be applicable to the varieties of spontaneous hypoglycemia observed under clinical conditions (e.g., insulinoma, Somoygi phenomenon) in which plasma glucose concentrations decrease less abruptly. Shown in Fig. 6 are the changes in rates of glucose production and utilization and the associated changes in plasma concentrations of glucagon, epinephrine, growth hormone, and cortisol following induction of hypoglycemia by an intravenous injection of insulin in normal human subjects. The decrease in plasma glucose concentration is due to both suppression of glucose production and stimulation of glucose utilization; but restoration of normoglycemia is wholly due to a compensatory increase in glucose production since glucose utilization remains at, or slightly above, basal rates throughout the period during which plasma glucose concentration increases (GARBER et al. 1976; RIZZA et al. 1979a). Thus, only those hormones capable of acute stimulation of glucose production would appear involved in the restoration of normoglycemia under these conditions. Of the major glucocounterregulatory hormones, only glucagon and epinephrine fulfill this requirement (GARBER et al. 1976; RIZZA and GERICH 1979; RIZZA et al. 1979a). As shown in Fig. 6, increases in plasma glucagon and plasma epinephrine concentrations are demonstrable within 30 min following administration of insulin, a time at which the first increase in glucose production is apparent. Increases in plasma growth hormone

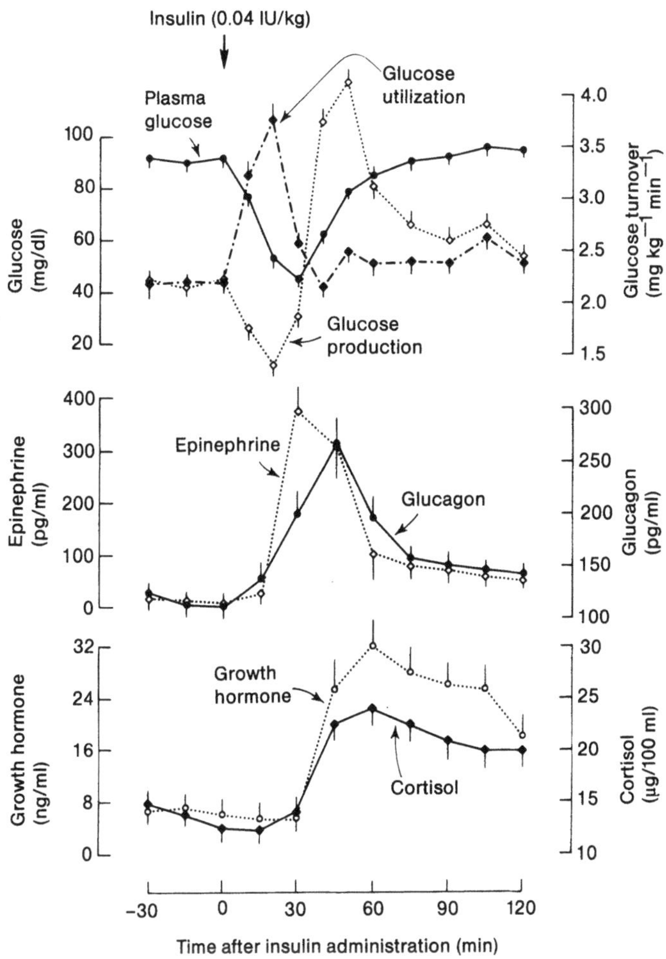


**Fig. 6.** Changes in plasma glucose, glucagon, epinephrine, cortisol, and growth hormone concentrations and rates of glucose production and utilization during insulin-induced hypoglycemia in 15 normal humans. Mean ± standard error. Adapted from RIZZA et al. (1979a)

and cortisol concentrations are not observed until after glucose production has increased. Although these observations appear to indicate that only glucagon and epinephrine could be involved in the initiation of counterregulation, there is evidence that increases in plasma cortisol concentrations such as those observed following insulin-induced hypoglycemia (and perhaps also in plasma growth hormone concentrations) may augment the actions of glucagon and epinephrine (SHAMOON et al. 1981). Thus, it is possible that, while not directly affecting the counterregulatory process by their intrinsic actions, these hormones may participate indirectly by augmenting or prolonging the effects of glucagon and epinephrine. Studies employing pharmacologic blockers in normal subjects and adrenalectomized patients, however, do not generally support such a role (GERICH et al. 1979; RIZZA et al. 1979a).

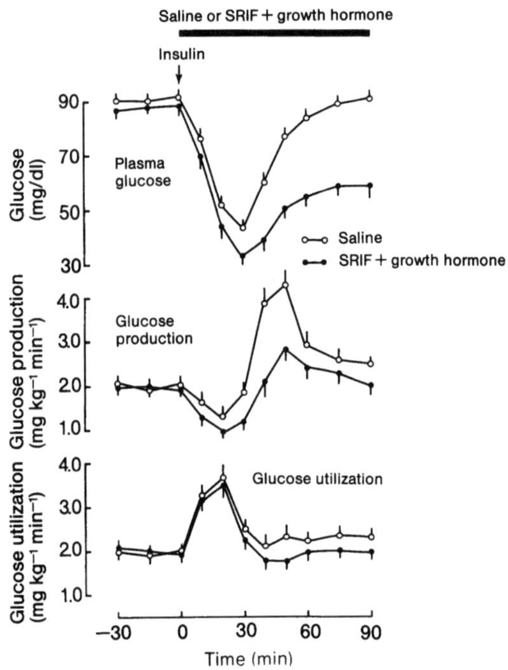


**Fig. 7.** Effect of isolated deficiency of glucagon on acute glucose counterregulation in 7 normal humans. Mean ± standard error. Adapted from RIZZA et al. (1979a)

Further direct evidence that glucagon is important in the recovery from hypoglycemia has come from studies in which inhibition of glucagon secretion has been shown to impair glucose counterregulation (GERICH et al. 1979; RIZZA et al. 1979a; CHIDECKEL et al. 1975; SACCA et al. 1977a; CHRISTENSEN et al. 1975). In Fig. 7 is shown an example of such a study: hypoglycemia was induced by an intravenous injection of insulin in normal subjects on two occasions, once during control studies and once during infusion of somatostatin (SRIF) which inhibits both glucagon and growth hormone secretion (GUILLEMIN and GERICH 1976). Growth hormone was infused to approximate plasma growth hormone concentrations observed in the control studies so that an isolated deficiency of glucagon was produced. Under these experimental conditions, glucagon deficiency markedly impaired glucose counterregulation as indicated by the failure of plasma glucose concentrations to return to normoglycemic levels; this was wholly due to a blunted compensatory increase in glucose production as might be expected in this impairment were due to deficiency of glucagon. It should be pointed out that some increase in glucose production did occur, despite the fact that plasma glucagon concentrations were suppressed. This residual counterregulation could, at least in part, have been due to adrenergic mechanisms since increased plasma epinephrine concentrations were observed under these experimental conditions. An intrinsic effect of hypoglycemia on the liver may also have contributed (SOSKIN et al. 1938; SACCA et al. 1979a). However, it is clear that neither adrenergic mechanisms nor glucose autoregulation were sufficient to maintain normal counterregulation in the absence of interact glucagon secretion.

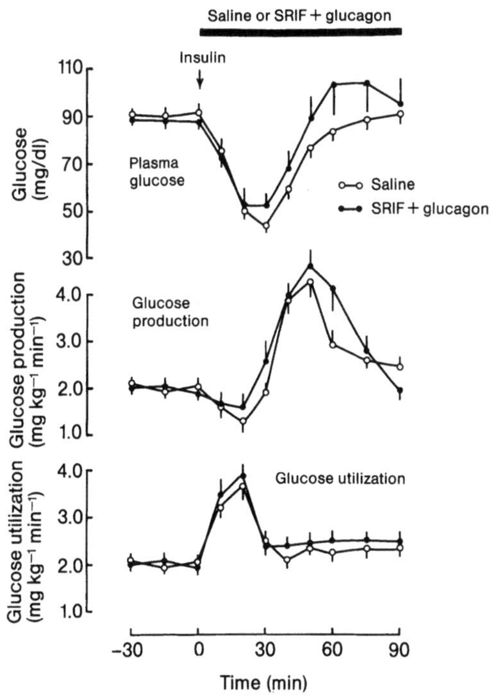


**Fig. 8.** Effect of isolated deficiency of growth hormone on acute glucose counterregulation in 7 normal humans. Mean ± standard error. Adapted from RIZZA et al. (1979 a)

In analogous studies, the consequences of an isolated deficiency of growth hormone and of a combined deficiency of growth hormone and glucagon have been examined (RIZZA et al. 1979 a). Although it is well established that chronic growth hormone deficiency can enhance the hypoglycemic action of insulin and impair glucose counterregulation (GREENWOOD and LONDON 1966), acute deficiency of growth hormone did not impair glucose counterregulation (Fig. 8). A similar lack of effect on acute glucose counterregulation has been observed in studies in which growth hormone secretion was inhibited by cyproheptadine (FELDMAN et al. 1975). As shown in Fig. 9, combined deficiency of growth hormone and glucagon produced no further impairment in glucose counterregulation than that observed during isolated deficiency of glucagon.

Taken together, these observations indicate that the acute increases in plasma growth hormone concentrations that are observed during hypoglycemia may not have an immediate role in the restoration of normoglycemia. Furthermore, they provide evidence against these changes in plasma growth hormone concentrations having synergistic effects on responses to glucagon and indicate that the impairment in glucose counterregulation observed during administration of somatostatin can be fully accounted for by inhibition of glucagon secretion. It should be pointed out, however, that these observations do not exclude the possibility that there may be delayed effects of these increases in plasma growth hormone which may, for example, be important during prolonged hyperinsulinemia or in the Somoygi phenomenon (posthypoglycemic hyperglycemia) (CAMPBELL 1976).

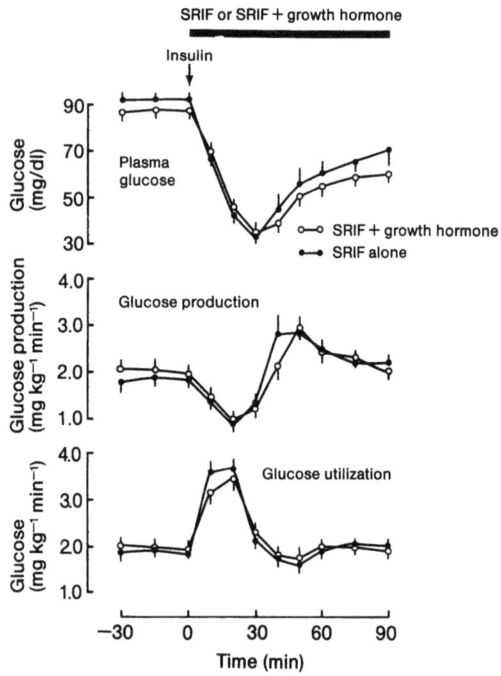


**Fig. 9.** Effects of combined deficiency of clucagon and growth hormone on acute glucose counterregulation in 7 normal humans. Mean ± standard error. Adapted from RIZZA et al. (1979a)

Although catecholamines can stimulate glucagon secretion in humans (GERICH et al. 1974), it now seems well established that the increases in plasma epinephrine and sympathetic nervous system activity play little if any role in modulating the A-cell secretory response to hypoglycemia: studies in several laboratories have demonstrated that plasma glucagon responses to insulin-induced hypoglycemia are unaffected by adrenergic blockade, adrenalectomy, and sympathectomy (WALTER et al. 1974; FRIER et al. 1981; PALMER et al. 1976; LILAVIVAT et al. 1981; GERICH et al. 1979; RIZZA et al. 1979a). Since plasma glucose profiles are generally also unaltered under such conditions (Fig. 10), the role of the adrenergic system in glucose counterregulation has been questioned. Nevertheless, recent studies suggest that the sympathetic nervous system may act as a secondary line of defense to compensate under conditions when glucagon secretion is inadequate, e.g., when the A-cell response to hypoglycemia is impaired such as in diabetes mellitus or when there is severe hypoglycemia (GERICH et al. 1979; RIZZA et al. 1979b; CRYER 1981; POPP et al. 1982).

In the studies illustrated in Fig. 7, the impaired glucose counterregulation associated with inhibition of glucagon secretion was accompanied by a further increase in circulating catecholamines. Since glucose counterregulation was not totally prevented under these conditions, the residual counterregulatory response could have been mediated in part at least by the sympathetic nervous system. The results of studies of the effect of combined adrenergic blockade and inhibition of glucagon secretion (Fig. 11) indicate that this indeed appears to be the case. The

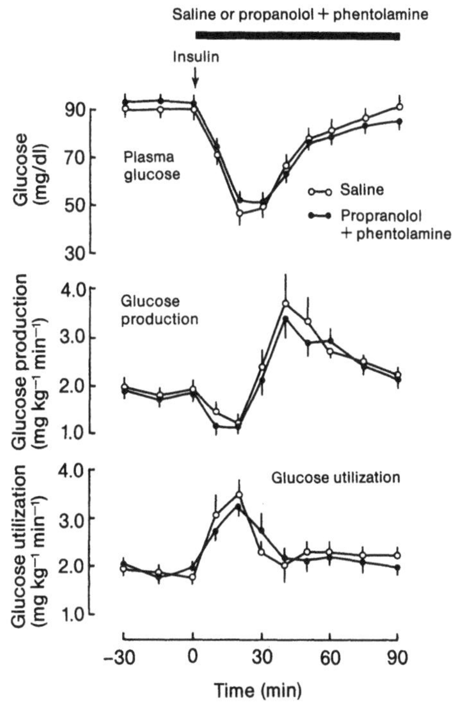


**Fig. 10.** Effects of $\alpha$- and $\beta$-adrenergic blockade on acute glucose counterregulation in 8 normal humans. Mean ± standard error. Adapted from RIZZA et al. (1979a)

impaired glucose counterregulation due to inhibition of glucagon secretion is further impaired by adrenergic blockade and this is due to further impairment of the compensatory increase in glucose production.

The results of studies in adrenalectomized patients (Fig. 12) suggest that the sympathetic nervous system participates in glucose counterregulation, primarily through release of adrenomedullary catecholamines. In these (Fig. 12a) and other studies (ENSINCK et al. 1976), adrenalectomy itself caused no impairment in glucose counterregulation when adrenalectomized patients were being treated with replacement doses of cortisol. These observations indicate that neither acute increases in the plasma concentration of cortisol nor of catecholamines of adrenomedullary origin were essential for normal recovery from hypoglycemia if glucagon secretion was intact. However, when lack of an adrenomedullary response was superimposed upon inhibition of glucagon secretion (Fig. 12b), near total paralysis of glucose counterregulation occurred. This should not have happened if neurally released norepinephrine rather than adrenomedullary catecholamines were primarily responsible for the sympathetic nervous system component.

Studies in the rat (SACCA et al. 1977b) also support a role for adrenomedullary catecholamines since it has been shown that, despite unaltered A-cell responses, combined reserpinization and adrenal demedullation augment insulin-induced hypoglycemia and markedly impair compensatory increases in glucose production. The lack of a compensatory increase in glucagon secretion under these conditions

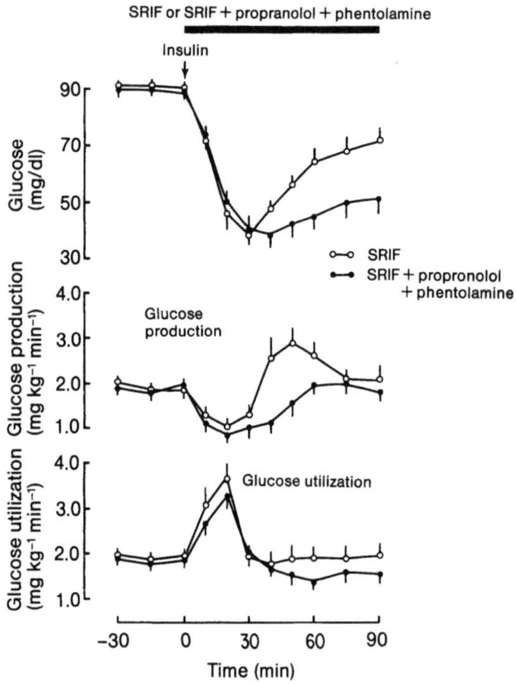


**Fig. 11.** Effects of glucagon deficiency and $\alpha$- and $\beta$-adrenergic blockade on acute glucose counterregulation in 6 normal humans. Mean ± standard error. Adapted from RIZZA et al. (1979a)

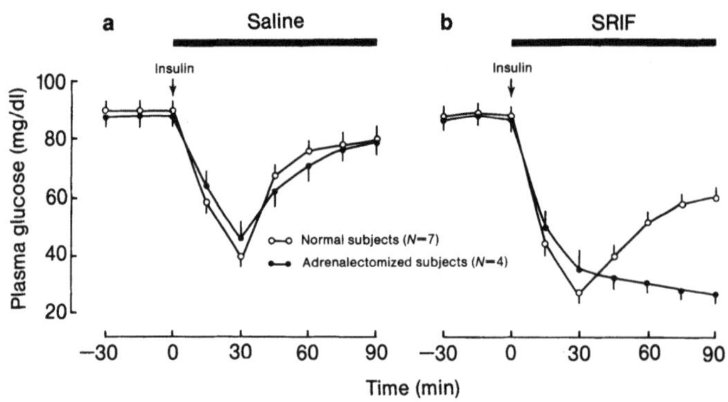


**Fig. 12.** Effects of adrenalectomy (**a**) and adrenalectomy plus glucagon deficiency (**b**) on plasma glucose concentrations after administration of insulin in 7 normal humans and 4 adrenalectomized subjects. Mean ± standard error. Adapted from GERICH et al. (1979)

is puzzling since A-cell responses appear to be proportional to hypoglycemia (see Chap. 23). Nevertheless, these observations indicate that, with severe hypoglycemia, both glucagon secretion and a sympathetically mediated adrenomedullary catecholamine response may be essential for restoration of normoglycemia.

Finally, studies in patients with diabetes mellitus, who have impaired A-cell responses to hypoglycemia believed to be due to a glucoreceptor defect (GERICH et

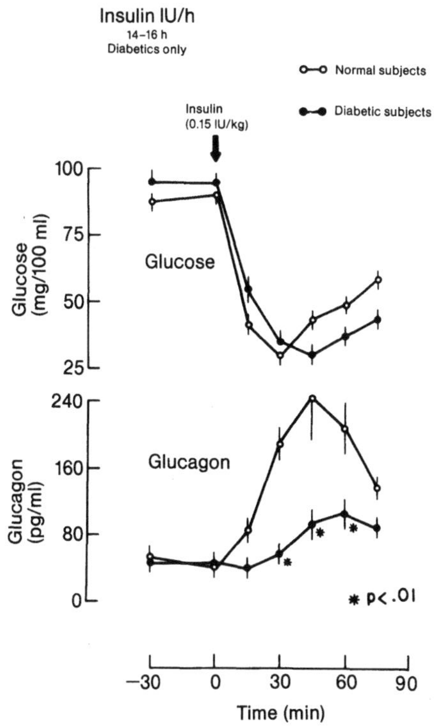


**Fig. 13.** Comparison of plasma glucose and glucagon reponses after administration of insulin in 8 normal humans and in 8 patients with insulin-dependent diabetes. Mean ± standard error. (J. GERICH et al. 1975, unpublished work)

al. 1973; REYNOLDS et al. 1977), have provided further evidence for a compensatory role of the sympathetic nervous system (POPP et al. 1982): following the attainment of normoglycemia by a prolonged intravenous infusion of insulin, further administration of insulin as bolus injection results in prolonged hypoglycemia compared with that observed in nondiabetic subjects administered the same dose of insulin (Fig. 13). This impaired glucose counterregulation, which can be explained on the basis of decreased glucagon secretion as well as decreased clearance of insulin due to the presence of insulin antibodies, is even further impaired by β-adrenergic blockade (POPP et al. 1982). This observation, while supporting the concept that secretion of glucagon is the primary hormonal mechanism involved in acute glucose counterregulation (since other mechanisms were not capable of fully compensating for glucagon deficiency), also indicates that the adrenergic system participates in this process via β-adrenergic mechanisms (RIZZA et al. 1980) as a secondary line of defense, the contribution of which increases when glucagon responses are impaired.

## D. Glucagon and the Somoygi Phenomenon

Clinically, one of the most commonly encountered types of hypoglycemia is that occurring in patients with diabetes being treated with insulin. Frequently, this hypoglycemia is followed by hyperglycemia and insulin resistance–the so-called Somoygi phenomenon (SOMOYGI 1959; BLOOM et al. 1969; CAMPBELL 1976). The immediate cause of the hypoglycemia is absolute or relative insulin excess (SCHMIDT et al. 1979; GALE et al. 1980). Thus, both administration of too much insulin on an absolute basis and administration of a normally appropriate amount of insulin followed by excessive exercise or by decreased food intake can lead to hypoglycemia. Since A-cell responses to hypoglycemia are impaired in patients with diabetes mellitus (GERICH et al. 1973) and since this results in impaired immediate glucose counterregulation (Fig. 13), it is possible that lack of appropriate glucagon secretion predisposes such patients to the development of hypoglycemia (REYNOLDS et al. 1977).

Multiple factors are probably responsible for the posthypoglycemic hyperglycemia. Plasma insulin concentrations, which may have been excessive prior to the development of hypoglycemia, are low during the initiation of hyperglycemia (SCHMIDT et al. 1979; GALE et al. 1980). This would tend to permit even normal counterregulatory mechanisms to produce excessive responses. Moreover, it should not be overlooked that, following an insulin reaction, it is quite common for patients to decrease or omit their subsequent insulin dose and to ingest or be administered carbohydrates.

The increases in circulating levels of counterregulatory hormones that are generally observed following such hypoglycemia (SCHMIDT et al. 1979; GALE et al. 1980; BRUCK and MACGILLIVRAY 1975; WINTER 1981; MOLNAR et al. 1971) are likely to play different roles in this situation from those after the acute induction of hypoglycemia following intravenous injections of insulin in normal individuals, especially in view of the differences in duration of their increases and the ambient plasma insulin concentrations. As indicated earlier, plasma glucagon responses to hypoglycemia are impaired in patients with diabetes mellitus (GERICH et al. 1973); thus one would not expect this hormone to play as important a counterregulatory role as it would in a nondiabetic individual. However, as shown in Fig. 3, trivial increases in plasma glucagon may have appreciable effects on glucose production if plasma insulin concentrations are low or are fixed.

There is substantial evidence that cortisol and growth hormone may participate in, but not be the sole mediators of the Somoygi phenomenon. First of all, the time scale of the hyperglycemic rebound (hours) is sufficiently long to permit both these hormones to exert anti-insulin actions (SHAMOON et al. 1980; GERICH et al. 1981). Second, although not all episodes of posthypoglycemic hyperglycemia are associated with increases in plasma concentrations of both hormones (BRUCK and MACGILLIVRAY 1975; MOLNAR et al. 1970; GALE et al. 1980; SCHMIDT et al. 1979; WINTER et al. 1981), situations in which their posthypoglycemic secretion has been inhibited or was not possible are generally associated with attenuated insulin resistance (MINTZ et al. 1968; OAKLEY et al. 1970). However, since the Somoygi phenomenon has been documented to occur in hypopituitarism (VESELY et al. 1976), it seems that secretion of neither of these hormones is essential.

Probably, as originally postulated by SOMOYGI (1959), epinephrine is the most important hormonal factor involved in hypoglycemic hyperglycemia. Its $\beta$-adrenergic actions which stimulate glucose production and inhibit glucose utilization (RIZZA et al. 1980) could readily account for the clinical picture of overproduction of glucose and insulin resistance. Moreover, it has recently been demonstrated that $\beta$-adrenergic blockade markedly attenuates posthypoglycemic hyperglycemia in patients with insulin-dependent diabetes (POPP et al. 1982). The failure to prevent such hyperglycemia by $\alpha$-adrenergic blockade (MINTZ et al. 1968) or to observe such hyperglycemia in some hypopituitary patients (MINTZ et al. 1968; OAKLEY et al. 1970) can be explained in this context by the fact that the effects of epinephrine are mediated through a $\beta$-adrenergic mechanism and that the cortisol deficiency associated with hypopituitarism may have decreased the production of epinephrine (FULLER 1973) as well as impaired tissue responses to the catecholamine (EXTON et al. 1970; GREEN et al. 1980).

## E. Summary

In conclusion, current evidence indicates that the liver is the main site at which moment-to-moment control of glucose homeostasis takes place and that in normal humans glucagon is the major glucose counterregulatory hormone. By antagonizing the suppressive effects of insulin on glucose production and by stimulating glucose production when appropriate, glucagon not only defends the organism against hypoglycemia, but also restores normoglycemia if hypoglycemia occurs. In this context, it plays an important role in glucose homeostasis not only in the postabsorptive state, but also in the postprandial state. The potency of glucagon as a counterregulatory hormone is such that a trivial imbalance in pancreatic A- and B-cell function can readily lead to hyperglycemia. Epinephrine normally plays a subsidiary role to glucagon in acute glucose counterregulation, but becomes critical when glucagon responses to hypoglycemia are impaired, as in diabetes mellitus, and may be the most important hormonal mediator of the Somoygi phenomenon. Acute changes in growth hormone and cortisol secretion did not directly affect moment-to-moment glucose counterregulation, but probably participate in glucose homeostasis through their long-term actions in modulating tissue responses to insulin, glucagon, and epinephrine.

## References

Alford F, Bloom S, Nabarro J, Hall R, Besser G, Coy D, Kosten A, Schally A (1974) Glucagon control of fasting glucose levels in man. Lancet 2:974–976

Altszuler N, Gottlieb B, Hampshire J (1976) Interaction of somatostatin, glucagon, and insulin on hepatic glucose output in the normal dog. Diabetes 25:116–121

Barling P, Beloff-Chain A (1973) Studies on the administration of glucagon and insulin antibodies to rats. Horm Metab Res 5:154–159

Baxter J, Funder J (1979) Hormone receptors. N Engl J Med 301:1149–1162

Bergman R, Bucolo R (1974) Interaction of insulin and glucose in the control of hepatic glucose balance. Am J Physiol 227:1314–1322

Blauth C, Sonksen P, Tompkins C, Bloom S (1977) The hypoglycemic action of somatostatin in the anesthetized dog. Clin Endocrinol (Oxf) 6:17–25

Bloom M, Mintz D, Field J (1969) Insulin-induced posthypoglycemic hyperglycemia as a cause of "brittle" diabetes. Am J Med 47:891–903

Bruck E, MacGillivray M (1975) Interaction of endogenous growth hormone, cortisol, and catecholamines with blood glucose in children with brittle diabetes mellitus. Pediatr Res 9:535–541

Byrne R, Nompleggi D, Ramey E, Penhos J (1977) Studies on the extrapancreatic metabolic effects of somatostatin. Proc Soc Exp Biol Med 155:507–510
Cahill G (1970) Starvation in man. N Engl J Med 282:668–675
Campbell I (1976) The Somoygi phenomenon: a short review. Acta Diabetol Lat 13:68–73
Chideckel E, Palmer J, Koerker D, Ensinck J, Davidson M, Goodner C (1975) Somatostatin blockade of acute and chronic stimuli of the endocrine pancreas and the consequences of this blockade on glucose homeostasis. J Clin Invest 55:754–762
Cherrington A, Chiasson J, Liljenquist J, Jennings A, Keller V, Lacy W (1976) The role or insulin and glucagon in the regulation of basal glucose production in the postabsorptive dog. J Clin Invest 58:1407–1418
Cherrington A, Caldwell M, Dietz M, Exton J, Crofford O (1977) The effect of somatostatin on glucose uptake and production by rat tissues in vitro. Diabetes 26:740–748
Cherrington A, Lacy W, Chiasson J (1978) Effect of glucagon on glucose production during insulin deficiency in the dog. J Clin Invest 62:664–677
Cherrington A, Liljenquist J, Shulman G, Williams P, Lacy W (1979) Importance of hypoglycemia-induced glucose production during isolated glucagon deficiency. Am J Physiol 236:263–271
Christensen N, Christensen S, Hansen A, Lundbaek K (1975) The effect of somatostatin on plasma noradrenaline and plasma adrenaline concentrations during exercise and hypoglycemia. Metabolism 24:1267–1272
Clark W, Santiago J, Thomas L, Haymond M, Ben-Galim E, Cryer P (1979) The role of adrenergic mechanisms in recovery from hypoglycemia in man: studies with adrenergic blockade. Am J Physiol 236:147–152
Cryer P (1981) Glucose counterregulation in man. Diabetes 30:261–264
Deri J, Williams P, Steiner K, Cherrington A (1981) Altered ability of the liver to produce glucose following a period of glucagon deficiency. Diabetes 30:490–495
Eisenberg S, Seltzer H (1962) The cerebral metabolic effects of acutely induced hypoglycemia in normal subjects. Metabolism 11:1162–1168
Ensinck J, Walter R, Palmer J, Brodows R, Campbell R (1976) Glucagon responses to hypoglycemia in adrenalectomized man. Metabolism 25:227–232
Epand R, Douglas R (1973) The effect of glucagon antibodies on plasma glucose and insulin levels. Biochim Biophys Acta 320:741–744
Exton J, Mallette L, Jefferson L, Wong E, Friedman N, Miller T, Park CR (1970) The hormonal control of hepatic gluconeogenesis. Recent Prog Horm Res 26:411–455
Feldman J, Plonk J, Bevens C (1975) The role of cortisol and growth hormone in the counter-regulation of insulin-induced hypoglycemia. Horm Metab Res 7:378–381
Felig P, Wahren J, Hendler R (1975) Influence of oral glucose ingestion on splanchnic glucose and gluconeogenic substrate metabolism in man. Diabetes 24:468–475
Frier B, Corrall R, Ratcliffe J, Ashby J, McClemont E (1981) Autonomic and neural control mechanisms of substrate and hormonal responses to acute hypoglycemia in man. Clin Endocrinol (Oxf) 14:552–558
Frohman L, Reichlin M, Sokal J (1970) Immunologic and biologic properties of antibodies to a glucagon-serum albumin polymer. Endocrinology 87:1055–1061
Fuller R (1973) Control of epinephrine synthesis and secretion. Fed Proc 32:1772–1781
Gale E, Kurtz A, Tattersall R (1980) In search of the Somoygi effect. Lancet 2:279–282
Garber A, Cryer P, Santiago J, Haymond M, Pagliara A, Kipnis D (1976) The role of adrenergic mechanisms in the substrate and hormonal response to insulin-induced hypoglycemia in man. J Clin Invest 58:7–15
Gerich J, Langlois M, Noacco C, Karam J, Forsham P (1973) Lack of glucagon response to hypoglycemia in diabetes: evidence for an intrinsic pancreatic alpha-cell defect. Science 182:171–173
Gerich J, Langlois M, Noacco C, Schneider V, Forsham P (1974) Adrenergic modulation of pancreatic glucagon secretion in man. J Clin Invest 53:1441–1446
Gerich J, Lorenzi M, Hane S, Gustafson G, Guillemin R, Forsham P (1975a) Evidence for a physiologic role of pancreatic glucagon in human glucose homeostasis: studies with somatostatin. Metabolism 24:175–182

Gerich J, Lorenzi M, Karam J, Schneider V, Forsham P (1975b) Abnormal pancreatic glucagon secretion and postprandial hyperglycemia in diabetes mellitus. JAMA 234:159–165

Gerich J, Lovinger R, Grodsky G (1975c) Inhibition by somatostatin of glucagon and insulin release from the perfused rat pancreas in response to arginine, isoproterenol, and theophylline: evidence for a preferential effect on glucagon secretion. Endocrinology 96:749–754

Gerich J, Davis J, Lorenzi M, Rizza R, Karam J, Lewis S, Kaplan R, Schultz T, Cryer P (1979) Hormonal mechanisms of recovery from insulin-induced hypoglycemia in man. Am J Physiol 236:380–385

Gerich J, Haymond M, Rizza R, Verdonk C, Miles J (1981) Hormonal and substrate determinants of hepatic glucose production in man. In: Veneziale C (ed) The regulation of carbohydrate formation and utilization in mammals. University Park Press, Baltimore, pp 419–457

Green G, Chenoweth M, Dunn A (1980) Adrenal glucocorticoid permissive regulation of muscle glycogenolysis: action on protein phosphatases and its inhibitors. Proc Natl Acad Sci USA 77:5711–5715

Greenwood F, Landon J (1966) Assessment of hypothalamic pituitary function in endocrine disease. J Clin Pathol 19:284–292

Grey N, McGuigan J, Kipnis D (1970) Neutralization of endogenous glucagon by high titer glucagon antiserum. Endocrinology 86:1383–1388

Guillemin R, Gerich J (1976) Somatostatin: physiologic and clinical significance. Annu Rev Med 27:379–388

Havel R (1972) Caloric homeostasis and disorders of fuel transport. N Engl J Med 287:1186–1192

Holst J, Galbo H, Richter E (1978) Neutralization of glucagon by antiserum as a tool in glucagon physiology. J Clin Invest 62:182–190

Huang S, Phelps M, Hoffman E, Sideris K, Selin C, Kühl D (1980) Noninvasive determination of local cerebral metabolic rate of glucose in man. Am J Physiol 238:E69–E82

Jennings A, Cherrington A, Liljenquist J, Keller V, Lacy W, Chiasson J (1977) The roles of insulin and glucagon in the regulation of gluconeogenesis in the postabsortive dog. Diabetes 26:847–856

Lickley H, Ross G, Vranic M (1979) Effects of selective insulin or glucagon deficiency on glucose turnover. Am J Physiol 236:255–262

Lilavivat U, Brodows R, Campbell R (1981) Adrenergic influence on glucocounterregulation in man. Diabetologia 20:482–488

Liljenquist J, Mueller G, Cherrington A, Keller V, Chiasson J, Perry J, Lacy W, Rabinowitz D (1977) Evidence for an important role of glucagon in the regulation of hepatic glucose production in normal man. J Clin Invest 59:369–374

Liljenquist J, Bloomgarden Z, Cherrington A, Perry J, Rabin D (1979a) Possible mechanism by which somatostatin-induced glucagon suppression improves glucose tolerance during insulinopenia in man. Diabetologia 17:139–143

Liljenquist J, Mueller G, Cherrington A, Perry J, Rabinowitz D (1979b) Hyperglycemia per se (insulin and glucagon withdrawn) can inhibit hepatic glucose production in man. J Clin Endocrinol Metab 48:171–175

Lund-Andersen H (1979) Transport of glucose from blood to brain. Physiol Rev 59:305–352

Molnar G, Fatourechi V, Ackerman E, Taylor W, Rosevear J, Gatewood L, Service F, Moxness K (1971) Growth hormone and glucose interrelationships in diabetes: studies of inadvertent hypoglycemic episodes during continuous blood glucose analysis. J Clin Endocrinol Metab 32:426–437

Mintz D, Finster J, Taylor A, Fefer A (1968) Hormonal genesis of glucose intolerance following hypoglycemia. Am J Med 45:187–197

Oakley N, Jacobs H, Turner R, Williams J, Aquino C, Nabarro J (1970) The effect of hypoglycemia on oral glucose tolerance in normal subjects and patients with pituitary and adrenal disorders. Clin Sci 36:663–674

Palmer J, Henry D, Bensen J, Johnson D, Ensinck J (1976) Glucagon response to hypoglycemia in sympathectomized man. J Clin Invest 57:522–525

Palmer J, Werner P, Hollander P, Ensinck J (1979) Evaluation of the control of glucagon secretion by the parasympathetic nervous system in man. Metabolism 28:549–552
Parrilla R, Goodman N, Toews C (1974) Effects of glucagon insulin ratios on hepatic metabolism. Diabetes 23:725–731
Popp D, Shah S, Cryer P (1982) The role of epinephrine-mediated beta adrenergic mechanisms in hypoglycemic glucose counterregulation and posthypoglycemic hyperglycemia in insulin-independent diabetes mellitus. J Clin Invest 69:315–326
Raskin P, Unger R (1978) Hyperglucagonemia and its suppression: importance in the metabolic control of diabetes. N Engl J Med 299:433–436
Reynolds C, Molnar G, Horwitz D, Rubenstein A, Taylor W, Jiang N (1977) Abnormalities of endogenous glucagon and insulin in unstable diabetes. Diabetes 26:36–45
Rizza R, Gerich J (1979) Persistent effects of sustained hyperglucagonemia on glucose production in man. J Clin Endocrinol Metab 48:352–354
Rizza R, Cryer P, Gerich J (1979a) Role of glucagon, catecholamines, and growth hormone in human glucose counterregulation: effects of somatostatin and combined $\alpha$- and $\beta$-adrenergic blockade on plasma glucose recovery and glucose flux rates following insulin-induced hypoglycemia. J Clin Invest 64:62–71
Rizza R, Miles J, Verdonk C, Gerich J (1979b) Effect of intermittent endogenous hyperglucagonemia on glucose homeostasis in normal and diabetic man. J Clin Invest 63:1119–1123
Rizza R, Verdonk C, Miles J, Service J, Haymond M, Gerich J (1979c) Somatostatin does not cause sustained hyperglycemia in man. Horm Metab Res 11:589–646
Rizza R, Cryer P, Haymond M, Gerich J (1980) Adrenergic mechanisms for the effects of epinephrine on glucose production and clearance in man. J Clin Invest 65:682–689
Rizza R, Mandarino L, Gerich J (1981a) Dose-response characteristics for the effects of insulin on production and utilization of glucose in man. Am J Physiol 240:630–639
Rizza R, Westland R, Hall L, Patton G, Haymond M, Clemens A, Gerich J, Service J (1981b) Effect of peripheral versus portal venous administration of insulin on postprandial hyperglycemia and glucose turnover in alloxan-diabetic dogs. Mayo Clin Proc 56:434–438
Sacca L, Perez G, Carteni G, Rengo F (1977a) Evaluation of the role of the sympathetic nervous system in the glucoregulatory response to insulin-induced hypoglycemia in the rat. Endocrinology 101:1016–1022
Sacca L, Perez G, Cartini G, Trimarco B, Rengo F (1977b) Role of glucagon in the glucoregulatory response to insulin-induced hypoglycemia in the rat. Horm Metab Res 9:209–212
Sacca L, Hendler R, Sherwin R (1978) Hyperglycemia inhibits glucose production in man independent of changes in glucoregulatory hormones. J Clin Endocrinol Metab 47:1160–1163
Sacca L, Cryer P, Sherwin R (1979a) Blood glucose regulates the effects of insulin and counterregulatory hormones on glucose production in vivo. Diabetes 28:533–536
Sacca L, Eigler N, Cryer P, Sherwin R (1979b) Insulin antagonistic effects of epinephrine and glucagon in the dog. Am J Physiol 237:487–492
Sacca L, Cicala M, Corso G, Ungaro B, Sherwin R (1981) Effect of counterregulatory hormones on kinetic response to ingested glucose in dogs. Am J Physiol 240:E465–E473
Schmidt M, Hadji-Georgopoulos A, Rendell M, Margolis S, Kowarski D, Kowarski A (1979) Fasting hyperglycemia and associated free insulin and cortisol changes in "Somoygi-like" patients. Diabetes Care 2:457–464
Shamoon H, Hendler R, Sherwin R (1980) Altered responsiveness to cortisol, epinephrine, and glucagon in insulin-infused juvenile-onset diabetes: a mechanism for diabetic instability. Diabetes 29:284–291
Shamoon H, Hendler R, Sherwin R (1981) Synergistic interactions among antiinsulin hormones in the pathogenesis of stress hyperglycemia in humans. J Clin Endocrinol Metab 52:1235–1241
Sherwin R, Hendler R, DeFronzo R, Wahren J, Felig P (1977) Glucose homeostasis during prolonged suppression of insulin and glucagon by somatostatin. Proc Natl Acad Sci USA 74:348–352

Shulman G, Liljenquist J, Williams P, Lacy W, Cherrington A (1978) Glucose disposal during insulinopenia in somatostatin-treated dogs. J Clin Invest 62:487–491
Sokal J, Sarcione E, Henderson A (1964) Relative potency of glucagon and epinephrine as hepatic glycogenolytic agents: studies with the isolated perfused rat liver. Endocrinology 74:930–938
Somoygi M (1959) Exacerbation of diabetes by excess insulin action. Am J Med 26:169–191
Soskin S, Essex H, Herrick J, Mann F (1938) The mechanism of the regulation of blood sugar by the liver. Am J Physiol 124:558–567
Unger R, Ohneda A, Aguilar-Parada E, Eisentraut A (1969) The role of aminogenic glucagon secretion in blood glucose homeostasis. J Clin Invest 48:810–822
Vesely C, Castro A, Levey G (1976) Somoygi effect in patient with hypopituitarism. Arch Intern Med 136:936–938
Wahren J, Felig P, Hagenfeldt L (1976) Effect of protein ingestion on splanchnic and leg metabolism in normal man and in patients with diabetes mellitus. J Clin Invest 57:987–999
Wahren J, Efendic S, Luft R, Hagenfeldt L, Björkman O, Felig P (1977) Influence of somatostatin on splanchnic glucose metabolism in postabsorptive and 60-hour fasted humans. J Clin Invest 59:299–307
Walter R, Dudl R, Palmer J, Ensinck J (1974) The effect of adrenergic blockade on the plasma glucagon responses to starvation and hypoglycemia in man. J Clin Invest 54:1214–1220
Winter R (1981) Profiles of metabolic control in diabetic children – frequency of asymptomatic nocturnal hypoglycemia. Metabolism 30:666–672

CHAPTER 38

# Glucagon and Its Relationship to Other Glucoregulatory Hormones in Exercise and Stress in Normal and Diabetic Subjects

H. L. A. LICKLEY, F. W. KEMMER, D. H. WASSERMAN, and M. VRANIC

## A. Introduction

Exercise and stress are two situations which have many features in common with respect to hormonal and neural responses to a changed metabolic environment. The interaction of glucagon with other glucoregulatory hormones is important both in stress and exercise. However, since there is a basic difference in fuel fluxes in the two states, we propose to address this topic in two separate sections, one dealing with exercise and the other with stress.

Although exercise may, in fact, be regarded as a special form of stress, it is also a normal physiologic condition. The cardiovascular, respiratory, hormonal, and metabolic adjustments which ensure adequate increases in oxygen supply to exercising muscle, depend to some extent, of course, on previous conditioning or training. The main metabolic aim of the individual during exercise is to ensure an adequate continuing increase in fuel supply for muscular activity, which is provided, in large part, through glycogenolysis and gluconeogenesis in the liver, and which glucagon appears to play a significant role. The increase in the flux of nutrients during exercise, particularly glucose and free fatty acids, is essentially unidirectional – always towards muscle. In normal individuals, there is a well-recognized drop in circulating plasma insulin levels (WAHREN et al. 1971; HARTLEY et al. 1972a; MURRAY et al. 1977), and this leads, in turn, to increased hepatic sensitivity to the counterregulatory hormones and presumably to the neural drive involved in regulation of glucose production and glycogenolysis in the liver. Another important feature of exercise is a marked increase in glucose clearance in muscle (CHAVEAU and KAUFMANN 1887). However, one striking feature of exercise, is that fuel utilization tends to be matched by fuel mobilization so that normoglycemia can be maintained until increased fuel resources are available. In diabetes mellitus, hormonal and metabolic responses to exercise are much more variable, depending as they do on such factors as the state of insulinization and the state of metabolic control. The temporal relationships between the performance of exercise and previous food intake and insulin treatment are also of prime importance.

Stress, on the other hand, is a physiologic to pathophysiologic continuum. Again, it shares many hormonal and metabolic features with exercise, and glucagon interplays in a multitude of ways, depending on the type of stress and the underlying condition of the individuals subjected to the stress situation. Fuel mobilization is important in stress, but this can lead to severe metabolic derangements. In stress, the fuel fluxes are not unidirectional as in exercise, but multidirectional; often directed, at least in the normal individual, toward repair processes.

As in exercise, glucose production rises in response to stress, but unlike the situation in exercise this is not matched by adequate increments in glucose utilization and hyperglycemia ensues. The insulin responses to stress are less clear-cut than those during exercise. There is generally thought to be a suppression of insulin release, an α-adrenergic response to the catecholamine release of stress (IVERSEN 1973; ROBERTSON and PORTE 1973; SAMOLS and WEIR 1979), but hyperglycemia can counterbalance this effect. Furthermore, it has been shown, in the dog, that there is a significant, though transient, release of insulin in response to an epinephrine infusion (GRAY et al. 1980), and indeed, β-adrenergic stimulation of insulin release has been described (GERICH et al. 1976). Thus, the two opposing effects of the catecholamines together with the insulinotropic effect of the prevailing hyperglycemia, can serve to prevent an absolute decrease in plasma insulin levels during stress. The insulin responses are both species dependent and dependent to some degree on the extent of elevation of serum catecholamine levels (GRAY et al. 1980). In addition to increased fuel mobilization during stress, there is a relative or, under some conditions, an absolute impairment in glucose clearance (ALTSZULER et al. 1967; GRAY et al. 1980; PEREZ et al. 1981; KEMMER et al. 1982). The hormonal and metabolic responses to stress depend not only upon the attendant catecholamine release, but in addition are influenced by the treatment modalities employed in countering the underlying causes of the stress condition. Finally, as in the responses to exercise, the responses to stress vary greatly between normal individuals and patients who are suffering from diabetes mellitus. In normal individuals, it is thought that the responses to stress are, by and large, favorable to the survival of the individual. In diabetes, and particularly in those patients with uncontrolled diabetes, the responses to stress may in fact be deleterious rather than favorable.

The development of isotope dilution techniques for determining glucose turnover (STEELE et al. 1956) has permitted a greater delineation of glucoregulation, by providing not only measurements of changes in glucose concentration, but also of glucose fluxes. This is important, for in the glucoregulatory system, hepatic glucose production and peripheral glucose utilization are controlled by different mechanisms. Thus, in situations in which glucose production and utilization change concurrently in the same direction, there will be changes in glucose turnover which measurements of circulating plasma glucose levels will not indicate. For example, when glucagon–insulin interactions were studied during the infusion of arginine in normal and depancreatized insulin-infused dogs (CHERRINGTON et al. 1974), the necessity of measuring tracer-determined glucose turnover became apparent. In both the normal and the depancreatized animals, arginine infusion failed to alter plasma glucose levels appreciably, but through entirely different mechanisms in the two experimental models. In the normal dogs, there was a simultaneous release of insulin and glucagon which maintained glycemia constant, although tracer techniques demonstrated a marked increase in glucose turnover. In the depancreatized insulin-infused dogs, glucose turnover did not change as the animals are incapable of altering either insulin or glucagon secretion in response to arginine. This was also well illustrated in exercise, where glucose fluxes increased 2–3-fold and yet normoglycemia was maintained (VRANIC et al. 1976a).

Extensive validation of the equations involved in the calculation of turnover has been carried out, including a specific application to the glucose system (COWAN and HETENYI 1971; HETENYI and NORWICH 1974; RADZIUK et al. 1974, 1978). The most widely used tracer for measuring glucose turnover today is glucose $^{3}$H-3. Because tritium is lost before three-carbon fragments are formed, it does not appear in recycled intermediates via the Cori cycle, as did the originally used label, glucose $^{14}$C. Thus, glucose $^{3}$H-3 can be used to measure "true glucose turnover" (ALTSZULER et al. 1975; KATZ and DUNN 1967). The tracer, glucose $^{3}$H-2, leads to a 13%–20% overestimation of glucose production under physiologic conditions, because the contribution of a "futile cycle" is incorporated in the calculation of turnover (ISSEKUTZ 1977; LICKLEY et al. 1979). Thus, by combining tracers, a measure of both recycling and "futile cycling" can be obtained. It has been reported that glucagon, the glucocorticoids, and exercise can all increase "futile cycling" as can diabetes, and as there are associated changed in glucose production, the rate of "futile cycling" may well play a part in sensitizing the liver to glycogenolytic and gluconeogenetic drives (SHAW et al. 1976; ISSEKUTZ 1977). In addition, rates of gluconeogenesis may also be assessed in exercise by infusing gluconeogenetic precursors (ISSEKUTZ et al. 1976); theoretical and experimental background for such approaches has been provided (HETENYI 1981).

In many of the studies described in this chapter, a technique has been used whereby the circulating levels of a hormone whose effects are under scrutiny are altered, while the other hormones are maintained at a fixed level, either by a constant infusion or by completely suppressing hormone release. The polypeptide, somatostatin suppresses the release of both insulin and glucagon from the endocrine pancreas (S. E. CHRISTENSEN et al. 1974), in addition to suppressing other hormones such as growth hormone, and thus provides an effective tool with which to study the role of these hormones and their interactions with each other and with other hormones in glucoregulation.

In the studies we will use to illustrate the interactions between glucagon and other hormones in exercise, and situations simulating stress, we will confine ourselves to a consideration of glucoregulation because hormonal interplays have been studied extensively. The importance of concomitant changes in protein and fat metabolism during exercise has been reviewed elsewhere (VRANIC and BERGER 1979).

## B. Exercise

### I. Metabolic Events During Exercise

With the onset of exercise, there is a rapid increase in energy consumption in working muscle. As energy sources in muscle itself are limited, there must be a precisely regulated flux of energy substrate from other sources, which is graded with respect to fuel availability and the type of muscular work being performed. The main fuels are glucose and free fatty acids, with amino acids and ketone bodies contributing to a minor degree.

The choice of metabolic substrates depends upon many variables including: (a) the duration and intensity of exercise; (b) cardiorespiratory status; (c) nutritional status; (d) endocrine status; (e) age; and (f) sex. In resting muscle, free fatty acids provide the main fuel source. With the onset of exercise, muscle glycogen is utilized first, then fuel from circulating glucose and plasma free fatty acids is added

(E. H. CHRISTENSEN and HANSEN 1939; AHLBORG et al. 1974). With heavy work, glucose requirements in muscle rise (ÅSTRAND and RODAHL 1970), but with prolonged physical training, the muscle can turn to a proportionately greater utilization of free fatty acids during strenuous exercise (ISSEKUTZ et al. 1965). However, free fatty acids can never replace glucose as an oxidative fuel, and thus increased hepatic glucose production is essential for the maintenance of glucose homeostasis in the face of continuing peripheral glucose utilization. However, with very prolonged exercise, hepatic glucose production fails to meet the continuing glucose oxidation in muscle, and presumably, the limiting factor is ultimately a decrease in glucose delivery to the brain (ISSEKUTZ et al. 1970).

The sudden changes in fuel demands in exercise of varying intensity and duration require a precise regulation of fuel fluxes which is provided by an interaction between hormonal and neural factors. During exercise, there is an increase in circulating levels of many of the hormones such as glucagon (BÖTTGER et al. 1972; FELIG et al. 1972; GALBO et al. 1975; VRANIC et al. 1976a), the catecholamines (GALBO et al. 1975; N. J. CHRISTENSEN et al. 1979), corticosteroids (HARTLEY et al. 1972a), and growth hormone (HANSEN 1970, 1971), all of which act to promote glucose and free fatty acid production. In addition, insulin levels decrease (WAHREN et al. 1971; HARTLEY et al. 1972a; VRANIC et al. 1976a; MURRAY et al. 1977), also serving to increase glucose and free fatty acid availability. In this chapter we will discuss these hormonal interactions, but would like to focus upon the particular role of glucagon during physical activity.

## II. Fuel Sources During Exercise

### 1. Carbohydrate-Derived Fuels

With the onset of exercise, muscle glycogen serves as the initial fuel source, particularly for short-term, high intensity activity. There is accelerated glycogenolysis in muscle within the first few minutes of exercise, which alows a high energy output independent of circulating substrate. As exercise continues, blood flow through the exercising muscle is increased and plasma glucose and free fatty acids become prominent fuel sources for the contractile process. Recent studies suggest that during prolonged exercise, both liver, through glucose production, and muscle glycogen, through muscle glycogenolysis to lactate and finally via the Cori cycle to glucose, contribute equally to provide a substrate for the oxidative process in exercising muscle (ISSEKUTZ 1981).

It was shown more than 90 years ago that glucose uptake in muscle is stimulated during exercise (CHAVEAU and KAUFMANN 1887). The mechanisms for increased glucose utilization are multifactorial and are not yet completely defined, but several theories have been advanced:

a. A "muscular activity factor", released by exercising muscle has been postulated (GOLDSTEIN et al. 1953; GOLDSTEIN 1961). This concept has been confirmed by some studies (R-CANDELA and R-CANDELA 1962; HAVITI and WERTHEIMER 1964), but other studies could not confirm this hypothesis (HELMREICH and CORI 1957; DULIN and CLARKE 1961).

b. Release of insulin or insulin-like materials from muscle has been suggested, for a rise in nonsuppressible insulin-like activity (NSILA) has been reported in the lymph from the exercising hindlimb in dogs (COUTURIER et al. (1971), and from the isolated forearm in humans (RENNI et al. 1976). This could of course represent merely a "washout phenomenon", rather than the release of insulin or insulin-like material from muscle.

c. The increased blood flow to exercising muscle (KJELLMER 1965) could result in increased insulin and fuel availability (VRANIC et al. 1976a; SCHULTZ et al. 1977).

d. An augmentation of insulin receptor sensitivity has also been postulated (KOIVISTO et al. 1979; PEDERSON et al. 1980).

e. Stimulation of glucose uptake by hypoxia was suggested by studies of the incubated rat diaphragm (RANDLE and SMITH 1958), however, other studies have shown that the degree of tissue hypoxia did not modulate the extent of glucose uptake during muscular contraction (BERGER et al. 1975).

The increased glucose requirements by muscle during exercise are met by increases in hepatic glucose production – initially glycogenolysis is the primary pathway for glucose production, then as exercise continues gluconeogenesis also becomes important. In short-term strenuous activity, plasma glucose may rise (E. H. CHRISTENSEN and HANSEN 1939), but ultimately, with prolonged exercise, hypoglycemia ensues (BERGSTRÖM et al. 1973). However, these are the extremes. Under most physiologic conditions, glucose utilization is matched by glucose production (VRANIC et al. 1976a; ZINMAN et al. 1977), and glucose homeostasis therefore can be maintained.

## 2. Fat-Derived Fuels

Although the adipose tissue provides the largest depot of storage energy, its relative contribution as an oxidative fuel during exercise depends upon the duration and intensity of the exercise as well as on preexisting conditions such as training and nutritional status. Adipose tissue triglycerides can provide free fatty acids (and subsequently ketone bodies) as well as glycerol, all of which can serve as gluconeogenic precursors. Fat becomes the predominent energy-yielding substrate during prolonged mild to moderate exercise (AHLBORG et al. 1974; FELIG and WAHREN 1975), and this is enhanced with a low carbohydrate diet (E. H. CHRISTENSEN and HANSEN 1939). However, fat oxidation can never fully replace the need for glucose as an energy source. The rate of free fatty acid turnover in the postabsorptive state is linearly related to arterial concentration, both at rest (HAGENFELDT 1975) and during established exercise (HAGENFELDT 1979). The regulation of free fatty acid availability is mediated through the activity of a hormone-sensitive lipase, which is inhibited by insulin, accounting for insulin's antilipolytic effects, and this lipase is stimulated by more than one of the lipolytic hormones (see Chap. 19).

The circulating levels and muscle uptake of acetoacetate and β-hydroxybutyrate, in normal humans, are quantitatively small except during starvation. Isolated muscle studies failed to demonstrate any significant utilization of these fuels in exercising muscle (HAGENFELDT and WAHREN 1968). However, the utilization of β-hydroxybutyrate by skeletal muscle is proportional to the plasma concentration (BERGER et al. 1978b), unless insulin deficiency prevails. Thus, as ketone body concentrations rise during prolonged physical activity (COURTICE and DOUGLAS 1936), they could become a significant fuel source under such conditions.

### 3. Protein-Derived Fuels

The protein of muscle itself is a potential source of fuel for exercise, although of limited usefulness if the structural integrity of the organism is to be maintained. Nitrogen balance does not change significantly during exercise (WILSON et al. 1925), except during very prolonged and strenuous activity (REFSUM and STRÖMME 1974), indicating sparing of muscle protein. The branched chain amino acids (valine, leucine, and isoleucine) can be delivered from the liver to exercising muscle, but they serve only as a minor energy source compared with carbohydrate- and fat-derived fuels.

Alanine fluxes from muscle to liver occur during exercise. As glucose is utilized by exercising muscle, alanine is synthesized from the branched chain amino acids, with a concomitant release of alanine, lactate, and pyruvate (FELIG and WAHREN 1975). Fractional hepatic extraction of alanine and other gluconeogenetic precursors increase as exercise proceeds, and the proportionate contribution of gluconeogenesis as compared with glycogenolysis to the whole of hepatic glucose production is increased. This increased flux of alanine from muscle to liver during exercise provides for the recycling of incompletely oxidized glucose (the glucose–alanine cycle), and for transport of amino nitrogen from muscle to liver (FELIG 1973).

## III. Hormonal Changes During Exercise

Exercise induces a decrease in plasma concentrations of insulin (WAHREN et al. 1971; HARTLEY et al. 1972a; MURRAY et al. 1977) and an increase in plasma concentrations of glucagon (BÖTTGER et al. 1972; FELIG et al. 1972; GALBO et al. 1975; VRANIC et al. 1976a), the catecholamines (GALBO et al. 1975; N. J. CHRISTENSEN et al. 1979), cortisol (HARTLEY et al. 1972a), growth hormone (HANSEN 1970, 1971), and gastroenteropancreatic hormones such as pancreatic polypeptide, vasoactive intestinal peptide, secretin, and somatostatin (HILSTED et al. 1980; FEURLE et al. 1980). The increases in counterregulatory hormones during exercise were intensified by preexercise fasting, suggesting that normal hormonal responses to exercise require normal insulin availability prior to exercise (GALBO et al. 1981a).

There was a marked drop in insulin both in dogs (VRANIC et al. 1976a) and in humans (HARTLEY et al. 1972a; MURRAY et al. 1977; WAHREN et al. 1971; KEMMER et al. 1979). The actual secretion of insulin is suppressed during exercise, as evidenced by a parallel decrease in circulating insulin and C-peptide levels (HILSTED et al. 1980; WIRTH et al. 1981). As catecholamine levels rise during exercise, α-adrenergic inhibition of insulin release could be the mechanism whereby insulin secretion decreases during exercise (IVERSEN 1973; ROBERTSON and PORTE 1973; SAMOLS and WEIR 1979; RIZZA et al. 1980b). In one study (HILSTED et al. 1980), the ratio of C-peptide to insulin fell during exercise, suggesting that insulin clearance decreases. This was not confirmed in other studies (WIRTH et al. 1981). With training, exercise gives rise to an increase in insulin-induced glucose uptake in the skeletal

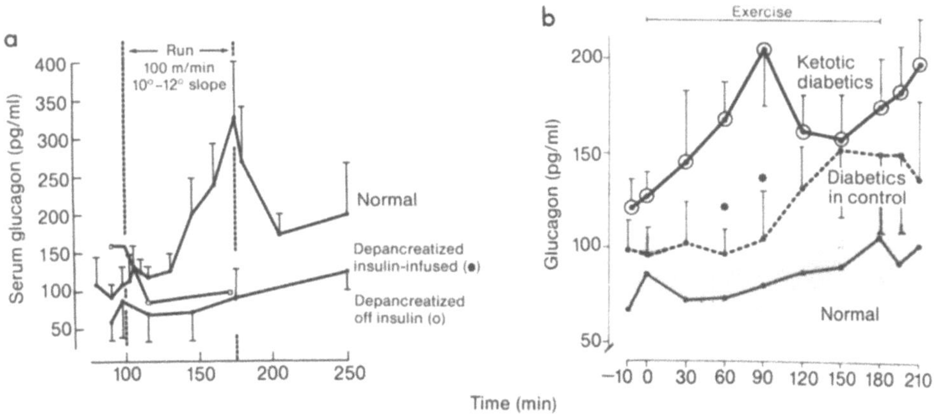


**Fig. 1. a** Serum levels of immunoreactive glucagon (mean ± standard error) during rest and exercise in five dogs before pancreatectomy (normal) and after pancreatectomy during basal infusions of insulin. Also shown are the average glucagon levels in three depancreatized dogs in which insulin infusion was discontinued for 60 min before the start of exercise. VRANIC et al. (1976a). **b** Serum levels of immunoreactive glucagon in healthy normal subjects, in diabetic patients in moderately good control (mean ± standard error), and diabetic patients who were ketotic following 24 h without insulin treatment (mean ± standard error). The *shaded area* indicated standard error about the mean control values. *Encircled values* are significantly different from corresponding values in the control group ($P<0.05$). *Asterisks* indicate statistically significant differences between corresponding values of the two groups of diabetic patients ($P<0.05$). Adapted from BERGER et al. (1977)

muscle, but glucose uptake by the liver decreases (MONDON et al. 1980). This suggests that training has primarily an affect on postreceptor events in tissue, and that changes in insulin receptor sensitivity may be of lesser importance. Also, the rise in plasma insulin in response to a glucose load is blunted in trained athletes and this has been ascribed mainly to increased insulin clearance (WIRTH et al. 1981). A decreased glucose sensitivity of the insulin-secretory mechanisms of chronically exercised rats has also been demonstrated (REAVEN and REAVEN 1981; GALBO et al. 1981b).

Both plasma epinephrine and norepinephrine can increase during both short-term and prolonged exercise (HARTLEY et al. 1972a, b; GALBO et al. 1975; N.J. CHRISTENSEN et al. 1979). If autonomic blockade is provided during exercise, the increase in circulating catecholamines is intensified (GALBO et al. 1975, 1976). If normoglycemia is maintained by means of a glucose infusion, the epinephrine response to exercise is blunted, but the norepinephrine increase is fully expressed (GALBO et al. 1977a). In poorly controlled diabetes, there is an exaggerated catecholamine release to exercise (N.J. CHRISTENSEN 1974).

In normal rats (LEFEBVRE et al. 1972; LUYCKX et al. 1975), dogs (BÖTTGER et al. 1972; VRANIC et al. 1976a), and humans (BÖTTGER et al. 1972; GALBO et al. 1975; LUYCKX et al. 1978), during exhaustive exercise, plasma glucagon concentrations increased gradually, the highest value being realized at the point of exhaustion. The magnitude of the glucagon response was greater in dogs than in humans. In one study illustrated in Fig. 1a, strenuous exercise in dogs induced a threefold increase in plasma immunoreactive glucagon (IRG) by 75 min (VRANIC et al. 1976a), whereas the increase in IRG to exhaustive exercise in humans was less pronounced. Mild to moderate exercise to 40% or 60% $\dot{V}O_{2\,max}$ caused a delayed and attenuated IRG response (FELIG 1973; AHLBORG et al. 1974; FELIG and WAHREN 1975; BERGER et al. 1977). As shown in Fig. 1 b, mild exercise induced a significant rise in IRG only after 3 h in normal subjects. However, in diabetics, the increase in IRG was more pronounced, particularly when insulin treatment was withheld for 24 h and the patients became ketotic.

The glucagon response to exercise also depends upon training and the prevailing metabolic and nutritional state. Training resulted in a markedly decreased glucagon response to exercise in normal volunteers (GYNTELBERG et al. 1977). Although the state of insulinization appears to affect the glucagon response in humans (Fig. 1 b; BERGER et al. 1977), it did not alter the glucagon response to exercise in depancreatized dogs (Fig. 1 a; VRANIC et al. 1976 a). However, in the depancreatized dogs, the glucagon originated from extrapancreatic A-cells which are not in juxtaposition to other endocrine secreting cells as are pancreatic A-cells. Thus, the regulation of secretion of extrapancreatic glucagon during exercise may well be different. Also, extrapancreatic A-cells are less sensitive to stimuli such as arginine (CHERRINGTON et al. 1974), growth hormone (SIREK et al. 1979), and to prevailing glycemia (OHNEDA et al. 1980), and altered regulation of the A-cell has been confirmed in systematic studies (LEFEBVRE and LUYCKX 1977, 1978) in the isolated canine stomach (see also Chap. 33).

In obese diabetic (MINUK et al. 1981) and nondiabetic (MINUK et al. 1980) patients, IRG did not rise during exercise. Also, starvation failed to affect the glucagon response to exercise in obese humans (MINUK et al. 1980). It is possible that the increase in glucagon secretion during exercise, is mediated, at least in part, by the sympathoadrenal system. Catecholamine levels rose shortly after the onset of exercise and epinephrine itself has been shown to stimulate glucagon release in dogs (GRAY et al. 1980), baboons (CHIDECKEL et al. 1977), and humans (GERICH et al. 1973, 1976). The question of whether epinephrine-stimulated glucagon release represents an α-adrenergic effect (HARVEY et al. 1974) of β-adrenergic effect (LUYCKX et al. 1975) has been argued. Recent studies in the isolated perfused canine pancreas (SAMOLS and WEIR 1979) have indicated that both α- and β-adrenergic mechanisms play a role, but the effect of the β-adrenoceptors on the glucagon-secreting cells is predominant. The release of glucagon was not affected by α- or β-adrenergic blockade, but either blocking agent enhanced the catecholamine response (GALBO et al. 1976, 1977). As the catecholamine responses were enhanced, the blocked receptor was stimulated by higher catecholamine concentrations than would have been present in the absence of blockade of the opposite receptor, thus accounting for similar glucagon responses. A β-adrenoceptor mechanism has been shown to mediate clearance of the catecholamines (CRYER 1980), thus explaining the markedly elevated epinephrine levels during β-blockade at rest (RIZZA et al. 1980 b).

Insulin can suppress glucagon secretion by a negative feedback mechanism which may be provided by changes in circulating insulin levels or by a more direct "paracrine" effect (see Chap. 31). It is therefore likely that exercise-induced suppression of insulin secretion facilitates glucagon release. However, in moderate exercise, insulin decreases are still apparent, whereas changes in plasma glucagon may be minimal or even absent. Also, in strenuous exercise, there is a rapid fall in plasma insulin levels, but plasma glucagon levels only reach their peak at the point of exhaustion (BÖTTGER et al. 1972).

## IV. Hormonal Interactions in the Control of Glucoregulation During Exercise

### 1. Role of Insulin

It was generally felt that insulin did not play a major role in regulating the response to exercise, because a lowering in plasma glucose concentration occurred in response to exercise, even in diabetics who had been deprived of insulin for more than 12 h. However, the presence of insulin antibodies in these patients precluded determination of plasma insulin levels, and it is likely that insulin continued to be made available to exercising muscles, in such patients, by dissociation of insulin–antibody complexes, and also because a small amount of circulating insulin is present for at least 48 h after the last injection of long-acting insulin.

The depancreatized dog is a useful model for examining the role of insulin during exercise, for treatment with porcine insulin to replace endogenous insulin does not lead to antibody formation for at least 1–2 months (VRANIC et al. 1976a). Therefore, reliable insulin measurements can be made, and also insulin cannot be released from hormone–antibody complexes. Basal intraportal insulin infusions, which maintain normoglycemia, allowed an assessment of the significance of basal insulin concentrations and the consequences of acute insulin deficiency can also be assessed during exercise (VRANIC and WRENSHALL 1969).

It became obvious that insulin plays a crucial role in glucoregulation during exercise when it was shown that, if depancreatized dogs were deprived of insulin for 48 h, glucose utilization did not increase during exercise (VRANIC and WRENSHALL 1969). As glucose production increased, there was a rapid rise in plasma glucose concentration. During acute insulin withdrawal, glucose uptake rose in proportion to glucose concentration as hyperglycemia developed, but clearance did not increase; thus, exercise accelerated the development of frank hyperglycemia (VRANIC et al. 1976a). Similarly, studies in isolated, perfused skeletal muscle of diabetic rats showed that glucose uptake in muscle was not increased by exercise, but abundant administration of insulin restored exercise-induced augmentation of glucose uptake (BERGER et al. 1975). Deficient glucose uptake has also been demonstrated in exercising calf muscle of juvenile-onset diabetic patients (STANDL et al. 1976).

Thus, although there is a decrease in plasma insulin concentration in exercise, the presence of a certain minimum amount of insulin is essential for increased glucose uptake during exercise. This increased glucose uptake occurs almost instantaneously, and it is possible that there is increased insulin delivery because of the increased blood flow and increased capillary surface areas.

There may also be changes in insulin receptor affinity, again providing increased insulin effectiveness, in spite of decreased plasma insulin concentrations. However, these changes cannot explain rapid changes in glucose uptake occurring at the onset of exercise. Studies have shown that, after 3 h mild exercise, there was increased insulin affinity to insulin receptors in normal humans (KOIVISTO et al. 1979) and insulin-dependent diabetics (PEDERSON et al. 1980). The site of increased insulin sensitivity was mainly in the skeletal muscle, whereas liver sensitivity was decreased (MONDON et al. 1980); thus, there could be a preferential increase in glucose uptake by muscle without a concomitant increase in the effectiveness of insulin in suppressing glucose production in the liver.

One study in which arteriovenous differences were measured across the working limb in humans (KALANT et al. 1978) demonstrated that there was increased insulin uptake by muscle during exercise. Also, studies in normal rats exercised on a treadmill, showed that there was an accumulation of $^3H$ in skeletal muscle following subcutaneous injection of insulin $^3H$ (BERGER et al. 1978a; HALBAN et al. 1978). Thus, there is some support for the concept that the increased uptake of glucose by muscle can be explained, at least in part, by increased binding of insulin to muscle, related to vascular changes in the exercising muscle. The vascular changes also bring about efflux of intravascular fluid into the interstitial space which could also facilitate delivery of insulin to this compartment (KJELLMER 1965).

The suppression of insulin secretion during exercise may be of greater importance in sensitizing the liver to the action of glycogenolytic and gluconeogenic factors (ALTSZULER et al. 1976; CHERRINGTON et al. 1976; LICKLEY et al. 1979; ISSEKUTZ 1980, 1981; RIZZA et al. 1981). This became apparent when it was shown in depancreatized dogs (VRANIC et al. 1974b, 1975; KAWAMORI and VRANIC 1977) and in diabetic humans (ZINMAN et al. 1977), that glucose production does not increase substantially during exercise when insulin levels are increased, either by insulin administration or exercise-induced increases in mobilization of injected insulin. It has been shown, using a glucose clamp technique to maintain normoglycemia, that splanchnic glucose production was more sensitive to changes in plasma insulin than was glucose uptake (RIZZA et al. 1981). This was supported by studies in exercising dogs, that showed a twofold increase in glycemia and glucose production in response to mannoheptulose-induced hypoinsulinemia, whereas glucose clearance did not change (ISSEKUTZ 1981). Thus, the main role of the drop in plasma insulin during exercise appears to be to increase hepatic glucose production (VRANIC et al. 1976a). During prolonged strenuous exercise, the fall in plasma insulin serves to limit peripheral glucose uptake, and thus preserve homeostasis. Low insulin levels could also be important in ensuring a similar contribution of liver glycogen to provide glucose directly, and muscle glycogen to provide glucose through the Cori cycle, as substrate for the working muscle. Under conditions of hyperinsulinemia, not only glucose, but also free fatty acid mobilization decreased (MURRAY et al. 1977; MARTINS et al. 1981), suggesting that even mild hyperinsulinemia interferes with the normal metabolic response to exercise, again confirming that the drop in insulin concentration normally seen during exercise is an important regulatory process, not merely a secondary consequence of decreasing plasma glucose levels.

## 2. Role of Catecholamines

Plasma epinephrine and norepinephrine levels both increased during exercise, the latter correlating with the increase in heart rate, vascular resistance, and pulmonary arterial saturation (N. J. CHRISTENSEN and BRANDSBORG 1973; N. J. CHRISTEN-

SEN et al. 1975c). Increases in circulating norepinephrine, when provided in the resting state, have little effect on plasma glucose or free faty acid concentrations (SCHADE and EATON 1979). It is possible that changes in norepinephrine concentrations at the nerve endings resulting from adrenergic stimulation may be of greater importance than circulating levels of norepinephrine. A rise in plasma epinephrine levels during exercise reflects increased sympathoadrenal activity. Increases in epinephrine have been shown to have marked stimulatory effects on glucose and free fatty acid production and an inhibitory effect on glucose uptake by muscle (ALTSZULER et al. 1976; MÜLLER et al. 1977; GALBO et al. 1978; SACCA et al. 1978; GRAY et al. 1980; PEREZ et al. 1980; RIZZA et al. 1980a; CHIASSON et al. 1981). Conversely, hypoglycemia stimulates epinephrine release (GALBO et al. 1977a; N.J. CHRISTENSEN et al. 1979; HARTLEY et al. 1972a).

Since the catecholamines exert their effects through both α- and β-adrenoceptor mechanisms, the effects of α- and β-blockade have been assessed during exercise (N.J. CHRISTENSEN et al. 1975c; GALBO et al. 1976, 1977b; ISSEKUTZ 1978). Studies during combined α- and β-blockade, and during isolated α-blockade, provide limited information because of the ensuing major circulatory disturbances. During β-blockade, in the dog (ISSEKUTZ 1978), peripheral glucose uptake during exercise was enhanced indirectly by decreasing muscle glycogenolysis and decreased free fatty acid production. However, glucose production in the liver was not suppressed by β-blockade. Experiments in rats indicate that the catecholamines probably promote glucose production during exercise through activation of α-adrenoceptors in the liver, whereas the peripheral stimulation of glycogenolysis and lipolysis are mediated through β-adrenergic mechanisms (CHAN and EXTON 1978).

Studies in adrenodemedullated and chemically sympathectomized rats showed a decrease in the glucagon response and a smaller drop in insulin in response to exercise, with an associated decrease in the breakdown of muscle and hepatic glycogen when these animals were compared with normal controls (GALBO et al. 1978). It was the adrenodemedullation that proved to be responsible for these changes (RICHTER et al. 1980, 1981a,c). Thus, during prolonged exercise, adrenomedullary hormones inhibit insulin secretion and stimulate glucagon secretion, both of which effects enhance muscle and hepatic glycogenolysis in the rat. In addition, it appears that the adrenal medulla, rather than sympathetic nerve endings, was responsible for most of the increased circulating norepinephrine during exercise.

Physiologic infusions of epinephrine (SACCA et al. 1978; GRAY et al. 1980; PEREZ et al. 1980) or glucagon (CHERRINGTON and VRANIC 1974; CHERRINGTON et al. 1974) have only transient effects on glucose production in resting dogs or humans, but the effects of both epinephrine and glucagon can be sustained if hypoinsulinemia prevails (SHERWIN et al. 1976; ROSS et al. 1978; RIZZA et al. 1979a; PEREZ et al. 1980; KEMMER et al. 1982). Recent experiments in the dog (EIGLER et al. 1979) have also demonstrated that the effects of epinephrine and glucagon become sustained if resting dogs have been exposed to elevated glucocorticoid levels. Thus, it is possible that insulin suppression is of major importance in increasing and prolonging the effectiveness of glucagon and the catecholamines during the initial stages of exercise, and that the glucocorticoids may play a similar role in exercise of prolonged duration.

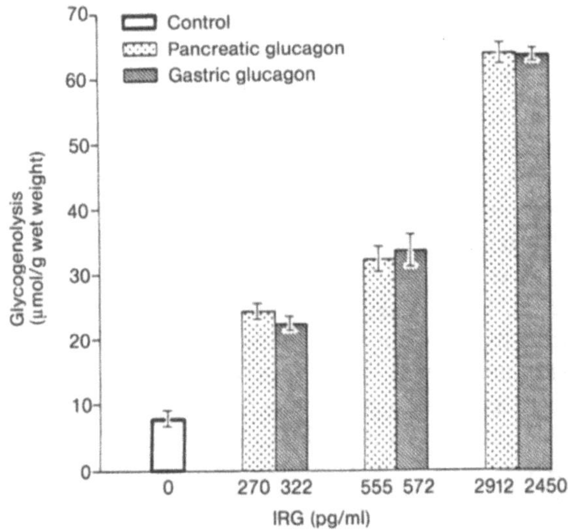


**Fig. 2.** The effects of pancreatic and gastric immunoreactive glucagon (IRG) on glycogenolysis in isolated rat hepatocytes in fed rats. The hepatocytes were incubated for 30 min in 1 ml Krebs–Ringer bicarbonate buffer (pH 7.4) that contained bovine charcoal-treated serum (15%), bovine serum albumin (1.5 g/100 ml), and glucose (100 mg/dl). DOI et al. (1979)

## 3. Role of Glucagon

Glucagon is released in response to strenuous exercise, in rats, dogs, and humans. As increased glycogenolysis and gluconeogenesis are essential in exercise, and as glucagon is known to stimulate both these metabolic processes (see Chap. 14 and 15), several studies have examined the role of glucagon in exercise. One study, in which antigen-stripped glucagon antibodies were administered to exercising rats, in order to examine the role of glucagon (RICHTER et al. 1981 b), demonstrated that glucagon appeared to enhance hepatic glycogen depletion. Also, glucagon's stimulatory effect on insulin secretion during exercise appeared to be countered by a sympathoadrenal inhibition of insulin secretion. The role of glucagon in exercise was also studied in normal and depancreatized insulin-infused dogs (VRANIC et al. 1976a; ISSEKUTZ and VRANIC 1980; VRANIC and ISSEKUTZ 1980), and it was felt from both studies, that the presence of some glucagon was essential to permit normal metabolic responses to exercise, but that it was not necessary for plasma glucagon levels to rise.

Normal and depancreatized insulin-infused dogs have the same resting glucagon levels. However, the glucagon in the normal animals is of pancreatic origin, whereas in the depancreatized dogs, the glucagon is of extrapancreatic origin. As reviewed in Chap. 33, the main extrapancreatic source is the fundus of the stomach (VRANIC et al. 1974a; BLASQUEZ et al. 1976; MORITA et al. 1976; MÜLLER et al. 1978). Both the pancreatic and stomach glucagons, when extracted and purified, have been shown to have the same molecular weight (3,500 daltons), immunologic, and physiochemical characteristics (MORITA et al. 1976; SRIKANT et al. 1977). The

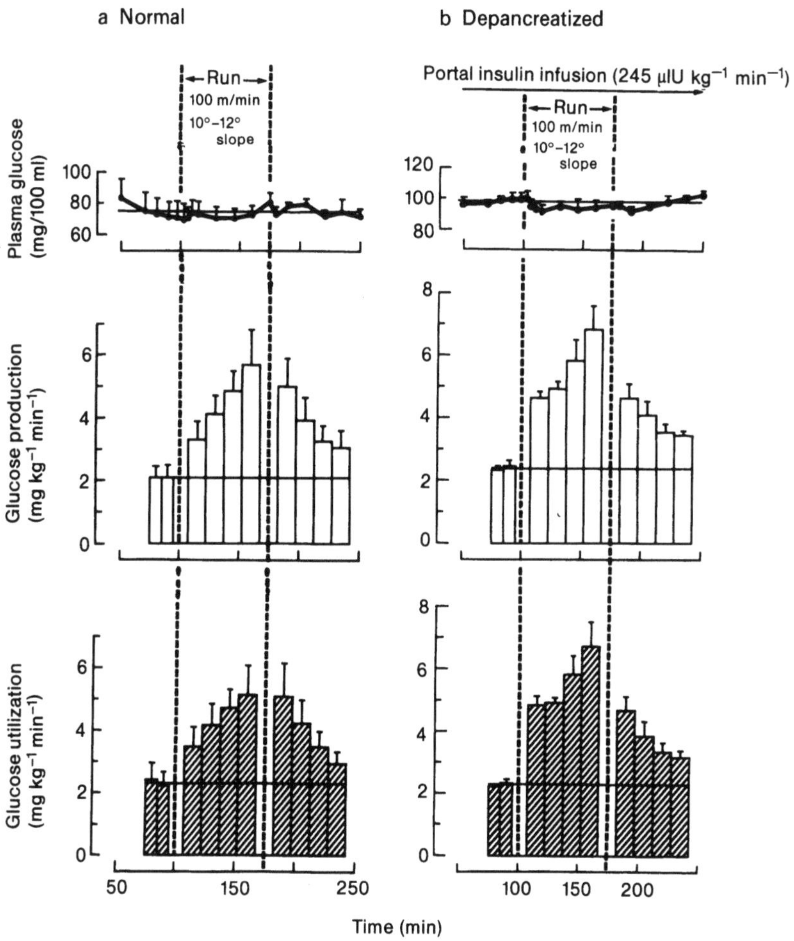


**Fig. 3 a, b.** The effects of exercise on plasma glucose concentration, rates of glucose production and utilization in (**a**) five normal dogs, (**b**) depancreatized insulin-infused dogs. In both groups of dogs, synchronized increments in glucose production and utilization occurred and glycemia remained unchanged. Adapted from VRANIC et al. (1976a)

glucagons from both sources also have the same biologic activity for, when stomach glucagon was purified to immunologic homogeneity and its effects were assessed in isolated rat liver cells with respect to glycogenolysis (Fig. 2), gluconeogenesis, concentrations of cyclic AMP, lactate, pyruvate, and urea (DOI et al. 1979), the effects of this peptide were the same as those of pancreatic glucagon. In both the normal and depancreatized dogs (VRANIC et al. 1976a), the rate of glucose production rose in response to exercise (Fig. 3). Plama immunoreactive insulin (IRI) levels were comparable during exercise, but in the normal dogs, plasma immunoreactive glucagon (IRG) levels rose threefold, whereas in the insulin-infused or acutely insulin-deprived depancreatized dogs, plasma IRG remained unchanged (see Fig. 1a). Thus, a glucagon surge did not appear to be an essential mediator of the increased hepatic glucose production during exercise.

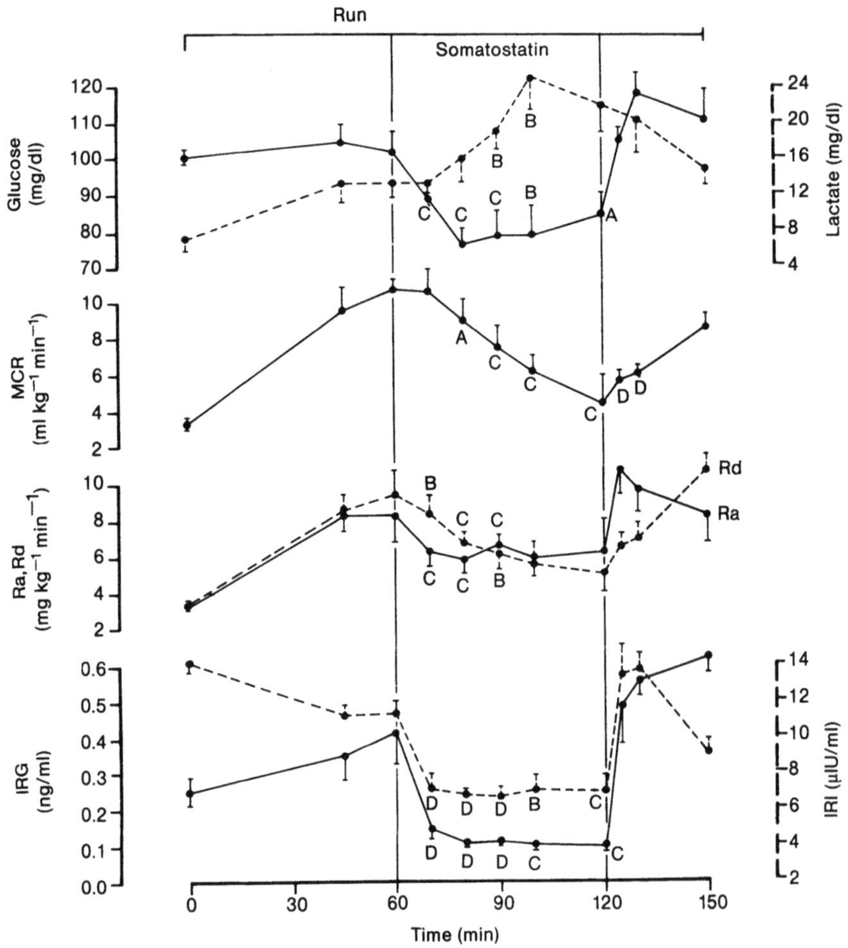


**Fig. 4.** The effects of somatostatin during exercise on plasma concentrations of glucagon (IRG), plasma insulin (IRI), glucose, and lactate, and on rates of glucose turnover, R (glucose appearance = Ra, glucose utilization = Rd) and metabolic glucose clearance (MCR) in steadily exercising dogs. *A–D* represent levels of significance of deviation from presomatostatin values (*A*) $P<0.05$; (*B*) $P<0.02$; (*C*) $P<0.01$; (*D*) $P<0.001$. ISSEKUTZ and VRANIC (1980)

Normal dogs were subjected to glucagon suppression by a 60-min somatostatin infusion during strenuous exercise, in order to determine if the presence of glucagon was essential for the immediate and/or maintained increase in glucose turnover during exercise (ISSEKUTZ and VRANIC 1980). As shown in Fig. 4, plasma IRG increased and IRI decreased during exercise. Both the rate of hepatic glucose production and the metabolic clearance rate (MCR) of glucose rose threefold and glucose homeostasis was maintained. There was also a slight increase in plasma lactate. When somatostatin infusion was commenced, plasma insulin levels were slightly further suppressed. Glucagon was also suppressed well below basal levels. Hepatic glucose production decreased faster than utilization. This imbalance led to a rapid fall in plasma glucose. Surprisingly, the metabolic clearance of glucose decreased

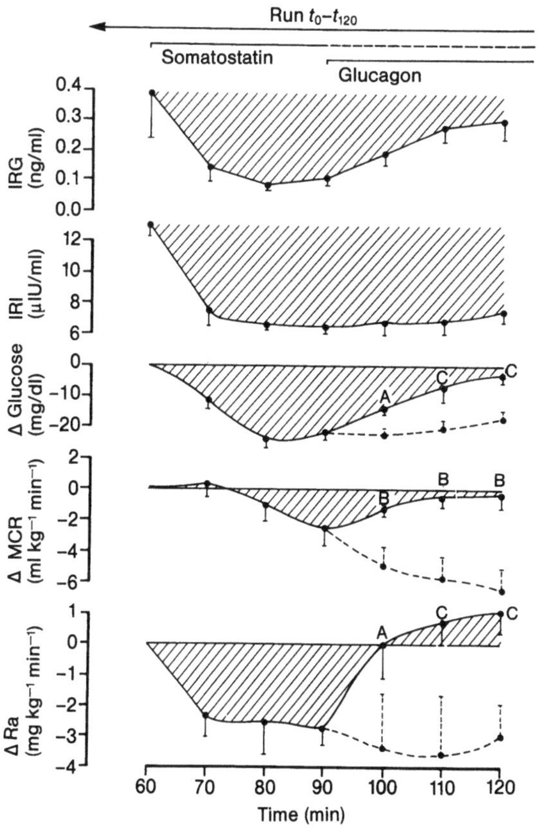


**Fig. 5.** The effects of somatostatin plus glucagon on concentrations of glucagon (IRG), insulin (IRI), and on glucose turnover in four exercising dogs. The run commenced at $t=0$ min, somatostatin was infused from $t=60$ to $t=120$ min. Three dogs served as controls without glucagon infusion (*broken lines*). The rates of hepatic glucose output (Ra), glucose clearance (MCR), and glucose concentrations are shown as deviations (Δ) from values measured at $t=60$ min. *A–D* represent the levels of significance of the differences from preglucagon values (see legend of Fig. 4 for key). ISSEKUTZ and VRANIC (1980)

as well; since glucagon does not control peripheral glucose uptake directly, an indirect effect, which is discussed later, has been suggested. Plasma lactate levels rose further during somatostatin infusion. Somatostatin also accentuated free fatty acid turnover during exercise. After the cessation of somatostatin infusion, there was a pronounced rebound effect on plasma IRI, IRG, glucose, and lactate levels and also a rebound effect on the rates of glucose production, utilization, and clearance. These studies suggest that glucagon plays an important role in the control of glucose production directly and glucose utilization by muscle indirectly during exercise.

As somatostatin has effects in addition to suppression of insulin and glucagon, another group of dogs were given glucagon at physiologic doses together with the somatostatin infusion during exercise. The infusion of glucagon at a time when insulin remained suppressed prevented all the metabolic effects of the somatostatin infusion and plasma concentrations of glucose and the rate of glucose production and clearance were normalized (Fig. 5). Thus, it was indeed the effect of the

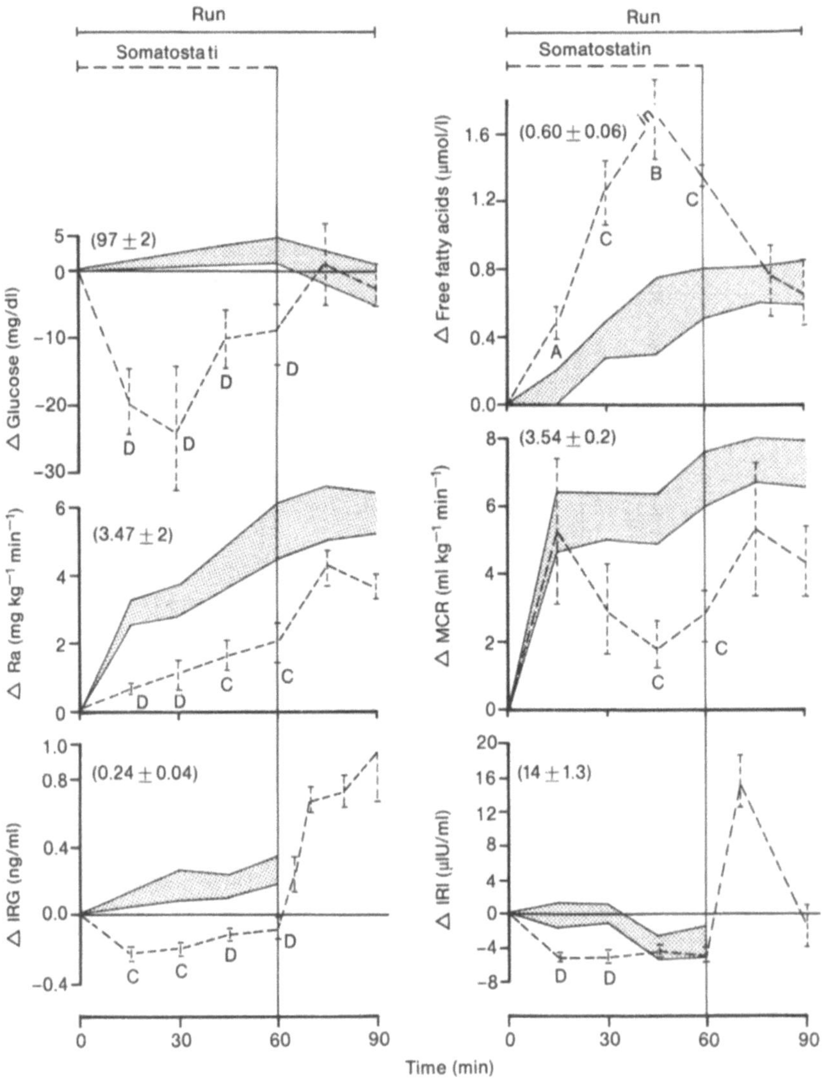


**Fig. 6.** The effect of somatostatin during rest-to-work transition. A priming dose of glucose $^{3}$H-3 was given at $t = 90$ min, followed by a constant tracer infusion. Baseline values were obtained between $t = -60$ and $t = 0$ min, when both exercise and somatostatin infusion (0.22 μg kg$^{-1}$ min$^{-1}$) were started. Values represent deviations ($\Delta$) from baseline (resting) values. The *shaded area* represents the mean ± standard error obtained in 8–12 control experiments (exercising dogs without somatostatin infusion). *Broken lines* represent the mean ± standard error of four experiments when somatostatin was given. *A–D* represent levels of significance of differences between the control and somatostatin-infused animals (see legend of Fig. 4 for key). Values in parentheses (mean ± standard error) are resting values. ISSEKUTZ and VRANIC (1980)

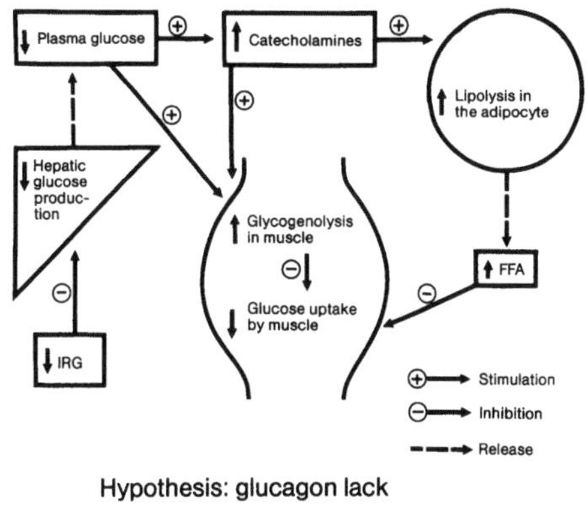


**Fig. 7.** A schema representing the hypothesis that glucagon suppression leads to a direct suppression of glucose production and an indirect suppression of glucose uptake by muscle, caused both by the increase in muscle glycogenolysis and free fatty acid release, both due to hypoglycemia-induced catecholamine release. Adapted from VRANIC and ISSEKUTZ (1980)

somatostatin on glucagon secretion, rather than other effects of somatostatin, which induced the perturbation in glucose homeostasis.

When somatostatin was infused starting concurrently with the onset of exercise, there was an early decrease in plasma IRI and IRG fell well below basal (Fig. 6). When the somatostatin-infused exercising dogs were compared with a group of control animals exercised without concomitant somatostatin infusion, somatostatin strikingly inhibited the normal increase in hepatic glucose production. Within 15 min, there was an 80% inhibition and, even at the end of the 60-min somatostatin infusion, the increase in the rate of appearance of glucose (Ra) was only one-half that occurring in the absence of somatostatin. This marked suppression of the hepatic response of Ra to exercise was responsible for the sharp drop in plasma glucose. Glucose clearance (MCR) initially rose to the same extent in the somatostatin-infused and in the control animals, however, after the first 15 min, MCR began to decline in the somatostatin-infused animals until it was 60%–70% lower than in the control group. This resulted in a partial recovery in plasma glucose levels. Free fatty acid levels rose more quickly to levels that were twice as high in the somatostatin-infused dogs as in the control group. There was a pronounced rebound in plasma IRI and IRG levels following cessation of the somatostatin infusions. Thus, again, it is apparent that glucagon plays an important role in glucoregulation during exercise.

The schema shown in Fig. 7 presents the hypothesis that hypoglucagonemia with its concomitant decrease in glucose production can induce a fall in circulating plasma glucose which may, in turn, trigger excessive release of catecholamines. This exercise-induced catecholamine release, in the presence of hypoinsulinemia, could accelerate both hepatic and muscle glycogenolysis (ISSEKUTZ 1978; RICHTER et al. 1980, 1981 a, b; RIZZA et al. 1980 b, CHIASSON et al. 1981; SACCA et al. 1979 a). This concept is supported by the increase in plasma lactate which was noted at a time when glucose utilization was decreased. Both the increased free fatty acid uptake into muscle and the increased glycogenolysis in muscle could result in a decrease in glucose uptake by muscle (NEWSHOLME and RANDLE 1964; COLOWICK 1973). Thus, glucagon could play a direct role in initiating and maintaining elevated hepatic glucose production and an indirect role in the control of glucose uptake during exercise.

The importance of basal glucagon levels during exercise has also been examined recently in humans (BJÖRKMAN et al. 1981). When somatostatin was given just prior to exercise, there was a sharp decrease in plasma glucose, as in the dog. However, when somatostatin was infused for 2 h prior to and during exercise, plasma glucose rose with the onset of exercise and the increase in net splanchnic glucose production was the same as in controls. Plasma glucose levels had not changed in a control group not given somatostatin, thus it was concluded that the rise in plasma glucose was related to decreased glucose utilization, which could be secondary to the prolonged hypoinsulinemia (VRANIC et al. 1976a). BJÖRKMAN et al. (1981) concluded that their findings contradict the view that glucagon always plays an important role in glucoregulation during exercise (ISSEKUTZ and VRANIC 1980). However, although plasma glucagon levels were suppressed, in exercising humans (BJÖRKMAN et al. 1981), circulating levels of glucagon were still approximately 70 pg/ml. As the studies in dogs (VRANIC et al. 1976a; ISSEKUTZ and VRANIC 1980) suggested that the presence of glucagon rather than a rise in glucagon was of importance in allowing normal responses to exercise, the studies in humans are not necessarily incompatible with those in dogs. It is also possible that, during chronic hypoinsulinemia caused by prolonged somatostatin infusion, the liver is even more sensitive to the glucagon or epinephrine released during exercise, so that IRG levels of the order of 70 pg/ml could be very effective indeed. However, this point has not been clarified as glucagon replacement studies were not carried out (BJÖRKMAN et al. 1981).

In another series of studies, glucagon–insulin interactions on exercise-induced splanchnic glucose output were studied during a glucose infusion. This prevented a rise in plasma glucagon during exercise and preexercise insulin levels were increased so that during exercise these elevated insulin levels returned to normal basal values (FELIG and WAHREN 1979). These experiments confirm the fact that increased glucagon secretion is not essential for stimulation of increased splanchnic glucose output during exercise. The studies also suggest that it is perhaps not the absolute insulin levels which are important in glucoregulation during exercise, but that the rate of fall in plasma insulin may also play an important role.

## V. The Response to Exercise in Diabetic Subjects

In insulin-dependent diabetes, neither conventional subcutaneous insulin therapy nor even the most sophisticated intravenous insulin pump device can ever restore normal physiologic insulin delivery. The lack of a normal portal–peripheral insulin gradient can lead to hepatic underinsulinization and/or peripheral overinsulinization. Previously, interest in the effects of exercise on diabetes has focused mainly on the role of insulin deficiency or treatment (VRANIC and BERGER 1979). We would like to examine the role of glucagon and its interactions with insulin and other glucoregulatory hormones in the exercising diabetic.

### 1. Exercise in Insulin-Dependent Diabetics

An increase in circulating plasma glucagon does not appear to be essential for an increase in hepatic glucose production in response to exercise, in either depancreatized, insulin-treated, or acutely insulin-deprived dogs (VRANIC et al. 1976a). The lack of glucagon response to exercise in depancreatized dogs is thought to be related to the lesser responsiveness to exercise of the gastric A-cell than the pancreatic A-cell (for further discussion, see Chap. 33). In humans, IRG increased only modestly when diabetes was well controlled, and such subjects usually have peripheral hyperinsulinemia. However, in diabetics who were poorly controlled, IRG rose substantially during exercise (BERGER et al. 1977). The role of the catecholamines during exercise in diabetes has not been adequately characterized, although it is known that in poorly controlled diabetics, the catecholamine response to exercise is excessive (N. J. CHRISTENSEN et al. 1979; GALBO et al. 1981a). Thus, an excessive hyperglycemic response to exercise in poorly controlled diabetic patients could reflect excessive responses of glucagon and/or the catecholamines and, in addition, hypoinsulinemia may sensitize the liver to these glucoregulatory hormones.

The importance of excessive glucagon and hypoinsulinemia in regulating glucose production by the liver has been substantiated in experiments in resting alloxan-induced diabetic (PEREZ et al. 1981) and depancreatized (KEMMER et al. 1982) dogs, respectively. When epinephrine was infused in the former, IRG rose to maximal levels, while in the latter, the IRG response was marginal. In both studies, however, glucagon's effects on the liver were substantial, because it was found that, with glucagon suppression by somatostatin, the liver's response to epinephrine was markedly attenuated. On the other hand, in normal dogs, the effect of epinephrine was not mediated by glucagon, since with or without glucagon suppression, the increase in hepatic glucose production in response to epinephrine was the same (GRAY et al. 1980). We feel that the key to understanding these observations is that, in normal dogs infused with epinephrine, insulin increased transiently and never fell below basal values, whereas in the diabetic dogs, insulin levels were always below basal values. These observations in resting dogs suggest that the role of IRG may be enhanced during exercise because of this suppressed insulin secretion.

The presence of some insulin is essential for facilitated glucose uptake by exercising muscle. This has been demonstrated in vitro in perfused rat muscle (BERGER et al. 1975) and in vivo in depancreatized dogs (VRANIC and WRENSHALL 1969).

In contrast to the insulin-deficient state in which there is a further deterioration of glucoregulation in response to exercise, diabetics who are relatively well controlled by subcutaneous regular or intermediate to long-acting insulin, may experience a lowering of plasma glucose levels in response to exercise. It has been shown that under some conditions, exercise can enhance mobilization of insulin from the injection site, resulting in an increase in circulating insulin levels. This phenomenon was first described in depancreatized dogs (KAWAMORI and VRANIC 1977), in whom insulin levels can be measured (VRANIC et al. 1976a) and in newly diagnosed insulin-dependent diabetics, before the development of insulin antibodies (ZINMAN et al. 1977, 1979) in which radioimmunoassay for insulin also gives interpretable results. Further evidence for insulin mobilization from the injection site during exercise was obtained from studies in which $^3$H-labeled regular insulin was injected

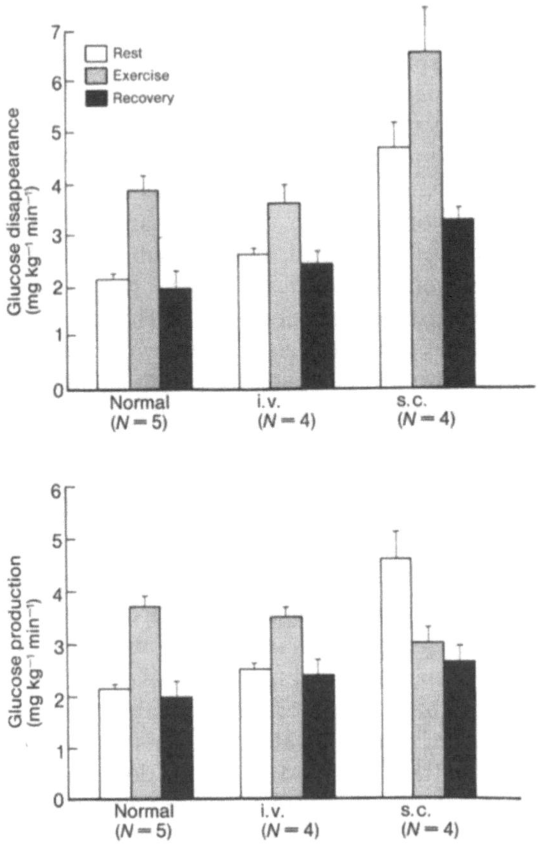


**Fig. 8.** Effects of exercise on glucose disappearance and production at rest during exercise, and during recovery from exercise in normal controls, intravenous insulin-infused (i.v.) diabetics, and diabetics treated with subcutaneous insulin injection (s.c.). ZINMAN et al. (1979)

in rats (BERGER et al. 1978 a; HALBAN et al. 1978), and its appearance in plasma determined. Also, studies in exercising insulin-dependent diabetics (KOIVISTO and FELIG 1978) in which external counting of insulin $^{125}$I over the injection site was carried out, suggested a similar mechanism. However, recent observations in exercising hyperinsulinemic insulin-dependent diabetics (KEMMER et al. 1979) demonstrated that the enhanced absorption of insulin from its injection site was not essential for plasma glucose levels to decrease. This is attributed to hyperinsulinemia which was already present before the start of exercise.

Insulin-dependent diabetics were investigated after injection of regular insulin both in the resting state and during exercise. After insulin injection, plasma insulin leves rose and this rise was not further enhanced by exercise initiated 35 min after the insulin injection. Although plasma IRI levels were similar in the resting and exercise states, plasma glucose levels dropped more dramatically during exercise. A group of normal controls who exercised without insulin administration, had lower IRI levels at the onset of exercise and IRI fell promptly. Thus, the initial hyperinsulinemia in the diabetic subjects, and their inability to lower plasma insulin are

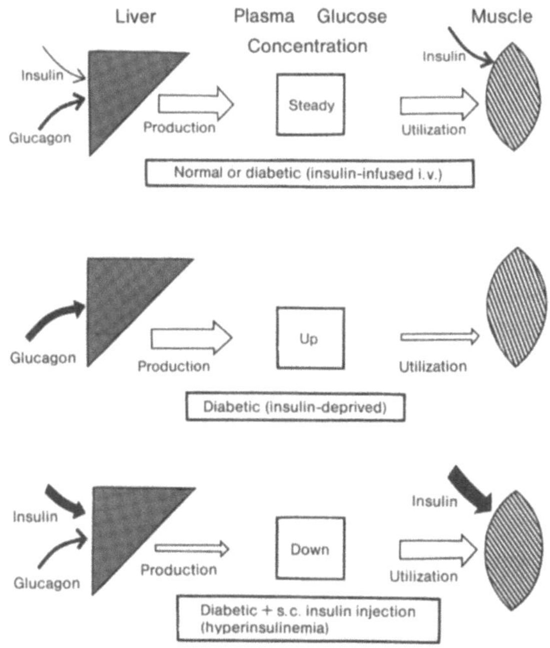


**Fig. 9.** A hypothesis concerning the role of insulin in the regulation of glucose production and utilization in normal subjects and diabetic subjects who are normoinsulinemic, hypoinsulinemic, or hyperinsulinemic. Adapted from VRANIC and BERGER (1979)

both responsible for the glucose-lowering effect of exercise (KEMMER et al. 1979). This does not preclude the possibility that, under different conditions, accelerated insulin absorption could be the main cause of hyperinsulinemia. The presence of suprabasal insulin levels prevents an adequate increase in hepatic glucose production in response to exercise as shown in Fig. 8, which was shown both in the depancreatized dogs (KAWAMORI and VRANIC 1977) and in insulin-dependent diabetics (ZINMAN et al. 1977), which leads in turn to a significant decrease in plasma glucose concentrations. There may also be decreased mobilization of glycerol and free fatty acids from adipose tissue (MARTIN et al. 1981). In addition to having increased insulin levels at the start of exercise, these insulin-dependent diabetics may also be paradoxically hyperglycemic, which will also inhibit hepatic glucose production. It has been shown that under prevailing hyperinsulinemia, hepatic glucose production is inversely correlated with the initial blood glucose levels (KAWAMORI and VRANIC 1977; ISSEKUTZ 1981). This is true only when suprabasal insulin levels are present to prevent an adequate rise in glucose production during exercise. If only basal insulin levels are provided by insulin infusion, glucose production increased with exercise (Fig. 8), whether or not the subjects were normoglycemic or hyperglycemic (VRANIC et al. 1976a; ZINMAN et al. 1977). The role of insulin levels in control of glycemia during exercise is summarized in Fig. 9.

The beneficial effects of exercise on glucoregulation in insulin-dependent diabetics has been well documented (LAWRENCE 1926; BÜRGER and KRAMER 1928; STRUWE 1977). A recent study (COCE et al. 1979) has examined this in insulin-dependent diabetics, given their usual insulin injection once daily, and who consumed five meals daily. The subjects were then exercised 30 min after breakfast, lunch, and dinner at a work load of 50% $\dot{V}O_{2\,max}$ for 30 min. The exercise periods resulted in consistently lower plasma glucose levels than were seen when the patients rested following their meals.

Regular exercise results in surprisingly prolonged improvement in glucoregulation. There may be increased sensitivity to insulin during exercise (KOIVISTO et al. 1979; MONDON et al. 1980; PEDERSON et al. 1980), which can persist far beyond the period of physical activity per se. Also, an enlarged capillary bed and increased blood flow in skeletal muscle has been described during endurance training (HOLLOSZY and NARAHARA 1967), which could provide increased insulin and substrate availability for prolonged periods. Also, exercise is known to decrease gastric mobility and could therefore decrease the absorption of food. This blunting of the food absorption peaks caused by physical training could account for some of the improved glycemic control in diabetics. Thus, regular exercise could lead to effects which are consistent with decreased insulin requirements in diabetics.

### 2. Exercise in Noninsulin-Dependent Diabetics

Although the majority of diabetics are noninsulin-dependent, it is the insulin-dependent diabetics who have been more extensively studied with respect to their responses to exercise. Certainly, exercise as a potential therapeutic modality in noninsulin-dependent diabetes requires further attention. Glucoregulation during exercise was examined in obese diabetics and obese nondiabetic control subjects (MINUK et al. 1981). They were exercised in the postabsorptive state for 45 min on a bicycle ergometer at 50% of maximum oxygen consumption. One-half of the diabetics were treated by diet alone and the remainder with diet plus chlorpropamide. During exercise, there was a significant decrease in plasma glucose levels in both groups of diabetics, while plasma glucose levels remained constant in the control subjects. Resting plasma insulin levels were elevated in all groups studied and decreased in the obese nondiabetic subjects, while they did not decrease in obese diabetics. Measurements of glucose turnover demonstrated that the decrease in glycemia in the diabetics was due to a failure to increase hepatic glucose production during exercise while peripheral glucose uptake was increasing. It is postulated that the sustained hyperinsulinemia and prevailing hyperglycemia in the diabetic subjects was responsible for preventing the normal increase in hepatic glucose output in response to exercise. The studies in obese diabetic and nondiabetic subjects (MINUK et al. 1981) demonstrated that exercise can improve hyperglycemia in obese noninsulin-dependent diabetics, and this could have therapeutic implications.

## C. Stress

Many diverse crisis situations give rise to what we have come to consider under the broad general heading of stress. In more primitive animals, and in humans in earlier times, the commonly quoted examples were *fight* and *flight*. In both, the instant and intense muscular work would lead to hypoglycemia if the glucose were not replaced by increased production of energy-yielding fuels. In present times, strenuous and extreme exercise corresponds most closely to these stress situations. However, some common stress situations also encountered at the present time, are the pathologic conditions such as hypovolemic shock, major surgery, major trauma, burns, severe infections, myocardial infarction, and severe hypoglycemia. After serious injury, shock can reduce cerebral blood flow and thereby diminish glucose delivery to the brain. Survival of the organism ultimately may depend on the ability of the glucoregulatory system to increase the arterial glucose concentration and compensate thus for the decreased cerebral circulation. There is a myriad of hormonal and metabolic sequelae of stress; only those directly related to glucoregulation will be examined in this section. We will also report the results of studies using several experimental models of stress.

## I. Hormonal Response to Stress

Although the underlying causes of stress can vary greatly, the common response which characterizes all stress situations is increased sympathoadrenal activity. Elevated epinephrine and norepinephrine levels have been suggested or demonstrated in such stress situations as major surgery (HALTER et al. 1977; HALTER and PFLUG 1980; CRYER 1980), burns (WILMORE et al. 1974a; WOLFE et al. 1979), severe infections (ROCHA et al. 1973), myocardial infarction (N. J. CHRISTENSEN and VIDEBAEK 1974; CRYER 1980), and severe hypoglycemia (CANNON et al. 1924; HOUSSAY et al. 1924; N. J. CHRISTENSEN et al. 1975a; GARBER et al. 1976a; GERICH et al. 1979; SACCA et al. 1979b; FRIER et al. 1981). In addition to increased adrenergic activity in response to hypoglycemia, activation of postganglionic cholinergic nerves, which, can stimulate, for example, glucagon release (BLOOM et al. 1974; FRIER et al. 1981) has been implicated. Markedly increased epinephrine and norepinephrine levels are also seen in diabetic ketoacidosis (CRYER 1980), which could be added to the list of stress situations, although often one of the previously mentioned stress conditions such as infection precedes the deterioration in diabetic control.

These increased catecholamine levels of stress may be associated with changes in insulin secretion. Although circulating insulin levels may be normal (RIZZA et al. 1979c) or even increased (ROCHA et al. 1973; N. J. CHRISTENSEN and VIDEBAEK 1974), the levels are inappropriately low with respect to the prevailing hyperglycemia. This is certainly related to the catecholamine excess. However, when considering surgical stress, it should be noted that several anesthetic agents also inhibit insulin release directly (HALTER and PFLUG 1980; HSU and HUMMEL 1981). The effect of the catecholamines on insulin secretion is therefore a relative suppression (PORTE et al. 1966; ROBERTSON and PORTE 1973; IVERSEN 1973; SAMOLS and WEIR 1979; HALTER and PFLUG 1980), which represents a balance between $\alpha$-adrenergic suppression and $\beta$-adrenergic stimulation of insulin release (GERICH et al. 1976; ATKINSON et al. 1981; NAKHOODA et al. 1981), and the effect of catecholamine-induced hyperglycemia on pancreatic B-cell function. The predominance of one of these effects over the other depends somewhat on the circulating catecholamine levels and temporal factors (GRAY et al. 1980).

Stress has also been associated with hyperglucagonemia (ROCHA et al. 1973; LINDSEY et al. 1974, 1975; WILMORE et al. 1974b) which occurs as a result of $\beta$- and possibly α-adrenergic stimulation of glucagon release (IVERSEN 1973; GERICH et al. 1976; SAMOLS and WEIR 1979; NAKHOODA et al. 1981), probably caused at least in part by catecholamine release, but possibly representing, in addition, enhanced A-cell sensitivity (HAMAJI et al. 1979). Elevated levels of growth hormone (ROTH et al. 1963; GREENWOOD et al. 1966; GARBER et al. 1976a; RIZZA et al. 1979b), β-endorphin (IPP et al. 1978), and cortisol (GREENWOOD et al. 1966; GARBER et al. 1976a; RIZZA et al. 1979b) have also been observed under hypoglycemic stress. Somatostatin also rises during hypoglycemia, but not in response to the stress of major surgery (WASS et al. 1980).

## II. Major Changes in Fuels or Energy Substrate During Stress

Stress can greatly affect the metabolism of carbohydrates, lipids, and proteins, mainly indirectly through changes in circulating hormones – chiefly the catecholamines, glucagon, and insulin.

### 1. Carbohydrate-Derived Fuels

Hyperglycemia has often been noted in many stress situations, which may be related to elevated catecholamine and glucagon levels and relative suppression of insulin release. The catecholamines induce hyperglycemia which is initiated by increased hepatic glucose production and maintained by a sustained impairment in glucose clearance (RIZZA et al. 1979c; GRAY et al. 1980). Both hepatic glycogenolysis (SOKAL et al. 1964) and gluconeogenesis (CHAN and EXTON 1978) are increased initially. Epinephrine rapidly activates phosphorylase in isolated rat hepatocytes, through α-adrenergic mechanisms (EXTON et al. 1978), whereas β-adrenergic mechanisms, increase cAMP and activate protein kinase only transiently (CHERRINGTON and EXTON 1976). However, the direct effects of epinephrine on tracer-determined glucose production and utilization, in humans, appear to be mediated by β-adrenergic mechanisms (RIZZA et al. 1980a). The α-adrenergic effects of epinephrine appear to be mediated mainly through suppression of insulin release. Glucagon levels which are also elevated during stress lead to increased hepatic glycogenolysis, possibly through rapid activation of cAMP-dependent protein kinase (CHERRINGTON and EXTON 1976), and gluconeogenesis. Catecholamines can also accelerate glycogenolysis in muscle and decrease muscle uptake of glucose (WALAAS and WALAAS 1950).

### 2. Fat-Derived Fuels

Both catecholamine excess, partly via β-adrenoceptors in adipose tissue (HIMMS-HAGEN 1970), and insulinopenia can stimulate lipolysis, which in turn, can increase ketogenesis indirectly through increased delivery of free fatty acids and glycerol to the liver (N. J. CHRISTENSEN et al. 1975a; MÜLLER et al. 1977) and directly through activation of liver carnityl transferase, which result, as reviewed in Chap. 17, from relative insulinopenia and hyperglucagonemia (MCGARRY and FOSTER 1977). The catecholamines may also stimulate ketogenesis, directly or indirectly, through release of hormones such as glucagon, growth hormone, and cortisol or suppression of insulin (N. J. CHRISTENSEN et al. 1979; SCHADE and EATON 1979, 1980). However, in shock, which represents extreme stress, there is decreased ability to oxidize free fatty acids (LIDDELL et al. 1979).

### 3. Protein-Derived Fuels

Catecholamines can decrease the release of alanine from intact rat skeletal muscle through $\beta$-adrenergic mechanisms (GARBER et al. 1976b). Short infusions of norepinephrine (SILVERBERG et al. 1978), but not epinephrine (CLUTTER et al. 1980) can decrease circulating alanine in humans. Nor does epinephrine decrease any of the gluconeogenetic amino acids: alanine, glutamate, or glutamine (N. J. CHRISTENSEN et al. 1975a). However, prolonged infusions of epinephrine can decrease amino acids, particularly the branched chain amino acids (SHAMOON et al. 1979). In shock, massive protein breakdown can occur (LIDDELL et al. 1979).

## III. Hormonal Interactions in Glucoregulation During Stress

### 1. Pathologic Stress States

Several severe stress situations, which are associated with release of catecholamines and glucagon together with relative insulin deficiency, such as shock (LIDDELL et al. 1979), severe burns (ALLSOP et al. 1978; BURKE et al. 1979; WOLFE et al. 1979; DURKOT and WOLFE 1981), sepsis (LONG et al. 1978; DURKOT and WOLFE 1981), and major trauma (LONG et al. 1978; ELWYN et al. 1979) have been studied in experimental animals and in injured patients. Although hyperglucagonemia is a finding common to all these forms of injury, changes in glucose kinetics vary. In burns, glucose production (Ra) and clearance (MCR) both increased (ALLSOP et al. 1978; DURKOT and WOLFE 1981), which appeared to involve both $\beta$- and $\alpha$-adrenoceptor mechanisms. Insulin resistance also occurred (DURKOT and WOLFE 1981). In sepsis, however, Ra and MCR both decreased (DURKOT and WOLFE 1981). Paradoxically, gluconeogenesis could not be suppressed by glucose administration (WOLFE et al. 1977; LONG et al. 1978). Gluconeogenesis was increased after severe trauma, but this could be suppressed by high infusion rates of glucose (ELWYN et al. 1979).

### 2. Experimental Stress Models

Several experimental models have been devised to study hormonal interactions in glucoregulation in mild, moderate, and severe stress.

#### a) The Injection of Pyrogen: A Model for Mild Stress

Pyrogen induces malaise and fever with rapid endogenous secretion of all stress hormones. This was studied in well-controlled insulin-dependent diabetics during basal insulin infusion (SCHADE and EATON 1980). The rise in plasma glucagon preceeded the rise in plasma glucose by more than 1 h. The rise in free fatty acids and ketone bodies, heralding the deterioration of the metabolic state in these subjects, was preceeded by a rise in catecholamines, cortisol, and growth hormone. Thus, it was suggested that the stress hormones initiate the metabolic deterioration in stressed, diabetic humans.

### b) Hypoglycemia: A Model for Moderate to Severe Stress

The role of the counterregulatory hormones in preventing severe hypoglycemia in diabetic humans was recently clearly demonstrated. It was reported (BODEN et al. 1981), that an insulin-dependent diabetic patient, who was capable of synthesizing and releasing all the major counterregulatory hormones, did not secrete them appropriately in response to insulin-induced hypoglycemia. As a result, this patient had many severe and protracted hypoglycemic episodes, and brain damage resulted. As reviewed in details in Chap. 37, several investigators have examined the mechanisms of recovery from insulin-induced hypoglycemia. In one study in humans, the maximum suppression in glucose production occurred when insulin levels were approximately 60 μIU/ml at an insulin receptor saturation of only 10%, whereas the maximal effect of insulin on glucose utilization occurred at insulin levels between 200 and 700 μIU/ml, at an insulin receptor saturation of 49% (RIZZA et al. 1981).

It was shown, in the rat (SACCA et al. 1977), that insulin-induced hypoglycemia initially decreased glucose production (Ra) and stimulated glucose uptake, but that recovery from hypoglycemia ultimately occurred owing to an increase in Ra which was not attenuated by demedullation plus reserpinization. As glucagon levels were the same in the normal and demedullated–reserpinized animals, it was concluded that glucagon was of greater importance than the catecholamines in the recovery from hypoglycemia. This reasoning, however, cannot be justified without a parallel study which examines the effects of glucagon suppression at a time when catecholamine levels are elevated during hypoglycemia, as will be discussed later in this chapter. It has been suggested that about 75% of hepatic glucose production in the postabsorptive state is glucagon-mediated (CHERRINGTON et al. 1979a). Naturally occurring glucagon deficiency has been reported (VIDNES and OYASAETER 1977), in the neonatal period, which resulted in severe hypoglycemia, suggesting that, as reviewed in Chap. 42, glucagon may be essential to maintain basal glucose production in the neonatal period (UNGER 1981) in humans. This is consistent with the observation that, in postabsorptive neonatal pups, plasma glucagon is three times higher than in adult dogs, and this parallels higher concentrations of glucagon, both in the pancreas and in the stomach of the pups. However, during fasting, plasma glucagon levels decreased paradoxically, and this was consistent with a decrease in gluconeogenesis. Possibly, the neonatal brain in less susceptible to hypoglycemia than the adult brain. Also, the responsiveness of the A-cell to various stimuli is not yet developed. Thus, glucagon did not increase with spontaneous hypoglycemia, and did not decrease in response to somatostatin. It was suggested that the unresponsiveness of the A-cell and decrease of glucagon during fasting might be useful in protecting the newborn from excessive changes in fuel mobilization at a time when fuel sources are scarce (HETENYI et al. 1976).

In studies in humans, the predominent role of glucagon in the recovery from insulin-induced hypoglycemia was again indicated, and it was suggested that the catecholamines assumed importance only during glucagon deficiency (RIZZA et al. 1979b; CHERRINGTON et al. 1979a). It should be emphasized that these observations with respect to hypoglycemia should be restricted to events after an insulin injection. Liver sensitivity to glucagon will be affected first by an inhibitory effect of insulin as it rises in the circulation, whereas, later, the abrupt fall of insulin could increase liver sensitivity to glucagon. This could explain a lesser role for glucagon during sustained hyperinsulinemia (GAUTHIER et al. 1980). Conversely, it was shown, in one study (GARBER et al. 1976b), that only the increase in catecholamines preceded the increase in glycogenolysis and gluconeogenesis from alanine after insulin-induced hypoglycemia. To reconcile these findings, it has been shown that the more profound the hypoglycemia, the greater is the increase in glucose production following hypoglycemia, and that this is associated with a greater catecholamine release (N. J. CHRISTENSEN et al. 1975a). Also, the recovery from hypoglycemia is not impaired by glucagon suppression by somatostatin (N. J. CHRISTENSEN et al. 1975b), and in this study, the hypoglycemia occurring in response to insulin and the ensuing catecholamine release were greater than when somatostatin was not given. Thus, the degree of hypoglycemia may determine the relative role of glucagon versus the catecholamines in the recovery from insulin-induced hypoglycemia. Also, whether insulinemia is transient or sustained may be important.

This was strongly supported by a series of studies from our laboratory. In order to explore the comparative roles of glucagon and epinephrine in the response to a drop in plasma glucose, conscious dogs were studied during the infusion of either phlorhizin or insulin (GAUTHIER et al. 1980). Phlorhizin caused a small decrease in plasma glucose levels from $109 \pm 5$ to $91 \pm 3$ mg/dl (Fig. 10), by increasing renal clearance of glucose. As plasma glucose fell, glucose production rose, and this rise appeared to be mediated entirely by an increase in plasma glucagon, for when somatostatin was infused, plasma IRG and Ra returned to basal, and profound rather than mild hypoglycemia resulted. The slight decrease in plasma insulin probably occurred as a result of the drop in plasma glucose, and hypoinsulinemia presumably increased liver sensitivity to glucagon.

Insulin infusion (7 mIU/$kg^{-1}$ $min^{-1}$) resulted in a profound drop in plasma glucose concentration to $43 \pm 3$ mg/dl (Fig. 11). Although IRG levels rose, and there was an accompanying rise in Ra, the addition of somatostatin to the insulin infusion restored IRG to basal levels, but Ra remained elevated, although a continuing rise in Ra was restrained. Thus, when profound hypoglycemia occurs, and liver sensitivity to glucagon is decreased because of sustained hyperinsulinemia, plasma IRG does not play an important role in increasing hepatic glucose production, whereas small changes about the normal "set point" for plasma glucose result in changes in circulating glucagon which act to control glucose production. During recovery from hypoglycemia, when insulin levels were normal, glucagon's role again became evident, illustrating the importance of insulin in controlling liver sensitivity to glucagon. When plasma glucose falls precipitously, other factors come into play, and glucose production rises, also, in part, because of the effects of the rate of change of glycemia on the liver. One study, in tetraplegic patients (FRIER et al. 1981), revealed a normal recovery from hypoglycemia which suggests that this recovery is not dependent solely on sympathetic innervation. Norepinephrine levels did not rise; however, epinephrine levels were not measured. Plasma glucagon levels rose comparably whether or not the tetraplegic patients were given atropine during insulin-induced hypoglycemia, yet recovery from hypoglycemia was impaired by the cholinergic blockade, and this appeared to be related to a delayed

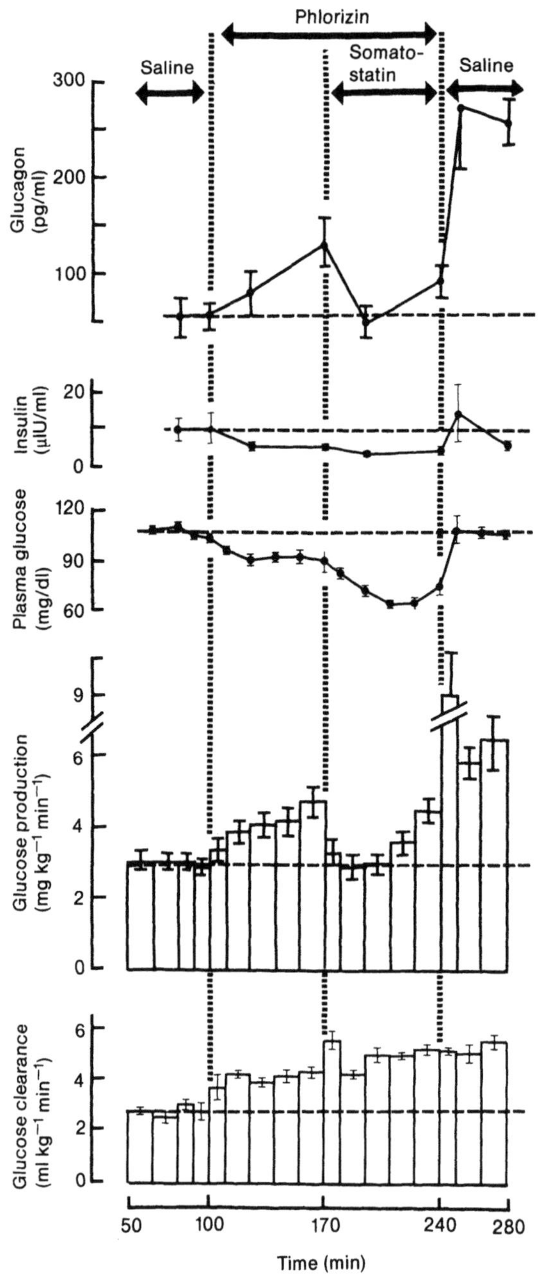


**Fig. 10.** The effects of the infusion of phlorhizin (50 μg $kg^{-1}$ $min^{-1}$), followed by the infusion of phlorhizin plus somatostatin (0.35 μg $kg^{-1}$ $min^{-1}$) on plasma concentrations of immunoreactive glucagon, immunoreactive insulin, and glucose, and on rates of glucose appearance (Ra) and glucose metabolic clearance (MCR), in five normal dogs. *Vertical bars* represent standard error. Adapted from GAUTHIER et al. (1980)

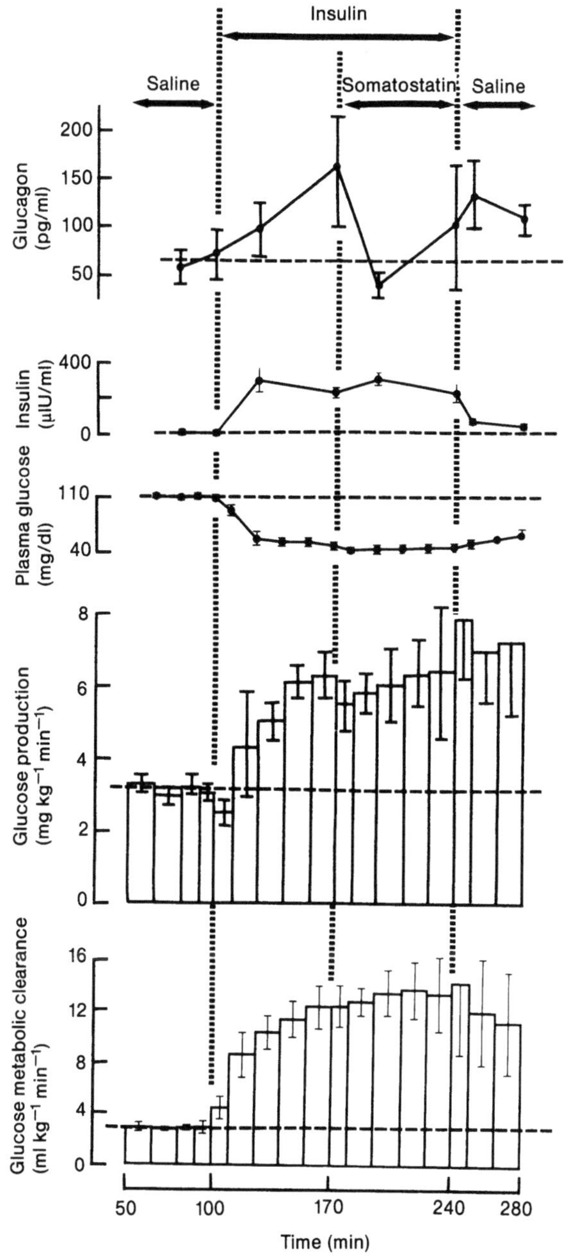


**Fig. 11.** The effects of the infusion of insulin (7 mIU $kg^{-1}$ $min^{-1}$), followed by the infusion of insulin plus somatostatin (0.35 μg $kg^{-1}$ $min^{-1}$) on plasma concentrations of immunoreactive glucagon, immunoreactive insulin, and glucose, and on rates of glucose appearance (Ra) and glucose metabolic clearance (MCR) in three normal dogs. *Vertical bars* represent standard error. Adapted from GAUTHIER et al. (1980)

adrenocorticotropic hormone and attenuated cortisol response. The mechanism of this attenuation, however, needs further exploration, as changes in cortisol do not usually have rapid metabolic effects.

c) Epinephrine Infusion: A Model for Moderate to Severe Stress

Epinephrine infusions simulate many of the hormonal changes which occur in response to stress, and have been extensively used to examine hormonal interactions in glucoregulation. Spontaneous stress situations and epinephrine infusions both result in increases in catecholamines and glucagon and induce a relative suppression of insulin release. The difference is that in normal subjects, the glucagon response to epinephrine is much less marked than in stress situations. We felt that epinephrine–glucagon–insulin interactions must be dissected in order to gain an understanding of the responses to stress.

## 3. Glucagon–Insulin Interactions in Glucoregulation and the Diabetogenic Role of Glucagon

Glucagon–insulin interactions play a predominant role in the acute regulation of glucose metabolism under all physiologic and pathophysiologic conditions, and catecholamines contribute to this interplay only in stress and exercise. Both insulin and glucagon change under conditions of stress, resulting in a relative excess of circulating glucagon and relative deficiency of circulating insulin. In order to analyze these interrelationships systematically, we will examine some glucagon–insulin interactions before proceeding to the more complex relationships between epinephrine and the two major pancreatic hormones. Both glucagon and insulin interact in regulating glycogenolysis and gluconeogenesis in the liver, whereas only insulin is effective in stimulating glucose utilization in peripheral tissues (CHERRINGTON et al. 1974). It has often been suggested that the insulin : glucagon ratio should be used to assess the combined effects of these two hormones. Although unquestionably, a relative excess of glucagon and deficiency of insulin will promote catabolic processes and vice versa, the numerical ratio can give misleading interpretations of metabolic processes, for the metabolic effects of a given ratio change with time after the initiation of a change in circulating insulin and/or glucagon levels. First, the effects of glucagon are transient, and second, when insulin and glucagon change concurrently, the effects of the change in circulating glucagon predominate initially, with the effects of insulin supervening eventually (CHERRINGTON and VRANIC 1974; CHERRINGTON et al. 1974). Third, glucagon and insulin interact differently in the regulation of glycogenolysis and gluconeogenesis respectively (CHERRINGTON et al. 1979b).

Situations which are characterized by, at least, relative hyperglucagonemia and hypoinsulinemia include, in addition to stress, fasting (CAHILL et al. 1966), exercise (KEMMER and VRANIC 1981), diabetes (UNGER et al. 1970), and also pancreatectomy where the glucagon released is of gastric origin (SUTHERLAND and DE DUVE 1948; VRANIC et al. 1974a). In spite of the similar changes in glucagon and insulin in these conditions, the changes in glucoregulation are not uniform. In exercise, we have seen that except when liver glycogen is depleted, both glucose production and utilization increase concurrently and normoglycemia is maintained. With fasting, glucose production decreases more than glucose utilization, and the net result is a fall in plasma glucose levels (see Chap. 34). In naturally occurring untreated diabetes and after pancreatectomy, glucose production increases and utilization in insulin-sensitive tissues is impaired, leading to hyperglycemia (see Chap. 44). Stress usually brings about an increase in glucose production and an associated hyperglycemia, but there can be a variable affect of stress on glucose utilization relating to the underlying cause of and intensity of the stress situation.

The infusion of glucagon in the normal dog (CHERRINGTON et al. 1972; CHERRINGTON and VRANIC 1974; SACCA et al. 1978) results in a minimal and transient increase in plasma glucose. However, the rise in glucose turnover is substantial, but glycemia is restrained by concomitant insulin secretion. If endogenous glucagon secretion is stimulation by arginine infusion (CHERRINGTON et al. 1974; RIZZA et al. 1979a), the rise in glucose production does not lead to hyperglycemia, because of the concomitant stimulation of insulin release.

Insulin–glucagon interactions can be best studied if one can vary the concentration of each hormone independently. Two models have been used to examine insulin–glucagon interactions in glucoregulation:

1. The depancreatized dog, with or without a subcutaneously implanted pancreatic autograft, can be suddenly rendered insulin deficient, either by interrupting the blood supply to the autograft or by maintaining the depancreatized dogs on insulin, then starting an intraportal insulin infusion at the start of an experimental study (VRANIC and WRENSHALL 1968; VRANIC et al. 1971; CHERRINGTON and VRANIC 1974) after which the insulin infusion rate can be manipulated at will (VRANIC et al. 1976a). The glucagon present in the depancreatized dogs is extrapancreatic glucagon secreted by the gastric fundus (VRANIC et al. 1974a; BLASQUEZ et al. 1976; MORITA et al. 1976; MÜLLER et al. 1978) and is identical to pancreatic glucagon in its physiochemical characteristics (MORITA et al. 1976; SRIKANT et al. 1977) and its biologic activity (DOI et al. 1979), as previously discussed.

2. Somatostatin infusion, which suppresses release of both glucagon and insulin (S. E. CHRISTENSEN et al. 1974; KOERKER et al. 1974), can be accompanied by infusion of either hormone to achieve the combination of insulin–glucagon to be scrutinized. Somatostatin studies provided strong evidence for the important role of glucagon in the hyperglycemia of diabetes mellitus (DOBBS et al. 1975; UNGER and ORCI 1975; UNGER 1978), and in the development of diabetic ketoacidosis (GERICH et al. 1975).

### a) Studies in Depancreatized Dogs

The administration of glucagon, in depancreatized dogs, results in hyperglycemia. Insulin infusion rates 12 times basal values were required to restore the responses to glucagon seen in normal dogs (CHERRINGTON and VRANIC 1974). The effects of marked hyperglucagonemia predominated until insulin levels were very high, and even then the effects of insulin were mainly on peripheral glucose uptake. In both normal and diabetic humans (FELIG et al. 1976), the effects of glucagon were seen to be transient, which cannot be attributed to increases in circulating insulin, as it

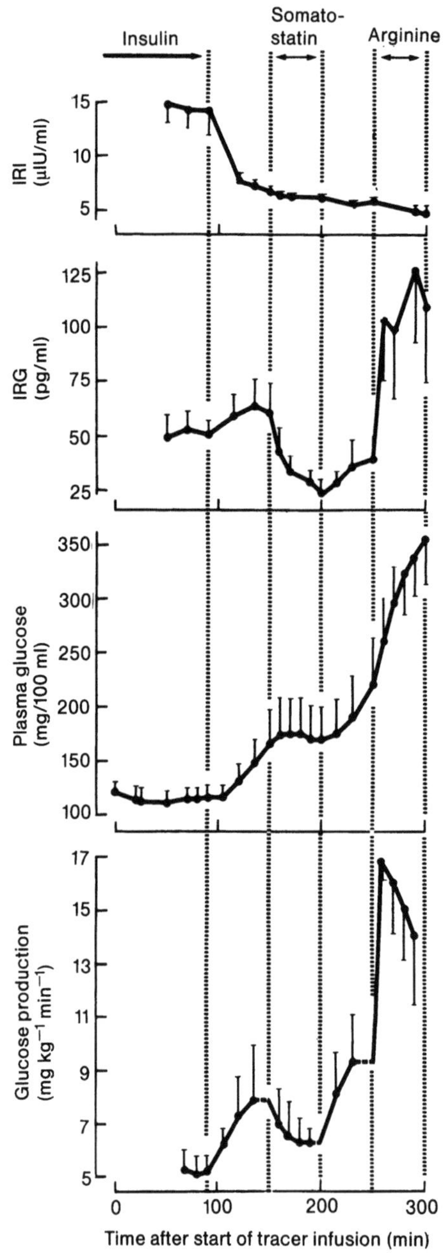


**Fig. 12.** Diabetogenic role of glucagon during acute insulin deficiency. Four depancreatized dogs were initially maintained normoglycemic with intraportal insulin (200–250 μIU $kg^{-1}$ $min^{-1}$). After insulin withdrawal, somatostatin (0.83 μg $kg^{-1}$ $min^{-1}$) was given, followed by a further period of insulin withdrawal, then arginine (12.5 mg $kg^{-1}$ $min^{-1}$) was administered. The plasma concentrations of immunoreactive insulin, immunoreactive glucagon, and glucose, and rates of glucose production are illustrated. *Vertical bars* represent standard error. Adapted from Ross et al. (1978)

**Table 1.** Mean basal values ± standard error and mean values ± standard error at point of maximal change from basal, during glucagon suppression by somatostatin

| | Alloxan-induced diabetic dogs | | Depancreatized insulin-deprived dogs | |
|---|---|---|---|---|
| | Basal | Value at point of maximal change | Basal | Value at point of maximal change |
| Plasma immunoreactive glucagon concentration (pg/ml) | 172 ± 7 | 61 ±27 | 271 ±48 | 43 ± 6 |
| Plasma immunoreactive insulin concentration (μIU/ml) | 5.6± 0.7 | 4.6± 0.6 | Unmeasurable | |
| Plasma glucose concentration (mg/dl) | 274 ±20 | 161 ±20 | 336 ± 7 | |
| Rate of glucose production (mg kg$^{-1}$ min$^{-1}$) | 5.9± 0.5 | 3.3± 0.4 | 8.7± 1.0 | 7.9± 1.1 |

was also observed in the depancreatized dogs with constant basal intraportal insulin infusions (CHERRINGTON and VRANIC 1974).

In recent studies in our laboratory, we have examined the effects of suppression and stimulation of extrapancreatic glucagon in depancreatized dogs (ROSS et al. 1978). As shown in Fig. 12, it can be seen that insulin withdrawal resulted in a sharp drop in plasma insulin and a slight increase in extrapancreatic IRG. There was also an increase in Ra and plasma glucose. The infusion of somatostatin suppressed IRG well below basal, Ra fell, and the rise in plasma glucose was curtailed. Arginine induced a marked hyperglucagonemia, and Ra and plasma glucose rose rapidly to extremely high values. Thus, in these depancreatized insulinopenic dogs, glucagon played a major role in stimulating glucose production, and contributing thus to the acute development of diabetes.

In a further series of experiments, we compared the effects of glucagon suppression in depancreatized dogs deprived of insulin for 3 days and in alloxan-induced diabetic dogs (LICKLEY et al. 1981 a). As shown in Table 1, both groups of animals had marked basal hyperglycemia and hyperglucagonemia. These depancreatized dogs were virtually totally insulin deficient, whereas the alloxan-induced diabetic dogs had low, but measurable plasma insulin concentrations. With the administration of somatostatin, there was a sustained decrease in plasma glucagon levels to near the sensitivity of the assay in both depancreatized and alloxan-induced diabetic animals. In the alloxan-induced diabetic dogs, glucagon suppression resulted in a profound and sustained drop in plasma glucose levels initially due to a 45% decrease in glucose production. However, in the depancreatized dogs, despite equivalent glucagon suppression, there was no significant improvement in glycemia. Thus, glucagon supression was effective in improving glucose homeostasis under conditions of partial, but not total, insulin deficiency.

These observations would indicate that: (1) glucagon suppression is antidiabetogenic only in the presence of some residual insulin; and (2) amelioration of gly-

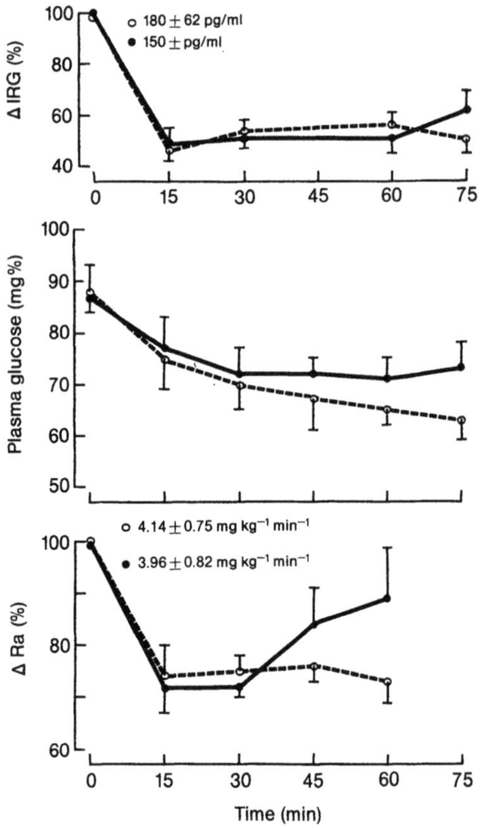


**Fig. 13.** The effect of somatostatin (0.5 μg $kg^{-1}$ $min^{-1}$, *full lines*) and somatostatin together with basal insulin replacement (200 μIU $kg^{-1}$ $min^{-1}$, *broken lines*) on plasma immunoreactive insulin (IRI) concentrations, change (*Δ*) in plasma immunoreactive glucagon (IRG) concentrations, plasma glucose concentrations, and change in rate of glucose production or appearance (Ra), in five normal dogs. *Vertical bars* represent standard error. Basal values for IRG and Ra are indicated. Adapted from LICKLEY et al. (1979)

cemia results only from decreased glucose production by the liver, and is accompanied by a further reduction of peripheral glucose utilization. This led to the hypothesis that suppression of glucagon in diabetic conditions, including those complicated by stress, should be contemplated as an adjuvant to the therapy of diabetes only when peripheral insulin concentrations are normal, and thus normal peripheral glucose metabolism is secured. Glucagon suppression might be desirable in insulin-treated patients, because it appears that conventional insulin treatment results in peripheral hyperinsulinemia. Peripheral hyperinsulinemia is presumably necessary to maintain adequate portal insulin levels when insulin is administered through a peripheral route. We hypothesize that partial glucagon suppression would sensitize the liver to insulin and thus a functional portal–peripheral insulin gradient would be established, and glucose homeostasis could therefore be maintained without elevated peripheral insulin levels (VRANIC et al. 1976b).

### b) Studies During Somatostatin Infusion

We have examined insulin–glucagon interactions in normal dogs under conditions of combined insulin and glucagon deficiency (LICKLEY et al. 1979). As shown in Fig. 13, both plasma insulin (IRI) and glucagon (IRG) levels fell during the infusion of somatostatin alone. There was an acute, but transient, decrease in Ra which led to a small decrease in plasma glucose concentrations, despite marked suppression of insulin release. Thus, glucagon lack prevailed over insulin lack, and Ra fell as has also been shown by other investigators (ALTSZULER et al. 1976; CHERRINGTON et al. 1976). It has also been shown that, during prolonged infusion of somatostatin, hyperglycemia will ultimately occur as insulin deficiency impairs glucose utilization and returns Ra toward normal (SHERWIN et al. 1977; CHERRINGTON et al. 1978).

During selective glucagon deficiency (Fig. 13), when basal insulin levels were provided during somatostatin infusion, the decrease in Ra was sustained rather than transient (CHERRINGTON et al. 1976; LICKLEY et al. 1979). Thus, glucagon is required to maintain basal glucose production, and is important in regulating glucose production about its normal set point.

During selective insulin deficiency, when glucagon was infused together with somatostatin to restore normal basal glucagon levels (CHERRINGTON et al. 1976, 1978), plasma glucose rose with an accompanying rise in Ra. When hyperglucagonemia was induced during insulin suppression by somatostatin (LICKLEY et al. 1979), glucose production and glycemia both increased to a greater extent (Fig. 14). This simulates the glucagon–insulin interactions occurring under stress situations, and indicates that glucagon can precipitate the acute onset of diabetes under conditions of hypoinsulinemia.

## 4. Glucagon–Insulin–Epinephrine Interactions in Glucoregulation

Epinephrine has been shown to increase hepatic glycogenolysis in vitro (SOKAL et al. 1964; CHERRINGTON and EXTON 1976) and to cause sustained hyperglycemia, owing to a transient increase in glucose production and a sustained failure to increase glucose utilization (ALTSZULER et al. 1967). Later studies have shown that the increase in hepatic glucose production and decrease in glucose clearance are directly mediated by $\beta$-adrenergic mechanisms and possibly also indirectly through $\alpha$-adrenergic inhibition of insulin release (RIZZA et al. 1980a, b). It should be noted that the effects of combined infusion of epinephrine, glucagon, and cortisol act to increase glucose production to a greater extent than the additive effect of the individual hormones. Thus, the hyperglycemia of stress may reflect synergism among the stress-related hormones (SHAMOON et al. 1980, 1981). It should be noted, however, that a synergistic effect of elevated cortisol levels becomes apparent only after 3 h cortisol infusion.

Epinephrine's effects can take place even in the absence of changes in insulin or glucagon secretion (RIZZA et al. 1980a; WOODSON et al. 1980). One study in normal baboons, however (CHIDECKEL et al. 1977), suggested that epinephrine-induced glucagon release also made a major contribution to the effects of epineph-

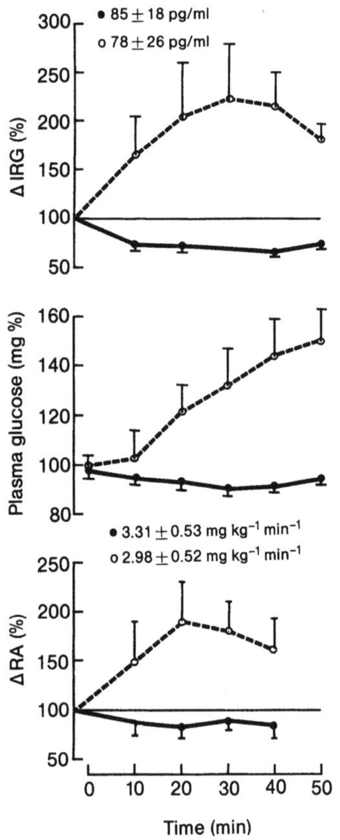


**Fig. 14.** The effects of somatostatin (0.8 μg $kg^{-1}$ $min^{-1}$, *full lines*) and somatostatin together with glucagon (0.13–0.43 μg $kg^{-1}$ $h^{-1}$, *broken lines*) on plasma immunoreactive insulin (IRI) concentrations, change (*Δ*) in serum immunoreactive glucagon (IRG) concentrations, plasma glucose concentrations, and change in rate of glucose production or appearance (Ra), in five normal dogs. *Vertical bars* represent standard error. Basal values for IRG and Ra are indicated. Adapted from LICKLEY et al. (1979)

rine on carbohydrate metabolism. However, in normal dogs (ALTSZULER et al. 1967; SACCA et al. 1979a), and humans (RIZZA et al. 1979c, 1980a, b) epinephrine stimulated hepatic glucose production and this could occur whether or not glucagon and insulin were suppressed by somatostatin (RIZZA et al. 1979a, 1980a), indicating that epinephrine can act independently of glucagon. In insulin-dependent diabetic subjects, however, epinephrine stimulated glucagon release which was shown to account for approximately one-half of the epinephrine-induced hyperglycemia (GERICH et al. 1976). The response to epinephrine in juvenile-onset diabetics was also more prolonged than that in normal subjects (SHAMOON et al. 1980). In order to assess this possible role of glucagon in epinephrine-induced hyperglycemia, we have examined the responses to epinephrine infusion with or without concomitant glucagon suppression by somatostatin, in normal, alloxan-induced diabetic, and depancreatized dogs.

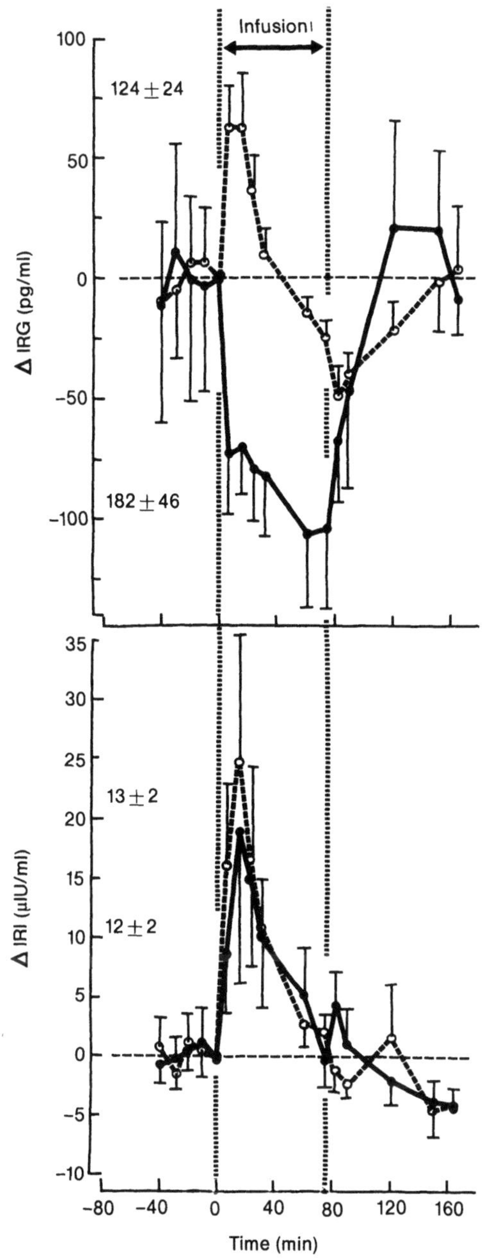


**Fig. 15.** The effects of epinephrine alone (0.1 μg $kg^{-1}$ $min^{-1}$, *broken lines*) or with somatostatin (0.05 μg $kg^{-1}$ $min^{-1}$, *full lines*), in normal dogs. Plasma immunoreactive glucagon (IRG) and insulin (IRI) levels are shown as deviations ($\Delta$) from the average control values indicated. The infusions were begun at $t=0$ after a 120-min equilibration period. Values are expressed as mean ± standard error. $N=6$ for infusion of epinephrine alone and $N=5$ for infusion of epinephrine plus somatostatin. Adapted from GRAY et al. (1980)

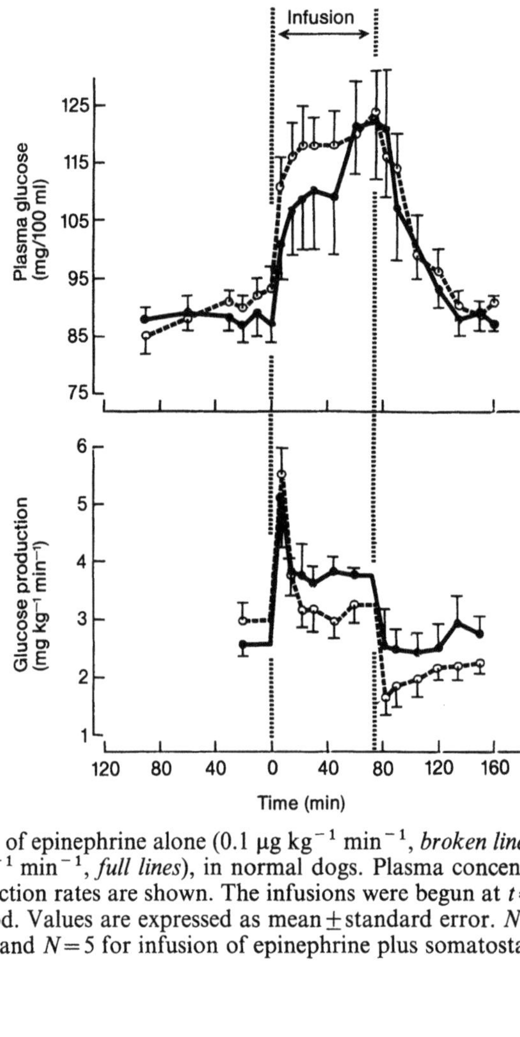


**Fig. 16.** The effects of epinephrine alone (0.1 μg $kg^{-1}$ $min^{-1}$, *broken lines*) or with somatostatin (0.05 μg $kg^{-1}$ $min^{-1}$, *full lines*), in normal dogs. Plasma concentrations of glucose and glucose production rates are shown. The infusions were begun at $t=0$ after a 120-min equilibration period. Values are expressed as mean ± standard error. $N=6$ for infusion of epinephrine alone and $N=5$ for infusion of epinephrine plus somatostatin. Adapted from GRAY et al. (1980)

### a) Epinephrine Infusion in Normal Dogs

The infusion rate of epinephrine (0.1 μg $kg^{-1}$ $min^{-1}$) induced a significant, but transient increase in plasma IRG (Fig. 15). When somatostatin (0.05 μg $kg^{-1}$ $min^{-1}$) was infused together with epinephrine, plasma glucagon levels were suppressed well below basal. Epinephrine also induced a transient rise in plasma IRI. Plasma IRI levels returned toward basal whether or not somatostatin was added to the epinephrine infusion, in spite of prevailing hyperglycemia, but never fell below basal. The low rate of infusion of somatostatin given, although capable of suppressing insulin release by 50% when given alone, did not significantly alter the insulin response to epinephrine, possibly because of as yet unclarified interactions between epinephrine and somatostatin at the level of the pancreatic B-cells. Thus, this combination of epinephrine and somatostatin enabled us to examine the effects of epinephrine during selective glucagon suppression. Glucose production increased significantly, but transiently, during the epinephrine infusions, but when somatostatin was given together with epinephrine, there was a similar maximum increase in Ra which was even more sustained than that seen with the infusion of epinephrine alone (Fig. 16). The infusion of epinephrine alone or together with somatostatin provided an equivalent glycemic response in these animals. There was also a slight decrease in the metabolic clearance rate of glucose when epinephrine was infused, with or without somatostatin. Thus, it appears that epinephrine can exert its full hyperglycemic effect independently of glucagon in the normal dog.

### b) Epinephrine Infusion in Alloxan-Induced Diabetic Dogs

Although epinephrine appears to produce its hyperglycemic effect independently of glucagon, in normal subjects, there is evidence that glucagon plays a major role in mediating the effects of epinephrine on the liver in diabetics (GERICH et al. 1976). We examined tracer-determined glucose kinetics in alloxan-induced diabetic dogs (PEREZ et al. 1981), which were infused with epinephrine (0.1 $\mu g\ kg^{-1}\ min^{-1}$) or epinephrine plus somatostatin (0.1 $\mu g\ kg^{-1}\ min^{-1}$), and in order to assure that the effects of somatostatin were due to somatostatin-induced glucagon suppression, and not to a direct effect of somatostatin on epinephrine-induced hepatic glucose overproduction (SACCA et al. 1979a), another group of alloxan-induced diabetic dogs received an infusion of epinephrine together with somatostatin plus glucagon (10 $ng\ kg^{-1}\ min^{-1}$). Basal peripheral insulin levels were low, and increased only slightly with epinephrine infusion, but remained at basal values when epinephrine was infused with somatostatin or somatostatin plus glucagon. As shown in Fig. 17, epinephrine infusion resulted in a 3–4-fold increase in plasma epinephrine concentrations. The increments in plasma epinephrine were not significantly different whether epinephrine was infused alone or together with somatostatin. There was a maximal increase to nearly five times basal in plasma glucagon during epinephrine infusion. The addition of somatostatin to the epinephrine infusion not only prevented the rise in plasma glucagon, but also induced a small decrease in circulating IRG. When glucagon was added to the infusion of epinephrine and somatostatin, the hyperglucagonemia attained was slightly greater than that noted during the infusion of epinephrine alone.

The infusion of epinephrine resulted in a 120% increase in Ra by 20 min (Fig. 17). Thereafter, Ra declined gradually, but was still 60% above basal by the end of the infusion. These changes in Ra resulted in a large and sustained increase in plasma glucose concentrations. In contrast, when somatostatin was given together with epinephrine, there was only a marginal change in Ra, and in plasma glucose levels. When hyperglucagonemia was restored by infusing glucagon together with epinephrine and somatostatin, the increases in Ra and plasma glucose were reestablished. Glucose clearance was unchanged during all three infusions. The apparent lack of suppression of glucose clearance by epinephrine is probably related to the increased renal clearance of glucose occasioned by the prevailing hyperglycemia. As renal clearance is included in the measurements of total glucose clearance, this factor may have obscured a decrease in the metabolic clearance of glucose. Since the curves for changes in glucose production in the animals receiving epinephrine alone and those given epinephrine together with somatostatin were

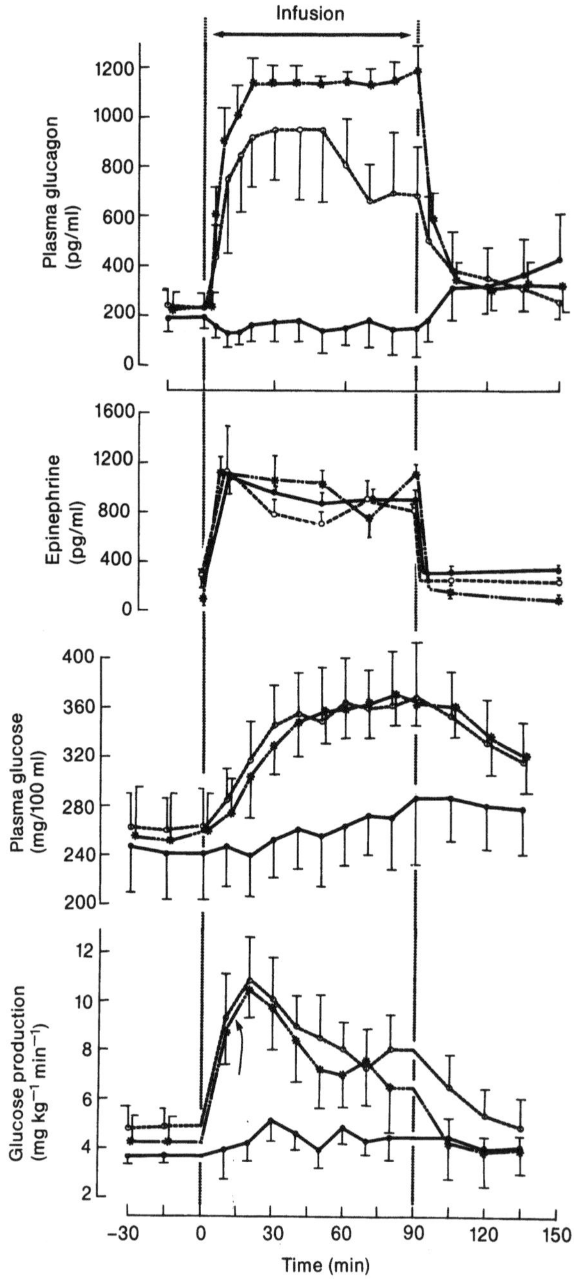


**Fig. 17.** The effect of epinephrine alone (0.1 μg $kg^{-1}$ $min^{-1}$, *broken line*), or epinephrine with somatostatin (0.1 μg $kg^{-1}$ $min^{-1}$, *full line*), or epinephrine with somatostatin plus glucagon (10 ng $kg^{-1}$ $min^{-1}$, *dash-dotted line*) on plasma immunoreactive glucagon, plasma epinephrine, and glucose concentrations, and rates of production of glucose in five alloxan-induced diabetic dogs. Values are expressed as mean ± standard error. Adapted from PEREZ et al. (1981)

virtually superimposable, the suggestion that somatostatin may sensitize the liver to epinephrine, as in normal dogs (SACCA et al. 1979a), was not confirmed in these diabetic dogs. Clearly, these studies suggest that, in contrast to the minimal role glucagon may play in the response to stress in normal subjects, it may well participate to a major extent in the stress-related deterioration of diabetes. Thus, both hyperglucagonemia and hypoinsulinemia may act in concert to increase hepatic sensitivity to epinephrine in diabetic subjects.

### c) Epinephrine Infusion in Depancreatized Dogs in Good or Poor Metabolic Control

Glucagon was shown to participate in epinephrine-induced hepatic glucose overproduction in alloxan-induced diabetic dogs (PEREZ et al., 1981), and it also contributes to epinephrine-induced hyperglycemia in diabetic humans (GERICH et al. 1976), but not in normal humans (RIZZA et al. 1979c) or dogs (GRAY et al. 1980). We therefore entertained the hypothesis that the degree of diabetic control might be pivotal in determining whether or not glucagon plays a major role in mediating the effects of epinephrine on glucose production. We examined this in depancreatized dogs which are capable of secreting extrapancreatic immunoreactive glucagon (eIRG) from the gastric fundus as previously described (VRANIC et al. 1974a; BLASQUEZ et al. 1976; SRIKANT et al. 1977; MÜLLER et al. 1978), and can release eIRG in response to some, but not all, glucagonatropic stimuli (ROSS et al. 1978).

The depancreatized dogs were studied under conditions of good metabolic control (normoglycemia, $95 \pm 5$ mg/dl) or poor metabolic control (hyperglycemia, $212 \pm 10$ mg/dl). The desired level of glycemia was achieved by means of constant intraportal basal or subbasal insulin infusions. Epinephrine (0.1 $\mu g\ kg^{-1}\ min^{-1}$) infusions were given to the animals, with or without somatostatin (0.1 $\mu g\ kg^{-1}\ min^{-1}$) and these studies were carried out on four separate days (KEMMER et al. 1982). When the animals were studied under normoglycemic conditions (Fig. 18), epinephrine induced a modest rise in eIRG, comparable to the glucagon response to epinephrine in normal dogs (GRAY et al. 1980), but considerably smaller than that seen in the alloxan-induced diabetic dogs (PEREZ et al. 1981). The rise in eIRG was abolished when somatostatin was given together with epinephrine. There was an epinephrine-induced increase in glucose production which was associated with a rise in plasma glucose concentration, and these changes were not attenuated by somatostatin.

When the animals were studied under hyperglycemic conditions (Fig. 19), the rise in eIRG was similar to that seen when the animals were given epinephrine under conditions of normoglycemia. The epinephrine-induced rise in eIRG was prevented by somatostatin, and eIRG levels actually fell slightly below basal. The striking feature in this study was the considerably greater rise in glucose production and plasma glucose levels in response to epinephrine when the animals were studied under hyperglycemic conditions than was seen under normoglycemic conditions. When somatostatin was added to epinephrine, when hyperglycemia prevailed, the rise in Ra and plasma glucose was equivalent to the rise seen under normoglycemic conditions with the infusions of either epinephrine or epinephrine plus somatostatin.

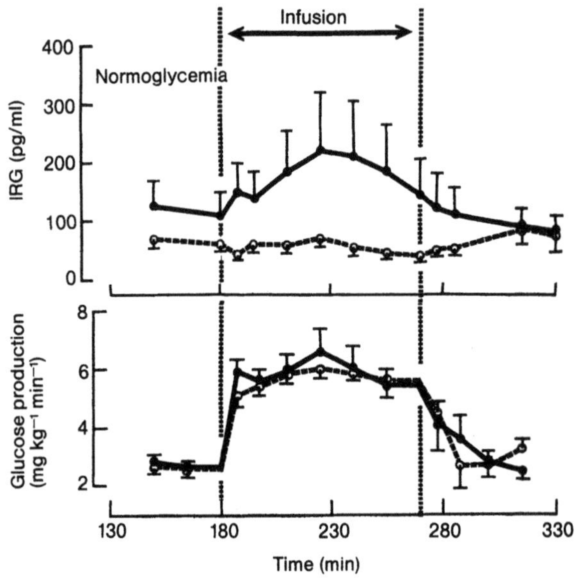


**Fig. 18.** The effects of epinephrine alone (0.1 μg $kg^{-1}$ $min^{-1}$, *full line*) or epinephrine plus somatostatin (0.1 μg $kg^{-1}$ $min^{-1}$, *broken line*), in six depancreatized dogs, studied under conditions of normoglycemia, on plasma glucagon concentrations and rates of glucose production. Values are expressed as mean ± standard error. Adapted from KEMMER et al. (1982)

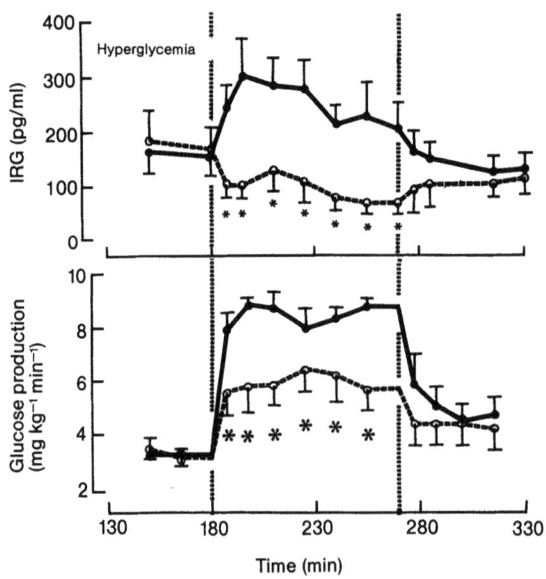


**Fig. 19.** The effects of epinephrine alone (0.1 μg $kg^{-1}$ $min^{-1}$, *full line*) or epinephrine plus somatostatin (0.1 μg $kg^{-1}$ $min^{-1}$, *broken line*), in six depancreatized dogs, studied under conditions of hyperglycemia, on plasma glucagon concentrations and rates of glucose production. Significant differences in the responses to epinephrine or epinephrine plus somatostatin are represented by *asterisks*. Values are expressed as mean ± standard error. Adapted from KEMMER et al. (1982)

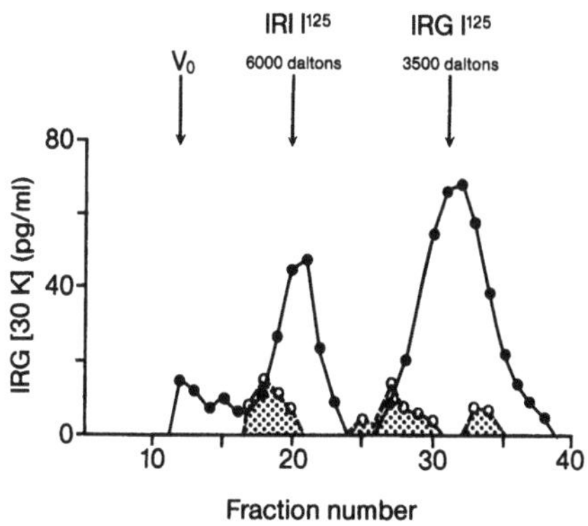


**Fig. 20.** Chromatographic profile on Biogel P-30 of the 30 K immunoreactivity in the plasma from depancreatized dogs under basal conditions (*shaded areas*) and after 15 min epinephrine infusion (0.2 μg $kg^{-1}$ $min^{-1}$,*full line*). *Arrows* indicate the void volume ($V_0$) and the elution values of the radioactive marker insulin $^{125}I$ (IRI $^{125}I$) of molecular weight 6,000 daltons and glucagon $^{125}I$ (IRG $^{125}I$) of molecular weight 3,500 daltons. Adapted from LICKLEY et al. (1981 b)

As there was no exaggerated release of glucagon under conditions of hyperglycemia in the depancreatized dogs, it appears that the gastric A-cell was less sensitive to hypoinsulinemia than the pancreatic A-cell of the alloxan-induced diabetic dogs. Thus, although glucagon was released in response to epinephrine under conditions of both good and poor metabolic control, it was only in the latter situation that glucagon was shown to contribute to epinephrine-induced hepatic glucose overproduction and hyperglycemia. Thus, the state of metabolic control does govern glucagon's participation in the responses to epinephrine. The two important characteristics of poor metabolic control are hyperglycemia and hypoinsulinemia. Hyperglycemia per se acts to decrease Ra (BERGMAN 1977), and thus cannot act to enhance the effectiveness of glucagon on the liver during epinephrine infusion. However, selective insulin deficiency has been shown to sensitize the liver to the effects of glucagon (ALTSZULER et al. 1976; CHERRINGTON et al. 1978; LICKLEY et al. 1979). It is therefore conceivable that hypoinsulinemia is the major factor acting to enhance the liver's sensitivity to glucagon during epinephrine infusion as well. Thus, even small increments in glucagon may be of major importance whenever hepatic sensitivity is increased by hypoinsulinemia (KEMMER et al. 1982).

In order to characterize the eIRG released in response to epinephrine in depancreatized dogs, plasma samples were taken before and during epinephrine infusion and subjected to fractionation using chromatography, and the eluates were assayed for IRG using the 30 *K* glucagon antiserum (LICKLEY et al. 1981 b). The chromatographic profile showed that the form of eIRG which was released in response to epinephrine, was, as in normal dogs, an eIRG of molecular weight 3,500 daltons, and a smaller peak of 9,000 daltons was also noted (Fig. 20).

## 5. Hormonal Regulation of "Futile Cycling" in the Liver

Alloxan-induced diabetes, administration of pharmacologic doses of glucagon, and exercise are all characterized by excess circulating glucagon and relative or absolute insulin deficiency, and have been shown to increase "futile cycling" in the liver (SHAW et al. 1976; ISSEKUTZ 1977). "Futile cycling" has been defined in Sect. A. It has been suggested that the increase in "futile cycling" mainly reflects the increased activity of glucose-6-phosphatase in the liver (ISSEKUTZ 1977). Since increases in "futile cycling" and increases in glucose production are directly correlated under many metabolic situations (ISSEKUTZ 1977), it is conceivable that this "futile cycling" may play a part in sensitizing the liver to glycogenolytic and gluconeogenetic drives such as encountered in stress, exercise, and diabetes. The importance of interactions between insulin and glucagon in this was suggested by the fact that combined suppression of insulin and glucagon did not affect "futile cycling" (LICKLEY et al. 1979), but relative insulin deficiency due to either infusion of mannoheptulose (ISSEKUTZ 1977) or to combined infusion of somatostatin and glucagon, increased "futile cycling" substantially (LICKLEY et al. 1979).

"Futile cycling" accounts for 48% of measured glucose production in well-controlled depancreatized dogs (VRANIC et al. 1981) as opposed to the 13%–20% previously reported in normal animals (ISSEKUTZ et al. 1977; LICKLEY et al. 1979). When the depancreatized dogs were studied under conditions of hyperglycemia and hypoinsulinemia, "futile cycling" increased to account for 60% of measured glucose production (VRANIC et al. 1981). When Ra was maximally stimulated by endogenous hyperglucagonemia, "futile cycling" increased even further (VRANIC et al. 1981), indicating that both insulin deficiency and glucagon excess can play a role, and that the increase in "futile cycling" might be a relevant factor in the glucoregulatory responses to stress. This is supported by evidence that pretreatment with prednisolone is the most potent potentiator of "futile cycling", both in the basal state and when dogs were subjected to hyperglucagonemia, hypoinsulinemia, or exercise (ISSEKUTZ 1977), and increased glucocorticoid activity is a prominent feature of stress.

## D. Summary

The energy requirements of working muscle are provided initially by glycogenolysis in muscle, then as physical activity proceeds, mainly from circulating glucose and ultimately circulating free fatty acids. Glycemia is maintained constant because of a precisely coordinated balance between glucose uptake by muscle and glucose production by the liver. Insulin, the catecholamines, and glucagon interact in the control of glucoregulation during exercise. Insulin decreases and the catecholamines rise during exercise, and glucagon also rises when exercise is strenuous and prolonged. Although a rise in plasma glucagon does not appear to be essential for increased hepatic glucose production during exercise, the presence of glucagon does appear to be necessary. The prompt decrease in plasma insulin levels during exercise not only acts to increase hepatic glucose production, but may also facilitate the glycogenolytic effect of glucagon and the catecholamines.

In insulin-dependent diabetics, the state of insulinization determines whether exercise leads to an improvement or deterioration of the metabolic state. If insulin levels are insufficient, hepatic glucose production increases and peripheral glucose uptake decreases leading to hyperglycemia. Also, the effect of the catecholamines and glucagon may be accentuated by hypoinsulinemia, and contribute to deterioration of the metabolic state. If hyperinsulinemia occurs because of excessive administration of insulin, lack of ability to control insulin delivery to peripheral tissues, or increased absorption of insulin from the injection site during exercise, hepatic glucose production does not rise to match the increased glucose utilization in working muscle and plasma glucose levels fall. Thus, unless overt hypoglycemia ensues, exercise exerts a beneficial effect on glucoregulation if sufficient circulating insulin is present.

Energy requirements during stress are provided mainly be circulating glucose and free fatty acids. Hyperglycemia is a prominent feature of stress, and is initiated by increased hepatic glycogenolysis and gluconeogenesis and maintained by a sustained impairment in glucose clearance. Again, as during exercise, insulin, the catecholamines, and glucagon interact in the control of glucoregulation during stress. Stress in characterized by increased sympathoadrenal activity and increased circulating catecholamine levels. There is a relative suppression of insulin secretion during stress which represents a balance between $\alpha$-adrenergic suppression and $\beta$-adrenergic stimulation of insulin secretion by the catecholamines. Hyperglucagonemia is also a feature of stress which also occurs mainly as a result of $\beta$-adrenergic stimulation. During mild hypoglycemia, glucagon appears to be the main determinant for increasing hepatic glucose production and restoring normoglycemia. However during more profound hypoglycemia, glucagon does not play a major role in mediating the restoration of normoglycemia. The return of plasma glucose levels toward normal probably occurs as a result of catecholamine release.

During epinephrine infusion, which simulates many of the hormonal responses to stress, glucagon does not play a major role in epinephrine-induced hyperglycemia in normal animals, but participates to a major degree in diabetes, particularly when the diabetic state is out of control. This can take place either because of an exaggerated release of glucagon or because of increased liver sensitivity to glucagon and possibly also to the catecholamines. This increased liver sensitivity is induced by portal hypoinsulinemia. Thus, glucagon suppression might prove to be effective in countering the diabetic instability of stress, but is unlikely to alter the responses to stress when normoglycemia prevails.

*Acknowledgements.* The work reported in this chapter has been supported by the Medical Research Council of Canada, The Canadian Diabetes Association, Women's College Hospital Research Fund, and Bayer Pharmaceuticals, Wuppertal, West Germany. Dr. KEMMER was a recipient of a postdoctoral fellowship from the Deutsche Forschungsgemeinschaft, West Germany, which he took up in the Department of Physiology, University of Toronto. Somatostatin was generously supplied by Ayerst Company, Montreal, Canada and by Dr. P. BRAZEAU at the Salk Institute, La Jolla, California. We are grateful to DONNA WILSON, LYN COOK, and DEBRA BILINSKI for their help in the preparation of the manuscript. Mr. DAVID WASSERMAN is a graduate student in the Department of Physiology, University of Toronto.

## References

Åstrand PO, Rodahl K (1970) Textbook of work physiology. McGraw-Hill, New York

Ahlborg G, Felig P, Hagenfeldt L, Hendler R, Wahren J (1974) Substrate turnover during prolonged exercise in man. J Clin Invest 39:1080–1090

Allsop JR, Wolfe RR, Burke JF (1978) Glucose kinetics and responsiveness to insulin in the rat injured by burn. Surg Gynecol Obstet 147:565–573

Altszuler N, Steele R, Rathgeb I, DeBodo RC (1967) Glucose metabolism and plasma insulin levels during epinephrine infusion in the dog. Am J Physiol 212:677–682
Altszuler N, Barkai A, Bjarknes L, Gottlieb B, Steele R (1975) Glucose turnover values in the dog obtained with various species of labeled glucose. Am J Physiol 229:1662–1669
Altszuler N, Gottlieb B, Hamshire J (1976) Interaction of somatostatin, glucagon and insulin on hepatic glucose output in the normal dog. Diabetes 25:116–121
Atkinson RL, Dahms WT, Bray GA, Sperling MA (1981) Adrenergic modulation of glucagon and insulin secretion in obese and lean humans. Horm Metab Res 13:249–253
Berger M, Hagg SA, Ruderman NB (1975) Glucose metabolism in perfused skeletal muscle. Interaction of insulin and exercise on glucose uptake. Biochem J 146:231–238
Berger M, Berchtold P, Cüppers HJ, Drost H, Kley HK, Müller WA, Wiegelmann W, Zimmermann-Telschow H, Gries FA, Krüskemper HL, Zimmermann H (1977) Metabolic and hormonal effects of muscular exercise in juvenile type diabetes. Diabetologia 13:355–365
Berger M, Halban PA, Müller WA, Offord RE, Vranic M, Renold AE (1978a) Mobilization of subcutaneously injected tritiated insulin in rats: effects of muscular exercise. Diabetologia 15:133–140
Berger M, Kemmer FW, Goodman MN, Zimmermann-Telschow H, Ruderman NB (1978b) Ketone body metabolism in isolated perfused muscle in various metabolic states. In: Söling HD, Seufert CD (eds) Biochemical and clinical aspects of ketone body metabolism. Thieme, Stuttgart, pp 193–203
Bergman RN (1977) Integrated control of hepatic glucose metabolism in the dog. Ann NY Acad Sci 148:441–468
Bergström J, Hultmann E, Saltin B (1973) Muscle glycogen consumption during cross country skiing (the Vasa ski race). Int Z Angew Physiol 31:71–75
Björkman O, Felig P, Hagenfeldt L, Wahren J (1981) Influence of hypoglucagonemia on splanchnic glucose output during leg exercise in man. Clin Physiol 1:43–57
Blasquez E, Muñoz-Barragan L, Patton GS, Orci L, Dobbs RE, Unger RH (1976) Gastric A-cell function in insulin deprived depancreatized dogs. Endocrinology 99:1182–1188
Bloom SR, Vaughan NJA, Russell RCG (1974) Vagal control of glucagon secretion in man. Lancet 2:546–549
Boden G, Reichard GA Jr, Hoeldtke RD, Rezvani J, Owen OE (1981) Severe insulin-induced hypoglycemia associated with deficiencies in the release of counterregulatory hormones. N Engl J Med 305:1200–1205
Böttger I, Schlein EM, Faloona GR, Knochel JP, Unger RH (1972) The effect of exercise on glugacon secretion. J Clin Endocrinol Metab 35:117–125
Bürger M, Kramer M (1928) Über die durch Muskelarbeit hervorgerufene Steigerung der Insulinwirkung auf den Blutzuckergehalt beim normalen und gestörten Kohlenhydratstoffwechsel und ihre praktische und theoretische Bedeutung. Klin Wochenschr 7:745–750
Burke JF, Wolfe RR, Mullany CJ, Mathews DE, Bier DM (1979) Glucose requirements following burn injury. Ann Surg 190:274–285
Cahill GR Jr, Herrera MG, Morgan AP, Soeldner JS, Steinke J, Levy PL, Reichard GA Jr, Kipnis DM (1966) Hormone fuel interrelationships during fasting. J Clin Invest 45:1751–1769
Cannon WB, McIver MA, Bliss SW (1924) Studies on the conditions of activity in endocrine glands. XIII. A sympathetic and adrenal mechanism for mobilizing sugar in hypoglycemia. Am J Physiol 69:46–66
Chan TM, Exton JH (1978) Studies on α-adrenergic activation of hepatic glucose output. J Biol Chem 253:6393–6400
Chaveau MA, Kaufmann M (1887) Experiences pour la détermination du coefficient de l'activité nutritive et respiratoire des muscles en repos et en travail. C R Acad Sci [D] (Paris) 104:1126–1132
Cherrington AD, Exton JH (1976) Studies on the role of cAMP-dependent protein kinase in the action of glucagon and catecholamines on liver glycogen metabolism. Metabolism 25 [Suppl 1]:1351–1354
Cherrington A, Vranic M (1974) Effect of interaction between insulin and glucagon on glucose turnover and FFA concentration in normal and depancreatized dogs. Metabolism 23:729–744

Cherrington A, Vranic M, Fono P, Kovacevic N (1972) Effect of glucagon on glucose turnover and plasma free fatty acids in depancreatized dogs maintained on matched insulin infusion. Can J Physiol Pharmacol 50:946–954
Cherrington AD, Kawamori R, Pek S, Vranic M (1974) Arginine infusion in dogs: model for the roles of insulin and glucagon in regulating glucose turnover and free fatty acid levels. Diabetes 23:805–815
Cherrington AD, Chiasson JL, Liljenquist JE, Jennings AS, Keller K, Lacy WW (1976) The role of glucagon in maintaining basal glucose production. J Clin Invest 58:1407–1418
Cherrington AD, Lacy WW, Chiasson J-L (1978) Then effects of glucagon on glucose production during insulin deficiency in the conscious dog. J Clin Invest 62:664–677
Cherrington AD, Liljenquist JE, Shulman GI, Williams PE, Lacy WW (1979a) Importance of hypoglycemia-induced glucose production during isolated glucagon deficiency. Am J Physiol 236:E263–E271
Cherrington AD, Williams PE, Liljenquist JE, Lacy WW (1979b) The control of glycogenolysis and gluconeogenesis in vivo by insulin and glucagon. In: Pierluissi J (ed) Endocrine pancreas in diabetes. Excerpta Medica, Amsterdam, pp 172–191
Chiasson J-L, Shikama H, Chu DTW, Exton JH (1981) Inhibitory effect of epinephrine on insulin-stimulated glucose uptake by rat skeletal muscle. J Clin Invest 68:706–713
Chideckel EW, Goodner CJ, Koerker DJ, Johnson DG, Ensinck JW (1977) Role of glucagon in mediating metabolic effects of epinephrine. Am J Physiol 232:464–470
Christensen EH, Hansen O (1939) Arbeitsfähigkeit und Ernährung. Skand Arch Physiol 81:160–171
Christensen NJ (1974) Plasma norepinephrine and epinephrine in untreated diabetics, during fasting and after insulin administration. Diabetes 23:1–8
Christensen NJ, Brandsborg O (1973) The relationship between plasma catecholamine concentration and pulse rate during exercise and standing. Eur J Clin Invest 3:299–306
Christensen NJ, Videbaek J (1974) Plasma catecholamines and carbohydrate metabolism in patients with acute myocardial infarction. J Clin Invest 54:278–286
Christensen NJ, Alberti KGMM, Brandsborg O (1975a) Plasma catecholamines and blood substrate concentrations: studies in insulin-induced hypoglycemia and after adrenaline infusions. Eur J Clin Invest 5:415–423
Christensen NJ, Christensen SE, Hansen AP, Lunder K (1975b) The effect of somatostatin on plasma noradrenaline and plasma adrenaline concentrations during exercise and hypoglycemia. Metabolism 24:1267–1272
Christensen NJ, Trap-Jensen J, Clausen JP (1975c) Effect of beta-receptor blockade on heart rate, hepatic blood flow and circulating noradrenaline during exercise in man. Acta Physiol Scand 95:62A–63A
Christensen NJ, Galbo H, Hansen JF, Hesse B, Richter EA, Trap-Jensen J (1979) Catecholamines and exercise. Diabetes 28 [Suppl 1]:58–62
Christensen SE, Hansen AP, Iversen J, Lundbaek K, Orskov H, Seyer-Hansen K (1974) Somatostatin as a tool in studies of basal carbohydrate and lipid metabolism in man: modification of glucagon and insulin. Scand J Clin Lab Invest 34:321–325
Clutter N, Bier D, Shah S, Cryer P (1980) Epinephrine plasma metabolic clearance rate and physiologic thresholds for metabolic and hemodynamic actions in man. J Clin Ivest 66:94–101
Coce R, Femenic R, Skrabalo A, Vranic M (1979) The effect of repetitive exercise in daily control of glycemia in insulin-dependent diabetics. In: IVth international symposium on early diabetes. Academic Press, New York
Colowick SP (1973) The hexokinases. In: Boyer PB (ed) The enzymes, 3rd edn, vol 9. Academic, New York, pp 1–48
Courtice FC, Douglas CG (1936) The effect of prolonged muscular exercise on the metabolism. Proc R Soc 13:381–439
Couturier E, Rasio E, Conard V (1971) Insulin in plasma and lymph and tissue glucose uptake in the exercised hind limb of the dog. Horm Metab Res 3:382–386
Cowan JS, Hetenyi G Jr (1971) Glucoregulatory responses in normal and diabetic dogs recorded by a new tracer method. Metabolism 20:360–372
Cryer PE (1980) Physiology and pathophysiology of the human sympathoadrenal neuroendocrine system. N Engl J Med 303:436–444

Dobbs RE, Sakurai H, Faloona GR, Valverde I, Baetens D, Orci L, Unger RH (1975) Glucagon: role in the hyperglycemia of diabetes mellitus. Science 187:544–547
Doi K, Prentki M, Yip C, Müller WA, Jeanrenaud B, Vranic M (1979) Identical biological effects of pancreatic glucagon and purified moiety of canine gastric glucagon. J Clin Invest 63:525–531
Dulin WE, Clarke JJ (1961) Studies concerning a possible humoral factor produced by working muscles. Its influence on glucose utilization. Diabetes 10:289–297
Durkot MJ, Wolfe RR (1981) Effects of adrenergic blockade on glucose kinetics in septic and burned guinea pigs. Am J Physiol 241:R222–R227
Eigler N, Sacca L, Sherwin RS (1979) Synergistic intractions of physiologic increments of glucagon, epinephrine and cortisol in the dog. A model for stress-induced hyperglycemia. J Clin Invest 63:114–123
Elwyn DH, Kinney JH, Jeevanandam M, Gump FE, Broell JR (1979) Influence of increasing carbohydrate intake on glucose kinetics in injured patients. Am Surg 190:117–127
Exton JH, Assimacopoulos-Jeannet FD, Blackmore PF, Cherrington AD, Cahn TM (1978) Mechanisms of catecholamine action on liver carbohydrate metabolism. Adv Nucleotide Res 9:441–452
Felig P (1973) The glucose-alanine cycle. Metabolism 22:179–207
Felig P, Wahren J (1975) Fuel homeostasis in exercise. N Engl J Med 293:1078–1084
Felig P, Wahren J (1979) Role of insulin and glucagon in the regulation of hepatic glucose production during exercise. Diabetes 28 [Suppl 1]:71–75
Felig P, Wahren J, Hendler R, Ahlborg G (1972) Plasma glucagon levels in exercising man. N Engl J Med 287:184–185
Felig P, Wahren J, Sherwin R, Hendler R (1976) Insulin, glucagon and somatostatin in normal physiology and diabetes mellitus. Diabetes 25:1091–1099
Feurle GE, Wirth A, Diehm C, Lorenzen M, Schlierf G (1980) Exercise-induced release of pancreatic polypeptide and its inhibition by propanolol: evidence for adrenergic stimulation. Eur J Clin Invest 10:249–251
Frier BM, Corrall RJM, Ratcliffe JG, Ashby JP, McClemont EJW (1981) Autonomic neural control mechanisms of substrate and hormonal responses to acute hypoglycemia in man. Clin Endocrinol (Oxf) 14:425–433
Galbo H, Holst JJ, Christensen NJ (1975) Glucagon and plasma catecholamine responses to graded and prolonged exercise in man. J Appl Physiol 38:70–75
Galbo H, Holst JJ, Christensen NJ, Hilsted J (1976) Glucagon and plasma catecholamines during beta-receptor blockade in exercising man. J Appl Physiol 40:855–863
Galbo H, Christensen NJ, Holst JJ (1977a) Glucose induced decrease in glucagon and epinephrine responses to exercise in man. J Appl Physiol 42:525–530
Galbo H, Christensen NJ, Holst JJ (1977b) Catecholamines and pancreatic hormones during autonomic blockade in exercising man. Acta Physiol Scand 101:428–437
Galbo H, Richter EA, Christensen NJ, Holst JJ (1978) Sympathetic control of metabolic and hormonal responses to exercise in rats. Acta Physiol Scand 102:441–449
Galbo H, Christensen NJ, Mikines KJ, Sonne B, Hilsted J, Hagen C, Fahrenkrug J (1981a) The effect of fasting on the hormonal response to graded exercise. J Clin Endocrinol Metab 52:1106–1112
Galbo H, Hedeskov CJ, Capito K, Vinten J (1981b) The effect of physical training on insulin secretion of rat pancreatic islets. Acta Physiol Scand 111:75–79
Garber AJ, Cryer PE, Santiago JV, Haymond MW, Pagliara AS, Kipnis DM (1976a) The role of adrenergic mechanisms in the substrate and hormonal responses to insulin-induced hypoglycemia. J Clin Invest 58:7–15
Garber AJ, Karl IE, Kipnis DM (1976b) Alanine and glutanine synthesis and release from skeletal muscle. IV. beta adrenergic inhibition of amino acid release. J Biol Chem 251:1851–1857
Gauthier C, Vranic M, Hetenyi G Jr (1980) Importance of glucagon in regulatory rather than emergency responses to hypoglycemia. Am J Physiol 238:E131–E140
Gerich JE, Karam JH, Forsham PH (1973) Stimulation of glucagon secretion by epinephrine in man. J Clin Endocrinol Metab 37:470–481
Gerich JE, Lorenzi M, Bier DM, Schneider V, Tsalikian E, Karam JH, Forsham PH (1975) Prevention of human diabetic ketoacidosis by somatostatin. Evidence for an essential role of glucagon. N Engl J Med 292:985–989

Gerich JE, Lorenzi M, Tsalikian E, Karam JH (1976) Studies on the mechanism of epinephrine-induced hyperglycemia in man. Diabetes 25:67–71
Gerich J, Davis J, Lorenzi M, Rizza R, Bohannon N, Karam J, Lewis S, Kaplan R, Schultz T, Cryer P (1979) Hormonal mechanisms of recovery from insulin-induced hypoglycemia in man. Am J Physiol 236:E380–E385
Goldstein MS (1961) Humoral nature of the hypoglycemic factor of muscular work. Diabetes 10:232–234
Goldstein MS, Mullick V, Huddlestun B, Levine R (1953) Action of muscular work on transfer of sugar across cell barriers: comparison with the action of insulin. Am J Physiol 173:212–216
Gray DE, Lickley HLA, Vranic M (1980) Physiologic effects of epinephrine on glucose turnover and plasma free fatty acid concentrations mediated independently of glucagon. Diabetes 29:600–609
Greenwood FC, Landon J, Stamp TCB (1966) The plasma sugar, free fatty acid, cortisol and growth hormone response to insulin. I. In control subjects. J Clin Invest 45:429–436
Gyntelberg F, Rennie MJ, Hickson RC, Holloszy JO (1977) Effect of training on the response of plasma glucagon to exercise. J Appl Physiol 43:302–305
Hagenfeldt L (1975) Turnover of individual free fatty acids. Fed Proc 34:2246–2249
Hagenfeldt L (1979) Metabolism of free fatty acids and ketone bodies during exercise in normal and diabetic man. Diabetes 28 [Suppl 1]:66–70
Hagenfeldt L, Wahren J (1968) Human forearm muscle metabolism during exercise. III. Uptake, release and oxidation of $\beta$-hydroxybutyrate and observations on the $\beta$-hydroxybutyrate/acetoacetate ratio. Scand J Clin Lab Invest 21:314–320
Halban P, Berger M, Gjinovici A, Renold A, Vranic M, Offord R (1978) Pharmacokinetics of subcutaneously injected semi-synthetic tritiated insulin in rats. In: Offord RE, Di Bello C (eds) Semisynthetic peptides and proteins. Academic Press, New York, pp 237–246
Halter JB, Pflug AE (1980) Relationship of impaired insulin secretion during surgical stress to anaesthesia and catecholamine release. J Clin Endocrinol Metab 51:1093–1098
Halter JB, Pflug AE, Porte D Jr (1977) Mechanism of plasma catecholamine increases during surgical stress in man. J Clin Endocrinol Metab 45:936–944
Hamaji M, Nakao K, Kiso K (1979) Pancreatic glucagon and insulin response during surgery. Horm Metab Res 11:488–489
Hansen AP (1970) Abnormal serum growth hormone response to exercise in juvenile diabetics. J Clin Invest 49:1467–1478
Hansen AP (1971) Normalization of growth hormone hyperresponse to exercise in juvenile diabetics after "normalization" of blood sugar. J Clin Invest 50:1806–1811
Hartley LH, Mason JW, Hogan RP, Jones LG, Kotchen TA, Mougey EH, Pennington LL, Ricketts PT (1972a) Multiple hormonal responses to graded exercise in relation to physical training. J Appl Physiol 33:602–606
Hartley LH, Mason JW, Morgan RP, Jones LG, Kotchen TA, Mougey EH, Wherry FE, Pennington LL, Ricketts PT (1972b) Multiple hormonal response to prolonged exercise in relation to physical training. J Appl Physiol 33:607–610
Harvey WD, Faloona GR, Unger RH (1974) The effect of adrenergic blockade on exercise-induced hyperglycemia. Endocrinology 94:1254–1258
Havivi E, Wertheimer HE (1964) A muscle activity factor increasing sugar uptake by rat diaphragms in vitro. J Physiol (Lond) 172:342–352
Helmreich E, Cori CF (1957) Studies of tissue permeability. II. Distribution of pentoses between plasma and muscle. J Biol Chem 224:663–679
Hetenyi G Jr (1981) Calculation of the rate of gluconeogenesis in vivo. In: Cobelli C, Bergman RN (eds) Carbohydrate metabolism. Wiley, New York, pp 201–219
Hetenyi G Jr, Norwich KH (1974) Validity of the rates of production and utilization of metabolites as determined by tracer methods in intact animals. Fed Proc 33:1841–1848
Hetenyi G Jr, Kovacevic N, Hall SE, Vranic M (1976) Plasma glucagon in pups, decreased by fasting, unaffected by somatostatin or hypoglycemia. Am J Physiol 231:R1377–R1382
Hilsted J, Galbo H, Sonne B, Schwartz T, Fahrenkrug J, Schaffalitzky de Muckadell OB, Lauritsen KB, Tronier B (1980) Gastroenteropancreatic hormonal changes during exercise. Am J Physiol 239:G136–G140

Himms-Hagen J (1970) Adrenergic receptors for metabolic responses in adipose tissue. Fed Proc 29:1388–1401
Holloszy JO, Narahara HT (1967) Enhanced permeability to sugar associated with muscle contraction. Studies on the role of $Ca^{++}$. J Gen Physiol 50:551–552
Houssay BA, Lewis JT, Molinelli EA (1924) Rôle de la sécrétion d'adrénaline pendant l'hypoglycémie produite par l'insuline. C R Soc Biol (Paris) 91:1011–1013
Hsu WH, Hummel SR (1981) Xylaxine-induced hypoglycemia in cattle: a possible involvement of α2-adrenergic receptors regulating insulin release. Endocrinology 109:825–827
Ipp E, Dobbs RE, Unger RH (1978) Morphine and β-endorphin influence the secretion of the endocrine pancreas. Nature 276:190–191
Issekutz B (1977) Studies on hepatic glucose cycles in normal and methylprednisolone treated dogs. Metabolism 26:157–170
Issekutz B Jr (1978) Role of β-adrenergic receptors in mobilization of energy sources in exercising dogs. J Appl Physiol 44:869–876
Issekutz B Jr (1980) The role hypoinsulinemia in exercise metabolism. Diabetes 29:629–635
Issekutz B Jr (1981) Effects of glucose infusion on hepatic and muscle glycogenolysis in exercising dogs. Am J Physiol 240:E451–E457
Issekutz B Jr, Vranic M (1980) Significance of glucagon in the control of glucose production during exercise in dogs. Am J Physiol 238:E13–E20
Issekutz B Jr, Miller HI, Paul P, Rodahl K (1965) Aerobic work capacity and plasma FFA turnover. J Appl Physiol 20:293–296
Issekutz B Jr, Issekutz AC, Nash D (1970) Mobilization of energy sources in exercising dogs. J Appl Physiol 29:691–697
Issekutz B Jr, Shaw WA, Issekutz AC (1976) Lactate metabolism in resting and exercising dogs. J Appl Physiol 40:312–319
Iversen J (1973) Adrenergic receptors and the secretion of glucagon and insulin from the isolated perfused canine pancreas. J Clin Invest 52:2102–2116
Kalant N, Leibovici T, Rohan I, McNeill K (1978) Effect of exercise on glucose and insulin utilization in the forearm. Metabolism 27:333–340
Katz J, Dunn A (1967) Glucose 2-T as a tracer for glucose metabolism. Biochemistry 6:1–5
Kawamori R, Vranic M (1977) Mechanism of exercise induced hypoglycemia in depancreatized dogs maintained on long-acting insulin. J Clin Invest 59:331–337
Kemmer FW, Vranic M (1981) The role of glucagon and its relationship to other glucoregulatory hormones in exercise. In: Unger RH, Orci L (eds) Glucagon physiology, pathophysiology, and morphology of the pancreatic A-cell. Elsevier, New York, pp 297–331
Kemmer FW, Berchtold P, Berger M, Cüppers HJ, Starke A, Gries FA, Zimmermann H (1979) A mechanism of the exercise-induced fall of blood glucose in insulin-treated diabetics. Diabetes [Suppl 2]:360
Kemmer FW, Lickley HLA, Gray DE, Perez G, Vranic M (1982) The state of metabolic control determines the role of epinephrine-glucagon interactions in glucoregulation in diabetes. Am J Physiol 242:E428–E436
Kjellmer J (1965) Studies on exercise hyperaemia. Acta Physiol Scand 64 [Suppl 244]:1–27
Koerker DJ, Ruch W, Chideckel G, Palmer J, Goodner CJ, Ensinck J, Gale CC (1974) Somatostatin: hypothalamic inhibition of the endocrine pancreas. Science 184:482–483
Koivisto V, Felig P (1978) Effects of leg exercise on insulin absorption in diabetic patients. N Engl J Med 298:77–83
Koivisto V, Soman V, Conard P, Hendler R, Nadel E, Felig P (1979) Insulin binding to monocytes in trained athletes: changes in the resting state and after exercise. J Clin Invest 64:1011–1015
Lawrence RD (1926) The effect of exercise on insulin action in diabetes. Br Med J 1:648–650
Lefèbvre PJ, Luyckx AS (1977) Factors controlling gastric-glucagon release. J Clin Invest 59:716–722
Lefèbvre PJ, Luyckx AS (1978) Glucose and insulin in the regulation of glucagon release from isolated perfused dog stomach. Endocrinology 103:1579–1582
Lefèbvre PJ, Luyckx AS, Federspil G (1972) Muscular exercise and pancreatic function in rats. Isr J Med Sci 8:390–398
Lickley HLA, Ross GG, Vranic M (1979) Effects of selective insulin or glucagon deficiency on glucose turnover. Am J Physiol 236:E255–E262

Lickley HLA, Doi K, Vranic M (1981 a) Glucagon suppression improves glycemia in partial but not in total insulin deficiency (Abstr 308). Diabetologia 21:297

Lickley HLA, Kemmer FW, Gray DE, Kovacevic N, Hatton TW, Perez G, Vranic M (1981 b) Chromatographic pattern of extrapancreatic glucagon and glucagon-like immunoreactivity before and during stimulation by epinephrine, and participation of glucagon in epinephrine-induced hepatic glucose overproduction. Surgery 90:186–194

Liddell MJ, MacLean LD, Shizgal HM (1979) The role of stress hormones in the catabolic metabolism of shock. Surg Gynecol Obstet 149:822–830

Lindsey CA, Santeusanio F, Braaten J, Faloona GR, Unger RH (1974) Pancreatic alpha-cell function in trauma. JAMA 227:757–761

Lindsey CA, Faloona GR, Unger RH (1975) Plasma glucagon levels during rapid exsanguination with and without adrenergic blockade. Diabetes 24:313–319

Long CL, Schiller WR, Geiger JW, Blakemore WS (1978) Gluconeogenic response during glucose infusion in patients following skeletal trauma or during sepsis. J Parent Ent Nutr 2:619–626

Luyckx AS, Dresse A, Cession-Fossion A, Lefèbvre PJ (1975) Catecholamines and exercise-induced glucagon and fatty acid mobilization in the rat. Am J Physiol 229:376–383

Luyckx AS, Pirnay F, Lefèbvre PJ (1978) Effect of glucose on plasma glucagon and free fatty acids during prolonged exercise. Eur J Appl Physiol 39:53–61

Martins MJ, Horwitz DL, Nattrass M, Granger JF, Rochman H, Ash S (1981) Effects of mild hyperinsulinemia on the metabolic response to exercise. Metabolism 30:688–694

McGarry JD, Foster DW (1977) Hormonal control of ketogenesis. Arch Intern Med 137:485–501

Minuk HL, Hanna AK, Marliss EB, Vranic M, Zinman B (1980) The metabolic response to moderate exercise in obese man during prolonged fasting. Am J Physiol 238:E322–E329

Minuk HL, Vranic M, Marliss EB, Hanna AK, Albisser AM, Zinman B (1981) The glucoregulatory and metabolic response to exercise in obese non-insulin dependent diabetes. Am J Physiol 240:E458–E464

Mondon CE, Dolkas CB, Reaven GM (1980) Site of enhanced insulin sensitivity in exercise trained rats at rest. Am J Physiol 239:E169–E177

Morita S, Doi K, Yip C, Vranic M (1976) Measurement and partial characterization of immunoreactive glucagon in gastrointestinal tissues of the dog. Diabetes 25:1018–1025

Müller WA, Aoki TT, Egdahl RH, Cahill GF Jr (1977) Effects of exogenous glucagon and epinephrine in physiological amounts on the blood levels of free fatty acids and glycerol in dogs. Diabetologia 13:55–58

Müller WA, Girardier L, Seydoux J, Berger M, Renold AE, Vranic M (1978) Extrapancreatic glucagon and glucagon-like immunoreactivity in depancreatized dogs: a quantitative assessment of secretion rates and anatomical delineation of sources. J Clin Invest 62:124–132

Murray FT, Zinman B, McClean PA, Denoga A, Albisser AM, Leibel BS, Nakooda AF, Stokes EF, Marliss EB (1977) The metabolic response to moderate exercise in diabetic man receiving intravenous and subcutaneous insulin. J Clin Endocrinol Metab 44:708–720

Nakhooda AF, Sole MJ, Marliss EB (1981) Adrenergic regulation of glucagon and insulin secretion during immobilization stress in normal and spontaneously diabetic BB rats. Am J Physiol 240:E373–E378

Newsholme EA, Randle PJ (1964) Regulation of glucose uptake by the muscle. Biochem J 93:641–651

Ohneda A, Kobayashi T, Nihei J (1980) Response of extrapancreatic glucagon to glycemic changes. Endocrinol Jpn 1:121–126

Pedersen O, Beck-Nielsen H, Heding L (1980) Increased insulin receptors after exercise in patients with insulin-dependent diabetes mellitus. N Engl J Med 302:886–892

Perez G, Ungaro B, Covelli A, Morrone G, Lombardi G, Scopacasa F, Rossi R (1980) Altered glucoregulatory response to physiological infusions of epinephrine and glucagon in hyperthyroidism. J Clin Endocrinol Metab 51:972–977

Perez G, Kemmer FW, Lickley HLA, Vranic M (1981) The importance of glucagon in mediating epinephrine-induced hyperglycemia in alloxan-diabetic dogs. Am J Physiol 241:E328–E335

Porte D, Graber AL, Kuzuya T, Williams RH (1966) The effects of epinephrine on IRI levels in man. J Clin Invest 45:228–236
Radziuk J, Norwich K, Vranic M (1974) Measurement and validation of nonsteady turnover rates with application to the insulin and glucose system. Fed Proc 33:1855–1864
Radziuk J, Norwich KH, Vranic M (1978) Experimental validation of measurements of glucose turnover in nonsteady state. Am J Physiol 234:E84–E93
Randle PJ, Smith GH (1958) Regulation of glucose uptake by muscle. II. The effect of insulin anaerobiosis and cell poisons on the penetration of isolated rat diaphragm by sugars. Biochem J 70:502–508
R-Candela R, R-Candela JL (1962) Possible factor produced during muscular contractions which influences the passage of glucose. Proc Soc Exp Biol Med 110:803–804
Reaven EP, Reaven GM (1981) Structure and function changes in the endocrine pancreas of aging rats with reference to the modulating effects of exercise and caloric restriction. J Clin Invest 68:75–84
Refsum HE, Strömme SB (1974) Urea and creatinine production and excretion in urine during and after prolonged heavy exercise. Scand J Clin Lab Invest 33:247–254
Rennie MJ, Park DM, Sulaiman WR (1976) Uptake and release of hormones and metabolites by tissues of exercising leg in man. Am J Physiol 231:967–973
Richter EA, Galbo H, Sonne B, Holst JJ, Christensen NJ (1980) Adrenal medullary control of muscular and hepatic glycogenolysis and of pancreatic hormonal secretion in exercising rats. Acta Physiol Scand 108:235–242
Richter EA, Galbo H, Christensen NJ (1981 a) Control of exercise-induced muscular glycogenolysis by adrenal medullary hormones in rats. J Appl Physiol 50:21–26
Richter EA, Galbo H, Holst JJ, Sonne B (1981 b) Significance of glucagon for insulin secretion and hepatic glycogenolysis during exercise in rats. Horm Metab Res 13:323–326
Richter EA, Sonne B, Christensen NJ, Galbo H (1981 c) Role of epinephrine for muscular glycogenolysis and pancreatic hormonal secretion in running rats. Am J Physiol 240:E526–E532
Rizza R, Verdonk C, Miles J, Service FJ, Gerich J (1979 a) Effect of intermittent endogenous hyperglucagonemia on glucose homeostasis in normal and diabetic man. J Clin Invest 63:1119–1123
Rizza RA, Cryer PE, Gerich JE (1979 b) Role of glucagon, catecholamines and growth hormone in human glucose counterregulation. Effects of somatostatin and combined α- and β-adrenergic blockade in plasma glucose recovery and glucose flux rate after insulin-induced hypoglycemia. J Clin Invest 64:62–70
Rizza R, Haymond M, Cryer P, Gerich J (1979 c) Differential effects of epinephrine on glucose production and disposal in man. Am J Physiol 237:E356–E362
Riza RA, Cryer PE, Haymond MW, Gerich JE (1980 a) Adrenergic mechanisms for the effects of epinephrine on glucose production and clearance in man. J Clin Invest 65:682–689
Rizza RA, Haymond MW, Miles JM, Verdonk CA, Cryer PE, Gerich JE (1980 b) Effect of α-adrenergic stimulation and its blockade on glucose turnover in man. Am J Physiol 238:E467–E472
Rizza RA, Mandarino LJ, Gerich JE (1981) Dose-response characteristics for effects of insulin on production and utilization of glucose in man. Am J Physiol 240:E630–E639
Robertson RP, Porte D Jr (1973) Adrenergic modulation of basal insulin secretion in man. Diabetes 22:1–8
Rocha DM, Santeusanio F, Faloona GR, Unger RH (1973) Abnormal pancreatic alpha-cell function in bacterial infection. N Engl J Med 288:700–703
Ross G, Lickley HLA, Vranic M (1978) Extrapancreatic glucagon in control of glucose turnover in depancreatized dogs. Am J Physiol 234:E213–E219
Roth J, Glick GM, Yalow RS, Berson SA (1963) Hypoglycemia: a powerful stimulus to secretion of growth hormone. Science 140:987–991
Sacca L, Perez G, Carteni G, Rengo G (1977) Evaluation of the role of the sympathetic nervous system in the glucoregulatory response to insulin-induced hypoglycemia in the rat. Endocrinology 101:1016–1022
Sacca L, Sherwin R, Felig P (1978) Effects of sequential infusions of glucagon and epinephrine on glucose turnover in the dog. Am J Physiol 235:E287–E290

Sacca L, Sherwin R, Felig P (1979a) Influence of somatostatin on glucagon- and epinephrine-stimulated hepatic glucose output in the dog. Am J Physiol 236:E113–E117
Sacca L, Sherwin R, Hendler R, Felig P (1979b) Influence of continuous physiologic hyperinsulinemia on glucose kinetics and counter-regulatory hormones in normal and diabetic humans. J Clin Invest 63:849–857
Samols E, Weir GC (1979) Adrenergic modulation of pancreatic A, B, and D cells. J Clin Invest 63:230–238
Schade DS, Eaton RP (1979) The regulation of plasma ketone body concentration by counter-regulatory hormones in man. III. Effects of norepinephrine in normal man. Diabetes 28:5–10
Schade DS, Eaton RP (1980) The temporal relationship between endogenously secreted stress hormones and metabolic decompensation in diabetic man. J Clin Endocrinol Metab 50:131–136
Schultz TA, Leweis SB, Westbie DK, Wallin JD, Gerich JE (1977) Glucose delivery: a modulation of glucose uptake in contracting skeletal muscle. Am J Physiol 233:E514–E518
Shamoon H, Jacob R, Sherwin RS (1979) Epinephrine-induced hypoaminoacidemia in man: a $\beta$-adrenergic effect. Clin Res 27:595A
Shamoon H, Hendler R, Sherwin R (1980) Altered responsiveness to cortisol, epinephrine and glucagon in insulin-infused juvenile onset diabetics: a mechanism for diabetic instability. Diabetes 29:284–291
Shamoon H, Hendler R, Sherwin RS (1981) Synergistic interaction among antiinsulin hormones in the pathogenesis of stress hyperglycemia in humans. J Clin Endocrinol Metab 52:1235–1241
Shaw WAS, Issekutz TB, Issekutz B Jr (1976) Gluconeogenesis from glycerol at rest and during exercise in normal diabetic and methylprednisolone treated dogs. Metabolism 25:329–339
Sherwin RS, Fisher M, Hendler R, Felig P (1976) Hyperglucagonemia and blood glucose regulation in normal, obese and diabetic subjects. N Engl J Med 294:455–461
Sherwin RS, Tamborlane W, Hendler R, Sacca L, de Fronzo RA, Felig P (1977) Influence of glucagon replacement on the hyperglycemic and hyperketonemic response to prolonged somatostatin infusion in normal man. J Clin Endocrinol Metab 45:1104–1107
Silverberg AB, Shah SD, Haymond MW, Cryer PE (1978) Norepinephrine: hormone and neurotransmitter in man. Am J Physiol 234:E252–E256
Sirek A, Vranic M, Sirek OV, Vigas M, Policova Z (1979) The effect of growth hormone on acute glucagon and insulin release. Am J Physiol 237:E107–E112
Sokal JE, Sarcione EJ, Henderson AM (1964) Relative potency of glucagon and epinephrine as hepatic glycogenolytic agents: studies with the isolated perfused rat liver. Endocrinology 74:930–938
Srikant CB, McKorkle K, Unger RH (1977) Properties of immunoreactive glucagon fractions of canine stomach and pancreas. J Biol Chem 252:1847–1851
Standl E, Janka HV, Dexel T, Kolb HJ (1976) Muscle metabolism during rest and exercise: influence on the oxygen transport system of blood in normal and diabetic subjects. Diabetes 25 [Suppl 2]:914–919
Steele R, Wall JS, deBodo RC, Altszuler N (1956) Measurement of size and turnover rate of body glucose pool by the isotope dilution method. Am J Physiol 187:15–24
Struwe FE (1977) Stoffwechselführung diabetischer Kinder unter körperlicher Belastung. In: Jahnke K, Mehnert H, Reis HD (eds) Muskelstoffwechsel, körperliche Leistungsfähigkeit und Diabetes mellitus. Schattauer, Stuttgart, pp 313–316
Sutherland EW, de Duve C (1948) Origin and distribution of hyperglycemic glycogenolytic factor of the pancreas. J Biol Chem 175:663–674
Unger RH (1978) Role of glucagon in the pathogenesis of diabetes: the status of the controversy. Metabolism 27:1691–1709
Unger RH (1981) The milieu interieur and the islets of Langerhans. Diabetologia 20:1–11
Unger RH, Orci L (1975) Hypothesis: the essential role of glucagon in the pathogenesis of diabetes mellitus. Lancet 1:14–16
Unger RH, Aguilar-Parada E, Müller W, Eisentraut A (1970) Studies of pancreatic alpha cell function in normal and diabetic subjects. J Clin Invest 49:837–848

Vidnes J, Oyasaeter S (1977) Glucagon deficiency causing severe neonatal hypoglycemia in a patient with normal insulin secretion. Pediatr Res 11:943–949
Vranic M, Berger M (1979) Exercise and diabetes mellitus. Diabetes 28:147–167
Vranic M, Issekutz B Jr (1980) The important roles of glucagon and insulin in the regulation of glucose fluxes during exercise in health and diabetes. In: Andreani D, Lefèbvre PJ, Marks V (eds) Current views on hypoglycemia and glucagon. Academic Press, New York, pp 57–70
Vranic M, Wrenshall GA (1968) Matched rates of insulin infusion and secretion and concurrent tracer determined rates of glucose appearance in fasting dogs. Can J Physiol Pharmacol 46:383–390
Vranic M, Wrenshall GA (1969) Exercise, insulin and glucose turnover in dogs. Endocrinology 85:165–171
Vranic M, Fono P, Kovacevic N, Lin BJ (1971) Glucose kinetics and fatty acids in dogs on matched insulin infusion after a glucose load. Metabolism 20:954–967
Vranic M, Pek S, Kawamori B (1974a) Increased "glucagon immunoreactivity" in plasma of totally depancreatized dogs. Diabetes 23:905–912
Vranic M, Kawamori R, Wrenshall GA (1974b) Mechanism of exercise-induced hypoglycemia in depancreatized insulin-treated dogs (Abstr). Diabetes 23 [Suppl 1]:353
Vranic M, Kawamori R, Wrenshall GA (1975) The role of insulin and glucagon in regulating glucose turnover in dogs during exercise. Med Sci Sports 7:27–33
Vranic M, Kawamori R, Pek S, Kovacevic N, Wrenshall GA (1976a) The essentiality of insulin and the role of glucagon in regulating glucose turnover during strenuous exercise. J Clin Invest 57:245–255
Vranic M, Ross GG, Doi K, Lickley HLA (1976b) The role of glucagon-insulin interactions in control of glucose turnover and its significance in diabetes. Metabolism 25 [Suppl 1]:1375–1380
Vranic M, Lickley HLA, Kemmer FW, Perez G, Hetenyi G Jr, Hatton TW, Kovacevic N (1981) Interaction between insulin and the counterregulatory hormones in the development of diabetes. In: Martin J, Ehrlich R (eds) Etiology and pathogenesis of diabetes. Raven, New York, pp 153–178
Wahren J, Felig P, Ahlbor G, Forfeldt L (1971) Glucose metabolism during leg exercise in man. J Clin Invest 50:2715–2725
Walaas O, Walaas E (1950) Effect of epinephrine on rat diaphragm. J Biol Chem 187:769–775
Wass JAH, Penman E, Medbaks S, Dawson AM, Tsiolakis D, Marks V, Besser GM, Rees LH (1980) Immunreactive somatostatin changes during insulin-induced hypoglycemia and operative stress in man. Clin Endocrinol (Oxf) 12:269–275
Wilmore DW, Long JM, Mason AD (1974a) Catecholamines: mediator of the hypermetabolic response to thermal injury. Ann Surg 180:653–669
Wilmore DW, Lindsey CA, Maylan JA, Faloona GR, Pruitt BA, Unger RH (1974b) Hyperglucagonemia after burns. Lancet 1:73–75
Wilson DW, Long WL, Thompson HC, Thurlow S (1925) Changes in the composition of the urine after muscular exercise. J Biol Chem 65:755–771
Wirth A, Diehm C, Mayer H, Mörl H, Vogel I, Björntorp P, Schlierf G (1981) Plasma C-peptide and insulin in trained and untrained subjects. J Appl Physiol 50:71–77
Wolfe RR, Allsop JR, Burke JF (1977) Experimental sepsis and glucose metabolism: time course of response. Surg Forum 28:42–43
Wolfe RR, Durkot MJ, Allsop JR, Burke JF (1979) Glucose metabolism in severely burned patients. Metabolism 28:1031–1039
Woodson LC, Bee DE, Potter DE (1980) Catecholamine-induced hyperglycemia in dogs: independent from alterations in pancreatic hormone release. Horm Metab Res 12:434–439
Zinman B, Murray FT, Vranic M, Albisser AM, Leibel BS, McClean PA, Marliss EB (1977) Glucoregulation during moderate exercise in insulin treated diabetes. J Clin Endocrinol Metab 45:641–652
Zinman B, Murray FT, Vranic M, Albisser M, Leibel BS, McClean PA, Marliss EB (1979) Glucoregulation during moderate exercise. Diabetes 28 [Suppl 1]:82–88

# Catabolism of Glucagon

CHAPTER 39

# The Metabolic Clearance Rate of Glucagon

K. S. POLONSKY, J. B. JASPAN, and A. H. RUBENSTEIN

## A. Introduction

Studies performed over the last 10 years have clearly shown that glucagon plays a pivotal role in the normal control of blood glucose as well as being extremely important in the pathogenesis of the metabolic abnormalities found in a number of disease states (JASPAN and RUBENSTEIN 1977). Thus, glucagon is important for the maintenance of normal blood glucose levels in the fasted state (MARLISS et al. 1970; ALFORD et al. 1974) as well as after protein feeding (UNGER et al. 1969) and during exercise (ISSEKUTZ and VRANIC 1980). Glucagon is the most important defense against the development of hypoglycemia induced by insulin (RIZZA et al. 1979). Although its precise role in the pathogenesis of diabetes is uncertain, there is substantial evidence that glucagon is important in aggravating the hyperglycemia due to insulin deficiency (UNGER 1976; UNGER and ORCI 1981; GERICH 1976) as well as being essential for the development of ketoacidosis in the insulin-deprived state (MCGARRY and FOSTER 1977; GERICH et al. 1974). Elevated levels of glucagon are found in cirrhosis (MARCO et al. 1973; SHERWIN et al. 1974), renal failure (KUKU et al. 1976), glucagonoma (JASPAN and RUBENSTEIN 1977), and stress conditions such as shock (UNGER 1971; WILLERSON et al. 1974), and the hyperglucagonemia probably contributes to the glucose intolerance found in all of these conditions.

To understand the economy of glucagon in the body, a knowledge of its secretion rate under differing circumstances and the factors controlling this rate as well as an appreciation of the nature of the regulation of glucagon catabolism is necessary. This chapter will deal with the metabolic clearance rate of glucagon in plasma, as well as its plasma half-life.

## B. Principles of Measurement

The metabolic clearance rate (MCR) of any substance, in this case glucagon, is the least amount of plasma totally cleared of the substance per unit time. It is a theoretical value since in reality the whole plasma volume is being partially cleared of glucagon continuously. The MCR is a value of importance, however, since it allows the whole body or organ metabolism of different substances to be compared in a meaningful way.

Although the MCR of peptide hormones can be assessed under non-steady-state conditions, such as after intravenous boluses or secretory stimuli, most studies of the MCR of glucagon have utilized the constant infusion technique of TAIT (1963). This technique involves the measurement of the plasma immunoreactive glucagon (IRG) level followed by a constant primed infusion of the hormone. When a new steady-state level is reached, plasma IRG is again measured and the MCR (usually expressed in ml $kg^{-1}$ $min^{-1}$) is calculated according to the formula

$$MCR = \text{Infusion rate of glucagon/Steady-state IRG} - \text{preinfusion IRG}$$

This formula is based on the principle that, under steady-state conditions the amount of hormone infused per unit time must equal the whole body disposal rate of the hormone. The calculation makes a number of assumptions:

1. The endogenous (preinfusion) IRG remains constant throughout the period of infusion. This consideration is mathematically unimportant if the postinfusion IRG level greatly exceeds the preinfusion IRG. In studies performed in our laboratory (Jaspan et al. 1981), glucagon MCR was measured after suppression of basal secretion of 3,500 dalton glucagon with somatostatin. In this situation, it can be safely assumed that the high molecular weight IRG fraction present during somatostatin infusion will be constant throughout the infusion of exogenous hormone.
2. Exogenously infused glucagon is handled in the same way as endogenously secreted glucagon. It is difficult to examine this question directly. However, we have found that both the kidney and liver handle exogenous and endogenous glucagon in a similar fashion (Jaspan et al. 1981), and since these are the two major sites of glucagon clearance, it is highly unlikely that the total metabolic clearance rate of exogenous and endogenous glucagon would be different.

This technique also has the significant advantage that, under steady-state conditions the MCR of glucagon is independent of the number of compartments into which the hormone is distributed and the nature of the interactions between these compartments. Furthermore, glucagon MCR is constant over a wide concentration range (Jaspan et al. 1981; Emmanouel et al. 1978; Alford et al. 1976). Thus, the constant infusion technique appears to be a satisfactory method for studying the metabolic clearance rate of glucagon.

## C. Glucagon Metabolic Clearance Rate in Laboratory Animals

The MCR of glucagon has been evaluated in the rat, the dog, and the pig and species differences have been found to exist. Thus, in the rat the glucagon MCR was found to be $31.8 \pm 1.2$ ml $kg^{-1}$ $min^{-1}$ (Emmanouel et al. 1978). In rats in which both ureters were ligated, the MCR fell to $22.9 \pm 3.3$ ml $kg^{-1}$ $min^{-1}$ and this rate was similar to that found in 70% nephrectomized animals ($22.3 \pm 1.6$ ml $kg^{-1}$ $min^{-1}$), as well as totally nephrectomized rats ($23.2 \pm 1.2$ ml $kg^{-1}$ $min^{-1}$). These data are summarized in Fig. 1.

In experiments performed in normal mongrel dogs, the MCR was found to be lower than in the rat, $12.5 \pm 0.8$ ml $kg^{-1}$ $min^{-1}$ (Table 1; Jaspan et al. 1981). This agrees closely with data obtained by Lefebvre and Luyckx (1976), and Muller et al. (1978) in the same species. Furthermore, the MCR is independent of arterial IRG concentrations over a concentration range from basal levels to values as high as 20 ng/ml (Jaspan et al. 1981). We also found that, when the MCR of insulin and

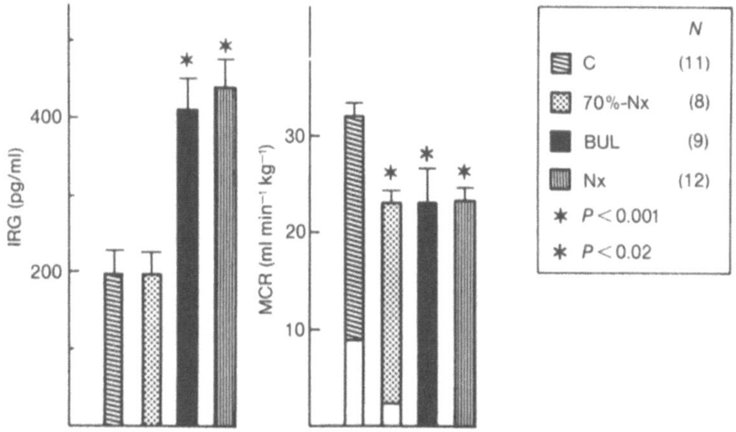


**Fig. 1.** Baseline IRG levels and the MCR of the hormone in normal rats (*C*) as well as rats with partial nephrectomy (*70%-Nx*), bilateral ureteric ligation (*BUL*), and nephrectomy (*Nx*). The simultaneously measured renal contribution to overall MCR in C and 70%-Nx rats is indicated by the *clear bars*. EMMANOUEL et al. (1978)

**Table 1.** The hepatic extraction, MCR, and plasma half-life $t_{\frac{1}{2}}$ of glucagon. Values for insulin are given for comparison. Values are mean ± standard error, numbers in parentheses are numbers of observations. JASPAN et al. (1981)

| | Hepatic extraction (%) | MCR (ml kg$^{-1}$ min$^{-1}$) | $t_{\frac{1}{2}}$ (min) |
|---|---|---|---|
| | (25) | (22) | (13) |
| Glucagon | 22.8 ± 1.8 | 12.6 ± 0.8 | 5.5 ± 0.5 |
| Insulin | 45.2 ± 3.3 | 19.5 ± 1.0 | 4.1 ± 0.3 |

glucagon were measured simultaneously, there appeared to be no significant regression relationship between these two parameters, suggesting that the metabolic clearance of these two hormones are independently regulated processes.

Although it has been suggested that somatostatin may influence glucagon metabolism (ISHIDA et al. 1980), we found that both the hepatic metabolism and MCR of glucagon (JASPAN et al. 1981) were unaffected by somatostatin. These observations were similar to those of VAN HOORN et al. (1978), who, in studies in the pig, found that the MCR of glucagon was not different in somatostatin-infused (46 ± 7 ml kg$^{-1}$ min$^{-1}$) and pancreatectomized animals (39 ± 8 ml kg$^{-1}$ min$^{-1}$).

## D. Glucagon Metabolic Clearance Rate in Humans

A number of investigators have also measured the MCR in humans. ALFORD et al. (1976) studied the MCR in normal subjects as well as seven insulin-dependent diabetics. The MCR was similar in both groups (9.0 ± 0.6 and 11.4 ± 1.0 ml kg$^{-1}$

$min^{-1}$, respectively). The concentration range over which the MCR was studied varied between 50 and 1,300 pg/ml. FISCHER et al. (1976) found that the MCR was comparable in the postabsorptive state in obese and nonobese subjects. However, as the obese subjects underwent prolonged fasting, there was a 20% reduction in MCR by 3 days and a 35% reduction by 3–4 weeks into the fasting period. BURMAN et al. (1980) found that the fall in glucagon MCR which occurs with fasting can be prevented by triiodothyronine administration. Since triiodothyronine levels fall with fasting, they suggested that this is the mechanism whereby glucagon levels are increased with fasting to maintain gluconeogenesis.

Since, as reviewed in Chaps. 40 and 48, the liver has been established as an important site of glucagon metabolism and hyperglucagonemia has been discovered in cirrhosis (MARCO et al. 1973; SHERWIN et al. 1974), the measurement of MCR in cirrhosis is of some interest. ALFORD et al. (1979) found glucagon MCR to be similar in controls ($13.0 \pm 1.3$ ml $kg^{-1}$ $min^{-1}$) and cirrhotics ($13.3 \pm 1{,}9$ ml $kg^{-1}$ $min^{-1}$) prior to undergoing a surgical portasystemic shunt procedure. After shunt surgery, however, the MCR declined significantly to $7.6 \pm 1.3$ ml $kg^{-1}$ $min^{-1}$. SHERWIN et al. (1978) reported that the MCR was no different in healthy controls, cirrhotics with and without portal systemic shunting, and cirrhotics with patent surgical end-to-site portacaval anastomosis. These authors therefore concluded that the hyperglucagonemia in cirrhosis was due to hypersecretion rather than diminished metabolism. In humans, reduced MCR has also been found in uremia (SHERWIN et al. 1976). The MCR of glucagon in uremics was reduced by 58% as compared with controls and the values were not increased by hemodialysis.

## E. Organ Contribution to Overall Glucagon Metabolic Clearance Rate

Although MCR is an important measurement of whole body metabolism (as discussed in Sect. B), the relative contribution of various organs to the metabolism of glucagon must be considered in evaluation of the overall metabolism of this peptide by the body. From the preceding discussion, it is evident that the liver and kidney are important in the degradation of glucagon. Thus, in our studies in the rat, renal clearance constituted 30% of the overall metabolic clearance of the hormone (EMMANOUEL et al. 1978). After 70% nephrectomy and bilateral ureteric ligation, this value fell to around 10% (see Fig. 1). Similarly, in the dog, we have found that the kidney contributes $28.7\% \pm 3.7\%$ to glucagon MCR (K. POLONSKY, unpublished work) and the liver $34.7 \pm 2.6\%$ (Fig. 2; JASPAN et al. 1981). The pattern of in vivo handling of glucagon is in fact remarkably similar to that of plasma somatostatin-like immunoreactivity (POLONSKY et al. 1981), since the liver and kidney also each account for approximately 30% of somatostatin clearance.

## F. Plasma Half-Life

The plasma half-life $t_{1/2}$ of glucagon has also been assessed in laboratory animals and humans. Since the $t_{1/2}$ of the hormone is dependent of its MCR as well as its volume of distribution, studies of this parameter are of limited value as a measure

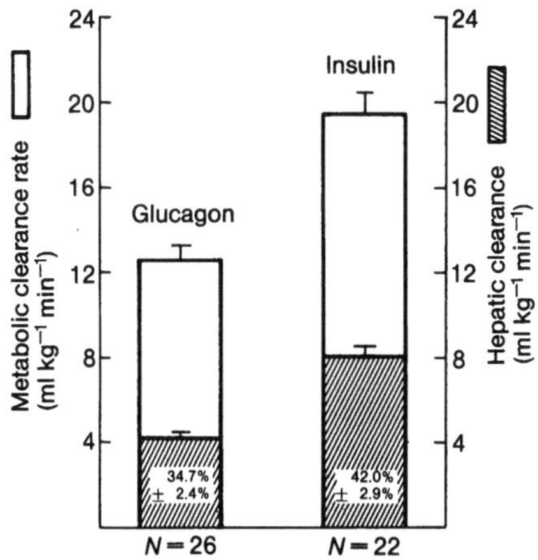


**Fig. 2.** The metabolic and hepatic clearance rates of glucagon in the dog. Simultaneously measured values for insulin are shown. The contribution of hepatic clearance to total MCR for each hormone is shown within the *hatched bars*. JASPAN et al. (1981)

of metabolism. In addition, the acute disappearance of glucagon does not follow a linear experimental decay and therefore cannot be used to predict the MCR of the hormone and as such is a somewhat questionable measure of metabolism.

In the dog, we found that the $t_{1/2}$ of glucagon was $5.5 \pm 0.5$ min (Table 1; JASPAN et al. 1981). This agrees closely with the values of $5.1 \pm 0.6$ and $5.3 \pm 0.3$ min, found in humans for exogenous and endogenous glucagon, respectively (ALFORD et al. 1976). In the latter study, although the MCR was similar in diabetics and non-diabetics, the $t_{1/2}$ was significantly greater in diabetics ($6.6 \pm 0.5$ min) than in non-diabetics ($4.8 \pm 0.2$ min). There was no correlation between $t_{1/2}$ and MCR in this study. A similar lack of correlation between $t_{1/2}$ and MCR has also been shown for growth hormone and insulin. The authors concluded that this difference was probably due to an increase in the apparent distribution space of the hormone. Thus, in diabetics, glucagon is distributed in an abnormal and larger pool.

## G. Summary and Conclusions

1. The metabolic clearance rate of glucagon in humans is approximately 10 ml kg$^{-1}$ min$^{-1}$. Similar values are found in the dog and pig and values approximately three times this level are found in the rat.
2. The MCR is independent of plasma glucagon concentration over a wide concentration range stretching into the pharmacologic range of glucagon concentrations.
3. The liver and kidney are major sites of glucagon clearance, each organ contributing 30% to overall MCR.
4. Glucagon MCR is unaffected by insulin and somatostatin. Reduced glucagon MCR is found in starvation, liver disease, and uremia, but is unchanged in diabetes.
5. Plasma half-life of glucagon is approximately 5 min and is prolonged in diabetes.

## References

Alford FP, Bloom SR, Nabarro JDN, Hall R, Berger GM, Coy DM, Kaskin AJ, Schally AV (1974) Glucagon control of fasting glucose in man. Lancet 2:974–976
Alford FP, Bloom SR, Nabarro JDN (1976) Glucagon metabolism in man. Studies on the metabolic clearance rate and the plasma acute disappearance time of glucagon in normal and diabetic subjects. J Clin Endocrinol Metab 43:830–838
Alford FP, Dudley FJ, Chisholm DJ, Findlay DM (1979) Effect of portasystemic venous shunt surgery on hyperglucagonemia in cirrhosis: paired studies of the pre- and post-shunt subjects. Gut 20:817–824
Burman KD, Smallridge RC, Jones L, Ramos EA, O'Brian JT, Wright FD, Wartofsky L (1980) Glucagon kinetics in fasting: physiological elevations in serum 3′5′3′ triiodothyronine increases the metabolic clearance rate of glucagon. J Clin Endocrinol Metab 51:1158–1165
Emmanouel DS, Jaspan JB, Rubenstein AH, Huen AHJ, Find E, Katz A (1978) Glucagon metabolism in the rat: contribution of the kidney to the metabolic clearance rate of the hormone. J Clin Invest 62:6–13
Fischer M, Sherwin RS, Hendler R, Felig P (1976) Kinetics of glucagon in man. Effects of starvation. Proc Natl Acad Sci USA 73:1735–1739
Gerich JE (1976) Alpha cell dysfunction in diabetes mellitus. Metabolism 25 [Suppl 1]:1513–1516
Gerich JE, Lorenzi M, Bier DM, Schneider V, Karam J, Rivier J, Guillemin R, Forsham P (1974) Effects of somatostatin on plasma glucose and glucagon levels in human diabetes mellitus. Pathophysiologic and therapeutic implications. N Engl J Med 291:544–547
Ishida T, Rojdmark S, Bloom G, Chou MCY, Field JB (1980) The effect of somatostatin on the hepatic extraction of insulin and glucagon in the anesthetized dog. Endocrinology 106:220–230
Issekutz B, Vranic M (1980) Role of glucagon in regulation of glucose production in exercising dogs. Am J Physiol 238:E13–20
Jaspan JB, Rubenstein AH (1977) Circulating glucagon plasma profiles and metabolism in health and disease. Diabetes 26:887–904
Jaspan JB, Polonsky KS, Lewis M, Pensler J, Pugh W, Moossa AR, Rubenstein AH (1981) Hepatic metabolism of glucagon in the dog. Am J Physiol 240:E233–244
Kuku SF, Zeidler A, Emmanouel DS, Katz AI, Rubenstein AH, Levin NW, Lello A (1976) Heterogeneity of plasma glucagon: patterns in patients with chronic renal failure and diabetes. J Clin Endocrinol Metab 42:173–176
Lefèbvre PJ, Luyckx AS (1976) Plasma glucagon after kidney exclusion: experiments in somatostatin infused and in eviscerated dogs. Metab Clin Exp 25:721–768
Marco J, Diego J, Villanueva M, Dray-Fierros M, Valverde I, Segovia J (1973) Elevated plasma of glucagon levels in cirrhosis of the liver. N Engl J Med 289:1107–1111
Marliss EB, Aoki TT, Unger RH, Soeldner JS, Cahill GF (1970) Glucagon levels and metabolic effects in prolonged fasted man. J Clin Invest 49:2256–2270
McGarry DJ, Foster DW (1977) Hormonal control of ketogenesis. Arch Intern Med 137:495–501
Müller WA, Girardier L, Seydoux J, Berger M, Renold AE, Vranic M (1978) Extrapancreatic glucagon and glucagonlike immunoreactivity in depancreatized dogs. A quantitative assessment of secretion rates and anatomical delineation of sources. J Clin Invest 62:124–132
Polonsky KS, Jaspan JB, Berelowitz M, Emmanouel DS, Dhorajiwala J, Moosa AR (1981) The hepatic and renal metabolism of somatostatin-like immunoreactivity: simultaneous assessment in the dog. J Clin Invest 68:1149–1157
Rizza R, Cryer P, Gerich J (1979) Role of glucagon, catecholamines and growth hormone in glucose counterregulation. J Clin Invest 64:62–71
Sherwin R, Joshi P, Hendler PR, Felig P, Cann HO (1974) Hyperglucagonemia in Laennec's cirrhosis. N Engl J Med 290:239–242

Sherwin RS, Bastl C, Finkelstein FO, Fischer M, Black H, Hendler R, Felig P (1976) Influence of uremia and hemodialysis on the turnover and metabolic effects of glucagon. J Clin Invest 57:722–733
Sherwin RS, Fisher M, Bessoff J, Snyder N, Hendler R, Conn HO, Felig P (1978) Hyperglucagonemia in cirrhosis. Altered secretion and sensitivity to glucagon. Gastroenterology 74:1224–1228
Tait PG (1963) The use of isotopic steroids for measurement of production rates in vivo. J Clin Endocrinol 23:1285–1297
Unger RH (1971) Glucagon and the insulin: glucagon ratio in diabetes and other catabolic illnesses. Diabetes 20:834–838
Unger RH (1976) Diabetes and the alpha cell. Diabetes 25:136–151
Unger RH, Orci L (1981) Glucagon in the pathogenesis of diabetes mellitus. The bihormonal hypothesis. In: Unger R, Orci L (eds) Glucagon. Elsiever/North-Holland Amsterdam, pp 369–376
Unger RH, Ohneda A, Aguila-Parada E, Eisentraut AM (1969) The role of aminogenic glucagon secretion in blood glucose homeostasis. J Clin Invest 48:810–822
Van Hoorn WA, Vinik A, Hickman A (1978) The metabolic clearance of endogenous immunoreactive glucagon in pancreatectomized and sham operated pigs. Endocrinology 103:1084–1089
Willerson JF, Hutcheson D, Leshin S, Faloona G, Unger RH (1974) Serum glucagon and insulin levels and their relationship to blood glucose values in patients with acute myocardial infarction and acute coronary insufficiency. Am J Med 57:747–753

CHAPTER 40

# Hepatic Handling of Glucagon

T. ISHIDA and J. B. FIELD

## A. Introduction

The anatomic relationship between the pancreas and the liver dictates that glucagon and other pancreatic hormones secreted by the islets must traverse the liver before reaching the systemic circulation. Thus, hepatic extraction of these hormones can modulate significantly their peripheral concentrations. Furthermore, glucagon has important effects on hepatic carbohydrate metabolism. The action of glucagon on hepatic carbohydrate metabolism is modulated by insulin since these two hormones have a number of opposing effects on this process (EXTON and PARK 1972; FELIG and WAHREN 1975; JENNINGS et al. 1977). Studies with isolated liver cells (WAGLE 1975) and perfused rat livers (MACKRELL and SOKAL 1969; EXTON et al. 1971; PARRILLA et al. 1974) indicate that regulation of hepatic glucose production depends upon the insulin: glucagon molar ratio rather than the absolute concentration of either hormone. Therefore, hepatic extraction of glucagon and insulin may determine their relative effects on hepatic carbohydrate metabolism. However, the relative role of these hormones on this process is still not completely defined. STEINER et al. (1981) reported that the portal vein insulin: glucagon molar ratio was not the primary regulator of glucose production and suggested that the absolute amount of each hormone is more important.

ASSAN (1972) reported the uptake of glucagon by the liver under various experimental conditions based on the portal–peripheral gradient for plasma glucagon concentrations and the excretion of both endogenous and exogenous glucagon in bile. However, in that review, the quantitation of hepatic extraction of glucagon was not provided because of lack of blood flow and glucagon measurements in appropriate vessels.

This chapter examines the hepatic extraction of glucagon and some of the factors which regulate it. Hepatic extraction of insulin is discussed when it is relevant to that of glucagon. The relationship between hepatic glucose metabolism and glucagon and insulin is also considered based on the absolute concentrations of these two hormones, their molar ratio in the peripheral circulation and the portal vein, and the molar ratio of the amount of each hormone extracted by the liver.

## B. Evidence for Glucagon Extraction by the Liver

### I. In Vitro Studies

The existence of a significant uptake of glucagon by the liver was suggested by GOLDNER et al. (1954) and LEFEBVRE and LUYCKX (1965) using bioassay procedure. Although in vitro experiments indicate that the exogenous glucagon $^{131}I$ is extensively degraded by liver homogenates (BERSON et al. 1957; KENNY 1956; WILLIAMS et al. 1959) and liver cell fractions (ANSORGE et al. 1971), they do not provide quantitation of hepatic extraction of the hormone. BUCHANAN et al. (1968) also demonstrated hepatic clearance of glucagon in isolated perfused rat liver and suggested the liver is a major site for degradation of the hormone.

## II. Relationship Between Portal and Peripheral Vein Glucagon Levels

The measurement of glucagon concentrations in the portal vein compared with those in peripheral vessels provided some evidence for hepatic extraction of the hormone. Thus BLACKARD et al. (1974) reported that the portal: peripheral glucagon ratio in eight fasting nondiabetic subjects was $1.7 \pm 0.5:1$. A similar ratio was reported by DENCKER et al. (1975) in three normal subjects after a meal. FELIG et al. (1974) found the portal: peripheral ratio for glucagon (1.1–1.5:1) during surgery was consistently less than that for insulin (1.8–4.0:1). These data thus suggest that, compared with insulin, smaller amounts of glucagon are removed by the liver during a single transhepatic passage. In rats, however, a substantial portal: peripheral venous glucagon ratio (2.8:1) was observed by JASPAN et al. (1977). In these experiments they estimated that hepatic glucagon extraction was $58\% \pm 3\%$, very similar to the fraction of insulin removed by the liver. While suggestive, such experimental results cannot be utilized quantitatively to assess hepatic extraction of glucagon because appropriate blood flows were not measured and the effect of dilution of the peripheral venous blood by nonhepatic blood flow was not considered. Measurement of the glucagon concentration in the hepatic vein would circumvent some, but not all, of these objections because the concentration of the hormone in the portal vein is significantly different from that in the hepatic artery.

## III. Studies in Subjects with Portacaval Shunts

Measurement of peripheral glucagon concentration in patients or animals with portacaval shunts has also provided evidence supporting hepatic extraction of that hormone. Patients with such shunts almost always have underlying hepatic disease, which may influence the results and make interpretation of the data difficult. However, a noncirrhotic subject with a mesocaval shunt had elevated fasting peripheral levels of glucagon and an exaggerated response after administration of arginine (MARCO et al. 1973). SHERWIN et al. (1974) reported that peripheral glucagon levels were highest in four patients with portacaval anastomosis, were less elevated in ten cirrhotic patients with spontaneous portal–systemic shunting, and were normal in four cirrhotic patients without portal–systemic shunting. However, these elevated peripheral glucagon levels could result from increased secretion of glucagon from the pancreas, decreased clearance of the hormone by the liver, or both. More recently, SHERWIN et al. (1978) reported similar metabolic clearance rates of glucagon in control and cirrhotic subjects, with and without surgically created portacaval shunts. They suggested that the elevated peripheral glucagon levels in cirrhotics were secondary to increased secretion of glucagon from the pancreas and that the liver did not play a significant role in the clearance of the hormone. This conclusion was supported by the studies of GRECO et al. (1979) who measured both portal and peripheral venous glucagon concentrations. Different results were obtained by ALFORD et al. (1979) who studied the effect of portal–systemic shunts on the basal peripheral glucagon levels, metabolic clearance rate, and basal systemic delivery rate of glucagon in paired studies in ten cirrhotic subjects before and after surgery. They demonstrated that in the presence of cirrhosis, the magnitude of portal–systemic shunting is important in determining the degree of hyperglucagonemia. In cirrhotics prior to shunting, the raised basal glucagon levels are principally due to pancreatic hypersecretion, whereas after surgery the elevated glucagon levels reflect both pancreatic hypersecretion and delayed clearance of glucagon by the liver. These results indicate that, in humans, the liver contributes significantly to the clearance of circulating glucagon. However, the heterogeneity of plasma glucagon immunoreactivity may contribute to the results of peripheral hyperglucagonemia in cirrhotics, as discusses in Sect. B.V

## IV. Direct Measurement of Basal Hepatic Extraction of Glucagon

Although in vitro studies indicated that exogenous glucagon is extensively degraded by the liver, they do not provide quantitation of hepatic extraction of the hormone. Direct determination of glucagon extraction by the liver requires measurement of the total amount of glucagon presented to and leaving the liver. This necessitates measurements of the glucagon concentrations in the portal and hepatic vein and artery as well as portal vein and hepatic artery blood flow. Portal vein and hepatic artery blood flows must be measured separately since the glucagon concentrations in these two vessels may be quite different. Such flow measurements permitted quantitative determination of glucagon flux across the liver, assuming that the glucagon concentration in the left common hepatic vein which was sampled was representative of all the hepatic veins.

We have examined hepatic extraction of glucagon and insulin in the basal state and after a variety of pertubations of the amount of these hormones presented to the liver in an attempt to understand and characterize these processes better. Mongrel adult dogs weighing about 20 kg were fasted overnight and anesthetized with pentobarbital (30 mg/kg body weight). The surgical procedure and experimental methods have been described in detail elsewhere (KADEN et al. 1973; HARDING et al. 1975; ROJDMARK et al. 1978a; ISHIDA et al. 1981).

Electromagnetic flow probes were placed on the portal vein immediately caudal to the superior pancreaticoduodenal vein, which was ligated, and on the hepatic artery. Blood sampling catheters were placed in the left common hepatic vein, the femoral artery, and the portal vein. The tip of the multiple-hole portal vein sampling catheter lay immediately before the hepatic portal vein bifurcation. When hormones were to be infused, another catheter was placed either in the superior mesenteric vein or the superior pancreaticoduodenal artery. The tip of the mesenteric vein catheter was approximately 10 cm caudal to the portal vein sampling catheter. After the completion of the surgery, a 60–120 min equilibration period was allowed. Multiple blood samples were obtained from the portal and hepatic veins and the artery. Blood flow measurements were obtained after each blood sampling and were converted to plasma flow based on hematocrits. The pancreatic hormone reaching the liver was the sum of the contributions of the portal vein (portal vein hormone concentration × portal vein plasma flow) and the hepatic artery (femoral artery hormone concentration × hepatic artery plasma flow). The pancreatic hormone leaving the liver was the product of the hepatic vein hormone concentration and the sum of the portal vein and hepatic artery plasma flows. The percentage hepatic extraction was determined by the formula:

Pancreatic hormone to the liver – pancreatic hormone from the liver/Pancreatic hormone to the liver × 100

The net hepatic glucose output has been calculated in a similar fashion:

Total glucose from the liver – total glucose to the liver/Body weight

Initially, experiments were done on anesthetized animals, but more recently with the use of chronically implanted Doppler flow probes, the hepatic extraction of pancreatic hormones has been measured in conscious, unrestrained animals. This has eliminated the possible effect of the stress of anesthesia and surgery on the hepatic extraction of pancreatic hormones.

During the control period, about 10%–20% of the 40–60 ng/min glucagon presented to the liver was extracted by that organ, considerably less than that of insulin (40%–60% of the 10–20 mIU/min insulin delivered to the liver was extracted by a single transhepatic passage; Table 1). However, JASPAN et al. (1977) reported a significantly greater hepatic glucagon extraction rate (58% ± 3%) in rats based on portal–peripheral venous glucagon gradients. As discussed previously, these experiments did not include blood flow measurements or take into consideration the dilution of hepatic venous blood by systemic venous blood. Recently, JASPAN et al. (1981) utilized an experimental design similar to ours and reported much lower basal hepatic glucagon extraction (23% ± 2%) in anesthetized dogs. BROCKMAN et al. (1976) observed hepatic glucagon extraction to be 7% in the conscious fed sheep. They utilized venoarterial plasma concentration differences and measured portal and hepatic vein plasma flow by a *p*-aminohippuric acid infusion into the mesenteric vein. The highest hepatic extraction rate for glucagon (61.5% ± 3.8%) was found by FISCHER et al. (1978a) utilizing an experimental preparation similar to ours. Since their report does not provide details, it is difficult to explain the discrepancy. It is possible that some of the variability could reflect the heterogeneity of glucagon immunoreactivity. This factor is discussed in Sect. B.V.

## V. Effect of Glucagon Heterogeneity on Hepatic Glucagon Extraction Rate

The heterogeneity of plasma glucagon immunoreactivity has been well documented (VALVERDE et al. 1975; KUKU et al. 1976; JASPAN et al. 1977), and can be an important factor in the determination of hepatic extraction of glucagon. The glucagon immunoreactivity determined with antibody 30K is composed of at least four different components of widely different molecular weights, as discussed in detail by VALVERDE in Chap. 11. The biologic activity and hepatic extraction of these fractions are quite different (VALVERDE et al. 1975; KUKU et al. 1976; JASPAN et al.

**Table 1.** Comparison of the mean (±standard error) basal hepatic extraction of glucagon and insulin in anesthetized dogs

| Reference | Glucagon | | Insulin | |
|---|---|---|---|---|
| | Basal hepatic extraction rate (%) | Total amount to liver (ng/min) | Basal hepatic extraction rate (%) | Total amount to liver (mIU/min) |
| RÖJDMARK et al. (1978a) | 19± 9 | 32± 9 | 62±6 | 7±1 |
| RÖJDMARK et al. (1978b) | 5± 3 | 51±10 | 63±5 | 16±7 |
| RÖJDMARK et al. (1979) | 15±11 | 52± 8 | 66±6 | 23±4 |
| ISHIDA et al. (1980) | 9± 3 | 65± 5 | 61±5 | 10±3 |
| ISHIDA et al. (1980) | 28± 6 | 44± 2 | 65±4 | 16±3 |
| ISHIDA et al. (1981) | 12± 8 | 52± 6 | 62±7 | 15±3 |
| ISHIDA et al. (1981) | 12± 8 | 76±21 | 49±8 | 11±3 |
| JASPAN et al. (1981) | 23± 2 | | 45±3 | |
| FISCHER et al. (1978a) | 62± 4 | | 30[a] | |
| BROCKMAN et al. (1976) | 7[b] | | | |

[a] Data from another report of FISCHER et al. (1978b)
[b] Data from conscious fed sheep

1977). Only the 3,500 daltons species appears to be removed by the liver (JASPAN et al. 1977). This species constitutes a variable fraction of the total plasma glucagon immunoreactivity and would account for an apparently low hepatic extraction if it represented only a small fraction of the total plasma glucagon. This does not explain the small hepatic extraction of glucagon which we have reported since, in our experiments, more than 75% of the portal vein glucagon immunoreactivity was due to the 3,500 daltons species (RÖJDMARK et al. 1978a, ISHIDA et al. 1980). However, VALVERDE (1977) reported that the 3,500 daltons component was a much smaller fraction of the total glucagon immunoreactivity. Big plasma glucagon comprised 55% and the 3,500 daltons fraction was 17% of the total in normal dogs respectively.

## VI. Relationship Between Hepatic Extraction of Glucagon and Insulin

Investigations utilizing a variety of techniques, from perfused livers to studies in vivo, indicated that about 50% of the insulin delivered to the liver was removed in a single transhepatic passage (Table 1; MORTIMORE et al. 1959; KADEN et al. 1973; HARDING et al. 1975; MONDON et al. 1975; CAMU 1975; TERRIS and STEINER 1976; MISBIN et al. 1976). However, some recent studies have reported significantly lower values for hepatic extraction of insulin. NAVALESI et al. (1976) determined insulin kinetics after portal and peripheral injection of insulin $^{125}$I in conscious dogs and reported a mean hepatic extraction of 19.6% (range 9.6%–36.2%). They estimated hepatic extraction of insulin based on the difference in the metabolic clearance rate of insulin $^{125}$I infused into the portal or peripheral vein. A total absence of hepatic extraction of insulin was observed by STRIFFLER and CURRY (1979) using an in situ perfused rat liver–pancreas preparation from fed rats. However, the hepatic insulin extraction rate was about 50% when rats were fasted for 24 h. Since the determination of hepatic insulin extraction rate in the liver perfusion system may depend on the flow rate, a precise figure from such in vitro studies is difficult to obtain. Thus, the wide discrepancies between the various results probably reflect the different experimental approaches used and the absence of blood flow and hormone measurements in the appropriate vessels.

Our previous results (RÖJDMARK et al. 1978 a and 1978 b; ISHIDA et al. 1980, 1981 a) and those of JASPAN et al. (1981) indicated that the basal hepatic glucagon extraction rate of 10%–20% was significantly less than that of insulin. However, the results of FISCHER et al. (1978 a, b) are in conflict with this conclusion. Utilizing an experimental procedure similar to ours in anesthetized dogs, they reported that hepatic extraction of insulin was 30% while that of glucagon was 62%. Although these were reported in different papers, the reason for this discrepancy is not clear. However, the higher portal–peripheral venous insulin gradient (between 2 and 4) compared with the glucagon gradient (< 2) in fasting subjects supports the conclusion that hepatic extraction of insulin exceeds that of glucagon (BLACKARD and NELSON 1970; BLACKARD et al. 1974; FELIG et al. 1974).

# C. Factors Regulating Hepatic Extraction of Glucagon

## I. Changes in Splanchnic Blood Flow

The effect of changes in splanchnic blood flow on hepatic extraction of glucagon has not been examined, while the effect on hepatic extraction of insulin is controversial. ROJDMARK et al. (1978 a, b) reported that increased portal vein plasma flow had no effect on the hepatic extraction of glucagon and insulin. In contrast, decreased hepatic extraction of insulin in the isolated perfused liver was associated with increased portal vein blood flow (MISBIN et al. 1976; BUCOLO 1978). However, HONEY and PRINCE (1979) found no effect of changes in blood flow on insulin extraction by the perfused liver. Recently, we found significant reduction of portal vein plasma flow in anesthetized and laparotomized dogs (ISHIDA and FIELD 1981 b). There are conflicting results about the effect of anesthesia and laparotomy on the hepatic vein and splanchnic blood flow. No difference between the anesthetized and conscious state (GILMORE 1958; EVRINGHAM et al. 1959; BOND et al. 1980), decreased blood flow with anesthesia (SHACKMAN et al. 1953; STRANDELL 1978; EADE and GINN 1978), and increased blood flow by chronic administration of pentobarbital (YATES et al. 1979) have all been reported. KATZ and BERGMAN (1969) compared the hepatic and portal venous blood flow in three conscious and eight anesthetized dogs by means of *p*-aminohippuric acid infusion into the mesenteric vein and reported a significant increase in the portal vein blood flow and decreased hepatic artery flow in conscious dogs; the hepatic vein blood flow was similar to that in anesthetized dogs. As shown in Table 2, anesthesia significantly decreased portal vein and hepatic vein plasma flow, but not the hepatic artery plasma flow. However, this had no effect on the hepatic extraction of glucagon and insulin. Thus, changes in splanchnic blood flow may not be an important factor in the regulation of these processes.

**Table 2.** Comparison of the mean (±standard error) basal hepatic extraction of insulin and glucagon in the anesthetized and conscious state

| | Anesthetized state ($N=30$) | Conscious state at 1 week after surgery ($N=17$) | Conscious state at 2 weeks after surgery ($N=9$) |
|---|---|---|---|
| Hepatic artery | | | |
| Plasma flow (ml/min) | 122 ±11 | 121 ±13 | 90 ± 8 |
| Plasma glucose (mg/100 ml) | 88 ± 3 | 77 ± 2 | 83 ± 4 |
| Plasma insulin (μIU/ml) | 17 ± 1 | 17 ± 2 | 16 ± 2 |
| Plasma glucagon (pg/ml) | 128 ± 9 | 173 ±19[a] | 144 ±13 |
| Portal vein | | | |
| Plasma flow (ml/min) | 299 ±20 | 487 ±41[a] | 426 ±66[a] |
| Plasma glucose (mg/100 ml) | 87 ± 3 | 77 ± 2 | 80 ± 3 |
| Plasma insulin (μIU/ml) | 56 ± 8 | 45 ± 6 | 41 ± 9 |
| Plasma glucagon (pg/ml) | 220 ±14 | 296 ±27[a] | 235 ±26 |
| Hepatic vein | | | |
| Plasma flow (ml/min) | 421 ±26 | 608 ±40[a] | 516 ±66 |
| Plasma glucose (mg/100 ml) | 93 ± 3 | 82 ± 2 | 89 ± 4 |
| Plasma insulin (μIU/ml) | 21 ± 2 | 22 ± 3 | 21 ± 4 |
| Plasma glucagon (pg/ml) | 160 ±11 | 232 ±23[a] | 194 ±15 |
| Total insulin to the liver (mIU/min) | 18 ± 2 | 25 ± 4 | 18 ± 4 |
| Total insulin from the liver (mIU/min) | 9 ± 1 | 14 ± 3 | 10 ± 2 |
| Hepatic insulin extraction (%) | 45 ± 4 | 43 ± 4 | 41 ± 7 |
| Total glucagon to the liver (ng/min) | 82 ± 7 | 167 ±18[a] | 104 ±15 |
| Total glucagon from the liver (ng/min) | 69 ± 7 | 144 ±18[a] | 94 ±11 |
| Hepatic glucagon extraction (%) | 15 ± 5 | 14 ±13 | 7 ± 6 |
| Hepatic glucose output (mg $kg^{-1}$ $min^{-1}$) | 1.3± 0.1 | 1.6± 0.3 | 2.1± 0.4 |

[a] Significant change from the anesthetized state ($P<0.01$)

## II. Anesthesia and Laparotomy

Most of the in vivo studies of hepatic glucagon extraction have involved anesthetized and laparotomized animals. Under these circumstances, anesthesia and laparotomy may exert an important effect on the results, which therefore may not reflect the normal physiologic state. However, in current experiments, we have used chronically implanted catheters and Doppler flow probes which permit us to compare the results in the anesthetized and conscious state. During the control period in anesthetized dogs, 15%±5% of the 82±7 ng/min glucagon presented to the liver was extracted in a single transhepatic passage (Table 2). In the conscious state 1 week after surgery, the basal amount of glucagon delivered to the liver increased significantly to 167±18 ng/min, concomitant with the significant increase of the portal vein plasma flow from 299±20 ml/min in anesthetized animals to 487±41 ml/min in conscious dogs. However, hepatic extraction of glucagon was 14%±13% and not different from that of anesthetized dogs, despite a significant increase of portal vein plasma flow. When the flow probes and catheters had been in situ for 2 weeks the basal amount of glucagon reaching the liver returned to the level observed in the anesthetized state, but the portal vein plasma flow still significantly exceeded that of the anesthetized state. At this time, the basal hepatic extraction of glucagon was 7% ± 6%.

Furthermore, hepatic extraction of insulin in the anesthetized state (45% ± 4%) was not different from that in conscious dogs (43% ± 4%), despite the significant difference in portal vein plasma flow. Thus, anesthesia and the stress of surgery did not influence the basal hepatic extraction of glucagon and insulin. These results also suggest that our previous findings in anesthetized dogs may be relevant to more normal physiology. However, it is obvious that they must be expanded and applied to other experimental conditions to exclude an effect of anesthesia and surgery on the hepatic extraction of these hormones.

## III. Changes in Portal Vein Glucagon Concentration

### 1. Increased Amount of Glucagon Presented to the Liver

#### a) Arginine and Cholecystokinin–Pancreozymin

During the infusion of arginine (10 g) and cholecystokinin–pancreozymin (CCK–PZ) (75 Ivy dog units CCK and 300 Crick, Haper and Raper units PZ), the portal vein glucagon and insulin concentrations increased from the control value of 126 ± 31 to about 1,000 pg/ml and from the basal level of 40 ± 8 to about 250 μIU/ml, respectively (ROJDMARK et al. 1978a). The basal hepatic extraction rate of glucagon (19% ± 9% of the 32 ± 9 ng/min glucagon presented to the liver) was unaffected by the infusion despite a 15-fold increased amount of that hormone reaching the liver. This increment is much larger than that of portal vein glucagon concentration because of the concomitant increased portal vein plasma flow. During the arginine–CCK–PZ infusion, the fraction of the insulin extracted by the liver significantly decreased from a control value of 62% ± 6% to a nadir of 22% ± 13% at 40 min (Fig. 1). More recently, we observed a similar effect of arginine–CCK-PZ on these processes in paired studies in seven anesthetized and conscious dogs (ISHIDA and FIELD 1981).

Thus, these findings indicate that glucagon and insulin extraction by the liver are quite different, are regulated independently, and are not affected by anesthesia.

#### b) Exogenous Glucagon and Insulin Infusion

Similar effects on hepatic extraction of glucagon and insulin were obtained when pharmacologic amounts of these hormones were infused through the pancreatic artery (RÖJDMARK et al. 1978b). During the infusion of glucagon (16 ng $kg^{-1}$ $min^{-1}$) and insulin 1.8 mIU $kg^{-1}$ min $^{1}$), the portal vein glucagon and insulin con-

**Fig. 1a–c.** Effect of arginine–cholecystokinin–pancreozymin on insulin (**a**) and glucagon ▶ (**b**) concentrations in portal vein and femoral artery and on hepatic extraction of insulin and glucagon (**c**) in seven anesthetized dogs. *Asterisks* indicate significant changes from control ($p \leq 0.05$). RÖJDMARK et al. (1978a)

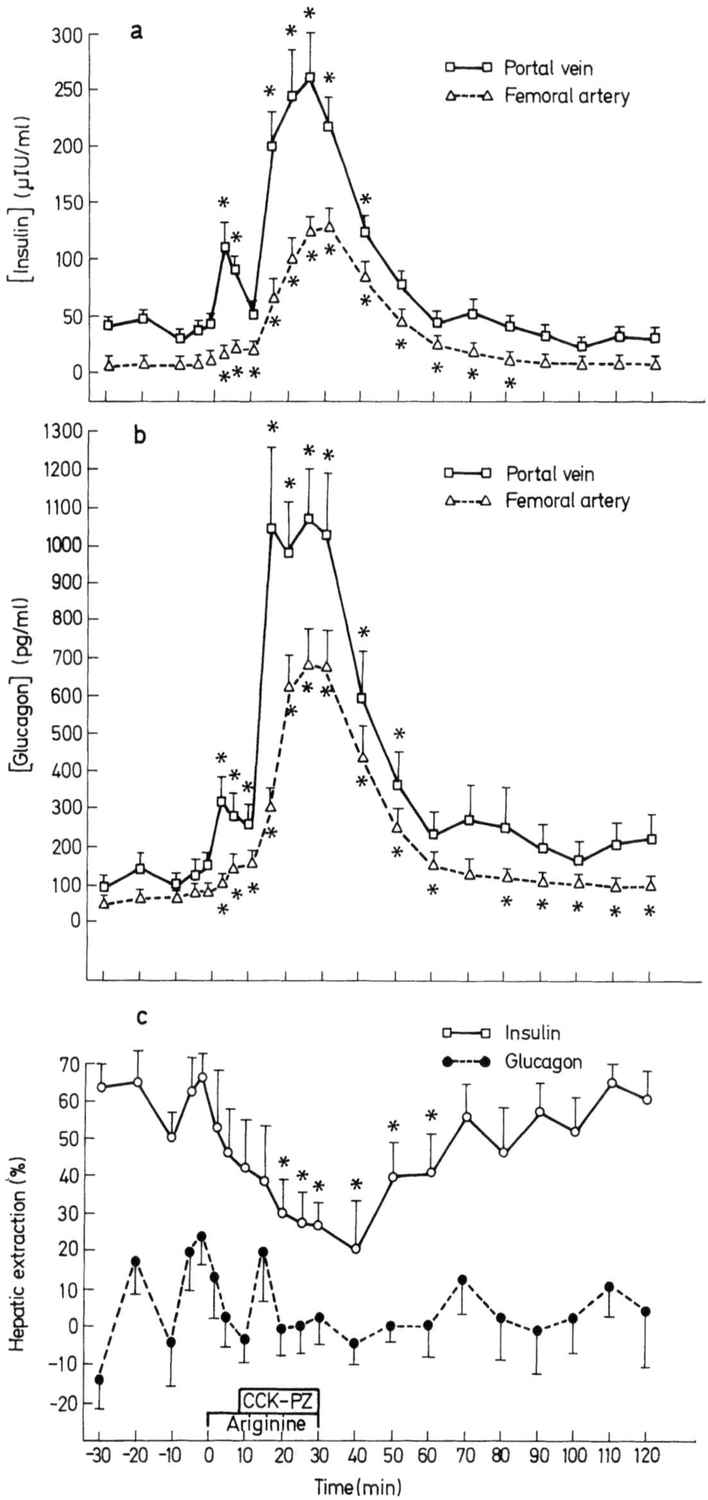
a
[Insulin] (μIU/ml)
300
250
200
150
100
50
0
Portal vein
Femoral artery
b
[Glucagon] (pg/ml)
1300
1200
1100
1000
900
800
700
600
500
400
300
200
100
0
Portal vein
Femoral artery
c
Hepatic extraction (%)
70
60
50
40
30
20
10
0
-10
-20
Insulin
Glucagon
CCK-PZ
Ariginine
-30 -20 -10 0 10 20 30 40 50 60 70 80 90 100 110 120
Time (min)

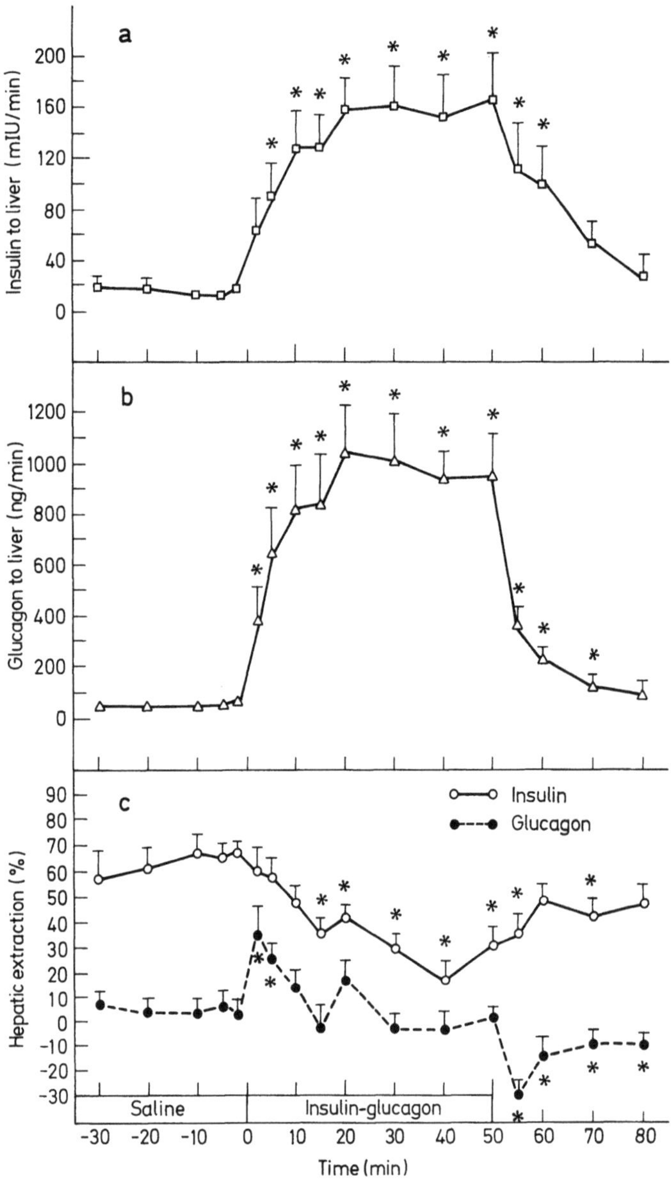


**Fig. 2 a–c.** Effect of intrapancreatic infusion of insulin and glucagon on the total amount of insulin (**a**) and glucagon (**b**) presented to the liver and on hepatic extraction of these hormones (**c**) in six anesthetized dogs. *Asterisks* indicate significant changes from control ($p \leqq 0.05$). RÖJDMARK et al. (1978 b)

centration increased from the basal level of 218±40 to about 3,300 pg/ml and from the control value of 70±21 to 640 μIU/ml, respectively. As shown in Fig. 2, the basal hepatic extraction of glucagon (5%±3%) showed no significant changes during the infusion period, while that of insulin deceased significantly from the control value of 63%±5% to a nadir of 18%±8% at 40 min. However, after the infusion was stopped, the hepatic glucagon extraction rate fell promptly and became negative while the hepatic insulin extraction rate changed in the opposite direction and returned to control levels. It is unlikely that the negative hepatic glucagon extraction values at the end of the experiment can be explained by laminar blood flow in the portal vein, inadequate mixing of portal vein blood, or nonrepresentative samples of the hepatic vein, since these factors should affect both glucagon and insulin in a similar fashion, inasmuch as they were infused together and measured in the same blood samples from the portal and hepatic veins.

Although hepatic release of glucagon has not been observed in vivo, it is possible that the liver, when exposed to high plasma glucagon concentrations, binds more glucagon than it can inactivate. The later decline in plasma glucagon concentrations may trigger a release of bound glucagon into the hepatic vein, and thus, at least in part, account for the negative hepatic glucagon extraction. We also confirmed these results utilizing pharmacologic amounts of glucagon (20 ng $kg^{-1}$ $min^{-1}$) and insulin (1.7 mIU $kg^{-1}$ $min^{-1}$) infused into the mesenteric vein instead of the pancreatic artery (ISHIDA et al. 1980). However, when glucagon (4 ng $kg^{-1}$ $min^{-1}$) and insulin (1.0 mIU $kg^{-1}$ $min^{-1}$) were infused into the mesenteric vein in both anesthetized and conscious dogs in amounts which produced more physiologic levels of glucagon and insulin in the portal vein, no change in hepatic extraction of glucagon or insulin was found (ISHIDA and FIELD 1981). The infusion of 4 ng $kg^{-1}$ $min^{-1}$ glucagon and 1 mIU $kg^{-1}$ $min^{-1}$ insulin raised portal vein hormone concentrations to apparently the same values which were obtained after infusion of arginine–CCK–PZ. These results indicate that the portal vein glucagon and insulin concentrations alone do not regulate their hepatic extraction and that arginine–CCK–PZ, independent of its elevation of glucagon, appears to decrease hepatic extraction of insulin. The reduction in hepatic extraction of insulin cannot be attributed to the concomitant increase in portal vein glucagon levels. It is possible that the amount of glucagon presented to the liver as well as the stimulus might be important.

We recently observed an increase in hepatic glucagon extraction rate during the infusion of somatostatin (250 ng $kg^{-1}$ $min^{-1}$), glucagon (20 ng $kg^{-1}$ $min^{-1}$), and insulin (1.7 mIU/kg/min) into the mesenteric vein (ISHIDA et al. 1980). Hepatic glucagon extraction rate increased significantly from 7% ± 12% to 35% ± 11%, while hepatic extraction of insulin was unchanged (Fig. 3). Thus, somatostatin increased the fraction of hepatic glucagon extraction and also abolished the decrease of hepatic insulin extraction rate induced by glucagon and insulin infusion. Since somatostatin increased both exogenous glucagon and insulin removal by the liver, it is possible that it has a common action on both processes, although we have previously presented evidence that hepatic glucagon and insulin extraction were separately regulated. In contrast to our results, JASPAN et al. (1981) reported no significant changes of hepatic extraction of glucagon and insulin during peripheral infusion of somatostatin (800 ng $kg^{-1}$ $min^{-1}$) and glucagon (1–300 ng $kg^{-1}$ $min^{-1}$) which increased the portal vein glucagon concentrations to 250–2,000 pg/ml, respectiveley. This discrepancy might reflect the different infusion route and the larger amounts of hormones infused compared with our studies. Both of these studies demonstrate a very large capacity of the liver to extract glucagon since saturation occurred only at glucagon delivery rates above 2,000 ng/min (JASPAN et al. 1981).

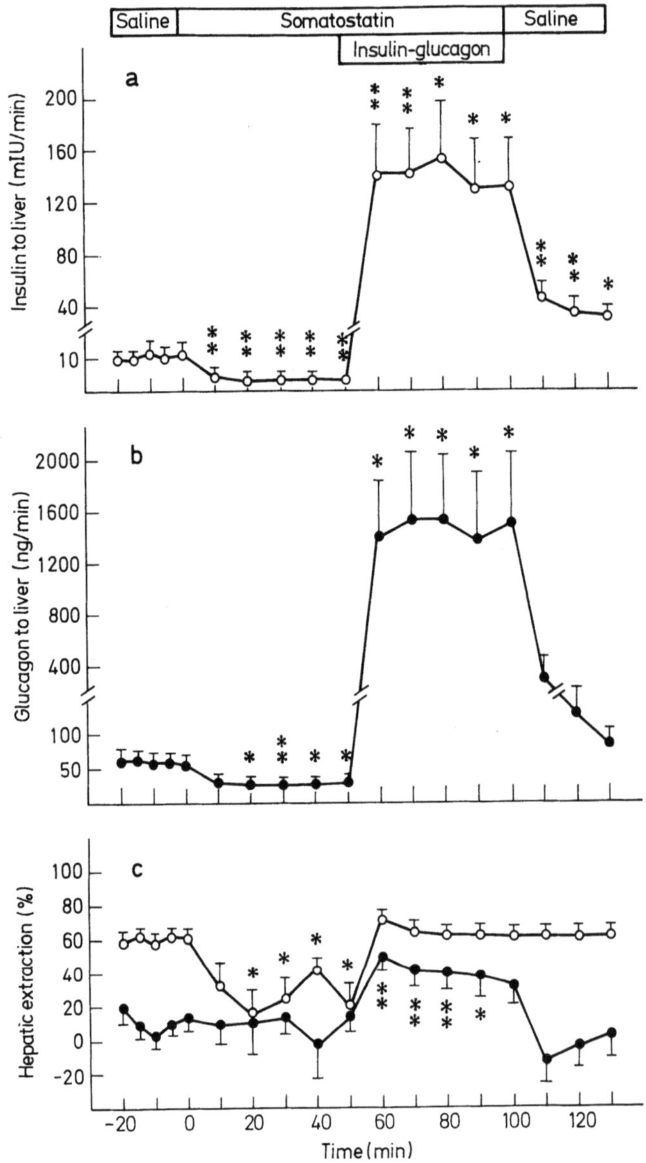


**Fig. 3a–c.** Effect of intraportal infusion of somatostatin, insulin, and glucagon on the total amount of insulin (**a**) and glucagon (**b**) presented to the liver and on hepatic extraction of these hormones (**c**) in eight anesthetized dogs. *Open circles* insulin extraction; *full cirles* glucagon extraction. *Asterisks* indicate significant changes from basal level ($p<0.01$); *double asterisks* $p<0.05$. ISHIDA et al. (1980)

---

**Fig. 4a–c.** Total amount of glucagon reaching the liver (*full circles*) and leaving the liver ▶ (*open circles*) in response to tolbutamide in 12 anesthetized dogs (**a**) and insulin (**b**) infusion in 6 anesthestized dogs. Hepatic extraction of glucagon after tolbutamide (*open circles*) and insulin (*triangles*) infusion (**c**). *Asterisks* indicate significant changes from basal level ($p<0.05$) ISHIDA et al. (1981)

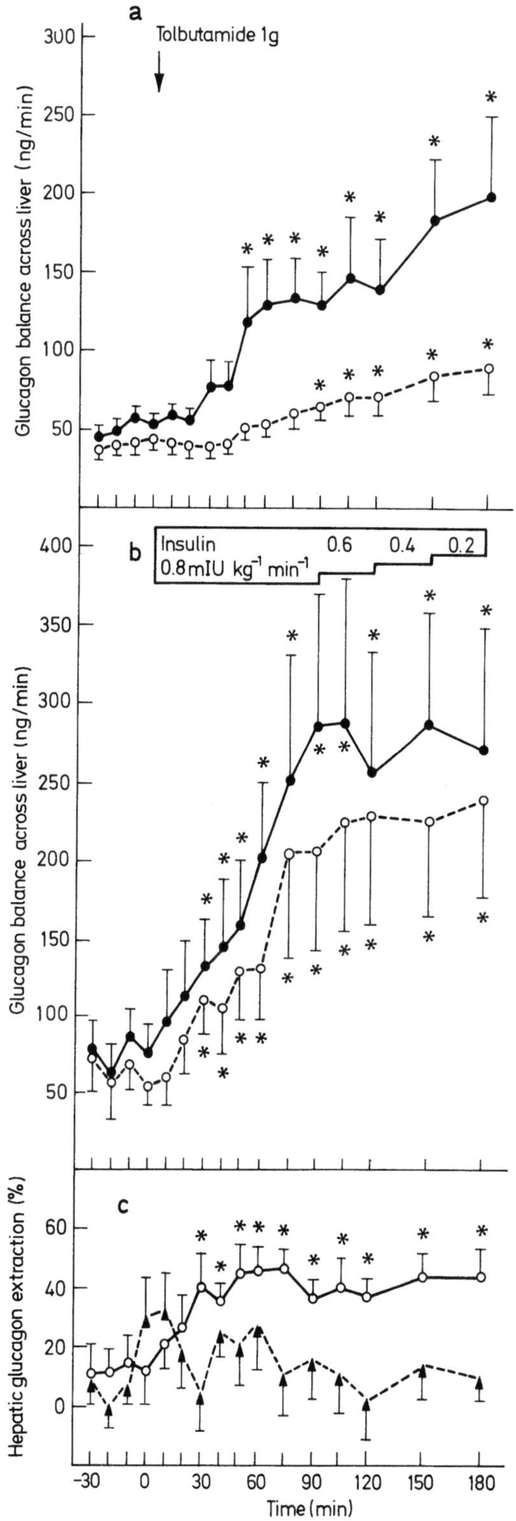
a
Tolbutamide 1g
Glucagon balance across liver (ng/min)
300
250
200
150
100
50
b
Insulin
0.8 mIU kg⁻¹ min⁻¹
0.6
0.4
0.2
400
350
300
250
200
150
100
50
c
Hepatic glucagon extraction (%)
60
40
20
0
-30
0
30
60
90
120
150
180
Time (min)

c) Tolbutamide

Administration of tolbutamide and its attendant hypoglycemia caused a significant increase in glucagon secretion and the amount of glucagon presented to the liver (ISHIDA et al. 1981 a). Prior to this increase in glucagon secretion, tolbutamide had no effect on hepatic extraction of the hormone (control value of 12% ± 8%). However, the hepatic extraction rate of glucagon rose to 41% ± 12% when the amount of glucagon presented to the liver increased (Fig. 4). Such augmented hepatic extraction of glucagon appeared to be a direct effect of tolbutamide and could not be attributed to either the hypoglycemia or the increased amounts of glucagon or insulin presented to the liver. Thus, the infusion of insulin to match the hypoglycemia and portal vein glucagon and insulin concentrations achieved by administration of tolbutamide had no effect on hepatic extraction of glucagon (Fig. 4). The failure of insulin extraction to change, despite increased amounts of both glucagon and insulin reaching the liver, provides additional support for the concept that physiologically elevated glucagon per se is not sufficient to alter hepatic extraction of insulin. Since hepatic extraction of glucagon and insulin can increase or decrease indpendent of each other, it is obvious that peripheral levels of these hormones may not accurately reflect their pancreatic secretion.

d) Intraportal Calcium Infusion

Although calcium administration was reported to alter glucagon and insulin concentrations in the peripheral vessels, there was no significant change in hepatic extraction of either glucagon or insulin, despite a 48% increase in glucagon production by the pancreas and a significant fall in the amount of insulin presented to the liver to 62% of the control values during the intraportal calcium infusion at a rate of 8 mg $kg^{-1}$ $h^{-1}$ (ROJDMARK et al. 1979).

### 2. Decreased Amount of Glucagon Presented to the Liver

a) Somatostatin

Infusion of somatostatin (250 ng $kg^{-1}$ $min^{-1}$) into the mesenteric vein decreased portal vein glucagon concentration from the basal value of 247 ± 37 to 111 ± 24 pg/ml (ISHIDA et al. 1980). Prior to infusion of somatostatin, the basal hepatic glucagon extraction rate was 9% ± 6% and during the infusion period it was 7% ± 12% (see Fig. 3). In these studies, somatostatin inhibited the 3,500 daltons fractions more than the total glucagon immunoreactivity, but did not abolish it completely. The 3,500 daltons fraction of glucagon comprised 74% ± 4% of the total glucagon immunoreactivity during the control period and 62% ± 5% during the somatostatin infusion period. However, JASPAN et al. (1981) observed complete suppression of the 3,500 daltons glucagon component by the intravenous infusion of somatostatin (800 ng $kg^{-1}$ $min^{-1}$) and obtained complete absence of hepatic extraction of glucagon, since the liver extracts the 3,500 daltons fraction almost exclusively. Hepatic extraction of endogenous insulin decreased from the basal level of 61% ± 5% of the 10 ± 3 mIU/min insulin presented to the liver to 29% ± 10% of the 3 ± 1 mIU/min insulin reaching the liver during the infusion of somatostatin (see Fig. 3). However, this decrease in hepatic extraction rate of endogenous insulin is based on the mean of the individual values of eight dogs and may be spurious because of the very low concentrations of insulin which were being measured and the fact that at such low concentrations, some dogs appeared to have negative hepatic insulin extraction.

b) Glucose

Oral administration of glucose (1 g/kg) to conscious dogs decreased the portal vein glucagon concentrations from 247±30 pg/ml to a nadir of 72±9 pg/ml at 105 min. However, the basal hepatic extraction of glucagon (18%±4%) did not change significantly. In contrast, the hepatic extraction of insulin increased from the basal level of 39%±4% to 57%±5% at 45 min and remained significantly elevated for 120 min. However, JASPAN et al. (1980) reported that glucose suppressed the portal vein glucagon level from 174±10 to 105±12 pg/ml and reduced the hepatic extraction of glucagon from 22.6%±2.6% to 8.3%±6.1% during the first 120 min in conscious dogs. They also demonstrated an increase in the fraction of hepatic insulin extraction from 33.5±12.3% to 58.9%±11.4% during this period. The reasons for the discrepancy in hepatic glucagon extraction are not known.

## IV. Partial Hepatectomy

The effect of glucagon on the regeneration of the remaining liver in response to a partial hepatectomy is now well documented, as discussed in detail by LEFFERT in Chap. 21. ITATSU et al. (1979) reported an 11-fold increase in portal vein glucagon and a 70% decrease in portal vein insulin concentrations 3 h after 80% hepatectomy, concomitant with a 50% decrease in peripheral glucose levels. They and CARUANA and GAGE (1980) suggested that increased hepatic extraction of glucagon might be the signal for hepatic regeneration. These results are consistent with the assumption by LEFFERT et al. (1976) that 70% hepatectomized liver accumulates and/or sequesters glucagon more effectively than insulin. Such an increase of hepatic extraction of glucagon with unchanged insulin extraction is similar to what we have observed after hypoglycemia induced by tolbutamide, but not by insulin infusion (ISHIDA et al. 1981 a). Although hypoglycemia is common to these two experimental situations as well as hepatectomy, there are other differences which might account for the discrepant results. Hepatectomy is associated with decreased insulin delivery to the liver while both tolbutamide and insulin infusion augmented the amount of insulin presented to the liver. It is thus clear that no single factor can be identified as being responsible for the increased hepatic extraction of glucagon. While CARUANA and GAGE (1980) also demonstrated a significant increase in glucagon uptake per unit weight of remnant liver after a 72% hepatectomy, they also reported a similar change in insulin extraction by the remnant liver. In contrast to the results of ITATSU et al. (1979), CARUANA and GAGE (1980) did not find any difference in the amount of glucagon or insulin reaching the liver after hepatectomy or in their peripheral concentrations. The reasons for this discrepancy are not known. CORNELL (1980) also found increased hepatic extraction of both glucagon and insulin as well as the elevated basal portal venous levels of the two hormones after 67% hepatectomy of fasted rats. They suggested the increased hepatic extraction might be a hepatotropic factor for regenerating liver.

## D. Relationship Between Net Hepatic Glucose Output and Glucagon and Insulin

### I. Peripheral Vein Insulin: Glucagon Molar Ratio

Although, surprisingly a much smaller fraction of glucagon is extracted by the liver compared with insulin, the liver is considered the primary site for glucagon activity (EXTON and PARK 1972; FELIG and WAHREN 1975; JENNINGS et al. 1977). Since BIRNBAUMER and POHL (1973) reported that binding of glucagon by only a small proportion of its receptors on the hepatocyte is sufficient to elicit a biologic effect, such a small fraction of hepatic glucagon extraction does not imply that the liver is unimportant in glucagon metabolism. It is well recognized that glucagon and insulin exert opposing actions on hepatic glucose metabolism, as discussed in detail by CHIASSON and CHERRINGTON in Chap. 16. MACKRELL and SOKAL (1969) suggested that the net hepatic glucose balance may be determined by the relative availability of both hormones, rather than the absolute concentration of either hormone alone. Furthermore, a variety of in vitro studies (PARRILLA et al. 1974; WAGLE 1975; SEITZ et al. 1976) have also supported this concept. UNGER (1971, 1972) measured the peripheral vein insulin: glucagon molar ratio in physiologic and pathologic states and suggested that the insulin: glucagon molar ratio might be the signal controlling the catabolic and anabolic state of the liver. These studies are based on the assumption that the peripheral insulin: glucagon molar ratio accurately reflects that in the portal vein. This assumption may be invalid because of the markedly different hepatic extraction of the two hormones and the fact that hepatic removal of each hormone can vary independent of the other. Furthermore, it is still uncertain whether glucagon has any effect on peripheral glucose disposal (CHARRINGTON and VRANIC 1974a; CHERRINGTON et al. 1974b). Therefore, if this ratio has any physiologic meaning, it would seem more appropriate to measure it in the portal vein. This concept of portal vein insulin: glucagon molar ratio is discussed in Sect.-D.II.

Furthermore, VRANIC et al. (1976) demonstrated a significant negative correlation between peripheral insulin: glucagon molar ratios and increments of hepatic glucose appearance. They concluded that a role for the insulin: glucagon molar ratio in the regulation of hepatic glucose balance cannot be demonstrated under most physiologic conditions. They felt that absolute levels of each hormone should be considered separately, and were more important in control of hepatic carbohydrate metabolism. BODEN et al. (1978) reported that the chronic elevation of endogenous glucagon in a patient with glucagonoma stimulated ketogenesis, but had no effect on splanchnic glucose production. They also observed a discrepancy between the peripheral vein insulin: glucagon molar ratio, based on biologically active insulin and glucagon, and splanchnic glucose production.

### II. Portal Vein Insulin: Glucagon Molar Ratio

The relationship between the changes in the amount of glucagon and insulin presented to the liver and the hepatic glucose output are presented in Figs. 5 and 6. The basal hepatic glucose output of $38 \pm 6$ mg/min was increased to approximately

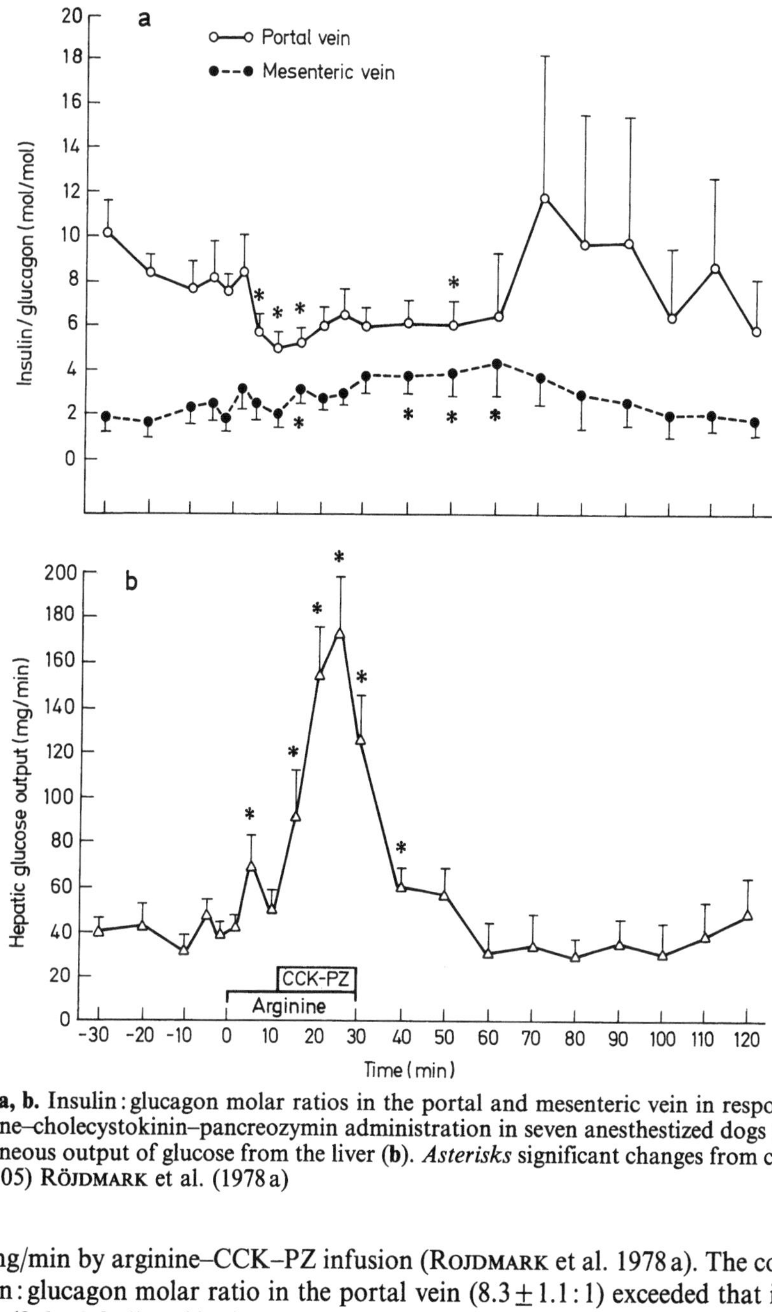


**Fig. 5a, b.** Insulin: glucagon molar ratios in the portal and mesenteric vein in response to arginine–cholecystokinin–pancreozymin administration in seven anesthetized dogs (**a**); simultaneous output of glucose from the liver (**b**). *Asterisks* significant changes from control ($p \leqq 0.05$) RÖJDMARK et al. (1978a)

150 mg/min by arginine–CCK–PZ infusion (ROJDMARK et al. 1978a). The control insulin: glucagon molar ratio in the portal vein (8.3 ± 1.1 : 1) exceeded that in the artery (3.6 ± 0.9 : 1) and in the mesenteric vein (2.1 ± 0.5 : 1) (Fig. 5). This reflects the greater hepatic extraction of insulin compared with that of glucagon. A similar higher insulin: glucagon molar ratio in the portal vein than in the peripheral vein was also demonstrated by FELIG et al. (1974). Arginine–CCK–PZ infusion increased hepatic glucose output and this correlated with a significant decrease in the portal vein insulin: glucagon molar ratio to a nadir of 5.0 ± 0.7 : 1. There was no

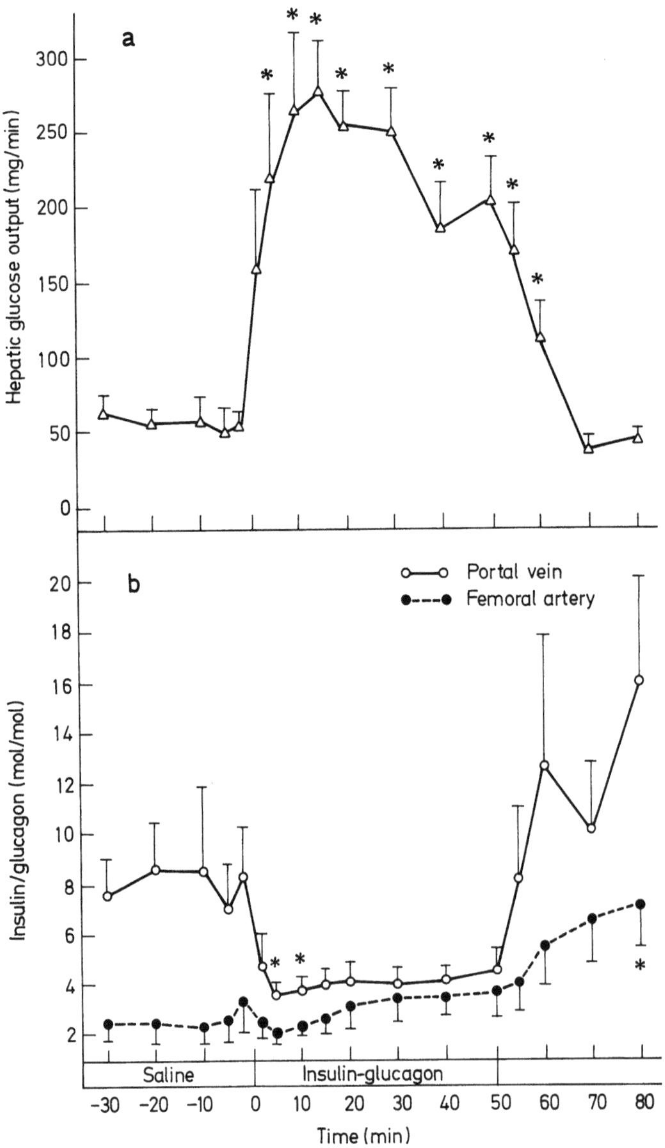


**Fig. 6 a, b.** Hepatic glucose output in response to intrapancreatic insulin and glucagon infusion in six anesthetized dogs (**a**); Concomitant insulin : glucagon molar ratios in portal vein and femoral artery (**b**). *Asterisks* indicate significant changes from control ($p \leq 0.05$). RÖJDMARK et al. (1978 b)

significant fall in the molar ratio in the artery, again emphasizing the discrepancy between these two ratios. After cessation of the infusion, hepatic glucose output returned to the basal levels. The portal vein insulin : glucagon molar ratio also returned to the control levels. Similar results were also demonstrated when exogenous glucagon (13 ng $kg^{-1}$ $min^{-1}$) and insulin (1.5 mIU $kg^{-1}$ $min^{-1}$) were infused (RÖJDMARK et al. 1978 b) (Fig. 6). The basal hepatic glucose output

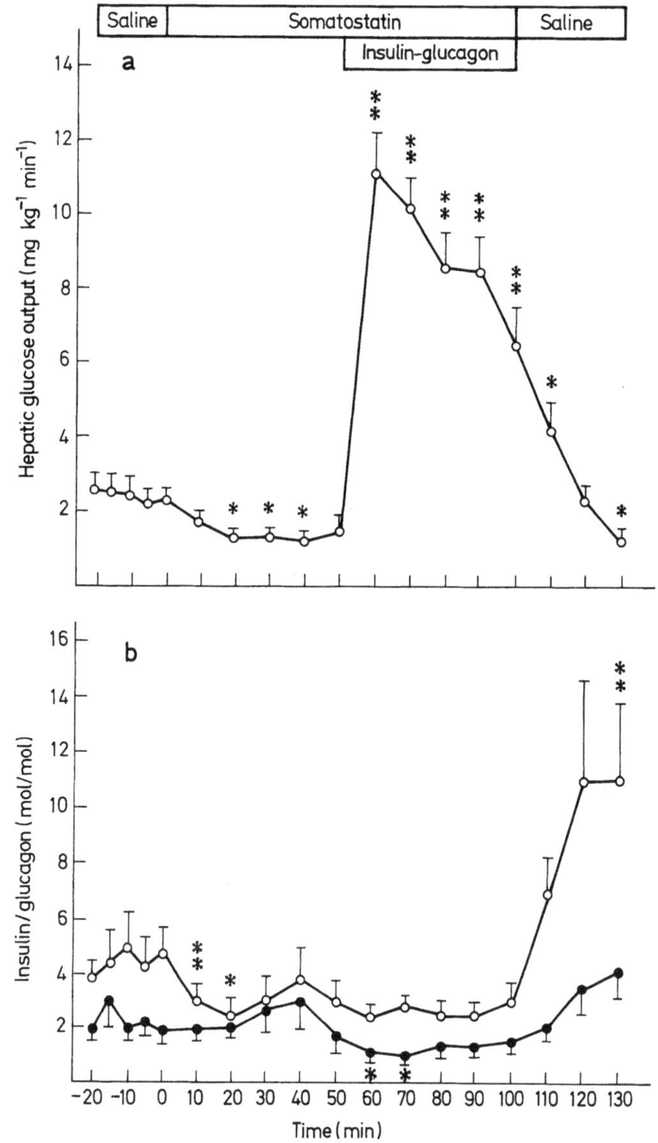


**Fig. 7 a, b.** Hepatic glucose output in response to intraportal somatostatin, insulin and glucagon infusions in eight anesthetized dogs (**a**); concomitant insulin : glucagon molar ratios in portal venous (*open circles*) and femoral arterial (*full circles*) blood (**b**). *Asterisks* indicate significant changes from basal level ($p<0.01$), *double asterisks* $p<0.005$. ISHIDA et al. (1980)

(56±11 mg/min) increased to 278±36 mg/min after infusion, associated with a corresponding decline in the portal vein insulin : glucagon molar ratio (from 7.2±2.2 : 1 to 3.5±0.3 : 1). There was no significant change in the ratio in the artery. When the infusion ceased, the portal vein insulin : glucagon molar ratio tended to rebound, associated with hepatic glucose output returning to the basal level.

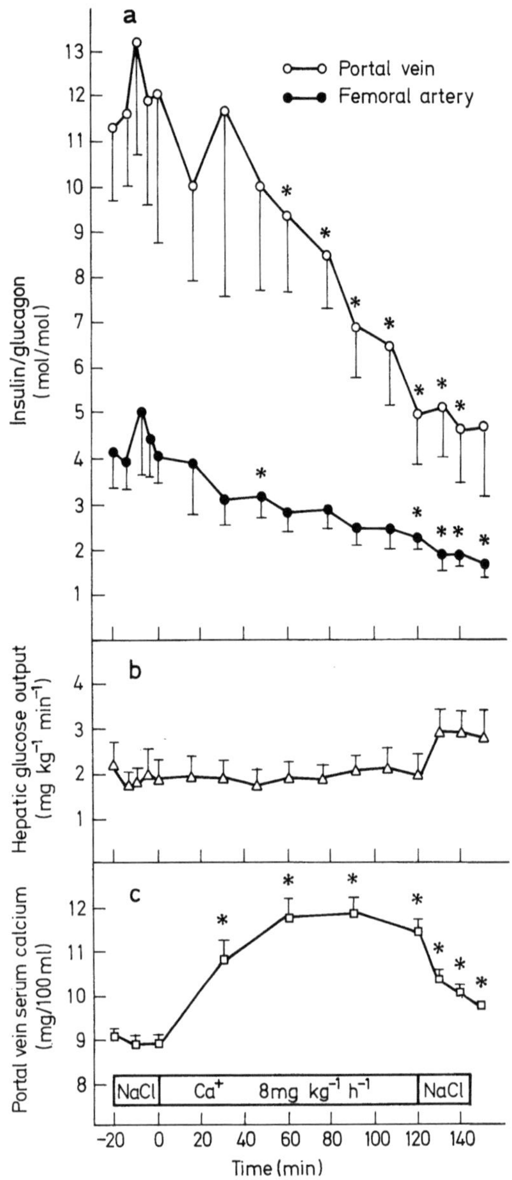


**Fig. 8 a–c.** Insulin : glucagon molar ratio in the portal vein and femoral artery in response to intraportal calcium infusion in six anesthetized dogs (**a**); concomitant hepatic glucose output (**b**) and calcium concentration (**c**) in the portal vein. *Asterisks* indicate significant changes from control ($p \leqq 0.05$). RÖJDMARK et al. (1979)

Although some studies supported the importance of the portal vein insulin : glucagon molar ratio in the regulation of hepatic carbohydrate metabolism, discrepancies have also been observed which raise questions about its physiologic importance. As shown in Fig. 7, somatostatin (250 ng kg$^{-1}$ min$^{-1}$) infusion into the mesenteric vein decreased the hepatic glucose output from the basal value of

$2.5 \pm 0.4$ to $1.4 \pm 0.3$ mg $kg^{-1}$ $min^{-1}$ (ISHIDA et al. 1980). This was associated with a significant decrease in the portal vein insulin: glucagon molar ratio from $4.4 \pm 1.0:1$ to $2.3 \pm 0.7:1$, rather than the expected increase if hepatic carbohydrate output was regulated by the ratio. It is possible that disparate effects of somatostatin on the various components of glucagon immunoreactivity could exert a significant effect on the ratio, if it is calculated based on the total glucagon immunoreactivity. However, the biologic insulin: glucagon molar ratio in the portal vein (based on the 3,500 dalton fraction for glucagon) did not increase. During the combined infusion of somatostatin, glucagon (20 ng $kg^{-1}$ $min^{-1}$) and insulin (1.7 mIU $kg^{-1}$ $min^{-1}$), the portal vein insulin: glucagon molar ratio did not change, despite an increase in hepatic glucose output to $11 \pm 1.1$ mg $kg^{-1}$ $min^{-1}$.

This increase of hepatic glucose output was similar to that achieved by the exogenous glucagon and insulin infusion without somatostatin, but this latter increase was associated with a significant decrease of the portal vein insulin: glucagon molar ratio. The lack of change in the portal vein insulin: glucagon molar ratio cannot be attributed to the heterogeneity of glucagon because exogenous glucagon represents only the 3,500 molecular weight fraction. Furthermore, changes in the portal vein insulin: glucagon molar ratio might be difficult to interpret since they assume parallel dose–response curves for both hormones. The interpretation is further confounded if the ratio is altered by the administration of pharmacologic amounts of glucagon and insulin which might not modify the maximal effect of either hormone. Another discrepancy between the portal vein insulin: glucagon molar ratio and hepatic glucose output was observed when calcium was infused into the portal vein (ROJDMARK et al. 1979). During the infusion of calcium, there was a significant fall in the portal vein insulin: glucagon molar ratio from $12.1 \pm 2.1:1$ to $4.9 \pm 1:1$ by 120 min, but this was not associated with any change in hepatic glucose output (Fig. 8). These results indicate that changes in hepatic glucose output do not always correlate with alteration in the portal vein insulin: glucagon molar ratio.

Other factors besides the portal vein insulin: glucagon molar ratio might be important determinants of hepatic carbohydrate metabolism. The studies of EL-REFAI and BERGMAN (1979) indicated that the timing of the increased insulin and glucagon rather than the ratio in the portal vein may be more important in regulating changes in hepatic glucose output in the isolated, perfused rat liver. When insulin levels in the portal vein were elevated before glucagon administration, a subsequent increase in glucagon concentrations overcame insulin inhibition of hepatic glucose output. Conversely, if glucagon concentrations in the portal vein were increased first, even though the final hormone concentrations and therefore their ratio were similar, an increase in insulin inhibited the effect of glucagon on hepatic glucose output. FRADKIN et al. (1980) also reported that, in humans, absolute hyperglucagonemia independent of portal vein insulin concentrations was associated with either a rise or fall in hepatic glucose output, depending on whether the glucagon concentration has been acutely raised or lowered. They suggested that the liver is thus responsive to changes in glucagon levels rather than their absolute concentrations and adapts to be new glucagon level. Recently, STEINER et al. (1981) also observed a discrepancy between the portal vein insulin: glucagon molar ratio and hepatic glucose production. They concluded that the hepatic glucose output is dependent on the absolute amounts of each hormone presented to the liver.

### III. Molar Ratio of Insulin and Glucagon Extracted by the Liver

Since the portal vein insulin: glucagon molar ratio did not always correlate with hepatic glucose output, it is worth considering other possible parameters. It is possible that rather than the ratio of the two hormones in the portal vein, a more physiologically meaningful parameter would be the ratio of the two hormones extracted by the liver. This would be especially true if the extraction was associated with a biologic effect and not only degradation. Since hepatic extraction of glucagon and insulin are different and vary independently, the ratio of the amounts of hormone extracted by the liver may be more relevant. However, it is not known whether all of the hormone that is extracted by the liver exerts an effect on carbohydrate metabolism or whether it is degraded independent of any physiologic effect. Recently, we examined the relationship between hepatic glucose output and the molar ratio of insulin and glucagon extracted by the liver in dogs given either tolbutamide or an insulin infusion to produce comparable hypoglycemia (ISHIDA et al. 1981). The initial fall in blood glucose after tolbutamide administration was associated with a significant reduction of hepatic glucose output from $1.8 \pm 0.5$ to $1.0 \pm 0.3$ mg $kg^{-1}$ $min^{-1}$ at 30 min (Fig. 9). However, despite a continuing fall in the glucose concentrations, hepatic glucose output returned to the basal level at 60 min and even increased to a peak of $2.6 \pm 0.4$ mg $kg^{-1}$ $min^{-1}$ at 105 min. The portal vein insulin: glucagon molar ratio rapidly rose from $7.3 \pm 1.1$:1 to $19 \pm 3$:1 at 20 min, as hepatic glucose output was falling. However, when glucagon secretion was increased and hepatic glucose output returned to normal, the portal vein ratio still remained elevated. The change in the arterial insulin: glucagon molar ratio was somewhat greater than that in the portal vein, reflecting the increased hepatic extraction of glucagon and unchanged removal of insulin. Although the molar ratio of hepatic extraction of insulin and glucagon increased from a basal value of $24.1 \pm 7.6$:1 to $47.6 \pm 14.4$:1 after tolbutamide administration, the changes were not significant and did not correlate with hepatic glucose output as well as did the portal vein insulin: glucagon molar ratio. Measurement and evaluation of the molar ratio of the hormones extracted by the liver may be very difficult because of the low and sometimes even negative hepatic extraction of glucagon.

## E. Fate of Glucagon Extracted by the Liver

Glucagon extraction by the liver may encompass several processes: binding to the hepatocyte receptors, degradation of glucagon, and excretion of that hormone into bile. However, it is unknown whether all of the glucagon extracted by the liver exerts a biologic effect as discussed in detail elsewhere in this book (see Chap. 13) or whether it is degraded independent of any physiological action.

**Fig. 9a–c.** Effect of tolbutamide (*open circles*) administration in 12 anesthetized dogs and ▸ insulin (*triangles*) infusion in six anesthetized dogs on hepatic glucose output (**a**); portal vein insulin: glucagon molar ratio (**b**); and hepatic insulin uptake: hepatic glucagon uptake molar ratio (**c**). Tolbutamide was injected intravenously over 1 min and insulin infusion rates were the same as in Fig. 4. *Asterisks* indicate significant changes from control ($p < 0.05$); *double asterisks* $p < 0.01$. ISHIDA et al. (1981)

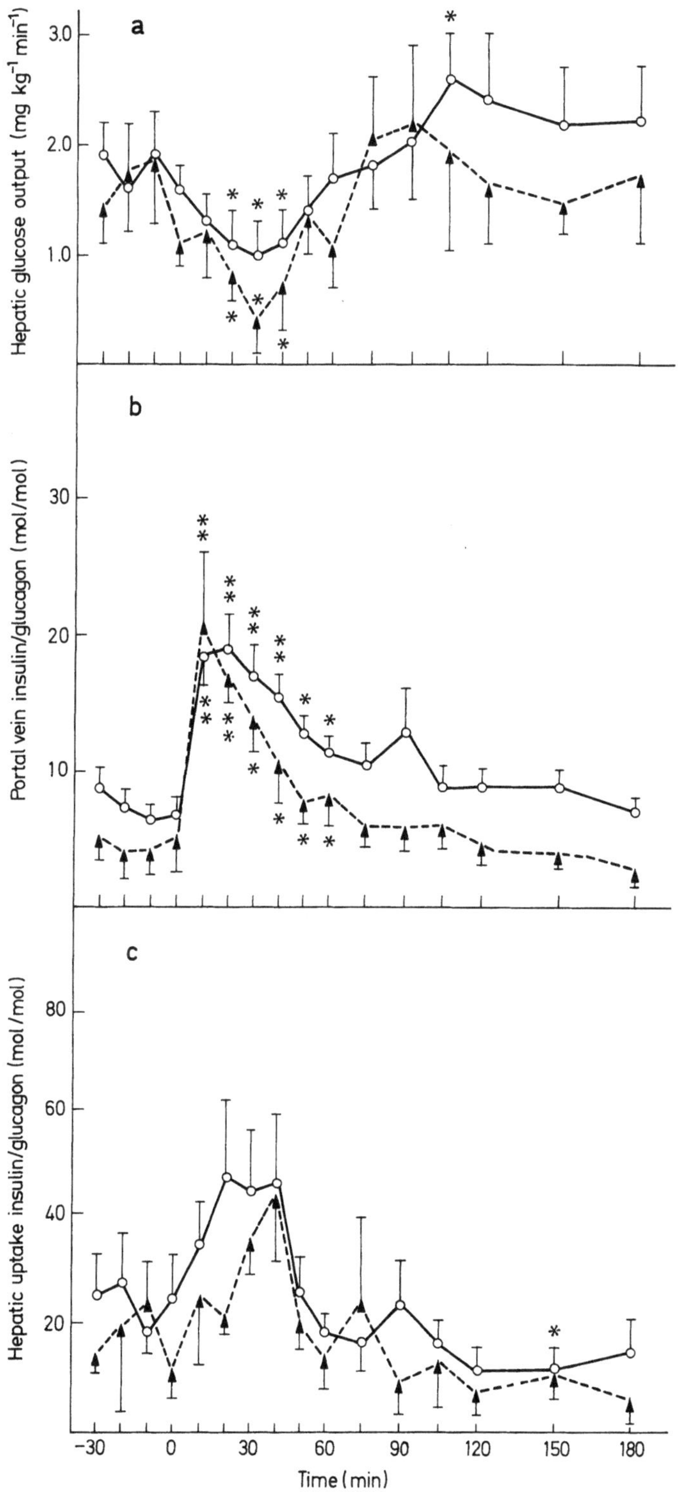
a
Hepatic glucose output (mg kg⁻¹ min⁻¹)
3.0
2.0
1.0
b
Portal vein insulin/glucagon (mol/mol)
30
20
10
c
Hepatic uptake insulin/glucagon (mol/mol)
80
60
40
20
-30
0
30
60
90
120
150
180
Time (min)

The biologic effect of glucagon is initiated by binding to receptors on the hepatocyte cell membrane, as discussed in detail by RODBELL in Chap. 13. In the liver cell, the site of glucagon degradation appears to be separate from the glucagon receptor (GIORGIO et al. 1974). Little information is available, however, on the nature of the subcellular site or the biochemical process involved in the inactivation of glucagon. Although isolated liver cell membranes possess components, functionally distinct from the receptors, that inactivate glucagon (DESBUQUOIS and CUATRECASAS 1972; POHL et al. 1972), intracellular sites of degradation may exist as well. ROUER et al. (1980) studied the functional relationship between the binding and the degradation of glucagon using isolated liver cells. They suggested that most of the degrading activity is unrelated to receptor binding and the degradation of cell-associated glucagon is a minor fraction of the total metabolism of that hormone, since cell-associated glucagon was degraded less rapidly than the unbound glucagon. This is in contrast to insulin (TERRIS and STEINER 1975). Insulin degradation may be one physiologic meachanism for termination of the cellular response to the hormone (CROFFORD 1968), but it is not clear whether this is also true for glucagon. Insulin and glucagon may be degraded by the same enzyme, the so-called insulin-specific protease, which was found in the soluble fraction of rat skeletal muscle homogenate (DUCKWORTH and KITABCHI 1974). Furthermore, the relationship of this enzyme to the glucagon-degrading system of liver cell membranes is unclear. Thus, the fate of the glucagon extracted by the liver is still uncertain, although it is possible that the catabolism of the hormone involves its cleavage into fragments, which are not detectable with the 30K antiserum. A fragment of glucagon encompassing its first six $NH_2$ terminal residues binds to the glucagon receptor and stimulates adenylate cyclase activity in rat liver plasma membranes, indicating that partial degradation does not cause complete biologic inactivation (WRIGHT and RODBELL 1979; see also Chap. 1 and 13).

## F. Conclusions

Hepatic extraction of glucagon and insulin plays an important role in the regulation of both peripheral venous hormone concentrations and hepatic glucose metabolism. In the basal state, approximately 10%–20% of the 40–60 ng/min glucagon delivered to the liver was extracted in a single transhepatic passage compared with about 40%–60% of the 10–20 mIU/min insulin presented to that organ. This considerably smaller fraction of hepatic glucagon extraction compared with that of insulin is somewhat surprising since the liver is the primary target for that hormone. This relatively low hepatic extraction rate for glucagon does not imply that the liver is unimportant in the regulation of glucagon homeostasis because binding of glucagon by only a small proportion of receptors on the hepatocyte is sufficient to elicit a biologic effect. Changes in splanchnic blood flow did not influence glucagon extraction. Alterations in the amount of glucagon reaching the liver did not regulate its hepatic extraction. The increased hepatic extraction of glucagon after administration of tolbutamide, but not after insulin-induced hypoglycemia appeared to be a direct effect of tolbutamide on the liver. Arginine–CCK–PZ infusion increased the amounts of glucagon and insulin presented to the liver. The infusion decreased insulin, but not glucagon extraction, suggesting a direct effect on the liver independent of the augmented amounts of the hormones.

Thus, hepatic glucagon and insulin extraction are quite different and regulated independently. Because of such changes in hepatic extraction of glucagon, glucagon concentrations in peripheral vessels may not accurately reflect pancreatic secretion of the hormone. The concept that the insulin: glucagon molar ratio regulates hepatic glucose output is clearly an oversimplification. The peripheral insulin: glucagon molar ratio does not always reflect that in the portal vein because of the different hepatic extraction of the two hormones. Even the portal vein insulin: glucagon molar ratio does not always correlate with hepatic glucose production. Although physiologically it might be more reasonable to relate changes in hepatic glucose output to alterations in the molar ratio of the glucagon and insulin extracted by the liver, rather than the portal vein ratio or the absolute concentrations in the portal vein, this parameter presents difficulties in its measurement because of the very low and sometimes even negative values for hepatic glucagon extraction. The fate of the glucagon extracted by the liver is also uncertain although it is possible that the catabolism of the hormone involves its cleavage into fragments, which are not detectable with the 30K antiserum, but still may be biologically active. It is still unknown whether all of the glucagon extracted by the liver has a biologic effect or whether it is degraded independent of any physiologic effect.

*Acknowledgments.* This work was supported by U.S. Public Health Service Grant AM 25253. The authors wish to thank SHELLEY DEARING for her outstanding assistance in the preparation of this manuscript.

## References

Alford FP, Dudley FJ, Chisholm DJ, Findlay DM (1979) Glucagon metabolism in normal subjects and in cirrhotic patients before and after portasystemic venous shunt surgery. Clin Endocrinol (Oxf) 11:413–424

Ansorge S, Bohley P, Kirschke H, Langner J, Hanson H (1971) Metabolism of insulin and glucagon: breakdown of radioiodinated insulin and glucagon in rat liver cell fractions. Eur J Biochem 19:283–288

Assan R (1972) In vivo metabolism of glucagon. In: Lefèbvre PL, Unger RH (eds) Glucagon. Molecular, physiology, clinical and therapeutic implications. Pergamon, Oxford, p 47

Berson SA, Yalow RS, Volk BW (1957) In vivo and in vitro metabolism of insulin-$I^{131}$ and glucagon-$I^{131}$ in normal and cortisone-treated rabbits. J Lab Clin Med 49:331–342

Birnbaumer L, Pohl SL (1973) Relation of glucagon-specific binding sites to glucagon-dependent stimulation of adenyl cyclase activity in plasma membranes of rat liver. J Biol Chem 248:2056–2061

Blackard WG, Nelson NC (1970) Portal and peripheral vein immunoreactive insulin concentration before and after glucose infusion. Diabetes 19:302–306

Blackard WG, Nelson NC, Andrews SS (1974) Portal and peripheral vein immunoreactive glucagon concentrations after arginine or glucose infusions. Diabetes 23:199–202

Boden G, Wilson RM, Owen OE (1978) Effects of chronic glucagon excess on hepatic metabolism. Diabetes 27:643–648

Bond JH, Prentiss RA, Levitt MD (1980) The effect of anesthesia and laparotomy on blood flow to the stomach, small bowel and colon on the dog. Surgery 87:313–318

Brockman RP, Manns JG, Bergman EN (1976) Quantitative aspects of secretion and hepatic removal of glucagon in sheep. Can J Physiol Pharmacol 54:666–670

Buchanan K, Solomon S, Vance J, Porter H, Williams RH (1968) Glucagon clearance by the isolated perfused rat liver. Proc Soc Exp Biol Med 128:620–623
Bucolo RJ (1978) Obesity, hyperinsulinemia and portal blood flow: a theoretic study of the effects of increased portal blood flow upon fasting peripheral insulin concentrations. Diabetes 27:840–848
Camu F (1975) Hepatic balance of glucose and insulin in response to physiological increments of endogenous insulin during glucose infusions in dogs. Eur J Clin Invest 5:101–108
Caruana JA, Gage AA (1980) Increased uptake of insulin and glucagon by the liver as a signal for regeneration. Surg Gynecol Obstet 150:390–394
Cherrington AD, Vranic M (1974) Effect of interaction between insulin and glucagon on glucose turnover and FFA concentration in normal and depancreatized dogs. Metabolism 23:729–744
Cherrington AD, Kawamori R, Pek S, Vranic M (1974) Arginine infusion in dogs. Model for the roles of insulin and glucagon in regulating glucose turnover and free fatty acid levels. Diabetes 23:805–815
Cornell RP (1980) Hyperinsulinemia and hyperglucagonemia in fasted rats during liver regeneration. Am J Physiol 240:E112–E118
Crofford OB (1968) The uptake and inactivation of native insulin by isolated fat cells. J Biol Chem 243:362–369
Dencker M, Hedner P, Holst J, Tranberg K-G (1975) Pancreatic glucagon response to an ordinary meal. Scand J Gastroenterol 10:471–474
Desbuquois B, Cuatrecasas P (1972) Independence of glucagon receptors and glucagon inactivation in liver cell membranes. Nature 236:202–204
Duckworth WC, Kitabchi AE (1974) Insulin and glucagon degradation by the same enzyme. Diabetes 23:536–543
Eade MN, Ginn RW (1978) The distribution of blood flow along the small intestine of the dog. Proc Soc Exp Biol Med 157:390–392
El-Refai M, Bergman RN (1979) Glucagon-stimulated glucogenolysis: time-dependent sensitivity to insulin. Am J Physiol 236:E246–E254
Evringham A, Brenneman EM, Horvath SM (1959) Influence of sodium pentobarbital on splanchnic blood flow and related functions. Am J Physiol 197:624–626
Exton JH, Park CR (1972) Interaction of insulin and glucagon in the control of liver metabolism. In: Steiner DF, Freinkel N (eds) Handbook of physiology, vol I, sect 7. American Physiological Society, Washington, DC, p 437
Exton JH, Lewis SB, Ho RJ, Robinson GA, Park CR (1971) The role of cyclic AMP in the interaction of glucagon and insulin in the control of liver metabolism. Ann NY Acad Sci 185:85–100
Felig P, Wahren J (1975) The liver as a site of insulin and glucagon action in normal, diabetic and obese humans. Isr J Med Sci 11:528–539
Felig P, Gusberg R, Hendler R, Gump FE, Kinney JM (1974) Concentrations of glucagon and the insulin: glucagon ratio in the portal and peripheral circulation. Proc Soc Exp Biol Med 147:88–90
Fischer U, Luyckx AS, Jutzi E, Hommel H, Lefèbvre PJ (1978a) Glucagon and the liver. In: Crepaldi G, Lefèbvre PJ, Alberti KGM (eds) Diabetes, obesity and hyperlipidemias. Academic, London, p 3
Fischer U, Jutzi E, Hommel H, Gottschling H-D, Salzsieder E, Fejeregyhazi I, Albrecht G (1978b) Pancreatic and hepatic blance of endogenous insulin and of glucose in the anesthetized dog during primed and non-primed infusions of glucose and of mannitol. Acta Biol Med Ger 37:1049–1064
Fradkin J, Shamoon H, Felig P, Sherwin RS (1980) Evidence for an important role of changes in rather than absolute concentrations of glucagon in the regulation of glucose production in humans. J Clin Endocrinol Metab 50:698–703
Gilmore JP (1958) Effect of anesthesia and hepatic sampling site upon hepatic blood flow. Am J Physiol 195–465–468
Giorgio NA, Johnson CB, Bleicher M (1974) Hormone receptors. IV. Properties of glucagon-binding proteins isolated from liver plasma membranes. J Biol Chem 249:428–437

Goldner M, Jauregui RH, Weisenfeld S (1954) Disappearance of HGF from insulin after liver perfusion. Am J Physiol 179:25–28

Greco AV, Crucitti F, Ghirlanda G, Manna R, Altomonte L, Rebuzzi AG, Bertoli A (1979) Insulin and glucagon concentrations in portal and peripheral veins in patients with hepatic cirrhosis. Diabetologia 17:23–28

Harding PE, Bloom G, Field JB (1975) Effect of infusion of insulin into portal vein on hepatic extraction of insulin in anesthetized dogs. Am J Physiol 228:1580–1588

Honey RN, Price S (1979) The determinations of insulin extraction in the isolated perfused rat liver. Horm Metab Res 11:111–117

Ishida T, Field JB (1981) Comparison of basal and stimulated insulin and glucagon extraction by the liver in conscious and anesthetized dogs (Abstr). Clin Res 29:540A

Ishida T, Röjdmark S, Bloom G, Chou MCY, Field JB (1980) The effect of somatostatin on the hepatic extraction of insulin and glucagon in the anesthetized dog. Endocrinology 106:220–230

Ishida T, Chou MCY, Lewis RM, Hartley CJ, Entman M, Field JB (1981) The effect of tolbutamide on hepatic extraction of insulin and glucagon and hepatic glucose output in anesthetized dogs. Endocrinology 109:443–450

Itatsu T, Kishimoto T, Shintani T, Ukai M (1979) The release and metabolism of pancreatic hormones after major hepatectomy in the dog. Endocrinol Jpn 23:319–324

Jaspan JB, Huen AH-J, Morley CG, Moossa AR, Rubenstein AH (1977) The role of the liver in glucagon metabolism. J Clin Invest 60:421–428

Jaspan J, Polonsky K, Dhorajiwala J, Pugh W, Moossa A (1980) Feeding alters hepatic extraction of insulin and glucagon (Abstr). Diabetes 29 [Suppl 2]:55A

Jaspan JB, Polonsky KS, Lewis M, Pensler J, Pugh W, Moossa AR, Rubenstein AH (1981) Hepatic metabolism of glucagon in the dog: contribution of the liver to overall metabolic disposal of glucagon. Am J Physiol 240:E233–E244

Jennings AS, Cherrington AD, Liljenquist JE, Keller U, Lacy WW, Chiasson JL (1977) The role of insulin and glucagon in the regulation of gluconeogenesis in the postabsorptive dog. Diabetes 26:847–856

Kaden M, Harding P, Field JB (1973) Effect of intraduodenal glucose administration on hepatic extraction of insulin in the anesthetized dog. J Clin Invest 52:2016–2028

Katz ML, Bergman EN (1969) Simultaneous measurements of hepatic and portal venous blood flow in the sheep and dog. Am J Physiol 216:946–952

Kenny AJ (1956) Inactivation of glucagon by tissues in vitro. Am J Physiol 186:419–426

Kuku SF, Jaspan JB, Emmanouel DS, Zeidler A, Katz AI, Rubenstein AH (1976) Heterogeneity of plasma glucagon. Circulating components in normal subjects and patients with chronic renal failure. J Clin Invest 58:742–750

Lefèbvre P, Luyckx A (1965) Modification par le foie isolé et perfusé de l'effect exercé par le glucagon sur la graisse epididymaire du rat. Ann Endocrinol (Paris) 26:369–374

Leffert HL, Koch KS, Rubalcava B (1976) Present paradoxes in the environmental control of hepatic proliferation. Cancer Res 36:4250–4255

Mackrell DJ, Sokal JE (1969) Antagonism between the effect of insulin and glucagon on the isolated liver. Diabetes 18:724–732

Marco J, Diego J, Villanueva ML, Diaz-Fierros M, Valverde I, Segovia JM (1973) Elevated plasma glucagon levels in cirrhosis of the liver. N Engl J Med 289:1107–1111

Misbin RI, Merimee TJ, Lowenstein JM (1976) Insulin removal by isolated perfused rat liver. Am J Physiol 230:171–177

Mondon CE, Dolkas CB, Olefsky JM, Reaven GM (1975) Insulin sensitivity of isolated perfused rat liver. Diabetes 24:225–229

Mortimore GE, Tietze F, Stetten D-W (1959) Metabolism of insulin-$I^{131}$. Studies in isolated, perfused rat liver and hind-limb preparations. Diabetes 8:307–314

Navalesi R, Pilo A, Ferrannini E (1976) Insulin kinetics after portal and peripheral injection of $^{125}$I-insulin. II. Experiments in the intact dog. Am J Physiol 230:1630–1636

Parrilla R, Goodman MN, Toews CJ (1974) Effect of glucagon: insulin ratios on hepatic metabolism. Diabetes 23:725–731

Pohl SL, Krans HMJ, Birnbaumer L, Rodbell M (1972) Inactivation of glucagon by plasma membranes of rat liver. J Biol Chem 247:2295–2301

Röjdmark S, Bloom G, Chou MCY, Jaspan JB, Field JB (1978a) Hepatic insulin and glucagon extraction after their augmented secretion in dogs. Am J Physiol 235:E88–E96
Röjdmark S, Bloom G, Chou MCY, Field JB (1978b) Hepatic extraction of exogenous insulin and glucagon in the dogs. Endocrinology 102:806–813
Röjdmark S, Ishida T, Bloom G, Chou MCY, Field JB (1979) Effect of intraportal calcium infusion on insulin and glucagon secretion and hepatic glucose output in anesthetized dogs. Endocrinology 104:814–821
Rouer E, Desbuquois B, Postel-Vinay M-C (1980) Interactions of glucagon with isolated rat-liver cells: fate and subcellular localization of cell-associated hormone. Mol Cell Endocrinol 19:143–164
Seitz HJ, Muller MJ, Nordmeyer P, Krone W, Tarnowski W (1976) Concentration of cyclic AMP in rat liver as a function of the insulin/glucagon ratio in blood under standardized physiological conditions. Endocrinology 99:1313–1318
Shackman R, Graber IG, Melrose DG (1953) Liver blood flow and general anesthesia. Clin Sci 12:307–315
Sherwin R, Joshi P, Hendler R, Felig P, Conn HO (1974) Hyperglucagonemia in Laennec's cirrhosis. The role of portal-systemic shunting. N Engl J Med 290:239–242
Sherwin RS, Fisher M, Bessoff J, Snyder N, Hendler R, Conn HO, Felig P (1978) Hyperglucagonemia in cirrhosis: altered secretion and sensitivity to glucagon. Gastroenterology 74:1224–1228
Steiner KE, Williams PE, Lacy WW, Cherrington AD (1981) Effects of insulin/glucagon molar ratio on glucose production in the dog (Abstr). Fed Proc 40:843
Strandell T (1978) Determination of hepatic blood flow. Acta Anaesthesiol Scand [Suppl] 70:19
Striffler JS, Curry DL (1979) Effect of fasting on insulin removal by liver of perfused liver-pancreas. Am J Physiol 237:E349–E355
Terris S, Steiner F (1975) Binding and degradation of $^{125}$I-insulin by rat hepatocytes. J Biol Chem 250:8389–8398
Terris S, Steiner DF (1976) Retention and degradation of $I^{125}$-insulin by perfused livers from diabetic rats. J Clin Invest 57:885–896
Unger RH (1971) Glucagon and the insulin: glucagon ratio in diabetics and other catabolic illnesses. Diabetes 20:834–838
Unger RH (1972) Insulin/glucagon ratio. Isr J Med Sci 8:252–257
Valverde I (1977) Quantification of plasma glucagon immunoreactive components in normal and hyperglucagonemic states. In: Foà PP, Bajaj JS, Foà NL (eds) Glucagon: its role in physiology and clinical medicine. Springer, Berlin Heidelberg New York, p 77
Valverde I, Dobbs R, Unger RH (1975) Heterogeneity of plasma glucagon immunoreactivity in normal, depancreatized, and alloxan-diabetic dogs. Metabolism 24:1021–1028
Vranic M, Kawamori R, Pek S, Kovacevic N, Wrenshall GA (1976) The essentiality of insulin and the role of glucagon in regulating glucose utilization and production during strenuous exercise in dog. J Clin Invest 57:245–255
Wagle SR (1975) Interrelationship of insulin and glucagon ratios on carbohydrate metabolism in isolated hepatocytes containing high glycogen. Biochem Biophys Res Commun 67:1019–1027
Williams RH, Hay JS, Tjaden MB (1959) Degradation of insulin-$I^{131}$ and glucagon-$I^{131}$ and factors influencing it. Ann NY Acad Sci 74:513–529
Wright DE, Rodbell M (1979) Glucagon 1–6 binds to the glucagon receptor and activates hepatic adenylate cyclase. J Biol Chem 254:268–269
Yates MS, Hiley CR, Challiner MR, Park BK (1979) Pentobarbitone effects on hepatic microsomal enzymes and liver blood flow in the guinea pig. Biochem Pharmacol 28:2856–2857

CHAPTER 41

# The Renal Handling of Glucagon

P. J. LEFEBVRE and A. S. LUYCKX

## A. Introduction

Until the mid-1970s, very little was known about the renal handling of glucagon. In their pioneer work published in 1958, NARAHARA et al., using glucagon $^{131}$I injections and autoradiographic techniques, demonstrated that rat kidney tubules reabsorb glucagon. In limited studies performed in four patients, ASSAN (1972) reported that the concentration of endogenous plasma glucagon was higher in renal arterial blood than in renal venous blood, thus suggesting that the kidney was involved in the clearance of *endogenous* glucagon. In addition, he observed that the plasma disappearance time of *exogenous* glucagon was significantly increased in patients with severe renal failure or after bilateral nephrectomy. Since he had been unable to detect significant amounts of glucagon in urine, he suggested that, under normal conditions, glucagon is filtered through the glomerulus and then reabsorbed by the tubules, after which it undergoes intrarenal catabolism. At about the same time, DAY (1972) reported, in contrast, that some material with the molecular weight and immunologic characteristics of glucagon was present in the urine of normal and diabetic subjects; he reported figures for urinary excretion of glucagon averaging $1{,}289 \pm 339$ and $1{,}194 \pm 264$ ng/24 h for normal and diabetic subjects, respectively.

Using dog kidneys, acutely transplanted to the neck vessels of a perfusing anesthetized dog, LEFEBVRE et al. (1974) were the first to quantify the renal uptake of glucagon precisely and to suggest that the kidney might be an important factor in determining the final concentration of circulating glucagon. Such a view has been rapidly confirmed by the demonstration by several groups that plasma glucagon levels rise rapidly after kidney exclusion (LEFEBVRE and LUYCKX 1975, 1976; EMMANOUEL et al. 1976) and are usually markedly elevated in renal failure (ASSAN 1972; BILBREY et al. 1974; KUKU et al. 1976; DAUBRESSE et al. 1976; SHERWIN et al. 1976; LEFEBVRE and LUYCKX 1977; BASTL et al. 1977; EMMANOUEL et al. 1976, 1978; MCEVOY and MCCAROLL 1978). Glucagon in renal insufficiency has been reviewed in detail elsewhere in this volume (see Chap. 47). In the present chapter, we will summarize the data available on the handling of glucagon by the kidney.

## B. Kidney Glucagon Uptake and Urinary Excretion

As already mentioned, experiments using dog kidneys acutely transplanted to the neck vessels of a perfusing anesthetized dog provided the first precise and quantitative data on the renal handling of glucagon (LEFEBVRE et al. 1974). As illustrated

**Table 1.** Handling of glucagon by the kidneys transplanted to the neck under basal conditions and during massive glucose infusion. Adapted from LEFEBVRE et al. (1974)

| | | | |
|---|---|---|---|
| Plasma glucose (mg/100 ml) | 136 ± 12 (12)[a] | 682 ± 36 (10) | $P<0.001$ |
| Plasma glucagon (pg/ml) | | | |
| Arterial | 193 ± 14 (14) | 136 ± 21 (10) | $P<0.05$[b] |
| Venous | 135 ± 16 (14) | 112 ± 22 (10) | N.S. |
| Renal plasma flow ($ml\ g^{-1} min^{-1}$) | 1.53 ± 0.07 (15) | 1.74 ± 0.13 (10) | N.S. |
| Glucagon uptake ($pg\ g^{-1} min^{-1}$) | 89 ± 14 (14) | 42 ± 5 (10) | $P<0.02$ |
| $Q$=quantity of glucagon entering the kidneys ($pg\ g^{-1} min^{-1}$) | 288 ± 37 (14) | 242 ± 48 (10) | N.S. |
| Glucagon uptake (% $Q$) | 37.1 ± 5.8 (14) | 23.5 ± 5.7 (10) | N.S. |
| Glucagon metabolic clearance ($ml\ g^{-1} min^{-1}$) | 0.482 ± 0.063 (14) | 0.315 ± 0.059 (9) | N.S. |

[a] The number of periods studied is indicated in parentheses. Results are expressed as mean ± standard error . The mean (± standard error) weight of both kidneys in this series of ten experiments was 51 ± 6.1 g (range 24.5–85)

[b] Paired comparison effected in eight dogs on arterial plasma glucagon before and after glucose infusion also revealed a highly significant reduction in the glucagon values: −46 ± 14 pg/ml ($P<0.01$). N.S. = not significant

**Table 2.** Urinary excretion of glucagon by the kidneys transplanted to the neck. Adapted from LEFEBVRE et al. (1974)

| | |
|---|---|
| Glomerular filtration rate ($ml\ g^{-1}\ min^{-1}$) | 0.611 ± 0.054[a] |
| Urine output ($ml\ g^{-1}\ min^{-1}$) | 0.021 ± 0.004 |
| Glucagon | |
| Arterial plasma concentration (pg/ml) | 169 ± 21 |
| Urinary concentration (pg/ml) | 160 ± 31 |
| Urinary excretion ($pg\ g^{-1}\ min^{-1}$) | 2.43 ± 0.63 |
| Urinary clearance ($ml\ g^{-1}\ min^{-1}$) | 0.013 ± 0.06 |
| Metabolic clearance ($ml\ g^{-1}\ min^{-1}$) | 0.315 ± 0.059 |
| Urinary clearance (% metabolic clearance) | 4.1 |

[a] Ten collection periods were studied in five dogs. Results are expressed as mean ± standard error ($N=10$). The weight of both transplanted kidneys was 27, 28, 39.5, 51, and 85 g in these five experiments

by Table 1, glucagon uptake by the dog kidney under basal conditions averaged 89 $pg\ g^{-1}\ min^{-1}$; it was significantly reduced by about 50% when the arterial plasma glucagon was decreased by massive glucose infusion to the perfusor dog. As shown in Table 2, the urinary clearance of glucagon represented only 4.1% of the total kidney glucagon clearance. In these experiments, no correlation was found between arterial plasma glucagon levels and kidney glucagon uptake; this was attributed to the fact that the range of arterial glucagon concentrations reached was relatively narrow (95–220 pg/ml).

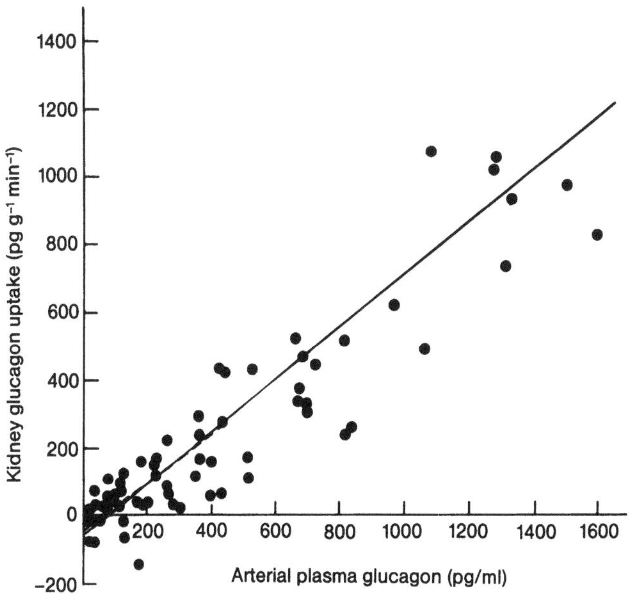


**Fig. 1.** Correlation between kidney glucagon uptake and arterial plasma glucagon in the isolated perfused dog kidney system. Adapted from LEFEBVRE et al. (1976)

Subsequent observations were made using the isolated dog kidneys perfused with whole canine blood (LEFEBVRE et al. 1976). In these experiments, exogenous glucagon was perfused in the system, thus permitting the investigation of kidney glucagon uptake at arterial plasma concentrations ranging from 26 to 1,184 pg/ml. In these conditions, renal vein plasma glucagon $V$ was proportional to renal artery plasma glucagon $A$. $V$ and $A$ were highly significantly correlated with $V = 0.733 \pm 0.034\ A$. As shown in Fig. 1, kidney glucagon uptake increased as a function of arterial plasma glucagon concentration ($r=0.940$; $P<0.01$). In order to compare renal glucagon uptake in this isolated perfused system with that obtained in the experiments on kidney transplanted to the neck, arterial plasma glucagon concentrations were selected to average $192 \pm 28$ pg/ml, a value similar to that reached under basal conditions with the kidney transplanted to the neck ($193 \pm 14$ pg/ml). Under both experimental conditions, renal glucagon uptake values were practically identical: $89 \pm 14$ versus $90 \pm 20$ pg $g^{-1}$ $min^{-1}$ for the kidneys transplanted to the neck and the isolated perfused kidneys, respectively. In the isolated system, the glucagon kidney clearance rate was similar at low, medium, or high plasma glucagon concentrations and averaged about 0.5 ml $g^{-1}$ $min^{-1}$. On average, kidney glucagon uptake in the isolated system represented 21.1% $\pm$2.7% of the quantity of glucagon entering the kidney in 1 min, thus confirming that the kidney may play an important role in determining the final concentration of circulating glucagon. In anesthetized normal rats, BASTL et al. (1977) observed a renal extraction rate of glucagon averaging $510 \pm 208$ pg $g^{-1}$ $min^{-1}$ of kidney with an extraction coefficient averaging 39% $\pm$5%. In that study, the urinary excretion of glucagon accounted for less than 2% of the amount of hormone extracted by the kidney. Renal extraction of glucagon and its extraction rate as a function of kidney weight were markedly reduced in experimentally induced renal failure.

EMMANOUEL et al. (1978), also working in the anesthetized rat, reported an overall metabolic clearance rate of glucagon averaging $31.8 \pm 1.2$ ml kg$^{-1}$ min$^{-1}$; they calculated that the kidney contributed to 30% of the overall metabolic clearance of glucagon; they confirmed that the renal extraction of endogenous and exogenous glucagon was similar, averaging $22.9\% \pm 1.6\%$ – a value extraordinarily similar to that reported in the dog by LEFEBVRE et al. (1976): $21.1\% \pm 2.7\%$ – and was independant of the plasma glucagon level over a wide range of arterial concentrations. Recent investigations of ROVIRA et al. (1982) have confirmed a high clearance of crystalline glucagon by the isolated perfused rat kidney; in contrast, their data suggest that the kidney may not play a major role in gut glucagon-like immunoreactivity (GLI) catabolism.

## C. Heterogeneity of Plasma Glucagon and Renal Uptake

As detailed elsewhere in this volume (see Chap. 11), circulating plasma glucagon exhibits marked heterogeneity. The studies of RUBENSTEIN, JASPAN and their collegues in Chicago have shown that the renal handling of the various circulating immunoreactive glucagon (IRG) components may involve different mechanisms. In particular, the metabolism of the 3,500 daltons fraction is dependent upon glomerular filtration, while the uptake of the 9,000 daltons material can proceed in its absence, as long as renal tissue is adequately perfused. This finding suggests that the 9,000 daltons component may be handled by peritubular uptake (EMMANOUEL et al. 1976). Further details on the heterogeneity of circulating glucagon in renal failure can be found in Chaps. 11 and 47, as well as in JASPAN et al. (1981).

## D. The Fate of the Glucagon Taken up by the Kidney

Since a large arteriovenous difference exists for the concentrations of glucagon accross the normal kidney (ASSAN 1972; LEFEBVRE et al. 1974, 1976; BASTL et al. 1977; EMMANOUEL et al. 1978) and since the urinary clearance of glucagon represents only a small fraction of the kidney glucagon uptake (LEFEBVRE et al. 1974; BASTL et al. 1977), glucagon accumulation and/or degradation in the kidney have to be considered. Kidney glucagon concentrations have been found to be very low by DUNBAR et al. (1977) and PEREZ-CASTILLO and BLAQUEZ (1980); therefore, kidney glucagon accumulation probably represents a very small part indeed of kidney glucagon uptake.

The early data of NARAHARA et al. (1958) suggest that glucagon is filtered through the glomerulus and reabsorbed at the tubular level. However, this does not exclude the possibility of a direct uptake from renal capillaries without previous filtration and reabsorption, as has been demonstrated for insulin (ZAHARKO et al. 1966; CHAMBERLAIN and STIMMLER 1967). This possibility, already mentioned by LEFEBVRE et al. (1974), has been supported by the finding that the renal clearance rate of glucagon may be higher than the glomerular filtration rate in the isolated perfused dog kidney (LEFEBVRE et al. 1976). Similarly, BASTL et al. (1977) reported that the glucagon extraction by the normal rat kidney significantly exceeded its filtered load, and therefore concluded that glucagon degradation by the renal tubuli involves both glomerular filtration and peritubular uptake. As already said, the observations of EMMANOUEL et al. (1978) suggest that the 3,500 daltons fraction of circulating IRG is handled primarily by glomerular filtration and tubular reabsorption, while peritubular uptake might be particularly important for the handling of the 9,000 daltons fraction.

The intimate mechanism by which the kidney degrades glucagon remains unknown. Extensive studies by DUCKWORTH (1976 a, b) have indicated that in rat kidney homogenates, the glucagon-degrading activity is located primarily in the $10^5$ *g* pellet and is apparently associated with the brush border. This glucagon-degrading activity can be solubilized by Triton X-100, it is inhibited by glutathione and EDTA, it is not affected by *N*-ethylmaleimide. These characteristics are identical with those of the neutral protease isolated from rabbit kidney brush border (GEORGE and KENNY 1973; KERR and KENNY 1974) and according to DUCKWORTH (1976 b) it appears that most of the glucagon-degrading activity seen in this preparation is due to this neutral protease. Another enzymatic activity is, however, also present in the cytosol; it seems to be responsible for insulin degradation by the kidney at physiologic insulin concentrations and also contributes to glucagon degradation in vitro (DUCKWORTH 1976 b). This enzyme has characteristics similar to those reported for the insulin–glucagon protease previously isolated from rat skeletal muscle (DUCKWORTH and KITABCHI 1974). On this muscle enzyme, it has been demonstrated that insulin served as a competitive inhibitor of glucagon degradation and that glucagon itself was a competitive inhibitor of insulin degradation (DUCKWORTH and KITABCHI 1974). Studies performed on the isolated perfused dog kidney have shown that, over a wide range of arterial concentrations, glucagon uptake by the kidney was not affected by the presence of insulin (Table 3); these observations led to the conclusion that insulin and glucagon are handled independently by the kidney (LEFEBVRE et al. 1976). Further studies by DUCKWORTH (1978) have examined the binding and degradation of insulin and glucagon by rat kidney cell membranes. Glucagon degradation by the kidney membrane preparation, which contained significant amounts of brush border membranes, was extremely active. Per unit weight of protein, the kidney membrane was over 20 times as active as the liver membrane. Even at 4 °C, significant glucagon degradation occurred. Because of this very active degradation, glucagon binding could not be accurately assessed. Again the glucagon-degrading activity was inhibited by glutathione and EDTA, but not affected by *N*-ethylmaleimide, adrenocorticotropic hormone, or insulin, all potent inhibitors of liver glucagon degradation. However, the apparent Michaelis constant $K_m$ for glucagon degradation by the kidney was similar to that for the liver, $2.4 \times 10^{-6}$ *M* (against $2.7 \times 10^{-7}$ *M* for insulin).

Taken as a whole, these findings tend to suggest that glucagon degradation by the kidney is primarily performed by the enzymatic activity associated with the brush border and that, under physiologic conditions, the bulk of the renal glucagon-degrading activity is carried out by mechanisms not affected by insulin.

PETERSON et al. (1982) recently compared the renal processing of glucagon and insulin. Iodinated glucagon or insulin was microinfused into surface nephrons of the rat kidney and the radiolabel recovered in the urine was quantified and char-

**Table 3.** Clearance rate of glucagon by the isolated perfused dog kidney at various glucagon or insulin concentrations. Adapted from LEFEBVRE et al. (1976)

A. Effect of glucagon concentrations (at all insulin levels)

| | | | |
|---|---|---|---|
| Arterial plasma glucagon (pg/ml) | Low 0 –200 | Medium 201 –500 | High >500 |
| Glucagon clearance[a] (ml g$^{-1}$ min$^{-1}$) | 0.53± 0.16 (7) | 0.52± 0.09 (12) | 0.57±0.05 (17) |

Low vs. Medium: N.S.; Medium vs. High: N.S.; Low vs. High: N.S.

B. Effect of insulin concentrations (at all glucagon levels)

| | | | |
|---|---|---|---|
| Arterial plasma insulin (μIU/ml) | Low 0 –10 | Medium 150 –300 | High 400 –500 |
| Glucagon clearance[a] (ml g$^{-1}$ min$^{-1}$) | 0.45± 0.06 (26) | 0.50±0.06 (14) | 0.43±0.12 (8) |

Low vs. Medium: N.S.; Medium vs. High: N.S.; Low vs. High: N.S.

[a] Results are expressed as mean ± standard error. The number of periods studied is indicated in parentheses. N.S. = not significant

acterized. Proximal, but not distal reabsorption of both peptides was observed, and uptake varied similarly as a function of tubular length. After proximal microinfusion, glucagon $^{125}$I was largely degraded to tyrosine $^{125}$I in the urine, but remained intact after distal infusion. In contrast, insulin $^{125}$I was recovered as intact peptide following both proximal and distal microinfusion. Prolonged tubular sequestration of $^{125}$I label was observed following proximal microinfusion of insulin $^{125}$I, with efflux requiring longer than 1 h for completion. No measurable sequestration of glucagon $^{125}$I was observed. Incubation of both peptides with rabbit renal microvilli membranes resulted in enzymatic hydrolysis of glucagon $^{125}$I. These data suggest that glucagon and insulin are processed in the proximal tubule by predominantly different mechanisms and are in agreement with the proposal that glucagon is mainly degraded by brush border enzymes of the proximal tubule.

## E. Conclusions

Of the glucagon which is entering the kidney, 20%–40% is taken up by this organ. Urinary glucagon excretion represents only 2%–4% of kidney glucagon uptake. Most experimental evidence suggest that glomerular filtration and tubular reabsorption are the main mechanisms involved in the renal handling of glucagon, particularly $IRG^{3500}$. Peritubular uptake can also occur and might be important for the handling of $IRG^{9000}$. Although not still fully understood, the intimate mechanism by which the kidney degrades glucagon probably involves a brush border enzyme, possibly identical to the "neutral protease" isolated from the rabbit kidney brush border.

## References

Assan R (1972) In vivo metabolism of glucagon. In: Lefèbvre PJ, Unger RH (eds) Glucagon: molecular physiology, clinical and therapeutic implications. Pergamon, Oxford New York, p 45

Bastl C, Finkelstein FO, Sherwin R, Hendler R, Felig P, Hayslett JP (1977) Renal extraction of glucagon in rats with normal and reduced renal function. Am J Physiol 233:F67–F71

Bilbrey GL, Faloona GR, White MG, Knochel JP (1974) Hyperglucagonemia of renal failure. J Clin Invest 53:841–847

Chamberlain MJ, Stimmler L (1967) The renal handling of insulin. J Clin Invest 46:911–919

Daubresse JC, Lerson G, Plomteux G, Rorive G, Luyckx AS, Lefèbvre PJ (1976) Lipids and lipoproteins in chronic uremia. A study of the influence of regular hemodialysis. Eur J Clin Invest 6:159–166

Day JL (1972) Detection of glucagon in urine: radioimmunoassay studies in normal and diabetic subjects. J Endocrinol 53:Xli

Duckworth WC (1976a) Insulin and glucagon degradation by the kidney. I. Subcellular distribution under different assay conditions. Biochim Biophys Acta 437:518–530

Duckworth WC (1976b) Insulin and glucagon degradation by the kidney. II. Characterization of the mechanisms at neutral pH. Biochim Biophys Acta 437:531–542

Duckworth WC (1978) Insulin and glucagon binding and degradation by kidney cell membranes. Endocrinology 102:1766–1774

Duckworth WC, Kitabchi AE (1974) Insulin and glucagon degradation by the same enzyme. Diabetes 23:536–543

Dunbar JC, Silverman H, Kirkman E, Foà PP (1977) Role of submaxillary gland and the kidney in the hyperglucagonemia of eviscerated rats. In: Foà PP, Bajaj JS, Foà N (eds) Glucagon: its role in physiology and clinical medicine. Springer, Berlin Heidelberg New York, p 157

Emmanouel DS, Jaspan JB, Kuhn SF, Rubenstein AH, Katz AI (1976) Pathogenesis and characterization of hyperglucagonemia in the uremic rat. J Clin Invest 58:1266–1279

Emmanouel DS, Jaspan JB, Rubenstein AH, Huen AH-J, Fink E, Katz AI (1978) Glucagon metabolism in the rat. Contribution of the kidney to the metabolic clearance rate of the hormone. J Clin Invest 60:6–13

George SG, Kenny AJ (1973) Studies on the enzymology of purified preparation of brush border from rabbit kidney. Biochem J 134:43–57

Jaspan JB, Polonsky KS, Rubenstein AH (1981) The heterogeneity of immunoreactive glucagon in plasma: clinical implications. In: Unger RH, Orci L (eds) Glucagon. Physiology, pathophysiology, and morphology of the pancreatic A-cells. Elsevier, New York, p 77

Kerr MA, Kenny AJ (1974) The molecular weight and properties of a neutral metallo-endopeptidase from rabbit kidney brush border. Biochem J 137:489–495

Kuku SF, Jaspan JB, Emmanouel DS, Zeidler A, Katz AI, Rubenstein AH (1976) Heterogeneity of plasma glucagon. Circulating components in normal subjects and patients with chronic renal failure. J Clin Invest 58:742–750

Lefèbvre PJ, Luyckx AS (1975) Effect of acute kidney exclusion by ligation of renal arteries on peripheral plasma glucagon levels and pancreatic glucagon production in the anesthetized dog. Metabolism 24:1169–1176

Lefèbvre PJ, Luyckx AS (1976) Plasma glucagon after kidney exclusion: experiments in somatostatin-infused and in eviscerated dogs. Metabolism 25:761–768

Lefèbvre PJ, Luyckx AS (1977) Glucagon and the kidney. In: Foà PP, Bajaj JS, Foà N (eds) Glucagon: its role in physiology and clinical medicine. Springer, Berlin Heidelberg New York, p 167

Lefèbvre PJ, Luyckx AS, Nizet AH (1974) Renal handling of endogenous glucagon in the dog: comparison with insulin. Metabolism 23:753–761

Lefèbvre PJ, Luyckx AS, Nizet AH (1976) Independence of glucagon and insulin handling by the isolated perfused dog kidney. Diabetologia 12:359–365

McEvoy J, McCarrol AM (1978) High circulating concentrations of glucagon in non-anaemic uraemic patients. Br Med J 2:17–18

Narahara HT, Everett NB, Simmons BS, Williams RH (1958) Metabolism on insulin-$I^{131}$ and glucagon-$I^{131}$ in the kidney of the rat. Am J Physiol 192:227–231

Perez-Castillo A, Blazquez E (1980) Tissue distribution of glucagon, glucagon-like immunoreactivity and insulin in the rat. Am J Physiol 238:E258–E266

Peterson DR, Carone FA, Oparil S, Christensen EI (1982) Differences between renal tubular processing of glucagon and insulin. Am J Physiol 242:F112–F118

Rovira A, Lopez-Novoa JM, Zubiaur M, Metasanz R, Ghiglione M, Pascual JM, Valverde I (1982) Renal metabolism of gut glucagon-like immunoreactivity. Endocrinology 110:2030–2036

Sherwin R, Bastl C, Finkelstein FO, Fisher M, Black M, Hendler R, Felig P (1976) Influence of uremia and hemodialysis on the turnover and metabolic effects of glucagon. J Clin Invest 57:722–731

Zaharko DS, Beck LV, Blankenbaker R (1966) Role of the kidney in the disposal of radioiodinated insulin in dogs. Diabetes 15:680–685

# Glucagon in Pathology

CHAPTER 42

# Glucagon Deficiency

G. BODEN

## A. Introduction

Endocrine deficiency states are of interest mainly for two reasons: (1) they can serve as experiments of nature, helping us to understand the physiologic roles of deficient hormones; and (2) they may be important disease entities. Chronic, severe glucagon deficiency has recently been shown to exist in human subjects after total duodenopancreatectomy and hemigastrectomy. Furthermore, pharmacologic approaches to suppress glucagon release have been employed successfully to create acute and transient states of glucagon deficiency. Data obtained with these techniques have greatly expanded our knowledge of the physiologic role of glucagon. In this chapter, we will review the literature on surgically and pharmacologically induced as well as on naturally occurring states of glucagon deficiency and on the effects of glucagon deficiency on protein and carbohydrate metabolism. Because of the lack of published work on the effects of selective glucagon deficiency on ketogenesis, lipolysis, and lipoprotein metabolism, these aspects of glucagon activity are not discussed. For the purpose of this discussion, glucagon deficiency is defined as an abnormally low basal plasma concentration of bioactive glucagon and/or abnormally low increases in plasma concentrations of bioactive glucagon after stimulation with one or more provocative agents. Glucagon refers to the 3,500 daltons moiety ($IRG^{3500}$) whereas IRG refers to total immunoreactive glucagon which includes various bioactive and bioinactive glucagon species.

## B. Chronic Glucagon Deficiency

### I. After Surgical Resection of A-cells

Production of glucagon deficiency by surgical resection is complicated by the presence of glucagon-secreting A-cells in different organs, the distribution of which varies from species to species (for details, see Chap. 33).

#### 1. In Humans

Glucagon-containing cells have been demonstrated by immunochemical techniques in the gastric fundus, the duodenum, and the colon of human subjects (ORCI 1976; RAVAZZOLA et al. 1981; SASAGAWA et al. 1974; KNUDSEN et al. 1975). It appears, however, that human gastric mucosa contains much less glucagon than canine gastric mucosa (see Chap. 33). For instance, in one study, A-like cells were found in only one of eight gastric fundi studied (MUÑOZ-BARRAGAN et al. 1977). These cells were few in number and were irregularly dispersed between parietal cells. Acid–ethanol-extractable IRG ranged from 13 to 200 pg/mg of fundic mucosal tissue, as compared with about 2 ng/mg of fundic mucosa of normal dogs (DOI et al. 1979). Unfortunately, the human tissues were studied 3–8 h postmortem and showed signs of advanced autolysis. This casts some doubt on the validity of the findings since glucagon is unstable and disappears rapidly from tissue when left at room temperature for several hours (MUÑOZ-BARRAGAN et al. 1977). In addition to the intestinal extrapancreatic glucagon, a high molecular weight glucagon has been extracted from human sublingual and submaxillary glands (LAWRENCE et al. 1976; BHATHENA et al. 1977). However, the role of this salivary glucagon remains unknown. Despite the presence of extrapancreatic glucagon, there is now convincing evidence that total pancreatectomy, which usually includes hemigastrectomy, duodenectomy, and splenectomy, leads to glucagon deficiency in human subjects. This deficiency, however, only becomes apparent when the bioactive $IRG^{3500}$ is measured as total IRG usually does not decrease after pancreatectomy. BODEN et al. (1980) found that total plasma IRG of nine totally pancreatectomized patients was not significantly different from that of ten normal controls ($286 \pm 101$ compared with $213 \pm 46$ pg/ml as measured with 30 K antiserum). When basal plasma from six of these nine patients was fractionated over Sephadex G-50, $81\% \pm 9\%$ of the total IRG eluted within the void volume. When rechromatographed on Sephadex G-200, this IRG moiety was found to have an apparent molecular weight of approximately 200,000 daltons. Only $18 \pm 9$ pg/ml plasma eluted in the glucagon ($IRG^{3500}$) region. This was significantly less than the $IRG^{3500}$ content in the plasma of normal controls which was $49 \pm 9$ pg/ml. Moreover, IRG concentrations did not rise after stimulation with intravenous arginine in most of these patients.

MÜLLER et al. (1979) studied eight duodenopancreatomized patients. Their basal IRG concentrations were normal ($162 \pm 68$ pg/ml) and did not change during arginine infusion. Using Biogel P-30 column chromatography, $IRG^{3500}$ was undetectable in seven of the eight patients. The same author (MÜLLER et al. 1974) had reported in a prior publication that between 75 and 90 pg/ml of total IRG (30 K antiserum) remained in the plasma of two totally pancreatectomized patients. Between 15 and 35 pg/ml of this could be bound to charcoal and, therefore, was suspected to be "true glucagon". BARNES and BLOOM (1976) detected no IRG in plasma of five pancreatectomized patients using their RCS-5 antiserum. However, these investigators added to each patient's standard curve plasma rendered "glucagon free" by pretreatment with an immune absorbant. The addition of "stripped plasma" which probably contained high molecular weight IRG components may have created an arbitrary zero standard and led to underestimation of plasma IRG. BODEN et al. (1980) using RCS-5 antiserum detected subnormal amounts of $IRG^{3500}$ in eight of their nine pancreatectomized patients. VILLANUEVA et al. (1976) and BOTHA et al. (1977) each studied one patient with total pancreatectomy and reported that $IRG^{3500}$ represented only a very small fraction of total IRG or was undetectable. BRINGER et al. (1981) recently reported a series of ten pancreatectomized patients in whom basal glucagon levels, using 30 K antiserum, averaged $69 \pm 8$ pg/ml. In two of these patients, a clear-cut rise in circulating glucagon was observed during intravenous arginine infusion. In another series of 21 pancreatectomized patients reported by TIENGO et al. (1982), basal plasma immunoreactive glucagon averaged 230 ± 26 pg/ml, but after ethanol extraction of plasma it became undetectable. In these patients, intravenous arginine did not stimulate glucagon. Thus, the presence of partial glucagon deficiency in human subjects after total duodenopancreatectomy and hemigastrectomy has been convincingly documented by several investigators. The origin of the IRG (total IRG and $IRG^{3500}$) remaining in human plasma after total duodenopancreatectomy and hemigastrectomy is uncertain. Possible sources include A-cells in human colon (KNUDSEN et al. 1975) and in vivo conversion of glucagon-like immunoreactivity (GLI), present in large amounts in the gut, into glucagon, as has been demonstrated in piglets by KORANYI et al. (1981).

## 2. In Dogs

In contrast to human subjects, pancreatectomy in dogs does not lead to glucagon deficiency, in fact, plasma IRG concentration usually increases after surgery owing to release of substantial amounts of extrapancreatic glucagon (MATSUYAMA and FOÀ 1974; VRANIC et al. 1974; MASHITER et al. 1975). Glucagon-containing A-cells are abundant in the canine fundic mucosa (LARSSON et al. 1975; BAETENS et al. 1976) and the quantity of extractable glucagon far exceeds that extractable from human mucosa. The increase in plasma glucagon concentration in pancreatectomized dogs has been reported to consist predominantly of $IRG^{3500}$ and $IRG^{9000}$ (VALVERDE et al. 1975). Some of the IRG which increased after pancreatectomy in response to arginine has been reported to have a molecular weight greater than 3,500 daltons (MASHITER et al. 1975). The release of extrapancreatic glucagon in the dog appears to be controlled by the concentration of circulating insulin and glucose. In normal dogs, the gastric fundus does not release significant quantities of glucagon during the basal state or after arginine stimulation (MUÑOZ-BARRAGAN et al. 1976). In contrast, insulin-deficient, hyperglycemic, pancreatectomized dogs release substantial amounts of glucagon which increases severalfold during arginine stimulation (BLAZQUEZ et al. 1977; LEFEBVRE and LUYCKX 1977; MÜLLER et al. 1978; ROSS et al. 1978). Thus, it appears that the main reasons why glucagon deficiency does not occur in pancreatectomized dogs are: (1) larger quantities of extrapancreatic glucagon in the intestinal tract of dogs compared with humans; and (2) the presence of insulin deficiency which facilitates the release of this glucagon.

## 3. In Other Experimental Animals

The presence of extrapancreatic A-cells and/or extractable IRG has been demonstrated in other experimental animals including cats, pigs, rats, and monkeys (LARSSON et al. 1975; CHISHOLM et al. 1978; HELMSTEADTER et al. 1977). Pancreatectomy results in increases in plasma IRG in cats and pigs (CHISHOLM et al. 1978; HOLST et al. 1978b). Whether or not pancreatectomized cats and pigs are deficient in bioactive glucagon has not been established unequivocally, mainly because few studies have been performed in these species.

A special situation seems to prevail in birds. Chickens, for instance, have been reported to possess two types of islets, one type consisting mainly of A- and D-cells, the other mainly of B- and D-cells. Selective removal of the A-cell-containing islets which are confined to special well-defined lobes of the pancreas, results in severe hypoglycemia and death within 12–36 h after resection (MIKAMI and KAZUYUKI 1962).

## II. Idiopathic Chronic Glucagon Deficiency in Neonates

Since the original description of presumed A-cell deficiency in babies with familial infantile hypoglycemia (MCQUARRIE et al. 1950), glucagon deficiency has been frequently discussed as a possible cause of neonatal hypoglycemia (GOTLIN and SILVER 1970; GROLLMAN et al. 1964; WAGNER et al. 1969). However, these reports were based on histologic evidence of a reduced number of pancreatic A-cells and glucagon determinations were not obtained. Furthermore, no differentiation was made between $A_1$- and $A_2$-cells and, it remained unclear whether the number of glucagon-producing $A_2$-cells or that of somatostatin-producing $A_1$-cells (now called D-cells) were reduced. Recently, however, glucagon deficiency has been reported in two newborns babies (VIDNES and OYASEATER 1977; KOLLEE et al. 1978). Both had very low basal plasma glucagon and normal plasma insulin concentrations and developed severe hypoglycemia associated with convulsions during the first 3 days after delivery. Both responded to injections of exogenous glucagon with dramatic clinical improvement and a rise in plasma glucose concentrations. In the case reported by VIDNES and OYASAETER (1977), additional studies revealed that plasma glucagon concentrations did not rise in response to alanine, that basal plasma concentrations of free fatty acids were low, that concentrations of several glucogenic amino acids, including asparagine, serine, glycine, and methionine, were high while the rate of gluconeogenesis from alanine was reduced. Urinary catecholamines were within normal limits. In a third case, reported in abstract form, glucagon deficiency was associated with hypoglycemia in an adult male (BLEICHER et al. 1970). This case was complicated by simultaneous deficiencies in the secretion of insulin and glucagon.

Several questions can be raised with respect to these reports. First, how well documented was the glucagon deficiency? In KOLLEE's case, plasma IRG concentrations (10 pg/ml, antiserum 30 K) seem to have been at the limit of assay detectability, indicating that basal plasma contained little or no pancreatic or extrapancreatic glucagon. Stimulatory tests were not performed. In the VIDNES case, basal IRG concentration was 55 pg/ml (antiserum 964, Novo Research Institute, Bagsvaerd, Denmark) which may have been within the normal range. $IRG^{3500}$ was not determined. However, there was no IRG response to intravenous administration of alanine. Thus, at least in this case, it is not clear whether or not basal glucagon concentrations were abnormally low. Second, what is the evidence that glucagon deficiency was responsible for the severe hypoglycemia in these two neonates? In adult patients with total duodenopancreatectomy, chronic glucagon deficiency is usually not associated with persistent hypoglycemia. Similarly, acute glucagon deficiency produced by infusion of somatostatin and insulin causes only a transient and mild hypoglycemia, followed by normal blood glucose levels (LILJENQUIST et al. 1977). This occurs despite a steep fall in the rate of hepatic glucose production because the falling plasma glucose concentration inhibits the release of insulin, which in turn raises plasma glucose concentration by increasing hepatic glucose production and by decreasing peripheral glucose utilization. Unfortunately, very little is known about the control of glucose production and utilization in neonates (PAGLIARA et al. 1973). It is possible, perhaps even likely, that because of their relatively greater glucose need and smaller gluconeogenetic potential, glucagon may play a greater role in neonates than in adults (see Chap. 36). However, of probably greater pathogenetic significance for the development of hypoglycemia than glucagon deficiency may have been the relative hyperinsulinemia present in these neonates. Both maintained normal insulin concentrations during severe hypoglycemia. For instance, in KOLLEE's case, plasma insulin concentrations were 9 and 15 μU/ml with corresponding glucose concentrations of 2.6 and 3.4 mmol/l, respectively. In VIDNES' case, a mean insulin value of 9.6 μU/ml was associated with a mean glucose concentration of 34.4 mg/dl. Under normal conditions, plasma insulin concentrations should be undetectable or certainly less than 5 μIU/ml for this degree of hypoglycemia. Therefore, the insulin levels observed in these neonates indicate disturbances in insulin release and suggest the presence of Beta-cell pathology. In summary, chronic or acute glucagon deficiency does not cause persistent hypoglycemia in adults. In one of the two reports on neonatal hypoglycemia, basal glucagon deficiency was not documented unequivocally, whereas in both cases, there was clear evidence of relative hyperinsulinemia. Thus, there is presently no convincing evidence that there is a well-defined neonatal glucagon deficiency syndrome and that *isolated* glucagon deficiency causes hypoglycemia.

## C. Acute Glucagon Deficiency

### I. After Administration of Anti-Glucagon Sera

Several investigators have attempted to use anti-glucagon sera to produce acute glucagon deficiency. The results have been equivocal. High titer antisera had no effects on glucose or insulin concentrations in resting fed or fasted rats (HOLST et al. 1978a). A glucose-lowering effect of glucagon antiserum in 48-h fasted adrenalectomized rats reported by GREY et al. (1970) could not be reproduced by another group (HOLST 1978).

### II. After Infusion of Somatostatin

Recently, somatostatin has been used successfully to inhibit the release of glucagon in human subjects and in experimental animals. A state of "isolated glucagon deficiency" can be produced by infusion of pharmacologic amounts of somatostatin together with replacement doses of other suppressed hormones such as insulin and/or growth hormone. During somatostatin infusion, plasma glucagon concentrations decrease by approximately 70%–80%. The majority of the remaining glucagon immunoreactivity consists of high molecular weight IRG of unknown biologic activity. This approach also has its problems inasmuch as somatostatin not only inhibits glucagon release, but in addition affects a large number of other hormonal peptides as well as intestinal and pancreatic exocrine secretion, splanchnic blood flow, and absorption of nutrients (WILSON et al. 1977; WAHREN et al. 1977).

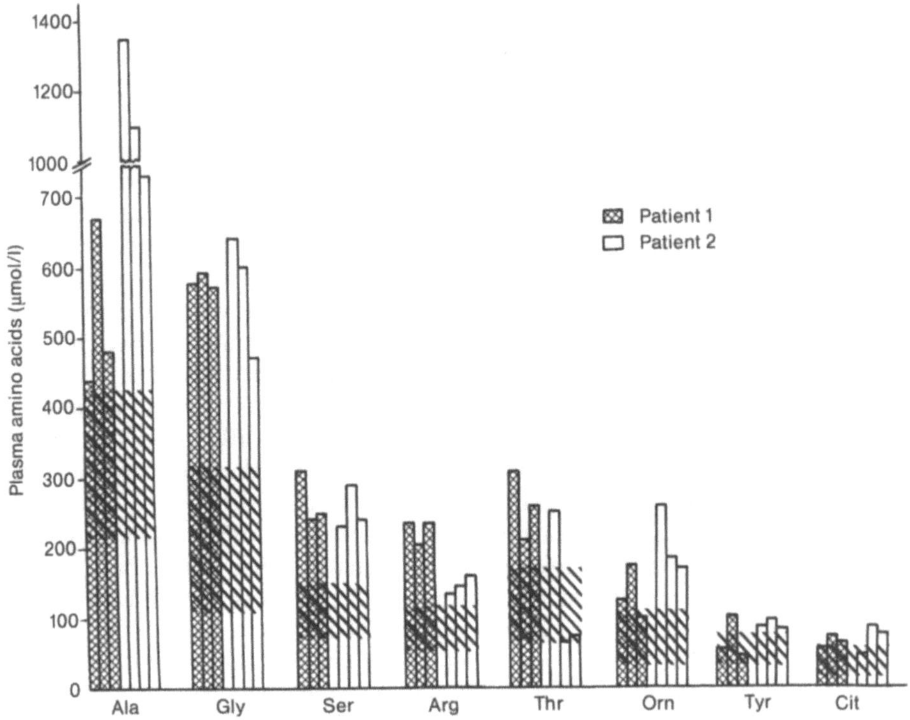


**Fig. 1.** Amino acids in two pancreatectomized patients. Shown are plasma concentrations of peripheral venous amino acids in Patient 1 (*crosshatched bars*) and Patient 2 (*open bars*) determined on 3 consecutive days. Blood was drawn after an overnight fast and before insulin administration. The *hatched areas* represent the normal range as determined in 20 normal subjects. BODEN et al. (1980)

Thus, "isolated glucagon deficiency" is never isolated and it is difficult to be sure that any observed metabolic effects were the exclusive consequence of glucagon deficiency. Despite these potential shortcomings, this technique has produced the best data available on the metabolic effects of acute glucagon deficiency.

## D. Metabolic Effects of Glucagon Deficiency

### I. On Protein and Amino Acid Metabolism

Strong support for an important role of basal glucagon concentrations on amino acid metabolism has recently been provided by the demonstration of distinct and reproducible effects of glucagon deficiency on plasma amino acid concentrations and on hepatic gluconeogenesis. BODEN et al. (1980) have studied 9 duodenopancreatectomized patients with documented glucagon deficiency and found significant elevations of postabsorptive plasma concentrations of serine, alanine, arginine, glycine, threonine, citrulline, α-aminobutyrate, and tyrosine compared with values obtained in 20 normal controls. Figure 1 shows elevations in plasma con-

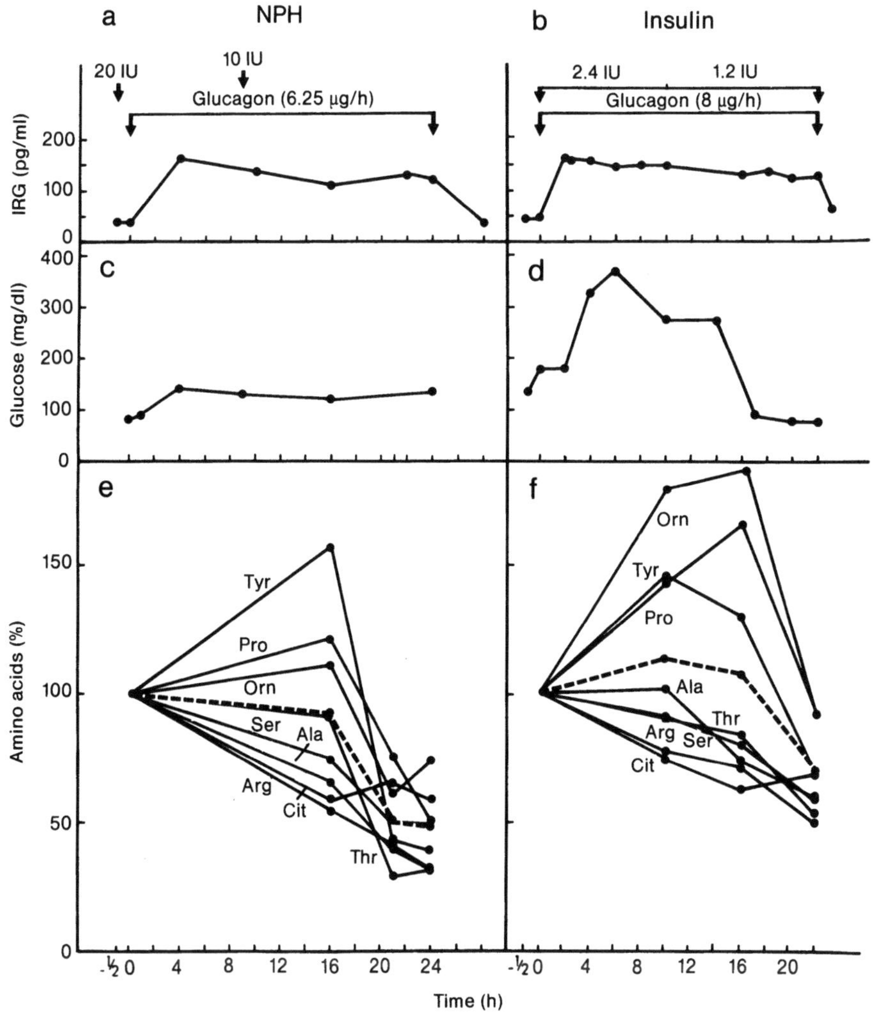


**Fig. 2 a–f.** Effect of glucagon infusion on glucagon, glucose, and amino acid levels. Shown are plasma IRG (**a, b**), plasma glucose (**c, d**), and individual (*full lines*) and mean (*broken lines*) amino acid concentration (**e, f**) in Patient 1 (**a, c, e**) and Patient 2 (**b, d, f**) during infusion of glucagon (6.25 μg/h for 24 h in Patient 1 and 8.0 μg/h for 22 h in Patient 2). Amino acids are expressed as percentages of the preinfusion (8 a.m.) concentrations. NPH = isophane insulin suspension NPH (Lilly, Indianapolis, Indiana). Boden et al. (1980)

centrations of 8 glucogenic amino acids determined on 3 consecutive days in 2 of these 9 patients. In the same two patients, replacement doses of glucagon (6.25 and 8 μg/h) were infused and resulted in normalization of the previously existing hyperaminoacidemia within 22 h (Fig. 2). Similarly, Müller et al. (1979) measured 14 amino acids in 5 patients with duodenopancreatectomy and found plasma concentrations of alanine, serine, ornithine, and arginine to be approximately twice the normal values. In both studies, plasma concentrations of the branched chain amino acids leucine, isoleucine, and valine were within normal limits, suggesting that insulin deficiency was not responsible for the observed changes.

The concept that glucagon deficiency was responsible for the hyperaminoacidemia in pancreatectomized patients was further strengthened by the observation that acute selective glucagon deficiency, produced in normal subjects by infusion of somatostatin, together with replacement doses of insulin, led to significant increases in plasma concentrations of glycine, alanine, arginine, glutamine, and lysine within 8 h (BODEN et al. 1979). Similarly, GERICH et al. (1976) observed that venous alanine concentrations rose markedly in a group of insulin-dependent diabetic patients 4–6 h after commencement of somatostatin infusion. These authors attributed the hyperalaninemia to glucagon deficiency. CHERRINGTON et al. (1979) produced selective glucagon deficiency in dogs by infusing somatostatin together with replacement doses of insulin. They observed a large increase in plasma alanine concentration (+102 μmol/l) when normoglycemia was maintained and a lesser rise (+48 μmol/l) when hypoglycemia was allowed to develop. As mentioned before, the validity of the conclusions drawn from these data rests on the assumption that somatostatin has no effects of its own on the liver other than those mediated by changes in glucagon and insulin release. This assumption is supported by a number of studies in dogs and studies with rat liver slices (CHERRINGTON et al. 1976, 1979; CHIDECKEL et al. 1975). On the other hand, findings by SACCA et al. (1979) that somatostatin reduced the stimulatory effect of epinephrine on hepatic glucose output, by OLIVER and WAGLE (1975) that somatostatin inhibited glucagon-stimulated glycogenolysis and gluconeogenesis in isolated hepatocytes, and by SACKS et al. (1977) that somatostatin suppressed glucagon-stimulated glucose release from isolated perfused rat livers emphasizes the need for caution in the interpretation of data based on somatostatin-induced suppression of glucagon release.

The studies provide no information on the mechanism by which glucagon deficiency leads to hyperaminoacidemia. There is, however, evidence in the literature to indicate that glucagon deficiency inhibits gluconeogenesis, suggesting alterations in the rate of glucose formation from amino acids as the mechanism by which glucagon controls plasma amino acid concentrations. JENNINGS et al. (1977) have demonstrated that selective glucagon deficiency produced in dogs severely inhibited gluconeogenesis from alanine, and RABIN et al. (1979) showed that selective glucagon deficiency in normal male subjects resulted in reduced uptake of alanine by the splanchnic bed. WAHREN et al. (1977) in a study on overnight and on 60-h fasted healthy volunteers showed that hypoglucagonemia inhibited glycogenolysis as well as gluconeogenesis; and that the glucagon-sensitive steps in gluconeogenesis affected by somatostatin involved primarily intrahepatic disposal rather than net hepatic uptake of the glucose precursors lactate, pyruvate, glycerol, and amino acids. Compatible with the thesis that glucagon controls amino acid concentration through actions on gluconeogenesis are the findings of several groups of investigators that: (a) the most prominent elevations occurred in plasma concentrations of those amino acids (alanine, glycine, serine, threonine) that are most avidly extracted by the splanchnic bed; and (b) acute selective glucagon deficiency in normal subjects significantly decreased urinary urea secretion. It is also noteworthy that those amino acids that are potent stimulators of glucagon release (particularly alanine and arginine) are also most affected by its metabolic actions. This suggests the existence of a feedback system between glucagon and certain glucogenic amino acids.

## II. On Hepatic Glucose Production

As reviewed in Chap. 16, selective glucagon deficiency produced by combined infusion of somatostatin and replacement doses of insulin has been shown to result in an approximately 30% decrease in the rate of hepatic glucose production and in hypoglycemia in overnight fasted anesthetized dogs (CHERRINGTON 1976, 1979) and in human subjects (LILJENQUIST et al. 1977). However, the true extent to which basal glucagon concentrations were responsible for the maintenance of basal glucose production became apparent only when hypoglycemia was prevented and normoglycemia maintained by infusion of glucose. Under these conditions, isolated glucagon deficiency led to a 67% decline in glucose production rates (CHERRINGTON et al. 1979). These studies illustrated that hypoglycemia significantly increased glucose production rates. The mechanisms for this effect are not well understood. Hypoglycemia mediated release of catecholamines, particularly of epinephrine, appears to be a possibility.

On the other hand, there are data to suggest that low glucose concentrations per se may stimulate hepatic glucose production (RUDERMAN and HERRERA 1968; GLINSMANN and MORTIMORE 1968). The decline in glucose production rates during isolated glucagon deficiency reflects primarily a decrease in glycogenolysis. The rate of gluconeogenesis from alanine and lactate has also been shown to decrease by about 20% (CHERRINGTON et al. 1979). However, this could reduce overall hepatic glucose production by not more than 5%, inasmuch as the maximal contribution of gluconeogenesis to glucose production in the postabsorptive state is only about 25% (FELIG 1973). The studies by CHERRINGTON et al. (1976, 1979) also provided evidence that the actions of basal glucagon concentrations on basal glucose production are not evanescent, as are the effects of absolute or relative hyperglucagonemia. Nevertheless, it remains unknown as to how glucagon can support two-thirds of basal hepatic glucose production for prolonged periods of time, but is unable to increase glucose production above basal levels for more than a few minutes or perhaps hours. In summary, most of the available evidence, obtained through studies of selective glucagon deficiency, supports the conclusion that in the presence of normal basal concentrations of insulin, glucagon is essential for the maintenance of most of the basal hepatic glucose production.

# E. Summary and Conclusions

Chronic glucagon deficiency develops in human subjects after duodenopancreatectomy–hemigastrectomy. Most of these patients have very low basal glucagon ($IRG^{3500}$) concentrations and absent glucagon responses to stimulation with arginine. In contrast, pancreatectomized dogs do not develop glucagon deficiency, probably because their gastrointestinal tract contains large quantities of extrapancreatic glucagon, the release of which is facilitated by insulin deficiency. In addition, acute glucagon deficiency can be produced in human subjects and in experimental animals by infusion of pharmacologic amounts of somatostatin together with replacement doses of insulin. Using this approach, it has been demonstrated that basal glucagon concentrations play an important role in amino acid and glucose metabolism. Acute glucagon deficiency raises plasma concentrations of most glucogenic amino acids, probably by reducing the rate of gluconeogenesis from amino acids. Moreover, studies in which acute glucagon deficiency was produced with somatostatin have shown that glucagon is responsible for about two-thirds of normal basal hepatic glucagon production.

Life-threatening hypoglyemia associated with abnormally low plasma glucagon concentrations has recently been described in two neonates. In both cases, treatment with exogenous glucagon resulted in prompt elevation of plasma glucose concentration and cessation of seizures. The cause of the glucagon deficiency was not elucidated. Moreover, it was not established that glucagon deficiency per se was responsible for the observed hypoglycemia as there was evidence of insulin hypersecretion in both neonates.

*Acknowledgments.* This work was supported by USPHS grants AM-19397 and 5 MOl RR 349.

## References

Baetens D, Rufener C, Srikant BC, Dobbs R, Unger R, Orci L (1976) Identification of glucagon-producing cells (A cells) in dog gastric mucosa. J Cell Biol 69:455–464
Barnes AJ, Bloom SR (1976) Pancreatectomised man: a model for diabetes without glucagon. Lancet 1:219
Bhathena SJ, Smith SS, Voyles NR, Penhos JC, Recant L (1977) Studies on submaxillary gland immunoreactive glucagon. Biochem Biophys Res Commun 74:1574–1581
Blazquez E, Muñoz-Barragan L, Patton GS, Dobbs RE, Unger RH (1977) Demonstration of gastric glucagon hypersecretion in insulin-deprived alloxan-diabetic dogs. J Lab Clin Med 89:971–977
Bleicher SJ, Levy LJ, Zarowitz H, Spergel G (1970) Glucagon-deficient hypoglycemia: a new syndrome (Abstr). Clin Res 18:355
Boden G, Rezvani I, Master RW, Trapp V, Schwartz M, Owen OE (1979) Glucagon deficiency causes hyperaminoacidemia (Abstr). Clin Res 27:482
Boden G, Master RW, Rezvani I, Palmer JP, Lobe TE, Owen OE (1980) Glucagon deficiency and hyperaminoacidemia after total pancreatectomy. J Clin Invest 65:706–716
Botha JL, Vinik AI, Child PT, Paul M, Jackson WPU (1977) Pancreatic glucagon like immunoreactivity in a pancreatectomized patient. Horm Metab Res 9:199
Bringer J, Mirouze J, Marchal G, Pham T-CH, Luyckx A, Lefèbvre P, Orsetti A (1981) Glucagon immunoreactivity and antidiabetic action of somatostatin in the totally duodeno-pancreatectomized and gastrectomized human. Diabetes 10:851–856
Cherrington AD, Chiasson JL, Liljenquist JE, Jennings AS, Keller U, Lacy WW (1976) The role of insulin and glucagon in the regulation of basal glucose production in the postabsorptive dog. J Clin Invest 58:1407–1418
Cherrington AD, Liljenquist JE, Shulman GI, Williams PE, Lacy WW (1979) Importance of hypoglycemia-induced glucose production during isolated glucagon deficiency. Am J Physiol 236:E263–E271
Chideckel E, Palmer J, Koerker DH, Ensinck J, Davidson MB, Goodner CJ (1975) Somatostatin blockade of acute and chronic stimuli of the endocrine pancreas and consequences of this blockade on glucose homeostasis. J Clin Invest 55:754–762
Chisholm DJ, Alford FP, Harewood MS, Findlay DM, Gray BN (1978) Nature and biologic activity of "extrapancreatic glucagon": studies in pancreatectomized cats. Metabolism 27:261–273
Doi K, Prentki M, Yip C, Müller WA, Jeanrenaud B, Vranic M (1979) Identical biological effects of pancreatic glucagon and a purified moiety of canine gastric immunoreactive glucagon. J Clin Invest 63:525–531
Felig P (1973) The glucose-alanine cycle. Metabolism 24:179–207

Gerich JE, Lorenzi M, Bier DM, Tsalikian E, Schneider V, Karam JH, Forsham PH (1976) Effects of physiologic levels of glucagon and growth hormone on human carbohydrate and lipid metabolism. J Clin Invest 57:875–884
Glinsmann WH, Mortimore GE (1968) Influence of glucagon and 3′,5′-AMP on insulin responsiveness of the perfused rat liver. Am J Physiol 215:553–559
Gotlin RW, Silver HK (1970) Neonatal hypoglycemia, hyperinsulinism and absence of pancreatic alpha-cells. Lancet 1:1346
Grey N, McGuigan JE, Kipnis DM (1970) Neutralization of endogenous glucagon by high titer glucagon antiserum. Endocrinology 86:1383–1388
Grollman A, McCaleb WE, White FN (1964) Glucagon deficiency as a cause of hypoglycemia. Metabolism 13:686–690
Helmstaedter J, Feurle GE, Forssmann WG (1977) Relationship of glucagon-somatostatin and gastrin-somatostatin cells in the stomach of the monkey. Cell Tissue Res 177:29–46
Holst JJ (1978) Extrapancreatic glucagons. Digestion 17:168–190
Holst JJ, Galbo H, Richter EA (1978a) Neutralization of glucagon by antiserum as a tool in glucagon physiology. J Clin Invest 62:182–190
Holst JJ, Kreutzfeldt M, Holm G, Jensen E, Poulsen JSD, Sparso B, Schmidt A (1978b) Absence of true pancreatic glucagon but persistence of circulating pancreatic glucagon-like immunoreactivity after pancreatectomy in pigs. Diabete Metab 4:75–79
Jennings AS, Cherrington AD, Liljenquist JE, Keller U, Lacy WW, Chiasson JL (1977) The roles of insulin and glucagon in the regulation of gluconeogenesis in the postabsorptive dog. Diabetes 26:847–856
Knudsen JB, Holst JJ, Asnoes S, Johanson AA (1975) Identification of cells with pancreatic-type and gut-type glucagon immunoreactivity in the human colon. Acta Pathol Microbiol Scand [A] 83:741–743
Kollee LA, Monnens LA, Ceika V, Wilms RH (1978) Persistent neonatal hypoglycemia due to glucagon deficiency. Arch Dis Child 53:422–424
Koranyi L, Peterfy F, Torok A, Guoth M, Tamas GY Jr (1981) Evidence for transformation of glucagon-like immunoreactivity of gut into pancreatic glucagon in vivo. Diabetes 30:792–794
Larsson L-I, Holst J, Håkanson R, Sundler F (1975) Distribution and properties of glucagon immunoreactivity in the digestive tract of various mammals: an immunohistochemical and immunochemical study. Histochemistry 44:281–290
Lawrence AM, Tan S, Hojvat S, Kirsteins L (1976) Salivary gland hyperglycemic factor: an extrapancreatic source of glucagon-like material. Science 195:70–72
Lefèbvre PJ, Luyckx AS (1977) Factors controlling gastric-glucagon release. J Clin Invest 59:716–722
Liljenquist JE, Mueller GL, Cherrington AD, Keller U, Chiasson JL, Perry JM, Lacy WW, Rabinowitz D (1977) Evidence for an important role of glucagon in the regulation of hepatic glucose production in normal man. J Clin Invest 59:369–374
Mashiter K, Harding PE, Chou M, Mashiter GD, Stout J, Diamond D, Field JB (1975) Persistent pancreatic glucagon but not insulin response to arginine in pancreatectomized dogs. Endocrinology 96:678–693
Matsuyama T, Foà PP (1974) Plasma glucose, insulin, pancreatic and enteroglucagon levels in normal and depancreatized dogs. Proc Soc Exp Biol Med 147:97–102
McQuarrie I, Bell ET, Zimmermann B, Wright WS (1950) Deficiency of alpha cells of pancreas as possible etiological factor in familial hypoglycemosis. Fed Proc 9:337
Mikami S-I, Kazuyuki O (1962) Glucagon deficiency induced by extirpation of alpha islets of the fowl pancreas. Endocrinology 71:464–473
Müller WA, Brennan MF, Tan MH, Aoki TT (1974) Studies of glucagon secretion in pancreatectomized patients. Diabetes 23:512–516
Müller WA, Girardier L, Seydoux J, Berger M, Renold AE, Vranic M (1978) Extrapancreatic glucagon and glucagonlike immunoreactivity in depancreatized dogs: a quantitative assessment of secretion rates and anatomical delineation of sources. J Clin Invest 62:124–132
Müller WA, Berger M, Suter P, Cuppers HJ, Reiter J, Wyss T, Berchtold P, Schmidt FH, Assal J-P, Renold AE (1979) Glucagon immunoreactivities and amino acid profile in plasma of duodenopancreatectomized patients. J Clin Invest 63:820–827

Muñoz-Barragan L, Blazquez E, Patton GS, Dobbs RE, Unger RH (1976) Gastric A-cell function in normal dogs. Am J Physiol 231:1057–1061
Muñoz-Barragan L, Rufener C, Srikant CB, Dobbs RE, Shannon WS Jr, Baetens D, Unger RH (1977) Immunocytochemical evidence for glucagon-containing cells in the human stomach. Horm Metab Res 9:37–39
Oliver JR, Wagle SR (1975) Studies on the inhibition of insulin release, glycogenolysis and gluconeogenesis by somatostatin in the rat islets of Langerhans and isolated hepatocytes. Biochem Biophys Res Commun 62:772–777
Orci L (1976) Cells containing somatostatin in pancreas and gastro-intestinal tract. Ciba Found Symp 41:344
Pagliara AS, Karl IE, Haymond M, Kipnis DM (1973) Hypoglycemia in infancy and childhood. J Pediatr 82:365–379, 558–577
Rabin D, Mueller GL, Lacy WW, Liljenquist JE (1979) Splanchnic metabolism of alanine in intact man: effects of somatostatin and somatostatin plus insulin. Diabetes 28:486–490
Ravazzola M, Unger RH, Orci L (1981) Demonstration of glucagon in the stomach of human fetuses. Diabetes 30:879–882
Ross G, Lickley L, Vranic M (1978) Extrapancreatic glucagon in control of glucose turnover in depancreatized dogs. Am J Physiol 234:E213–E219
Ruderman NB, Herrera MG (1968) Glucose regulation of hepatic gluconeogenesis. Am J Physiol 241:1346–1351
Sacca L, Sherwin R, Felig P (1979) Influence of somatostatin on glucagon- and epinephrine-stimulated hepatic output in the dog. Am J Physiol 236:E113–E117
Sacks H, Waligora K, Matthews J, Pimstone B (1977) Inhibition by somatostatin of glucagon-induced glucose release from the isolated perfused rat liver. Endocrinology 101:1751–1759
Sasagawa TS, Kobayashi S, Fujita T (1974) Electron microscope studies on the endocrine cells of the human gut and pancreas. In: Fujita T, Igahu S (eds) Gastro-entero-pancreatic endocrine systems. A cell biological approach. Williams and Wilkins, Baltimore, p 17–38
Tiengo A, Bessioud M, Valverde I, Tabbi-Anneni A, Delprato S, Alexandre J, Assan R (1982) Absence of islet alpha cell function in pancreatectomized patients. Diabetologia 22:25–32
Valverde I, Dobbs R, Unger RH (1975) Heterogeneity of plasma glucagon immunoreactivity in normal, depancreatized and alloxan-diabetic dogs. Metabolism 24:1021–1028
Vidnes J, Oyasaeter S (1977) Glucagon deficiency causing severe neonatal hypoglycemia in a patient with normal insulin secretion. Pediatr Res 11:943–949
Villanueva ML, Hedo JA, Marco J (1976) Plasma glucagon immunoreactivity in a totally pancreatectomized patient. Diabetologia 12:613–616
Vranic M, Pek S, Kawamori R (1974) Increased "glucagon immunoreactivity" in plasma of totally depancreatized dogs. Diabetes 23:905–912
Wagner T, Spranger J, Brunch HJ (1969) Kongenitaler α-Zellmangel als Ursache einer chronischen infantilen Hypoglykämie? Monatsschr Kinderheilkd 117:236–238
Wahren J, Efendic S, Luft R, Hagenfeldt L, Bjørkman O, Felig P (1977) Influence of somatostatin on splanchnic glucose metabolism in postabsorptive and 60-hour fasted humans. J Clin Invest 59:299–307
Wilson RM, Boden G, Shore LS, Essa-Koumar N (1977) Effect of somatostatin on meal-stimulated pancreatic exocrine secretion in dogs. Diabetes 26:7–10

CHAPTER 43

# The Glucagonoma Syndrome

S. M. WOOD, J. M. POLAK, and S. R. BLOOM

## A. Introduction

The glucagonoma syndrome is now well characterised by a number of distinctive clinical and biochemical features associated with glucagon-producing islet cell tumours of the pancreas. In retrospect the first clear report of such a syndrome appeared in 1942. A patient, who was found to have at post mortem a malignant pancreatic tumour of unknown cell type, had suffered from an unusual erythematous migratory rash which was resistant to treatment. Diabetes mellitus, anaemia, weight loss and severe depression were other features of the illness and the patient had died after an acute thrombosis of the left iliac vein (BECKER et al. 1942). In 1956 similar features were reported in a patient with a pancreatic tumour containing $A_2$-cells (ZHDANOV 1956); and in 1960 an extract of a tumour from another patient with a migratory rash, was found to be hyperglycaemic when injected into animals (GOSSNER and KORTING 1960). It was not however until 1966 that it became possible to identify glucagon in both tumour and plasma of a patient with the now classical features of the disease (MCGAVRAN et al. 1966). Despite these reports it took some years before the full syndrome was characterised and the first series of nine patients published in 1974 (MALLINSON et al. 1974b).

The discovery of the glucagonoma syndrome has given a unique insight into the long-term effects of glucagon elevation. Most if not all, of the effects can be explained by an action on the liver, i.e. stimulation of gluconeogenesis. No "escape" from the continued elevation of glucagon occurs, indeed if secretion is temporarily blocked by somatostatin, hypoglycaemia immediately occurs. In addition, as will be detailed later, diabetes, if it occurs is usually mild and free of complications or ketoacidosis.

## B. Incidence

There have now been a number of recent reviews of the glucagonoma syndrome, these include those by: LOKICH et al. (1977), BINNICK et al. (1977), BLOOM and POLAK (1978), HIGGINS et al. (1979), LEICHTER (1980) and LUYCKX and LEFÈBVRE (1981). Glucagon-producing tumours are rare, 47 cases were reviewed by HIGGINS et al. in 1979; and 41 by LEICHTER in 1980. As recently reviewed by GUILLAUSSEAU et al. (1982), the literature now contains more than 90 reported cases of the syndrome.

### I. Age

Patients range from 19 to 73 years, with a mean age of about 55 years at the time of diagnosis (SOLER et al. 1976; HIGGINS et al. 1979; RIDDLE et al. 1978; PROYE et al. 1980). Many patients have a long history of symptoms, suggesting existence of the tumour at a much earlier stage.

### II. Sex

Amongst the patients reported before 1975 the majority were middle-aged women, as more cases have been reported the female preponderance has lessened, the ratio reported by HIGGINS et al. (1979) was 28 women to 19 men.

## C. Clinical Features

The majority of patients with glucagon-secreting pancreatic tumours have cutaneous manifestations including a necrolytic migratory erythematous rash, glossitis and angular stomatitis, and indeed many patients owe their initial diagnosis to a dermatologist. In association with the rash there is mild diabetes, weight loss, anaemia, depression and a susceptibility to deep vein thrombosis (CHURCH and CRANE 1967; BIANCHI et al. 1968; SHIMA et al. 1970; HOLST et al. 1975). There are however a small number of patients with no cutaneous features, who appear to have the same degree of hyperglucagonaemia as those with skin lesions. However, their diabetes tends to be much more severe (YOSHINAGA et al. 1966) and all those reported have had malignant tumours (MALLINSON et al. 1973; RUTTMAN et al. 1980). A few glucagonomas have been found in association with the pluriglandular syndrome, none with skin lesions (CROISIER et al. 1971; CROUGHS et al. 1972; WOODTLI and HEDINGER 1978).

### I. Skin Lesions

Perhaps the most distinctive feature of the glucagonoma syndrome is the skin rash, named necrolytic migratory erythema (NME) by WILKINSON (1973). This description incorporates the clinical feature of the figurate erythema and the histological changes of toxic epidermal necrolysis which are felt to be characteristic (SWEET 1974).

#### 1. Pathogenesis

The role of hyperglucagonaemia in the pathogenesis of the skin lesions is unknown. Hypoaminoacidaemia, can be induced in normal humans by infusion of glucagon (MARLISS et al. 1970), and is found in patients with glucagonomas (MALLINSON et al. 1974a, b; BINNICK et al. 1977). Kwashiorkor is associated with a rash with some similarities to the necrolytic migratory erythema and these patients have low plasma amino acid levels (HENINGTON et al. 1958), it has therefore been suggested that amino acid deficiency may contribute to skin changes. The skin lesions of kwashiorkor have been cured by vitamin-free casein or a mixture of pure amino

a
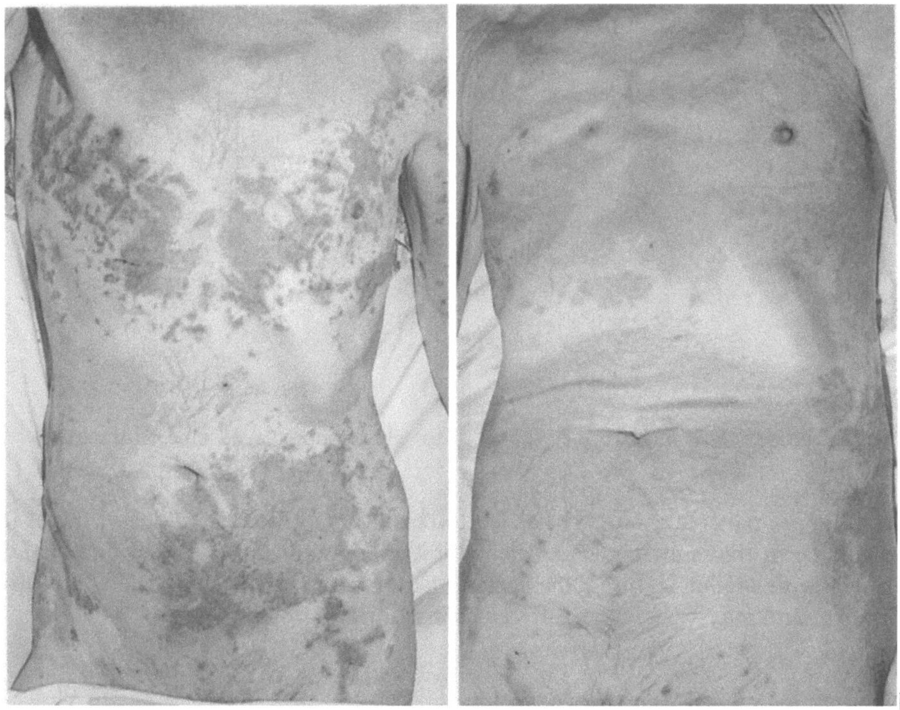
b

**Fig. 1a, b.** The skin rash of the glucagonoma syndrome before (**a**) and after treatment with topical and oral zinc (**b**) showing healing with pigmentation

acids (TRUSWELL et al. 1962), while there are reports of dramatic improvement of NME by intravenous amino acids (NORTON et al. 1979; MALLINSON et al. 1974a). Fatty acid deficiency in humans is associated with a dermatitis similar to NME (PROTTEY et al. 1975), however fatty acid abnormalities have not been well documented in the glucagonoma syndrome.

Glucagon has been shown to stimulate the production of "acute phase" proteins (MCMILLAN 1975). Raised levels of one of these, fibrinogen, could possibly alter skin blood flow, leading to skin changes. Zinc deficiency has been reported in patients with the glucagonoma syndrome (AMON et al. 1976; MALLINSON et al. 1977; HORROBIN and CUNNANE 1980). The rash has been found to respond in some cases to intravenous and topically administered zinc (Fig. 1) (MALLINSON et al. 1977). Acrodermatitis enteropathica and intravenous feeding are both associated with low plasma zinc and skin lesions resembling NME (HORROBIN and CUNNANE 1980; KAY and TASMAN JONES 1974). Rapid improvement in the skin lesions of the glucagonoma syndrome has been reported soon after surgical removal of the islet cell tumour (HIGGINS et al. 1979; JOHNSON et al. 1981), and with treatment with intravenous somatostatin (SOHIER et al. 1980; KAHN et al. 1981). In both cases this has been thought to be secondary to the fall in circulating glucagon levels.

There are a number of cases of NME in the literature not associated with A-cell tumours, but with other pancreatic abnormalities (THIVOLET 1980; RAMIERZ ARIAS

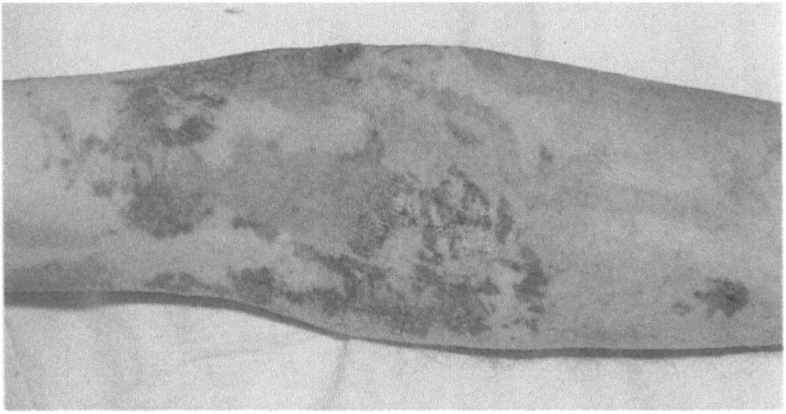

**Fig. 2.** The necrolytic migratory erythema of the glucagonoma syndrome on the leg

et al. 1979), pancreatic surgery has led to resolution of the rash in one of these. It would seem that nutritional deficiency is the common link between NME and rashes with similar features, the specific factors causing the skin lesions are unknown, and may vary under different conditions.

## 2. Distribution and Appearance

The rash most commonly involves the buttocks, groin, perineum, thighs and distal extremities (FREEDBERG and GALDABINI 1976; BINNICK et al. 1977; GATRELLI-BELTZER 1980). Lesions fall into two types: either pale brown macules and papules with superficial scaling which gives an eczematous or psoriasiform appearance; alternatively erythematous areas appear and enlarge to form blisters at their centre, these break down and become encrusted, eventually healing with pigmentation, but without scarring, each lesions evolving over about 14 days. Crops of lesions tend to appear and coalesce, resulting in extensive circinate, gyrate or serpiginous patterns (Fig. 2). The rash characteristically waxes and wanes, but trauma plays an important part in its severity and site. Bacterial and fungal superinfection is common, however, antibiotics, antifungal agents and corticosteroids have little or no long-term effect (WILKINSON 1973; MALLINSON et al. 1974b; BINNICK et al. 1977; SWENSON et al. 1978; LEWIS 1979). Perioral and paranasal crusting also involving the cheeks and forehead is frequent, as are atrophic glossitis, stomatitis and angular cheilitis. Nail dystrophy and thinning of hair are also associated features (BINNICK et al. 1977).

## 3. Histology

The characteristic histological features may be missed unless biopsies are taken from the edge of a fresh lesion (WILKINSON 1973). Microscopically, focal parakeratosis, with vesicles in the superficial epidermis, leading to subcorneal separation is characteristic. In the superficial granular layer and upper half of the epidermis there are swollen pale vacuolated cells with pyknotic nuclei. Acantholysis is

Table 1. Differential diagnosis of necrolytic migratory erythema

| |
|---|
| Pemphigus foliaceous |
| Chronic mucocutaneous candidiasis |
| Seborrhoeic dermatitis |
| Psoriasis vulgaris |
| Pustular psoriasis |
| Subcorneal pustular dermatosis (Sneddon – Wilkinson disease) |
| Chronic benign familial pemphigus (Hailey – Hailey disease) |
| Acrodermatitis enteropathica |
| Erythema annulare centrifugum |
| Chronic erythema multiforme |
| Toxic epidermal necrolysis |

notably absent (WILKINSON 1973; BINNICK et al. 1977). Only a mild perivascular inflammatory infiltrate is seen in the dermis. Immunofluorescence studies for immunoglubulin and complement are negative (WILKINSON 1973; KAHAN et al. 1977).

### 4. Differential Diagnosis of Necrolytic Migratory Erythema

NME has similarities to a number of skin diseases (Table 1), it can usually be differentiated from these by either clinical or histological appearances; the lack of acantholysis is particularly useful (KAHAN et al. 1977).

## II. Diabetes Mellitus

Most patients have abnormal glucose tolerance, there is however a poor correlation between hyperglucagonaemia and hyperglycaemia. This latter observation may be explained by such factors as the secretion of several abnormal molecular forms of glucagon by the tumour with reduced biological activity, the variable reactive hyperinsulinaemia in these patients and the possible effect of hepatic metastases in moderating the gluconeogenesis stimulated by high glucagon levels (LEICHTER 1980; JOHNSON et al. 1981; WEILLER et al. 1981). The diabetes of these patients is mild, often requiring only oral hypoglycaemic treatment. Ketoacidosis rarely occurs, perhaps because of the compensatory hyperinsulinaemia seen in these patients (DOMEN et al. 1980; GANDA 1980). Diabetic complications are not reported. Surgical removal of the tumour has led to cure of the diabetes (LIGHTMAN and BLOOM 1974; HIGGINS et al. 1979) as long as sufficient pancreatic tissue remains.

## III. Anaemia

Many patients have a normochromic, normocytic anaemia at some stage in the disease, though the cause is unknown [1]. The fall in haemoglobin seems to correlate

1 NAETS and GUNS (1980) have recently demonstrated a marked inhibitory effect exerted by glucagon on erythropoiesis in mice and rats, this effect seems to be exerted at the level of erythroid stem cell differentiation

with advanced disease and severe skin involvement. Although some patients have low plasma iron concentrations, these improve on oral iron without normalisation of the haemoglobin; no abnormalities have been found in the marrow with respect to cellular turnover, and iron stores (Mallinson et al. 1974b; Valverde et al. 1976).

### IV. Other Features

Loss of weight is common, perhaps due to the catabolic action of glucagon. Although about 20% of patients have diarrhoea, malabsorption is not part of the syndrome. Barium studies have demonstrated coarse folds of the duodenum in a number of cases, while in one case large jejunal villi were found similar to those seen in the case reported as an enteroglucagonoma (Gleeson et al. 1971; Mallinson et al. 1974b). Venous thrombosis has been an associated feature in about 30% of cases (Mallinson et al. 1974b; Pedersen et al. 1976). Routine tests of coagulation however are normal, although there is an unconfirmed finding of increased platelet adhesion. Psychiatric disturbances, particularly depression have been reported. There has been one case of widespread neurological involvement in association with the glucagonoma syndrome, with remission of symptoms after chemotherapy (Khandekar et al.1979).

## D. Biochemical Findings

### I. Plasma Glucagon

The fasting concentration of plasma glucagon in patients with the glucagonoma syndrome greatly exceeds that in normal subjects (Unger et al. 1963). In our assay, for example, glucagon concentrations in healthy young adults fasted overnight, lie between 5 and 20 pmol/l, while glucagonoma patients are always over 50 pmol/l and often over 500 pmol/l. Four molecular weight forms of glucagon have been characterised in plasma by gel filtration techniques (Fig. 3). These correspond to "big" glucagon (molecular weight >90,000 daltons), proglucagon (>9,000 daltons), glucagon (3,500 daltons) and a low molecular weight immunoreactive form (3,000 daltons) (Perrino et al. 1975; Danforth et al. 1976; Recant et al. 1976; Weir et al. 1976; Villar et al. 1981). These forms also occur in normal plasma (see Chap. 11), the 3,500 daltons form greatly predominating, but the amount of larger and smaller forms is much greater in the plasma of tumour patients. The biological activity of each molecular species may vary and would explain why the concentration of total glucagon measured in the plasma of these patients often does not correlate with their clinical or metabolic abnormalities. These molecular forms have been shown to change after surgical excision (Villar et al. 1981) and chemotherapy (Danforth et al. 1976).

#### 1. Hyperglucagonaemia: Its Differential Diagnosis

Disorders involving hyperglucagonaemia include diabetes mellitus, burn injury, acute trauma, septicaemia, cirrhosis, renal failure and Cushing's syndrome (Doyle

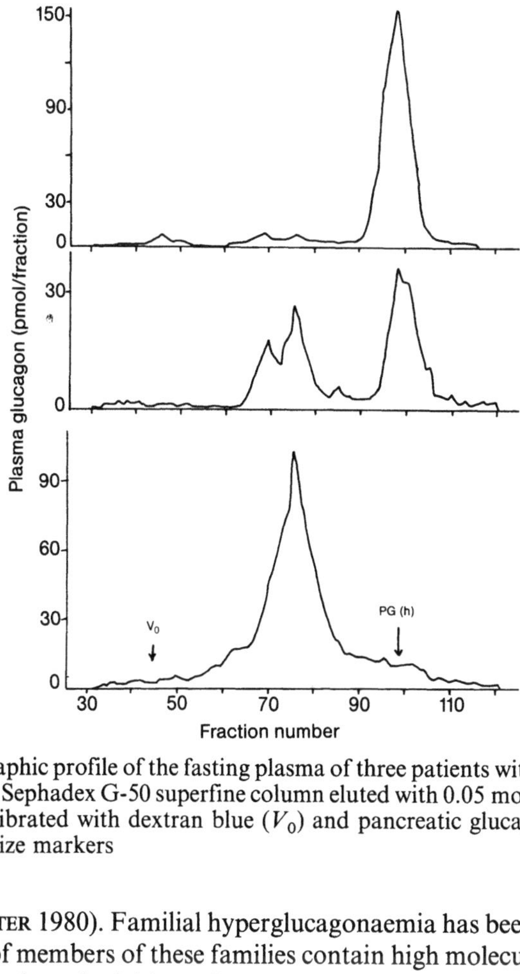


**Fig. 3.** Chromatographic profile of the fasting plasma of three patients with glucagon-secreting tumours; 60-cm Sephadex G-50 superfine column eluted with 0.05 mol/l phosphate buffer (pH 7.4) and calibrated with dextran blue ($V_0$) and pancreatic glucagon standard (PG Std) as molecular size markers

et al. 1979; LEICHTER 1980). Familial hyperglucagonaemia has been reported to occur. The plasma of members of these families contain high molecular weight forms of glucagon; none have had islet cell tumours (PALMER et al. 1976; BODEN and OWEN 1977). This group obviously requires to be differentiated from the familial glucagonoma syndrome (CROUGHS et al. 1972; YOUKER and RIDDLE 1980; STACPOOLE et al. 1981), which exists as part of the multiple endocrine adenomatosis.

## 2. Glucagon Secretory Patterns

The secretory behaviour of glucagon from tumours is often abnormal, as demonstrated by glucagon hypersecretion in response to intravenous arginine (HOLST 1979; BERGER et al. 1980; KRAMER et al. 1976) or adrenaline. Intravenous glucose causes an appropriate, but subnormal decrease in plasma glucagon; in contrast oral carbohydrate results in a paradoxical rise (VON SCHENCK 1979; HOLST 1979; HOLST et al. 1979; KHANDEKAR 1979; LEICHTER 1980; BERGER et al. 1980). Somatostatin, a potent inhibitor of endocrine secretion, is reported to lower the circulating concentration of glucagon in some patients with tumours (MALLINSON et al. 1977; LONG et al. 1979; HOLST et al. 1979; KAHN et al. 1981) and has improved the skin lesions in a number of cases (SOHIER et al. 1980).

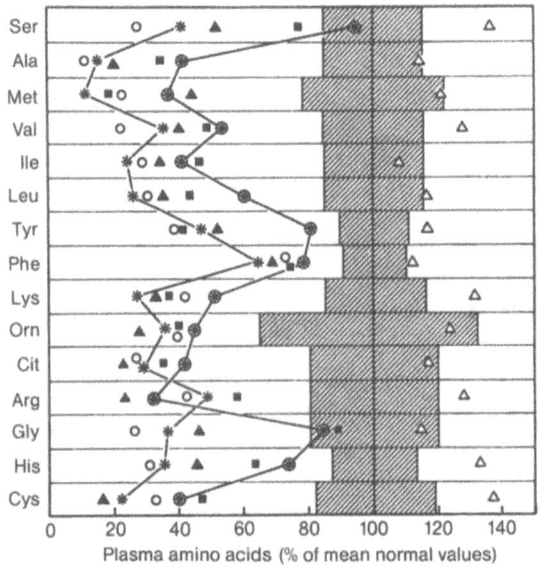


**Fig. 4.** Fasting plasma amino acid levels in four patients with the glucagonoma syndrome (Patient 1: preoperative *full triangles*, postoperative *open triangles*; 2: *squares*; 3: *circles*; 4: *asterisks*, in remission *circled asterisks*). *Hatched area* is mean ± standard deviation of normal values

## II. Plasma Insulin

Plasma levels of endogenous insulin are often raised (MALLINSON 1974b) and may explain the mildness of the carbohydrate intolerance and rarity of ketoacidosis.

## III. Glucose Tolerance

The glucose intolerance of glucagonomas is mild despite very high levels of circulating glucagon (LAWRENCE and DORSCH 1980), suggesting that other factors modify the glycaemic response, such as the biological activity of the glucagon produced by the tumour, the compensatory hyperinsulinaemia, the influence of hepatic metastases on glycogenolysis and gluconeogenesis, and the effect of other peptides released by the tumour.

## IV. Other Peptides Secreted by Glucagonomas

Many other peptides have been found to be elaborated by islet cell tumours and released into the circulation, these include pancreatic polypeptide (in some series as many as 50% have high plasma pancreatic polypeptide levels), insulin, adrenocorticotropic hormone, vasoactive intestinal peptide, calcitonin, gastrin, neurotensin and somatostatin (GREIDER et al. 1970; BELCHETZ et al. 1973; FRIESEN et al. 1973; POLAK et al. 1976; BODEN et al. 1977; SCHWARTZ 1979; HIGGINS 1979; OHNEDA et al. 1979). Each of these in turn may modify the clinical and biochemical presentation of the patient.

### V. Plasma Amino Acids

The concentrations of glucogenic, ketogenic, as well as the combined glucogenic and ketogenic amino acids have been found to be reduced in glucagonoma patients (Fig. 4). Alanine and glutamine have been shown to be particularly low (MALLINSON et al. 1974a, b; HOLST et al. 1979, HOLST 1979). This is an expected finding in view of the effect of exogenous infusions of glucagon in lowering plasma amino acids (MARLISS et al. 1970). Parenteral replacement of amino acids in the presence of a caloric source such as dextrose has been found to improve the skin lesions of the glucagonoma syndrome (MALLINSON et al. 1974a; NORTON et al. 1979). This deficiency may also contribute to both the cachexia and anaemia of the syndrome.

## E. Tumour Characteristics

The glucagonomas can be divided into two main groups on clinical and pathological grounds, i.e. those associated with the glucagonoma syndrome (BORDI et al. 1979; RUTTMAN et al. 1980) and tumours which do not secrete.

### I. Tumours Associated with the Glucagonoma Syndrome

#### 1. Site and Spread

The tumours are usually single and large in size, 78% are reported to be greater than 5 cm in diameter at diagnosis (BORDI et al. 1979). Most of these are found in the body and tail of the pancreas which conforms with the distribution of the A-cells in the normal gland; and rarely at other sites (ROGGLI et al. 1979; HIGGINS et al. 1979). The majority of tumours have malignant characteristics, either by virtue of local infiltration into adjacent tissues or their metastatic spread, which most commonly occurs in the liver, bone, adrenal glands and lymph nodes. The rate of malignancy reported by different reviewers varies from 59% to 81.6% of cases (BINNICK et al. 1977; BORDI et al. 1979; HIGGINS et al. 1979; LEICHTER 1980). Glucagonomas are therefore more frequently malignant than the other islet cell tumours, although this may purely relate to the longer delay that often occurs before diagnosis of the glucagonoma syndrome.

#### 2. Light Microscopy

Haematoxylin and eosin staining of tumours shows no common cellular pattern. There is enormous variation between tumours; trabecular, solid or diffuse arrangements are the most frequent patterns of growth. Individual tumour cells are often multinucleated, with granular and abundant cytoplasm. The degree of cellular anaplasia tends to vary from one area to another in the same tumour (BORDI et al. 1979; LUBETZKI et al. 1980).

#### 3. Immunofluorescence

The development of specific immunohistochemical techniques has allowed identification of the peptide content of islet cells and their tumours (STERNBERGER 1979).

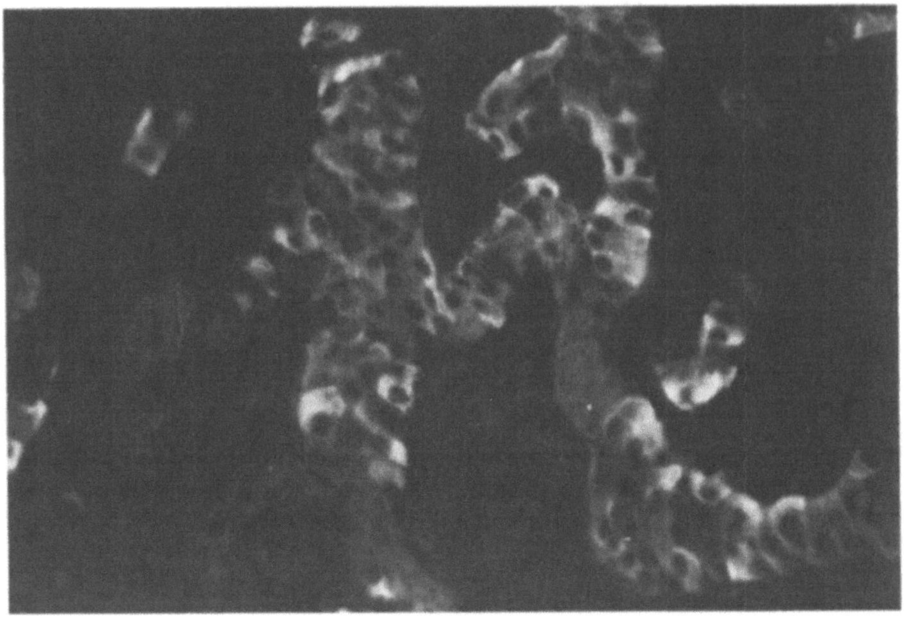

**Fig. 5.** Glucagon immunostaining in a glucagonoma (indirect immunofluorescence). ×625

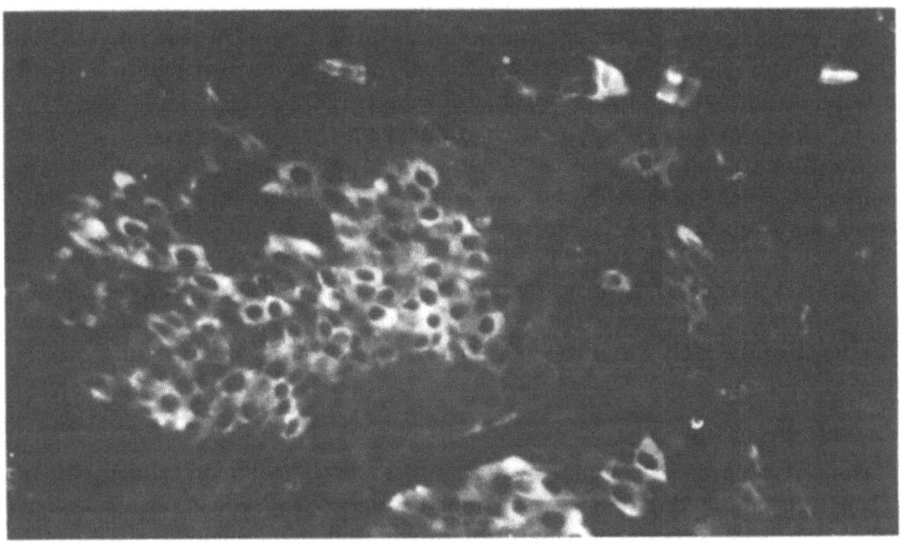

**Fig. 6.** Pancreatic polypeptide in a glucagonoma (indirect immunofluorescence). ×625

Indirect immunofluorescence is used to demonstrate glucagon-containing cells in glucagonomas and has shown these to be unevenly distributed, with clusters of positive cells scattered amongst predominantly unreactive cells (Fig. 5). Enteroglucagon, which is presumed to be a precursor of glucagon (see Chap. 6), may be produced in these tumours in a greater amount than glucagon. In addition, other en-

**Fig. 7.** Neuron specific enolase in a glucagonoma (peroxidase-antiperoxidase method). ×625

docrince cell types are often identified in individual tumours; for example, cells containing pancreatic polypeptide are the most frequently found nonglucagon cells (Fig. 6) (LARSSON et al. 1973; BORDI et al. 1979).

## 4. Neuron Specific Enolase

This protein, which is a neuron-specific form of the glycolytic enzyme enolase has been found to be present in the peptide-containing cells of the diffuse neuroendocrine system. It is a useful marker of these cells and their tumours; and islet cell tumours stain heavily with antibodies against this enzyme (Fig. 7; TAPIA et al. 1981).

## 5. Electron Microscopy

Glucagonoma tumour cells contain a number of populations of secretory granules as shown by electron microscopy. The majority of these cells have small to medium, dense, round granules (Fig. 8) which do not closely resemble these found either in the A-cell or the other B-, D- and PP islet cell types (Fig. 9; BORDI et al. 1979). The typical α-granule of the adult human islet which has a round dense core surrounded by an eccentric lighter halo, is only seen in a few tumour cells (Fig. 10). Some granules have a clear centre with a dense periphery similar to humal foetal α-granules. Such nonspecific granules are also found in other pancreatic endocrine tumours including insulinomas, vipomas and gastrinomas (KOSTIANOVSKY 1980).

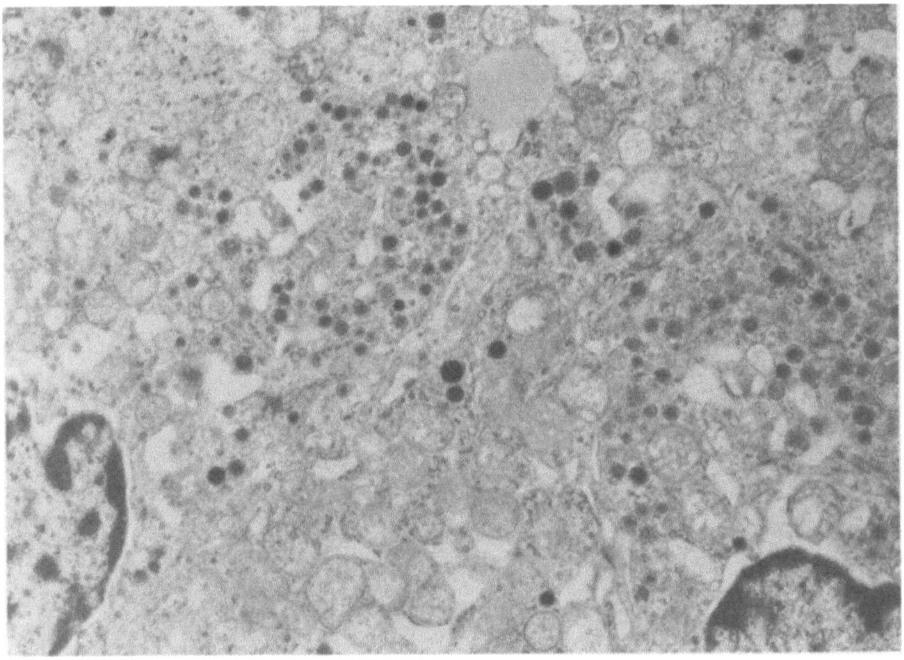

**Fig. 8.** Electron micrograph of a glucagonoma with atypical secretory granules (non A-cell type). × 17,500

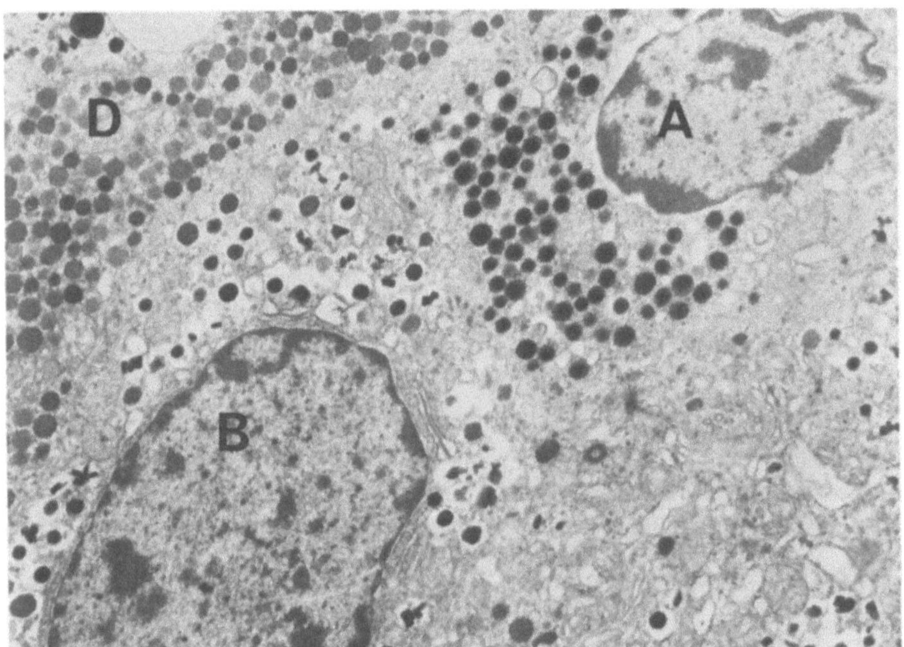


**Fig. 9.** Electron micrograph of a human pancreatic islet showing A-cell (glucagon), B-cell (insulin) and D-cell (somatostatin) with their characteristic secretory granules. × 12,500

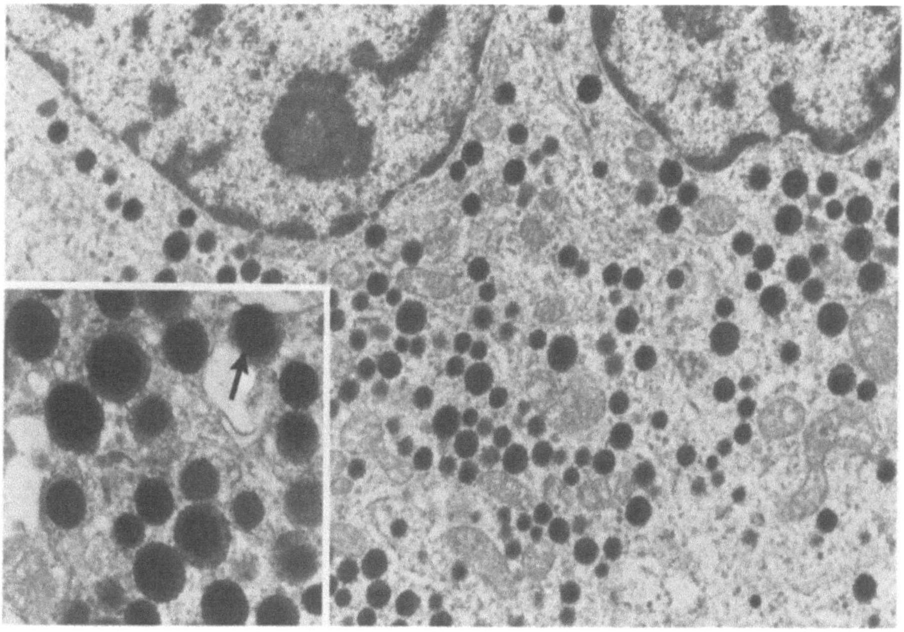

**Fig. 10.** Electron micrograph of a glucagonoma with typical A-cell granules. ×17,500. *Inset* shows characteristic acentric core of the A-cell granules. ×25,000

Despite the lack of similarity between tumour and normal islet cell granules, evidence from immunofluorescence and peptide extraction of tumours indicate that these abnormal granules must indeed store and release the specific peptides (BORDI et al. 1979; LOKICH et al. 1977).

### II. Tumours not Associated with the Glucagonoma Syndrome

The pathology of glucagon cell tumours not associated with the glucagonoma syndrome is relatively characteristic. The tumours are small and benign, immunofluorescence shows most cells to stain for glucagon and typical α-type secretory granules are seen at electron microscopy (CROISIER et al. 1971; BORDI et al. 1979). These features are all in sharp contrast to those of tumours associated with the clinical syndrome, as is the lack of glucagon hypersecretion, which may indicate that the cells function like normal cells and are under normal regulation. Many of these "nonfunctioning" adenomas have been found in patients which multiple endocrine adenomatosis. In such cases some of the larger adenomas have some characteristics in common with functioning glucagon tumours, suggesting possible evolution to this type (CROUGHS et al. 1971; WOODTLI and HEDINGER 1978).

## F. Localisation of Tumours

Experience with a number of techniques now available has greatly increased the success in accurate localisation of pancreatic tumours.

### I. Arteriography

Selective coeliac and superior mesenteric arteriography has proved to be the most reliable way of detecting the primary tumour. The tumour is identified by its pathological circulation or displacement of major vessels. The method gives valuable information on the site, size, resectability of the tumour and the presence of metastases (STANLEY and LEICHTER 1975; HIGGINS et al. 1979; SHUPACK et al. 1978; REHFELD 1979; KAMIMURA et al. 1980).

### II. Percutaneous Transhepatic Portal Venous Sampling

The percutaneous transhepatic approach to the portal vein has made selective catheterisation of the pancreatic veins feasible, and allowed localisation of tumours in the pancreas by arteriovenous hormone gradients. This technique is usually only performed after angiography has failed to localise the primary tumour. It is not without complications (perforation of the gallbladder, hepatic haematomas), and localisation may be difficult owing to aberrant venous drainage (INGEMANSSON et al. 1977; REICHARDT et al. 1979).

### III. Other Techniques

Ultrasound, and isotope scanning of pancreas and liver have proved more useful in the detection of metastases rather than the primary tumour. Computerised axial tomography (emission and transmission) and nuclear magnetic resonance techniques need further assessment, but could prove to be useful noninvasive investigative tests (HIGGINS et al. 1979).

## G. Diagnosis and Treatment

Diagnosis of the glucagonoma syndrome is made on the basis of the presenting clinical features, supported by an elevated fasting plasma glucagon concentration. Localisation of the tumour is then important to assess resectability and the presence of metastases, on which the choice of treatment will depend.

### I. Surgery

In the small proportion of patients that have benign tumours, surgical resection is the treatment of choice (MALLINSON et al. 1974b; BINNICK et al. 1977). Ealier diagnosis could well increase the number of patients appropriate for surgery. Of the 47 patients reviewed by HIGGINS et al. (1979), 15 underwent apparently successful surgical resection, but at least 3 of these developed recurrences postoperatively. When surgical cure is impossible, considerable improvement in symptoms has been achieved by reduction in tumour mass by "surgical debulking" (MONTENEGRO et al. 1980). This should always be considered as glucagonomas, like other islet cell tumours, are very slow growing, even when malignant.

## II. Chemotherapy

When surgery is not feasible, a certain number of patients will prove to be responsive to chemotherapy as judged by improvement in symptoms, reduction in circulating glucagon concentrations and regression of tumour and metastases. The agents that have been found to be effective in some cases include streptozotocin, either alone (MURRAY-LYON et al. 1968; LECLERE et al. 1977; YOSHINO et al. 1979; MOERTEL et al. 1980) or in combination with 5-fluorouracil (KHANDEKAR et al. 1979, MOERTEL et al. 1980). The Eastern Cooperative Oncology Group (ECOG) have recently compared the effectiveness of streptozotocin alone and together with 5-fluorouracil in the treatment of all islet tumours. They report an increased rate of response (63% compared with 36%) with combination therapy (MOERTEL et al. 1980). This group is in the process of investigating other chemotherapeutic agents including chlorozotocin which is structurally related to streptozotocin, and the combination of streptozotocin with doxirubicin (MOERTEL et al. 1980). In addition, there have been several recent reports of good clinical responses to dimethyltriazenoimidazole carboxamide (dacarbazine) (MARYNICK et al. 1980).

## III. Antisecretory Therapy

Somatostatin, a tetradecapeptide found naturally in the D-cells of the islet of Langerhans, inhibits the release of a large number of peptides including glucagon, and therefore has potential as a therapeutic agent in the glucagonoma syndrome, particularly in the long-acting octapeptide form (LONG et al. 1979). In a small number of patients, infusion of somatostatin has resulted in rapid improvement in their skin rash (SOHIER et al. 1980). Recently analogues of somatostatin, such as [des $Trp^8$, des $Cys^{14}$] somatostatin, have been developed which, under certain circumstances, may produce a selective inhibition of glucagon release (KAHN et al. 1981). However somatostatin does not inhibit hormone synthesis and a large rebound release of hormone often occurs after its cessation because of the remaining large stores of peptide. Diphenylhydantoin has been found to inhibit glucagon and insulin secretion in vitro, it was therefore thought to be of possible use in treatment of the glucagonoma syndrome. A reduction in plasma glucagon concentration has been reported in two cases (KRAMER et al. 1976; MACHINA et al. 1980).

## IV. Hepatic Artery Embolisation

In cases where the tumour has metastasized to the liver, resulting in peptide production from both primary and secondary tumours, the therapeutic alternative to chemotherapy or surgery has been hepatic artery embolisation. This procedure obliterates the arterial blood supply to tumour tissue, reducing its viability and peptide production, in many cases allowing temporary alleviation of symptoms. As a procedure it has the advantage of being relatively noninvasive, being performed under local anaesthesia, and being repeatable if necessary. The most serious complication includes hepatic abscess formation in necrotic tissue. It is, of course only a palliative procedure, but since the tumours are slow growing and their effects mainly secondary to the effects of the secreted peptide, quite long periods of good remission can be achieved (ALLISON 1978).

### V. Symptomatic Treatment

Symptomatic treatment of NME with topical and systemic antibiotics appears helpful only when secondary infection is a problem. Steroids have been tried with only short-term beneficial effect. As discussed in Sect. C.I.1 there have been some good responses to both oral and topical zinc preparations, and an occasional report of remission with intravenous amino acids. The effect of intravenous somatostatin has been discussed in Sects. C.I.1 and G.III and, though occasionally effective, is unlikely to be useful in the chronic situation. Treatment with dipyridamole and aspirin has been used as prophylaxis against the increased incidence of venous thrombosis and pulmonary embolus occurring in these patients (Mallinson et al. 1977; Kessinger et al. 1977) and may be especially helpful when the patient has to remain in bed (as when admitted).

## H. Prognosis

The islet cell tumours are slow growing, and patients present with significant clinical features often many years after the start of the disease (Binnick et al. 1977; Lokich et al. 1977). For those with benign tumours surgical cure is possible (Lightman and Bloom 1974; Higgins et al. 1979; Proye et al. 1980) and there are now reports of survival for over 9 years following surgery (Molini et al. 1981). The prognosis for those patients with metastases at diagnosis, is obviously less favourable, the median survival is approximately 3 years, which is comparable for both functioning and nonfunctioning tumours (Moertel et al. 1980). The effect of chemotherapy or hepatic artery embolisation on the length of survival is unknown as it has not been assessed in a sufficient number of patients.

## References

Allison DH (1978) Therapeutic embolization. Br J Hosp Med 20:707–715
Amon RH, Swenson KH, Hanifin JM, Hambidge RM (1976) The glucagonoma syndrome (necrolytic migratory erythema) and zinc (letter). N Engl J Med 295:962
Becker SW, Kahn D, Rothman S (1942) Cutaneous manifestations of internal malignant tumours. Arch Dermatol Syphilol 45:1069
Belchetz PE, Brown CL, Makin HLJ, Trafford DJH, Stuart-Mason A, Bloom SR, Ratcliffe JG (1973) ACTH, glucagon and gastrin production by a pancreatic islet cell carcinoma and its treatment. Clin Endocrinol (Oxf) 2:307–316
Berger M, Teuscher A, Halban P, Trimble E, Studer PP, Wollheim CB, Zimmermann-Telschow H, Müller WA (1980) In vitro and in vivo studies on glucagonoma tissue. Horm Metab Res 12:144–150
Bianchi C, Macor M, Zar E (1968) Dermatosi paraneoplastica in carcinoma di tipo endocrino del pancreas. Riv Pinato Patol Oncol 33:319
Binnick AN, Spencer SK, Dennison WL Jr, Horton ES (1977) Glucagonoma syndrome. Report of two cases and literature review. Arch Dermatol 113:749–754
Bloom SR, Polak JM (1978) The glucagonoma syndrome. Adv Exp Med Biol 106:183–194
Boden G, Owen OE (1977) Familial hyperglucagonaemia – an autosomal dominant disorder. N Engl J Med 296:534–538
Boden G, Owen OE, Rezvani I, Elfenbein BI, Quickel KE (1977) An islet-cell carcinoma containing glucagon and insulin. Chronic glucagon excess and glucose homeostasis. Diabetes 26:128–137

Bordi C, Ravazzola M, Baetens D, Gorden P, Unger RH, Orci L (1979) A study of glucagonomas by light and electron microscopy and immunofluorescence. Diabetes 28:925–936
Church RE, Crane WA (1967) A cutaneous syndrome associated with an islet-cell carcinoma of the pancreas. Br J Dermatol 79:284–286
Croisier JC, Lehy T, Zeitoun P (1971) $A_2$-cell pancreatic microadenomas in a case of multiple endocrine adenomatosis. Cancer 28:707–713
Croughs RJM, Hulsmans HA, Israel DE, Hackeng WHL, Schopman W (1972) Glucagonoma as a part of the polyglandular syndrome. Am J Med 52:690–698
Danforth DN, Triche T, Doppman JL, Beazley RM, Perrino PV, Recant L (1976) Elevated plasma proglucagon-like component with a glucagon-secreting tumour. N Engl J Med 295:242–245
Domen RE, Shaffer MB Jr, Finke J, Sterin WK, Hurst CB (1980) The glucagonoma syndrome. Report of a case. Arch Intern Med 140:262–263
Doyle JA, Schroeter AL, Rogers RS (1979) Hyperglucagonaemia and necrolytic migratory erythema in cirrhosis – possible pseudo-glucagonoma syndrome. Br J Dermatol 100:581–587
Freedberg IM, Galdabini JJ (1976) Dermatitis, weight loss and filling defects in the liver. Case records of the Massachusetts General Hospital. N Engl J Med 292:1117–1123
Friesen SR, Hermreck AS, Mantz FA (1973) Glucagon, gastrin and carcinoid tumours of the duodenom, pancreas and stomach, polypeptide "apudomas" of the foregut. Am J Surg 127:90–101
Ganda OP (1980) Glucagonoma and diabetic ketoacidosis? Arch Intern Med 140:1397–1398
Gatrelli-Beltzer E (1980) Lesions cutanées rèvelatrices d'un glucagonome. Sem Hop Paris: 579–582
Gleeson MH, Bloom SR, Polak JM, Henry K, Dowling RH (1971) Endocrine tumour in kidney affecting small bowel structure, motility and absorptive function. Gut 12:773–782
Gossner W, Korting GW (1960) Metastasierendes Inselzellen-Karzinom vom A-Zell-Typ bei einem Fall von Pemphigus foliaceous mit Diabetes renalis. Dtsch Med Wochenschr 85:434–440
Greider MH, Bencosme JA, Lechago J (1970) The human pancreatic islet cells and their tumours. I. The normal pancreatic islet. Lab Invest 22:344–354
Guillausseau PJ, Guillausseau C, Villet R, Kaloustian E, Valleur P, Hautefeuille P, Lubetzki J (1982) Les glucagonomes. Aspects cliniques, biologiques, anatomopathologiques et thérapeutiques (Revue générale de 130 cas). Gastroenterol Clin Biol 6:1029–1041
Henington UM, Caroe E, Derbes V (1958) Kwashiorkor: report of 4 cases from Louisana. Arch Dermatol 78:157–168
Higgins GA (1979) Pancreatic islet cell tumours: insulinoma, gastrinoma, and glucagonoma. Surg Clin North Am 59:131–141
Higgins GA, Recant L, Rischman AB (1979) The glucagonoma syndrome: surgically curable diabetes. Am J Surg 137:142–148
Holst JJ (1979) Possible entries to the diagnosis of a glucagon-producing tumour. Scand J Gastroenterol [Suppl] 14:53–56
Holst JJ, Jonsson J, Pedersen NB, Thomsen K (1975) The glucagonoma syndrome. Ugeskr Laeger 137:2631–2636
Holst JJ, Helland S, Ingemannson S, Pedersen NB, von Schenck H (1979) Functional studies in patients with the glucagonoma syndrome. Diabetologia 17:151–156
Horrobin DF, Cunnane SC (1980) Interactions between zinc, essential fatty acids and prostaglandins: relevance to acrodermatitis enteropathica, total parenteral nutrition, the glucagonoma syndrome, diabetes, anorexia nervosa and sickle cell anaemia. Med Hypotheses 6:277–296
Ingemansson S, Holst J, Larsson LI, Lunderquist A (1977) Localisation of the glucagonomas by catheterisation of the pancreatic veins and with glucagon assay. Surg Gynecol Obstet 145:509–516

Johnson RD, Larkins RG, Iser JH, Roberts-Thomson IC, Kune GA (1981) The glucagonoma syndrome: an unusual skin rash associated with diarrhoea and diabetes. Aust NZ J Med 11:59–63

Kahan RS, Perez-Figarado RA, Neimanis A (1977) Necrolytic migratory erythema. Distinctive dermatosis of the glucagonoma syndrome. Arch Dermatol 113:792–797

Kahn CR, Bathena SJ, Recant L, Rivier J (1981) Use of somatostatin and somatostatin analogs in a patient with a glucagonoma. J Clin Endocrinol Metab 53:543–549

Kamimura R, Matsul O, Kadoya M, Kitagawa K, Saito Y, Takashima T (1980) Arteriography in islet cell tumor – including one case of glucagonoma (in Japanese). Rinsho Hoshanen 25:329–336

Kay RG, Tasman-Jones C (1974) Acute zinc deficiency in man during intravenous alimentation. Aust NZ J Surg 45:325–330

Khandekar JD (1979) Islet-cell tumours of the pancreas: clinico-biochemical correlations. Ann Clin Lab Sci 9:212–218

Khandekar JD, Oyer D, Miller HJ, Vick NA (1979) Neurologic involvement in glucagonoma syndrome: response to combination chemotherapy with 5-fluorouracil and streptozotocin. Cancer 44:2014–2016

Kessinger A, Lemon HM, Foley JF (1977) The glucagonoma syndrome and its management. J Surg Oncol 9:419–424

Kostianovsky M (1980) Endocrine pancreatic tumours: ultrastructure. Ann Clin Lab Sci 10:65–75

Kramer S, Machina T, Marcus J (1976) Metabolic studies in the malignant glucagonoma syndrome. Diabetes 25:370

Larsson LI, Sundler F, Grimelius L, Håkanson R, Holst J (1973) Immuno-histochemical demonstration of glucagon in an $A_2$-cell carcinoma. Experientia 29:698–699

Lawrence AM, Dorsch T (1980) The glucagonoma syndrome. In: Podolsky S, Viswanathan M (eds) Secondary diabetes: the spectrum of the diabetic syndrome. Raven, New York, pp 287–295

Leclere J, Vicari F, Laurent J, Jeanpierre R, Ploquet J, Grignon G, Hartemann P (1977) Islet-cell tumour with diarrhoea and diabetes (glucagonoma?) associated with hyperparathyroidism. Long-term results of local regional treatment with streptozotocin. Ann Endocrinol (Paris) 38:153–154

Leichter SB (1980) Clinical and metabolic aspects of glucagonoma. Medicine 59:100–113

Lewis AE (1979) The glucagonoma syndrome. Int J Dermatol 18:17–22

Lightman SL, Bloom SR (1974) Cure of insulin-dependent diabetes mellitus by removal of a glucagonoma. Br Med J 1:367–368

Lokich J, Anderson N, Rossini A, Hadley W, Federman M, Legg M (1977) Pancreatic alpha cell tumours. Cancer 45:2675–2683

Long RG, Barnes AJ, Adrian TE, Mallinson CN, Brown MR, Vale W, Rivier JE, Christofides ND, Bloom SR (1979) Suppression of pancreatic endocrine tumour secretion by long acting somatostatin analogue. Lancet 2:764–767

Lubetzki J, Grupper G, Malbec D, Warnet A, Guillausseau PJ, Luyckx A, Hautefeuille P, Galian A, Scotto J, Rault P, Eskenazi A, Mathieu M (1980) Clinical, biological, histological, ultrastructural and therapeutic studies in one case of glucagonoma. Nouv Presse Med 9:1565–1569

Luyckx AS, Lefèbvre PJ (1981) Les glucagonomes. Diab Metab 7:289–300

Machina T, Marcus R, Levin SR (1980) Inhibition of glucagon secretion by diphenylhydantoin in a patient with glucagonoma. West J Med 132:357–360

Mallinson CN, Salmon PR, Barrowman J, Bloom SR (1973) The association of a specific skin lesion with islet-cell tumours of the pancreas. Gut 14:827

Mallinson CN, Cox B, Bloom SR (1974a) Plasma levels of amino acids and glucagon in patients with pancreatic glucagonomas. Gut 15:340

Mallinson CN, Bloom SR, Warin AP, Salmon PR, Cox B (1974b) A glucagonoma syndrome. Lancet 2:1–5

Mallinson CN, Adrian TE, Hanley J, Bryant M, Bloom SR (1977) Metabolic and clinical responses in patients with pancreatic glucagonomas. Ir J Med Sci 146:37

Marliss EB, Aoki TT, Unger RH (1970) Glucagon levels and metabolic effects in fasting man. J Clin Invest 49:2256–2270
Marynick SP, Fagadau WR, Duncan LA (1980) Malignant glucagonoma syndrome: response to chemotherapy. Ann Intern Med 93:453–454
McGavran MH, Unger RH, Recant L, Polk HC, Kilo C, Levin ME (1966) A glucagon secreting alpha-cell carcinoma of the pancreas. N Engl J Med 274:1408–1413
McMillan DE (1975) Deterioration of the microcirculation in diabetes. Diabetes 24:944–957
Moertel C, Hanley JA, Johnson LA (1980) Streptozotocin alone compared with streptozotocin plus fluorouracil in the treatment of advanced islet-cell carcinoma. N Engl J Med 303:1189–1194
Molini C, Naudan P, Daly JP, Assan R, Laverdant C (1981) Glucagonoma: 9 years' survival after exeresis (letter). Nouv Presse Med 10(3):117–118
Montenegro F, Lawrence GD, Macon W, Pass C (1980) Metastatic glucagonoma. Improvement after surgical debulking. Am J Surg 139:424–427
Murray-Lyon IM, Eddleston ALWF, Williams R (1968) Treatment of multiple hormone producing malignant islet cell tumour with streptozotocin. Lancet 2:895–898
Naets JP, Guns M (1980) Inhibitory effect of glucagon on erythropoiesis. Blood 55:997–1002
Norton JA, Kahn CR, Shiebinger R, Gonschboth C, Brennan MF (1979) Amino acid deficiency and the skin rash associated with glucagonoma. Ann Intern Med 91:213–215
Ohneda A, Otsuki M, Fujiva H, Yaginuma N, Kokubo T, Ohtani H (1979) A malignant insulinoma transformed into a glucagonoma syndrome. Diabetes 28:962–969
Palmer JP, Werner L, Benson JW, Ensinck JW (1976) Dominant inheritance of large molecular weight immunoreactive glucagon. Diabetes [Suppl 1] 25:326
Pedersen NB, Jonsson L, Holst JJ (1976) Necrolytic migratory erythema and glucagon cell tumour of the pancreas: the glucagonoma syndrome. Report of two cases. Acta Derm Venereol (Stockh) 56:391–395
Perrino PP, Lavine RL, Bhathena SJ, Burns WA, Recant L (1975) Big glucagon in glucagonoma. Clin Res 37:446A
Polak JM, Bloom SR, Adrian TE, Heitz P, Bryant MG, Pearse AGE (1976) Pancreatic polypeptide in insulinomas, gastrinomas, vipomas and glucagonomas. Lancet 1:328–330
Prottey C, Hartop PJ, Press M (1975) Correction of the cutaneous manifestations of essential fatty acid deficiency in man by application of sunflower seed oil to the skin. J Invest Dermatol 64:228–234
Proye C, Lefèbvre J, Lelievre G, Mazzuca M, Bergoend H, Patoir A, Lagache G, Linquette M (1980) A case of glucagonoma in a 72 year old woman. Cephalic duodenopancreatectomy. Result after 1 year and a half. Chirurgie 106:600–604
Ramierz Arias JL, Anzures ME, Mercado Perez C, Rodriguez L, Albores J, Crotte A (1979) Cancer of the alpha cells of the pancreas. Rev Interam Radiol 4:33–36
Recant L, Perrino PV, Bhathena SJ, Danforth DN Jr., Lavine RL (1976) Plasma immunoreactive glucagon fractions in four cases of glucagonoma: increased "large glucagon-immunoreactivity". Diabetologia 12:319–326
Reichardt W, Ericsson M, Holst JJ, Ingemansson S, Lunderquist A (1979) Glucagon producing endocrine pancreas tumours. Symptoms, diagnosis, localization, therapy and follow up. Chirurg 50:754–758
Rehfeld JF (1979) Radioimmunoassay in diagnosis, localization and treatment of endocrine tumours in gut and pancreas. Scand J Gastroenterol [Suppl] 14:33–38
Riddle MC, Golper TA, Fletcher WS, Ensinck JW, Smith PH (1978) Glucagonoma syndrome in a 19 year old woman. West J Med 129:68–72
Roggli VL, Judge DM, McGavran MH (1979) Duodenal glucagonoma: a case report. Hum Pathol 10:350–353
Ruttman E, Kloppel G, Bommer G, Kiehn M, Heitz PV (1980) Pancreatic glucagonoma with and without syndrome. Virchows Arch [Pathol Anat] 388:51–67
Schwartz TW (1979) Pancreatic polypeptide (PP) and endocrine tumours of the pancreas. Scand J Gastroenterol 14:93–100
Shima K, Tsujii T, Sasano N, Dol O, Kyoba S (1970) Glucagonoma. Jpn J Cancer Clin 16:866–870

Shupack JL, Berczeller PH, Stevens DM (1978) The glucagonoma syndrome. J Dermatol Surg Oncol 4:242–247
Sohier J, Jeanmougin M, Lombrail P, Passa Ph (1980) Rapid improvement of skin lesions in glucagonomas with intravenous somatostatin infusion. Lancet 1:40
Soler NG, Oatis GD, Malins JM, Cassar J, Bloom SR (1976) Glucagonoma syndrome in a young man. Proc R Soc Med 69:429–431
Stacpoole PW, Jaspan J, Kasselberg AG, Halter SA, Polonsky K, Gluck FW, Liljenquist JE, Rabin D (1981) A familial glucagonoma syndrome: genetic, clinical and biochemical features. Am J Med 70:1017–1026
Stanley RJ, Leichter S (1975) Pancreatic glucagonoma: a clinical-angiographic correlation. Case report. Mo Med 72:113–119
Sternberger L (1979) Immunocytochemistry, 2nd edn. Wiley, New York, pp 24–59
Sweet RD (1974) A dermatosis specifically associated with a tumour of pancreatic alpha cells. Br J Dermatol 90:301–308
Swenson KH, Amon RB, Hanifin JM (1978) The glucagonoma syndrome. A distinctive cutaneous marker of systemic disease. Arch Dermatol 114:224–228
Tapia FJ, Polak JM, Barbosa AJA, Bloom SR, Marangos PJ, Dermody C, Pearse AGE (1981) Neuron specific enolase is produced by neuroendocrine tumours. Lancet 1:808–811
Thivolet J (1980) Necrolytic migratory erythema without glucagonoma. Arch Dermatol 117:4
Truswell AS, Liadsley C, Wittman W (1962) Are the skin lesions of kwashiorkor pellagrous. S Afr Med J 36:965–966
Unger RH, Eisentraut AM, Lochner JR (1963) Glucagon producing tumours of the islets of Langerhans. J Clin Invest 42:987–988
Valverde I, Lemon HM, Kessinger A, Unger RH (1976) Distribution of plasma glucagon immunoreactivity in a patient with suspected glucagonoma. J Clin Endocrinol Metab 42(5):804–808
Villar HV, Johnson DG, Lynch PJ, Pond GD, Smith PH (1981) Pattern of immunoreactive glucagon in portal, arterial and peripheral plasma before and after removal of glucagonoma. Am J Surg 141:148–152
von Schenck H, Thorell JI, Berg J, Bojs C, Dymling JF, Hallengren B, Ljungberg O, Tibblin S (1979) Metabolic studies and glucagon gel filtration pattern before and after surgery in a case of glucagonoma syndrome. Acta Med Scand 205:155–162
Weiller PJ, Weiller M, Lam-My S, Pizzi M, Sayag J, Sarles JC, Mongin M (1981) Glucagonoma without diabetes: a case report. Ann Med Interne (Paris) 132:41–43
Weir GC, Horton E, Aoki TT, Slovik DM (1976) Increased large glucagon immunoreactivity in the glucagonoma syndrome. Diabetes 25:326.
Wilkinson DS (1973) Necrolytic migratory erythema with carcinoma of the pancreas. Trans St Johns Hosp Dermatol Soc 59:244–248
Woodtli W, Hedinger C (1978) Pancreatic islet-cell tumours and their syndromes. II. Zollinger-Ellison syndrome, glucagonoma syndrome, multiple endocrine adenomatosis and islet-cell tumours not obviously provable. Schweiz Med Wochenschr 108:1997–2007
Yoshinaga T, Okuno G, Shinji Y, Tsujii T, Nismikawa M (1966) Pancreatic A cell tumour associated with severe diabetes mellitus. Diabetes 15:709–713
Yoshino G, Kasumi T, Morita S, Kobayashi N, Terashi K, Baba S (1979) Glucagon secretion during the development of insulin-secreting tumours induced by streptozotocin and nicotinamide. Endocrinol Jpn 26:655–660
Youker GD, Riddle MC (1980) High big plasma glucagon (BPG) in a glucagonoma patient's family. Clin Res 28:55 A
Zhdanov VC (1956) Diabetes and malignancy of islet-cell in pancreas (in Russian). Arch Pathol (Moskva) 92:306–309

CHAPTER 44

# Glucagon in Diabetes Mellitus

R. H. UNGER and L. ORCI

## A. A-cell Function in Human Diabetes

A-cell function is abnormal in all known forms of overt diabetes mellitus. This includes all varieties of human diabetes, including that produced by total pancreatectomy and experimental diabetes in animals. It is conceptually convenient and perhaps physiologically correct to separate the functional abnormalities of the diabetic A-cell into two categories: (1) loss of glycemic control of glucagon secretion; and (2) hyperresponsiveness of glucagon secretion to stimulation. The first abnormality constitutes a loss of the normal reciprocal relationship between glucose and glucagon concentrations. This physiologic relationship seems to be partly or entirely mediated by the concomitant glucose-induced secretion of insulin or at least be an insulin-requiring response. In contrast to nondiabetics, in whom hyperglycemia suppresses and hypoglycemia stimulates glucagon secretion, in diabetics the plasma levels of immunoreactive glucagon are at all times high, both in relation to the steady-state level of fasting hyperglycemia (UNGER 1976) and to an increase in hyperglycemia resulting from the ingestion or infusion of glucose (Fig. 1, UNGER et al. 1970); in fact, glucagon levels may rise paradoxically during hyperglycemia produced by a carbohydrate meal (BUCHANAN and MCCARROLL 1972). Nor does hypoglycemia elicit in diabetic patients the rise in glucagon secretion that occurs in normal subjects (GERICH et al. 1978). This loss of the normal A-cell response to changes in glucose concentration appears to be a selective one inasmuch as glucagon suppression by increased levels of free fatty acids is perfectly normal in diabetics (GERICH et al. 1976a).

The second abnormality observed in most forms of diabetes in which it has been looked for is hyperresponsiveness to stimulation by arginine (UNGER et al. 1970), alanine (MULLER et al. 1971), and protein (MULLER et al. 1970). This response, which in nondiabetics is under the dominant influence of the ambient glucose level (RASKIN et al. 1978), in diabetics is independent of changing levels of glycemia (RASKIN et al. 1978). For example, in nondiabetics, hyperglycemia induced by intravenous glucose completely prevents the protein-induced rise in plasma glucagon that occurs during normoglycemia (MULLER et al. 1970). In contrast, in diabetics, hyperglycemia does not prevent the protein-induced rise in immunoreactive glucagon (IRG); glucagon rises just as much as in normoglycemic nondiabetics (Fig. 2). This fact may be of clinical importance in the management of diabetic hyperglycemia, inasmuch as it signifies that a carbohydrate-free protein meal can raise plasma glucose levels in the diabetic. Inappropriate postprandial rises in glucagon will cause postprandial hyperglycemia in diabetics, even if carbohydrate has

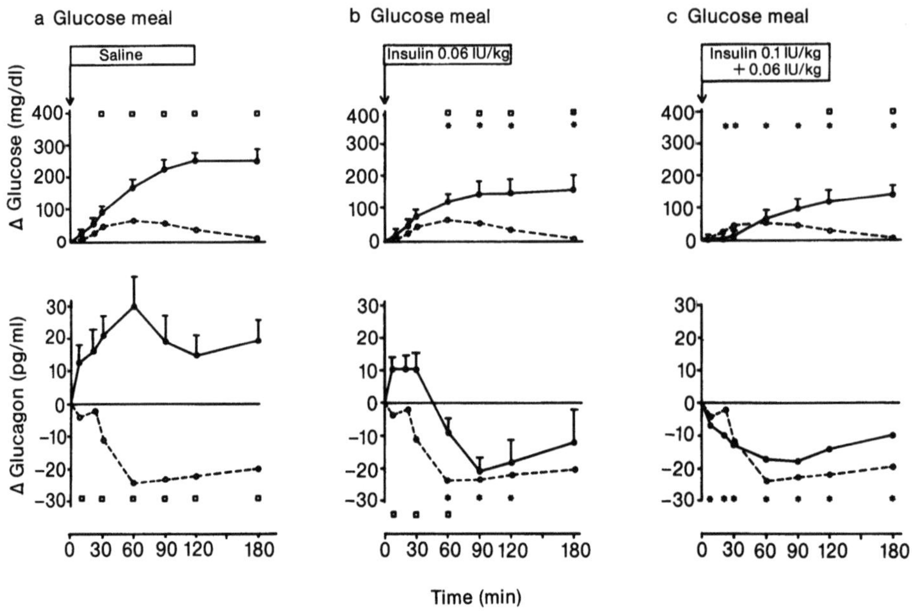


**Fig. 1 a–c.** The plasma glucose and IRG responses to a glucose meal in ten type I diabetics (*full line*) compared with nine nondiabetic controls (*broken line*). **a** the response during an infusion of saline, i.e., no exogenous insulin has been provided for the diabetics; **b** a constant intravenous infusion of insulin has been given, there is marked improvement in the glucagon response, but it remains somewhat above normal in terms of total IRG decrement; **c** the diabetics have been given an initial bolus of insulin in addition to the constant insulin infusion. The total decrement is now no different from normal. *Squares* indicate $P<0.05$ diabetics compared with nondiabetics; *asterisks* indicate $P<0.05$ diabetics without insulin compared with diabetics with insulin. AYDIN et al. (1977)

been eliminated from the diet unless, of course, there is a compensatory protein-induced rise in insulin.

## B. Relationships of A-cell Malfunction to Insulin

While there is general agreement with respect to the existence of A-cell malfunction in overt diabetes, considerable disagreement prevails concerning the pathogenesis of the anomaly. It has been suggested that the A-cell dysfunction in human diabetes is a discrete primary abnormality entirely separate from the B-cell disorder (UNGER and ORCI 1977a), and KIRK et al. (1975) have put forth evidence in support of this possibility. On the other hand, it has been argued that the A-cell abnormalities in some cases of human diabetes are the passive consequences of insulin deficiency and can, therefore, be restored to normal by repletion with insulin (CHISHOLM and ALFORD 1977). However, accumulating evidence tends to suggest that neither of these views is strictly correct and that the issue is more complex.

First, it would appear that, while both juvenile-onset and adult-onset diabetes exhibit overlapping abnormalities of A-cell function, the degree to which administered insulin can correct the A-cell malfunction is different in these two forms of

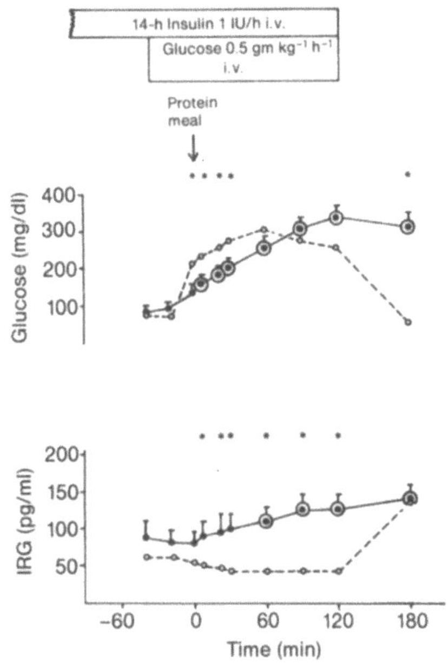


**Fig. 2.** The effect of hyperglycemia on the plasma glucose and IRG responses to a protein meal in 10 type I diabetics (*full line*) compared with 12 hyperglycemic nondiabetics (*broken line*). IRG rises in response to a protein meal much as it does in nondiabetics during normoglycemia. *Asterisks* indicate $P<0.05$ diabetics compared with nondiabetics; *circled dots* indicate $P<0.05$ compared with baseline. RASKIN et al. (1978)

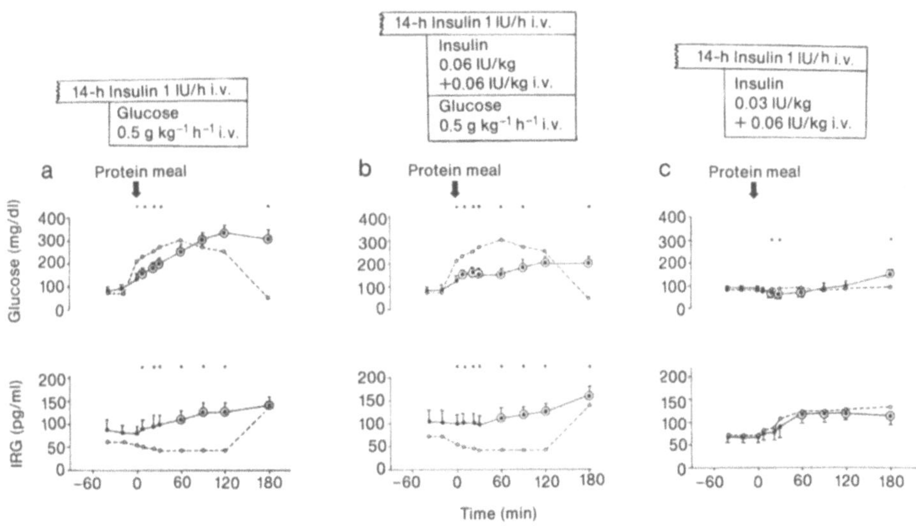


**Fig. 3a–c.** The plasma glucose and IRG responses to a protein meal in ten subjects with type I diabetes during induced hyperglycemia (**a**), hyperglycemia plus hyperinsulinemia (**b**), and hyperinsulinemia (**c**). *Full circles* diabetic subjects; *open circles* nondiabetic subjects ($N=12$); *asterisks* indicate $P<0.05$ diabetic compared with nondiabetic subjects; *circled dots* indicate $P<0.05$ compared with baseline

the disease. Second, it must be recognized that reduction of the increased concentrations of glucagon by insulin does not necessarily signify that A-cell function has been returned to normal, that is, that the A-cell has regained the normal ability to secrete glucagon appropriately in response to changes in the concentration of glucose. For example, insulin infused at a constant rate will reduce the glucagon levels of any diabetic patient, but this does not necessarily restore to normal the ability of the A-cell to vary plasma glucagon concentrations appropriately in response to changes in nutrient levels (UNGER 1978). Moreover, even a completely "normal" glucagon response in a diabetic with fixed insulin levels will induce clinically significant hyperglycemia (Fig. 3) – thus, A-cell function is endocrinologically "normal", but metabolically inappropriate. This issue may have important therapeutic implications: if A-cell malfunction can be corrected by simple replacement of insulin, a novel therapeutic agent capable of correcting glucagon secretion would be unnecessary. If, on the other hand, insulin treatment, at least by conventional delivery systems, is incapable of correcting the A-cell malfunction, a search for a glucagon-suppressing agent or for a glucagon antagonist would become a potentially important objective of pharmaceutical research (see Chap. 32).

## C. The Islets in Diabetes: A-cell Relationships Within the Islets

In Chap. 31, the normal relationships of islets cells have been considered in detail. The intercellular relationships in the islets are considered essential for normal function of the A-cell and in all probability other components of the endocrine pancreas. Intraislet signals include unidentified messages that may be transmitted via connexons, the individual units that form gap junctions and link islet cells to one another (ORCI et al. 1975). In addition, paracrine signals have been proposed (UNGER and ORCI 1977 b) in which a hormone secreted by one cell may influence the function of nearby cells through passive contact on the interstitium. In the cases of insulin and glucagon, a positive–negative feedback, first proposed in classic studies by SAMOLS et al. (1965, 1972), appears to be operative and extremely important. Similar feedbacks may exist between other cells, particularly between somatostatin-secreting D-cells and other islet cells (see also Chap. 31). The normal A-cell–B-cell morphofunctional relationships are schematically depicted in Fig. 4 a.

Anything that interferes with this insulin–glucagon feedback relationship may alter A-cell function. That hyperglucagonemia is present in the type I diabetic is hardly surprising considering morphological alterations that have been described by ORCI et al. (1976). In the insulin-deficient islet (type I diabetes, streptozotocin- or alloxan-induced diabetes in animals) the central mass of B-cell is largely absent. This results in a relative increase in the percentage of non-B-cells that make up the endocrine pancreas. However, in addition, there appears to be an absolute increase in the number of A-cells, which make up ~30% of the nondiabetic islet and 70% of the diabetic islet. The total volume of A-cells per pancreas is approximately doubled. Obviously, the normal pattern of A-cell–B-cell contacts is completely lost and a chaotic, seemingly random arrangement of the surviving cells replaces the nonrandom pattern of the normal islet in which the A-cells occupy a peripheral po-

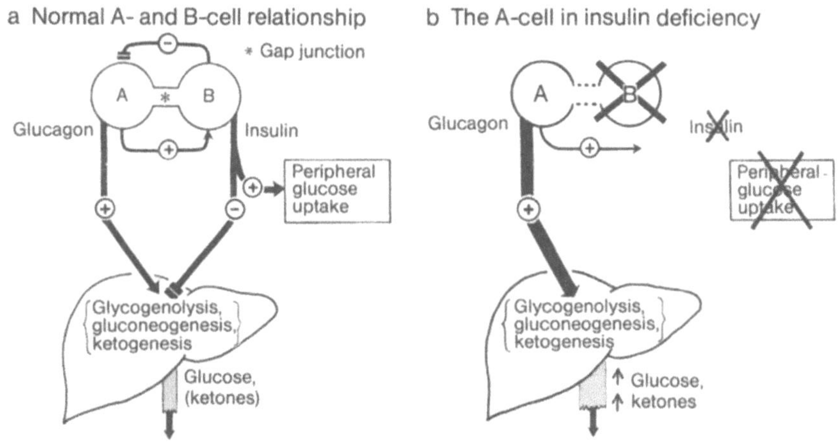


**Fig. 4. a** Postulated mechanisms of "functional coupling" of A- and B-cells via paracrine and gap junctional connections. The stimulatory effects of glucagon on the B-cells and the inhibitory effect of insulin on the A-cells provide a short "paracrine" loop by which simultaneous responses of the two hormones might protect against perturbations of nutrient concentration that would result from unilateral action of one of them. **b** The absence of a sufficient number of B-cells eliminates the normal inhibitory effect of insulin on the A-cells (see **a**) and both paracrine and gap junctional coupling with B-cells is lacking. The resulting hyperglucagonemia stimulates glucose and ketone production in the liver without opposition from insulin. The resulting hyperglycemia can neither be corrected by compensatory insulin-mediated uptake in tissues nor reduced by inhibition of glucagon secretion, since this too is an insulin-requiring process. UNGER (1981)

sition which permits them to surround and make contact with B-cells (ORCI and PERRELET 1981). The functional changes of the A-cells already described could well be the consequence of insufficient cell–cell contacts with insulin-secreting cells. Function can be restored to normal by simulation of normal insulin secretory patterns (RASKIN et al. 1975, 1976, 1978, 1979; GERICH et al. 1976b; AYDIN et al. 1977; YAMAMOTO et al. 1979). The morphofunctional relationships of A- and B-cells in this form of diabetes are depicted schematically in Fig. 4b.

Studies of the morphological changes in type II diabetes are few. Some such pancreata have been examined by L. ORCI and co-workers (1980, unpublished work), but striking differences from normal in the A-cell–B-cell relationships comparable to those observed in type I diabetes were not identified. It is, therefore, difficult at present to attribute abnormalities in A-cell function to identifiable changes in the cell–cell relationships. It may, however, be relevant to point out that in the ob/ob mouse, considered by many to constitute an animal model analogous to type II diabetes associated with obesity, the islets are enormously enlarged as the result of hyperplasia of the B-cells (BAETENS et al. 1978). The increase in the B-cell mass "stretches" the peripheral layers of cells, which are comprised largely of A-cells, and simultaneously separates A-cells from D-cells. Whether or not this has functional implications in the ob/ob mouse is unknown, and its relevance, if any, to human type II diabetes is not now apparent.

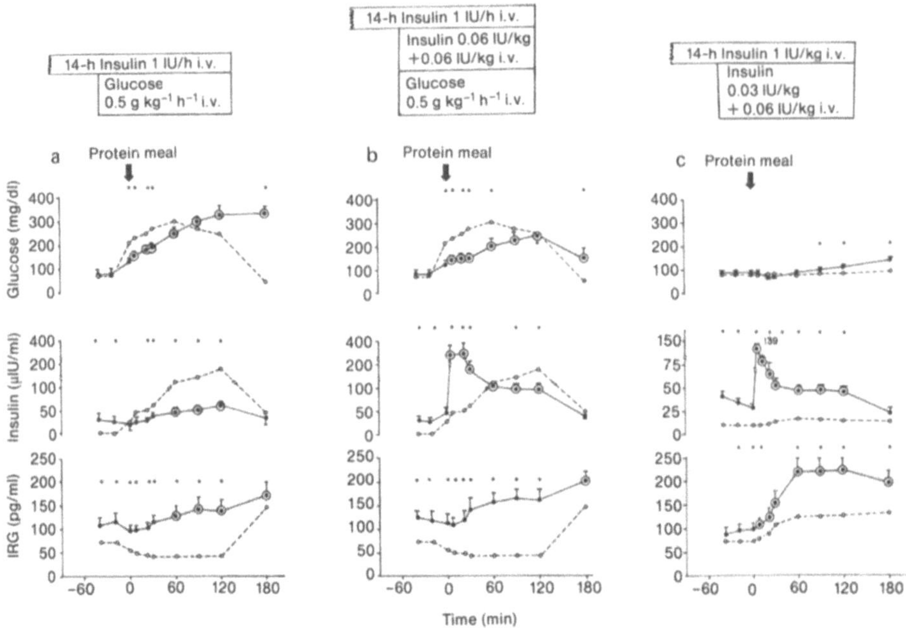


**Fig. 5a–c.** The plasma glucose, insulin, and glucagon (IRG) responses to a protein meal in ten subjects with type II diabetes during induced hyperglycemia (**a**), hyperglycemia plus hyperinsulinemia (**b**), and hyperinsulinemia (**c**). *Full circles* diabetic subjects; *open circles* nondiabetic subjects ($N = 12$); *asterisks* indicate $P < 0.05$ diabetics compared with nondiabetics; *circled dots* indicate $P < 0.05$ compared with baseline. RASKIN et al. (1978)

## D. The Effects of Insulin on the A-cell Abnormalities of Type I Diabetics

In juvenile-onset (type I) diabetics in whom B-cells are markedly reduced, plasma glucagon levels do not exhibit a reciprocal relationship to the blood glucose level. In fact, it is when such patients are most severely hyperglycemic, as in ketoacidosis or nonketotic hyperosmolar coma, that their hyperglucagonemia is most marked (ASSAN et al. 1969; MÜLLER et al. 1973; LINDSEY et al. 1974). Such acutely ill patients are markedly dehydrated and the effect of hypersecretion of glucagon upon plasma glucagon levels is exaggerated by decreased glucagon clearance resulting from hypoperfusion of the kidney, shown by LEFÈBVRE and LUYCKX (1976) to be a major site of glucagon degradation.

It seems that the A-cell does not have an intrinsic glucose-sensing capability of its own in the absence of appropriate glucose-induced change in insulin secretion. That normal glucose sensing by the A-cell is either mediated by or dependent upon insulin is suggested by the fact that diabetic A-cell responses to glycemia can be restored to normal if exogenous insulin is delivered with the glucose in a manner that simulates the normal insulin response to changes in glycemia (GERICH et al. 1976b; AYDIN et al. 1977; YAMAMOTO et al. 1979). When an insulin-deprived type I diabetic is made normoglycemic by a constant rate insulin infusion, and glucagon

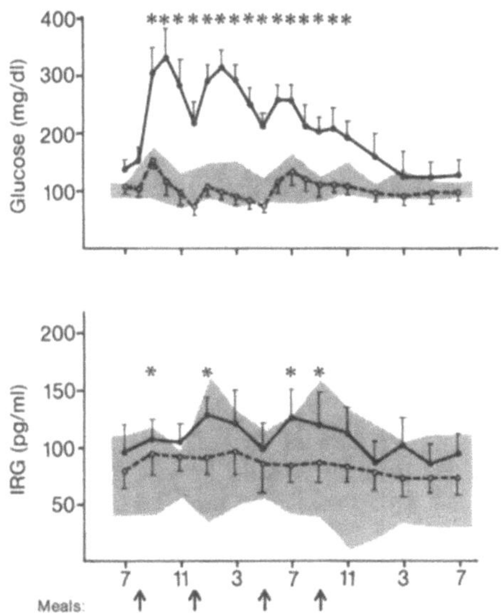


**Fig. 6.** Mean (±standard error) plasma glucose and IRG profiles during 24 h conventional insulin treatment (*full circles*) and after 4–5 weeks continuous subcutaneous insulin infusion (*open circles*) with portable insulin infusion pumps in five type I diabetics. *Asterisks* indicate $P<0.05$ in the diabetics with insulin infusion compared with conventional therapy. The *shaded area* represents the mean ±2 standard deviations of values from nine nondiabetics on a similar diet included for comparison. Times of meals are indicated by *arrows*. RASKIN et al. (1979)

levels thereby lowered into the so-called normal range, hyperglycemia induced by intravenous glucose will suppress IRG levels by only about one-half as much as in nondiabetics made similarly hyperglycemic (RASKIN et al. 1975). However, if when the constant rate insulin infusion is supplemented by a bolus injection of insulin at the time of the intravenous administration of glucose, thereby simulating the first phase of insulin secretion that occurs in nondiabetics, the glucagon response is reduced (Fig. 5; YAMAMOTO et al. 1979). Delivery of insulin by portable subcutaneous insulin pumps also seems to restore A-cell function to normal and/or overcome the metabolic consequences of any uncorrected abnormalities (Fig. 6; RASKIN et al. 1979).

## E. The Effects of Insulin on A-cell Abnormalities of Type II Diabetics

Adult-onset (type II) diabetics, in whom B-cells are, of course, present, also respond to the infusion of insulin with a lowering of fasting IRG levels (RASKIN et al. 1975). The response of glucagon levels of such patients to a rapid intravenous infusion of glucose will depend on the adequacy of their endogenous insulin response to glucose (YAMAMOTO et al. 1979). The mean decrement in IRG after such a challenge may be virtually identical to that of nondiabetics if the insulin response

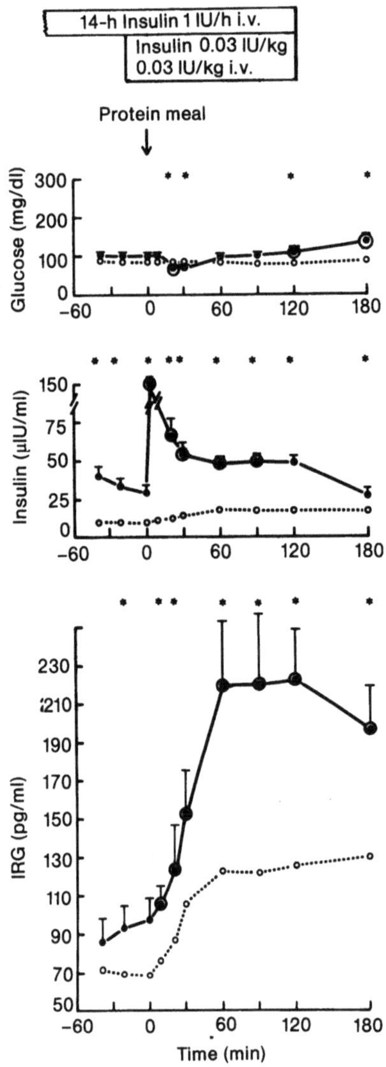


**Fig. 7.** The effect of exogenous insulin upon the glucagon response to a protein meal in a group of 10 type II diabetics (*full line*) compared with the response of 12 nondiabetics (*broken line*). Insulin in quantities that result in supraphysiologic plasma levels fail to diminish the exaggerated IRG response. *Asterisks* indicate $P<0.05$ diabetics compared with nondiabetics; *circled dots* indicate $P<0.05$ compared with baseline. RASKIN et al. (1978)

is brisk, or markedly obtunded if the insulin response is obtunded (HATFIELD et al. 1977). Thus, the negative A-cell response to glucose appears to be related to the positive B-cell response to glucose. The IRG response to ingested glucose or carbohydrate is often abnormal, or even paradoxical (BUCHANAN and MCCARROLL 1972). The fact that there is a difference between the IRG response to infused and ingested glucose suggests that in such diabetics there may be an exaggerated response to a glucagon-stimulating gut factor released during glucose absorption, rather than to an impaired response to hyperglycemia itself, and that supraphysiologic quantities of insulin cannot restore it to normal (UNGER et al. 1972; AYDIN et al. 1977).

Studies in obese white American type II diabetics have led to the conclusion that neither a constant insulin infusion nor a bolus of insulin sufficient to raise insulin levels well above the physiologic range, correct the exaggerated response to arginine and protein (Fig. 7), in striking contrast to type I diabetics. This claim should probably be reevaluated with more sophisticated techniques of insulin delivery in view of the report by KAWAMORI et al. (1980) that, in lean Japanese type II diabetics, A-cell function was restored to normal by insulin delivered via the artificial pancreas. Whether this discrepancy reflects differences in the type II diabetic subpopulations studied or in the methods of insulin delivery has not been determined. Nevertheless, at this time it seems fair to conclude that at least in obese diabetics in whom B-cells are presumably present, there is an exaggerated glucagon response to certain stimuli that is not corrected by insulin and which must therefore be ascribed to a cause other than insulin or B-cell deficiency.

## F. The Bihormonal Abnormality Hypothesis

The classical view that insufficiency of insulin or of insulin activity is the direct cause of the metabolic abnormalities of diabetes mellitus is derived from two historic experiments, those of VON MERING and MINKOWSKI (1889), in which removal of the canine pancreas was observed to produce this metabolic syndrome, and those of BANTING and BEST (1921), in which the injection of insulin-containing pancreatic extract was found to reverse the metabolic abnormalities resulting from pancreatectomy. It was logical to conclude from those two landmark events that all of the metabolic abnormalities of diabetes were the *direct* consequence of a single abnormality, insulin lack. This "unihormonal deficiency" concept had gone unchallenged for more than half a century.

Recently, however, an alternative concept, the "bihormonal abnormality hypothesis", was proposed. It is equally consistent with the findings of VON MERING and MINKOWSKI (1889) and of BANTING and BEST (1921). It is based on the recent recognition (VRANIC et al. 1974; MATSUYAMA and FOÀ 1975; MASHITER et al. 1975; DOBBS et al. 1975) that total pancreatectomy in dogs, as performed by VON MERING and MINKOWSKI (1889), is rapidly followed by an increase in glucagon secretion originating in extrapancreatic A-cells located for the most part in the gastric fundus (BLAZQUEZ et al. 1976; DOBBS et al. 1975), and that the administration of insulin, as performed by BANTING and BEST, promptly suppresses this extrapancreatic hyperglucagonemia (BLAZQUEZ et al. 1976). Because glucagon is biologically capable of increasing hepatic production of glucose and ketones, particularly in the presence of insulinopenia, and because relative hyperglucagonemia is present in all other forms of spontaneous and experimental diabetes (MÜLLER et al. 1970, 1971; UNGER et al. 1970; DOBBS et al. 1975), it was hypothesized that, in the absence of glucagon, insulin deficiency cannot directly produce all of the metabolic derangements that occur in the presence of glucagon; rather *both* glucagon and insulinopenia are essential to the development of the complete syndrome (UNGER and ORCI 1975).

**Table 1.** The "bihormonal abnormality" hypothesis: postulated contribution of insulin deficiency and of relative glucagon (IRG) excess to the metabolic abnormalities of diabetes mellitus

| Abnormalities | Insulin deficiency | Relative IRG excess |
|---|---|---|
| 1. Decreased glucose utilization | ++++ | 0 |
| 2. Increased hepatic glucose production from glycogen | + | ++++ |
| 3. Increased mobilization of amino acids | ++++ | 0 |
| 4. Increased gluconeogenesis | ++ | ++++ |
| 5. Increased lipolysis | ++++ | +(?) |
| 6. Increased ketogenic capacity of liver | +(+) | ++++ |

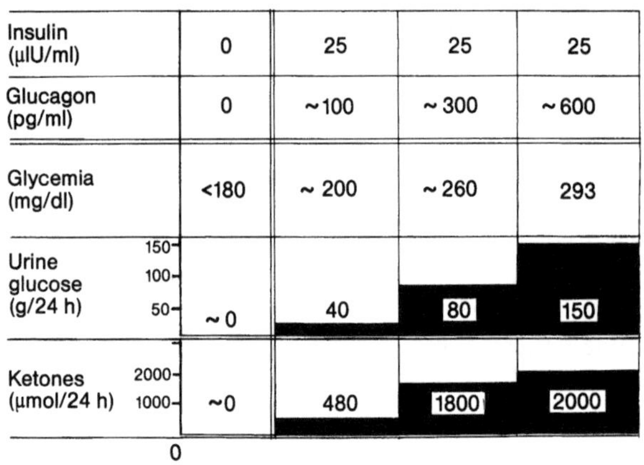


**Fig. 8.** A résumé of clinical data demonstrating the roles of insulin and glucagon in hepatic fuel overproduction. When both insulin and glucagon are absent, the massive hyperglycemia and hyperketonemia observed in the presence of glucagon does not occur. In patients in whom insulin was clamped at approximately 25 μU/ml for 3 days, hyperglycemia, glycosuria, and ketonuria increased progressively as glucagon levels rose as the result of a constant low dose glucagon infusion. Thus, in the absence of glucagon, deficiency of insulin does not result in massive overproduction of fuels by the liver. UNGER (1981)

According to this theory, insulin lack is the cause of impairment in glucose utilization by insulin-sensitive tissues, such as muscle, fat, and liver, of increased lipolysis in adipocytes, and of increased amino acid release from muscle (Table 1). But the marked hepatic overproduction of glucose and ketones is secondary to the combination of the relative excess of glucagon and insulinopenia and does not occur when both hormones are lacking (Fig. 8). The direct evidence favoring this hypothesis includes demonstrations in normal animals and humans that when plasma glucagon levels are reduced by somatostatin, blood glucose levels do not rise even though insulin is concomitantly suppressed (KOERKER et al. 1974; SAKURAI et al. 1974, 1975). Thus, insulin deficiency does not cause severe hyperglycemia so long as glucagon is absent, but when exogenous glucagon is infused during suppression of endogenous glucagon by a somatostatin infusion, hyperglycemia appears (Fig. 9; SAKURAI et al. 1974, 1975; RASKIN and UNGER 1977, 1978; GERICH et al.

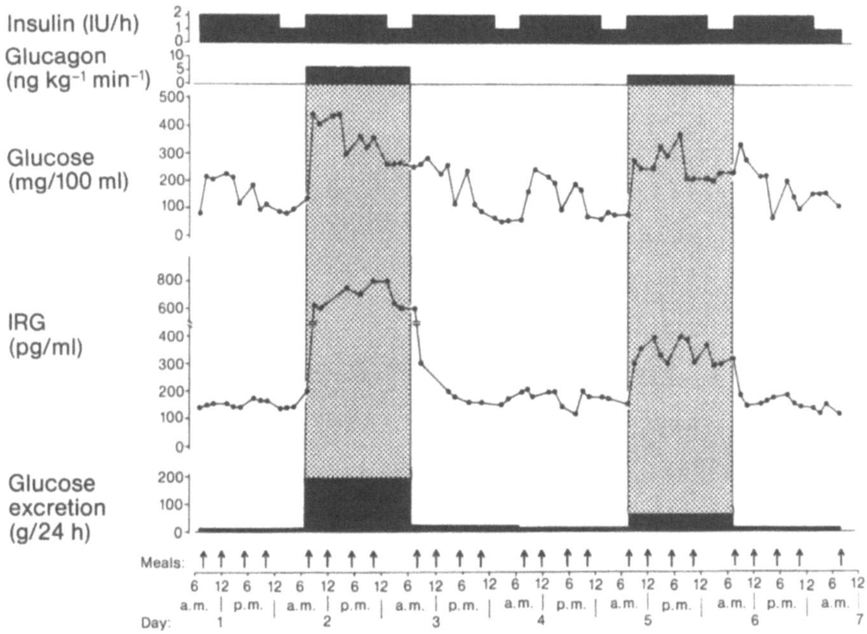


**Fig. 9.** Demonstration that relatively modest increments in plasma glucagon increase the hyperglycemia and glycosuria in a type I diabetic (Patient 2 of Table 2) controlled by insulin infused intravenously at a relatively constant rate. RASKIN und UNGER (1977)

1975). Somatostatin-induced suppression of glucagon also lowers plasma glucose levels in severely insulin-deficient alloxan-induced diabetic dogs (SAKURAI et al. 1974), both in the fasting state and during intravenous alanine infusion. In humans, GERICH et al. (1975) similarly observed that the suppression of glucagon by somatostatin infusion prevents both the severe endogenous hyperglycemia and the increase in blood ketones that otherwise occur in type I diabetics when insulin is discontinued. They also reported dramatic reduction in postprandial hyperglycemia in somatostatin-treated patients, but this may in part have been due to reduced entry of ingested glucose (WAHREN and FELIG 1976). In totally depancreatized dogs, suppression of extrapancreatic glucagon secretion similarly prevents the development of severe endogenous hyperglycemia (SAKURAI et al. 1975), and it appears promptly when replacement doses of glucagon are given with the somatostatin.

Evidence of the essential role of glucagon in the pathogenesis of severe endogenous hyperglycemia and hyperketonemia of insulin-deficient states has also been provided in more chronic circumstances. Hyperglycemia and ketosis is frequently absent in diabetics with pituitary insufficiency and this may in large part be the consequence of concomitant hypoglucagonemia. NAKABAYASHI et al. (1978) reported that hypophysectomy in dogs results in marked hypoglucagonemia. Total pancreatectomy in such hypophysectomized dogs (Houssay preparation) is associated with normal or low glucose levels, despite the absence of any endogenous or exogenous insulin, whereas sham-hypophysectomized dogs, in which hypergluca-

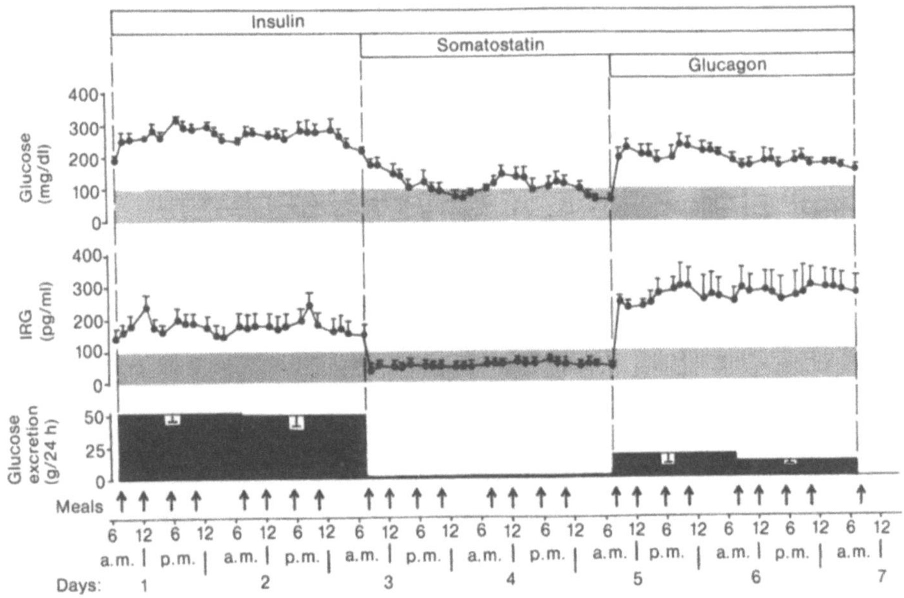


**Fig. 10.** Effects of somatostatin with glucagon on the daily profiles of mean (±standard error) levels of glucose and IRG and on glucose excretion in four patients with type I diabetes receiving a continuous insulin infusion and a diet containing 150 g/day carbohydrate. RASKIN and UNGER (1978)

gonemia follows removal of the pancreas, develop severe hyperglycemia. Glucagon given to the hypoglucagonemic Houssay dogs in replacement doses is followed by prompt appearance of endogenous hyperglycemia and glycosuria (NAKABAYASHI et al. 1978). Thus, hypoglucagonemia may in large part account for the normoglycemia so frequently present in the Houssay animal.

A second example of a chronic bihormonal deficiency state, hypoinsulinemia and hypoglucagonemia, may be observed in patients with somatostatinoma. In the first two such cases to be reported, severe endogenous hyperglycemia or ketoacidosis was absent, despite the insulin deficiency (LARSSON et al. 1977; GANDA et al. 1977).

A third example of protracted bihormonal deficiency can be experimentally induced in type I diabetics suboptimally treated with insulin and given the glucagon-suppressing hormone, somatostatin (RASKIN and UNGER 1978). In such patients, hyperglycemia and glycosuria disappeared and ketone and urea nitrogen excretion was reduced. Subsequent glucagon replacement was associated with a prompt worsening of all four parameters (Fig. 10; RASKIN and UNGER 1978). This provided strong evidence that hyperglucagonemia is a significant deleterious factor whenever insulin levels are fixed and unable to rise to counter the hyperglycemic effects of glucagon.

The foregoing experimental support for the bihormonal abnormality concept has been strengthened by recent work at the cellular level, suggesting that insulin's direct action on the liver is to oppose glucagon actions, i.e., that in the absence of glucagon it may have no action (BOYD et al. 1981).

## G. The "Glucagon Controversy"

The bihormonal abnormality concept was subjected to careful scrutiny and criticism. Doubt as to the importance of glucagon in the pathogenesis of the diabetic abnormalities was based on the following three main issues: (1) glucagon infusion was reported not to worsen the hyperglycemia or ketonemia of insulin-treated diabetics (SHERWIN et al. 1976); (2) the glycogenolytic action of glucagon was found to be transient and it was therefore argued that glucagon could not be of long-term importance in the excess hepatic glucose production of diabetes; (3) circulating glucagon was reported to be absent in totally depancreatized humans who are well known to develop at least mild endogenous hyperglycemia and ketoacidosis. As reviewed elsewhere (UNGER 1978; LEFEBVRE and LUYCKX 1979; UNGER and ORCI 1981a, b), none of these arguments can be regarded as invalidating the bihormonal hypothesis.

1. Glucagon does not worsen diabetes if insulin is present. The inability of SHERWIN et al. (1976) to demonstrate worsening of the hyperglycemia of insulin-treated type I diabetics during exogenously induced hyperglucagonemia has not been confirmed and may have been the consequence of the experimental design of their study. A restudy of this question using identical amounts of glucagon has revealed that in type I diabetics receiving a constant dose of insulin, a modest increase in plasma glucagon produced by infusing glucagon at the same rates used by SHERWIN and coworkers causes marked worsening of hyperglycemia, glycosuria, and ketone and urea nitrogen excretion (Table 2; Fig. 9; RASKIN and UNGER 1977).

2. The action of glucagon is transient. It is now recognized that the so-called transiency of the glucagon-mediated increase in glucose production reflects a decline in the glycogenolytic contribution to hepatic glucose production, but not in gluconeogenesis (JENNINGS et al. 1977) or ketogenesis (KELLER et al. 1977). Moreover, the evanescence is observed only when glucagon is infused at a constant rate; when the glucagon is infused at a variable rate or in pulses which simulate the secretory pattern of endogenous glucagon secretion, the glycogenolytic effect of glucagon does not wane, each new increment in glucagon finding a responsive liver (RIZZA and GERICH 1979; FRADKIN et al. 1979). The evanescence of glycogenolytic activity observed during constant glucagon infusions is believed to be mediated by hyperglycemia, which in the presence of insulin inhibits glycogenolysis (LILJENQUIST et al. 1979; FERRANNINI et al. 1980). But evancescence of glycogenolysis notwithstanding, when insulin is deficient, the glucagon-induced hyperglycemia resulting from glucagon-induced glycogenolysis persists long after the glycogenolytic action of glucagon has waned (CHERRINGTON et al. 1978). The issue of evanescence is, therefore, irrelevant for many reasons to the potential importance of glucagon in diabetic hyperglycemia.

3. Glucagon is absent in totally depancreatized patients who nevertheless develop endogenous hyperglycemia and ketoacidosis. Initially it was believed that insulin-deprived, totally depancreatized patients were aglucagonemic, a model of diabetes without glucagon (BARNES and BLOOM 1976; BARNES et al. 1977). It is now believed that although their glucagon levels are far less than in type I diabetics similarly deprived of insulin, some true glucagon is present in sufficient concentration

**Table 2.** Values of 24-h urinary glucose, urea nitrogen, and ketone excretion before, during, and after intravenous administration of glucagon to diabetics at a rate of 6 $ng/kg^{-1}/min^{-1}$ (A) or 3 $ng/kg^{-1}/min^{-1}$ (B). RASKIN and UNGER (1977)

| Patient | Control day | | | Glucagon day | | | Control day | | |
|---|---|---|---|---|---|---|---|---|---|
| | Glucose (g/24 h) | Urea nitrogen (g/24 h) | Ketones (μmol/24 h) | Glucose (g/24 h) | Urea nitrogen (g/24 h) | Ketones (μmol/24 h) | Glucose (g/24 h) | Urea nitrogen (g/24 h) | Ketones (μmol/24 h) |
| (A) 1 | 44 | 5.4 | 340 | 158 | 9.2 | 690 | 33 | 5.7 | 320 |
| 2 | 5 | 9.6 | 553 | 188 | 15.4 | 2,445 | 21 | 10.7 | 1,286 |
| 3 | 69 | 4.3 | 296 | 163 | 7.6 | 3,265 | 51 | 4.7 | 753 |
| 4 | 26 | 8.9 | 738 | 99 | 14.7 | 1,331 | 19 | 12.2 | 609 |
| Mean | 36 ± 14 | 7.0 ± 1 | 482 ± 102 | 152 ± 19[a] | 11.7 ± 2[a] | 1,933 ± 573[a] | 31 ± 7[a] | 8.3 ± 2[a] | 742 ± 202[a] |
| (B) 1 | 5 | 5.6 | 307 | 18 | 9.1 | 708 | 6 | 5.5 | 346 |
| 2 | 3 | 10.2 | 936 | 59 | 13.7 | 1,308 | 16 | 9.8 | 867 |
| 3 | 44 | 10.2 | 506 | 168 | 12.6 | 3,248 | 36 | 5.7 | 616 |
| Mean | 17 ± 13 | 8.7 ± 1.5 | 583 ± 186 | 82 ± 45 | 11.8 ± 1.4 | 1,755 ± 766 | 19 ± 9 | 7.0 ± 1.4 | 550 ± 104 |

[a] $P < 0.05$ vs, preceding day

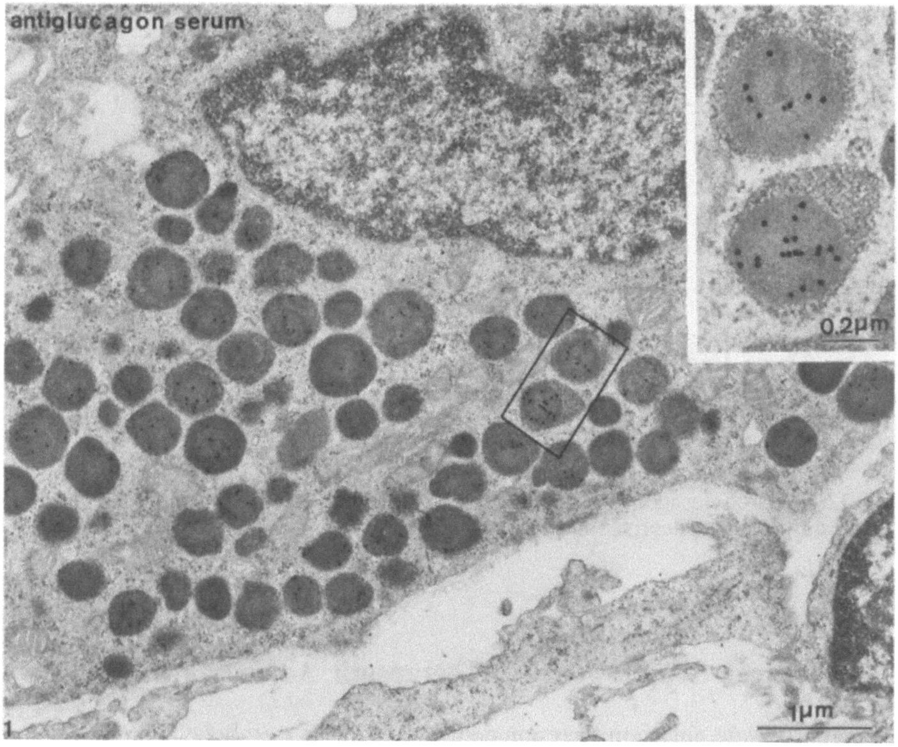


**Fig. 11.** Identification of an A-cell in the gastric fundus of a human fetus by means of COOH terminally directed anti-glucagon and the protein A gold technique. The grains represent anti-glucagon binding sites. Note the typical appearance of human A-cells with a glucagon-containing central "core" and the surrounding "halo" which contains glucagon precursor (or precursors). RAVAZZOLA et al. (1981)

to account for the hyperglycemia and ketonemia (which tends to be less severe than in type I diabetics). IRG of molecular weight 3,500 daltons was present in the totally depancreatized patients studied by BODEN et al. (1980), the mean level averaging 18 pg/ml. The source of this glucagon is uncertain since gastric A-cells comparable to those identified in dogs (BAETENS et al. 1976), have not been identified in adult humans (MUÑOZ-BARRAGAN et a. 1977), although their presence in the better preserved fetal stomach has been established (Fig. 11; RAVAZZOLA et al. 1981). A-cells have also been identified in the human colon (KNUDSEN et al. 1975), and conversion in circulating plasma of enteric glucagon-like immunoreactivity to glucagon has recently been demonstrated (KORANYI et al. 1981).

Figure 12 demonstrates apparent aglucagonemia in a pancreatectomized patient in whom endogenous hyperglycemia and hyperketonemia were absent during insulin deprivation until glucagon was administered (SANTEUSANIO et al. 1981). In summary, there is no clear-cut current evidence that renders untenable the bihormonal abnormality hypothesis.

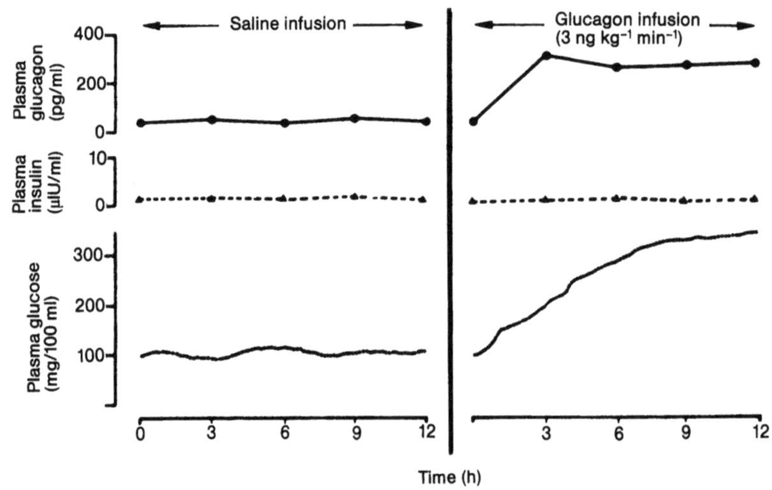


**Fig. 12.** Effect of glucagon infusion upon blood glucose in comparison with saline infusion alone, during insulin deprivation, in a pancreatectomized patient. SANTEUSANIA et al. (1981)

# H. Glucagon Suppression as a Therapeutic Adjunct in Diabetes

If inappropriate glucagon secretion contributes to the metabolic abnormalities of diabetes, as now seems highly likely, suppression of glucagon would seem a reasonable therapeutic approach in the management of diabetic hyperglycemia. Studies of RIZZA and GERICH (1981) and of RASKIN and UNGER (1978) suggest that somatostatin-induced glucagon suppression markedly improves hyperglycemia throughout the day, eliminates glycosuria (Fig. 10), and reduces ketone and urea nitrogen excretion. This form of therapy could make possible the attainment of continuous normoglycemia throughout the day by means of a dose of insulin that does not cause hypoglycemia between meals. While some of the somatostatin-induced improvement in hyperglycemia can be attributed to blockade of glucose absorption, recent studies demonstrate that glucagon suppression by itself is an important cause of the improvement (RIZZA and GERICH 1981). Such findings raise the hope that constant slow delivery of insulin supplemented by glucagon suppression could maintain relative normoglycemia by blocking inappropriate secretion of glucagon and slowing the rate of glucose absorption from the gut. However, newer techniques of insulin delivery via portal subcutaneous infusion pumps have made possible a far more physiologic pattern of insulin delivery than can be attained by conventional insulin delivery methods. As shown in Fig. 6, such a system of insulin delivery restores glucagon as well as glucose levels to the normal range (RASKIN et al. 1979) and would, therefore, seem to reduce the need for glucagon suppression.

An alternative to glucagon suppression is inhibition of glucagon action at its receptor site; promising data in this matter have recently been reported by JOHNSON et al. (1982).

*Acknowledgments.* This work was supported by the Veterans Administration Institutional Research Support Grand 549-8000-1; National Institutes of Health Grant AM 02700 and Contract NO1-AM-62219; and the Swiss National Science Foundation Grant 3.668.80.

## References

Assan R, Hautecouverture G, Guillemant S, Dauchy F, Protin P, Dérot M (1969) Evolution de paramètres hormonaux (glucagon, cortisol, hormone somatotrope) et énergètiques (glucose, acides gras libre, glycerol) dans dix acido-cétoses diabétiques graves traitées. Pathol Biol (Paris) 17:1095–1098

Aydin I, Raskin P, Unger RH (1977) The effect of insulin on the glucagon response to a carbohydrate meal in adult-onset and juvenile-type diabetes. Diabetologia 13:629–636

Baetens D, Rufener C, Srikant CB, Dobbs R, Unger RH, Orci L (1976) Identification of glucagon-producing cells (A-cells) in dog gastric mucosa. J Cell Biol 69:455–464

Baetens D, Stefan Y, Ravazzola M, Malaisse-Lagae F, Coleman DL, Orci L (1978) Alteration of islet cell populations in spontaneously diabetic mice. Diabetes 21:1–7

Banting FG, Best CH (1921) The internal secretion of the pancreas. J Lab Clin Med 7:251–266

Barnes AJ, Bloom SR (1976) Pancreatectomized man: a model for diabetes without glucagon. Lancet 1:219–221

Barnes AJ, Bloom SR, Mashiter K, Alberti KGMM, Smythe P, Turnell D (1977) Persistent metabolic abnormalities in diabetes in the absence of glucagon. Diabetologia 13:71–75

Blazquez E, Muñoz-Barragan L, Patton GS, Orci L, Dobbs RE, Unger RH (1976) Gastric A-cell function in insulin-deprived depancreatized dogs. Endocrinology 99:1182–1188

Boden G, Master RW, Rezvani I, Palmer JP, Lobe TE, Owen OE (1980) Glucagon deficiency and hyperaminoacidemia after total pancreatectomy. J Clin Invest 65:706–716

Boyd ME, Albright EB, Foster DW, McGarry JD (1981) In vitro reversal of the fasting state of liver metabolism in the rat. Reevaluation of the roles of insulin and glucose. J Clin Invest 68:142–152

Buchanan KD, McCarroll AM (1972) Abnormalities of glucagon metabolism in untreated diabetes mellitus. Lancet 2:1394–1395

Cherrington AD, Lacy WW, Chiasson JL (1978) Effect of glucagon on glucose production during insulin deficiency in the dog. J Clin Invest 62:664–677

Chisholm DJ, Alford FP (1977) The nature of the A-cell abnormality in human diabetes mellitus. In: Foà PP, Bajaj JS, Foà NL (eds) Glucagon: its role in physiology and clinical medicine. Springer, Berlin Heidelberg New York, pp 651–661

Dobbs RE, Sakurai H, Faloona GR, Valverde I, Baetens D, Orci L, Unger RH (1975) Glucagon: role in the hyperglycemia of diabetes mellitus. Science 187:544–547

Ferrannini E, DeFronzo R, Sherwin R (1980) Mechanism of the transient response of the liver to glucagon in man. Clin Res 28:519A

Fradkin J, Sherwin RS, Shamoon H, Felig P (1979) Effects of glucagon on hepatic glucose output: a consequence of changes in rather than absolute concentrations of plasma glucagon. Endocrinology 102 [Suppl 1]:167

Ganda OP, Weir GC, Soeldner JS, Legg MA, Chick WL, Patel YC, Ebeid AM, Gabbay JM, Reichlin S (1977) Somatostatinoma: a somatostatin containing tumor of the endocrine pancreas. N Engl J Med 296:963–967

Gerich JE, Lorenzi M, Bier DM, Schneider V, Tsalikian E, Karam JH, Forsham PH (1975) Prevention of human diabetic ketoacidosis by somatostatin: evidence for an essential role of glucagon. N Engl J Med 292:985–989

Gerich J, Langlois M, Noacco C, Lorenzi M, Karam J, Forsham P (1976a) Comparison of the suppressive effects of elevated plasma glucose and free fatty acid levels on glucagon secretion in normal and insulin-dependent diabetic subjects; evidence for a selective alpha-cell insensitivity to glucose in diabetes mellitus. J Clin Invest 58:320–325

Gerich J, Lorenzi M, Tsalikian E, Bohanon N, Noacco C, Karam J, Forsham P (1976b) Effects of acute insulin withdrawal and administration on plasma glucagon response to intravenous arginine in insulin-dependent diabetic subjects. Diabetes 25:955–960

Gerich J, Langlois M, Noacco C, Karam J, Forsham P (1978) Lack of glucagon response to hyperglycemia in diabetes: evidence for an intrinsic pancreatic alpha cell defect. Science 182:171–173
Hatfield HH, Banasiak MF, Driscoll T, Kim H-J, Kalkhoff RK (1977) Glucose suppression of glucagon: relationship to pancreatic beta cell function? J Clin Endocrinol Metab 44:1080–1087
Jennings AS, Cherrington AD, Liljenquist JE, Keller U, Lacy WW, Chiasson JL (1977) The roles of insulin and glucagon in the regulation of gluconeogenesis in the postabsorptive dog. Diabetes 26:847–856
Johnson DG, Goebel CU, Hruby VJ, Bregman MD, Trivedi D (1982) Hyperglycemia of diabetic rat decreased by a glucagon receptor antagonist. Science 215:1115–1116
Kawamori R, Shichiri M, Kikuchi M, Yamasaki Y, Abe H (1980) Perfect normalization of excessive glucagon responses to intravenous arginine in human diabetes mellitus with the artificial beta cell. Diabetes 29:762–765
Keller U, Chiasson JL, Liljenquist JE, Cherrington AD, Jennings AS, Crofford OB (1977) The roles of insulin, glucagon and free fatty acids in the regulation of ketogenesis in dogs. Diabetes 26:1040–1051
Kirk RD, Dunn PJ, Smith JR, Beaven DW, Donald RA (1975) Abnormal pancreatic alpha-cell function in first-degree relatives of known diabetics. J Clin Endocrinol Metab 40:913–916
Knudsen JB, Holst JJ, Asnaes S, Johansen A (1975) Identification of cells with pancreatic-type and gut-type glucagon immunoreactivity in human colon. Acta Pathol Microbiol Scand [A] 83:741
Koerker DJ, Ruch W, Chideckel E, Palmer J, Goodner CJ, Ensinck J, Gale CC (1974) Somatostatin: hypothalamic inhibitor of the endocrine pancreas. Science 184:482–489
Koranyi L, Peterfy F, Szabo J, Torok A, Guoth M, Tamas GY Jr (1981) Evidence for transformation of glucagon-like immunoreactivity of gut into glucagon in vivo. Diabetes 30:722–794
Larsson LI, Hirsch MA, Holst JJ, Ingemansson S, Kühl C, Lindbaer-Jensen S, Lundquist G, Rehfeld JF, Schwartz TW (1977) Pancreatic somatostatinoma: clinical features and physiological implications. Lancet 1:666–668
Lefèbvre PJ, Luyckx AS (1976) Plasma glucagon after kidney exclusion: experiments in somatostatin-infused and in eviscerated dogs. Metabolism 25:761–768
Lefèbvre PJ, Luyckx AS (1979) Glucagon and diabetes: a reappraisal. Diabetologia 16:347–354
Liljenquist JE, Mueller GL, Cherrington AD, Perry JM, Rabinowitz D (1979) Hyperglycemia per se (insulin and glucagon withdrawn) can inhibit hepatic glucose production in man. J Clin Endocrinol Metab 48:171–175
Lindsey CA, Faloona GR, Unger RH (1974) Plasma glucagon in nonketotic hyperosmolar coma. JAMA 229:1771–1773
Mashiter K, Harding PE, Chou M, Mashiter GD, Stout J, Diamond ND, Field JB (1975) Persistent pancreatic glucagon but not insulin response to arginine in pancreatectomized dogs. Endocrinology 96:678–693
Matsuyama T, Foà P (1975) Plasma glucose, insulin, pancreatic, and enteroglucagon levels in normal and depancreatized dogs. Proc Soc Exp Biol Med 147:97–102
Müller WA, Faloona GR, Aguilar-Parada E, Unger RH (1970) Abnormal alpha cell function in diabetes: response to carbohydrate and protein ingestion. N Engl J Med 283:109–115
Müller WA, Faloona GR, Unger RH (1971) The effect of experimental insulin deficiency on glucagon secretion. J Clin Invest 50:1992–1999
Müller WA, Faloona GR, Unger RH (1973) Hyperglucagonemia in diabetic ketoacidosis: its prevalence and significance. Am J Med 54:52–57
Muñoz-Barragan L, Rufener C, Srikant CB, Dobbs RE, Shannon WA Jr, Baetens D, Unger RH (1977) Immunocytochemical evidence for glucagon-containing cells in the human stomach. Horm Metab Res 9:37–39
Nakabayashi H, Dobbs RE, Unger RH (1978) The role of glucagon deficiency in the Houssay phenomenon of dogs. J Clin Invest 61:1355–1362

Orci L, Perrelet A (1981) The morphology of the A-cell. In: Unger RH, Orci L (eds) Glucagon. Physiology, pathophysiology, and morphology of the pancreatic A-cells. Elsevier North-Holland, New York, pp 3–36
Orci L, Malaisse-Lagae F, Ravazzola M, Rouiller D, Renold AE, Perrelet A, Unger RH (1975) A morphological basis for intercellular communication between A- and B-cells in the endocrine pancreas. J Clin Invest 56:1066–1070
Orci L, Baetens D, Rufener C, Amherdt M, Ravazzola M, Studer P, Malaisse-Lagae F, Unger RH (1976) Hypertrophy and hyperplasia of somatostatin-containing D-cells in diabetes. Proc Natl Acad Sci USA 73:1338–1342
Raskin P, Unger RH (1977) Effects of exogenous hyperglucagonemia in insulin-treated diabetics. Diabetes 26:1034–1039
Raskin P, Unger RH (1978) Hyperglucagonemia and its suppression: importance in the metabolic control of diabetes. N Engl J Med 299:433–436
Raskin P, Fujita T, Unger RH (1975) Effect of insulin-glucose infusion in plasma glucagon levels in fasting diabetics and nondiabetics. J Clin Invest 56:1132–1138
Raskin P, Aydin I, Unger RH (1976) Effects of insulin on the exaggerated glucagon response to arginine stimulation in diabetes mellitus. Diabetes 25:227–229
Raskin P, Aydin I, Yamamoto T, Unger RH (1978) Abnormal alpha cell function in human diabetics. The response to oral protein. Am J Med 64:988–997
Raskin P, Pietri A, Unger RH (1979) Changes in glucagon responses after four to five weeks of glucoregulation by portable insulin infusion pumps. Diabetes 28:1033–1035
Ravazzola M, Unger RH, Orci L (1981) Demonstration of glucagon in the stomach of human fetuses. Diabetes 30:879–882
Rizza R, Gerich J (1979) Effect of intermittent endogenous hyperglucagonemia on glucose metabolism in normal and diabetic man. J Clin Invest 63:1119–1123
Rizza R, Gerich J (1981) Treatment of diabetic hyperglucagonemia. In: Unger RH, Orci L (eds) Glucagon. Physiology, pathophysiology, and morphology of the pancreatic A-cells. Elsevier North-Holland, New York, pp 377–398
Sakurai H, Dobbs R, Unger RH (1974) Somatostatin-induced changes in insulin and glucagon secretion in normal and diabetic dogs. J Clin Invest 54:1395–1402
Sakurai H, Dobbs RE, Unger RH (1975) The role of glucagon in the pathogenesis of the endogenous hyperglycemia of diabetes mellitus. Metabolism 24:1287–1297
Samols E, Marri G, Marks V (1965) Promotion of insulin secretion by glucagon. Lancet 2:415–416
Samols E, Tyler JM, Marks V (1972) Glucagon-insulin interrelationships. In: Lefèbvre PJ, Unger RH (eds) Glucagon. Molecular physiology, clinical and therapeutic implications. Pergamon, Oxford, pp 151–173
Santeusanio F, Massi-Benedetti M, Angeletti G, Calabrese G, Bueti A, Brunetti P (1981) Glucagon and carbohydrate disorder in a totally pancreatectomized man (a study with the aid of an artificial endocrine pancreas). J Endocrinol Invest 4:93–96
Sherwin RS, Fisher M, Hendler R, Felig P (1976) Glucagon and glucose regulation in normal, obese and diabetic subjects. N Engl J Med 294:455–461
Unger RH (1976) Diabetes and the alpha cell. Diabetes 25:135–151
Unger RH (1978) Role of glucagon in the pathogenesis of diabetes: the status of the controversy. Metabolism 27:1691–1709
Unger RH, (1981) The mileu interieur and the islets of Langerhans. Diabetologia 20:1–11
Unger RH, Orci L (1975) Hypothesis: the essential role of glucagon in the pathogenesis of diabetes mellitus. Lancet 1:14–16
Unger RH, Orci L (1977a) Role of glucagon in diabetes. Arch Intern Med 137:482–491
Unger RH, Orci L, (1977b) Possible roles of the pancreatic D-cell in the normal and diabetic states. Diabetes 26:241–244
Unger RH, Orci L (1981a) Glucagon and the A-cell (Part 1). N Engl J Med 304:1518–1524
Unger RH, Orci L (1981b) Glucagon and the A-cell (Part 2). N Engl J Med 304:1575–1580
Unger RH, Aguilar-Parada E, Müller WA, Eisentraut AM (1970) Studies of pancreatic alpha cell function in normal and diabetic subjects. J Clin Invest 49:837–848
Unger RH, Madison LL, Müller WA (1972) Abnormal alpha cell function in diabetics: response to insulin. Diabetes 21:301–307

von Mering J, Minkowski O (1889) Aus dem Laboratorium der med. Klinik von Straßburg i.e. Diabetes mellitus nach Pankreasexstirpation. Arch Exp Pathol Pharmacol 26:371–378
Vranic M, Pek S, Kawamori R (1974) Increased "glucagon immunoreactivity" (IRG) in plasma of totally depancreatized dogs. Diabetes 23:905–912
Wahren J, Felig P (1976) Influence of somatostatin on carbohydrate disposal and absorption in diabetes mellitus. Lancet 2:1213–1216
Yamamoto T, Raskin P, Aydin I, Unger RH (1979) The effects of insulin on the response of immunoreactive glucagon to an intravenous glucose load in human diabetes. Metabolism 28:568–574

CHAPTER 45

# Glucagon in Human Endocrine and Exocrine Disorders

A. M. LAWRENCE

## A. Introduction

In 1972, I prepared an article entitled "Pancreatic alpha-cell function in miscellaneous clinical disorders" (LAWRENCE 1972). In that chapter it was noted that, despite the enthusiastic search for and desire to implicate abnormalities of pancreatic glucagon secretion in the cause of certain human diseases, with the exception of the glucagonoma syndrome, glucagon's causal role in other diseases remained elusive. A decade later, it is still uncertain whether abnormally high levels of measured glucagon in the blood of some individuals with, for example, acromegaly, hyperparathyroidism, or pheochromocytoma, reflect a general endocrine hypersecretory state akin to what has been described in the multiple endocrine neoplasia (MEN) syndrome or whether glucagon contributes causally to such endocrinopathies. Interpretive uncertainties arise, in some part, from our less than perfect ability to measure pancreatic glucagon in bodily fluid or from our inability to interpret accurately what such measurements actually reflect in terms of the pathogenesis of associated disease entities in humans. As an example, radioimmunoassay measurements of glucagon from peripheral blood provide glucagon measurements after substantial hepatic extraction (see Chap. 40) and, in general, measure glucagon of several origins (see Chap. 11). In what follows, however, an attempt has been made to update previous observations which appear to involve glucagon in disordered endocrine and exocrine physiology.

## B. Human Endocrine Metabolic Disorders

### I. Hypoglycemia

There is little question that glucagon plays a critical physiologic role in the maintenance of adequate hepatic glucose output in humans (LILJENQUIST et al. 1977; GERICH et al. 1980; see Chaps. 16, 37, and 38). Thus, it is not surprising that glucagon levels have been measured in a variety of hypoglycemic states, including idiopathic hypoglycemia in children, ketotic hypoglycemia of childhood, reactive or functional postprandial hypoglycemia, in patients with unusually hypoglycemia-prone type I diabetes mellitus, in pancreatectomized patients and in patients with severe endstage chronic pancreatic insufficiency with recurring hypoglycemia. In addition, recent attention has been drawn to an association between the use of $\beta$-adrenergic blockers and a susceptibility to spontaneous hypoglycemia in patients receiving such medication. Chapter 42 in this book has dealt with the issue of glucagon and hypoglycemia in greater detail.

### 1. Hypoglycemia in Infancy and Childhood

Although absolute glucagon deficiency has been suspected to be responsible for some forms of infant hypoglycemia, it is remarkable that no more than a few cases have had this suspicion documented by measurement of blood glucagon levels. Although GOTLIN and SILVER (1970) and others before them described cases of neonatal hypoglycemia where examination of autopsy specimens showed an apparent absence of pancreatic A-cells, confirmatory blood glucagon measurements have been lacking. In 1976, VIDNES and OYASAETER described a newborn with hypoglycemia in whom low basal glucagon levels could not be stimulated by either hypoglycemia or by an infusion of alanine. A brother and sister had died in infancy, perhaps of hypoglycemia. There was a dramatic response in this patient to the administration of zinc glucagon. The Pakistani parents were close relatives and when they were given an infusion of alanine, there was no rise in circulatory glucagon levels. Later, in 1979, KOLLEE et al. described a neonate with profound hypoglycemia and barely measurable levels of glucagon in the blood. As in the former case, this neonate also evidenced a dramatic response to zinc protamine glucagon. Follow-up studies in these children have not been reported.

### 2. Reactive Hypoglycemia

The role of inadequate glucagon secretion in reactive or functional postprandial symptomatic hypoglycemia remains controversial (LEFÈBVRE et al. 1976). Whereas there are reports of increased levels of glucagon-like immunoreactivity in the plasma of some patients suffering from alimentary hyperglycemia followed by severe symptomatic postprandial hypoglycemia (REHFELD et al. 1970), and particularly in gastric operated patients (MARCO et al. 1972; BREUER et al. 1975), others have demonstrated an apparent "glucagon deficiency" in such patients (FOÀ et al. 1980). The differences in these observations stem, in part, from the use of different antisera to glucagon in the radioimmunoassay.

In earlier studies, such antisera almost always measured glucagon of mixed origins. FOÀ et al. (1980) attempted to study this issue in their subjects using an antiserum cross-reactive with extracts of intestinal mucosa and pancreatic glucagon and compared such measurements with those obtained using an antiserum specific for the carboxyterminus of glucagon, an antiserum believed to measure A-cell or pancreatic islet glucagon. In seven subjects who developed symptomatic reactive hypoglycemia following a standard glucose challenge, there was a significantly greater rise in total and estimated gut glucagon in the control subjects compared with what was observed in individuals in whom symptomatic neurohypoglycemia developed. For two subjects who developed plasma glucose levels of 28 and 24 mg%, respectively, the rise in total glucagon and in estimated gut glucagon was significantly lower than for the remainder of the hypoglycemic group in that study. Using an antiserum, now believed to be relatively nonspecific for pancreatic islet glucagon, LAWRENCE (1966b) reported that certain gastrectomy patients, with symptomatic dumping syndrome and hypoglycemia, exhibited a blunted glucagon rise following oral glucose challenge when compared with a normal control group. Thus, it would appear that there may be a subpopulation of individuals who do, in fact, suffer from reactive hypoglycemia in some way related to defective glucagon counterregulation. However, in a series of patients investigated by GIUGLIANO et al. (1979), no abnormalities in plasma glucagon were observed.

### 3. Hypoglycemia in Diabetes and in Chronic Pancreatitis

GERICH et al. (1974, 1980) were among the first to draw attention to the frequent finding of inadequate glucagon responses to hypoglycemia in type 1 diabetics. Since that phenomenon was originally reported, others have suggested that the degree of unresponsiveness of glucagon to induced hypoglycemia can be correlated with poor diabetic control and/or the presence of diabetic neuropathy. REYNOLDS et al. (1973) noted that the plasma glucagon rise following insulin-induced hypoglycemia was highest in nondiabetic controls, less in stable diabetics, and least or even absent in poorly controlled diabetics. Later, MAHER et al. (1977) observed a twofold increase in circulating glucagon levels following insulin-induced hypoglycemia in eight diabetics without evident neuropathy compared with a threefold increase obtained in nondiabetic controls. In nine diabetic subjects with documented neuropathy, five of whom were judged to be unstable diabetics, no glucagon rise followed the hypoglycemic stimulus. These workers postulate a wider involvement of disordered glucoregulatory mechanisms in the neuropathic unstable diabetic than just deficient glucagon secretory responsiveness. Some have postulated that instability or brittleness of certain type I diabetics, including those with endstage fibrotic pancreatitis, are particularly prone to recurring hypoglycemia because of deficient glucagon release (AGUILAR-PARADA et al. 1969; LAWRENCE and ABRAIRA 1980). Recent data reported by BOLLI et al. (1983) and WHITE et al. (1983) have confirmed the theory that abnormalities in glucagon secretion are involved in the abnormal glucose counterregulation found in some type 1 diabetic patients.

### 4. Glucagon Antibodies

Antibodies to contaminants in commercial insulin, including glucagon antibodies, have been reported (CRESTO et al. 1974; VILLALPANDO and DRASH 1979; FITZPATRICK and PATEL 1981). Even despite improved purity of currently available commercial insulin preparations, about 14% of insulin-dependent diabetics have glucagon antibodies that bind approximately 10–30 ng/100 IU insulin (FITZ-PATRICK and PATEL 1981). It is not clear, however, whether such antibodies to circulating glucagon pose any serious clinical problem. CRESTO et al. (1974) are of the opinion that there was essentially no difference in the clinical course of patients with insulin-dependent diabetes, whether glucagon antibodies were present or not. VILLALPANDO and DRASH (1979), however, studying 66 adolescents who had had diabetes mellitus for 6–12 years, determined that significant nocturnal hypoglycemia developed in 6 of their young patients with detectable titers of glucagon antibodies. From their experience, they believe rather strongly that antibodies to glucagon can significantly alter the clinical course of diabetes, particularly as regards nocturnal hypoglycemia. In their six patients, a relative deficiency of free glucagon was thought to be largely responsible for the lack of counterregulatory mechanisms to prevent hypoglycemia in these children. In a somewhat different vein, DELPRETE et al. (1978) studied a subset of individuals in whom autoantibodies to pancreatic

A-cells had been demonstrated. In several, a variety of autoimmune disorders coexisted but, in two instances, subjects were found to have islet cell antibodies that reacted only with glucagon-producing A-cells. No abnormalities of glucagon secretory dynamics could be demonstrated in these two patients although, as the authors note, it may be necessary to repeat such pharmacologic tests at a future time in order to assess accurately whether the presence of autoantibodies to glucagon-secreting cells poses any serious threat to the long-term functional integrity of the pancreatic A-cell.

### 5. Beta-Adrenergic Blockade

The use of β-adrenergic blockade has been associated with serious hypoglycemia in a variety of patients, particularly those receiving insulin or sulfonylurea therapy for their diabetes. This may be secondary to impaired glucagon secretory dynamics (LAWRENCE et al. 1972; see also Chap. 32) or may be secondary to an intrinsic glucose-lowering effect of the first generation noncardioselective β-blockers, such as propranolol. In patients with cirrhosis, there may be some increased susceptibility to β-blocker-associated hypoglycemia. In cirrhotic individuals with increased hepatic arterial resistance, hepatic blood flow may be diminished by 30% or more, allowing drugs such as propranolol, with a high first-pass hepatic extraction, to accumulate (HÖKFELT et al. 1978). Some have suggested, however, that the newer cardioselective β-blocking agents, such as metoprolol, are free of the hypoglycemia association previously described for propranolol and similar drugs (HÖKFELT et al. 1978). It is likely that additional studies will have to be undertaken before it is adequately determined whether the cardioselective β-blockers are indeed free of risk of hypoglycemia in diabetics who are receiving insulin and sulfonylurea. In any event, it is reasonable that glucagon be used for the treatment of β-blocker-associated hypoglycemia as it has been demonstrated that such therapy will normalize the blood glucose in hypoglycemic individuals with inappriopriately low immunoassayable glucagon (ROBSON 1980).

## II. Hyperglucagonemia

### 1. Glucagonoma

Although first accurately described and documented with immunoassays for glucagon (MCGAVRAN et al. 1966), glucagon-secreting adenomas and adenocarcinomas have probably been recognized since the original report, by ROTHMAN (1925), of a patient with a necrotizing dermopathy and autopsy findings of a pancreatic carcinoma metastatic to the liver which microscopically resembled an islet cell carcinoma. As discussed in greater detail in Chap. 43, the glucagonoma syndrome is now a well-established entity and has been described in several recent reviews (MALLINSON et al. 1974; LAWRENCE and DORSCH 1980; STACPOOLE 1981). Patients as young as 16 and as old as 80 years have been described. A majority of patients presented with one or more of the features described, including the distinctive dermopathy, necrolytic migratory erythema, marked weight loss with prostration, often without loss of appetite, stomatitis, glossitis, cheilosis, sparse hair, and crumbly, pliable fingernails. Inexorable weight loss, prostration, and ultimately death have ensued, at a variable course, unless the patient responds to therapeutic intervention. Most, but not all patients have mild to overt glucose intolerance although some with perfectly normal carbohydrate homeostasis have been reported (LAWRENCE and DORSCH 1980; PRINZ et al. 1981; STACPOOLE 1981). As with other islet cell carcinomas, insulinomas, and gastrinomas, functional or nonfunctional, tumor growth, including metastases, may develop very slowly and certain of these patients may be able to live in reasonably good health for many years, despite local and hepatic metastases.

#### 2. Nonislet Glucagon-Secreting Tumors

To the best of my knowledge, only two nonpancreatic glucagon-secreting malignancies have been described. These have contained cells which appear histologically identical to those of pancreatic A-cell origin. Extraction of both the duodenal tumor and the renal carcinoma revealed the presence of large amounts of a high molecular weight glucagon-like material (GLEESON et al. 1971; BLOOM 1972). Both patients were clinically characterized by effects on the gastrointestinal tract, including colonic and jejunal stasis and malabsorption, similar to what has been observed in animals receiving large amounts of glucagon chronically.

#### 3. The Multiple Endocrine Neoplasia

Glucagonomas are now recognized in association with the familial multiple endocrine neoplasia type I (MEN-I) (CROUGHS et al. 1977; STACPOOLE et al. 1981). Several with other endocrine adenomas have exhibited hyperglucagonemia without evidence of a functioning glucagonoma (VANCE et al. 1969). Frequently, such patients present with the classical features of primary hyperparathyroidism, and one family has been described in which three members with hyperparathyroidism also presented with features of the glucagonoma syndrome, two of whom had metastatic disease at the time of initial presentation (STACPOOLE et al. 1981). One patient has been described with primary hyperparathyroidism and calcific pancreatitis with coexistent hyperglucagonemia and striking A-cell hyperplasia on histologic examination of the surgically resected pancreas as treatment for intractable pain (PALOYAN et al. 1967a).

### III. Pheochromocytoma

In 1966, it was reported that several patients with pheochromocytoma exhibited significantly increased levels of circulating glucagon immunoreactivity (LAWRENCE 1966a). The antiserum used for those measurements was nonspecific for pancreatic glucagon and it must be concluded that those measurements assayed glucagon of mixed origins (LAWRENCE 1972). More recently, HAMAJI (1979) reported data on seven patients with pheochromocytoma where he found a low glucagon response to intravenous arginine infusion; interestingly enough, the glucagon response to arginine was normalized after removal of the tumor. Further studies are needed to clarify this issue. It should be remembered that glucagon, given intravenously, has been used as a tool for the diagnosis of pheochromocytoma; this aspect is considered in detail in Chap. 51.

## IV. Pituitary Interrelationships

### 1. Acromegaly

A variety of experimental and clinical observations has long suggested that there exists a special relationship between pancreatic glucagon and pituitary growth hormone secretion (see also Chap. 52). DORNER and STAHL (1957) reported that removal of the pituitary from dogs ultimately led to A-cell hypoplasia, and KRACHT (1953) reported that growth hormone administration to such animals would result in restoration of A-cell morphology. In 1951, BORNSTEIN et al. had described the presence of a material in pancreatic duodenal vein blood, which appeared after growth hormone administration and would cause hyperglycemia in recipient animals. Studies of this kind were, of course, marred by lack of any available glucagon measurements at that time.

With the advent of the glucagon immunoassay, a brief flurry of clinical studies was undertaken in individuals with a variety of pituitary disorders. However, as in the several other clinical circumstances discussed, there is great need for building a data base from which concrete conclusions can be drawn. In acromegaly, it was reported that an arginine infusion elicited an exaggerated serum glucagon rise (GOLDFINE et al. 1972). Although it was then known that the glucagon antiserum used was relatively nonspecific, it was believed that a glucagon response to an arginine infusion represented a rise in pancreatic glucagon. Since that time, however, it has been shown that arginine will also elicit a rise in glucagon species of gastrointestinal origin (see Chap. 33). More recently, TRIMBLE et al. (1980) were unable to confirm earlier findings of an exaggerated glucagon rise to an arginine infusion. As in GOLDFINE's report, basal glucagon levels in patients with acromegaly did not differ from normal controls, but unlike the findings of GOLDFINE et al., the latter group could show no exaggerated glucagon rise in the 16 acromegalic subjects they studied. It must be remembered, however, that the earlier study had utilized a totally nonspecific glucagon antiserum and measured nonextracted serum samples whereas the more recent report used an antiserum known to react only which the COOH terminal portion of the glucagon molecule and plasma samples were extracted with ethanol before assay so that all larger molecular forms of glucagon-like immunoreactivity were removed. Thus, these disparate results may simply reflect on the different assay systems where one measured total plasma glucagon of several sources and the latter determined glucagon primarily of pancreatic islet origin. Alternatively, the variegated endocrinologic makeup of acromegalic patients may be responsible for these differing results. As noted earlier, this area of clinical investigation is also in need of more thorough and careful scrutiny. Of interest, however, are three reports said to show increased glucagon secretion in normal volunteers receiving human growth hormone administration (PEK et al. 1971; LAWRENCE 1972; GOLDFINE et al. 1972).

### 2. Hypopituitarism

If total immunoreactive glucagon levels in serum of acromegalics are raised to either basal or higher than normal levels after suitable provocation, it could be anticipated that levels would be low in patients with hypopituitarism. Although this also is an area that has not been thoroughly studied, some have reported that basal glucagon levels are low in hypopituitary individuals and that glucagon levels do not rise significantly in response to an arginine infusion in such patients (LAWRENCE 1972). Several intriguing clinical observations support the suspicion that pancreatic glucagon secretion may be subnormal in states of hypopituitarism. Such individuals commonly exhibit marked reactive hypoglycemia following a standard oral glucose challenge (GOLD et al. 1968). Perhaps by analogy, glucagon levels may also be low in ateliotic dwarfism, where a unitropic deficiency of growth hormone exists. These patients demonstrate impaired glucose tolerance, presumably due to insufficient and delayed insulin release. These abnormalities can be reversed after several months of growth hormone treatment (GOLD et al. 1968; MERIMEE 1970). These relationships with regard to glucagon, however, need to be better quantified in many more hypopituitary patients before any exact interdependence between the pancreatic islet A-cell and secretions of the anterior pituitary can be firmly established. As reviewed in Chap. 52, glucagon is useful, diagnostically, in screening short-statured children for growth hormone deficiency.

## V. Glucagon and the Parathyroid Glands

### 1. Hyperparathyroidism

The role, if any, of glucagon hypersecretion in the pathogenesis of hyperparathyroidism also has yet to be determined. Based primarily on the work of PALOYAN and co-workers, the suspicion of such an association was raised. That group showed that the chronic administration of glucagon caused marked hyperplasia of the parathyroid glands in rabbits (PALOYAN et al. 1966; PALOYAN 1967) and that infused glucagon would significantly lower the prevailing serum calcium levels in rabbits (PALOYAN et al. 1967a) and in humans (PALOYAN 1967). They also describe certain patients with hyperparathyroidism and chronic calcific pancreatitis where hyperglucagonemia was found preoperatively and normalized following subtotal resection of the diseased pancreas for intractable pain (PALOYAN et al. 1967b). Histologic examination of the pancreas from one such patient demonstrated giant islets, many larger than 250 μm, composed almost exclusively of A-cells. It has also been reported by the same group that there is an increased frequency of islet cell hyperplasia in the pancreas of patients with hyperparathyroidism who have come to autopsy from an unrelated problem (PALOYAN 1967). Further, serum glucagon levels have been found to be significantly elevated in patients with chronic pancreatitis (KANNAN et al. 1979). As glucagon has been shown to promote significant lowering of the prevailing plasma calcium concentration (PICKLEMAN et al. 1969; AVIOLI et al. 1969), some have theorized that the association of hypersecreting parathyroid adenomas in patients with chronic calcific pancreatitis stems directly from this hypocalcemic effect of glucagon.

Parathyroid hyperfunction may exist in variants of the MEN syndrome, and the presence of hyperglucagonemia and even glucagonomas, in some of these patients, has been of intriguing interest. As noted earlier, VANCE et al. (1969) simultaneously measured insulin, glucagon, and gastrin in blood samples drawn from several relatives of a patient with hyperinsulinemia, islet cell carcinoma, and adrenal cortical hyperplasia. All relatives demonstrated excessively elevated levels of one or more of these peptide hormones whether they were symptomatic or not. Glucagonomas, as part of the MEN-I syndrome, have been reported in patients who have presented with the symptoms and features of primary hyperparathyroidism (CROUGHS et al. 1977; STACPOOLE 1981), and sporadically in patients thought to have idiopathic primary hyperparathyroidism uncomplicated by diffuse endocrine adenomatosis (PALOYAN et al. 1967b). Unfortunately, no large prospective series has yet been reported where basal and provoked pancreatic glucagon levels have been measured in patients with clear-cut idiopathic hyperparathyroidism versus MEN-I, preoperatively and following surgical resection of hyperfunctioning parathyroid tissue. Whether glucagon of pancreatic origin truly contributes to a chain of events promoting parathyroid hyperplasia remains to be proved. The small amount of information available to date only hints at such an association.

### 2. Hypoparathyroidism

TORELLA et al. (1982) recently reported that the glucagon response to intravenous arginine was modest and delayed. The acute correction of hypocalcemia produced a striking increase in basal glucagon levels and restored to normal the glucagon response to arginine. They concluded that glucagon secretion in humans is critically dependent on plasma calcium concentration (see Chap. 26).

## VI. Glucagon and Other Endocrine Diseases

### 1. Cushing's Syndrome

Excessive glucagon response to arginine has been reported in patients with Cushing's syndrome (SEINO et al. 1977) as it is after glucocorticoid administration in healthy controls (see Chaps. 29 and 32). However, oral glucose substantially suppresses plasma glucagon in patients with Cushing's syndrome (IMURA et al. 1977).

### 2. Hyperthyroidism

The plasma glucagon response to arginine infusion has been reported to be slightly, but not significantly reduced in hyperthyroidism (SEINO et al. 1974).

### 3. Hypothyroidism

The glucagon response to arginine is usually exaggerated in hypothyroidism (SEINO et al. 1977).

### 4. Addison's Disease

OKUNO and TAKENAKA (1977) investigated A-cell function in a single patient with primary adrenocortical insufficiency. Basal plasma glucagon levels and the glucagon response to hypoglycemia were normal. The glucagon response to arginine, before treatment, was about one-half that observed in normal controls. This response became supranormal, but markedly prolonged after cortisol treatment.

### VII. Comment

What new has really been learned in the past decade with regard to glucagon's role in human endocrine and metabolic disorders? The answer is that very little clinically relevant information, not previously recognized, has actually emerged in the past 10 years. Glucagon's clinical usefulness in combating a variety of hypoglycemic disorders remains unchallenged (see Chap. 55). Its utility in screening children with suspected growth hormone deficiency continues to enjoy clinical applicability (see Chap. 52). In patients with equivocal laboratory results in whom an insulinoma (Chap. 55) or a pheochromocytoma (Chap. 51) is suspected, glucagon provocation may assist in documenting the preoperative diagnosis, and in 10 years' time these procedures have been sufficiently refined so they no longer pose significant hazard to the patient under study. What is missing form the past decade's literature are serious and in-depth prospective analyses of whether serum glucagon levels are abnormal in all patients with acromegaly, pheochromocytoma, or hyperparathyroidism, and if so, whether raised levels are altered by surgical resection of the offending hypersecreting endocrine organ. It would be of considerable interest to know for certain whether defective glucagon kinetics contribute to the poor glucoregulation of hypopituitary patients and to have some solid information on glucagon profiles in states of altered thyroid and adrenal function in humans. Perhaps, now that the earlier preoccupation with glucagon's role in diabetes mellitus has been placed in a more rational perspective, it will be possible to return to many of these sketchily probed areas in clinical endocrinology. There can only be hope that disappointment stemming from finding that glucagon does not play a pivotal role in the pathogenesis of human diabetes will not divert interest from pursuing a more methodical assessment of glucagon's place in other disorders of endocrine physiology.

## C. Glucagon and the Exocrine Pancreas

### I. Interplay with the Islets of Langerhans

Embryologically and phylogenetically the insulin- and glucagon-secreting islets of Langerhans originate from the primitive ductular system of the developing exocrine pancreas (LIKE and ORCI 1972; FALKMER and OSTBERG 1977). Although these islets ultimately become isolated from pancreatic exocrine ducts, many intriguing relationships between exocrine and endocrine tissue of the pancreas persist and even suggest that this intimacy is necessary for the proper function of both the islets and the pancreatic digestive organ. Although mammalian islets separate from their mother ducts, several close arrangements, histologic and functional, between the pancreatic endocrine and exocrine tissues persist. For example, the major blood supply to the exocrine pancreas is largely derived from efferent capillaries draining the individual islets and radiating into adjacent acinar tissue. Exocrine tissue close to the islets has larger cells which are more densely packed with zymogen granules, and enzyme content of acinar tissue close to the islets differs from that of enzyme content further away from endocrine tissue (HENDERSON 1969; MALAISSE-LAGAE et al. 1975; BAETENS et al. 1979). Furthermore, acinar tissue immediately surrounding the islets of Langerhans can easily be distinguished from more distant exocrine tisue as those in close proximity contain significantly larger cells with larger Golgi and are more densely packed with zymogen granules (MALAISSE-LAGAE et al. 1975). These peri-insular "halos" contain distinctly different amylase, lipase, and chymotrypsin content from acinar tissue located further from the islets.

It has previously been suggested that, if there is any important relationship between the exocrine and endocrine tissues of the pancreas in mammals, neural and hormonal factors known to affect one should affect the other's activity (HENDERSON 1969). This is certainly so with regard to the gastrointestinal hormones, secretin, gastrin, and pancreozymin, as well as with stimulation of the distal end of the cut vagus, each of which will result in pancreatic exocrine flow as well as release of insulin and/or glucagon (FALKMER and PATENT 1972; FUJITA and KOBAYASHI 1974; FALKMER and OSTBERG 1977; PFEIFFER et al. 1977).

In both human diabetes and experimentally induced diabetes, for example, marked anatomic and functional changes in pancreatic exocrine anatomy and function occur. These changes include diminished excretory pancreatic flow in response to administered secretin, epithelial duct vacuolization, acinar fibrosis, and fatty lymphocytic infiltration. HENDERSON (1969) has theorized that these anatomic and functional derangements are a result of impaired infusion of the pancreatic exocrine tissue with adequate amounts of insular hormones. Others have shown that pancreatic glucagon has a marked pancreozymin-like activity upon the exocrine pancreas and that amylase production by the exocrine pancreas drops significantly when islets are impaired or destroyed following administration of streptozotocin (HENDERSON 1969; DANIELSON and SEHLEN 1974; FUJITA and WATANABE 1974; KANNO and SAITO 1976; KORC et al. 1978; SHAPIRO and LUDEWIG 1978). Insulin release in response to secretin administration is seriously impaired in nondiabetic patients with pancreatic insufficiency (HENDERSON 1969; PFEIFFER et al. 1977). Patients with chronic pancreatic insufficiency fail to show the high insulin levels generated during the course of an oral glucose challenge and others have shown that nondiabetic patients with cystic fibrosis and pancreatic insufficiency also fail to release insulin following secretin administration and have an impaired insulin response to oral glucose as well (PFEIFFER et al. 1977). From such observations, it has been inferred that insulin release in response to signals from gastrointestinal hormones and from hyperglycemia is severely blunted in patients with chronic pancreatitis and, in some, is virtually absent.

## II. Insulin and Glucagon in Pancreatic Exocrine Fluid

From the foregoing considerations, it is not altogether surprising that there have been reports of the presence of immunoreactive insulin and glucagon-like materials in pancreatic exocrine fluid (SATAKE et al. 1972; TEWSKBURY et al. 1973; BAILY et al. 1975; PRINZ et al. 1980). SATAKE et al. (1972) reported immunoreactive insulin-like substances in pancreatic juice obtained from nonanesthetized dogs with indwelling cannulae and showed that the infusion of pancreozymin/secretin had no apparent effect on insulin content of exocrine fluid, but that levels rose, late in the course of an intravenously administered glucose load. More recently, the glucagon immunoassay has been applied to carefully collected pancreatic fluid from fistulized dogs and humans (PRINZ et al. 1978, 1980; LAWRENCE et al. 1979). When such fluid is freshly obtained in ice-chilled tubes we have not been able to find any immunoassayable glucagon. However, following brief treatment with trypsin, significant amounts of a proglucagon-like immunoreactive material are generated with a molecular weight of approximately 13,000 daltons. Careful analysis of the nonimmunoreactive precursor, or preproglucagon, reveals a molecular weight of approximately 14,000 daltons. Carboxypeptidase B treatment of the immunoreactive proglucagon generated by trypsin results in the appearance of an A-cell-sized glucagon (A. M. LAWRENCE 1982, unpublished observation). These glucagon precursors are similar to compounds described by others, extracted either from islets or from the exocrine pancreas (PATZELT et al. 1979; LUND et al. 1980).

## III. Pancreatitis

For some time it has been known that glucagon suppresses gastric acid secretion in humans and in animals and that its acute administration results in diminished exocrine secretion. Its chronic administration in animal studies has revealed widespread degranulation of pancreatic acinar cells. In addition, glucagon significantly reduces motility of the gastrointestinal tract (KANNAN et al. 1979; CHRISTIANSEN 1980). For these and other reasons, it was thought that parenteral administration of glucagon in patients with acute pancreatitis could help to diminish the severity of the disease and its morbid complications.

Initially, isolated case reports did purport to show benefit from glucagon administration in such patients, but, more recently, more carefully designed, prospective, randomized double-blind trials of glucagon in acute pancreatitis have failed to demonstrate such an effect. In a 1980 report from Denmark, 22 patients presenting with a first attack of severe acute pancreatitis were randomized with the glucagon-treated group receiving 1 mg glucagon as a bolus intravenous injection followed by an infusion of 6 mg in 1,500 ml saline, every 24 h for 3 days. The number of severe complications, duration of the acute disease, and death were essentially no different between the control and glucagon-treated patients (KRONBORG et al. 1980). In a cooperative clinical trial designed by the Working Party on the Treatment of Acute Pancreatitis of the Medical Research Council of Great Britain, 257 patients with acute pancreatitis were studied in a randomized double-blind fashion where standard therapy was compared with patients randomized into a group receiving glucagon or aprotinin. All patients received nasogastric aspiration, intravenous fluids, and analgesics as required. Glucagon was administered as a 2 mg intravenous bolus followed by 2 mg every 6 h for 5 days and thereafter 8 mg every 24 h. At the conclusion of the study, the resulting data showed no benefit derived from glucagon therapy over standard conservative therapy alone (MEDICAL RESEARCH COUNCIL MULTICENTRE TRIAL 1980). DEBAS et al. (1980) also concluded, contrary to expectations, that the use of glucagon therapy in acute pancreatitis in a series of 66 patients assigned to conventional therapy or to glucagon treatment failed to reveal any beneficial results accruing from the administration of glucagon. More recently, REGAN et al. (1981), from the mayo Clinic, came to similar conclusions after studying 20 patients assigned to either nasogastric suction, nasogastric suction and glucagon, or nasogastric suction and cimetidine either alone or with glucagon. Their results found neither drug approach superior over standard nasogastric suction.

## References

Aguilar-Parada E, Eisentraut AM, Unger RH (1969) Pancreatic glucagon secretion in normal and diabetic subjects. Am J Med Sci 257:415–419

Avioli LV, Birge SJ, Scott S, Shieber W (1969) Role of the thyroid gland during glucagon-induced hypocalcemia in the dog. Am J Physiol 216:939–945

Baetens D, Malaisse-Lagae F, Perrelet A, Orci L (1979) Endocrine pancreas: three-dimensional reconstruction shows two types of islets of Langerhans. Science 206:1323–1325

Baily CJ, Flatt PR, Atkins TW, Best LC, Matty AJ (1975) Immunoreactive insulin in pancreatic juice and bile of rats. Diabetologia 11:330–331

Bloom SR (1972) An enteroglucagon tumor. Gut 13:520–523

Bolli G, De Feo P, Compagnucci P, Cartechini MG, Angeletti G, Santeusanio F, Brunetti P, Gerich JE (1983) Abnormal glucose counterregulation in insulin-dependent diabetes mellitus. Interaction of anti-insulin antibodies and impaired glucagon and epinephrine secretion. Diabetes 32:134–141

Bornstein J, Reid E, Young FG (1951) The hyperglycaemic action of blood from animals treated with growth hormone. Nature 168:903–905

Breuer RI, Zuckerman L, Hauch TW, Green W, O'Gara P, Lawrence AM, Foà PP, Matsuyama T (1975) Gastric operations and glucose homeostasis. II. Glucagon and secretin. Gastroenterology 69:598–606

Christiansen J (1980) Pancreatic glucagon and gastric acid secretion. Scand J Gastroenterol 15:257–258

Cresto J, Lavine R, Perrino P, Recant L, August G, Hung W (1974) Glucagon antibodies in diabetic patients. Lancet 1:1165

Croughs RJM, Hulsman HAM, Israel DE, Hackeng WHZ, Schopman W (1977) Glucagonoma as part of the polyglandular adenoma syndrome. Am J Med 52:690–695

Danielson E, Sehlen R (1974) Transport and oxidation of amino acids and glucose in isolated exocrine mouse pancreas: effects of insulin and pancreozymin. Acta Physiol Scand 91:557–565

Debas HT, Hancock RJ, Soon-Shiong P, Smyth HA, Cassim MM (1980) Glucagon therapy in acute pancreatitis: prospective randomized double-blind study. Can J Surg 23:578–580

DelPrete GF, Tiengo A, Nosadini R, Bottazzo GF, Betterle C, Bersani G (1978) Glucagon secretion in two patients with auto-antibodies to glucagon secreting cells. Horm Metab Res 10:260–261

Dorner M, Stahl J (1957) Recherches sur le glucagon. II. Relations entre l'hypophyse antérieure et le système alpha-cellulaire du pancréas chez le chien. Rev Fr Etud Clin Biol 2:118–128

Falkmer S, Ostberg Y (1977) Comparative morphology of pancreatic islets in animals. In: Volk BW, Wellmann KF (eds) The diabetic pancreas. Plenum, New York, p 15

Falkmer S, Patent GJ (1972) Comparative and embryological aspects of the pancreatic islets. In: Steiner DF, Freinkel N (eds) The endocrine pancreas. Williams & Wilkins, Baltimore (Handbook of physiology, Vol 1, p 1)

Fitz-Patrick D, Patel YC (1981) Antibodies to insulin, pancreatic polypeptide, glucagon, and somatostatin in insulin-treated diabetics. J Clin Endocrinol Metab 52:948–952

Foà PP, Dunbar JC, Klein SP, Levy SH, Malik MA (1980) Reactive hypoglycemia in alpha-cell ('pancreatic') glucagon deficiency in the adult. JAMA 244:2281–2285

Fujita T, Kobayashi S (1974) The cells and hormones of the GEP endocrine system: the current of studies. In: Fujita T (ed) Gastro-entero-pancreatic endocrine system. Williams & Wilkins, Baltimore, p 1

Fujita T, Watanabe Y (1974) The effects of islet hormones upon the exocrine pancreas. In: Fujita T (ed) Gastro-entero-pancreatic endocrine system. Williams & Wilkins, Baltimore, p 164

Gerich JE, Schneider V, Dippe SE, Langlois M, Noacco C, Karam JH, Forsham PH (1974) Characterization of the glucagon response to hypoglycemia in man. J Clin Endocrinol Metab 38:77–82
Gerich P, Cryer P, Rizza R (1980) Hormonal mechanisms in acute glucose counter-regulation: the relative roles of glucagon, epinephrine, growth hormone and cortisol. Metabolism 29:1164–1175
Giugliano D, Luyckx A, Binder D, Lefèbvre P (1979) Comparative effects of metformin and indanorex in the treatment of reactive hypoglycemia. Int J Clin Pharmacol 17:76–81
Gleeson MH, Bloom SR, Polak JM, Henry K, Dowling RH (1971) Endocrine tumour in kidney affecting small bowel structure, motility and absorptive function. Gut 12:773–777
Gold H, Spector S, Samaan NA, Pearson OH (1968) Effect of growth hormone on carbohydrate metabolism in hypopituitary dwarfs. Metabolism 17:74–83
Goldfine ID, Kirsteins L, Lawrence AM (1972) Excessive glucagon responses to arginine in active acromegaly. Horm Metab Res 4:97–100
Gotlin RW, Silver HK (1970) Neonatal hypoglycemia, hyperinsulinism and absence of pancreatic alpha cells. Lancet 1:1346
Hamaji M (1979) Pancreatic $\alpha$- and $\beta$-cell function in pheochromocytoma. J Clin Endocrinol Metab 49:322–325
Henderson JR (1969) Why are the islets of Langerhans? Lancet 2:469–470
Hökfelt B, Hansson BG, Heding LG, Nilsson KO (1978) Effect of insulin-induced hypoglycemia on the blood levels of catecholamines, glucagon, growth hormone, cortisol, C-peptide and proinsulin before and during medication with the cardioselective beta-receptor blocking agent metoprolol in man. Acta Endocrinol (Copenh) 87:659–667
Imura H, Ikeda M, Seino Y (1977) Significance of the measurement of plasma insulin and glucagon in the diagnosis of diabetes mellitus. Jpn J Med 16:55–56
Kannan V, Nabarro JDN, Cotton PB (1979) Glucagon secretion in chronic pancreatitis. Horm Res 11:203–212
Kanno T, Saito A (1976) The potentiating influences of insulin on pancreozymin-induced hyperpolarization and amylase release in the pancreatic acinar cell. J Physiol (Lond) 261:505–521
Kollee LA, Monnens LA, Cecjka V, Wilms RM (1978) Persistent neonatal hypoglycemia due to glucagon deficiency. Arch Dis Child 53:422–424
Korc M, Sankaran H, Wong KY, Williams JA, Goldfine ID (1978) Insulin receptors in isolated mouse pancreatic acini. Biochem Biophys Res Commun 84:293–299
Kracht J (1953) Wirkung von Wachstumhormon auf die Langerhans'schen Inseln des Ratten Pankreas. Naturwissenschaften 40:607–608
Kronborg O, Bulow S, Joergensen PM, Svendsen LB (1980) A randomized double-blind trial of glucagon in treatment of first attack of severe acute pancreatitis without associated biliary disease. Am J Gastroenterol 73:423–425
Lawrence AM (1966a) Radioimmunoassayable glucagon levels in man: effects of starvation, hypoglycemia, and glucose administration. Proc Natl Acad Sci USA 55:316–320
Lawrence AM (1966b) Failure of normal glucagon response in the dumping syndrome. Proc annual meeting endocrine society, June, 1966
Lawrence AM (1972) Pancreatic alpha-cell function in miscellaneous clinical disorders. In: Lefèbvre PJ, Unger RH (eds) Glucagon, molecular physiology, clinical and therapeutic implications. Pergamon, New York, p 259
Lawrence AM, Abraira C (1977) Glucagon and its clinical significance. Lab Med 8:21–26
Lawrence AM, Abraira C (1980) Glucagon and diabetes mellitus. In: Podolsky S (ed) Clinical diabetes: modern management. Appleton-Century-Crofts, New York, p 589
Lawrence AM, Dorsch T (1980) The glucagonoma syndrome. In: Podolsky S, Viswanathan M (eds) Secondary diabetes: the spectrum of the diabetic syndrome. Raven, New York, p 287
Lawrence AM, Hagen TC, Kirsteins L (1972) Propranolol hypoglycemia and impaired glucagon secretion. Clin Res 20:770
Lawrence AM, Prinz RA, Paloyan E, Kokal WA (1979) Glucagon and insulin in pancreatic exocrine secretions. Lancet 1:1354

Lefèbvre P, Luyckx AS, Lecomte MJ (1976) Studies on the pathogenesis of reactive hypoglycemia: role of insulin and glucagon. Horm Metab Res 6 (Suppl.):91–98
Like AA, Orci L (1972) Embryogenesis of the human pancreatic islets: a light and electron microscopic study. Diabetes 21:511–534
Liljenquist JE, Mueller GL, Cherrington AD, Keller U, Chiasson JL, Perry JM, Lacy WW, Rabinowitz D (1977) Evidence for an important role of glucagon in the regulation of hepatic glucose production in normal man. J Clin Invest 59:369–374
Lund PK, Goodman RH, Jacobs JW, Habener JF (1980) Glucagon precursors identified by immunoprecipitation of products of cell-free translation of messenger RNA. Diabetes 29:583–586
Maher TD, Tanenberg RJ, Greenberg BZ, Hoffman JE, Dol RP, Goetz FC (1977) Lack of glucagon response to hypoglycemia in diabetic autonomic neuropathy. Diabetes 26:196–200
Malaisse-Lagae F, Ravazzola M, Robberecht P, Vandermeers A, Malaisse WJ, Orci L (1975) Exocrine pancreas: evidence for topographic partition of secretory function. Science 90:795–797
Mallinson CN, Bloom SR, Warin AP, Salmon PR, Cox B (1974) A glucagonoma syndrome. Lancet 2:1
Marco J, Baroja IM, Diaz-Fierros ML (1972) Relationship between insulin and gut glucagon immunoreactivity (GLI) in normal and gastrectomized subjects. J Clin Endocrinol Metab 34:188–191
McGavran MH, Unger RH, Recant L, Polk HC, Kilo C, Levin ME (1966) A glucagon-secreting alpha cell carcinoma of the pancreas. N Engl J Med 274:1408–1411
Medical Research Council Multicentre Trial (1980) Morbidity of acute pancreatitis: the effect of aprotinin and glucagon. Gut 21:334–339
Merimee TJ, Fineberg SE, McKusick VA, Hall J (1970) Diabetes mellitus and sexual ateliotic dwarfism: a comparative study. J Clin Invest 49:1096–1102
Okuno G, Takenaka H (1977) Pancreatic A-cell function in a patient with Addison's disease. In: Foà PP, Bajaj JS, Foà NL (eds) Glucagon: its role in physiology and clinical medicine. Springer, New York Heidelberg Berlin, p 777
Paloyan E (1967) Recent developments in the early diagnosis of hyperparathyroidism. Surg Clin North Am 47:61–69
Paloyan E, Lawrence AM, Straus FH II, Paloyan D, Harper PV, Cummings D (1966) Alpha cell hyperplasia in calcific pancreatitis associated with hyperparathyroidism. JAMA 200:757–761
Paloyan E, Paloyan D, Harper PV (1967 a) Glucagon-induced hypocalcemia. Metabolism 16:35–39
Paloyan E, Paloyan D, Harper PV (1967 b) The role of glucagon hypersecretion in the relationship of pancreatitis and hyperparathyroidism. Surgery 62:167–173
Patzelt C, Tager HS, Carroll RJ, Steiner DF (1979) Identification and processing of proglucagon in pancreatic islets. Nature 282:260–266
Pek S, Fajans SS, Floyd JC, Knopf RF, Prchkow VK, Sherman RM, Weisman PN, Conn JW (1971) Effects upon plasma glucagon of human growth hormone and of dexamethasone in man. Clin Res 19:482
Pfeiffer FF, Raptis S, Fussganger R (1977) Gastrointestinal hormones and islet function. In: Pfeiffer EF (ed) Gastrointestinal hormones. Springer, Berlin Heidelberg New York
Pickleman JR, Ernst K, Brown S, Paloyan E (1969) Glucagon-induced hypocalcemia: effect of the thyroid gland. Surg Forum 20:85
Prinz R, Kokal W, Kirsteins L, Smith M, Lawrence AM, Paloyan E (1978) Presence of high concentrations of glucagon and insulin in pancreatic exocrine secretions. Surg Forum 29:460–461
Prinz RA, Kirsteins L, Connick E, Paloyan E, Lawrence AM (1980) Insulin and glucagon in human pancreatic exocrine fluid. Horm Metab Res 12:38–39
Prinz RA, Dorsch TR, Lawrence AM (1981) Clinical aspects of glucagon-producing islet-cell tumors. Am J Gastroenterol 76:125–131
Regan PT, Malagelada JR, Go LLW, Wolf AM (1981) A prospective study of the antisecretory and therapeutic effects of cimetidine and glucagon in human acute pancreatitis. Mayo Clin Proc 56:499–503

Rehfeld JF, Heding LG, Holst JJ (1973) Increased gut glucagon as a pathogenetic factor in reactive hypoglycemia. Lancet 1:116–118
Reynolds C, Molnar GD, Jiang N, Jones JD, Taylor WF (1973) Abnormal glucagon response to hypoglycemia in unstable diabetics. Diabetes 22:327
Robson RH (1980) Glucagon for beta-blocker poisoning. Lancet 1:1357–1358
Rothman S (1925) Über Hauterscheinungen bei bösartigen Geschwulsten Innerer Organe. Ann Dermatol Syphiligr 149:99–123
Satake K, Pairent FW, Romero F, Appert HE, Howard JM (1972) The concentration of insulin in pancreatic juice. Surg Gynecol Obstet 134:589–592
Seino Y, Goto Y, Taminato T, Ikeda M, Imura H (1974) Plasma insulin and glucagon responses in patients with thyroid dysfunction. J Clin Endocrinol Metab 38:1136–1140
Seino Y, Goto Y, Kurahachi H, Sakurai H, Ikeda M, Kadowaki S, Inoue Y, Mori K, Taminato T, Imura H (1977) Alteration of plasma glucagon response to arginine after treatment in patients with diabetes mellitus, Cushing's syndrome and hypothyroidism. Horm Metab Res 9:28–32
Shapiro H, Ludewig RM (1978) The effect of glucagon on the exocrine pancreas. Am J Gastroenterol 70:274–281
Stacpoole PW (1981) The glucagonoma syndrome: clinical features, diagnosis and treatment. Endocrinol Rev 2:347–361
Stacpoole PW, Jaspan J, Kasselberg AG, Halter SA, Polonsky K, Gluck FW, Liljenquist JE, Rabin D (1981) A familial glucagonoma syndrome. Am J Med 70:1017–1026
Tewskbury DA, Wyman AJB, Mulholland D, Dumas D (1973) Immunoreactive insulin in human duodenal aspirates. Acta Endocrinol (Copenh) [Suppl] 177:329–331
Torella R, Giugliano D, Scognamiglio G, Passariello N, Tirello A (1982) Glucagon secretion in patients with hypoparathyroidism: effect of serum calcium on glucagon release. J Clin Endocrinol Metab 54:229–232
Trimble ER, Atkinson AB, Buchanan KD, Hadden DR (1980) Plasma glucagon and insulin concentrations in acromegaly. J Clin Endocrinol Metab 51:626–631
Vance E, Stoll RW, Kitabchi AE, Williams RH, Wood FC (1969) Nesidioblastosis in familial endocrine adenomatosis. JAMA 207:1682–1689
Vidnes J, Oyasaeter S (1976) Glucagon deficiency causing severe neonatal hypoglycemia in a patient with normal insulin secretion. Pediatr Res 11:943–949
Villalpando S, Drash A (1979) Circulating glucagon antibodies in children who have insulin dependent diabetes mellitus. Diabetes 28:294–299
White NH, Skor DA, Cryer PE, Levandoski LA, Bier DM, Santiago JV (1983) Identification of type 1 diabetic patients at increased risk for hypoglycemia during intensive therapy. N Engl J Med 308:485–491

CHAPTER 46

# Glucagon and Hyperlipoproteinemias

R.P. EATON

## A. Introduction

In humans, glucagon has a potent stimulating action upon hepatic glucose production (Chaps. 14, 15 and 16), ketogenesis (Chap. 17), and lipolysis. These metabolic effects have been shown to be sensitive to the simultaneous secretion of insulin and/or the hepatic exposure to insulin which counteracts the lipolysis, ketogenesis, and gluconeogenesis. Insulin also exerts a well-characterized effect upon the liver to enhance the production of triglyceride-bearing lipoprotein (REAVEN et al. 1967; STREGA et al. 1977). In this setting, a counteractive action of glucagon has been observed, suggesting a hypotriglyceridemic response to this hormone, which could participate in the pathophysiology of endogenous hyperlipoproteinemic states (see Chap. 20).

## B. The Glucagon-Insulin Environment

It is known that treatment with pharmacologic doses of glucagon induces a reduction in triglyceride levels in normal humans as well as patients with acquired and genetic hypertriglyceridemia (AMATUZIO et al. 1962; PALOYAN and HARPER 1961; AUBRY et al. 1974). A variety of factors may modify endogenous glucagon secretion, and thus potentially influence the triglyceride "set" of a patient. Among these are exercise, fasting, dietary carbohydrate composition, obesity, and oral contraceptive hormones, as well as the coexistence of the diabetic state. In each case, in a given patient in whom a change in one of these factors may take place, a change in the relative glucagon: insulin ratio will secondarily evolve. Such a change in this hormone equilibrium may be expected to alter the relative rate of hepatic triglyceride (very low density lipoprotein, VLDL) production, and result in a new metabolic steady state of plasma triglyceride concentration. This sequence of events has been elegantly demonstrated with prolonged fasting in obese humans (STREGA et al. 1977). In this setting, a linear correlation exists between the elevation in the glucagon: insulin ratio and the reduction in hepatic VLDL production, with a resulting reduced steady-state triglyceride concentration (STREGA et al. 1977). Similarly, in the rat, the augmentation of endogenous glucagon secretion resulting from diazoxide administration has been shown to result in reduced hepatic VLDL production as the glucagon: insulin ratio increases (EATON and SCHADE 1980). Comparable evalution in humans of "physiologic" elevations in glucagon concentration have been difficult to achieve experimentally, exclusive of a counteracting rise in insulin concentration (SHERWIN et al. 1976; SCHADE and EATON 1977).

Recognizing the ability of an altered glucagon-insulin environment to influence the state of hepatic lipoprotein production, and potentially, peripheral lipoprotein removal, it has been proposed that this hormonal axis may participate in the pathophysiology of endogenous hypertriglyceridemia (EATON et al. 1974b). In humans, this clinical diagnosis includes a spectrum of interacting disease states, including obesity, type I diabetes, type II diabetes, aging, carbohydrate sensitivity, and a host of "genetic" hyperlipoproteinemic states of mixed genetic description. In such diverse company, an altered glucagon-insulin environment may take on differing metabolic significance, depending upon coexistent physiologic events such as fatty acid availability or receptor sensitivity.

## C. Studies in the Zucker Genetic Hyperlipemic Rat

The Zucker strain of obese rat represents an animal model of genetic hyperlipemia associated with obesity, hyperinsulinemia, and mild glucose intolerance (BARRY and GRAY 1969; SCHONFELD et al. 1974; SCHONFELD and PFLEGER 1971; ZUCKER 1965, 1972; EATON et al. 1976a). It has been suggested that thyroid hypofunction may be a contributing factor in the lipemia and/or obesity in these rats (YORK et al. 1972), as has been reported in human genetic hyperlipemia (HAZZARD and BIERMAN 1972). However, while induction of the hyperthyroid state will improve the cholesterol level, the hypertriglyceridemia is only minimally influenced in this animal model of type IV hyperlipemia (ENGELKEN and EATON 1981). In addition, the animals demonstrate unusual sensitivity to the hyperlipemic response to estrogen (WILSON et al. 1982) similar to that observed in human hyperlipemic states (DAVIDOFF et al. 1973).

### I. Glucagon Secretion

Impaired glucagon secretion is characteristic of this genetic animal model. Basal plasma glucagon concentration is reduced in the fed state, and unexpectedly declines during food deprivation (EATON et al. 1976b). This observation contrasts with the behavior of plasma glucagon concentration in thin littermates, which demonstrate both a higher basal concentration and a rise in hormone concentration with fasting as reported in other mammals (EATON 1973a; and reviewed in Chap. 20). Moreover, in response to a potent substrate secretagogue (arginine), the obese animals fail to demonstrate an appropriate secretion of glucagon as compared with thin littermate animals. In contrast to the apparent hormone secretory defect, physiologic suppression of glucagon release by glucose administration remains intact. Thus, the genetic hyperlipemic obese Zucker rat appears to be characterized by a hormonal abnormality of reduced glucagon secretion and/or concentration.

### II. Glucagon Regulation

Three factors, insulin, glucose, and free fatty acids (FFA), may be visualized as contributing to a physiologic suppression of normal pancreatic A-cell function.

First, insulin has been reported to suppress glucagon secretion in vitro in isolated pancreatic islets (BUCKMAN and MAWHINNEY 1973). Similarly, in diabetic humans (MÜLLER et al. 1973), insulin administration results in a reduction in endogenous glucagon levels. The markedly elevated basal insulin levels in the obese hyperlipemic rat, and the hypersecretion of insulin in response to glucose and arginine, would suggest that excess insulin could participate in a hormonal suppression of endogenous glucagon secretion. The reduction in insulin secretion with exercise training (SIMONELLI and EATON 1978 b) resulting in a relative glucagon excess and reduction in triglyceride production is consistent with this hypothesis. Second, hyperglycemia is reported to suppress glucagon secretion both in vitro in the perfused pancreas (GERICH et al. 1974a), and in vivo in normal humans (GERICH et al. 1974 b). The persistent mild hyperglycemic state of the obese hyperlipemic rat may thus potentiate the tendency for glucagon suppression. Third, the elevated concentration of plasma FFA might also exert a suppressive action on endogenous glucagon secretion (see review in Chap. 25). In both humans (GERICH et al. 1974b) and dogs (LUYCKX and LEFEBVRE 1970; MADISON et al. 1968), the increase in plasma FFA induced by infusion of triglycerides and heparin or heparin alone is associated with an immediate reduction in glucagon concentration. Similarly, in the isolated islet preparation (EDWARDS and TAYLOR 1970), FFA are reported to reduce glucagon secretion.

The possibility that the combined interaction of elevated insulin, glucose, and FFA may participate in the relative reduction in glucagon secretion is supported by the response to pharmacologic correction of these parameters. Reduction in plasma FFA, glucose, and insulin with halofenate in the Zucker obese hyperlipemic rat is reported to augment arginine-stimulated glucagon secretion (EATON et al. 1976b). In this setting, plasma triglyceride concentration is simultaneously reduced in parallel with the change in glucagon-insulin environment. Detailed study of clofibrate, a lipid-lowering drug similar in action to halofenate, has shown that enhanced peripheral clearance of triglyceride may participate in the hypolipemic response (SIMONELLI and EATON 1978 a).

A primary genetic abnormality of the pancreatic islet cells may also be present which results in reduced hormone production independent of circulating factors. In support of this possibility is the observation that in vitro perifusion of the isolated islet from the Zucker obese rat is characterized by persistent B-cell hypersecretion of insulin (SCHADE and EATON 1975d; STERN et al. 1975). A related abnormality in the A-cell with persistent reduction in glucagon secretion would be compatible with these observations, but has not been investigated. The possibility that the obesity itself may predispose to reduced glucagon secretion independent of recognized regulatory factors is suggested by some, but not all studies in human obesity (SCHADE and EATON 1974; WISE et al. 1973).

## III. Effect of Reduced Glucagon

One of the suggested consequences of decreased glucagon secretion in the Zucker rat may be hyperlipemia. Glucagon is an effective lipid-lowering hormone in all species studied, including the rat (EATON 1973b; PALOYAN and HARPER 1961). A reduction in this hyperlipemic hormonal effect might be expected to result in hyperlipemia, as has been suggested for other forms of hyperlipemia in humans and animals (EATON et al. 1974b). A similar reduction in glucagon secretion has been reported in carbohydrate-induced hyperlipemia (EATON et al. 1974a), while a resistance to the lipid-lowering actions of glucagon is observed in cobalt-induced hyperlipemia (EATON 1973c). Thus, a genetic predisposition to a limitation of glucagon secretion suggests that this hormonal alteration may represent a common abnormality shared by diet-induced, drug-induced, and familial lipemic states.

### IV. Effect of Elevated Insulin

The elevated insulin secretion may also participate in hormonally induced hyperlipemia as has been suggested from both human and rat investigations (BAGDADE et al. 1971; BECK et al. 1975; BERNSTEIN et al. 1975; BJÖRNTORP et al. 1971; EATON and KIPNIS 1969; EATON and NYE 1973; OLEFSKY et al. 1974; REAVEN et al. 1967; ROBERTSON et al. 1973). Accelerated triglyceride secretion is reported in association with excess insulin, occurring both in the presence and absence of obesity (STREGA et al. 1977; BAGDADE et al. 1971; EATON and NYE 1973; OLEFSKY et al. 1974; ROBERTSON et al. 1973). In the experimental induction of obesity in the desert sand rat, *Psammomys obesus* (ROBERTSON et al. 1973), increased rates of triglyceride secretion are reported in correlation with progressive elevation in circulating insulin levels. Reduction in insulin secretion by diazoxide administration results in a rapid reduction in circulating lipid (EATON and NYE 1973). Moreover, insulin deficiency induced with alloxan in the rat (EATON and KIPNIS 1969) results in a marked depression of hepatic lipoprotein and secretion. Thus, the combined hormonal abnormality of excess insulin and reduced glucagon in the genetic hyperlipemic rat may participate in the endogenous hyperlipemia characteristic of this animal.

## D. Studies in Human Hyperlipemia

### I. Immunoassayable Glucagon

While there is no doubt that glucagon acts to reduce hepatic triglyceride production and to lower plasma triglyceride concentration in normal human physiology, the potential role in clinical hyperlipemia has been difficult to quantitate. The initial report of radioimmunoassay levels of plasma glucagon by MARKS et al. (1971), demonstrated elevated basal hormone levels in patients with type IV and type V hyperlipemia. This observation has been confirmed by numerous investigators, demonstrating not only elevated basal levels of plasma glucagon (MARKS et al. 1971; EATON and SCHADE 1973; TIENGO et al. 1975, 1977; RATZMAN et al. 1980), but also an inappropriately high response to arginine stimulation (EATON and SCHADE 1973; TIENGO et al. 1975, 1977), and a failure to suppress glucagon with glucose administration (RATZMANN et al. 1980).

### II. Hypolipemic Response to Glucagon

The presence of elevated plasma glucagon in hyperlipemic subjects, seems inappropriate in view of lipid-lowering action of the hormone. Resistance to this action of glucagon was thus postulated (EATON and SCHADE 1973), and studies were carried out to determine the response to exogenously administered glucagon. While a physiologic dose-response pharmacokinetic evaluation has not been reported, studies with pharmacologic doses of glucagon given both intravenously (ELKES and HAMBLEY 1976; ANDEL et al. 1980; NOSADINI et al. 1978) and intramuscularly (AUBRY et al. 1974; TOMMASO et al. 1977) have demonstrated supranormal reductions in plasma triglyceride concentrations compared with results in control patients. Studies were also performed in patients with "primary" hypertriglyceridemia and compared with patients with "secondary" hypertriglyceridemia in association with a diabetic state or with obesity (RATZMANN et al. 1980; NOSADINI et al. 1978; TOMMASO et al. 1977). No conclusive differences in response could be demonstrated, but a clear hypotriglyceridemic response to glucagon was readily observed in all hyperlipemic subjects. These studies represented heterogeneous lipemic populations, with large differences in the absolute basal triglyceride concentration. Much confusion exists as to whether the relative reduction or absolute fall in plasma lipid concentration represents the most appropriate method of comparison. Nevertheless, a normal to exaggerated lipid-lowering response to exogenous glucagon has consistently been reported in the five investigations cited.

## III. Glucagon Structure-Function Relationships

To reconcile the elevated immunoassayable plasma glucagon levels in the presence of a supranormal lipid-lowering response to exogenous glucagon, it may be necessary to exclude the possibility of biologically inactive hormone secretion in hyperlipemic patients. While the radioimmunoassay of glucagon has focused upon pancreatic extractable glucagon of molecular weight 3,500 daltons, it is known that three or four distinct molecular forms of the hormone circulate in human plasma (see Chap. 11). In normal subjects, most of the immunoreactive glucagon is in a high molecular weight form (greater than 15,000 daltons) (KUKU et al. 1976; VALVERDE et al. 1976; WEIR et al. 1975). However with A-cell stimulation (arginine infusion), it is the 3,500 daltons form or "true pancreatic glucagon" which changes most dramatically (VALVERDE et al. 1976). It has also been shown that in decompensated diabetic ketoacidosis, the predominant circulating form of glucagon is the true hormone of a molecular weight 3,500 daltons. A different pattern of molecular forms of glucagon is recognized in patients with chronic renal failure (BILBREY et al. 1974; KUKU et al. 1976). In this state, an intermediate molecular weight form of glucagon is circulating (9,000 daltons) with lesser amounts of molecular weight 3,500 daltons and still less of a high molecular weight form of over 15,000 daltons. The molecular weight of 9,000 daltons in the plasma of chronic renal failure patients is similar to a component isolated from the pancreas (RIGOPOULOU et al. 1970) consistent with the molecular weight of proglucagon (TAGER and STEINER 1973). Unfortunately, detailed investigation of the biologic activity of these differing molecular forms of glucagon has not been reported with regard to the lipid-lowering action. Thus, studies in hyperlipemic patients with elevated immunoreactive glucagon need to be examined for the presence of a molecular species of glucagon which may have lost hypolipemic actions.

### IV. Response to Therapy

Consistent with the probability of participation by glucagon in hyperlipemic patients has been the response to therapy. In such subjects (TIENGO et al. 1975) as in nonlipemic subjects (EATON et al. 1977), treatment with clofibrate resulted in a significant elevation in the glucagon: insulin bihormonal ratio, in concert with a reduction in plasma triglyceride concentration. This altered hormonal environment of reduced insulin relative to augmented glucagon, is consistent with a critical role for both hormones in lipid regulation, providing some biologically active glucagon can be secreted. A similar correlation was reported in patients with lipemia induced by diets with different carbohydrate content (FUJITA et al. 1974). While this hormonal response to clofibrate administration may be implicated in the triglyceride-lowering actions of the drug, it is recognized that the agent is also reported to improve oral glucose tolerance (EATON and SCHADE 1974; FENDERSON et al. 1974), which may independently participate in reduced serum lipid concentrations. It can be concluded that a putative role for a loss of glucagon activity in hyperlipidemic states remains attractive, but at present still unsubstantiated.

## E. The Mechanism of the Hypolipemic Response to Glucagon

The effect of glucagon on plasma triglyceride concentration remains a controversial issue in contrast to glucagon's accepted role in enhancing plasma nonesterified fatty acid and ketone body concentration. A recent in-depth review of the current controversy has been published (EATON 1977). HEIMBERG et al. (1969a, b) examined the pharmacologic effects of glucagon on the hepatic secretion of triglycerides. In these studies, it was observed that this hormone suppressed the hepatic secretion of triglyceride at all concentrations of nonesterified fatty acids within the physiologic range. These studies have since been extended to physiologic glucagon concentrations utilizing the isolated hepatocyte and have confirmed glucagon's ability to decrease hepatic triglyceride formation (CHRISTIANSEN 1977; WITTERS and TRASKO 1978). In these studies, WITTERS and TRASKO demonstrated that in the fed animal, physiologic concentrations of glucagon decreased hepatic triglyceride synthesis by 20%–50% (WITTERS and TRASKO 1978). Of interest was their observation that this activity was only observed in the nonfasting state when the liver's rate of triglyceride synthesis was inherently high. Glucagon's hypolipemic activity in vivo has been difficult to demonstrate, primarily because of the difficult of controlling all of the variables present in vivo that alter hepatic triglyceride formation (EATON 1977; SHERWIN et al. 1976). These variables include glucagon's alteration of plasma nonesterified fatty acid concentration (SCHADE and EATON 1975a, b), insulin (SCHADE and EATON 1975c; WOODSIDE and HEIMBERG 1976), growth hormone (VANDERSCHUEREN-LODEWEYCKX et al. 1974), and catecholamines (LEFEBVRE and LUYCKX 1972). This difficulty may account for a lack of hypolipemic response observed in some studies (SHERWIN et al. 1976).

In patients with glucagon-secreting tumors, hepatic catheterization studies have demonstrated that endogenous glucagon suppresses the hepatic output of triglycerides (BODEN et al. 1978). Other studies utilizing physiologic infusions of glucagon into both normal and diabetic humans have shown that under appropriate conditions, glucagon may induce a hypolipemic response (SCHADE and EATON 1977). The intimate mechanism or mechanisms by which glucagon may exert its inhibitory effect on triglyceride synthesis by the liver have been analyzed in detail in Chap. 20.

## References

Amatuzio DS, Grande F, Wada S (1962) Effect of glucagon on the serum lipids in essential hyperlipemia and in hypercholesterolemia. Metabolism 11:1240–1249

Andel M, Brodan V, Kucerova L, Grafnetter D, Kuhn E, Veselkova A (1980) Cholesterol and triglyceride plasma levels in patients with primary hyperlipoproteinemia. Cas Lek Cesk 119:606–621

Aubry F, Marcel YL, Davignon L (1974) Effects of glucagon on plasma lipids in different types of primary hyperlipoproteinemia. Metabolism 23:225–238

Bagdade JD, Biermann EL, Porte D (1971) Influence of obesity on the relationship between insulin and triglyceride levels in endogenous hypertriglyceridemia. Diabetes 20:664–666

Barry WS, Gray GA (1969) Plasma triglycerides in genetically obese rats. Metabolism 18:833–839

Beck P, Eaton RP, Arnett DM, Alsever RN (1975) Effect of contraceptive steroids on arginine-stimulated glucagon and insulin secretion in women. I. Lipid physiology. Metabolism 24:1055–1065

Bernstein RS, Grant N, Kipnis DM (1975) Hyperinsulinemia and enlarged adipocytes in patients with endogenous hyperlipoproteinemia without obesity or diabetes mellitus. Diabetes 24:207–214

Bilbrey GL, Faloona GR, White MG, Knochel JP (1974) Hyperglucagonemia of renal failure. J Clin Invest 53:841–847

Björntorp P, Gustafson A, Persson B (1971) Adipose tissue fat cell size and number in relation to metabolism in endogenous hypertriglyceridemia. Acta Med Scand 190:363–366

Boden G, Wilson RM, Owen OE (1978) Effects of chronic glucagon excess on hepatic metabolism. Diabetes 27:643–648

Buckman KD, Mawhinney WA (1973) Insulin control of glucagon release from insulin-deficient rat islets. Diabetes 22:801–803

Christiansen RZ (1977) Regulation of palmitate metabolism by carnitine and glucagon in hepatocytes isolated from fasted and carbohydrate-refed rats. Biochim Biophys Acta 488:249–262

Davidoff F, Tisbler S, Rosoff C (1973) Marked hyperlipidemia and pancreatitis associated with oral contraceptive therapy. N Engl J Med 289:552–555

Eaton RP (1973 a) Effect of clofibrate on arginine-induced insulin and glucagon secretion. Metabolism 22:763–767

Eaton RP (1973 b) Hypolipemic action of glucagon in experimental endogenous lipemia in the rat. J Lipid Res 14:312–318

Eaton RP (1973 c) Glucagon secretion and activity in the cobalt-chloride treated rat. Am J Physiol 225:67–73

Eaton RP (1977) Glucagon and lipoprotein regulation in man. In: Foà PP, Bajaj JS, Foà NL (eds) Glucagon: its role in physiology and clinical medicine. Springer, Berlin Heidelberg New York, p 533

Eaton RP, Kipnis DM (1969) Effect of glucose feeding on lipoprotein synthesis in the rat. Am J Physiol 217:1153–1159

Eaton RP, Nye WHR (1973) The relationship between insulin secretion and triglyceride concentration in endogenous lipemia. J Lab Clin Med 81:682–695

Eaton RP, Schade DS (1973) Hypothesis: glucagon resistance: an hormonal basis for endogenous hyperlipemia. Lancet 2:973–974

Eaton RP, Schade DS (1974) The effect of clofibrate on arginine stimulated glucagon and insulin secretion in man. Metabolism 23:445–454

Eaton RP, Schade DS (1980) The effect of diazoxide-induced hormonal secretion on plasma triglyceride concentration in the rat. Diabetologia 18:301–306

Eaton RP, Kipnis DM, Karl I, Eisenstein AB (1974a) Effects of glucose feeding upon insulin and glucagon secretion and hepatic gluconeogenesis in the rat. Am J Physiol 227:101–104
Eaton RP, Schade DS, Conway M (1974b) Hypothesis: decreased glucagon activity: a mechanism for both genetic and acquired endogenous hyperlipemia. Lancet 2:1545
Eaton RP, Conway M, Schade DS (1976a) Endogenous glucagon regulation in genetically hyperlipemic obese rats. Am J Physiol 230:1336–1341
Eaton RP, Oase R, Schade DS (1976b) Altered insulin and glucagon secretion in treated genetic hyperlipemia: a mechanism of therapy. Metabolism 25:245–249
Eaton RP, Schade DS, Lueker R (1977) Clofibrate-induced changes in glucagon and insulin secretion in patients with angiographically documented coronary artery disease. Am J Clin Nutr 30:2068–2077
Edwards JC, Taylor KW (1970) Fatty acids and the release of glucagon from isolated guinea-pig islets of Langerhans incubated in vitro. Biochim Biophys Acta 215:310–315
Elkes RS, Hambley J (1976) Glucagon resistance as a cause of hypertriglyceridemia. Lancet 2:18–20
Engelken SF, Eaton RP (1981) The effects of altered thyroid status on lipid metabolism in the genetic hyperlipemic Zucker rat. Atherosclerosis 38:177–188
Fenderson RW, Sekowski I, Mohan NC, Duetsch S, Benjamin Fred, Samuel P (1974) Effect of clofibrate on plasma clucose and serum immunoreactive insulin in patients with hyperlipoproteinemia. Am J Clin Nutr 27:22–28
Fujita Y, Gotto A, Unger RH (1974) Insulin, glucagon, and I/G ratio in carbohydrate-induced hyperlipemia. Diabetes 23:372
Gerich J, Charles A, Grodsky GM (1974a) Characterization of the effects of arginine and glucose on glucagon and insulin release from the perfused rat pancreas. J Clin Invest 54:833–841
Gerich JE, Langlois M, Schneider V, Karam JH, Noacco C (1974b) Effects of alterations of plasma free fatty acid levels in pancreatic glucagon secretion in man. J Clin Invest 53:1283–1289
Hazzard WR, Bierman EL (1972) Aggravation of broad-beta disease (type 3 hyperlipoproteiemia) by hypothyroidism. Arch Intern Med 130:822–828
Heimberg M, Van Harken DR, Weinstein I (1969a) Regulatory factors in ketogenesis and in the metabolism of triglycerides by liver. Adv Exp Biol Med 4:185–200
Heimberg M, Weinstein I, Kohout M (1969b) The effects of glucagon, dibutyryl cyclic and adenosine 3′:5′-monophosphate and concentration of free fatty acid on hepatic lipid metabolism. J Biol Chem 244:5131–5139
Kuku SF, Zeidler DS, Emmanouel DS, Katz AI, Rubenstein AH (1976) Heterogeneity of plasma glucagon: patterns in patients with chronic renal failure and diabetes. J Clin Endocrinol Metab 42:173–176
Lefèbvre P, Luyckx A (1972) Glucagon and catecholamines. In: Lefèbvre PJ, Unger RH (eds) Glucagon. Molecular physiology, clinical and therapeutic implications. Pergamon of Oxford, New York, pp 175–180
Luyckx AS, Lefèbvre PJ (1970) Arguments for a regulation of pancreatic glucagon secretion by circulating plasma free fatty acids. Proc Soc Exp Biol Med 133:524–528
Madison LL, Seyffert WA, Unger RH, Barker B (1978) Effect of plasma free fatty acids on plasma glucagon and serum insulin concentration. Metabolism 17:301–304
Marks V, Frizel S, Twycross RG, Buckman KD (1971) Effect of B-pyridylcarbinol on glucose tolerance, plasma glucagon, insulin and growth hormone in man. In: Metabolic effects of nicotinic acid and derivatives. Gey, Gey KF, Carlson LA (eds) Huber, Berne, pp 961–976
Müller WA, Faloona GR, Unger RH (1973) Hyperglucagonemia in diabetic ketoacidosis. Am J Med 54:52–57
Nosadini N, Solda G, Biasi F, Tiengo A (1978) Metabolic effects of glucagon in endogenous hypertriglyceridemia. Acta Diabetol Lat 15:251
Olefsky JM, Farquhar JW, Reaven GM (1974) Reappraisal of the role of insulin in hypertriglyceridemia. Am J Med 57:551–555

Paloyan E, Harper PV Jr (1961) Glucagon as a regulation factor of plasma lipids. Metabolism 10:315–323
Ratzmann KP, Schulz B, Witt S, Ziegler M, Heinke P (1980) Pancreatic glucagon response to glucose in hyperlipoproteinaemia with and without abnormalities in carbohydrate metabolism. Acta Endocrinol (Copenh) 94:76–83
Reaven GM, Lerner RL, Stern MP, Farquhar JW (1967) Role of insulin in endogenous hypertriglyceridemia. J Clin Invest 46:1744–1756
Rigopoulou D, Valverde I, Marco J, Faloona G, Unger RH (1970) Large glucagon immunoreactivity (LGI) in extracts of pancreas. J Biol Chem 245:496–501
Robertson RP, Gavareski DJ, Henderson JD, Porte D Jr, Bierman EL (1973) Accelerated triglyceride secretion: a metabolic consequence of obesity. J Clin Invest 52:1620–1626
Schade DS, Eaton RP (1974) The role of insulin and glucagon in obesity. Diabetes 23:657–661
Schade DS, Eaton RP (1975a) Modulation of fatty acid metabolism by glucagon in man. I. Effects in normal subjects. Diabetes 24:502–509
Schade DS, Eaton RP (1975b) Modulation of fatty acid metabolism by glucagon in man. II. Effects in insulin-deficient diabetics. Diabetes 24:510–515
Schade DS, Eaton RP (1975c) The contribution of endogenous insulin secretion to the ketogenic response to glucagon in man. Diabetologia 11:555–559
Schade DS, Eaton RP (1975d) Insulin secretion by perifused islets from the obese Zucker rat. Proc Soc Exp Biol Med 149:311–314
Schade DS, Eaton RP (1977) The effect of short-term physiological elevations of plasma glucagon concentration on plasma triglyceride concentration in normal and diabetic man. Horm Metab Res 9:253–257
Schonfeld G, Pfleger B (1971) Overproduction of very low density lipoproteins by livers of genetically obese rats. Am J Physiol 220:1178–1181
Schonfeld G, Felski C, Howald MA (1974) Characterization of the plasma lipoproteins of the genetically obese hyperlipoproteinemic Zucker fatty rat. J Lipid Res 15:457–465
Sherwin RS, Fisher M, Hendler R, Felig P (1976) Hyperglucagonemia and blood glucose regulation in normal, obese and diabetic subjects. N Engl J Med 294:455–461
Simonelli C, Eaton RP (1978a) Effect of clofibrate on in-vitro triglyceride production and clearance in genetically hyperlipidemic rats. Atherosclerosis 29:269–275
Simonelli C, Eaton RP (1978b) Reduced triglyceride secretion: a metabolic consequence of chronic exercise. Am J Physiol 243:E111–E227
Stern JS, Johnson PR, Batchelor BR, Zucker LM, Hirsch J (1975) Pancreatic insulin release and peripheral tissue resistance in Zucker obese rats fed high and low carbohydrate diets. Am J Physiol 228:543–548
Strega DA, Marliss EB, Steiner G (1977) The effects of prolonged fasting on plasma triglyceride kinetics in man. Metabolism 26:505–516
Tager HS, Steiner DF (1973) Isolation of a glucagon-containing peptide: primary structure of a possible fragment of proglucagon. Proc Natl Acad Sci USA 70:2321–2325
Tiengo A, Muggeo M, Assan R, Fedele D, Crepaldi G (1975) Glucagon secretion in primary endogenous hypertriglyceridemia before and after clofibrate treatment. Metabolism 24:901–914
Tiengo A, Nosadini R, Garotti MC, Fedele D, Muggeo M, Crepaldi G (1977) Glucagon and hypertriglyceridemia. In: Foà PP, Bajaj JS, Foà ML (eds) Glucagon: its role in physiology and clinical medicine. Springer, Berlin Heidelberg New York, p 735
Tommaso G, Arrigo F, Trifiri A, Consolo F (1977) Hyperlipemia and carbohydrate metabolism: effects of glucagon in primary and secondary hypertriglyceridemia. Boll Soc Ital Cardiol 22:363–371
Valverde I, Lemon HM, Kessinger A, Unger RH (1976) Distribution of plasma glucagon immunoreactivity in a patient with suspected glucagonoma. J Clin Endocrinol Metab 42:804–808
Vanderschueren-Lodeweyckx M, Wolter R, Malvaux P, Eggermont E, Eeckels R (1974) The glucagon stimulation test: effect on plasma growth hormone and on immunoreactive insulin, cortisol, and glucose in children. J Pediatr 85:182–187

Weir GC, Knowlton SD, Martin DB (1975) High molecular weight glucagon-like immunoreactivity in plasma. J Clin Endocrinol Metab 40:296–302
Wilson JN, Wilson SP, Eaton RP (1982) Influence of genetic hyperlipemia in the Zucker rat upon the lipemic response to graded estradiol exposure Atherosclerosis 41:99–114
Wise JK, Hendler R, Felig P (1973) Evaluation of alpha-cell function by infusion of alanine in normal diabetic and obese subjects. N Engl J Med 288:487–490
Witters LA, Trasko CS (1978) Hormonal regulation of hepatic ketogenesis and triglyceride synthesis. Diabetes 27:331
Woodside WF, Heimberg M (1976) Effects of anti-insulin serum, insulin and glucose on output of triglycerides and on ketogenesis by the perfused rat liver. J Biol Chem 251:12–23
York DA, Hershman JM, Utiger RD, Bray GA (1972) Thyrotropin secretion in genetically obese rats. Endocrinology 90:67–72
Zucker LM (1965) Hereditary obesity in the rat associated with hyperlipemia. Ann NY Acad Sci 131:447–458
Zucker LM (1972) Fat mobilization in vitro and in vivo in the genetically obese Zucker rat "fatty". J Lipid Res 13:234–243

CHAPTER 47

# Glucagon and Renal Insufficiency

J. B. Jaspan, K. S. Polonsky, and A. H. Rubenstein

## A. Introduction

Alterations in the endocrine environment which are prominent features of the uremic syndrome are associated with a variety of hormonal and metabolic disturbances (Feldman and Singer 1974; Wilson 1971; Emmanouel et al. 1980). In some instances, these derangements persist despite effective dialysis and adequate biochemical control, improving only with successful kidney transplantation, implicating a nonfunctioning kidney rather than the uremic environment in the pathogenesis of the metabolic disturbances. However, in other cases striking improvements occur following dialysis, suggesting that the disturbed metabolic environment of uremia and not loss of a particular renal action is responsible. In either case, the kidney clearly plays a central role in determining the prevailing endocrine milieu. There are a number of possible mechanisms by which the kidney may be responsible for disturbances of the endocrine and metabolic status. These include: (a) producing a hormone (erythropoeitin); (b) activating a prohormone (1,25-dihydroxyvitamin D); (c) stimulating release of a hormone (renin → angiotensin → aldosterone); (d) subserving an important catabolic role in the metabolic disposal of hormones (glucagon, insulin, parathyroid hormone, calcitonin, prolactin, and a number of anterior pituitary and gastrointestinal hormones); and (e) uremia-associated interference with the extrarenal degradation of hormones. In addition, the kidney is an organ of intermediary metabolism by virtue of its role in gluconeogenesis.

The endocrine disorders of uremia may therefore result from altered concentrations of active hormones presented to the target organs, from abnormal responses of target organs to appropriate hormone levels, or a combination of these two factors. This general classification of disorders is particularly relevant to the pathogenesis of derangements in carbohydrate metabolism that are associated with uremia. With respect to glucagon, the kidney is a major site of metabolic degradation of the peptide, and impairment of kidney function is associated with disturbances in the kinetics of glucagon metabolism, which may contribute to the metabolic abnormalities associated with uremia. Altered responses of target tissues to the action of glucagon have also been implicated in the disordered carbohydrate metabolism of renal failure.

In this chapter, the relationship of glucagon and renal insufficiency will be considered in three parts:

1. The effects of renal failure on circulating glucagon levels and molecular profiles

2. The effects of renal failure on glucagon secretion, biologic action, and receptor binding
3. The contribution of abnormalities in glucagon levels and action to the metabolic disturbances of uremia.

## B. The Effects of Renal Failure on Circulating Glucagon Levels and Molecular Profiles

In the period 1974–1976, plasma glucagon disturbances in uremia were described by a number of investigators (BILBREY et al. 1974, 1975; SHERWIN et al. 1976 a; LEFÈBVRE and LUYCKX 1975; KATSILAMBROS and HEDING 1975). BILBREY et al. (1975) reported that the hyperglucagonemia of renal failure was corrected by renal transplantation, but not by hemodialysis. However, these investigators showed that the elevated immunoreactive glucagon (IRG) levels in uremic patients were neither stimulated by arginine, nor suppressed by glucose administration to the same extent as in normal subjects. MARUMO et al. (1979) confirmed the observations with respect to arginine.

The explanation for these findings was provided by chromatographic analysis of plasma. As reviewed in Chap. 11, a number of groups have reported that IRG exists in several forms in the plasma of normal subjects, diabetic patients, patients with glucagonoma, and various experimental animals (WEIR et al. 1975, 1977; VALVERDE et al. 1974, 1975, 1976; MASHITER et al. 1975; KUKU et al. 1976 a, b; JASPAN et al. 1976, 1977, 1981 b; JASPAN and RUBENSTEIN 1977; FLANAGAN et al. 1980). Except for 3,485 daltons true pancreatic glucagon, little is known about the nature and biologic significance of these fractions. In this chapter, we will designate these fraction A-, B-, C-, and D-peaks, in order of their appearance in the eluates when plasma is subjected to gel filtration, according to terminology previously used by us.

The A-peak component, which coelutes with IgG (VALVERDE et al. 1974; WEIR et al. 1975; KUKU et al. 1976 a; JASPAN and RUBENSTEIN 1977) in the void volume of the column, and which is thought to have a molecular weight of approximately 150,000 daltons, is believed to result from plasma protein binding of the 3,485 daltons IRG component or to be an immunologically cross-reacting immunoglobulin, variably present in plasma. The B-peak (~ 9,000 daltons), is generally believed to be a precursor or an intermediate in the biosynthesis of glucagon (NOE and BAUER 1971, 1975; TAGER and STEINER 1973; NOE et al. 1975; HELLERSTRÖM et al. 1974; TRAKATELLIS et al. 1975). The C-peak (3,485 daltons), represents the fully active "true" pancreatic glucagon, and the fourth or D-peak (< 2,000 daltons), is probably a degradative product of glucagon in plasma, which retains some immunoreactivity (JASPAN and RUBENSTEIN 1977). In uremia, it was found that circulating IRG is also heterogeneous, with a significant proportion of the total plasma IRG level being present in the B-peak fraction (KUKU et al. 1976 a). Further analysis showed that plasma IRG in chronic renal insufficiency was composed of three distinct components: 9% of total plasma IRG existed in the A-peak component, 68% in the B-peak, and 21% in the C-peak fraction; a variable but small (< 5%) D-peak fraction was sometimes detected. By comparison, in normal subjects, these components comprised 54%, 0%, 46%, and 0% of plasma glucagon, respectively

**Table 1.** Plasma glucagon responses and distribution of glucagon immunoreactivity in the fasting state and after oral glucose administration in five patients with chronic renal failure (CRF) and four normal subjects

| Subject | Fasting | | | | After oral glucose administration | | | | |
|---|---|---|---|---|---|---|---|---|---|
| | IRG | IRG in fraction (pg/ml) | | | Time (min) | IRG | IRG in fraction (pg/ml) | | |
| | | A | B | C | | | A | B | C |
| CRF | | | | | | | | | |
| 1 | 480 | 20 | 263 | 257 | 180 | 220 | 24 | 206 | 12 |
| 2 | 740 | 52 | 499 | 267 | 180 | 520 | 40 | 442 | 85 |
| 3 | 580 | 67 | 523 | 158 | 180 | 400 | 59 | 311 | 40 |
| 4 | 440 | 155 | 179 | 55 | 180 | 400 | 158 | 190 | 0 |
| 5 | 600 | 62 | 358 | 253 | 180 | 310 | 32 | 295 | 60 |
| Mean | 568 | 71.2 | 364.4 | 198 | | 370 | 62.6 | 288.8 | 39.4 |
| Standard error | 52.4 | 22.5 | 66.3 | 40.8 | | 50.2 | 24.5 | 45.1 | 15.5 |
| Normal | | | | | | | | | |
| 1 | 180 | 60 | 0 | 128 | 90 | 135 | 76 | 0 | 0 |
| 2 | 250 | 85 | 0 | 79 | 120 | 180 | 105 | 0 | 47 |
| 3 | 120 | 8 | 0 | 59 | 90 | 80 | 5 | 0 | 18 |
| 4 | 155 | 34 | 0 | 79 | 60 | 54 | 34 | 0 | 0 |
| Mean | 176.3 | 46.8 | | 86.3 | | 112.3 | 55.0 | | 16.3 |
| Standard error | 27.5 | 16.6 | | 14.7 | | 28.2 | 22.1 | | 11.1 |

(KUKU et al. 1976b). The dominant B-peak component in chronic renal failure patients fails to increase after arginine stimulation or to decrease after oral glucose administration, despite quantitatively normal responses of the C-peak to both of these stimuli (KUKU et al. 1976b). The impaired stimulation and suppression of the B-peak component thus accounted for the abnormal responses of plasma IRG to arginine and glucose in uremic patients (BILBREY et al. 1974; MARUMO et al. 1979). These observations are illustrated in Table 1 and Fig. 1 (oral glucose) and Table 2 and Fig. 2 (intravenous arginine).

The hyperglucagonemia or renal failure has been shown by a number of workers to be due to impairments of glucagon catabolism normally effected by the healthy kidney (LEFÈBVRE et al. 1974; KUKU et al. 1976b; EMMANOUEL et al. 1976, 1978; KATZ and EMMANOUEL 1978; ARDAILLOU and PAILLARD 1980). The role of the kidney in glucagon metabolism and the contribution of the kidney to the abnormalities in circulating glucagon profiles are discussed in Chap. 41.

## C. The Effects of Renal Failure on Glucagon Secretion, Biologic Action, and Receptor Binding

### I. Glucagon Secretion

Theoretically, the hyperglucagonemia of uremia may be attributable to increased secretion in addition to the recognized decreased catabolism of glucagon in chronic

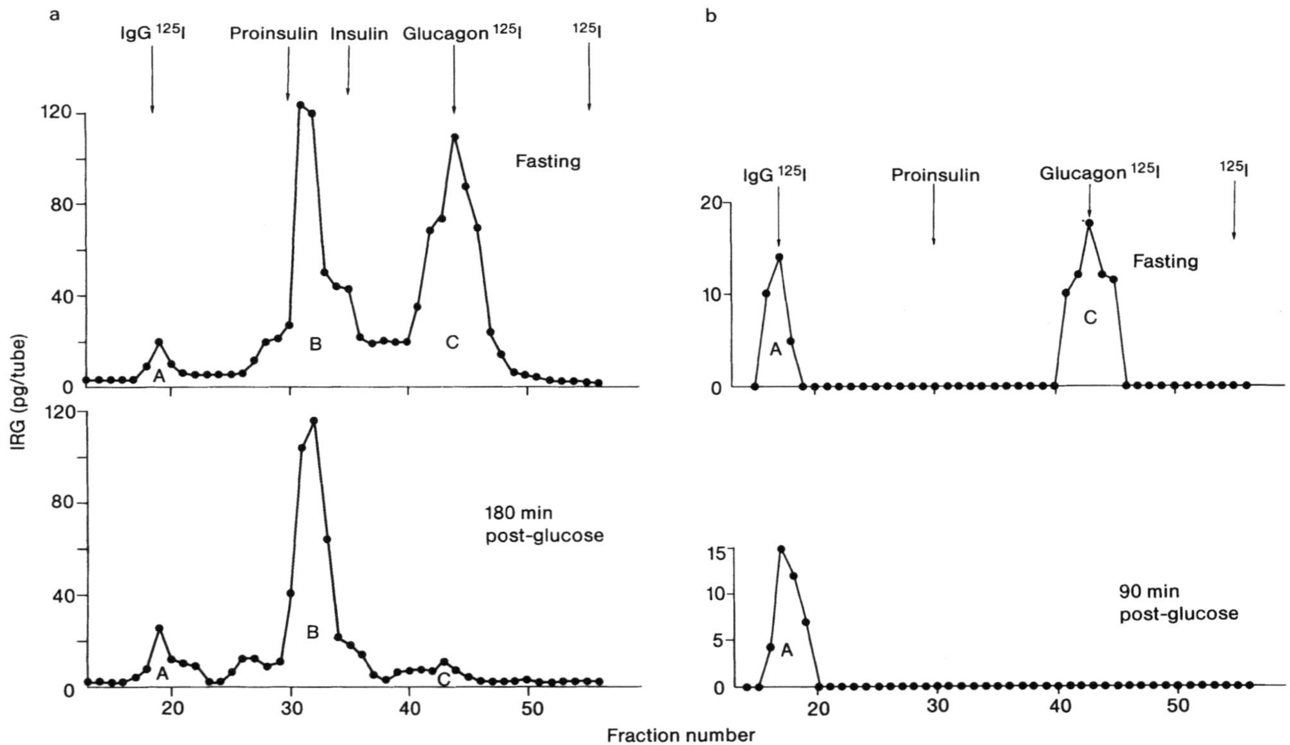


**Fig. 1 a, b.** Representative elution profiles of plasma IRG on 1 × 50-cm Biogel P-30 columns before and after administration of 100 g oral glucose in a patient with chronic renal failure (**a**) and in a healthy subject (**b**). KUKU et al. (1976b)

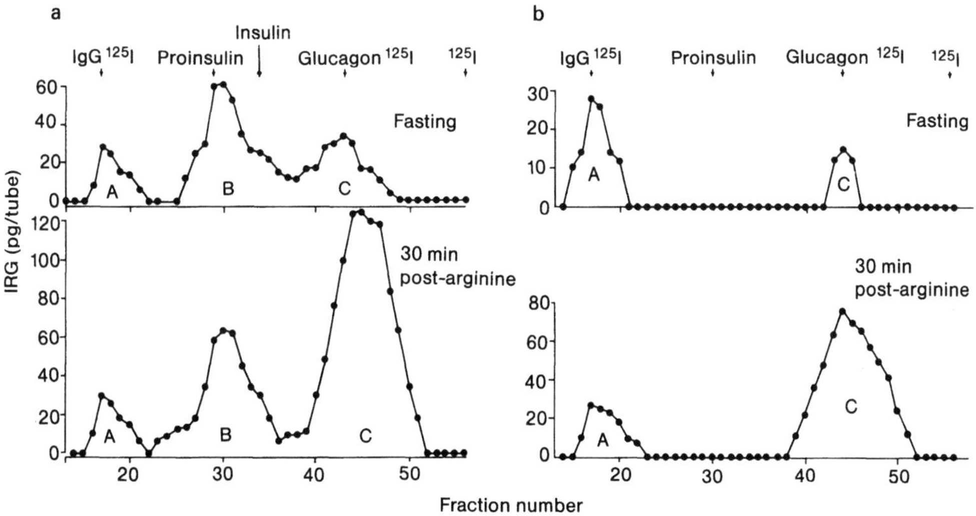


**Fig. 2 a, b.** Representative elution profile of plasma IRG on 1 × 50-cm Biogel P-30 columns before and after arginine infusion in a patient with chronic renal failure (CRF) (**a**) and in a healthy subject (**b**). The arginine was infused over 30 min at a dose of 250 mg/kg in the CRF patient and 500 mg/kg in the healthy subject. KUKU et al. (1976b)

**Table 2.** Plasma glucagon responses and distribution of glucagon immunoreactivity in the fasting state and after intravenous arginine administration in five patients with chronic renal failure (CRF) and four normal subjects

| Subject | Fasting | | | | After arginine infusion[a] | | | | |
|---|---|---|---|---|---|---|---|---|---|
| | IRG | IRG in fraction (pg/ml) | | | Time (min) | IRG | IRG in fraction (pg/ml) | | |
| | | A | B | C | | | A | B | C |
| CRF | | | | | | | | | |
| 1 | 600 | 42 | 383 | 312 | 30 | 1250 | 74 | 371 | 1233 |
| 2 | 680 | 91 | 388 | 199 | 30 | 1250 | 99 | 389 | 952 |
| 3 | 700 | 51 | 663 | 134 | 30 | 1100 | 48 | 580 | 381 |
| 4 | 500 | 173 | 180 | 56 | 30 | 540 | 135 | 206 | 237 |
| 6 | 420 | 131 | 247 | 69 | 30 | 700 | 135 | 263 | 449 |
| Mean | 580.0 | 97.6 | 372.2 | 154 | | 968.0 | 98.2 | 361.8 | 650.4 |
| Standard error | 53.3 | 24.6 | 82.9 | 47.0 | | 146.9 | 17.1 | 64.2 | 189.0 |
| Normal | | | | | | | | | |
| 4 | 95 | 45 | 0 | 20 | 30 | 460 | 63 | 0 | 294 |
| 5 | 80 | 80 | 0 | 16 | 30 | 420 | 105 | 0 | 375 |
| 6 | 100 | 56 | 0 | 54 | 45 | 800 | 25 | 0 | 604 |
| 7 | 50 | 53 | 0 | 0 | 45 | 400 | 36 | 0 | 210 |
| Mean | 81.3 | 58.5 | | 22.5 | | 520.0 | 57.3 | | 370.8 |
| Standard error | 11.3 | 7.5 | | 11.4 | | 94.2 | 17.8 | | 84.7 |

[a] 250 mg/kg was infused in the CRF patients and 500 mg/kg in the normal subjects

renal failure. However, in studies directed towards this question, SHERWIN et al. (1976a) demonstrated that basal plasma levels of glucagon in uremia were increased 3–4-fold above those in control subjects. Increments in plasma glucagon after alanine infusion in uremic patients were also 3–4-fold greater than in controls, and glucagon concentrations after oral glucose administration remained similarly elevated above control levels. Plasma glucagon increments after exogenous glucagon infusion were 2–3-fold greater in uremia. The metabolic clearance rate of glucagon in uremics was reduced by 58% compared with controls, although the basal systemic (posthepatic) delivery rate of glucagon in these patients was not different from controls. It was suggested that elevated basal plasma glucagon levels and increased plasma glucagon concentrations after alanine, glucose, or glucagon administration, were primarily the consequence of decreased glucagon catabolism in uremia rather than increased glucagon secretion (SHERWIN et al. 1976a). However, as they pointed out, the methods employed in these studies do not exclude the possibility of increased hepatic extraction of glucagon and associated hypersecretion of glucagon in uremic patients, since basal systemic delivery rates do not directly measure pancreatic secretion rates. This possible error would have added significance if the liver extracted glucagon to any substantial degree. In this regard, and as reviewed in Chap. 40, the liver has been demonstrated to be an important site of glucagon metabolism, with a high, nonsaturable capacity for glucagon extraction (JASPAN et al. 1981a). Furthermore, increased delivery of glucagon is accompanied by increased extraction such that increased glucagon secretion may not be reflected in the periphery.

## II. Biologic Activity

The biologic activity of glucagon may be increased in uremia. It was shown (SHERWIN et al. 1976a) that during glucagon infusion, increments of plasma glucose in nondialyzed uremic patients were 3–4 times greater than those observed in dialyzed patients or in normal subjects (Fig. 3). In controls, with glucagon infusion to plasma levels comparable to those found in uremic subjects, the glycemic response was still approximately twofold less than in nondialyzed uremic patients. Notably, the maximal plasma glucose response to infused glucagon showed a direct linear correlation with peak glucose concentration after oral glucose administration ($r = 0.82$, $P < 0.05$). The enhanced glucagon sensitivity in these uremic subjects was diminished by dialysis. These findings suggested that uremia is characterized by increased sensitivity to the hyperglycemic effects of physiologic increments in glucagon (SHERWIN et al. 1976a).

## III. Receptor Binding

The foregoing observations were extended by studies of glucagon and insulin binding to liver membranes of control and uremic rats (SOMAN and FELIG 1977). Specific receptor binding of glucagon was shown to be increased in uremia in parallel with increased glucagon-stimulated adenylate cyclase activity, while that of insulin was reduced. These alterations were suggested to be due to changes in receptor

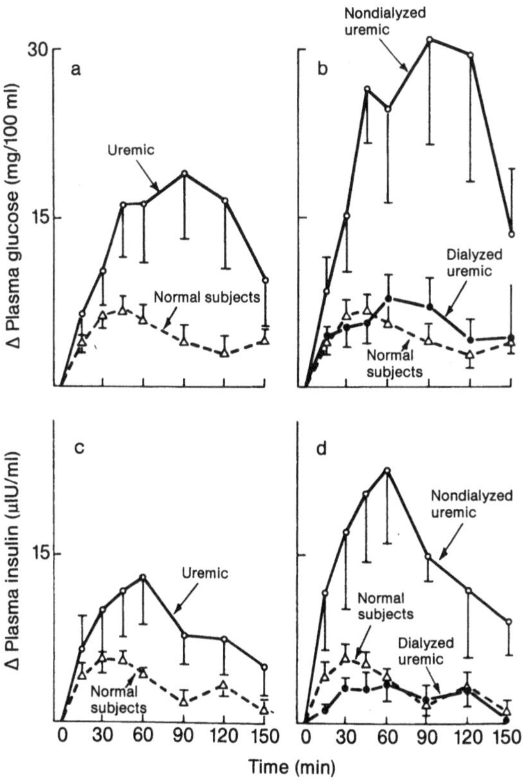


**Fig. 3a–d.** Changes in plasma glucose and insulin in response to the infusion of glucagon (3 ng $kg^{-1}$ $min^{-1}$) in normal and uremic subjects. In **a** and **c**, the responses of the dialyzed ($N=5$) and nondialyzed ($N=5$) uremics are combined, while in **b** and **d**, the two uremic groups are shown separately. The increments in plasma glucose were greater in the non-dialyzed uremics than in the dialyzed uremics or healthy controls at 45 ($P<0.005$), 60 ($P<0.05$), 90 ($P<0.05$), and 120 min ($P<0.05$). Similarly, the insulin response was greater in nondialyzed uremics than in dialyzed uremics or normal controls at 45 ($P<0.02$), 60 ($P<0.02$), 90 ($P<0.01$), and 150 min ($P<0.02$). SHERWIN et al. (1976a)

binding capacity for these hormones, rather than altered affinity of receptors. Opposite changes of glucagon and insulin binding excluded nonspecific interference with the binding of these peptide hormones to liver receptors, in the uremic state. Although it may be inferred that alterations in receptor binding of glucagon and insulin in uremia are attributable to the accumulation of a dialyzable factor, these changes may also be due to elevated plasma concentrations of each hormone. Thus, in the case of insulin, it is well recognized that exposure to high concentrations of this hormone results in loss of receptors or "downregulation" of its receptors in target cells (GAVIN et al. 1974). Although it has been suggested that increased prolactin levels induce or "upregulate" prolactin receptors (POSNER et al. 1975), there are conflicting data on the regulation of glucagon receptors by ambient glucagon concentrations (SOMAN and FELIG 1978; BHATHENA et al. 1978).

## D. The Role of Abnormalities in Glucagon Levels and Action in the Metabolic Disturbances of Uremia

It is well recognized that chronic renal failure is characterized by an increased incidence of glucose intolerance (HUTCHINGS et al. 1966; CERIETTY and ENGBRING 1967; HORTON et al. 1968; HAMPERS 1966; WESTERVELT 1969; SPITZ et al. 1970; LOWRIE et al. 1970; DEFRONZO et al. 1973; SWENSON et al. 1973). Less well known, however, is that the reverse situation may also be encountered where uremic patients are subject to hypoglycemic insult. This section will consider the role of glucagon in these diverse derangements of carbohydrate metabolism in uremia.

### I. Glucose Intolerance

#### 1. Hyperglucagonemia and Hepatic Glucose Production

Consideration must first be given to the possibility that the elevated glucagon levels associated with renal failure may result in glucose intolerance. Elevated glucagon levels in uremia may result in increased hepatic glucose production, leading to a state of insulin resistance, characterized by increased insulin levels in response to intravenous or oral glucose (SHERWIN et al. 1976a). However, this appears not to be the case, since hepatic glucose production has been reported to be normal in uremics, despite the hyperglucagonemia (DEFRONZO et al. 1973). That elevated glucagon levels are not the cause of carbohydrate intolerance in uremia is suggested by the observation that, despite marked improvement in carbohydrate tolerance and insulin sensitivity after dialysis, plasma glucagon levels are unaffected by this procedure (BILBREY et al. 1974, 1975). Furthermore, the presence and degree of hyperglucagonemia in uremia is not correlated with the degree of glucose intolerance (DEFRONZO et al. 1973). In part, these observations may be explained by the molecular heterogeneity of plasma glucagon as discussed in Sect. A of this chapter and in Chap. 11. Thus, although absolute increases in the levels of the biologically active, "true" pancreatic glucagon (C-peak) are observed in uremia, the functional importance of the hyperglucagonemia is less striking than might appear from direct assay of plasma glucagon levels which are contributed to, in major degree, by elevations in the biologically inactive A- and B-Peak components (KUKU et al. 1976b).

#### 2. Increased Tissue Sensitivity to Glucagon

a) Glycogenolysis

As discussed previously, an enhanced sensitivity of target organs to physiologic increments in glucagon has been demonstrated (SHERWIN et al. 1976a). In these studies which were performed after an overnight fast, and after glucose, glucagon, or alanine administration, the increased glucose levels reflected primarily an increased sensitivity of glycogenolysis to glucagon. These findings, together with the observations of others (CERIETTY and ENGBRING 1967; HORTON et al. 1968; HAMPERS et al. 1966; WESTERVELT 1969; SPITZ et al. 1970; SWENSON et al. 1973), suggest that the carbohydrate intolerance of uremia is the consequence of opposite changes in tissue responsiveness to glucagon and insulin. Sensitivity to the former appears to be increased, while that to the latter is blunted. In addition, it is apparent that these abnormalities in tissue responsiveness can be reversed by dialysis, a finding consistent with the marked improvement in carbohydrate tolerance in dialyzed uremics (DeFronzo et al. 1973).

The mechanisms of increased glucagon sensitivity in renal failure are unknown. Although increased sensitivity to glucagon with resultant hyperglycemia has been shown to occur in acute insulin deficiency induced by somatostatin infusion in normal humans (Alford et al. 1974) and in diabetics following insulin withdrawal (Sherwin et al. 1976b), physiologic increments of glucagon do not alter glucose tolerance in healthy subjects, in chemical diabetics, or in insulin-deficient (type-I) diabetics who are receiving adequate insulin therapy (Sherwin et al. 1976b). However, the presence of hyperinsulinemia in undialyzed uremic patients (Sherwin et al. 1976a) indicates that hypersensitivity to glucagon in uremia cannot be attributed to insulin deficiency. Thus, impairment of glucose tolerance associated with hyperinsulinemia in undialyzed uremic patients points to insulin resistance as a primary cause (Sherwin et al. 1976a), as previously suggested by others (Cerietty and Engbring 1967; Horton et al. 1968; Hampers et al. 1966; Westervelt 1969; Spitz et al. 1970; DeFronzo et al. 1973; Swenson et al. 1973). However, it has not been elucidated whether the insulin resistance of uremia is the cause or effect of the increased tissue sensitivity to glucagon already dicussed. It has been suggested that uremia may result in disturbances in hepatic glucagon receptors or postreceptor events unrelated to altered tissue responsiveness to insulin (Sherwin et al. 1976a).

### b) Gluconeogenesis

Glucagon infusion results in a decrease in the majority of plasma amino acid levels in uremic patients, but not in normal subjects (Sherwin et al. 1976b). In fact, the reduction in plasma alanine levels was greatest in uremics, suggesting the there is also increased sensitivity to the gluconeogenic effects of glucagon in renal failure. This is consistent with the observations of increased amino acid incorporation into glucose in perfused rat livers following bilateral nephrectomy (Frohlich et al. 1974). Nevertheless, despite glucagon-induced augmentation of gluconeogenesis, DeFronzo (1978), using glucose clamp techniques, has shown that the amount of glucose metabolized and the ratio of glucose metabolized to serum insulin levels (a measure of tissue sensitivity to insulin) are decreased in uremic subjects, both parameters improving with hemodialysis. Hepatic glucose production rates did not differ in controls, undialyzed, and dialyzed uremics. These observations support the contention that peripheral resistance to insulin in uremia is related to a defect in the cellular uptake or metabolism of glucose rather than the glucagon-induced augmentation of gluconeogenesis. However, the marked plasma alanine decreases in response to glucagon infusion in uremics are not corrected by dialysis, suggesting that uremia may exert differential effects on glucagon-induced glycogenolysis and gluconeogenesis (Sherwin et al. 1976a).

## II. Hypoglycemia

A number of reports have described spontaneously occurring, fasting hypoglycemia in chronic renal insufficiency (BLOCK and RUBENSTEIN 1970; WHITE and KUTZMAN 1971; RABAU et al. 1973; FRIZZEL et al. 1973; MATAVERDE et al. 1974; GARBER et al. 1974). This characteristic hypoglycemic syndrome appears to be unrelated to increased insulin levels and as such must be differentiated from other uremia-associated syndromes of hypoglycemia due to hyperinsulinism. These may occur following peritoneal dialysis in nondiabetic uremics (GREENBLATT 1972), or in uremic diabetics on insulin, where decreased or absent renal metabolism of insulin leads to hyperinsulinism (O'BRIEN and SHARPE 1967). Nor could increased insulin sensitivity be documented in these patients (FRIZZELL et al. 1973). The possible role of glucagon in this hypoglycemic syndrome of uremia must be considered. Decreases in gluconeogenesis might conceivably be implicated, owing to either glucagon deficiency or resistance to glucagon action. The former can be excluded since persistently elevated glucagon levels have been documented in these patients (GARBER et al. 1974). Although the latter possibility also seems unlikely in view of the observations of enhanced sensitivity to glucagon-induced gluconeogenesis (SHERWIN et al. 1976a), it is noteworthy that blood glucose unresponsiveness to hypoglycemia itself and to epinephrine and glucagon infusion has been reported in renal failure (COHEN 1962).

However, in an elegant study, GARBER et al. (1974) indicated that substrate limitation of gluconeogenesis is probably causative. Glucose and alanine turnover studies by tracer methodology, revealed markedly reduced glucose production rates, alanine turnover, and glucose production from alanine. Their data suggested that inadequate delivery of the prime gluconeogenic substrate, alanine, may occur in uremia and exert clinically significant rate limitation of glucose production. Inadequate glycogenolysis may also contribute to the hypoglycemia in these patients. However, in view of the findings of GARBER et al. (1974) and the delayed onset of hypoglycemia in relation to food intake, defective alanine metabolism unrelated to glucagon disturbances appears to be the dominant cause of hypoglycemia in these patients. This tendency to hypoglycemia is unaffected by dialysis. However, the mechanism of this uncommon metabolic disturbance in renal failure has not yet been established with certainty.

# E. Summary and Conclusions

1. Renal failure is accompanied by elevated plasma glucagon levels, which respond subnormally to glucose or arginine administration. This is accounted for by a markedly elevated 9,000 daltons component ($\sim 68\%$ of plasma IRG levels) which is biologically inactive and unresponsive to suppressive or stimulatory maneuvers. 3,485 daltons, "true" pancreatic glucagon is also elevated ($\sim 21\%$ of plasma IRG). An elevated void volume component and $<2,000$ daltons component are also variably present in the circulation in uremia.

2. The hyperglucagonemia and characteristic molecular distribution of plasma glucagon are the result of loss of peripheral degradative capacity rather than hypersecretion of glucagon.

3. The biologic activity of glucagon may be increased in uremia, as evidenced by increased sensitivity to the hyperglycemic effects of physiologic increments of glucagon. Increased receptor binding of glucagon, not related to insulin disturbances, has also been demonstrated.

4. The above derangements are related to both the uremic environment and the nonfunctioning kidney as evidenced by the correction of certain abnormalities by dialysis and others only by successful renal transplantation. Thus, the hyperglucogonemia and altered plasma glucagon profiles are corrected only by renal transplantation while enhanced glucagon-induced glycogenolysis and receptor binding are reversed by dialysis. On the other hand, the increased sensitivity to glucagon-induced glycogenolysis of uremia is not improved by dialysis.

5. These disturbances in plasma levels, molecular profiles, and biologic activity of glucagon in renal insufficiency contribute to the disordered carbohydrate metabolism associated with uremia. Increased tissue responsiveness to glucagon, rather than hyperglucagonemia, appears to be largely responsible for the glucagon-mediated disturbances of renal failure. On the other hand, spontaneous fasting hypoglycemia of chronic renal failure is unrelated to the glucagon abnormalities.

## References

Alford FP, Bloom SR, Nabarro JDN, Hall R, Besser GM, Coy DH, Kastin AJ, Schally AV (1974) Glucagon control of fasting glucose in man. Lancet 2:974–977

Ardaillou R, Paillard F (1980) Metabolism of polypeptide hormones by the kidney. Adv Nephrol 9:247–269

Bhathena SJ, Voyles NR, Smith S, Recant L (1978) Decreased glucagon receptors in diabetic rat hepatocytes. Evidence for regulation of glucagon receptors by hypergluca-gonemia. J Clin Invest 61:1488–1497

Bilbrey GL, Faloona GR, White MG, Knochel JP (1974) Hyperglucagonemia of renal failure. J Clin Invest 53:841–847

Bilbrey GL, Faloona GR, White MG, Atkins C, Hull AR, Knochel JP (1975) Hyperglucagonemia in uremia: reversal by renal transplantation. Ann Intern Med 82:525–528

Block MD, Rubenstein AH (1970) Spontaneous hypoglycemia in diabetes with renal insufficiency. JAMA 213:1963–1965

Cerietty JM, Engbring NH (1967) Azotemia and glucose intolerance. Ann Intern Med 66:1097–1108

Cohen BD (1962) Abnormal carbohydrate metabolism in renal disease. Blood glucose unresponsiveness to hypoglycemia, epinephrine, and glucagon. Ann Intern Med 57:204–213

DeFronzo RA (1978) Pathogenesis of glucose intolerance in uremia. Metabolism 27:1866–1880

DeFronzo RA, Andres AR, Edgar P, Walker WG (1973) Carbohydrate metabolism in uremia: a review. Medicine (Baltimore) 52:469–481

Emmanouel DS, Jaspan JB, Kuku SF, Rubenstein AH, Katz AI (1976) Pathogenesis and characterization of hyperglucagonemia in the uremic rat. J Clin Invest 58:1266–1271

Emmanouel DS, Jaspan JB, Rubenstein AH, Huen A, Fink E, Katz AI (1978) Glucagon metabolism in the rat: contribution of the kidney to the metabolic clearance rate of the hormone. J Clin Invest 62:6–13

Emmanouel DS, Lindheimer MD, Katz AI (1980) Pathogenesis of endocrine abnormalities in uremia. Endocr Rev 1:28–44

Feldman HA, Singer I (1974) Endocrinology and metabolism in uremia and dialysis: a clinical review. Medicine (Baltimore) 54:345–354

Flanagan RW, Murphy RF, Buchanan KD (1980) Circulating forms of glucagon and related peptides in normal subjects and uremic patients. Biochem Soc Trans 8:426–427

Frizzell M, Larsen PR, Field JB (1973) Spontaneous hypoglycemia associated with chronic renal failure. Diabetes 22:493–495
Frohlich J, Schölmerich J, Hoppe-Seyler G, Maier KP, Talke H, Schollmeyer P, Gerok W (1974) The effect of acute uraemia on gluconeogenesis in isolated perfused rat livers. Eur J Clin Invest 4:453–458
Garber AJ, Bier DM, Cryer PE, Pagliara AS (1974) Hypoglycemia in compensated chronic renal insufficiency. Substrate limitation of gluconeogenesis. Diabetes 23:982–986
Gavin JR, Roth J, Neville DM Jr, DeMeyts P, Buell DN (1974) Insulin-dependent regulation of insulin receptor concentrations: a direct demonstration in cell culture. Proc Natl Acad Sci USA 71:84–88
Greenblatt DJ (1972) Fatal hypoglycemia occurring after peritoneal dialysis. Br Med J 1:270–271
Hampers CL (1966) Effect of chronic renal failure and hemodialysis on carbohydrate metabolism. J Clin Invest 45:1719–1731
Hellerström C, Howell SL, Edwards JC, Anderson A, Östenson (1974) Biosynthesis of glucagon in isolated pancreatic islets of guinea pigs. Biochem J 140:13–23
Horton ES, Johson C, Lebovitz HE (1968) Carbohydrate metabolism in uremia. Ann Intern Med 68:63–74
Hutchings RH, Hegstrom RM, Scribner BH (1966) Glucose intolerance in patients on long-term intermittent dialysis. Ann Intern Med 65:275–285
Jaspan JB, Rubenstein AH (1977) Circulating glucagon: plasma profiles and metabolism in health and disease. Diabetes 26:887–904
Jaspan JB, Kuku S, Locker J, Huen A, Emmanouel DS, Katz AI, Rubenstein AH (1976) Heterogeneity of plasma glucagon in man. Metabolism 25 [Suppl 1]:1397–1401
Jaspan JB, Huen A, Gonen B, Rubenstein AH (1977) Circulating glucagon components: significance in health and disease. In: Foà PP, Bajaj JS, Foà NL (eds) Glucagon: its role in physiology and clinical medicine. Springer, Berlin Heidelberg New York, pp 93–112
Jaspan JB, Polonsky KS, Lewis M, Pensler J, Pugh W, Moossa AR, Rubenstein AH (1981 a) Hepatic metabolism of glucagon in the dog: contribution of the liver to overall metabolic disposal of glucagon. Am J Physiol 240:E233–244
Jaspan JB, Polonsky KS, Rubenstein AH (1981 b) The heterogeneity of immunoreactive glucagon in plasma: clinical implications. In: Current endocrinology. Unger RH, Orci L (eds) Elsevier/North-Holland, New York, pp 77–96
Katsilambros N, Heding LG (1975) Plasma glucagon levels in acute experimental uremia resulting from bilateral nephrectomy and bilateral ureter ligation in the rat. Int Res Commun Syst Med Sci 3:388
Katz AI, Emmanouel DS (1978) Metabolism of polypeptide hormones by the normal kidney and in uremia. Nephron 22:69–80
Kuku SF, Zeidler A, Emmanouel DS, Katz AI, Rubenstein AH, Levin NW, Tello A (1976 a) Heterogeneity of plasma glucagon: patterns in patients with chronic renal failure and diabetes. J Clin Endocrinol Metab 42:173–176
Kuku SF, Jaspan JB, Emmanouel DS, Zeidler A, Katz AI, Rubenstein AH (1976 b) Heterogeneity of plasma glucagon: circulating components in normal subjects and patients with chronic renal failure. J Clin Invest 58:742–750
Lefèbvre PJ, Luyckx AS (1975) Effect of acute kidney exclusion by ligation of renal arteries on peripheral plasma glucagon levels and pancreatic glucagon production in the anesthetized dog. Metab Clin Exp 24:1169–1176
Lefèbvre PJ, Luyckx AS, Nizet AH (1974) Renal handling of endogenous glucagon in the dog: comparison with insulin. Metabolism 23:753–761
Lowrie EG, Soeldner JS, Hampers CL, Merrill JB (1970) Glucose metabolism and insulin secretion in uremic, prediabetic, and normal subjects. J Lab Clin Med 76:603–615
Marumo F, Sakai T, Sato S (1979) Response of insulin, glucagon and growth hormone to arginine infusion in patients with chronic renal failure. Nephron 24:81–84
Mashiter K, Harding PE, Chou M, Mashiter GD, Stout J, Diamond D, Field JB (1975) Persistent pancreatic glucagon but not insulin response to arginine in pancreatectomized dogs. Endocrinology 96:678–693

Mataverde A, El-Ravi R, Cohen M (1974) Spontaneous hypoglycemia in chronic renal failure (Abstr). Clin Res 22:475A
Noe BD, Bauer GE (1971) Evidence for glucagon biosynthesis involving a protein intermediate in islets of the angler fish (Lophius americanus). Endocrinology 89:642–651
Noe BD, Bauer GE (1975) Evidence for sequential metabolic cleavage of proglucagon to glucagon in glucagon biosynthesis. Endocrinology 97:868–877
Noe BD, Bauer GE, Steffes MW, Sutherland DER, Najarian JS (1975) Glucagon biosynthesis in human pancreatic islets: preliminary evidence for a biosynthetic intermediate. Horm Metab Res 7:314–322
O'Brien JP, Sharpe AR (1967) The influence of renal disease on the insulin I-131 disappearance curve in man. Metabolism 16:76–83
Posner BI, Kelly PA, Friesen HG (1975) Prolactin receptors in rat liver: possible induction by prolactin. Science 188:57–59
Rabau M, Dor J, Adar R, Walden R, Mozes M (1973) Spontaneous hypoglycemia in a diabetic patient with renal failure. Isr J Med Sci 9:1036–1040
Sherwin RS, Bastl C, Finkelstein FO, Fisher M, Black H, Hendler R, Felig P (1976a) Influence of uremia and hemodialysis on the turnover and metabolic effects of glucagon. J Clin Invest 57:722–731
Sherwin RS, Fisher M, Hendler R, Felig P (1976b) Hyperglucagonemia and blood glucose regulation in normal, obese and diabetic subjects. N Engl J Med 294:455–461
Soman V, Felig P (1977) Glucagon and insulin binding to liver membranes in a partially nephrectomized uremic rat model. J Clin Invest 60:224–232
Soman V, Felig P (1978) Glucagon binding and adenylate cyclase activity in liver membranes from untreated and insulin-treated diabetic rats. J clin Invest 61:552–560
Spitz IM, Rubenstein AH, Bersohn C, Abrahams C, Lowy C (1970) Carbohydrate metabolism in renal disease. Q J Med 39:201–226
Swenson RS, Peterson DT, Eshleman M, Reaven GM (1973) Effect of acute uremia on various aspects of carbohydrate metabolism in dogs. Kidney Int 4:267–272
Tager HS, Steiner DF (1973) Isolation of a glucagon-containing peptide. Primary structure of a possible fragment of proglucagon. Proc Natl Acad Sci USA 70:2321–2325
Trakatellis AC, Tada K, Yamaji K, Gardik-Kouidou P (1975) Isolation and partial characterization of anglerfish proglucagon. Biochemistry 14:1508–1512
Valverde I, Villaneuva ML, Lozano I, Marco J (1974) Presence of glucagon immunoreactivity in the globulin fraction of human plasma ("big plasma glucagon"). J Clin Endocrinol Metab 39:1090–1098
Valverde I, Dobbs R, Unger RH (1975) Heterogeneity of plasma glucagon immunoreactivity in normal, depancreatized, and alloxan-diabetic dogs. Metab Clin Exp 24:1021–1028
Valverde I, Lemon HM, Kessinger A, Unger RH (1976) Distribution of plasma glucagon immunoreactivity in a patient with suspected glucagonoma. J Clin Endocrinol Metab 42:804–808
Weir GC, Knowlton SD, Martin DB (1975) High molecular weight glucagon-like immunoreactivity in plasma. J Clin Endocrinol Metab 40:296–302
Weir GC, Horton ES, Aoki T, Slovik DM, Jaspan JB, Rubenstein AH (1977) Secretion by glucagonomas of a possible glucagon precursor. J Clin Invest 59:325–330
Westervelt FB Jr (1969) Insulin effect in uremia. J Lab Clin Med 74:79–84
White MG, Kutzman NA (1971) Hypoglycemia in nondiabetics with renal failure. JAMA 215:117–120
Wilson DM (1971) Metabolic abnormalities in uremia. Med Clin North Am 55:1381–1392

CHAPTER 48

# Glucagon in Cirrhosis of the Liver

J. Marco

## A. Introduction

In patients with cirrhosis of the liver, a high frequency of abnormal carbohydrate tolerance, along with a state of insulin resistance, has been demonstrated (Creutzfeldt et al. 1970). The cause of these alterations is unknown, and the possible implication of elevated circulating growth hormone (Conn and Daughaday 1970), potassium depletion (Podolsky et al. 1973), and increased plasma levels of free fatty acids (Felber et al. 1967) has been considered.

The observation of hyperglucagonemia in human diabetes (Aguilar-Parada et al. 1969) as well as in some experimental conditions with impaired glucose metabolism (Müller et al. 1971; Marco et al. 1973a) led to the investigation of glucagon secretion in cirrhotic patients. On the other hand, glucagon is cleared during its passage through the liver (see Chap. 40) and abnormalities of hepatocyte function and liver blood flow might affect its concentration in peripheral plasma. This chapter reviews the studies on the behavior of circulating glucagon in cirrhotics, and the possible pathogenic role of this hormone as a diabetogenic factor in these patients.

## B. Plasma Glucagon Levels in Cirrhotic Patients

The existence of hyperglucagonemia in cirrhosis of the liver has been amply demonstrated, although this finding is not present in all patients. The abnormal elevation of plasma glucagon levels has been observed in the basal state and after A-cell stimulation. The suppressant effect of glucose administration on circulating glucagon concentrations has also been evidenced in cirrhotic subjects.

### I. In the Basal State

We reported that cirrhotic patients, as a group (Table 1), exhibit elevated fasting plasma glucagon levels as compared with normal subjects (Marco et al. 1973b). Basal glucagon values were significantly higher in cirrhotics with surgical portacaval anastomoses than in those patients without anastomoses. It is of note that the mean fasting plasma glucose levels of the control and cirrhotic groups were not significantly different. The mean plasma α-amino nitrogen level was elevated in the patients with cirrhosis as compared with the controls; the presence of a surgical portacaval anastomosis did not demonstrate a significant influence on this level.

**Table 1.** Basal plasma glucagon, insulin, amino nitrogen, and glucose concentrations in cirrhotic patients with and without surgical portacaval shunts and in control subjects; mean ± standard error (MARCO et al. 1973b)

| Group | Glucose (mg/100 ml) | Amino nitrogen (mg/100 ml) | Insulin (μIU/ml) | Glucagon (pg/ml) |
|---|---|---|---|---|
| Controls (15) [a] | 82±2.6 | 4.7±0.12 | 8±1 | 146±10 |
| Cirrhotics without shunts (28) | 87±2.1 | 5.3±0.13 [c] | 22±2 [d] | 217±23 [b] |
| Cirrhotics with shunts (8) | 79±3.7 | 5.1±0.12 [b] | 26±4 [d] | 455±63 [d,e] |

[a] Figures in parentheses represent number in group
[b] $P<0.05$ (with respect to control value)
[c] $P<0.01$ (with respect to control value)
[d] $P<0.001$ (with respect to control value)
[e] $P<0.001$ (with respect to cirrhotics without portacaval shunt)

No statistically significant difference could be detected between basal plasma insulin values in the cirrhotic patients with and without portacaval anastomoses and both were higher than the controls.

The elevation of fasting plasma glucagon concentrations in patients with cirrhosis of the liver has been confirmed by several authors (SHERWIN et al. 1974; GRECO et al. 1974; SHURBERG et al. 1977; PEREZ et al. 1978; MARCHESINI et al. 1979a; DUDLEY et al. 1979). HOLST et al. (1980) reported that, in a group of 11 cirrhotics, the basal plasma glucagon values did not differ from those of 10 matched subjects hospitalized for minor surgical disorders, without suspected disease of liver or pancreas; however, compared with 12 normal subjects, their cirrhotic patients had borderline, but significant fasting hyperglucagonemia.

## II. After A-cell Stimulation

We examined the behavior of plasma glucagon during stimulation of A-cell secretion by intravenous arginine in 25 cirrhotic patients (3 with surgical portacaval shunts) and in 15 control individuals (Fig. 1; MARCO et al. 1973b). The mean basal plasma glucagon level in the cirrhosis group was 239 ± standard error 32 pg/ml, while that of the controls was 146 ± 10 pg/ml ($P<0.05$). The glucagon response to arginine was also more intense in the cirrhotic subjects; the differences oscillated between 178 and 418 pg/ml ($P<0.05$ at 10 min and $<0.025$ at the other points). Both the basal and post-arginine insulin levels were significantly higher in the patients with cirrhosis than in the controls ($P<0.01$). The basal glucose levels of the two groups overlapped. However, the hyperglycemia occurring after arginine administration was less pronounced (6–10 mg/dl) in the cirrhotic group, with a statistically significant difference at 20 min ($P<0.05$).

Several authors have demonstrated an increased glucagon response in cirrhotic patients, using different stimuli of the A-cell, such as intravenous arginine (GRECO et al. 1974), intravenous alanine (SHERWIN et al. 1974), a protein-rich meal (DUDLEY et al. 1979), and insulin-induced hypoglycemia (PEREZ et al. 1978). In the previously mentioned study of HOLST et al. (1980), these authors found no difference

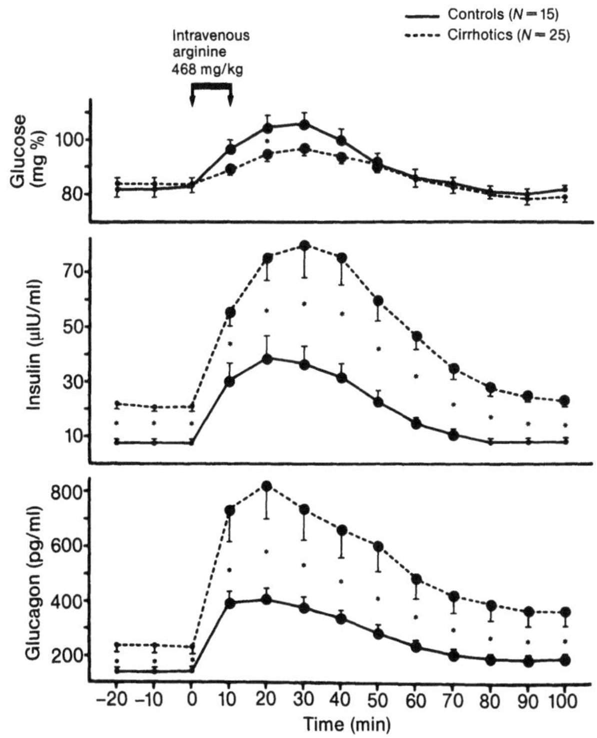


**Fig. 1.** Plasma glucose, insulin, and glucagon responses to intravenous arginine in cirrhotic and normal subjects (mean ± standard error). The *large dots* represent statistically significant differences from the basal value (average of three determinations), and the *asterisks* statistically significant differences between the control and the cirrhotic groups at a given time. MARCO et al. (1973b)

between the glucagon response to a mixed meal of hospitalized cirrhotic and noncirrhotic patients; in these two groups, however, the postprandial glucagon elevation was greater than in healthy individuals.

## III. After A-cell Inhibition

In the same group of cirrhotic patients who participated in the arginine experiments, we found that glucose ingestion was followed by a significant decline (about 40%) of circulating plasma glucagon levels that was more marked than in the control subjects (Fig. 2; MARCO et al. 1973b). Exaggerated hyperinsulinism, both basal and after glucose administration, was observed in these patients. This effect was accompanied by a more prolonged elevation of plasma glucose.

The suppressant effect of oral glucose on plasma glucagon concentrations in cirrhotic patients has also been observed by MARCHESINI et al. (1979a), DUDLEY et al. (1979), MCDONALD et al. (1979), and SMITH-LAING et al. (1980). The administration of glucose by the intravenous route has also been shown to elicit a reduction of basal glucagonemia in cirrhotics (GRECO et al. 1974; MCDONALD et al.

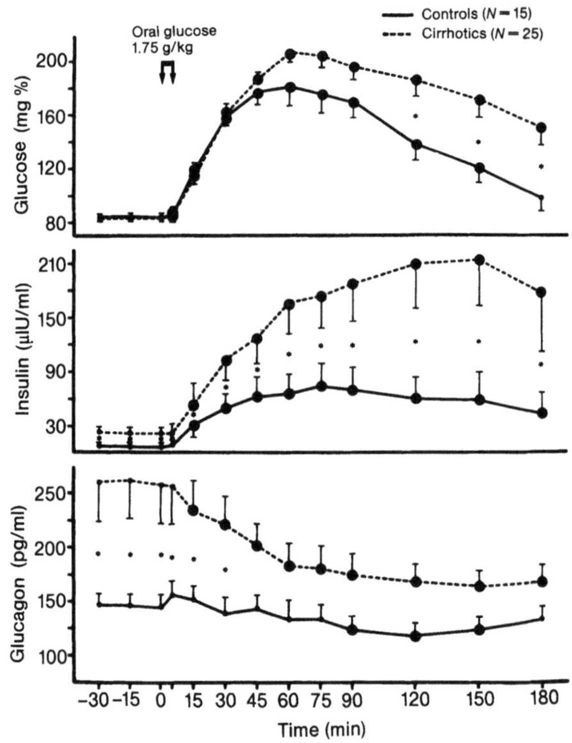


**Fig. 2.** Plasma glucose, insulin, and glucagon responses to oral glucose in cirrhotic and normal subjects (mean ± standard error). The *large dots* represent statistically significant differences from the basal value (average of three determinations), and the *asterisks* statistically significant differences between the control and the cirrhotic groups at a given time. MARCO et al. (1973 b)

1979). In a second report, GRECO et al. (1979 a) found no significant suppressant effect of intravenous glucose on either portal or peripheral plasma glucagon levels in these patients. Finally, the intravenous infusion of somatostatin causes an intense reduction of fasting glucagon values in cirrhotic subjects (GRECO et al. 1980).

## C. The Cause of Hyperglucagonism in Cirrhotic Patients

The investigation of the cause of hyperglucagonism in cirrhosis of the liver has been focused on two main aspects: the presence of spontaneous or surgical portasystemic shunts, and the implication of hepatocellular damage. It is obvious that diversion of the portal blood flow into the systemic circulation, and thus the escape of glucagon from the hepatic filter, has been considered as a factor in impaired breakdown of this hormone. The handling of glucagon by the liver is the subject of Chap. 40. On the other hand, several authors have presented data indicating increased glucagon secretion rather than diminished glucagon degradation in cirrhotic patients. It has been postulated that the augmented A-cell secretion, gener-

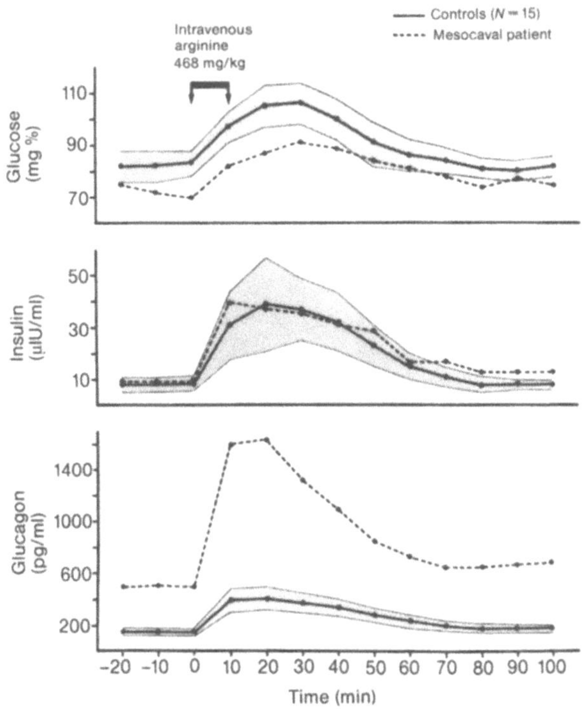


**Fig. 3.** Plasma glucose, insulin, and glucagon responses to intravenous arginine in a noncirrhotic subject with a mesocaval shunt and in 15 normal controls. *Shaded areas* represent calculated intervals of confidence (95% level) for the mean values of the control group. MARCO et al. (1973 b)

ated through a hypothetical signal, would be a consequence of the refractoriness of the cirrhotic liver to glucagon; this hepatic resistance to glucagon has been attributed to the diversion of the portal blood flow and/or to the liver cell damage.

## I. The Role of Portasystemic Shunting

As previously mentioned, we found that in the patients with surgical portacaval anastomoses, the mean fasting glucagon level was significantly higher than that of the cirrhotics without these shunts (Table 1). Moreover, we observed a marked hyperglucagonemia in a subject with a mesocaval anastomosis performed for bypass of a portal thrombosis, who had no clinical signs of liver ailment and normal tests of liver function (Fig. 3 and 4). On the basis of these data, we suggested that the hyperglucagonemia of cirrhotic patients might be in part due to impaired glucagon degradation by the liver, as a consequence of the diversion of portal blood flow through spontaneous or surgical shunts.

In agreement with our data, BILHEIMER et al. (1975) reported that, in a girl with the homozygous form of familial hypercholesterolemia, the creation of an end-to-side portacaval anastomosis resulted in marked increases of basal and arginine-stimulated plasma glucagon levels; however, this patient was receiving propanolol

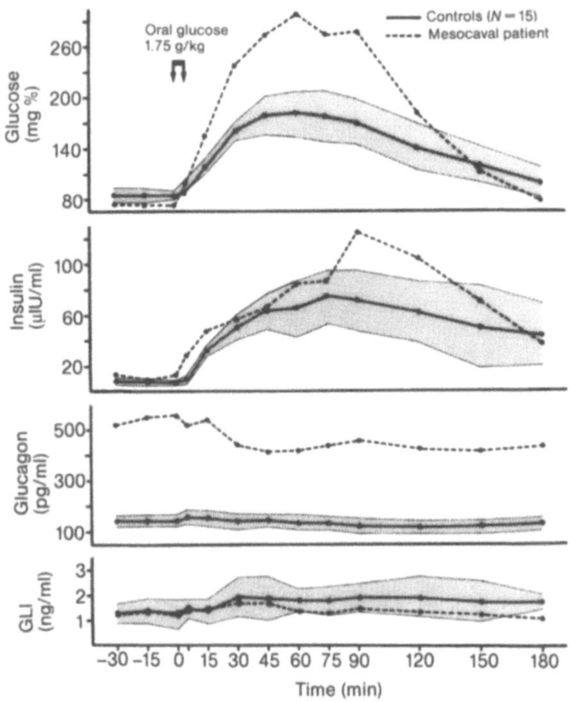


**Fig. 4.** Plasma glucose, insulin, glucagon and intestinal GLI responses to oral glucose in a noncirrhotic subject with a mesocaval shunt and in 15 control subjects. *Shades areas* represent calculated intervals of confidence (95% level) for the mean values of the control group. MARCO et al. (1973b)

and had slightly abnormal indices of liver function. Conversely, WEGLICKI et al. (1977) found no changes in glucagon values (basal and after oral or intravenous glucose administration) after portacaval diversion in a patient affected by the same lipid disorder, and MCDONALD et al. (1979) observed normal fasting glucagon values in some subjects after portacaval anastomosis.

Data obtained from experimental animals are also conflicting. SOETERS et al. (1977) observed that, in normal dogs, portasystemic shunting resulted in an increase of basal and arginine-stimulated plasma glucagon levels; the glucagon rise became most marked when hepatic failure supervened. However, LICKLEY et al. (1975) found no statistically significant difference between fasting glucagon concentrations in intact dogs and those with portacaval anastomoses.

The role of portasystemic shunting in the hyperglucagonemia of liver cirrhosis has been emphasized by several studies. SHERWIN et al. (1974) reported that the fasting plasma glucagon levels as well as the glucagon response to intravenous alanine were normal in four cirrhotics without evidence of portasystemic shunting, were elevated in ten cirrhotics with spontaneous portasystemic shunting, and were highest in four cirrhotics with surgical end-to-side portacaval anastomoses. DUDLEY et al. (1979) demonstrated that in cirrhotic patients before shunting, basal plasma glucagon levels as well as the glucagon response to a protein meal were significantly greater than in control subjects; this glucagon elevation increased further af-

**Table 2.** Plasma glucagon levels and glucagon kinetics in normal and cirrhotic subjects; mean ± standard error[a] (SHERWIN et al. 1978)

| Group[b] | Basal glucagon (pg/ml) | MCR (ml $m^{-2}$ $min^{-1}$) | EBSDR (ng $m^{-2}$ $min^{-1}$) |
|---|---|---|---|
| Healthy controls (15) | 90 ± 12 | 503 ± 37 | 45 ± 7 |
| Total with cirrhosis (17) | 320 ± 55<br>$P<0.005$[c] | 487 ± 55 | 170 ± 43<br>$P<0.01$ |
| Cirrhotics without PSS (4) | 77 ± 21 | 408 ± 60 | 34 ± 13 |
| Cirrhotics with PSS (7) | 255 ± 50<br>$P<0.01$ | 523 ± 104 | 138 ± 41<br>$P<0.05$ |
| Cirrhotics with PCA (6) | 557 ± 118<br>$P<0.001$ | 497 ± 99 | 297 ± 90<br>$P<0.02$ |

[a] Abbreviations: MCR, metabolic clearance rate; EBSDR, estimated basal systemic delivery rate; PSS, spontaneous portasystemic shunt; PCA, surgical portacaval anastomosis
[b] Figures in parentheses represent number in group
[c] Significance of difference as compared with control group

ter shunt surgery. SHURBERG et al. (1977) also found that fasting plasma glucagon was elevated to a greater extent in shunted than in nonshunted alcoholic cirrhotic patients.

In a study on glucagon turnover in liver cirrhosis, SHERWIN et al. (1978) confirmed their earlier observation of elevated fasting plasma glucagon levels in cirrhotics with spontaneous or surgical portasystemic shunts, but normal glucagon values in those patients without evidence of shunting (Table 2). The metabolic clearance rate (MCR) of exogenously infused glucagon was comparable in each of the subgroups of cirrhotics and in control individuals; however, the estimated basal systemic delivery rate (EBSDR) of glucagon was 2–6-fold greater in the hyperglucagonemic cirrhotic patients. Accordingly, they postulated that hyperglucagonism in cirrhosis is a consequence of hypersecretion rather than decreased glucagon catabolism. Furthermore, these authors found that the increment of plasma glucose concentration elicited by glucagon infusion (3 ng $kg^{-1}$ $min^{-1}$) progressively decreased as the severity of the shunting increased. In addition, for the entire cirrhotic group, they obtained an inverse correlation between the glycemic response to injected glucagon and the basal output of endogenous glucagon. In the opinion of SHERWIN et al. (1978), the decreased hepatic responsiveness to the glycemic effect of glucagon (due to portasystemic shunting and/or hepatocellular damage) would generate a signal for glucagon release. Since, in cirrhotic patients without shunting, the glucagon-induced hyperglycemia was not reduced, they suggested that parenchymal damage may be less important than shunting in determining hepatic sensitivity to glucagon. However, liver cell refractoriness may be the consequence rather than the cause of hyperglucagonism in cirrhosis. In fact, chronic hyperglucagonemia has been shown to cause downregulation of the hepatic glucagon receptors along with a reduced cyclic AMP response to glucagon (BHATHENA et al. 1978), and several groups have demonstrated desensitization to glucagon after treatment of hepatocytes with this hormone (PLAS and NUNEZ 1975; DERUBERTIS and CRAVEN 1976; GURR and RUH 1980). This topic is discussed in Chap. 13.

In a study similar to that of SHERWIN et al. (1974), ALFORD et al. (1979) investigated the metabolism of glucagon in cirrhotic patients with proven esophageal varices, before and after portasystemic (side-to-side) venous shunt surgery. Fasting plasma glucagon levels were significantly higher in the cirrhotic group before shunt surgery (213 ± 27 pg/ml) than in the normal control group (53 ± 13 pg/ml), with further elevation occurring after shunt surgery (382 ± 73 pg/ml). The metabolic clearance rate of glucagon was similar in control and cirrhotic patients before shunt surgery, but was significantly decreased after shunt surgery. The EBSDR of glucagon was significantly increased in the cirrhotics, both before (3,042 ± 454 pg $kg^{-1}$ $min^{-1}$) and after shunt surgery (2,518 ± 535 pg $kg^{-1}$ $min^{-1}$), as compared with controls (750 ± 244 pg $kg^{-1}$ $min^{-1}$); there was no significant difference between the mean glucagon EBSDR values before and after surgery. From their data, ALFORD et al. (1979) concluded that the hyperglucagonemia of cirrhotic patients before shunt surgery is principally due to A-cell hypersecretion, while after surgery it is the consequence of both glucagon hypersecretion and delayed clearance of glucagon; thus, only surgical acute diversion of portal hepatic blood flow would result in diminished clearance of glucagon in cirrhotic subjects.

There is a discrepancy between the data obtained by SHERWIN et al. (1978) and those of ALFORD et al. (1979). In the former report, the glucagon MCR of patients with surgical portacaval anastomoses was not significantly lower than that of controls or of subjects with only spontaneous portasystemic shunting. ALFORD et al. performed their study of portacaval patients shortly (2–8 weeks) after the operation while although not specifically stated in the cases of SHERWIN et al. a longer period of time presumably elapsed between the intervention and their study. According to ALFORD et al., it is possible that with time the glucagon clearance mechanism of the liver may adapt to the situation created by the surgical portacaval diversion. These authors also pointed out that paired data were employed in their comparisons of glucagon MCR between cirrhotic groups before and after surgery, whereas SHERWIN et al. studied two different series of patients. In support of the concept of glucagon hypersecretion in cirrhotics, GRECO et al. (1979a) observed that the hyperglucagonemia of these patients is demonstrable not only in peripheral, but also in portal plasma.

## II. The Role of Hepatocellular Damage

The implication of portasystemic shunting in the hyperglucagonism of liver cirrhosis has been challenged by SMITH-LAING et al. (1980). These authors observed that in patients with long-standing portal venous block, but normal or minimal alterations in liver histology and/or liver function, plasma glucagon levels (both basal and after glucose ingestion) were superimposable on those of a control group formed by hospitalized patients with no evidence of digestive disease. The portal venous block subjects with elevated aspartate transaminase levels had significantly higher fasting glucagon concentrations than those with normal aspartate transaminase levels (79.4 ± 13.5 pg/ml compared with 41.5 ± 10.5 pg/ml, $P < 0.05$), although neither group differed significantly from the control value (51 ± 8 pg/ml). They found an elevation of plasma glucagon in cirrhotic patients with extensive spontaneous or surgical portasystemic shunts (167 ± 61 pg/ml, $P < 0.05$ compared with controls). In these patients, there was no significant correlation between fasting glucagon values and the degree of portasystemic shunting. A correlation, however, was found between basal glucagon levels and the aspartate transaminase levels in a group formed by cirrhotics and portal vein block subjects with elevated transaminase ($r = 0.68$, $P < 0.01$). From their data, SMITH-LAING et al. concluded that the hyperglucagonemia of cirrhosis is related to hepatocellular damage and not to portasystemic shunting. The refractoriness of hepatic tissue to glucagon as a cause of glucagon hypersecretion has been previously discussed. According to SMITH-LAING et al., such glucagon resistance would be due to alterations of the hepatic parenchyma and not, as postulated by SHERWIN et al. (1978), to the diversion of the portal blood flow. However, it can be thought that glucagon breakdown in the cirrhotic liver might be impaired as a consequence of the development of intrahepatic shunts (POPPER et al. 1952) and a decrease of the known degrading capacity of the hepatocyte (ANSORGE et al. 1972).

Finally, HOLST et al. (1980) reported that in 11 cirrhotic patients, liver function (as evaluated by the albumin: bilirubin ratio) correlated inversely with the integrated portal glucagon response to a mixed meal. SHERWIN et al. (1978) also observed a direct correlation between the basal systemic delivery rate of glucagon and aspartate transaminase levels in cirrhotic subjects.

### III. Other Factors

As shown in Table 1, in the cirrhotic subjects who participated in our study the plasma levels of amino nitrogen were elevated as compared with control values, a finding already described for this kind of patient (RUDMAN et al. 1971), and, as previously suggested (MARCO et al. 1973a), endogenous hyperaminoacidemia may represent a stimulant factor of the A-cell. It should also be considered that in cirrhotics, impaired glucose utilization at the level of the A-cell could impede the normal suppressor effect of this sugar, thus allowing glucagon hypersecretion (UNGER 1968). The elevation of circulating ammonia does not seem to stimulate glucagon secretion, since in cirrhotic patients without shunts the rise of this cation in blood evoked by ammonium chloride ingestion failed to alter plasma glucagon levels (SHERWIN et al. 1978). In contrast, however, STROMBECK et al. (1978) reported that in dogs, intravenous ammonia stimulated glucagon release from pancreatic and extrapancreatic sources. Finally, McDONALD et al. (1979) have observed an exaggerated gastric inhibitory peptide response to oral glucose in cirrhotics, and have speculated that this peptide may contribute to stimulation of glucagon secretion in liver disease.

## D. Plasma Glucagon-Like Immunoreactivity Levels in Cirrhotic Patients

We examined the plasma levels of glucagon-like immunoreactivity (GLI) of intestinal origin in a group of eight cirrhotic patients – three of them with surgical portacaval shunts – in whom, as measured with 30 K antiserum, high basal glucagon concentrations (245–662 pg/ml) and a frank glucagon hyperresponse to arginine (peaks 1,070–3,200 pg/ml) had previously been detected (Fig. 5). The control group

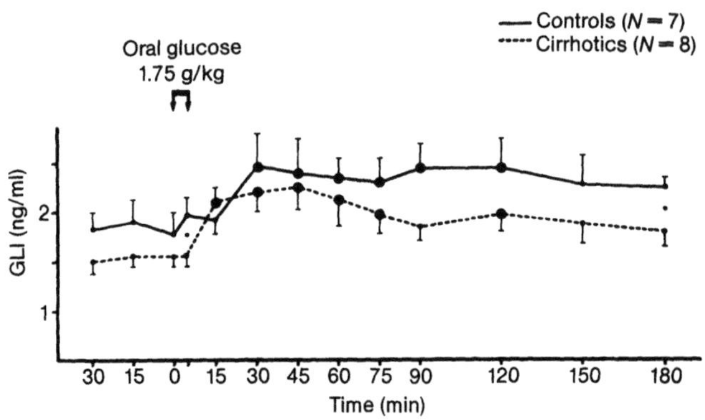


**Fig. 5.** Intestinal GLI responses to oral glucose in cirrhotic and normal subjects (mean ± standard error). The *large dots* represent statistically significant differences from the basal value (average of three determinations) and the *asterisks* statistically significant differences between the control and the cirrhotic groups at a given time

consisted of seven healthy subjects. GLI determinations were performed using 78J antiserum, which exhibits great avidity for this intestinal material (Harris et al. 1979).

The mean fasting plasma GLI concentration of the cirrhotic patients (1.04 ± 0.08 ng/ml) was not significantly different from that of the controls (1.34 ± 0.20 ng/ml). The elevation after oral administration of glucose was comparable in both groups (~0.7 ng/ml). At only two points was the GLI level significantly lower in the patients with cirrhosis: at 5 min (1.05 ± 0.07 ng/ml compared with 1.29 ± 0.22 ng/ml, $P < 0.05$) and at 180 min (1.3 ± 0.13 ng/ml compared with 1.74 ± 0.12 ng/ml, $P < 0.05$). The basal and post-glucose GLI plasma levels of the cirrhotics fell within the range found in our laboratory for normal subjects (Marco et al. 1977).

In a noncirrhotic patient with a mesocaval shunt (see Fig. 4), basal GLI levels and the GLI response to glucose ingestion also fell within normal values. In agreement with these results, Bilheimer et al. (1975) found no change in plasma GLI concentrations after portacaval shunt surgery in a patient with homozygous familial hypercholesterolemia.

The foregoing data indicate that circulating GLI does not appear to be affected by the liver to the same extent as glucagon, since neither the existence of liver damage nor portacaval shunts were accompanied by an elevation of its concentration. In this context, Gutman et al. (1973) found that intestinal GLI was not degraded when perfused through the isolated rat liver. In contrast, however, Buchanan et al. (1968) found a similar clearance for intestinal rat GLI and for rat pancreatic glucagon in the perfused rat liver.

## E. The Nature of Circulating Immunoreactive Glucagon in Cirrhotic Patients

The heterogeneity of plasma immunoreactive glucagon (IRG) is the subject of Chap. 11. The elevation of plasma glucagon observed in cirrhotic patients is mainly due to the increase of the 3,500 daltons component ($IRG^{3500}$) (VALVERDE and VILLANUEVA 1976; DUDLEY et al. 1979; MCDONALD et al. 1979). DUDLEY et al. found that, in cirrhotic subjects before and after shunt surgery, the rise in plasma IRG induced by a protein meal as well as the plasma IRG decline subsequent to glucose ingestion could almost entirely be accounted for by changes in the $IRG^{3500}$ fraction. These authors observed that after shunt surgery, the relative proportion of the $IRG^{9000}$ component decreased significantly and this fall was associated with a corresponding increase in the $IRG^{3500}$ moiety. MCDONALD et al. (1979) also found that in cirrhotics, the suppressant effect of glucose administration on circulating glucagon was due to major decreases of $IRG^{3500}$. According to JASPAN et al. (1981) the liver has a predilection for removal of the $IRG^{3500}$ fraction, with virtually no extraction of the other IRG components. Thus, it can be thought that both diversion of portal blood flow and hepatocellular damage may contribute to the elevation of $IRG^{3500}$ in peripheral plasma.

## F. Pathogenic Implications of Hyperglucagonemia in Cirrhosis of the Liver

The metabolic consequences of hyperglucagonemia in cirrhosis may not be the same as in conditions in which the liver is intact. Since this organ is the main target for glucagon, the diabetogenic effect of this hormone, in terms of its ability to stimulate gluconeogenesis and glycogenolysis, could be restricted by abnormalities of hepatocyte function and liver blood flow. In fact, a correlation between the elevation of plasma glucagon and the degree of abnormal glucose tolerance in these patients has not been observed.

As shown in Fig. 1, we found that the rise of blood glucose after arginine infusion was less pronounced in cirrhotics than in control subjects. SHERWIN et al. (1974) also found that the hyperglucagonemic response to alanine in cirrhosis was not accompanied by an augmented elevation of plasma glucose. Similarly, DUDLEY et al. (1979) observed that in cirrhotics, the plasma glucose levels corresponding to a protein meal test were unaltered by shunt surgery, despite the resultant increase in glucagon levels in the presence of unchanged insulin concentrations. Furthermore, in hyperglucagonemic cirrhotic patients, basal hepatic glucose output has been estimated to be normal (PEREZ et al. 1978) or even reduced (PROIETTO et al. 1980). GRECO et al. (1979b) observed that the insulin resistance of cirrhotic patients persisted during glucagon suppression by a concomitant infusion of glucose, insulin, and somatostatin.

In connection with these observations, FRANCAVILLA et al. (1978) found that in advanced cirrhotics, unlike healthy subjects, plasma cyclic AMP and glucose levels did not increase after intravenous glucagon infusion; they also observed that the adenylate cyclase from the livers of these patients was less responsive to glucagon stimulation than that from noncirrhotic subjects. MAEKUBO et al. (1980) observed that, in cirrhotics, while basal plasma cyclic AMP levels were elevated, the cyclic AMP response to exogenous glucagon was markedly blunted. DAVIES et al. (1976) found a reduced increase of plasma cyclic AMP after intravenous glucagon in six of eight cirrhotics with surgical portacaval shunts. As already discussed,

SHERWIN et al. (1978) obtained an inverse correlation between the glycemic response to glucagon infusion and the basal glucagon output in cirrhotics. Conversely, STRANGE et al. (1977) observed no significant differences between plasma cyclic AMP and glucose concentrations after glucagon injection in four patients with alcoholic cirrhosis and five healthy controls.

In cirrhotic patients, hyperglucagonism has been hypothesized to favor the accumulation of aromatic amino acids in the circulation, which in turn might be implicated in the pathogenesis of hepatic encephalopathy (SOETERS and FISCHER 1976); under the catabolic stimulus of glucagon, the large amounts of aromatic amino acids released from endogenous protein would not be catabolized efficiently by the failing liver. MARCHESINI et al (1979 b) found a correlation between glucagon and aromatic amino acid plasma levels in cirrhotics.

The possible influence of glucagon on ammonia production in cirrhosis has been examined by WALKER et al. (1974). These authors reported that in stable cirrhotics, supplementation of the diet with glucose, while increasing insulin plasma levels and lowering circulating glucagon, reduced blood ammonia concentrations. They suggested that this could reflect diminished ammonia generation from hepatic ureagenesis and from amino acids, secondary to the increase of the insulin: glucagon molar ratio induced by glucose. Finally, extreme hyperglucagonemia ($\sim$1,000 pg/ml) has been considered as the possible cause of the necrolytic migratory erythema found in a patient with liver cirrhosis and without evidence of glucagonoma (DOYLE et al. 1979).

## G. Summary

In patients with cirrhosis of the liver, an elevation of plasma glucagon, both in the basal state and after A-cell stimulation, has been demonstrated. The rise in IRG can be accounted for by the increase of the $IRG^{3500}$ fraction. Circulating levels of GLI of intestinal origin, under basal conditions or after stimulation by oral glucose, are in the normal range in these patients.

The cause of hyperglucagonism in liver cirrhosis has not been definitely established. The escape of glucagon from the hepatic filter through spontaneous or surgical portasystemic anastomoses may result in reduced degradation of this hormone. On the other hand, there is evidence indicating glucagon hypersecretion in cirrhotics. It has been hypothesized that hepatic resistance to the action of glucagon, as a consequence of diversion of portal blood flow and/or hepatocellular damage, would generate a signal for glucagon release. Hyperaminoacidemia, and impaired glucose utilization at the level of the A-cell, might also contribute to glucagon hypersecretion.

The metabolic consequences of hyperglucagonism in cirrhosis may not be the same as in conditions in which the liver is intact. Since this organ is the main target for glucagon, the diabetogenic effect of this hormone in terms of its ability to stimulate gluconeogenesis and glycogenolysis could be restricted by the abnormalities of hepatocyte function and liver blood flow. In fact, refractoriness to the hyperglycemic effect of both endogenous and exogenous glucagon has been observed in cirrhotic subjects, and, in hyperglucagonemic cirrhotics, basal hepatic glucose output has been estimated to be normal or reduced. Moreover, a relationship between glucagon levels and the degree of abnormal glucose tolerance has not been observed in these patients.

In cirrhosis, hyperglucagonemia has been hypothesized to represent a catabolic factor which would contribute to plasma amino acid imbalance and in turn to the genesis of hepatic encephalopathy. It has also been considered that in cirrhosis, oral glucose administration, by increasing the insulin: glucagon molar ratio, would reduce ammonia generation from hepatic ureagenesis and from amino acids. Necrolytic migratory erythema has been described in a single cirrhotic patient with extreme hyperglucagonemia and without evidence of glucagonoma.

*Acknowledgments.* This work was supported by research grants from the Instituto Nacional de Sanidad and from the Eugenio Rodríguez Pascual Foundation, Spain

## References

Aguilar-Parada E, Eisentraut AM, Unger RH (1969) Pancreatic glucagon secretion in normal and diabetic subjects. Am J Med Sci 257:415–419

Alford FP, Dudley FJ, Chisholm DJ, Findlay DM (1979) Glucagon metabolism in normal subjects and in cirrhotic patients before and after portasystemic venous shunt surgery. Clin Endocrinol (Oxf) 11:413–424

Ansorge S, Kaiser B, Senger U, Bohley P, Langner J, Kirschner H (1972) Metabolismus von Insulin und Glukagon: über die Heterogenität und zur Charakterisierung der Insulin und Glukagon abbauenden Enzymsysteme in Rattenleberzellfraktionen. Acta Biol Md Ger 28:761–777

Bhathena SJ, Voyles NR, Smith S, Recant L (1978) Decreased glucagon receptors in diabetic rat hepatocytes. Evidence for regulation of glucagon receptors by hyperglucagonemia. J Clin Invest 61:1488–1497

Bilheimer DW, Goldstein JL, Grundy SM, Brown MS (1975) Reduction in cholesterol and low density lipoprotein synthesis after portacaval shunt surgery in a patient with homozygous familial hypercholesterolemia. J Clin Invest 56:1420–1430

Buchanan KD, Solomon SS, Vance JE, Porter HP, Williams RH (1968) Glucagon clearance by the isolated perfused rat liver. Proc Soc Exp Biol Med 128:620–623

Conn HO, Daughaday WH (1970) Cirrhosis and diabetes. V. Serum human growth hormone levels in Laennec's cirrhosis. J Lab Clin Med 76:678–688

Creutzfeldt W, Frerichs H, Sickinger K (1970) Liver disease and diabetes mellitus. Prog Liver Dis 3:371–407

Davies TF, Prudhoe K, Douglas AP (1976) Plasma cyclic adenosine-3′,5′-monophosphate response to glucagon in patients with liver disease. Br Med J 1:931–933

DeRubertis FR, Craven P (1976) Reduced sensitivity of the hepatic adenylate cyclase-cyclic AMP system to glucagon during sustained hormonal stimulation. J Clin Invest 57:435–443

Doyle JA, Schroeter AL, Rogers RS III (1979) Hyperglucagonaemia and necrolytic migratory erythema in cirrhosis: possible pseudoglucagonoma syndrome. Br J Dermatol 100:581–587

Dudley FJ, Alford FP, Chisholm DJ, Findlay DM (1979) Effect of portasystemic venous shunt surgery on hyperglucagonaemia in cirrhosis: paired studies of pre- and post-shunted subjects. Gut 20:817–824

Felber JP, Magnenat P, Vannoti A (1967) Tolérance au glucose diminuée et réponse insulinique élevée dans la cirrhose. Schweiz Med Wochensch 97:1537–1539

Francavilla A, Jones AF, Starzl TE (1978) Cyclic AMP metabolism and adenylate cyclase concentration in patients with advanced hepatic cirrhosis. Gastroenterology 75:1026–1032

Greco AV, Ghirlanda G, Patrono C, Fedeli G, Manna R (1974) Behavior of pancreatic glucagon, insulin and HGH in liver cirrhosis, after arginine and i. v. glucose. Acta Diabetol Lat 11:330–339

Greco AV, Crucitti F, Ghirlanda G, Manna R, Altomonte L, Rebuzzi AG, Bertoli A (1979a) Insulin and glucagon concentrations in portal and peripheral veins in patients with hepatic cirrhosis. Diabetologia 17:23–28

Greco AV, Rebuzzi AG, Altomonte L, Manna R, Bertoli A, Ghirlanda G (1979b) Glucose, insulin and somatostatin infusion for the determination of insulin resistance in liver cirrhosis. Horm Metab Res 11:547–549

Greco AV, Altomonte L, Ghirlanda G, Rebuzzi AG, Manna R, Bertoli A (1980) Somatostatin infusion in liver cirrhosis: glucagon control of glucose homeostasis. Diabetologia 18:187–191

Gurr JA, Ruh TA (1980) Desensitization of primary cultures of adult rat liver parenchymal cells to stimulation of adenosine 3′,5′-monophosphate production by glucagon and epinephrine. Endocrinology 107:1309–1319

Gutman RA, Fink G, Voyles N, Selawry H, Penhos JC, Lepp A, Recant L (1973) Specific biologic effects of intestinal glucagon-like materials. J Clin Invest 52:1165–1175

Harris V, Faloona GR, Unger RH (1979) Glucagon. In: Jaffe BM, Behrman HR (eds) Methods of hormone radioimmunoassay, 2nd edn. Academic Press, New York, p 643

Holst JJ, Burcharth F, Kühl C (1980) Pancreatic glucoregulatory hormones in cirrhosis of the liver: portal vein concentrations during intravenous glucose tolerance test and in response to a meal. Diabete Metab 6:117–127

Jaspan JB, Polonsky KS, Röjdmark S, Ishida T, Field JB, Rubenstein AH (1981) Hepatic extraction of plasma immunoreactive glucagon components: predilection for 3500-daltons glucagon metabolism by the liver. Diab 30:767–772

Lickley HLA, Chisholm DJ, Rabinovitch A, Wexler M, Dupre J (1975) Effects of portacaval anastomosis on glucose tolerance in the dog: evidence of an interaction between the gut and the liver in oral glucose disposal. Metabolism 24:1157–1168

Maekubo H, Matsushima T, Okada F, Honma M, Ui M (1980) Anomalous plasma cyclic AMP responses to glucagon in patients with liver disease. Dig Dis Sci 25:700–704

Marchesini G, Forlani G, Angiolini A, Zoli M, Scolari MP, Bianchi FB, Pisi E (1979a) Oral glucose in cirrhotics. Effects on plasma amino acid patterns and the role of insulin and glucagon. Diab Metab 5:135–139

Marchesini G, Forlani G, Zoli M, Angiolini A, Scolari MP, Bianchi FB, Pisi E (1979b) Insulin and glucagon levels in liver cirrhosis. Relationship with plasma amino acid imbalance of chronic hepatic encephalopathy. Dig Dis Sci 24:594–601

Marco J, Calle C, Román D, Díaz-Fierros M, Villanueva ML, Valverde I (1973a) Hyperglucagonism induced by glucocorticoid treatment in man. N Engl J Med 288:128–131

Marco J, Diego J, Villanueva ML, Díaz-Fierros M, Valverde I, Segovia JM (1973b) Elevated plasma glucagon levels in cirrhosis of the liver. N Engl J Med 289:1107–1111

Marco J, Hedo JA, Villanueva ML, Calle C, Corujedo A, Segovia JM (1977) Effect of food ingestion on intestinal glucagon-like immunoreactivity (GLI) secretion in normal and gastrectomized subjects. Diabetologia 13:131–135

McDonald TJ, Dupre J, Caussignac Y, Radziuk J, Van Vhiet S (1979) Hyperglucagonemia in liver cirrhosis with portal-systemic venous anastomoses: responses of plasma glucagon and gastric inhibitory polypeptide to oral or intravenous glucose in cirrhotics with normal or elevated fasting plasma glucose levels. Metabolism 28:300–307

Müller WA, Faloona GR, Unger RH (1971) The effect of experimental insulin deficiency on glucagon secretion. J Clin Invest 50:1992–1999

Perez G, Trimarco B, Ungaro B, Rengo F, Sacca L (1978) Glucoregulatory response to insulin-induced hypoglycemia in Laennec's cirrhosis. J Clin Endocrinol Metab 46:778–783

Plas C, Nunez J (1975) Glycogenolytic response to glucagon of cultured fetal hepatocytes. J Biol Chem 250:5304–5311

Podolsky S, Zimmerman HJ, Burrows BA, Cardarelli JA, Pattavina CG (1973) Potassium depletion in hepatic cirrhosis: a reversible cause of impaired growth-hormone and insulin response to stimulation. N Engl J Med 288:644–648

Popper H, Elias H, Petty DE (1952) Vascular pattern of the cirrhotic liver. Am J Clin Pathol 22:717–729

Proietto J, Alford FP, Dudley FJ (1980) The mechanism of the carbohydrate intolerance of cirrhosis. J Clin Endocrinol Metab 51:1030–1036

Rudman D, Vogler WR, Howard CH, Gerron GG (1971) Observations on the plasma amino acids of patients with acute leukemia. Cancer Res 31:1159–1165
Sherwin R, Joshi P, Hendler R, Felig P, Conn HO (1974) Hyperglucagonemia in Laennec's cirrhosis. The role of portal-systemic shunting. N Engl J Med 290:239–242
Sherwin RS, Fisher M, Bessoff J, Snyder N, Hendler R, Conn HO, Felig P (1978) Hyperglucagonemia in cirrhosis: altered secretion and sensitivity to glucagon. Gastroenterology 74:1224–1228
Shurberg JL, Resnick R, Koff RS, Ros E, Baum RA, Pallota JA (1977) Serum lipids, insulin, and glucagon after portacaval shunt in cirrhosis. Gastroenterology 72:301–304
Smith-Laing G, Orskov H, Gore MBR, Sherlock S (1980) Hyperglucagonemia in cirrhosis. Diabetologia 19:103–108
Soeters PB, Fischer JE (1976) Insulin, glucagon, aminoacid imbalance, and hepatic encephalopathy. Lancet 2:880–882
Soeters PB, Weir G, Ebeid AM, Fischer JE (1977) Insulin, glucagon, portal systemic shunting, and hepatic failure in the dog. J Surg Res 23:183–188
Strange RC, Mjos OD, Henden T, Jynge P (1977) The effect of glucagon on plasma cyclic AMP and glucose concentrations in patients with alcoholic cirrhosis. Acta Med Scand 202:87–88
Strombeck DR, Rogers Q, Stern JS (1978) Effects of intravenous ammonia infusion on plasma levels of amino acids, glucagon and insulin in dogs. Gastroenterology 74:1165
Unger RH (1968) New ideas concerning the physiologic roles of glucagon. Am J Med Sci 255:273–276
Valverde I, Villanueva ML (1976) Heterogeneity of plasma immunoreactive glucagon. Metabolism [Suppl] 1:1393–1395
Walker C, Peterson W, Unger R (1974) Blood ammonia levels in advanced cirrhosis during therapeutic elevation of the insulin: glucagon ratio. N Engl J Med 291:168–171
Weglicki WB, Ganda OP, Soeldner JS, Murawski BJ, Cohn LH, Couch NP (1977) Portacaval diversion for severe hypercholesterolemia. Report of a case with measurements of glucose tolerance, insulin and glucagon levels. Arch Surg 112:634–636

CHAPTER 49

# Glucagon in Obesity

J. J. HOLST

## A. Introduction

Glucagon is capable of influencing many pathways of lipid metabolism (see Chap. 17, 19, 20, 25, 46) and although some uncertainty remains about the importance of glucagon in the regulation of lipid metabolism under physiologic circumstances in humans (SCHADE et al. 1979; MÜLLER et al. 1977; SCHADE and EATON 1976; GERICH et al. 1976; POZEFSKY et al. 1976; ANDREWS et al. 1975; LILJENQUIST et al. 1974; SHERWIN et al. 1976), glucagon's potential influence has been sufficiently conspicuous to justify several investigations of its possible role in human and experimental obesity.

This chapter will review results concerning derterminations of the concentration of glucagon in plasma in obesity as well as aspects of the biologic effects of glucagon in human nondiabetic obesity and in experimental obesity.

## B. Glucagon in Human Obesity

There is little doubt that obesity in humans is a multifactorial condition (see SIMS 1979 for detailed review) and it would be naive to expect a single abnormality e. g., in glucagon secretion. However, it is well established that most forms of obesity are associated with abnormalities in glucose metabolism, hyperinsulinism, and insulin resistance (SIMS 1979; JARRETT and BAKIR 1978; BECK-NIELSEN 1980; OLEFSKY 1981; JEANRENAUD 1979; GLASS et al. 1981); and since glucagon secretion and action are intimately coupled with insulin and glucose metabolism it is of interest to analyze the possible disturbances in glucagon secretion in obesity also (see Chaps. 23, 44).

### I. Plasma Concentrations of Glucagon

The available data concerning plasma glucagon in human obesity are listed in Table 1, which summarizes reported basal levels of glucagon in obese subjects compared with controls and the results of various tests of A-cell function.

#### 1. Basal State

Most authors (Table 1) have reported normal values of glucagon in obesity, but decreased levels (GÖSCHKE 1977; FALLUCA et al. 1975) as well as increased levels (PAULSEN and LAWRENCE 1968; GARLASCHI et al. 1975; TIENGO et al. 1972; SCHNEIDER et al. 1975; SAVAGE et al. 1979; MINUK et al. 1980) have been described.

**Table 1.** Plasma glucagon in human nondiabetic obesity

| Reference | Number of patients | Basal levels | Test of A-cell function | Response in obese subjects |
|---|---|---|---|---|
| TIENGO et al. (1972) | 7 | Insignificantly increased | i.v. Arginine | Increased |
| KALKHOFF et al. (1973) | 6+8 | Normal | i.v. Arginine ±weight loss | Increased |
| GOSSAIN et al. (1974) | 9 | Normal | i.v. Arginine<br>i.v. Glucose[a]<br>Meal | Increased<br>Normal |
| GERICH et al. (1974) | 15 | Normal | i.v. Arginine<br>i.v. Insulin<br>Oral glucose[a] | Increased<br>Normal |
| SCHADE and EATON (1974) | 10 | Normal | i.v. Arginine | Decreased |
| FALLUCA et al. (1975) | 7 | Insignificantly decreased | i.v. Arginine | Decreased |
| SCHNEIDER et al. (1975) | 13 infants | Normal or increased | i.v. Arginine | Increased |
| SANTIAGO et al. (1977) | 10 | Normal | i.v. Arginine<br>i.v. Glucose[a]<br>Oral glucose[a] | Normal |
| MESCHI et al. (1977) | 12 infants | Normal | i.v. Arginine | Normal |
| SAVAGE et al. (1979) | 10 | Increased | i.v. Arginine<br>i.v. Glucose[a]<br>±weight loss | Increased<br>Normal |
| WALTER et al. (1980) | 5 | Normal | i.v. Arginine | Normal |
| WISE et al. (1973a) | 5 | Normal | i.v. Alanine ±Dexamethasone | Normal |
| WISE et al. (1973b) | 10 | Normal | i.v. Alanine ±Fasting | Decreased |
| MÜLLER et al. (1975) | 15 | Normal | i.v. Alanine ±Fasting | Normal |
| PAULSEN and LAWRENCE (1968) | 13 infants | Increased | Oral glucose[a] | Increased |
| FELIG et al. (1974) | 13 | Normal | i.v. Glucose[a] | Normal |
| VRANIC et al. (1980) | 5 | Normal | Oral glucose[a] | Increased |
| GARLASCHI et al. (1975) | 18 infants | Elevated | Exercise | Normal |
| MINUK et al. (1980) | 5 | Elevated | Exercise<br>Fasting | Normal |
| TROVATI et al. (1981) | 6 | Decreased | Exercise | Decreased (reversed) |
| MARLISS et al. (1970) | 15 | Normal | Fasting | Normal |
| FISHER et al. (1976) | 7 | Normal | Fasting | Normal |
| GÖSCHKE (1977) | 14 | Decreased | Fasting | Decreased |
| TURPIN et al. (1978) | 5 | Normal | Fasting | Probably normal |
| ATKINSON et al. (1981) | 5 | Not reported | Adrenergic modulations | Normal |
| SHERWIN et al. (1976) | 7 | Normal | Not reported | |
| LASSMANN et al. (1980) | 13 | Normal | Meal | Normal |

[a] With glucose this response is a suppression below basal levels

## 2. Tests of A-cell Function

The most widely used test has been intravenous arginine infusion (see Chap. 24) with radioimmunologic determination of the plasma glucagon response. Again, several investigators have found normal responses to arginine (SANTIAGO et al. 1977; MESCHI et al. 1977; WALTER et al. 1980), whereas others have reported decreased (SCHADE and EATON 1974; FALLUCA et al. 1975) or increased responses (TIENGO et al. 1972; KALKHOFF et al. 1973; GOSSAIN et al. 1974; SCHNEIDER et al. 1975; SAVAGE et al. 1979).

Other tests have included intravenous infusion of alanine which in normal subjects is a potent stimulus for glucagon secretion (WISE et al. 1973 b). With this test, normal (WISE et al. 1973 a; MÜLLER et al. 1975) as well as decreased responses have been reported (WISE et al. 1972, 1973 b). Also, the inhibitory effect of glucose on glucagon secretion has been studied in obese subjects. The depression elicited by oral or intravenous glucose has been reported to be normal (GERICH et al. 1974; SANTIAGO 1977; FELIG et al. 1974; SAVAGE et al. 1979) or increased (PAULSEN and LAWRENCE 1968; VRANIC et al. 1980). Likewise, the glucagon response to exercise was found to be normal by some (GARLASCHI et al. 1975; MINUK et al. 1980) and decreased by others (TROVATI et al. 1981). Exercise, however, is less suitable as a test of A-cell function, since the response to short periods of moderate exercise, consists, if anything, of a depression of peripheral glucagon concentrations (GALBO et al. 1975) whereas more intense exercise leads to a moderate increase, probably due to sympathoadrenal activity (GALBO et al. 1975). Only prolonged exercise causes a reproducible, conspicuous rise in plasma glucagon, probably as a consequence of cerebral or A-cell glucose deprivation (GALBO et al. 1977, 1979). The effect of administration of adrenergic agonists or antagonists was normal in obese subjects (ATKINSON et al. 1981) and so was the A-cell response to intravenous dexamethasone (WISE et al. 1973 a).

The glucagon response to short-term fasting has also been studied in obese subjects. As a test of A-cell function this procedure is difficult to evaluate since it has been suggested that the very reproducible hyperglucagonemia elicited by fasting (AGUILAR-PARADA et al. 1969; MARLISS et al. 1970) is caused by decreased clearance of glucagon rather than increased secretion (SHERWIN et al. 1976). Furthermore, in some of the studies the response to fasting was not compared with a control group proper, but rather with results obtained in normal subjects on previous occasions. However, "normal" responses to a 3-day fast were found by MARLISS et al. (1970), GERICH et al. (1974), FISHER et al. (1976), and TURPIN et al. (1978), whereas WISE et al. (1972), as well as GÖSCHKE (1977) and MINUK et al. (1980), were unable to measure any increase in glucagon concentration in response to short-term fasting. Decreased glucagon responses to amino acid stimulation in obese subjects were reported to persist after short-term fasting by GÖSCHKE (1977), and WISE et al. (1972).

Rather than comparing obese subjects with controls, KALKHOFF et al. (1973) and SAVAGE et al. (1979) restudied their obese patients after weight loss, and both groups reported decreased (although insignificantly in the former study) basal levels and decreased responses to amino acid stimulation after weight reduction, indicating that obesity had been associated with inappropriately elevated levels of glucagon.

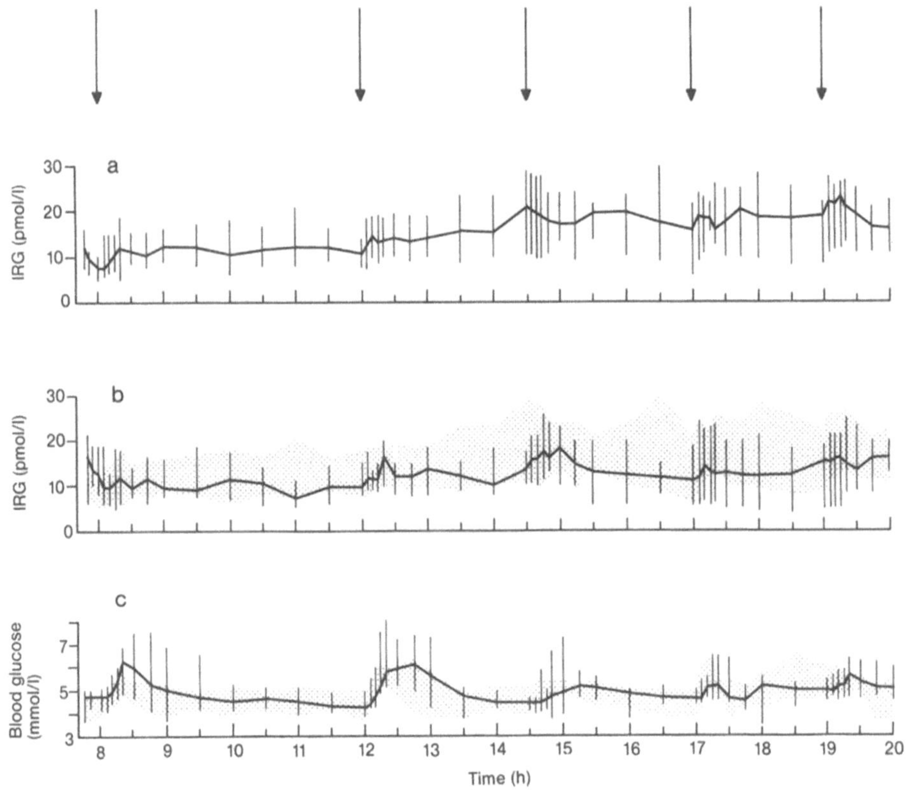


**Fig. 1 a–c.** Glucagon and glucose concentrations during the day in obese and normal subjects. Glucagon concentrations in plasma in five obese subjects (194% ± 14% of ideal body weight) and six age-matched controls (102 ± 4%) were followed from 7.30 a.m. to 8 p.m. The *arrows* indicate meals: breakfast (bread, cheese, butter, milk, 8 a.m.), lunch (beef, potatoes, gravy, 12 a.m.), tea (same as breakfast, 2.30 p.m.), dinner (protein-rich sandwiches, 5 p.m.), supper (same as breakfast, 7 p.m.). The diagrams show plasma concentrations (median and range) of COOH terminal glucagon in normal subjects (**a**) and the glucagon concentrations in obese subjects (**b**); they also demonstrate (**c**) that none of the obese subjects was diabetic (normal glucose values). The *stippled areas* in **b** and **c** indicate normal ranges

Thus, there are abundant conflicting data concerning glucagon secretion in obesity. Probably several factors are responsible for this. First, the data have been collected over a time span of about 13 years, during which period the technique of glucagon radioimmunoassay has undergone some development (see Chap. 10). Second, many of the differences observed between controls and obese subjects are not very conspicious, and many of the numbers have been rather small, so that statistical type I as well as type II errors may be suspected. Also, the degree of obesity has varied. The approach of studying obese subjects before and after weight loss is attractive (KALKHOFF et al. 1973; SAVAGE et al. 1979) because of its implicit solution to the problem of selecting appropriate controls, but it also illustrates another problem which is very difficult to get around, namely the fact that glucagon secretion is greatly influenced by the diet (MÜLLER et al. 1971; GALBO et al. 1979; SA-

**Table 2.** Effect of glucagon in human obesity

| Reference | Test system | Response in obese subjects vs controls |
|---|---|---|
| ARKY et al. (1970) | Insulin and glucose response to glucagon, 1 mg i.v. | Fluctuating responses during prolonged fasting in obese subjects |
| SCHADE and EATON (1975a) | Free fatty acid and $\beta$-hydroxybutyrate response to glucagon 1 µg/kg i.v. | Decreased or paradoxical response |
| SCHADE and EATON (1977) | As above, but 1–2 µg/kg | Decreased response to 1 µg/kg normal response to 2 µg/kg |
| SHERWIN et al. (1976) | i.v. Glucagon infusion, 120 ng $m^2 \times min^{-1}$ + oral glucose | Glucose and insulin response normal |
| FISHER et al. (1976) | Glucose responses to i.v. glucagon 3 ng $kg^{-1} min^{-1}$ | Normal responses |
| TURPIN et al. (1978) | Glucose, insulin, and plasma cyclic AMP responses to i.v. glucagon 1 mg | Responses unchanged before and after 3 days fasting |
| VRANIC et al. (1980) | Hepatic glucose production during graded doses of glucagon i.v. | Normal glucose production |

VAGE et al. 1979). Not only are the basal levels dependent on the preceeding diet, but also the responses to tests like protein meals (MÜLLER et al. 1971) or exercise (GALBO et al. 1979) vary with the diet. Diet-dependent changes in glucagon secretion may therefore explain the majority of differences observed between obese and nonobese subjects.

It therefore seems that differences in glucagon secretion between obese and nonobese normal subjects are likely to be small. In a recent study in this laboratory (HOLST et al. 1983), we studied peripheral glucagon levels in grossly obese nondiabetic subjects and age-matched controls subjected to identical diets and followed them for 12 h throughout all the meals of the day (Fig. 1). Consistent with most other observations, our obese subjects had significant hyperinsulinism (not shown), but all median glucagon values were clearly within the normal range, and the responses to the different daily meals were not different from those of the controls. Similar results were presented by SANTIAGO et al. (1977).

It thus seems unlikely that human obesity in itself has any major influence on A-cell secretion. A recent study of portal and peripheral vein glucagon concentrations in normal subjects and obese diabetics indicated that hepatic extraction of glucagon (and insulin) did not differ between the two groups (WALTER et al. 1980). Differences in glucagon secretion between obese diabetics and normal controls are therefore more likely to be due to diabetes than to obesity (see Chap. 44).

## II. Effect of Glucagon in Human Obesity

While glucagon secretion in humans seems to be little or very moderately influenced by obesity, this does not exclude the possibility that its biologic activities might be modulated in this condition, and this has been the subject of study in a number of reports (Table 2). Indeed, SCHADE and EATON (1975a) found a decreased free fatty acid and $\beta$-hydroxybutyrate response to a bolus injection of glucagon (1 μg/kg) in obese subjects, and, in a subsequent study Schade and Eaton (1977), suggested that the obese subjects were resistant to the lipolytic activity of glucagon since normal responses could be elicited by a larger dose of glucagon (2 μg/kg). The authors themselves suggested (Schade and Eaton 1975b; Schade et al. 1979) that the difference, rather than being due to different tissue sensitivity to glucagon, was due to the differences in insulin secretion elicited by glucagon secretion. The importance of the prevailing insulin concentrations in plasma for the lipolytic response to glucagon is apparent from other studies on nonobese subjects (Gerich et al. 1976; Marks 1973; Liljenquist et al. 1974).

However, with respect to other metabolic effects of glucagon, obese subjects seem to be similar to normal subjects. Thus, fasting neither affected the glucose, insulin (Arky et al. 1970) nor cyclic AMP response to glucagon in obese patients (Turpin et al. 1978); intravenous glucagon infusions with subsequent oral glucose tolerance tests gave similar results in obese subjects and controls (Sherwin et al. 1976). When measured with an isotope technique, the hepatic glucose production in response to infusion of graded doses of glucagon was also reported to be similar in obese subjects and normal controls (Vranic et al. 1980) and infusions of "physiologic" doses of glucagon in obese subjects gave normal glucose responses (Fisher et al. 1976).

Thus, it is not clearly established that obese subjects respond differently to the metabolic activity of glucagon, but it may be that the hyperinsulinemia, characteristic of human obesity, influences glucagon action in these patients. On the other hand insulin resistance, equally characteristic of obesity (Olefsky 1981) might to some extent level out this difference.

## C. Experimental Obesity

Experimental models of obesity include genetically transmitted obesity (Bray and York 1971), obesity resulting from bilateral ventromedial hypothalamic (VMH) lesion (Brobeck 1964), and obesity following dietary (Schemmel et al. 1970) or endocrine manipulations (Bray and York 1971). Aspects of the role of glucagon in the pathogenesis of obesity have been studied in several of these models (Tables 3 and 4).

Among the genetically obese animals, at least three patterns of inheritance have been described:

1. Transmission as an autosomal dominant trait, as seen in yellow obese mice
2. Transmission as an autosomal recessive trait, as demonstrated in ob/ob, db/db, and ad/ad mice and fa/fa rats
3. As polygenic inheritance

Data concerning glucagon secretion or effect are available from studies of the obese (ob/ob) mouse (Ingalls et al. 1950) and the fatty (fa/fa) rat described by Zucker and Zucker (1961).

**Table 3.** Glucagon secretion in experimental obesity

| Reference | Basal plasma levels | Test of A-cell function | Response in obesity |
|---|---|---|---|
| FINDLAY et al. (1973) | | Pancreatic glucagon content (radioimmunoassay) | Increased in ob/ob mice |
| LEMONNIER et al. (1974) | | Pancreatic glucagon content (radioimmunoassay) | Normal in Zucker rats |
| LAUBE et al. (1974) | | Glucagon secretion after arginine or glucose from isolated perfused pancreas | Increased in ob/ob mice |
| LAVINE et al. (1975) | Increased | Glucagon response to fasting | Decreased in ob/ob mice |
| LABURTHE et al. (1975) | Normal | Nature and content of pancreatic glucagon (radioimmuno-, radioreceptor-, and bioassay) | Normal in Zucker rats |
| EATON et al. (1976) | Reduced | Response to fasting, i.v. arginine and glucose | Decreased, Increased } in Zucker rats |
| BRYCE et al. (1977) | Normal | Release of glucagon from isolated islets | Decreased in Zucker rats |
| KARAKASH et al. (1977) | Increased | | Increased in VMH rats |
| INOUE et al. (1977a) | Decreased<br>Normal | | Decreased in VMH rats<br>Normal in high fat diet rats |
| HAYEK and WOODSIDE (1979) | | Release of glucagon from isolated pancreas; arginine and cyclic AMP | Decreased in Zucker rats |
| ROHNER-JEANRENAUD and JEANRENAUD (1980) | | Secretion from isolated perfused pancreas; basal and after arginine and epinephrine | Increased in VMH rats |
| KARAKASH et al. (1980) | | Secretion from isolated perfused pancreas; basal and after arginine | Increased in VMH rats |
| CHIKAMORI et al. (1980) | Decreased | Response to starvation or i.v. arginine | Decreased in VMH rats |
| BOBBIONI and COSCELLI (1980) | Normal | Portal vein response to i.v. arginine | Increased in VMH rats |
| ROHNER-JEANRENAUD and JEANRENAUD (1981) | | Secretion from isolated perfused pancreas; basal and after arginine and methacholine | Increased in VMH rats |
| SEINO et al. (1981) | Increased | Secretion from isolated perfused pancreas in response to arginine | Normal in VMH rats and Zucker rats |
| NISHIKAWA et al. (1981) | Normal | Response to epinephrine and cold exposure | Normal in Zucker rats |
| | | Arginine | Increased in VMH rats and Zucker rats |
| | | Insulin hypoglycemia | Decreased in Zucker rats |

**Table 4.** Glucagon action in experimental obesity

| Reference | Test system | Response |
|---|---|---|
| MANGANIELLO and VAUGHAN (1972) | Fat cell adenylate cylate response to glucagon | Decreased in large fat cells |
| KAHN et al. (1973) | Binding of glucagon $^{125}$I to liver cell membranes | Decreased in ob/ob mice |
| LIVINGSTON et al. (1974) | Binding of glucagon $^{125}$I to adipocytes | Decreased in large rat cells |
| DE SANTIS et al. (1974) | Phosphodiesterase in glucagon-treated adipocytes | Increased in large rat cells |
| CHANG et al. (1975) | Binding of glucagon $^{125}$I to liver cell membranes | Normal in ob/ob mice |
| | Adenylate cyclase response to glucagon | Increased in ob/ob mice |
| BROER et al. (1977) | Binding of glucagon $^{125}$I to isolated hepatocytes | Normal in Zucker rats |
| | Cyclic AMP response in isolated hepatocytes | Increased in Zucker rats |
| LE MARCHAND et al. (1977) | Binding of glucagon $^{125}$I to liver cell membranes | Decreased in ob/ob mice |
| MA et al. (1978) | Inhibition by glucagon of lipid biosynthesis | Decreased in ob/ob mice |
| | Insulin antagonism of glucagon-induced glycogenolysis | Decreased in ob/ob mice |
| | Glycogenolysis; all in perfused livers | Normal in ob/ob mice |
| MAHMOUD et al. (1978) | Binding of glucagon $^{125}$I to liver cell membranes | Normal in Zucker rats |
| | Adenylate cyclase response to glucagon | Increased in Zucker rats |
| | Free fatty acid response to glucagon | Decreased in Zucker rats |
| LE MARCHAND et al. (1978) | Binding of glucagon $^{125}$I to liver cell membranes | Normal in gold thioglucose treated obese rats |
| LOCKWOOD et al. (1979) | Binding of glucagon $^{125}$I to liver cell membranes | Insignificantly decreased in obese monkeys |
| YEN et al. (1980) | Hepatic cyclic AMP and glycogenolysis after i.v. glucagon | Decreased in ob/ob mice |

## I. The Obese Mouse

In the ob/ob mouse, the basal levels of glucagon were found to be elevated (LAVINE et al. 1975) and the pancreatic content of glucagon (as measured by radioimmunoassays) increased (FINDLAY et al. 1973). In addition LAUBE et al. (1974) reported increased glucagon responses to arginine from isolated perfused pancreases of ob/ob mice. The glucagon response to fasting was decreased in ob/ob mice (LAVINE et al. 1975), probably because the basal levels were already elevated (LAVINE et al. 1975). Furthermore, KAHN et al. (1973) and LE MARCHAND et al. (1977) found decreased binding of glucagon ($^{125}$I-labeled) to isolated liver cell membranes from ob/ob mice, whereas CHANG et al. (1975) found normal binding of glucagon, but an increased adenylate cyclase activity in response to glucagon with such membranes. MA et al. (1978) found a normal glycogenolytic response to glucagon, whereas glucagon-induced inhibition of lipid biosynthesis and insulin antagonism of glucagon-induced glycogenolysis was decreased in livers from ob/ob mice. These data seem to indicate that glucagon secretion and/or activity may be disturbed in the ob/ob mouse; however, the ob/ob mouse is, apart from its obesity, also characterized by hyperglycemia, hyperinsulinemia, and marked insulin resistance (see BRAY and YORK 1971), features which characterize certain forms of diabetes. Abnormalities of glucagon secretion in diabetes are well known (see Chap. 44), and it remains to be seen whether the ob/ob mouse is not a better model of diabetes (or at least diabetic obesity) than of obesity as such (BRYCE et al. 1977).

## II. The Zucker Rat

In the fatty rat (Zucker rat, fa/fa) which is not hyperglycemic, the basal levels of glucagon in plasma were reported to be normal (LABURTHE et al. 1975, BRYCE et al. 1977) or decreased (EATON et al. 1976). The pancreatic glucagon content as determined by radioimmunoassay (LEMONNIER et al. 1974) or radioreceptorassay and bioassay (LABURTHE et al. 1975) was also normal, whereas the secretion of glucagon from isolated islets (BRYCE et al. 1977) or from the intact rat in response to arginine (EATON et al. 1976) was found to be reduced. However, normal secretion from isolated perfused pancreata has also been reported (SEINO et al. 1981). Furthermore, the binding of $^{125}$I-labeled glucagon to isolated liver cell membranes (BROER et al. 1977; MAHMOUD et al. 1978) was normal, and the adenylate cyclase response (BROER et al. 1977) or the cyclic AMP response to glucagon isolated hepatocytes (MAHMOUD et al. 1978) was increased in Zucker rats. On the other hand, the lipolytic response to glucagon was apparently decreased in Zucker rats (MAHMOUD et al. 1978).

The Zucker rats resemble obese humans inasmuch as they are also hyperinsulinemic (BRAY and YORK 1971; EATON et al. 1976; BRYCE et al. 1977; MAHMOUD et al. 1978), and it is conceivable that the hyperinsulinism may modulate the secretion as well as the biologic effects of glucagon in this condition as well as in human obesity, as already discussed. However, the differences between the obese (fa/fa) rats and their lean littermates (fa/−) with respect to glucagon are either small or variable (Table 3), and on the whole it seems unlikely that glucagon is essential for the development of obesity in this condition.

## III. Rats with Dietary Obesity

In rats made obese by high fat diets, plasma levels were normal (INOUE et al. 1977a), but, as reviewed in Chap. 13, the adenylate cyclase response to glucagon and the binding of $^{125}$I-labeled glucagon was decreased and the phosphodiesterase activity in glucagon-treated adipocytes was increased in large fat cells from rats allowed to develop obesity by overeating (MANGANIELLO and VAUGHAN 1972; LIVINGSTON et al. 1974; DE SANTIS et al. 1974). Glucagon binding to liver cell membranes from obese monkeys has also been studied, and was reported to be insignificantly decreased (LOCKWOOD et al. 1979).

### IV. VMH Rats

Ventromedial hypothalamic (VMH) lesions in rats result in hyperphagia, hyperinsulinemia, hypertriglyceridemia, reduced fatty acid oxidation, and increased lipogenesis, all leading to obesity (ASSIMACOUPOULOS-JEANNET and JEANRENAUD 1976). The obesity is not dependent on overeating and changes in the function of the endocrine pancreas also occur if hyperphagia is prevented (ROHNER-JEANRENAUD and JEANRENAUD 1980). However, the disturbed pancreatic activity may be essential for the metabolic changes after endocrine hypothalamic lesions (KARAKASH et al. 1977; INOUE and BRAY 1979). Plasma glucagon levels after VMH lesions have been reported to be normal (BOBBIONI and COSCELLI 1980) as well as decreased (INOUE et al. 1977a; CHIKAMORI et al. 1980) or increased (KARAKASH et al. 1977; SEINO et al. 1981). In addition, the glucagon response to intravenous arginine was found to be decreased (CHIKAMORI et al. 1980) as well as increased (BOBBIONI and COSCELLI 1980; NISHIKAWA et al. 1981) in rats with hypothalamic lesions. The basal as well as the arginine-stimulated secretions of glucagon from isolated perfused pancreata were reported to be increased when the pancreata were obtained from rats with hypothalamic lesions (ROHNER-JEANRENAUD and JEANRENAUD 1980, 1981; KARAKASH et al. 1980; JEANRENAUD et al. 1981), or the values were similar to those in sham-operated controls (SEINO et al. 1981). The binding of $^{125}$I-labeled glucagon to isolated liver cell membranes was normal in membranes from rats with hypothalamic lesions due to goldthioglucose administration (LE MARCHAND et al. 1978).

It seems well documented that abnormalities in insulin secretion are intimately associated with the development of obesity in rats with VMH lesions (KARAKASH et al. 1977) and that the abnormalities are transmitted to the pancreas via the autonomic nervous system (INOUE et al. 1977b). In spite of convincing demonstration of increased glucagon responses to arginine from the pancreas of VMH rats in vitro, it is as yet not clear that such abnormalities of glucagon secretion are also present in vivo. Furthermore, although the mechanism worked out for experimental hypothalamic obesity may apply to human hypothalamic obesity as well (INOUE and BRAY 1979) it is not yet established that abnormal autonomic activity is involved in the remaining syndromes of obesity (GOLD et al. 1980). The role of glucagon in obesity, therefore, has not as yet emerged in greater clarity from the study of experimental hypothalamic obesity.

## D. Extrapancreatic Glucagon in Obesity

The first reported study on plasma levels of glucagon in human obesity (PAULSEN and LAWRENCE 1968) was performed with the use of an assay of the so-called cross-reacting type, i.e., the antiserum used was directed against a part of the sequence of the glucagon molecule which is also accessible for antibodies in the peptides from the gut mucosa which contain the glucagon sequence: the gut glucagons or glut GLIs (glucagon-like immunoreactants) (see Chaps. 7, 12). In addition, PAULSEN and LAWRENCE (1968) studied the response to oral glucose in their patients (obese children) whereby it may be concluded that the GLI studied was derived from the gut rather than the pancreas. In these obese children (but not in obese diabetic children), the GLI levels were abnormally elevated.

Also, TIENGO et al. (1972) studied GLI determined with a cross-reacting antiserum, but these authors were unable to confirm the increased levels of GLI. GLI secretion of intestinal origin in morbidly obese subjects was recently studied by BARRY et al. (1977), BESTERMAN et al. (1978), and HOLST et al. (1979). The most important finding in these papers was the demonstration of greatly elevated GLI levels after jejunoileal bypass operations, whereas little was said about unoperated obese subjects. However, normal levels of GLI were noted by BESTERMAN et al. (1978) and the meal response of morbidly obese subjects reported by HOLST et al. (1979) and more recently by SØRENSEN et al. (1983) indicate that GLI secretion is not elevated in obesity. On the contrary, data from our laboratory (HOLST et al. 1983) indicate that GLI secretion may be significantly decreased in obese subjects.

Intestinal glucagon has repeatedly been implicated in lipid metabolism. The two main forms: peak I, probably glicentin; and peak II, ("basic GLI") have both been assigned lipolytic activity (LEFEBVRE et al. 1969; GUTMAN et al. 1973), but impurities may have been responsible for much of the activity (GUTMAN et al. 1973; LANGSLOW 1973). More recent studies indicate that intestinal GLIs may be inactive or weakly lipolytic (HORIGOME et al. 1977; CONLON et al. 1979; KRUG 1978), or even inhibitory to lipolysis (KRUG and MIALHE 1977).

Oral or intraintestinal administration of triglycerides stimulate the secretion of intestinal GLI in humans and dogs (BÖTTGER et al. 1973; HOLST et al. 1976; CHRISTIANSEN et al. 1979). Medium chain triglycerides and intravenously administered triglycerides failed to influence GLI secretion (BÖTTGER et al. 1973). At present, therefore, there is little evidence to support suppositions about a key role for intestinal GLI in the pathogenesis of obesity.

## References


Aguilar-Parada E, Eisentraut AM, Unger RH (1969) Effects of starvation on plasma pancreatic glucagon in normal man. Diabetes 18:717–723

Andrews AS, Lopez-S A, Blackard WG (1975) Effect of lipids on glucagon secretion in man. Metabolism 24:35–44

Arky RA, Finger M, Veverbrants E, Braun AP (1970) Glucose and insulin response to intravenous glucagon during starvation. Am J Clin Nutr 23:691–695

Assimacopoulos-Jeannet F, Jeanrenaud B (1976) The hormonal and metabolic basis of experimental obesity. Clin Endocrinol Metab 5:337–365

Atkinson RL, Dahms WT, Bray GA, Sperling MA (1981) Adrenergic modulation of glucagon and insulin secretion in obese and lean humans. Horm Metab Res 13:249–253

Barry RE, Barisch J, Bray A, Sperling MA, Morin RJ, Benfield J (1977) Intestinal adaption after jejunoileal bypass in man. Am J Clin Nutr 30:32–42

Beck-Nielsen H (1980) Insulin receptors in man. The monocyte as model for insulin receptor studies. Thesis, Laegeforeningens, Copenhagen

Besterman HS, Sarson DL, Blackburn HM, Cleary J, Pilkington TRE, Bloom SR (1978) The gut-hormone profile in morbid obesity and following jejunoileal bypass. Scand J Gastroenterol 13 [Suppl 49]: 15

Bobbioni E, Coscelli C (1980) Portal levels of glucagon and insulin in VMH-lesioned rats. Horm Metab Res 12:480–481

Böttger I, Dobbs R, Faloona GR, Unger RH (1973) The effects of triglyceride absorption upon glucagon, insulin, and gut glucagon like immunoreactivity. J Clin Invest 52:2532–2541

Bray GA, York DA (1971) Genetically transmitted obesity in rodents. Physiol Rev 51:598–646

Brobeck JR (1974) Mechanism of the development of obesity in animals with hypothalamic lesions. Physiol Rev 26:541–559
Broer Y, Freychet P, Rosselin G (1977) Insulin and glucagon-receptor interactions in the genetically obese Zucker rat: studies of hormone binding and glucagon-stimulated cyclic AMP levels in isolated hepatocytes. Endocrinology 101:236–249
Bryce GF, Johnson PR, Sullivan AC, Stern JS (1977) Insulin and glucagon plasma levels and pancreatic release in the genetically obese Zucker rat. Horm Metab Res 9:366–370
Chang K-J, Huang D, Cuatrecasas P (1975) The defect in insulin receptors in obese-hyperglycemic mice: a probable accompaniment of more generalized alterations in membrane glycoproteins. Biochem Biophys Res Commun 64:566–573
Chikamori K, Nishimura N, Suehiro F, Sato K, Mori H, Saito S (1980) Alterations in glucagon secretion in obese rats with hypothalamic lesions. Horm Metab Res 12:56–59
Christiansen J, Bech A, Fahrenkrug J, Holst JJ, Lauritsen K, Moody AJ, Schaffalitzky de Muckadell O (1979) Fat-induced jejunal inhibition of gastric acid secretion and release of pancreatic glucagon, gastric inhibitory polypeptide, and vasoactive intestinal polypeptide in man. Scand J Gastroenterol 14:161–166
Conlon JM, Murphy RF, Buchanan KD (1979) Physicochemical and biological properties of glucagon-like polypeptides from porcine colon. Biochim Biophys Acta 577:229–240
De Santis RA, Gorenstein T, Livingston JN, Lockwood DH (1974) Role of phosphodiesterase in glucagon resistance of large adipocytes. J Lipid Res 15:33–38
Eaton RP, Conway M, Schade DS (1976) Endogenous glucagon regulation in genetically hyperlipemic obese rats. Am J Physiol 230:1336–1341
Falluca F, Menzinger G, Gambardella S, Tamburrano S, Andreani D (1975) Glucagon, insulin and growth hormone response in obese women. Acta Diabetol Lat 12:239–243
Felig P, Wahren J, Hendler R, Brundin T (1974) Splanchnic glucose and amino acid metabolism in obesity. J Clin Invest 53:582–590
Findlay JA, Rookledge KA, Beloff-Chain H, Lever JD (1973) A combined biochemical and histological study of the islets of Langerhans in the genetically obese hyperglycaemic mouse and in the lean mouse, including observations on the effects of Streptozotocin treatment. J Endocrinol 56:571–583
Fisher M, Sherwin RS, Hendler R, Felig P (1976) Kinetics of glucagon in man: effects of starvation. Proc Natl Acad Sci USA 73:1735–1739
Galbo H, Holst JJ, Christensen NJ (1975) Glucagon and plasma catecholamine responses to graded and prolonged exercise in man. J Appl Physiol 38:70–75
Galbo H, Richter EA, Hilsted J, Holst JJ, Christensen NJ, Henriksson J (1977) Hormonal regulation during prolonged exercise. Ann NY Acad Sci 301:72–80
Galbo H, Holst JJ, Christensen NJ (1979) The effect of different diets and of insulin on the hormonal response to prolonged exercise. Acta Physiol Scand 107:19–32
Garlaschi C, di Natale B, del Guercio MJ, Caccamo A, Gargantini L, Chiumello G (1975) Effect of physical exercise on secretion of growth hormone, glucagon, and cortisol in obese and diabetic children. Diabetes 24:758–761
Gerich JE, Schneider V, Langlois M, Karam JH, Noacco C (1974) Glucagon secretion in obesity. Clin Res 22:190A
Gerich JE, Lorenzi M, Bier DM, Tsalikian E, Schneider V, Karam JH, Forsham PH (1976) Effects of physiologic levels of glucagon and growth hormone on human carbohydrate and lipid metabolism. J Clin Invest 57:875–884
Glass AR, Burman KD, Dahms W, Boehm TM (1981) Endocrine function in human obesity. Metabolism 30:89–104
Göschke H (1977) Mechanism of glucose intolerance during fasting: differences between lean and obese subjects. Metabolism 26:1147–1153
Gold RM, Sawchenko PE, DeLuca C, Alexander J, Eng R (1980) Vagal mediation of hypothalamic obesity but not of supermarket dietary obesity. Am J Physiol 238:R447–R453
Gossain VV, Matute ML, Kalkhoff RK (1974) Relative influence of obesity and diabetes on plasma alpha-cell glucagon. J Clin Endocrinol Metab 38:238–243
Gutmann RA, Fink G, Voyles N, Selawry H, Penhos JC, Lepp A, Recant L (1973) Specific biologic effects of intestinal glucagon-like materials. J Clin Invest 52:1165–1175

Hayek A, Woodside W (1979) Correlation between morphology and function in isolated islets of the Zucker rat. Diabetes 28:565–569

Holst JJ, Christiansen J, Kühl C (1976) The enteroglucagon response to intrajejunal infusion of glucose, triglycerides, and sodium chloride, and its relation to jejunal inhibition of gastric acid secretion in man. Scand J Gastroenterol 11:297–304

Holst JJ, Sørensen TIA, Andersen AN, Stadil F, Andersen B, Lauritsen KB, Klein HC (1979) Plasma enteroglucagon after jejunoileal bypass with 3:1 or 1:3 jejunoileal ratio. Scand J Gastroenterol 14:205–207

Holst JJ, Schwartz TW, Løvgren NA, Pedersen O, Beck-Nielsen H (1983) Diurnal profile of pancreatic polypeptide, pancreatic glucagon, gut glucagon and insulin in human morbid obesity. Int. J Obesity (in press)

Horigome K, Ohneda A, Maruhama Y, Abe R, Kai Y (1977) Heterogeneity of extractable gut glucagon-like immunoreactivity (GLI) and its lipolytic activity. Horm Metab Res 9:370–374

Ingalls AM, Dickie MM, Snell GP (1950) Obese, new mutation in the mouse. J Hered 41:317–318

Inoue S, Bray GA (1979) An autonomic hypothesis for hypothalamic obesity. Life Sci 25:561–566

Inoue S, Campfield A, Bray GA (1977a) Comparison of metabolic alterations in hypothalamic and high fat diet-induced obesity. Am J Physiol 233:R162–R168

Inoue S, Bray GA, Mullen Y (1977b) The effect of transplantation of the pancreas on the development of hypothalamic obesity. Nature 266:742–744

Jarrett RJ, Bakir SM (1978) Insulin and obesity. Eur J Clin Invest 8:359–360

Jeanrenaud B (1979) Insulin and obesity. Diabetologia 17:133–138

Jeanrenaud B, Berthoud H-R, Bereiter DA, Rohner-Jeanrenaud F (1981) The role of the central nervous system in the regulation of the secretory activity of endocrine pancreas: a viewpoint from our laboratories. In: Enzi G, Crepaldi G, Pozza G, Renold AE (eds) Obesity: pathogenesis and treatment, vol 28. Academic, London, pp 23–35

Kahn CR, Neville DM, Roth J (1973) Insulin-receptor interaction in the obese-hyperglycemic mouce. J Biol Chem 248:244–250

Kalkhoff RK, Gossain VV, Matute ML (1973) Plasma glucagon in obesity. N Engl J Med 289:465–467

Karakash C, Hustvedt BE, Løvø A, le Marchand Y, Jeanrenaud B (1977) Consequences of ventromedial hypothalamic lesions on metabolism of perfused rat liver. Am J Physiol 232(3):E286–E293

Karakash C, Rohner-Jeanrenaud F, Hustvedt BE, Jeanrenaud B (1980) Nitrogen handling in adult hypothalamic obese rats. Am J Physiol 238:E32–E37

Krug E (1978) Antilipolytic nature of gut GLI, and mode of action of two highly potent intestinal lipolytic species in birds. Horm Metab Res 10:505–509

Krug E, Mialhe P (1977) A possible role for gut GLI: an inhibitor of lipolysis. Horm Metab Res 6:465–469

Laburthe M, Rancon F, Freychet P, Rosselin G (1975) Glucagon and insulin from lean rats and genetically obese fatty rats: studies by radioimmunoassay, radioreceptorassay and bioassay. Diabetologia 11:517–526

Langslow DR (1973) The action of gut glucagon-like immunoreactivity and other intestinal hormones on lipolysis in chicken adipocytes. Horm Metab Res 5:428–432

Lassmann V, Vague P, Vialettes B, Simon M-C (1980) Low plasma levels of pancreatic polypeptide in obesity. Diabetes 29:428–430

Laube H, Fussgänger RD, Pfeiffer EF (1974) Paradoxical glucagon release in obese hyperglycemic mice. Horm Metab Res 6:426

Lavine RL, Voyles N, Perrino PV, Recant L (1975) The effect of fasting on tissue cyclic cAMP and plasma glucagon in the obese hyperglycemic mouse. Endocrinology 97:615–620

Lefèbvre PJ, Unger RH, Valverde I, Rigopoulou D, Luyckx AS, Eisentraut A (1969) Effect of dog jejunum "glucagon-like immunoreactive material" on adipose tissue metabolism. Horm Metab Res 1:143–144

LeMarchand Y, Loten EG, Assimacopoulos-Jeannet F, Forgue M-E, Freychet P, Jeanrenaud B (1977) Effect of fasting and streptozotocin in the obese-hyperglycemic (ob/ob) mouse. Apparent lack of a direct relationship between insulin binding and insulin effects. Diabetes 26:582–590
LeMarchand Y, Freychet P, Jeanrenaud B (1978) Longitudinal study on the establishment of insulin resistance in hypothalamic obese mice. Endocrinology 102:74–85
Lemonnier D, Aubert R, Suquet J-P, Rosselin G (1974) Metabolism of genetically obese rats on normal or high-fat diet. Diabetologia 10:697–701
Liljenquist JE, Bomboy JD, Lewis SB, Sinclair-Smith BC, Felts PW, Lacy WW (1974) Effects of glucagon on lipolysis and ketogenesis in normal and diabetic men. J Clin Invest 53:190–197
Livingston JN, Cuatrecasas P, Lockwood DH (1974) Studies of glucagon resistance in large rat adipocytes: 125-I-labeled glucagon binding and lipolytic capacity. J Lipid Res 15:26–38
Lockwood DH, Hamilton CL, Livingston JN (1979) The influence of obesity and diabetes in the monkey on insulin and glucagon binding to liver membranes. Endocrinology 104:76–80
Ma GY, Gove CD, Hem DA (1978) Effects of glucagon and insulin on fatty acid synthesis and glucagon degradation in the perfused liver of normal and genetically obese (ob/ob) mice. Biochem J 174:761–768
Mahmoud HA, Wood PJ, Marks V (1978) The effect of induced hyperglucagonaemia on the Zucker fatty rat. Diabetologia 14:405–412
Manganiello V, Vaughan M (1972) Selective loss of adipose cell responsiveness to glucagon with growth in the rat. J Lipid Res 13:12–16
Marliss EB, Aoki TT, Unger RH, Soeldner JS, Cahill GF (1970) Glucagon levels and metabolic effects in fasting man. J Clin Invest 49:2256–2270
Marks V (1973) Glucagon and lipid metabolism in man. Postgrad Med J 49:615–619
Meschi F, Garlaschi C, di Natale B, del Guercio MJ, Gargantini L, Chiumello G (1977) Glucagon response to arginine stimulation in obese and diabetic children. Diabetes 26:558–560
Minuk HL, Hanna AK, Marliss EB, Vranic M, Zinman B (1980) Metabolic response to moderate exercise in obese man during prolonged fasting. Am J Physiol 238:E322–E329
Müller WA, Faloona GR, Unger RH (1971) The influence of the antecedent diet upon glucagon and insulin secretion. N Engl J Med 285:1450–1454
Müller WA, Aoki TT, Cahill GF (1975) Effect of alanine and glycine on glucagon secretion in postabsorptive and fasting obese man. J Clin Endocrinol Metab 40:418–425
Müller WA, Aoki TT, Egdahl RH, Cahill GF (1977) Effects of exogenous glucagon and epinephrine in physiological amounts on the blood levels of free fatty acids and glycerol in dogs. Diabetologia 13:55–58
Nishikawa K, Ikeda H, Matsuo T (1981) Abnormal glucagon secretion in Zucker fatty rats. Horm Metab Res 13:259–263
Olefsky JM (1981) Insulin resistance and insulin action. An in vitro and in vivo perspective. Diabetes 30:148–162
Paulsen EP, Lawrence AM (1968) Glucagon hypersecretion in obese children. Lancet 2:110
Pozefsky T, Tancredi RG, Moxley RT, Dupre T, Tobin JD (1976) Metabolism of forearm tissues in man. Studies with glucagon. Diabetes 25:128–135
Rohner-Jeanrenaud F, Jeanrenaud B (1980) Consequences of ventromedial hypothalamic lesions upon insulin and glucagon secretion by subsequently isolated perfused pancreases in the rat. J Clin Invest 65:902–910
Rohner-Jeanrenaud F, Jeanrenaud B (1981) Possible involvement of the cholinergic system in hormonal secretion by the perfused pancreas from ventromedial-hypothalamic lesioned rats. Diabetologia 20:217–222
Santiago JV, Haymond MW, Clarke WL, Pagliara AS (1977) Glucagon, insulin, and glucose responses to physiologic testing in normal and massively obese adults. Metabolism 26:1115–1122
Savage PJ, Bennion LJ, Flock EV, Nagule-Sparan M, Mott D, Roth J, Unger RH, Bennett PH (1979) Diet-induced improvement of abnormalities in insulin and glucagon secretion

and in insulin receptor binding in diabetes mellitus. J Clin Endocrinol Metab 48:999–1007
Schade DS, Eaton RP (1974) Role of insulin and glucagon in obesity. Diabetes 23:657–661
Schade DS, Eaton RP (1975a) Altered tissue response to glucagon in obesity. J Clin Endocrinol Metab 40:732–735
Schade DS, Eaton RP (1976) Modulation of fatty acid metabolism by glucagon in man. IV. Effects of a physiologic hormone infusion in normal man. Diabetes 25:978–983
Schade DS, Eaton RP (1977) Modulation of the catabolic activity of glucagon by endogenous insulin secretion in obese man. Acta Diabetol Lat 14:62–72
Schade DS, Eaton RP (1975b) The contribution of endogenous insulin secretion to the ketogenic response to glucagon in man. Diabetologia 11:555–559
Schade DS, Woodside W, Eaton RP (1979) The role of glucagon in the regulation of plasma lipids. Metabolism 28:874–885
Schemmel R, Mickelsen O, Gill JL (1970) Dietary obesity in rats: body weight and fat accretion in seven strains of rats. J Nutr 100:1041–1048
Schneider P, Rabinovitch A, Paunier L, Sizonenko PC (1975) Alpha cell function in short, obese, and diabetic children. Pediatr Res 9:733
Seino S, Seino Y, Takemura J, Tsuda K, Kuzuya H, Ishikawa K, Shimazu T, Imura H (1981) Somatostatin, insulin and glucagon secretion from isolated perfused pancreas of obese rats. Am J Physiol 241:E146–E150
Sherwin RS, Fisher M, Hendler R, Felig P (1976) Hyperglucagonemia and blood glucose regulation in normal, obese and diabetic subjects. N Engl J Med 294:455–461
Sims EAH (1979) Syndromes of obesity. In: Degroot JL, Cahill GF, Odell WD, Martini L, Potts, JT, Nelson DH, Steinbergher E, Winegrad AT (eds) Endocrinology 3. Grune and Stratton, New York, pp 1941–1962
Sørensen TIA, Nyboe Andersen A, Lauritsen KB, Holst JJ, Stadil F, Andersen B (1983) Gut and pancreatic hormone profile after jejuno-ileal bypass with 3:1 or 1:3 jenuno-ileal ratio. Digestion 66:137–145
Tiengo A, Assan R, Tchobroutsky G (1972) Metabolic and hormonal patterns after three days of total fasting in 27 obese and non-obese subjects. Isr J Med Sci 8:821–822
Trovati M, Lorenzatio R, Cassader M, Pisu E, Martiny W, Pagano G (1981) Metabolic effects of short-term treadmill exercise in moderately obese subjects. In: Enzi G, Crepaldi G, Pozza G, Renold AE (eds) Obesity: pathogenesis and treatment, vol. 28. Academic Press, London, pp 289–296
Turpin BP, Austin MW, Solomon SS (1978) Urinary and plasma cyclic AMP levels during short term starvation in obese man: response to glucagon stimulation. Horm Metab Res 10:36–37
Vranic M, Morita S, Steiner G (1980) Insulin resistance in obesity as analyzed by the response of glucose kinetics to glucagon infusion. Diabetes 29:169–176
Walter RM, Gold EM, Michas CA, Ensinck JW (1980) Portal and peripheral vein concentrations of insulin and glucagon after arginine infusion in morbidly obese subjects. Metabolism 29:1037–1041
Wise JK, Hendler R, Felig P (1972) Obesity: evidence of decreased secretion of glucagon. Science 178:513–514
Wise JK, Hendler R, Felig P (1973a) Influence of glucocorticoids on glucagon secretion and plasma amino acid concentrations in man. J Clin Invest 52:2774–2782
Wise JK, Hendler R, Felig P (1973b) Evaluation of alpha-cell function by infusion of alanine in normal, diabetic and obese subjects. N Engl J Med 288:487–490
Yen TT, Stamm NB, Fuller RW, Root MA (1980) Hepatic insensitivity to glucagon in ob/ob mice. Res Commun Chem Pathol Pharmacol 30:29–40
Zucker TF, Zucker LM (1961) Fatty, a new mutation in the rat. J Hered 52:275–278

# Pharmacological Effects of Glucagon and the Use of Glucagon for Diagnosis and in Therapeutics

CHAPTER 50

# Influence of Glucagon on Water and Electrolyte Metabolism

J. KOLANOWSKI

## A. Introduction

Since the original reports of STAUB et al. (1956, 1957) demonstrating that the injection of glucagon induced a strong diuretic effect in dogs, with an enhancement in renal excretion of sodium, chloride, phosphate, and potassium, several studies have been devoted to the possible role of glucagon in the regulation of water and electrolyte balance. Despite numerous investigations in this field, it still remains uncertain whether under physiologic conditions glucagon really plays a role in the maintenance of salt and water homeostasis. It has nevertheless been convincingly demonstrated that in some ketogenic states associated with insulinopenia, such as starvation or uncontrolled insulin-dependent diabetes, the increase in plasma glucagon levels promotes salt and water depletion.

The influence of glucagon on salt and water metabolism seems to be secondary to its direct effect on kidney function, thus on an organ which is involved to an important extent in the extraction and metabolic inactivation of glucagon. The degradation of glucagon by the kidney and the postulated influence of the hormone on renal function are in all likelihood unrelated and occur at different segments of the nephron. It seems advisable, however, to discuss both of these aspects of the relationship between glucagon physiology and kidney function in an attempt to delineate the possible role which glucagon may play in changes of water and mineral balance in some pathophysiologic situations.

## B. The Kidney as a Target Site for Glucagon Action on Water and Mineral Metabolism

### I. Renal Extraction, Handling, and Metabolism of Glucagon

Simultaneously with the first reports suggesting a direct influence of glucagon on kidney function (STAUB et al. 1957; ELRICK et al. 1958; DALLE et al. 1959) it was demonstrated that glucagon is readily reabsorbed across the luminal membrane of tubular cells and degraded, predominantly in proximal portions of the tubule (NARAHARA et al. 1958). The mechanism of glucagon inactivation by the kidney was further elucidated by DUCKWORTH (1976) who demonstrated that kidney tissue contains specific glucagon-degrading enzymes in both the proximal tubular cytosol and brush border.

A detailed analysis of the available data regarding the renal handling of glucagon can be found in Chap. 41. Additional information on plasma glucagon values in renal insufficiency is provided in Chap. 47. Since the kidney contributes significantly to the overall metabolism of glucagon, an impairment of renal function can induce a rise in circulating glucagon levels which may in turn modify renal hemodynamics and/or tubular function. In addition, it is of interest to note that hyperglucagonemia of early fast is due to a significant reduction in the metabolic clearance rather than to increased production of the hormone (FISHER et al. 1976; BURMAN et al. 1980). Although the contribution of the kidney to the fast-induced reduction in glucagon metabolism has not been appreciated in these studies, the decline in metabolic clearance rate of glucagon may represent one of several aspects of altered renal function that characterize the early phase of total starvation.

## II. Direct Effect of Glucagon on Renal Function

That glucagon may exert a direct influence on the kidney, stimulating mineral and water excretion, has been suggested by several investigators who demonstrated a unilateral increase in urinary flow and ion excretion when the hormone was infused directly into the renal artery. In all these experiments, however, highly pharmacologic amounts of glucagon were infused, and the question whether physiologic changes in glucagon secretion can really modify the renal function still awaits elucidation. From a theoretical point of view, the influence of glucagon on water and mineral handling by the kidney may result from an increase in glomerular filtration or renal blood flow as well as from a direct effect of the peptide on tubular function. As will appear from the review of the available data, all of these aspects of kidney function may be altered by glucagon, at least when large unphysiologic amounts of the hormone are administered systemically or directly into the renal artery.

### 1. Glucagon-Induced Increase in the Glomerular Filtration Rate

A significant increase in glomerular filtration rate (GFR) following an intravenous injection of glucagon was first demonstrated in dogs by STAUB et al. (1957) and in humans by ELRICK et al. (1958). This effect of glucagon was reproduced by PULLMAN et al. (1967) who infused the peptide into the dog renal artery and suggested that the resulting increase in GFR represented a direct effect of glucagon on the kidney, resulting in an enhancement in water and salt excretion. Using a comparable experimental approach, LEVY and STARR (1972) evidenced that either intravenous or intrarenal infusion of glucagon may increase GFR by 30% without any change in the ratio of end-proximal tubular fluid to plasma inulin and concluded that glucagon is devoid of any effect on the proximal tubule. Since the increase in GFR induced by glucagon cannot be prevented by renal denervation or by the blockade of adrenergic, cholinergic, dopaminergic, or histaminergic receptors (LEVY 1975a) it has been suggested that glucagon may stimulate glomerular filtration through an action on specific receptors. This conclusion has not been substantiated by the more recent study of URANGA et al. (1979) which demonstrated that the glucagon-induced rise in GFR is due to an increase in the hepatic production of a substance called glomerulopressin. These authors observed a significant concomitant increase in glomerulopressin release by the liver and in GFR when glucagon was infused into the portal vein, while the same amount of the hormone administered directly into the renal artery was devoid of effect on glomerular filtration. However, this interesting observation must be taken with caution since the physiologic role of glomerulopressin has not been clearly established as yet.

As to the mechanism of the glucagon-induced increase in GFR it seems to result from selective dilatation of afferent arterioles with subsequent increase in the filtration pressure (UEDA et al. 1977). KIRSCHENBAUM and ZAWADA (1980) have recently demonstrated that the increase in GFR induced in dogs by glucagon infused intravenously at the rate of 1 μg/min was not prevented by inhibitors of prostaglandin synthesis. On the other hand, at fivefold lower dosage, glucagon was without any measurable effect on GFR while still inducing a clear-cut natriuretic effect (KIRSCHENBAUM and ZWADA 1980) which was apparently related to the increase in prostaglandin synthesis by the kidney. It appears therefore that, at least in the dog, the effect of glucagon on GFR is rather a pharmacologic one and cannot explain entirely the natriuretic properties of this hormone.

The recent studies of PARVING et al. (1977, 1980) suggest nevertheless that hyperglucagonemia comparable to that observed in some pathologic states such as poorly controlled insulin-dependent diabetes mellitus, may significantly increase the rate of glomerular filtration. These authors demonstrated that, upon glucagon infusion into diabetic subjects (at the rate of 20–40 μg/h which increased the plasma glucagon level from 254 to 440 pg/ml), there occurred a significant rise in GFR highly correlated with the degree of induced hyperglucagonemia. These interesting observations suggest that the increase in glomerular filtration effectively occurring in the early phase of decompensated insulin-dependent diabetes may be partly due to the concomitant increase in glucagonemia (PARVING et al. 1980).

## 2. Influence of Glucagon on Renal Hemodynamics

That the glucagon-induced increase in water and ion excretion results from a direct vasodilatatory effect of the hormone on the kidney with an increase in renal plasma flow (RPF) was first suggested by PULLMAN et al. (1967). The essential role of the renal vasodilatatory influence of glucagon in the mechanism of its natriuretic effect was further underlined by the observation that the glucagon-induced natriuresis can be completely abolished by previous maximal renal vasodilatation (STOWE and HOOK 1970).

The influence of glucagon on renal hemodynamics was further substantiated by the infusion of the hormone into the renal artery in dogs (LEVY and STARR 1972; LEVY 1975b; UEDA et al. 1977) and directly evidenced with renal arteriography (DANFORD 1970). The pathophysiologic implications of this postulated effect of glucagon on kidney hemodynamics have been recently suggested by PARVING et al. (1980) who observed a slight, but significant increase in RPF in diabetic patients given intravenous glucagon in amounts raising plasma glucagon to levels encountered in poorly controlled diabetics. A direct relaxant effect of glucagon on renal vasculature has been demonstrated in strips of rabbit renal artery previously contracted with norepinephrine (GAGNON et al. 1978); it is of interest to note that this vasodilator effect of glucagon can be reproduced by cyclic AMP and is markedly potentiated by phosphodiesterase inhibitors. On the other hand, the vascular effect of glucagon can be inhibited by calcium antagonists such as verapamil (GAGNON et al. 1980). It has been suggested, therefore, that glucagon may exert its vasodilator effect by promoting calcium extrusion from the vascular smooth muscle cells by a mechanism involving an increased generation of cyclic AMP (GAGNON et al. 1980).

An alternative mechanism by which glucagon may produce its vasodilatatory effect on the kidney could be related to the stimulatory influence of glucagon on renal prostaglandin synthesis. It has been demonstrated recently (KIRSCHENBAUM and ZAWADA 1980) that the glucagon-induced increase in urine flow and sodium excretion rates is associated with an enhancement in prostaglandin E excretion, and all these renal effects of glucagon can be reversed by the administration of inhibitors of prostaglandin synthesis. Since prostaglandins stimulate water and sodium excretion by the kidney, at least in part as a consequence of their vasodilatatory effect on renal circulation (TANNENBAUM et al. 1975; ANDERSON et al. 1976; GLASSON and VALLOTTON 1980), it can be speculated that glucagon stimulates water and salt excretion by a prostaglandin-mediated mechanism. In favor of this interpretation is the recent observation of LAMEIRE et al. (1980) that other vasodilators with natriuretic properties, such as acetylcholine, also act on the kidney by inducing the synthesis of prostaglandins; furthermore, as is the case for glucagon, the effect of acetylcholine on the kidney can be prevented by inhibitors of prostaglandin synthesis. On the other hand, prostaglandins antagonize the effect of antidiuretic hormone (ADH) on the collecting tubule (ANDERSON et al. 1976), which may explain the observation that the stimulatory effect of glucagon on water diuresis cannot be completely suppressed by the administration of ADH (DAVID et al. 1960).

### 3. Postulated Direct Effect of Glucagon on Tubular Function

A direct depressing influence of glucagon on tubular reabsorption of sodium and other ions has been suggested by several investigators who reported that the hormone could promote salt and water excretion in the absence of a concomitant increase in GFR (STAUB et al. 1956, 1957; ELRICK et al. 1958; DALLE et al. 1959). PULLMAN et al. (1967) came to a similar conclusion when they noted that, following a unilateral infusion of glucagon into the renal artery, the excretion of Na, Cl, K, P, and Ca, as well as urinary flow, were significantly increased only in the perfused kidney, despite a comparable bilateral rise in GFR and RPF. Other investigators, who emphasized the essential role of renal vasodilatation in the natriuretic response to glucagon administration (STOWE and HOOK 1970; LEVY and STARR 1972) also suggested that, besides its vascular effect, the hormone may directly interfere with salt reabsorption by distal segments of the nephron. On the basis of these observations, AVIOLI (1972) concluded that the effect of glucagon on urinary electrolyte excretion is due primarily to a direct influence on tubular function while the concomitant changes in RPF and GFR have little, if any, additional effect on ion excretion. The possibility of a direct action of glucagon on renal tubules has also been emphasized by KATZ and LINDHEIMER (1977) in their review of the action of hormones on kidney function. Such conclusion was also supported by the recent work of BAILLY and AMIEL (1982) who clearly demonstrated in the rat an increase in the magnesium renal reabsorption. Since this effect was observed whatever the filtered load and independently of parathormone and calcitonin, it was suggested that this increased magnesium reabsorption may reflect a direct tubular effect of glucagon.

It must be stressed, however, that the evidence for a direct action of glucagon on tubular transport processes for sodium (or other ions) is still lacking. Since other hormones which influence sodium reabsorption by the mammalian kidney, such as aldosterone and insulin, also modify sodium transport by amphibian epithelia, of interest is the fact that glucagon is devoid of any effect on sodium transport across the toad bladder (CRABBÉ 1973). Moreover, glucagon does not influence the adenylate cyclase activity in intestinal epithelium and is without effect on sodium reabsorption by mouse intestine (KAUFMAN et al. 1980). It seems, therefore, that the reported inhibitory effect of glucagon on sodium and chloride reabsorption by the human jejunum (HICKS and TURNBERG 1974) results essentially, if not exclusively, from the influence of the hormone on the splanchnic blood flow (see Chap. 54).

If glucagon really acts directly on the nephron, one could expect the existence of specific glucagon receptors within the kidney tissue and the ability of the peptide to activate the adenylate cyclase system in tubular cells. While there is no evidence for a specific binding of glucagon to tubular cells as yet, it has been demonstrated that glucagon does stimulate the adenylate cyclase activity in cortical tubules isolated from rat kidney (MARCUS and AURBACH 1969; MELSON et al. 1970) and in human renal medulla (MULVEHILL et al. 1976; KIM et al. 1977). BAILLY et al. (1980) have demonstrated recently that the stimulatory effect of glucagon on cyclic AMP generation is most pronounced in the thick medullary portion of the ascending limb, thus in the portion of the nephron presumably involved in the effect of glucagon on sodium reabsorption (STOWE and HOOK 1970; LEVY and STARR 1972). The physiologic link between the stimulation by glucagon of cyclic AMP generation in the ascending limb and the postulated inhibitory influence of the hormone on sodium reabsorption by this diluting segment of the nephron still remains unclear. According to GILL et al. (1971), cyclic AMP can effectively block sodium reabsorption by the proximal tubule, but it does not completely prevent the reabsorption of the increased load of sodium reaching the distal nephron, hence there is a lack of a net natriuretic effect of this nucleotide. On the other hand, urinary excretion of cyclic AMP remains unchanged in fasting subjects (TURPIN et al. 1978), despite a significant increase in plasma glucagon levels associated with a clear-cut natriuresis.

While the ability of glucagon to stimulate the adenylate cyclase in tubular cells has been convincingly demonstrated, the mechanism by which glucagon may modify the reabsorption of sodium and other minerals by the nephron still awaits elucidation. The recent observations of KIRSCHENBAUM and ZAWADA (1980) suggest that the glucagon-induced increase in renal prostaglandin synthesis can explain not only the influence of the hormone on renal plasma flow, but also the direct tubular effect of the peptide reducing sodium reabsorption. It has been demonstrated indeed, that following the infusion of minute amounts of glucagon in the dog, there occurs a threefold increase in sodium excretion without any change in RPF, an effect that can be prevented by inhibitors of prostaglandin synthesis

(KIRSCHENBAUM and ZAWADA 1980). This observation is reminiscent of the significant natriuretic effect of small amounts of arachidonic acid infused directly into the renal artery (TANNENBAUM et al. 1975) which, in contrast to the vasodilatatory effect of large amounts of this precursor of prostaglandins, significantly stimulated renal sodium handling in the absence of a detectable increase in renal blood flow. It cannot be excluded, therefore, that the direct effect of glucagon on tubular sodium reabsorption implies a prostaglandin-mediated mechanism. Whether this glucagon-induced increase in prostaglandin synthesis interferes with sodium reabsorption exclusively as the result of changes in renal hemodynamics or also implies an inhibition of sodium transport by tubular cells cannot be definitely settled as yet. Whatever the mechanism may be, it seems unlikely that physiologic changes in glucagon secretion exert a significant influence on kidney function in normal conditions. Indeed, no changes in renal salt and water handling can be demonstrated upon physiologic increments in plasma glucagon levels either in humans (SHERWIN et al. 1977) or in dogs (FORREST et al. 1976).

## C. The Influence of Glucagon on Natriuresis of Starvation

The role of glucagon in renal salt and water loss attending the early phase of total starvation was suggested on the basis of a parallel increase in plasma glucagon levels and in renal sodium excretion during the first days of fasting; furthermore, glucagon can enhance fasting natriuresis and completely prevent the antinatriuresis of carbohydrate refeeding (KOLANOWSKI and CRABBÉ 1971; KOLANOWSKI et al. 1972; SAUDEK et al. 1973). The natriuretic properties of glucagon in fasting subjects have been confirmed by several additional studies (KOLANOWSKI et al. 1977, 1978; KOLANOWSKI 1980, 1981 b), but it seems that changes in glucagonemia cannot be considered as an essential hormonal mechanism influencing the salt and water balance in fasting conditions. Since this particular aspect of glucagon action has been discussed in detail recently (KOLANOWSKI 1979, 1981 a, b), the present review will be limited to certain key observations in an attempt to delineate the role which may be attributed to hyperglucagonemia in the fast-induced losses of sodium and other minerals.

The transient increase in plasma glucagon levels during the first days of fast, reported first by UNGER et al. (1963) and confirmed by numerous further studies (AGUILAR-PARADA et al. 1969; MARLISS et al. 1970; SAUDEK et al. 1973; AOKI et al. 1974; KOLANOWSKI et al. 1977) effectively parallels the progressive increase in natriuresis. The maximal increment in glucagonemia and in renal sodium excretion occurs on days 3 or 4 of starvation, and both parameters return progressively to control values after a few more days of fast. This apparent parallelism does not indicate, however, an obligatory causal relationship between changes in glucagonemia and in renal sodium handling. While the enhancement in natriuresis during the initial phase of fast seems to result essentially from increased excretion of organic acids (KOLANOWSKI et al. 1978) and from reduced insulin secretion (KOLANOWSKI 1979, 1981 b), the increase in glucagonemia is due to the reduction in metabolic clearance of the hormone without any significant change in its secretion rate (FISHER et al. 1976; BURMAN et al. 1980). Since the kidneys contribute significantly to glucagon disposal, it cannot be excluded that fasting hyperglucagonemia represents a consequence of the fast-induced alteration in renal function rather than an essential hormonal mechanism of the metabolic adaptation to starvation. In fact, correction of fasting hyperglucagonemia by means of somatostatin infusion does not reduce ketogenesis and natriuresis in the fasted state (KOLANOWSKI 1981 b) and suggests that glucagon can no longer be considered as the natriuretic hormone of fasting.

Nevertheless, it remains clear that glucagon infusion, even in doses increasing the plasma level of the hormone to within the pathophysiologic range, significantly enhances the excretion of sodium and other ions in fasting subjects (SAUDEK et al. 1973; KOLANOWSKI et al. 1977, 1978; KOLANOWSKI 1980, 1981 b). The previous suggestion that glucagon may exert its natriuretic properties in fasting subjects by inducing a tubular refractoriness to mineralocorticoids (O'BRIAN et al. 1974) cannot be considered as a satisfactory explanation, since the renal sensitivity to aldosterone is maintained in fasting subjects (KOLANOWSKI et al. 1976). Moreover, the infusion of glucagon increases not only the excretion of sodium, but also produces a clear-cut kaliuresis (KOLANOWSKI et al. 1977), and the blockade of mineralocorticoid receptors by spironolactone enhances considerably the renal sodium handling during starvation (KOLANOWSKI et al. 1976).

As to the mechanism of the natriuretic effect of even moderate hyperglucagonemia (a 2–3-fold increase from baseline) induced by glucagon infusion into fasting subjects, it seems to involve an increased fractional excretion of ketone bodies. It has, in fact, been observed that during a 10–24-h infusion of 0.1 mg glucagon into starving obese subjects, the plasma levels of ketone bodies remained unchanged or were slightly decreased, while the natriuresis and ketonuria were significantly enhanced (MARLISS et al. 1970; AOKI et al. 1974; KOLANOWSKI 1980, 1981 b). The GFR estimated from the clearance of endogenous creatinine remained unchanged in these conditions, but in the absence of simultaneous measurements of renal blood flow or prostaglandin excretion, the mechanism of this apparent effect of glucagon on tubular ketone body and sodium reabsorption has not been elucidated. Nevertheless, it can be postulated that in ketogenic states such as starvation, poorly controlled diabetes, or serious trauma, the increase in glucagonemia may promote the renal excretion of ketone bodies, hence resulting in an enhancement of natriuresis. The natriuretic role of hyperglucagonemia in insufficiently treated diabetics was in fact previously emphasized by SAUDEK et al. (1974).

The infusion of large amounts (1 mg over 6–10 h) of glucagon to fasting subjects, increasing at least 10–20-fold the already elevated plasma levels of the hormone, significantly enhances not only the renal handling of ketones and sodium, but also the plasma levels of ketone bodies (KOLANOWSKI 1980, 1981 b), probably as a consequence of the ketogenic effect of the hormone on the liver (MCGARRY et al. 1975; MERIMEE et al. 1978; see also Chap. 17). It is noteworthy that the same amounts of glucagon administered together with an oral glucose load, failed to prevent the reduction in ketonemia due to carbohydrate refeeding, but completely suppressed the fall in ketonuria and the antinatriuretic effect of glucose (KOLANOWSKI 1980, 1981 a, b). This observation provides an additional argument supporting the conclusion that hyperglucagonemia of the fasted state may contribute to the magnitude of sodium losses during starvation by reducing the tubular reabsorption of ketone bodies.

## D. The Effect of Glucagon on Plasma Electrolyte Levels

### I. Hypocalcemic Effect of Glucagon

AVIOLI et al. (1968) and BIRGE and AVIOLI (1969) have observed that the glucagon-induced increase in urinary calcium excretion (ELRICK et al. 1958; PULLMAN et al. 1967) was associated with a slight, but significant hypocalcemia. The total increment in urinary calcium excretion observed in normal subjects given glucagon was nevertheless insufficient to account for the observed hypocalcemia (BIRGE and AVIOLI 1969), which suggested that the enhancement in renal calcium loss cannot be considered as the sole factor involved in the hypocalcemic effect of glucagon. This conclusion was further substantiated by the observation that, even in nephrectomized dogs or rats, glucagon caused a significant fall in plasma calcium levels (AVIOLI et al. 1968, 1969, TANZER et al. 1969). AVIOLI et al. (1968, 1969) came to the conclusion that glucagon may lower calcemia through the stimulatory effect of this peptide on calcitonin release since glucagon was devoid of its hypocalcemic effect in thyroidectomized dogs. The calcitonin-mediated mechanism of glucagon-induced hypocalcemia was more directly suggested by the observations indicating a release of calcitonin by thyroid slices incubated in the presence of glucagon (STERN and BELL 1970; MELVIN et al. 1970). It seems that glucagon can affect calcitonin release by stimulating the adenylate cyclase activity, resulting in an increased generation of cyclic AMP (CARE et al. 1970). However, a glucagon concentration as high as 5 ng/ml was required to elicit this effect. Calcitonin release by glucagon must therefore be considered as a pharmacologic rather than physiologic event, and there are no available data in favor of an influence of glucagon on calcium metabolism in physiologic conditions. Indeed, no change in calcium excretion has been observed in human subjects given glucagon in amounts such that plasma glucagon levels remained within a pathophysiologic range (SAUDEK et al. 1973; SHERWIN et al. 1977).

### II. Glucagon-Induced Decrease in Plasma Phosphate Levels

The hypophosphatemic effect of glucagon was reported first by DE VENANZI (1955), who considered it as the result of increased phosphate utilization for glycogen breakdown. On the other hand, STAUB et al. (1957) and ELRICK et al. (1958) attributed the decrease in phosphatemia to the increased urinary excretion of phosphate ion resulting from a direct effect of glucagon on the kidney, either by increasing the filtered load of phosphate or decreasing its tubular reabsorption. Later on, TANZER et al. (1969) suggested that the glucagon-induced fall in phosphatemia may result from a direct action of glucagon on bone metabolism in a manner similar to calcitonin. It cannot be excluded, therefore, that hypophosphatemia which follows the injection of pharmacologic amounts of glucagon is at least partly due to the release of calcitonin. As for the hypocalcemic action of glucagon, it seems unlikely that glucagon influences phosphate balance under physiologic conditions. Since there are no available data on changes in plasma phosphate under physiologic alterations in glucagonemia without concomitant changes in insulin secretion, which are known to influence phosphatemia, further studies are required to appreciate the postulated effect of glucagon on plasma phosphate concentration.

## III. Influence of Glucagon on Plasma Potassium Levels

Ellis et al. (1957) have demonstrated that, following administration of high amounts of glucagon, there occurs a transient, but significant increase in plasma potassium levels. This increase in extracellular potassium levels cannot be entirely attributed to the glucagon-induced increase in glycogenolysis and in potassium efflux from the liver (Ellis and Becket 1963) since glucagon can increase plasma potassium concentration, even after liver glycogen depletion. These authors have postulated, therefore, that glucagon-induced hyperkalemia is mediated via a cellular mechanism independent of that related to glycogenolysis.

The ability of glucagon to increase extracellular potassium concentration is of particular interest since physiologic changes in insulinemia influence plasma potassium levels in the opposite way (De Fronzo et al. 1978). It must be stressed, however, that in the latter studies the concomitant decline in plasma insulin and glucagon levels induced by somatostatin infusion was associated with a significant hyperkalemic effect. It appears therefore that changes in insulin rather than in glucagon secretion influence potassium homeostasis in physiologic conditions.

More recently, Massara et al. (1980) have demonstrated that in the absence of hyperinsulinemia, the physiologic increase in plasma glucagon levels from 200 to 400 pg/ml produced a significant hyperkalemia. Therefore, these authors postulated that elevated plasma potassium levels occurring in serious stress or in decompensated diabetic patients may result not only from insulin deficiency, but also from a concomitant increase in glucagonemia. It seems unlikely that the hyperkalemic effect of glucagon could be secondary to the renal effect of this hormone, since the urinary excretion of potassium remains unchanged after glucagon administration (Saudek et al. 1973; Sherwin et al. 1977; De Fronzo et al. 1978) or even increases (Elrick et al. 1958; David et al. 1969; Pullman et al. 1967).

# E. Summary and Conclusions

While it still remains uncertain whether physiologic changes in glucagon secretion truly influence salt and water homeostasis, an increase in glucagonemia such as is observed in some catabolic states may contribute to the ensuing mineral and water losses by a direct influence of the hormone on the kidney. It appears from analysis of available data that a moderate hyperglucagonemia can promote renal loss of sodium and other ions, even in the absence of detectable changes in renal blood flow or filtration rate. This effect of glucagon on renal tubular function may be related to a stimulatory effect of glucagon on prostaglandin synthesis by the kidney. Following the administration of large, pharmacologic amounts of the hormone, there occurs in addition an increase in renal blood flow and in glomerular filtration which represents an additional mechanism, possibly responsible for the glucagon-induced enhancement in water and mineral handling by the kidney. The changes in glucagonemia during starvation or in ketogenic states such as insulin deficiency may also promote natriuresis, as a consequence of the inhibitory effect of glucagon on tubular ketone body reabsorption. The resulting increase in keto acid excretion is associated with a significant urinary loss of sodium, which may contribute to the postulated inhibition of tubular sodium reabsorption by glucagon. The additive nature of these two glucagon-dependent natriuretic effects may explain why glucagon is clearly more natriuretic after a few days of a total fast than in the fed state.

## References

Aguilar-Parada E, Eisentraut AM, Unger RH (1969) Effects of starvation on plasma pancreatic glucagon in normal man. Diabetes 18:717–723
Anderson RJ, Berl T, McDonald KM, Schrier RW (1976) Prostaglandins: effects on blood pressure, renal blood flow, sodium and water excretion. Kidney Int 10:205–215
Aoki TT, Müller WA, Brennan MF, Cahill GF Jr (1974) Effect of glucagon on aminoacid and nitrogen metabolism in fasting man. Metabolism 23:805–814
Avioli LV (1972) The effects of glucagon on mineral and electrolyte metabolism. In: Lefèbvre PJ, Unger RH (eds) Glucagon, molecular physiology, clinical and therapeutic implications. Pergamon, Oxford, p 181
Avioli LV, Birge SJ, Kanagawa H, Shieber W (1968) Glucagon-induced hypocalcemia in man and dog. J Clin Invest 47:3a
Avioli LV, Birge SJ, Scott S, Shieber W (1969) Role of the thyroid gland during glucagon-induced hypocalcemia in the dog. Am J Physiol 216:939–945
Bailly C, Imbert-Teboul M, Chabardès D, Hus-Citharel A, Montégut M, Clique A, Morel F (1980) The distal nephron of rat kidney: a target site for glucagon. Proc Natl Acad Sci USA 77:3422–3424
Bailly C, Amiel C (1982) Effect of glucagon on magnesium renal reabsorption in the rat. Pflügers Arch 392:360–365
Birge SJ, Avioli LV (1969) Glucagon-induced hypocalcemia in man. J Clin Endocrinol Metab 29:213–218
Burman KD, Smallridge RC, Jones L, Ramos EA, O'Brian JT, Wright FD, Wartofsky L (1980) Glucagon kinetics in fasting: physiological elevations in serum 3,5,3'-triiodothyronine increase the metabolic clearance rate of glucagon. J Clin Endocrinol Metab 51:1158–1165
Care AD, Bates RFL, Gitelman HJ (1970) A possible role for adenyl cyclase system in calcitonin release. J Endocrinol 48:1–15
Crabbé J (1973) Insulin, glucagon and active sodium transport: from man to amphibia – and back. In: Ussing HH, Thorn NA (eds) Transport mechanisms in epithelia. Munksgaard, Copenhagen, p 173
Dalle X, Tanghe J, Gryspeerdt W (1959) Influence du glucagon sur l'excrètion rénale des électrolytes. Arch Int Pharmacodyn Ther 120:505–507
Danford RO (1970) The effect of glucagon on renal hemodynamics and renal arteriography. Am J Roentgenol 108:665–673
David VMA, Horwath IW, Kovacs K (1960) Über die Wirkung des Glucagons auf den Wasser- und Elektrolytstoffwechsel. Endokrinologie 39:138–149
De Fronzo RA, Sherwin RS, Dillingham M, Hendler R, Tamborlane WV, Felig P (1978) Influence of basal insulin and glucagon secretion on potassium and sodium metabolism. J Clin Invest 61:472–479
De Venanzi F (1955) Comparison between changes in serum inorganic phosphorus induced by glucose and glucagon in diabetics. Proc Soc Exp Biol Med 90:112–115
Duckworth WC (1976) Insulin and glucagon degradation by the kidney. I. Subcellular distribution under different assay conditions. Biochim Biophys Acta 437:518–530
Ellis S, Beckett SB (1963) Mechanism of the potassium mobilizing action of epinephrine and glucagon. J Parmacol Exp Ther 142:318–328
Ellis S, Beckett SB, Boutwell JH (1957) Dibenamide blockade of epinephrine and glucagon hyperkalemias. Proc Soc Exp Biol Med 94:343–345
Elrick H, Huffman ER, Hlad CL Jr, Whipple N, Staub A, Smith AE, Yearwood-Drayton V (1958) Effects of glucagon on renal function in man. J Clin Endocrinol Metab 18:813–824
Fisher M, Sherwin RS, Hendler R, Felig P (1976) Kinetics of glucagon in man: effects of starvation. Proc Natl Acad Sci USA 73:1735–1739

Forrest JN, Fisher M, Hendler R, Soman V, Sherwin R, Felig P (1976) Contrasting roles of the kidney in the disposal and hormonal action of physiological concentrations of glucagon. Clin Res 24:400A

Gagnon G, Regoli D, Rioux F (1978) A new bioassay for glucagon. Br J Pharmacol 64:99–108

Gagnon G, Regoli D, Rioux F (1980) Studies on the mechanism of action of glucagon in strips of rabbit renal artery. Br J Pharmacol 69:389–396

Gill JR Jr, Casper AGT, Tate J (1971) Renal effects of adenosine 3′,5′-cyclic monophosphate and dibutyryl adenosine 3′,5′-cyclic monophosphate. J Clin Invest 50:1231–1240

Glasson P, Vallotton MB (1980) Interrelation between renal prostaglandins and the hormones involved in volume homeostasis by the kidney. Clin Exp Hypertens 2:761–775

Hicks T, Turnberg LA (1974) Influence of glucagon on the human jejunum. Gastroenterology 67:1114–1118

Katz AI, Lindheimer MD (1977) Actions of hormones on the kidney. Ann Rev Physiol 39:97–134

Kaufman ME, Dinno MA, Huang KC (1980) Effect of glucagon on ion transport in mouse intestine. Am J Physiol 238:G491–G494

Kim JK, Frohnert PP, Hui YSF, Barnes LD, Farrow GM, Dousa TP (1977) Enzymes of cyclic 3′,5′-nucleotide metabolism in human renal cortex and renal adenocarcinoma. Kidney Int 12:172–183

Kirschenbaum MA, Zawada ET (1980) The role of prostaglandins in glucagon-induced natriuresis. Clin Sci 58:393–401

Kolanowski J (1979) Influence of glucose, insulin, and glucagon on sodium balance in fasting obese subjects. Perspect Biol Med 22:366–376

Kolanowski J (1980) Le rôle du glucagon dans la natriurèse du jeûne. Ann Endocrinol (Paris) 41:237–238

Kolanowski J (1981 a) Associations between sodium retention and carbohydrate metabolism. In: Björntorp P, Cairella M, Howard AN (eds) Recent advances in obesity research III. Libbey, London, p 25

Kolanowski J (1981 b) Influence of insulin and glucagon on sodium balance in obese subjects during fasting and refeeding. Int J Obes [Suppl 1] 5:105–114

Kolanowski J, Crabbé J (1971) Rôle du glucagon dans le rétention du sodium observée en cas d'administration de glucose à l'obèse soumis au jeûne total. J Physiol (Paris) 63:243A

Kolanowski J, De Gasparo M, Desmecht P, Crabbé J (1972) Further evaluation of the role of insulin in sodium retention associated with carbohydrate administration after a fast in the obese. Eur J Clin Invest 2:439–444

Kolanowski J, Desmecht P, Crabbé J (1976) Sodium balance and renal tubular sensitivity to aldosterone during total fast and carbohydrate refeeding in the obese. Eur J Clin Invest 6:75–83

Kolanowski J, Salvador G, Desmecht P, Henquin JC, Crabbé J (1977) Influence of glucagon on natriuresis and glucose-induced sodium retention in the fasting obese subjects. Eur J Clin Invest 7:167–175

Kolanowski J, Bodson A, Desmecht P, Bemelmans S, Stein F, Crabbé J (1978) On the relationship between ketonuria and natriuresis during fasting and upon refeeding in obese patients. Eur J Clin Invest 8:277–282

Lameire N, Vanholder R, Ringoir S, Leusen I (1980) Role of medullary hemodynamics in the natriuresis of drug-induced renal vasodilatation in the rat. Circ Res 47:839–844

Levy M (1975 a) Further observations on the response of the glomerular filtration rate to glucagon: comparison with secretion. Can J Physiol Pharmacol 53:81–85

Levy M (1975 b) The effect of glucagon on glomerular filtration rate in dogs during reduction of renal blood flow. Can J Physiol Pharmacol 53:660–668

Levy M, Starr NL (1972) The mechanism of glucagon-induced natriuresis in dogs. Kidney Int 2:76–84

Marcus R, Aurbach GD (1969) Bioassay of parathyroid hormone in vitro with a stable preparation of adenyl cyclase from rat kidney. Endocrinology 85:801–810

Marliss EB, Aoki TT, Unger RH, Soeldner JS, Cahill GF Jr (1970) Glucagon levels and metabolic effects in fasting man. J Clin Invest 49:2256–2270

Massara F, Martelli S, Cagliero E, Camanni F, Molinatti GM (1980) Influence of glucagon on plasma levels of potassium in man. Diabetologia 19:414–417
McGarry JD, Wright PH, Foster DW (1975) Hormonal control of ketogenesis. J Clin Invest 55:1202–1209
Melson GL, Chase LR, Aurbach GD (1970) Parathyroid hormone-sensitive adenyl cyclase in isolated renal tubules. Endocrinology 86:511–518
Melvin KEW, Tashjian AH, Voelkel EF (1970) Medulary carcinoma of the thyroid: stimulation of calcitonin secretion by glucagon and calcium. In: Taylor S (ed) Calcitonin. Heinemann, London, p 487
Merimee TJ, Misbin RI, Pulkkinen AJ (1978) Sex variations in free fatty acids and ketones during fasting: evidence for a role of glucagon. J Clin Endocrinol Metab 46:414–419
Mulvehill JB, Hui YS, Barnes LD, Palumbo PJ, Dousa TP (1976) Glucagon-sensitive adenylate cyclase in human renal medulla. J Clin Endocrinol Metab 42:380–384
Narahara HT, Everett NB, Simmons BS, Williams RH (1958) Metabolism of insulin $I^{131}$ and glucagon $I^{131}$ in the kidney of the rat. Am J Physiol 192:227–231
O'Brian JT, Saudek C, Spark RF, Arky RA (1974) Glucagon induced refractoriness to exogenous mineralocorticoid. J Clin Endocrinol Metab 38:1147–1149
Parving HH, Noer J, Kehlet H, Mogensen CE, Svendsen PA, Heding L (1977) The effect of short-term glucagon infusion on kidney function in normal man. Diabetologia 13:323–325
Parving HH, Sandhal-Christiansen J, Noer I, Tronier B, Mogensen CE (1980) The effect of glucagon infusion on kidney function in short-term-insulin-dependent juvenile diabetics. Diabetologia 19:350–354
Pullman TN, Lavender AR, Aho I (1967) Direct effects of glucagon on renal hemodynamics and excretion of inorganic ions. Metabolism 16:358–383
Saudek CD, Boulter PR, Arky RA (1973) The natriuretic effect of glucagon and its role in starvation. J Clin Endocrinol Metab 36:761–765
Saudek CD, Boulter PR, Knopp RH, Arky RA (1974) Sodium retention accompanying insulin treatment of diabetes mellitus. Diabetes 23:240–246
Sherwin RS, Hendler R, Felig P (1977) Influence of physiologic hyperglucagonaemia on urinary glucose, nitrogen and electrolyte excretion in diabetes. Metabolism 26:53–58
Staub A, Springs V, Elrick H (1956) Effect of glucagon on renal excretion of electrolytes. Fed Proc 15:361
Staub A, Springs V, Stoll F, Elrick H (1957) A renal action of glucagon. Proc Soc Exp Biol Med 94:57–60
Stern PH, Bell NH (1970) Effects of glucagon on serum calcium in the rat and on bone resorption in tissue culture. Endocrinology 87:111–117
Stowe NT, Hook JB (1970) Role of alterations in renal hemodynamics in the natriuretic action of glucagon. Arch Int Pharmacodyn Ther 183:65–74
Tannenbaum J, Splawinski JA, Oates JA, Nies AS (1975) Enhanced renal prostaglandin production in the dog. I. Effect on renal function. Circ Res 36:197–203
Tanzer FS, Kennedy JW III, Talmage RV (1969) A comparison of the effects of thyrocalcitonin and glucagon on plasma calcium and phosphate. Proc Soc Exp Biol Med 133:500–505
Turpin BP, Austin MW, Solomon SS (1978) Urinary and plasma cyclic AMP levels during short term starvation in obese man: response to glucagon stimulation. Horm Metab Res 10:36–37
Ueda J, Nakanishi H, Miyazaki M, Abe Y (1977) Effects of glucagon on the renal hemodynamics of dogs. Eur J Pharmacol 41:209–212
Unger RH, Eisentraut AM, Madison LL (1963) The effects of total starvation upon the levels of circulating glucagon and insulin in man. J Clin Invest 42:1031–1039
Uranga J, Fuenzalida R, Rapoport AL, del Castillo E (1979) Effect of glucagon and glomerulopressin on the renal function of the dog. Horm Metab Res 11:275–279

CHAPTER 51

# Glucagon and Catecholamines

P. J. LEFEBVRE and A. S. LUYCKX

## A. Introduction

Glucagon and catecholamines share numerous metabolic effects. In addition, catecholamines are potent stimulators of glucagon secretion and, under certain conditions, glucagon stimulates catecholamine release from the adrenal medulla. Finally, the involvement of catecholamines has been invoked to explain some of the metabolic actions of glucagon. The stimulation of glucagon release by catecholamines has been reviewed in detail in Chap. 30 while the possible involvement of catecholamines in some effects of glucagon has been considered in Chaps. 37 and 38. In this short contribution, we will summarize the in vitro and in vivo studies dealing with the effects of glucagon on catecholamine release and review the potential of these findings for the diagnosis of pheochromocytoma.

## B. Glucagon and Catecholamine Release In Vitro

SCIAN et al. (1960) were the first to demonstrate that, at doses of 0.01–1.0 mg/ml, glucagon stimulated the release of epinephrine and norepinephrine form the isolated perfused dog adrenals. In the isolated perfused rat adrenals, a "pulse" injection of 0.01–0.1 mg glucagon in the arterial cannula did not stimulate catecholamine release (CESSION-FOSSION et al. 1966). At a dose of 0.5 mg, as an intraarterial pulse, a slight and inconstant increase in catecholamine release was noted. In perfusion experiments (0.1 mg/min for 9 min) an unequivocal increase in catecholamine release was observed in about one-half of the trials (CESSION-FOSSION et al. 1966; LEFEBVRE et al. 1968).

## C. Glucagon and Catecholamine Release In Vivo

### I. Rats

In the rat, the intraperitoneal injection of glucagon at a dose of 100 μg/kg of body weight significantly lowered the epinephrine and norepinephrine content of the adrenals, a return to basal values, or even a rebound effect being observed after 30 min (LEFEBVRE and DRESSE 1961). Such pharmacologic doses of glucagon induced a progressive decline in blood pressure which was preceded by a 10–30-s period of hypertension. The blood pressure drop reached 3–4 cmHg after 4–5 min and persisted for 45–60 min; after 90 min, recovery was complete. In the eviscerated rat, glucagon failed to induce a drop in arterial blood pressure whereas in the adrenalectomized rat, the fall in blood pressure was more pronounced and a complete cardiovascular collapse was often observed (CESSION-FOSSION et al. 1966; LEFEBVRE et al. 1968). The glucagon-induced fall in peripheral blood pressure has been attributed to marked visceral vasodilatation (see Cap. 54). The stimulation of the adrenergic system which followed the administration of pharmacologic doses of glucagon in rats was, in part, of reflex origin resulting from the hepatosplanchnic vasodilatation and the ensuing secondary drop in blood pressure and, in part, from direct stimulation of the adrenal medulla (LEFEBVRE et al. 1968). Injections to the rat of zinc-glucagon (1 mg) stimulated the adrenal medullary enzyme, phenylethanolamine-*N*-methyltransferase, which catalyzes the conversion of norepinephrine to epinephrine. The effect occurred within 2 h and was enhanced when insulin (6 IU) was injected simultaneously (KITABCHI et al. 1968).

## II. Guinea Pigs

In guinea pigs, administration of estrogens and growth hormone induces hyperplasia of the chromaffin tissue. Injections of glucagon reduced the norepinephrine content of the hyperplastic tissue, but had no effect on its epinephrine content (LUPULESCU et al. 1966).

## III. Dogs

In anesthetized dogs, SARCIONE et al. (1963) demonstrated that a single intravenous injection of 50 μg/kg glucagon rapidly induced an elevation of peripheral plasma epinephrine levels; a 10-min intravenous infusion of glucagon produced a marked elevation of both epinephrine and norepinephrine in adrenal venous blood.

## IV. Cats

In the cat, the intraperitoneal injection of 200 μg/kg glucagon induced after 1 h a 29% decrease in the epinephrine and norepinephrine adrenal content. These changes were not observed 3 h after the injection (LLORENS et al. 1973).

## V. Normal Humans

### 1. Plasma Catecholamines

KUSCHKE et al. (1966 a, b) initially reported a 35%–185% rise in peripheral plasma catecholamines between 10 and 30 min after the intravenous injection of 1 mg glucagon. LEFEBVRE et al. (1968) found a plasma catecholamine rise in two subjects 1–4 min after 1 mg glucagon given intravenously, but not in seven others after 5 and 15 min. SIMARD (1974) reported a plasma epinephrine rise 3 min after glucagon injection, but no rise after saline administration in six normal subjects. SIQUEIRA-FILHO et al. (1975) reported a significant increase of plasma catecholamine levels in 95% of a group of 124 control subjects 30–60 s after the intravenous administration of 1 mg glucagon. In contrast, no rise in plasma catecholamines after glucagon administration was found by CREMER et al. (1968) and SEBEL et al. (1974). As emphasized by YOUNG et al. (1976), some of the disparities may be due to sampling time following glucagon injection but, also and more probably, to a lack of specificity of the earlier plasma catecholamine assay procedures.

### 2. Urinary Catecholamines and Metabolites

ZICHA et al. (1965) reported a significant 52% increase in the urinary excretion of vanillin-mandelic acid after administration of 1 mg glucagon in humans. HINTERBERGER and WILCKEN (1974) observed a significant increase in norepinephrine and 4-hydroxy-3-methoxymandelic acid urinary excretion in 14 patients with congestive renal failure treated with pharmacologic doses of glucagon (5–7.5 mg/h). Later on, YOUNG et al. (1976) reported that the intravenous injection of a huge bolus of glucagon (4 mg) produced a 2–3-fold increase in urinary epinephrine that was, however, indistinguishable from the 2–3-fold rise also observed in the saline controls. They concluded that the observed increment had to be attributed more to the stress of the procedure than to glucagon itself, but admitted that the integrative nature of the urinary method did not completely rule out a transient catecholamine response to glucagon.

## D. Glucagon and Pheochromocytoma

### I. Glucagon as a Provocative Test in Pheochromocytoma

LAWRENCE and FORLAND (1964) and LAWRENCE (1965, 1967) were the first to describe the interest of glucagon administation as a provocative test for the diagnosis of pheochromocytoma: a marked pressor response was observed 15–60 s after the intravenous administration of glucagon in three documented cases of pheochromocytoma and in one suspected case; no response was observed in 126 patients not having pheochromocytoma, including 36 with other causes of hypertension. LEFEBVRE et al. (1966) described, in one patient with pheochromocytoma, a very pronounced pressor response to intravenous glucagon which disappeared after removal of the tumor (Fig. 1); the pressor response was accompanied by a dramatic rise in plasma catecholamine before the operation and a lack of response after it. SHEPS and MAHER (1968) compared the glucagon test with the standard cold pressure–histamine test in 12 patients with pheochromocytoma and 35 without tumor. There was no false positive response to glucagon while one patient without tumor had a positive pressor response to histamine. Of 11 patients who received 1 mg glucagon, 6 had a positive pressor reponse to glucagon (4 of whom had also a positive response to histamine). In one patient with pheochromocytoma, the pressor response was positive to histamine, but not to 0.5 mg glucagon. The effects of glucagon, 0.01 mg/kg body weight intravenously, were investigated in two patients with pheochromocytoma by SEBEL et al. (1974). There was a marked rise in blood pressure in one of the patients and a significant increase in plasma catecholamines in both patients. The authors considered that the glucagon test should not be used for screening patients for the diagnosis of pheochromocytoma, but that it may be useful in the rare patients in whom clinical features strongly suggest pheochromocytoma, but no increase in urinary excretion of catecholamines and their metabolites is detected; they also concluded that the diagnostic value of

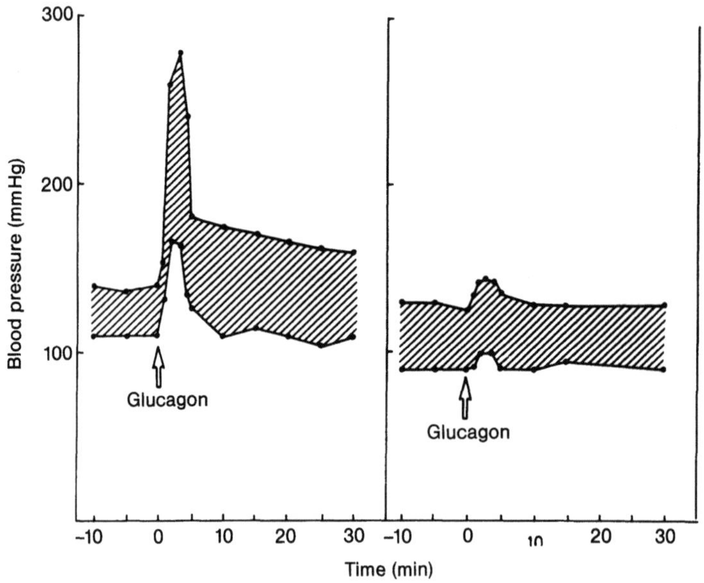


**Fig. 1.** Changes in blood pressure after administration of 1 mg glucagon intravenously in a patient with pheochromocytoma; lack of significant changes 6 weeks after removal of the tumor. LEFEBVRE et al. (1966)

the test can be enhanced by measuring plasma catecholamines as well as blood glucose during the procedure, as previously described by LEFÈBVRE et al. (1966). SIQUEIRA-FILHO et al. (1975) considered the glucagon test as negative in six patients with pheochromocytoma, including one with multiple endocrine neoplasia of type 2.

WAHLEN et al. (1983) reported an impressive rise in blood pressure accompanied by a rise in circulating catecholamines in a case of pheochromocytoma associated with neurofibromatosis.

BERNHEIM et al. (1978) have reported a case of extraadrenal pheochromocytoma in which the intravenous administration of 1 mg glucagon did not induce significant blood pressure changes, but was responsible for an impressive increase in plasma catecholamines; 13 months after removal of the tumor, intravenous glucagon did not significantly modify plasma catecholamines. MANDAI et al. (1980) have evaluated the role of the sympathic nervous system in essential hypertension by examining the response of urinary catecholamines to the intramuscular injection of glucagon (0.03 μg/kg body weight) in both young and elderly normal subjects and in both young and elderly patients with essential hypertension. The increments in urinary epinephrine and norepinephrine after glucagon injection were significantly higher in young hypertensive than in young normotensive subjects or normotensive and hypertensive elderly subjects. They concluded that this hyperreactivity of the sympathetic nervous system lends support to the hypothesis that that system is more important in the maintenance of hypertension in the young than in the elderly. In more than one-half of 67 patients suspected of having pheochromocytoma, and investigated by KUCHEL et al. (1981), glucagon stimulation (1 mg intravenously) increased plasma free norepinephrine (NE) and epinephrine (E) by 50% or more within 3 min, with rising blood pressure or pulse rate; however, only three patients had a proven pheochromocytoma. A low degree of catecholamine conjugation accounted for most of the false positive results. In patients with low conjugated NE+E levels, there was a greater rise in free NE+E and free E as well as in pulse rate after glucagon stimulation than in those with normal levels of conjugated NE+E. After sham administration of glucagon, there were rises in blood pressure, but not in free NE or E in four patients. They concluded that the glucagon-induced catecholamine test can be false positive in hyperadrenergic essential hypertensive patients with abnormally low conjugated NE+E. Saline alone in a sham glucagon test, in susceptible patients, raises systolic blood pressure and pulse rate, and therefore, if plasma free NE and E are measured and found not to rise, this type of false positive result can be eliminated.

UDA et al. (1982) reported that, in 9 patients with essential hypertension and normal or high plasma renin activity, intravenous injection of 1 mg glucagon caused a rise in both plasma epinephrine and norepinephrine while an increase in plasma renin activity was detected in 4 of 9 patients. In contrast, in 7 patients with low renin hypertension, plasma epinephrine failed to increase after glucagon whereas plasma norepinephrine increased significantly; in these patients, there was no discernible response of plasma renin activity to glucagon. It was concluded that the glucagon test may be useful for the simultaneous assessement of the reserve of both renin secretion and sympathico-adreno-medullary function in essential hypertension.

The glucagon provocative test in pheochromocytoma is usually well tolerated, except for occasional slight headache, mild flushing, nausea, and palpitations, all of a temporary nature (SHEPS and MAHER 1968). In view of the excessive blood pressure response observed in some patients (LEFEBVRE et al. 1966; SEBEL et al. 1974) the procedure is not without hazard.[1] Initially, 0.5 mg should be tried, and, if the pressor response is absent or equivocal, then 1.0 mg may be used (GALLOWAY 1972). In any case, a syringe filled with phentolamine should be at hand.

## II. Mode of Action of Glucagon in Pheochromocytoma

DEXTER and ALLEN (1970) were the first to show that glucagon activated adenylate cyclase in a homogenate of pheochromocytoma from a patient demonstrating a hypertensive response to glucagon, while, in contrast, glucagon did not activate adenylate cyclase in a pheochromocytoma from another patient not having a hypertensive response to administered glucagon. SCHOOR et al. (1972) found that glucagon, like adrenocorticotropic and thyroid-stimulating hormones stimulated the adenylate cyclase in both homogenates from a bilateral pheochromocytoma, but

1 As emphasized in Chap. 54, glucagon is often used for its hypotonic effect on the gut to facilitate oral use of gastrointestinal contrast and endoscopy. GEELHOED (1980) recently reported the case of a patient with pheochromocytoma who had had the only hypertensive attack of her life triggered during a glucagon-augmented computerized axial tomography scanning which revealed a large left adrenal mass. Similarly, MCLOUGHLIN et al. (1981) reported a life-threatening reaction in a 32-year-old woman with pheochromocytoma who had received 1 mg glucagon intramuscularly prior to computerized axial tomography. A similar case had been reported previously by BEGGS (1978)

two unilateral tumors were not stimulated by these hormones. LEVEY et al. (1975) demonstrated that glucagon activated adenylate cyclase in a homogenate of a pheochromocytoma over the concentration range $1 \times 10^{-8}$–$1 \times 10^{-6}$ $M$. The tumor receptor for glucagon was characterized and found to be similar in several ways to the glucagon receptor previously reported in normal tissues such as liver and heart. However, both glucagon $^{125}$I binding and adenylate cyclase activity were much more labile than in heart or liver. It was suggested that the effect of glucagon on pheochromocytoma was exerted through adenylate cyclase-mediated cyclic AMP synthesis, since cyclic AMP may induce tyrosine hydrolase activity as has been observed in some pheochromocytomas (ROTH et al. 1968). The presence of a glucagon-sensitive adenylate cyclase in pheochromocytomas has recently been confirmed by KUCHEL et al. (1981).

## E. Conclusions

Under most experimental conditions and in most species so far investigated, pharmacologic amounts of glucagon stimulate the release of catecholamine from the adrenal medulla. This property has been used as the basis of a provocative test for the diagnosis of pheochromocytoma. Such a test should not be performed routinely in screening patients for the presence of pheochromocytoma since: (1) the procedure is not without hazard; (2) estimations of urinary catecholamines and their metabolites are easier to perform and, often more reliable. In questionable cases however, the injection of glucagon accompanied by frequent, or continuous, measurements of blood pressure and by determination of plasma catecholamines may be a useful adjunct to the diagnostic strategy of pheochromocytoma.

## References

Beggs I (1978) Pheochromocytoma diagnosed during barium meal. Br J Radiol 51:918

Bernheim C, Jaillon P, Graisely B, Cloarec M, Debray J, Cheymol G (1978) Diagnostic d'un phéochromocytome par dosage des catécholamines plasmatiques au cours du test au glucagon. Nouv Presse Med 7:553–555

Cession-Fossion A, Lecomte J, Lefèbvre P (1966) Sur la stimulation médullo-surrénalienne induite chez le rat par le glucagon. CR Soc Biol (Paris) 160:723–726

Cremer GM, Molnar GD, Moxness KE, Sheps SG, Maher FT, Jones JD (1968) Hormonal and biochemical response to glucagon administration in patients with pheochromocytoma and in control subjects. Mayo Clin Proc 43:161–176

Dexter RN, Allen DO (1970) A glucagon-sensitive adenylcyclase system in pheochromocytoma. Clin Res 18:601

Galloway JA (1972) The pharmacology and clinical use of glucagon. In: Lefèbvre PJ, Unger RH (eds) Glucagon: molecular physiology, clinical and therapeutic implications. Pergamon, Oxford, p 299

Geelhoed GW (1980) CAT scans and catecholamines. Surgery 87:719–720

Hinterberger H, Wilcken DEL (1974) The effect of prolonged glucagon infusions on the urinary excretion of catecholamines and 3-hydroxy-3-methoxymandelic acid and on adrenal medullary tissue levels of catecholamines in patients with severe cardiac disease. Clin Chim Acta 52:153–161

Kitabchi AE, Solomon SS, Williams RH (1968) Stimulatory effect of insulin and glucagon on phenylethanolamine-*N*-methyl transferase of rat adrenal. Proc Soc Exp Biol Med 127:296–300

Kuchel O, Hamet P, Bun NT, Larochelle P (1981) Basis of falsepositive glucagon tests for pheochromocytoma. Clin Pharmacol Ther 29:687–694

Kuschke HJ, Klusmann H, Schölkens B (1966a) Über die Wirkung von Glucagon auf den Katecholamin-Plasmaspiegel beim Menschen. Arch Exp Pathol Pharmakol 253:65
Kuschke HJ, Klusmann N, Schölkens B (1966b) Untersuchungen über die Wirkung von Glukagon auf des sympatho-adrenale System des Menschen. Klin Wochenschr 44:1297–1300
Lawrence AM (1965) A new provocative test for pheochromocytoma. Ann Intern Med 63:905–906
Lawrence AM (1967) Glucagon provocative test for pheochromocytoma. Ann Intern Med 66:1091–1096
Lawrence AM, Forland M (1964) Glucagon provocative test for pheochromocytoma. J Lab Clin Med 64:878
Lefèbvre P, Dresse A (1961) Influence du glucagon sur le taux des catécholamines surrénaliennes chez le rat. CR Soc Biol (Paris) 155:412–414
Lefèbvre PJ, Cession-Fossion A, Luyckx AS (1966) Glucagon test for pheochromocytoma. Lancet 2:1366
Lefèbvre PJ, Cession-Fossion A, Luyckx AS, Lecomte J, Van Cauwenberge H (1968) Interrelationships glucagon-adrenergic system in experimental and clinical conditions. Arch Int Pharmacodyn 172:393–404
Levey GS, Weiss SR, Ruiz E (1975) Characterization of the glucagon receptor in a pheochromocytoma. J Clin Endocrinol Metab 40:720–723
Lloréns I, Borrell J, Borrell S (1973) Effects of insulin, glucagon and adrenocorticotrophin on the levels of corticosteroids, noradrenaline, adrenaline and ascorbic acid in the adrenal glands of cat. Horm Res 4:321–330
Lupulescu A, Chivu V, Petrovici A (1966) Effect of aldosterone, glucagon and growth hormone on the catecholamine content and the evolution of chromaffin hyperplasia in guinea pigs. Experientia 22:222–223
Mandai T, Ogihara T, Hata T, Okada Y, Ogasahara S, Mikami H, Nakamaru M, Kumahara Y (1980) Urinary catecholamine response to glucagon in young and elderly patients with essential hypertension. J Am Geriatr Soc 28:462–465
McLoughlin M, Langer B, Wilson DR (1981) Lifethreatening reaction to glucagon in a patient with pheochromocytoma. Radiology 140:841–842
Roth RH, Stjårne L, Levine RJ, Giarman NJ (1968) Abnormal regulation of catecholamines synthesis in pheochromocytoma. J Lab Clin Med 72:397–403
Sarcione EJ, Back N, Sokal JE, Mehlman B, Knoblock E (1963) Elevation of plasma epinephrine levels produced by glucagon in vivo. Endocrinology 72:523–526
Scian LF, Westermann CD, Verdesca AS, Hilton JG (1960) Adreno-cortical and medullary effects of glucagon, Am J Physiol 199:867–870
Schoor I, Hinshaw HT, Cooper MA, Mahaffee D, Ney RL (1972) Adenylcyclase hormone responses of certain human endocrine tumors. J Clin Endocrinol Metab 34:447–451
Sebel EF, Hull RD, Kleerekoper M, Stokes GS (1974) Response to glucagon in hypertensive patients with and without pheochromocytoma. Am J Med Sci 267:337–343
Sheps SG, Maher FT (1968) Histamine and glucagon tests in diagnosis of pheochromocytoma. JAMA 205:895–899
Simard SJ (1974) Valeur diagnostique des catécholamines plasmatiques dans le test au glucagon pour le phéochromocytome. Union Med Can 103:460–465
Siqueira-Filho AG, Sheps SG, Maher FT, Nai-Siang J, Elveback LR (1975) Glucagon-blood catecholamine test. Arch Intern Med 135:1227–1231
Uda M, Sakamoto N, Tsuchiya M, Ito K, Ikeda M, Matsukura S, Fujita T (1982) Responses of plasma catecholamine and plasma renin activity after glucagon stimulation in essential hypertension. Acta Endocr 101:242–247
Wahlen Ch, Appeltants A, Limet R (1983) Neurofibromatose et phéochromocytome. Histoire d'un cas. Rev. Méd. Liège 38:56–57
Young JB, Candsberg L, Knopp RH (1976) Effect of intravenous glucagon on urinary catecholamine excretion in normal man. Metabolism 25:233–237
Zicha L, Schmid E, Weist F, Winkler E (1965) Der Einfluß von Glukagon, Insulin, Adrenalin und Sulfonylharnstoffen auf Blutzucker und Vanillinmandelsäure im Harn bei gesunden Kontrollpersonen, Leberkranken und anderen Patienten mit Dysproteinämie. Verh Deutsch Ges Inn Med 71:499–501

CHAPTER 52

# Glucagon and Growth Hormone

T. J. MERIMEE

## A. Introduction

It has been speculated that small increments of insulin in the portal blood, counterbalanced by small increases in glucagon, could result in an enhanced protein anabolic effect with minimal shifts in glucose homeostasis. Such a hypothesis makes it reasonable to postulate that glucagon itself might be under pituitary control (see also Chap. 29).

Numerous claims have been made that the administration of growth hormone increases pancreatic islet tissue both in humans and animal models. These older reports have ben reviewed in detail in a classic examination of the effects of the pituitary gland on pancreatic function (KETTERER et al 1957). FOÀ et al. (1953) initially presented evidence for an effect of growth hormone on the secretion of a hyperglycemic factor into pancreatic blood: growth hormone administered to a donor dog caused a rise of blood glucose in a recipient dog when cross-circulation was established between the pancreatic and femoral veins, but no change in blood glucose in the recipient dog when the connection was made between mesenteric and femoral veins. Growth hormone administered to normal cats was found by others to cause an increase in portal blood of a substance which could bring about a rise in blood glucose in the alloxan-induced diabetic, hypophysectomized, adrenalectomized rat, as well as a decrease in liver glycogen (BORNSTEIN et al. 1951). Since these and other studies were performed prior to the development of accurate immunoassays for the measurements of growth hormone and glucagon, no definitive statement could be made on the hormone or hormones involved in these actions, although they strongly supported an effect of growth hormone on glucagon secretion.

## B. Acute and Chronic Effects of Growth Hormone on Glucagon Secretion

After the development of immunoassays for peptide hormones, multiple studies indicated that growth hormone modulated the secretion of insulin, although acute, direct effects on insulin release appeared unlikely (CERASI and LUFT 1964; MERIMEE et al. 1969; ELLKELES et al. 1969). As discussed in Chap. 29, the question of whether growth hormone could similarly modulate or stimulate the secretion of glucagon was not as easily answered. This was due in part to the presence of several sources of endogenous glucagon, multiple molecular species of glucagon, and cross-reactivity between many antisera for these molecular entities. Perfusion studies, which one might anticipate giving more suitable results, were also inconsistent.

Perfusion of the isolated rat pancreas with either bovine or ovine growth hormone was reported to cause a transient release of glucagon (TAI and PEK 1976); other studies utilizing isolated rat islets and highly purified human growth hormone failed to demonstrate this effect (LARSON et al. 1978). In most in vivo whole body studies, no conclusive effect of injecting pharmacologic doses of various growth hormone preparations was noted on glucagon release either in human or in animal studies (ADAMSON 1975; DAUGHADAY and KIPNIS 1966; FARMER et al. 1971; FROHMAN et al. 1971).

It should be pointed out that the secretion of insulin and glucagon may be reflected poorly in peripheral blood, however, portal samples may well show such a phenomenon. An examination of both insulin and glucagon secretion in this manner with a sufficient number of study groups to permit interpretation of the data was performed by SIREK et al. (1979). In a study of 23 dogs, with one group rendered diabetic by an intravenous injection of alloxan and another group surgically pancreatectomized, a clear-cut effect of growth hormone on glucagon secretion was demonstrated. In one set of experiments, four pancreatectomized, four alloxan-induced diabetic, and three normal dogs had the portal vein cannulated. Growth hormone was then injected into the cephalic vein at a single dose of 10 mg/kg body weight. In a second set of experiments, comparable studies were done with portal and peripheral blood being sampled after administering doses of growth hormone that gave physiologic concentrations of growth hormone in the blood reaching the pancreas (6 μg/kg body weight).

In portal, but not in peripheral blood, immunoreactive plasma glucagon and glucagon-like activity were significantly elevated 10 min after growth hormone was injected in normal and alloxan-treated dogs; no effect was seen in pancreatectomized animals. When a physiologic dose of growth hormone producing an ambient concentration of growth hormone of 40 ng/ml was studied, glucagon increased 70% within 2 min in portal blood. Figure 1, illustrates the effect of a single bolus of 10 mg/kg growth hormone on the portal concentration of glucagon in three animals as compared with the mean peripheral concentration in these same animals. It is evident that large increases of portal glucagon (up to 700 pg/ml) are reflected by much smaller changes in the peripheral circulation. When smaller, but significant increases of glucagon were noted in portal blood, no increase could be detected in the peripheral circulation (SIREK et al. 1979). The results summarized probably indicate why earlier investigators failed to observe the effect of ovine, bovine, and human growth hormone when they sampled only peripheral blood.

In addition to an acute effect of growth hormone on glucagon secretion, a tropic or modulating effect is suggested from clinical studies. Support for an effect of growth hormone on glucagon secretion in this manner was obtained from both the acromegalic and hypopituitary states, although data from the latter may be difficult to interpret for reasons that will be given (see also Chap. 45).

GOLDFINE et al. (1972) reported excessive glucagon responses to arginine in active acromegaly, but the antiserum used for the glucagon assay was not specific for pancreatic glucagon. SEINO et al. (1978a) investigated the ability of growth hormone to modulate arginine-induced glucagon release in acromegalic subjects. In

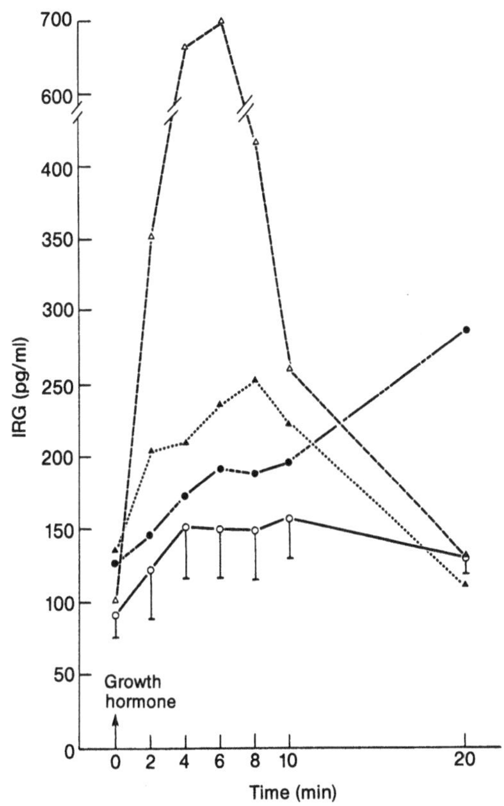


**Fig. 1.** The effect of a single injection of growth hormone on plasma glucagon concentration in portal blood is shown for three dogs (*full circles, full triangles, open triangles*). The mean (±standard error) glucagon response in peripheral blood is also indicated (*open circles*). In additional experiments with smaller increments of portal glucagon concentration, no increase of glucagon was seen in peripheral blood. SIREK et al. (1979)

these patients, basal and peak values of glucagon after arginine administration were greater than normal. In yet another study in acromegalics, oral glucose loads failed to suppress basal levels of glucagon as in normal subjects (SEINO et al. 1978a, b).

It has also been observed that fasting glucagon levels may be elevated in patients with an isolated deficiency of growth hormone. In one study, fasting glucagon levels were reported as 175±21 pg/ml in hypopituitary dwarfs compared with 85±8 pg/ml in normal children. The peak response to arginine in the growth hormone-deficient patients was 529±57 pg/ml, a value significantly greater than that observed in normal subjects: 308±27 pg/ml (SEINO et al. 1978a). BLACKARD et al. (1973) and SIZONENKO et al. (1977) independently observed slightly higher glucagon responses to arginine in growth hormone-deficient patients. In reviewing these reports, it appears that the elevated plasma glucagon in isolated growth hormone deficiency probably relates to fasting glucose values, which tend to be low in the growth hormone-deficient state. LARON et al. (1977) have shown that mean glucose values in blood are decreased in most growth hormone-deficient patients, and the level of blood glucose clearly affects the glucagon response to several stimuli. For example, in the portal vein of healthy pigs, the glucagon response to arginine correlates inversely with the plasma glucose concentration before the start of the arginine infusion (KÜHL et al. 1976).

In evaluating both acute and chronic studies with growth hormone, it would appear that growth hormone can cause an acute rise of portal glucagon concentration, usually reflected 2–10 min after the rise of growth hormone concentration in the portal circulation. Early attempts to detect such rises of glucagon probably failed by concentrating on peripheral samples. Glucagon levels may be elevated in acromegaly, but likewise may be elevated in the hypopituitary state. In the latter condition, this probably reflects exposure of the A-cell to consistently lower blood glucose concentrations.

## C. Effect of Glucagon on Growth Hormone Secretion

Most, if not all, stimuli causing growth hormone secretion act via a neural mechanism. It is thought that secretion of growth hormone may be mediated via any of three centers – the ventromedial nucleus (VMN), the arcuate nucleus (AN), and the limbic system. Although probably an oversimplification, norepinephrine is thought to be neurotransmittor for stimuli reaching the VMN, dopamine for the AN, and serotonin for stimuli mediated via the limbic system. Glucagon, like vasopressin and L-arginine has been postulated to act at the VMN [1] (MERIMEE and RABIN 1973).

The earliest attempts to detect elevation of growth hormone after administration of glucagon were unsuccessful. In one of the earliest reports, no significant change of venous growth hormone concentration was noted for periods up to 90 min following the injection of glucagon (CREMER et al. 1968). Later, several workers reported that growth hormone did increase after glucagon administration. In one study, 1 mg glucagon was administered subcutaneously to eight men and seven women and growth hormone levels were measured in plasma every 15 min for 240 min. The most common pattern of change seen in the serum concentration of glucagon was an elevation at the second hour, followed by a rapid decline in serum concentration. No subject showed a significant increase in glucagon within the first 30 min and control studies to evaluate the spontaneous secretion of growth hormone were not performed (MITCHELL et al. 1969). In a later study by the same group, growth hormone increased in 33 of 34 subjects after 1 mg glucagon was given either subcutaneously or intramuscularly (MITCHELL et al. 1970). However, the mean peak concentration of glucagon in serum was substantially lower than that

1 However, a recent investigation performed in healthy subjects by HAINER et al. (1981) has shown that thyrotropin-releasing hormone (TRH) and cyproheptadine significantly reduced the growth hormone response to glucagon (1 mg intramuscularly); in contrast pimozide, a dopamine-blocking agent, was much less effective. These findings prompted HAINER et al. to suggest that glucagon-induced growth hormone secretion is at least partly mediated via serotoninergic mechanisms while significant dopaminergic involvement does not seem probable. It was further suggested that TRH plays a substantial inhibitory role in glucagon-stimulated growth hormone secretion

seen after other provocative tests of growth hormone secretion, and the response was seen only 2 h after administration of glucagon. These same workers reported that propanolol (40 mg given 2 h before glucagon injection) augmented the growth hormone responses to glucagon, but this was not a consistent experience (MITCHELL et al. 1971).

CAIN et al. (1970) found only small increases of growth hormone following bolus injections of glucagon. In response to glucagon (1 or 2 mg intravenously), 50% of tested subjects did not exhibit a growth hormone rise greater than 2 ng/ml. In response to a 30-min infusion of glucagon, growth hormone increased to 20 ng/ml at 120 min in the latter study. The plasma glucose response to the glucagon infusion is worth noting. During the first 30 min of the infusion glucose concentration increased 40–50 mg/dl and subsequently decreased 80 mg/dl prior to any detectable increase of plasma growth hormone. PARKS et al. (1973) evaluated the sequential administration of oral propranolol and intravenous glucagon as a test of growth hormone secretion in 73 children. Side effects were noted in 13 of the subjects. Mean growth hormone levels in 49 children who responded to propranolol–glucagon were 5 ng/ml at the beginning of the test, 10.9 ng/ml following propranolol administration, and 28.8 ng/ml following glucagon administration. Of 20 patients failing to obtain growth hormone levels greater than 6 ng/ml, none had a normal response to insulin or to arginine. It is not possible in reviewing this latter study to dissociate the growth hormone response to glucagon from the concomitant decrease of plasma glucose.

One of the few studies of the effect of glucagon on growth hormone secretion in which levels of glycemia were regulated throughout was performed by EDDY et al. (1970). Growth hormone secretion was studied following administration of glucagon after subcutaneous injection, after intravenous injection, and after intravenous injection during constant glucose infusion. Only 10 of 24 subjects had growth hormone secretion in response to glucagon alone, with inconsistent timing in magnitude of the response. In no case did glucagon stimulate growth hormone secretion during constant glucose infusion. These results are consistent with our own (T. J. MERIMEE and S. E. FEINBERG 1975) unpublished work. While evaluating the ability of several agents to stimulate growth hormone secretion, we noted that glucagon initiated growth hormone secretion only after decreases occurred in the plasma glucose concentration. In five studies, we found the growth hormone response to glucagon totally obliterated when plasma glucose concentration was equal to or greater than 120 mg/dl. In essence, evidence that glucagon itself can stimulate growth hormone secretion is not conclusive. In vivo studies are inconsistent and the best show no growth hormone response when the "late" decrease of plasma glucose consequent to glucagon administration is prevented.

## D. Summary

There is evidence for both acute and chronic effects of growth hormone on the secretion of glucagon. Following the acute administration of growth hormone, increases of pancreatic glucagon can be detected consistently in portal blood, although changes in peripheral blood may be undetectable or small. In pathologic conditions with chronic elevation of growth hormone (acromegaly), both basal and stimulated secretion of glucagon appear increased. A direct effect of glucagon on growth hormone secretion cannot be convincingly substantiated. If such an effect occurs, it would appear to be largely dependent on glucagon-induced changes of the circulating glucose concentration.

## References


Adamson V (1975) Insulin-like and diabetogenic actions of human growth hormone in man. PhD dissertation, University of Stockholm, Stockholm

Blackard WG, Andrews SS, Lazarus EJ (1973) Effect of growth hormone deficiency on glucagon secretion. Proc Soc Exp Biol Med 143:1042–1044

Bornstein J, Reid E, Young FG (1951) The hyperglycemic action of blood from animals treated with growth hormone. Nature 168:903–905

Cain JP, Williams GH, Dluhy RD (1970) Glucagon stimulation of human growth hormone. J Clin Endocrinol Metab 31:222–224

Cerasi E, Luft R (1964) Insulin response to glucose loading in acromegaly. Lancet 2:769–771

Cremer GM, Molnar GD, Moxness KE, Sheps SG, Maher FT, Jones JD (1968) Hormonal and biochemical response to glucagon in patients with pheochromocytoma and in control subjects. Proc Mayo Clin 43:161–176

Daughaday WH, Kipnis DM (1966) The growth promoting and insulin actions of somatotropin. Recent Prog Horm Res 22:49–99

Eddy RL, Jones AL, Hirsch RM (1970) Effect of exogenous glucagon on pituitary polypeptide hormone release. Metabolism 19:904–911

Ellkeles RS, Wright AD, Lowy C, Frazer TR (1969) Serum insulin in acromegaly. Lancet 2:615–618

Foà PP, Magid EB, Glassman MD, Weinstein HR (1953) Anterior pituitary growth hormone and pancreatic secretion of glucagon. Proc Soc Exp Biol Med 83:758–761

Farmer RW, Pellizari ED, Fabre LF, Nonaka K, Sugase T, Foà PP (1971) Failure of growth hormone to stimulate glucagon secretion. Proc Soc Exp Biol Med 138:491–493

Frohman LA, MacGillivray MH, Aceto T Jr (1971) Acute effects of human growth hormone in normal and growth hormone deficient subjects. J Clin Endocrinol Metab 27:561–567

Goldfine ID, Kirstens L, Lawrence AM (1972) Excessive glucagon response to arginine in active acromegaly. Horm Metab Res 4:97–100

Hainer V, Urbanek J, Malec B, Krejcik L (1981) The effect of TRH, cyproheptadine and pimozide on the growth hormone response to intramuscular glucagon. Horm Metab Res 13:451–453

Ketterer B, Randle PJ, Young FG (1957) The pituitary growth hormone and metabolic processes. Ergeb Physiol Biol Chem Exp Pharmakol 49:121–211

Kühl C, Jensen SL, Nielson OV, Holst JJ (1976) Influence of basal plasma glucose concentration on the glucagon response to arginine. Scand J Gastroenterol [Suppl 37] 11:53–56

Laron Z, Mimoun M, Josephberg Z, Zadik Z, Doron M (1977) The effect of growth hormone deficiency on the insulin response to glucagon after oral glucose loading. Diabetologia 13:447–450

Larson BA, Williams TL, Lewis VJ, Vanderloon WP (1978) Insulin secretion from pancreatic islets: effect of growth hormone and related proteins. Diabetologia 15:129–132

Merimee TJ, Rabin D (1973) A survey of growth hormone secretion and action. Metabolism 22:1235–1251

Merimee TJ, Burgess JA, Rabinowitz D (1969) Influence of growth hormone on insulin secretion: studies of growth hormone deficient patients. Diabetes 16:478–482

Mitchell ML, Byrne MJ, Silver J (1969) Growth hormone release by glucagon. Lancet 1:289–290

Mitchell ML, Byrne MJ, Sanchez Y, Sawin CT (1970) Detection of growth hormone deficiency. The glucagon stimulation test. N Engl J Med 282:539–541

Mitchell ML, Suvunrungsi P, Sawin CT (1971) Effect of propanolol on the serum growth hormone response to glucagon. J Clin Endocrinol Metab 32:470–475
Parks JS, Amrhei JA, Vaidya V, Moshang T, Bogiovanni AM (1973) Growth hormone responses to propanolol-glucagon stimulation: a comparison with other tests of growth hormone reserve. J Clin Endocrinol Metab 37:85–92
Seino Y, Taminato T, Goto Y (1978 a) Growth hormone modulation of arginine-induced glucagon release: studies of isolated growth hormone deficiency and acromegaly. Clin Endocrinol (Oxf) 9:563–570
Seino Y, Taminato T, Goto Y, Inoue Y, Kadowaki S, Mari K, Imura H (1978 b) Acromegaly: insensitivity of pancreatic alpha cell to hyperglycemia. Clin Endocrinol (Oxf) 9:577–581
Sirek A, Vranic M, Sirek OV, Vigas M, Policova Z (1979) Effect of growth hormone on acute glucagon and insulin release. Am J Physiol 237:E107–E112
Sizonenko P, Rabinowitz A, Schneider P, Paunier L, Wollheim CB, Zahnd G (1977) Plasma growth hormone, insulin, and glucagon responses to arginine infusion. Pediatr Res 9:733–738
Tai TY, Pek S (1976) Direct stimulation by growth hormone of glucagon and insulin release from isolated rat pancreas. Endocrinology 99:669–677

CHAPTER 53

# Glucagon and the Heart

A. E. FARAH

## A. Introduction

The polypeptide hormones glucagon, secretin, and vasoactive intestinal peptide have homologous structures in the amino acid terminal region (JORPES 1968) and are known to release insulin, inhibit intestinal smooth muscle, increase blood glucose, mobilize free fatty acids, and increase cardiac rate and contractile force (FARAH and TUTTLE 1960; ROSS 1970; CHIBA 1976; SAID et al. 1972). CHATELAIN et al. (1979) have shown in rat heart homogenates that these polypeptides activate the adenylate cyclase in the presence of guanosine triphosphate. These findings suggest that these hormones act via similar mechanisms although they may not activate the same receptor (CHATELAIN et al. 1979a).

In this chapter, emphasis will be placed on the effects of glucagon on cardiac contractility, heart rate, and the electrophysiologic parameters controlling these functions. The receptor–adenylate cyclase relationship will be discussed and the data for heart muscle will be emphasized. Many of the membrane mechanisms of glucagon have been worked out in a variety of cells and have been discussed in this chapter wherever they may contribute to our understanding of the effects of glucagon on the heart. In a later section we discuss the clinical implications and the use of glucagon in human heart disease.

A hyperglycemic factor in pancreatic extracts was described by MURLIN et al. (1923) and COLLIP (1923) and glucagon was isolated by STAUB et al. (1953). Some insulin preparations contained impurities which had pharmacologic effects on both intestinal and cardiac muscle. Thus, VISSCHER et al. (1927) described an epinephrine-like effect on the heart of certain commercial insulin preparations and suggested that this effect was caused by contaminants of the insulin preparation. TADA (1929), TRENDELENBERG (1934), and FARAH (1938) described epinephrine-like effects of some insulin preparations in the isolated intestinal smooth muscle. The data of FARAH (1938) suggested that these effects of insulin preparations were not due to insulin per se, but to a contaminant or contaminants of these preparations. Crystalline insulin did not inhibit the movements of isolated rabbit intestines.

## B. Effects of Glucagon on Mechanical and Electrophysiologic Properties

### I. Cardiac Contractility and Rate

Experiments conducted in the dog heart–lung preparation have shown positive inotropic and chronotropic effects of amorphous insulin preparations; however,

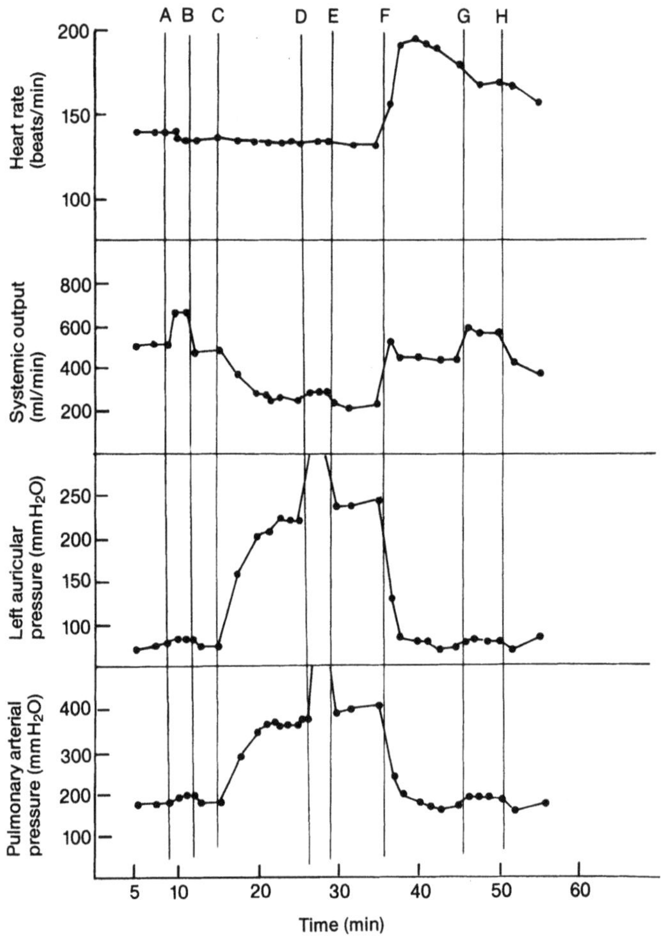


**Fig. 1.** The effect of glucagon on heart failure produced in a heart–lung preparation of the dog (Method of PATTERSON and STARLING, as modified by KRAYER and MENDEZ 1942). Male dog: 10.9 kg; heart weight 92 g; sodium pentobarbital anesthesia; temperature 37.5 °C; blood volume 780 ml; arterial resistance 80 mmHg. *A* raise inflow vessel by 5 cm; *B* lower inflow vessel by 5 cm; *C* infuse sodium pentobarbital 125 mg; *D* increase inflow vessel by 5 cm; *E* decrease inflow vessel by 5 cm; *F* inject 100 µg crystalline glucagon (Lilly); *G* increase inflow vessel by 5 cm; *H* decrease inflow vessel by 5 cm

crystalline insulin had no visible hemodynamic effects (FARAH 1938) and a quantitative assessment of the intestinal and cardiac effects of a number of these insulins did not correlate with the hypoglycemic effects of these preparations (FARAH 1938).

Following the isolation of glucagon from the pancreas, the hormone was tested in the dog heart–lung preparation and showed inotropic and chronotropic effects which were directly related to the hyperglycemic effects of a number of glucagon preparations (made available by Eli Lilly and Company, Indianapolis, Indiana). Crystalline glucagon had strong positive inotropic and chronotropic effects in the heart–lung preparation which, on a molar basis, were at least equal to those observed with epinephrine (FARAH and TUTTLE 1960). These inotropic effects of glu-

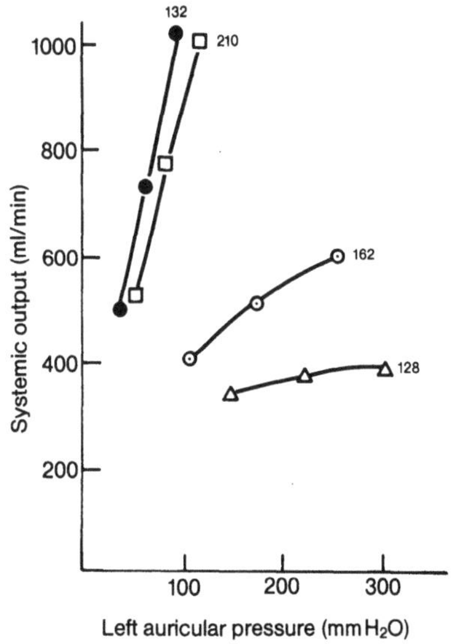


**Fig. 2.** Effect of glucagon on the mammalian heart–lung preparation. Male dog: 11.6 kg; sodium pentobarbital anestesia; temperature 37.5 °C; blood volume 800 ml; arterial resistance 78 mmHg; systemic output plotted against left auricular pressure. *Full circles* control; *full triangles* heart failure (pentobarbital 150 mg); *circled dots* after 15 µg glucagon; *squares* after 100 µg glucagon. Numbers are maximal heart rates observed (beats/min)

cagon could be observed in both spontaneous and barbiturate-induced failure and produced in most instances a complete reversal of these types of heart failure. Figure 1 depicts an experiment in a dog heart–lung preparation where a maximally effective dose of glucagon was used. A sufficiency test was conducted by raising the inflow vessel 5 cm, thus increasing the flow of blood to the right heart. Under control conditions, the procedure increased the cardiac output and produced minimal increases in the left auricular and pulmonary arterial pressures and heart rate. The administration of pentobarbital produced an increase in the auricular and pulmonary artery pressures, a reduction in the systemic output, and minimal changes in the heart rate; direct observation of the heart indicated a dilatation of the ventricles. When a repeat sufficiency test was conducted, cardiac output increased slightly and both auricular and pulmonary artery pressures increased significantly. The administration of 100 µg glucagon increased cardiac output, decreased cardiac volume and auricular and pulmonary artery pressures to control levels, and the sufficiency test was normalized to levels comparable to those seen during the control period. In Fig. 2 we have plotted the left auricular pressure (equivalent to the diastolic left ventricular pressure) against systemic output (cardiac output minus coronary blood flow). This relationship was profoundly changed by a barbiturate-induced heart failure and effective doses of glucagon reversed the auricular pressure–systemic output relationship and with 100 µg this was reversed to control conditions. Heart rate was dose dependent and increased by about 35–80 beats/min. That the heart rate increase was not responsible for the positive inotropic effect was demonstrated by SIMAAN and FAWAZ (1976), who administered the alkaloid, veratramine and reduced the glucagon-induced increase in heart rate without affecting

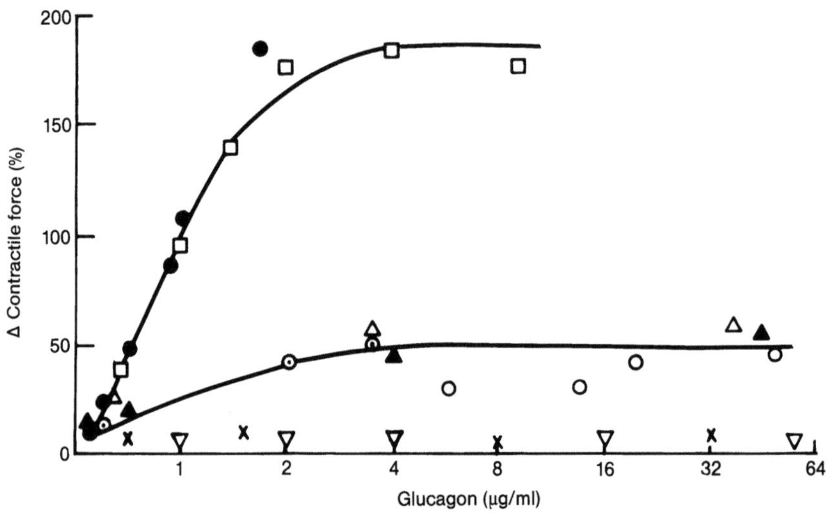


**Fig. 3.** The dose–response curve of glucagon in isolated heart tissue of various species. *Squares* dog trabeculae (A. FARAH 1967, unpublished work); *full circles* dog papillary muscle (PRASAD 1975a); *upward open triangles* cat papillary muscle (MARCUS et al. 1971); *upward full triangles* cat papillary muscle (GLICK et al. 1968); *open circles* cat heart (SATTLER and VAN ZWIETEN 1972); *downward open triangles* rabbit ventricular and auricular strips (A. FARAH 1967, unpublished work); *crosses* guinea pig papillary muscle (MACLEOD et al. 1981)

the positive inotropic effect of glucagon significantly (see Fig. 6). Veratramine is known to prevent or reverse the heart rate increases produced by a number of $\beta$-adrenergic agonists (KRAYER 1949, 1950). Furthermore, MARSIGLIA et al. (1970) kept the heart rate constant and demonstrated an increase in cardiac contractility following glucagon administration. Similar results were observed in isolated ventricular preparations where heart rate was kept constant. The effects of glucagon on cardiac contractility and rate were confirmed and extended by REGAN et al. (1964), WHITEHOUSE and JAMES (1966), GLICK et al. (1968), LUCCHESI (1968), and MATSUURA et al. (1971), who demonstrated the cardiac effects of glucagon in the intact anesthetized dog and in isolated cardiac preparations. BODER and JOHNSON (1972) observed both positive inotropic and chronotropic effects in cultured mouse heart cells and MOURA and SIMPKINS (1975) observed an increased content of cyclic AMP in cultured rat heart cells following the addition of 10 m*M* glucagon.

The minimally effective inotropic dose of glucagon was about $2–10 \times 10^{-9}$ *M* for isolated guinea pig auricles, dog, rat, and cat isolated ventricular muscle and maximal effects were obtained with a concentration of $10^{-6}$ *M* (Fig. 3). A reduction of the maximal contractile response of rat ventricular muscle was obtained with $10^{-5}$ *M* glucagon (MACLEOD et al. 1981; RODGERS et al. 1981). This reduced effect of a high dose of glucagon was not observed in heart–lung preparations or intact dogs.

The use of repeated doses of glucagon produced a reduction in the inotropic and chronotropic response (FARAH and TUTTLE 1960) and cumulative addition of glucagon produced significantly lower maximal responses in both guinea pig and rat atria and in rat ventricular muscle than when given as a single dose (MacLeod et al. 1979, 1981; Rodgers et al. 1981). These data suggest that the reduction in the response to cumulative doses of glucagon was due to a desensitization phenomenon (see Sect. K). In the isolated heart, glucagon increased the maximum ventricular systolic pressure and the rate of rise in pressure (MacLeod et al. 1981) and similar findings have been described for the intact dog (Glick et al. 1968).

Time to peak force in isolated heart muscle was reduced by catecholamines and cyclic AMP derivatives and this effect was related to cyclic AMP production (Reuter 1974b; Meinertz et al. 1976). Glick et al. (1968) and Spilker (1970) did not observe a decrease in time to peak tension in cat papillary and guinea pig auricular tissue following maximally effective doses of glucagon. Marcus et al. (1971) and Gaide et al. (1981) studied cat papillary muscle and reported that glucagon alone did not decrease the time to peak tension. However, Marcus et al. (1971) observed a reduction of the time to peak tension when glucagon was added in the presence of a phosphodiesterase inhibitor. Greeff (1976) and MacLeod et al. (1981) have observed a decrease in the time to peak tension, as well as the relaxation time following the addition of glucagon. Our own findings in dog isolated ventricular trabeculae confirm these observations and show that the decrease in time to peak tension is both dose dependent and inversely proportional to the increase in peak tension; it could not be clearly observed until the increase in contractile force exceeded 40%–50% of control values and was best observed at relatively low calcium ion concentrations. In many of the experiments where negative results were reported, contractility changes did not exceed 50%.

Heart rate increases following glucagon administration have been observed in all the hearts of responsive species. In the isolated heart, the increase in heart rate per unit increment in contractile force was greater for glucagon than for norepinephrine (Fig. 4; Chiba 1975; Kimura et al. 1974). The increase in heart rate produced by glucagon was dependent on the control heart rate and this increase lasted for about 25–40 min following the administration of 100–200 μg to the dog heart–lung preparation and intact dog (Glick et al. 1968; Lucchesi 1968).

Regan et al. (1964), Whitehouse and James (1966), Glick et al. (1968), and Lucchesi (1968) have conducted the definitive experiments on intact anesthetized dogs and have demonstrated an increase in the contractile force and heart rate and minimal changes in blood pressure when glucagon was administered intravenously. Similar results were obtained by Hammer et al. (1973) and Fricke et al. (1971). Glucagon in a dose of 0.5–16 mg produced a dose-dependent increase in heart rate and contractile force, reduced ventricular and diastolic pressure, and increased the rate of pressure development in the ventricles. These effects were demonstrated with bolus injections, as well as continuous infusions of glucagon. In contrast to the isolated heart, no tachyphylaxis was observed in the intact dog following repeated injections of glucagon (Lucchesi 1968). The contractile response to glucagon was most marked when the heart was in failure (Farah and Tuttle 1960; Glick et al. 1968). In intact dogs, the effects of 40–50 μg/kg glucagon lasted about 25–35 min on both heart rate and contractility (Lucchesi 1968; Glick et al. 1968).

Lydtin et al. (1972) and Smitherman et al. (1978) have determined a dose–response curve of glucagon in humans. The doses of glucagon infused were rather small. However, dose-related effects on heart rate, ejection fraction and systolic

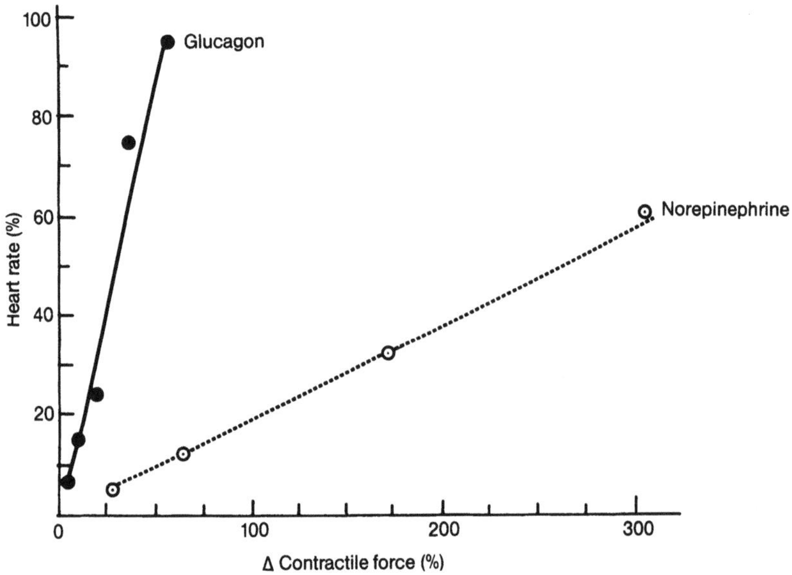


**Fig. 4.** The relation of increases in contractile force and heart rate in the canine isolated atria; comparison of glucagon and norepinephrine. After CHIBA (1975)

volume, plasma glucose, and insulin could be demonstrated. In heart failure cases, the increase in heart rate was less marked than in normal subjects (PARMLEY et al. 1968; ARMSTRONG et al. 1971). This difference was probably caused by the glucagon-induced improvement in cardiac activity, increased blood flow and pulse pressure, and the resulting reduction in sympathetic tone.

## II. Atrioventricular Conduction and Idioventricular Rhythms

WHITSITT and LUCCHESI (1968) were the first to show that glucagon could antagonize the propranalol-induced decrease in atrioventricular conduction velocity. This observation was extended by STEINER et al. (1969), who recorded His bundle potentials in dogs and showed that the interval between the pacing impulse and the His bundle spike was reduced by glucagon. When the auricles were driven at a high rate, the second-degree heart block produced was improved by the intravenous administration of glucagon (STEINER et al. 1969; LIPSKI et al. 1972; LAVARENNE et al. 1974) and these effects were not blocked by propranolol. This increase in atrioventricular conduction was also demonstrated by HAWTHORNE and HINDS (1972) and CURRY et al. (1972), who observed that glucagon increased the ventricular rate during an episode of auricular flutter or fibrillation without a change in auricular rate. Figure 5 illustrates this point and it can be seen that the glucagon injection, although it had no effect on auricular rate, increased the ventricular rate in a preparation where auricular flutter had been produced by the method of ROSENBLUETH and GARCIA RAMOS (1947). Using a perfused canine heart preparation, IIJIMA et al. (1974) were able to show that a dose of glucagon that had no effect on the sinus rate decreased the atrioventricular conduction time. Similar findings have been re-

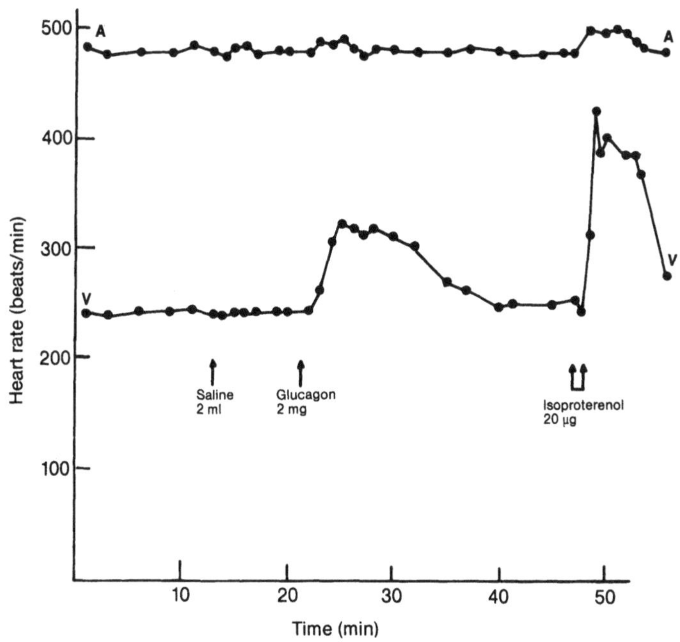


**Fig. 5.** Effect of glucagon and isoproterenol on the auricular (**A**) and ventricular (**V**) rate of a canine auricular flutter preparation. ROSENBLUETH and GARCIA RAMOS (1947)

ported in human patients with atrioventricular block (PARMLEY et al. 1969; KONES 1971; SANNA et al. 1975). However, clinical experience with glucagon in atrioventricular block is quite limited and negative results have also been reported in patients with various types of nodal rhythms (AVENHAUS et al. 1971; NISHIMURA et al. 1972; KONES et al. 1972; GAVRILESCU et al. 1972).

Atrioventricular pacemaker activity was studied by crushing the sinoatrial node (LUCCHESI et al. 1969), or by injecting a cholinergic agent into the sinus node of dogs (URTHALER et al. 1974) and thus stopping the discharge of the sinus node. Both groups showed that glucagon increased the rate of discharge of the atrioventricular pacemaker. URTHALER et al. (1974) compared the sinus rate changes with the rate of discharge of the atrioventricular node and found a direct relation between these pacemakers. Their data indicate that glucagon does not significantly change this relationship and thus it is possible that the increase in atrioventricular discharge rate is related to the increased sinus rate produced by glucagon. The effects of glucagon on the atrioventricular discharge rate were not blocked by propranolol (LUCCHESI et al. 1969). LUCCHESI et al. (1969) observed an idioventricular rhythm after the destruction of the sinus node and this rhythm did not respond to a relatively small dose of glucagon (4 µg/kg). Following the production of heart block by cauterization or by the injection of formalin into the atrioventricular area of dogs, it was shown by WOODS et al. (1970), DANIELL et al. (1970), and HURWITZ (1971) that during the acute phase of these experimental blocks, glucagon had no effect on the idioventricular rhythm, but during the chronic phase, glucagon increased the idioventricular discharge rate. URTHALER et al. (1974) injected physostigmine into the artery supplying the atrioventricular node, thus producing an idioventricular rhythm which was responsive to glucagon. The differences in results are possibly due to the amount of glucagon used in these experiments. In the negative experiments (LUCCHESI et al. 1969), 4 μg/kg were used, while doses up to 50 μg/kg were used in the positive experiments (WOODS et al. 1970; DANIELL et al. 1970; URTHALER et al. 1974).

Vagal stimulation was used to produce an idioventricular rhythm by STEINER et al. (1969), who were unable to observe a significant change in the vagal escape time or the idioventricular rate produced by glucagon. When WILKERSON et al. (1971) took into account the rate-dependent suppression of ventricular automaticity, glucagon produced a statistically significant increase in the number of ventricular beats and a decrease in the time of asystole. This was supported by the observation that glucagon increased the discharge rate in spontaneously beating Purkinje fibers (PRUETT et al. 1971; STEWART et al. 1969). LIPSKI et al. (1972) concluded from their studies that glucagon in relatively large dosage (50 μg/kg) increased the sinus rate, atrioventricular conduction, and ventricular automaticity, and decreased the vagal escape time during vagal stimulation. In a few dogs, glucagon changed the site of escape pacemaker to a higher focus and converted a discharge originating below the atrioventricular node to a rhythm originating at or above the nodal region.

The idioventricular rhythm that was observed after coronary artery ligation in dogs was studied by LUCCHESI et al. (1969), who administered relatively small doses of glucagon (12 μg/kg) and did not observe an increase in the irregularity rate of this experimental preparation, while small doses of epinephrine markedly increased the idioventricular rhythm. Using a similar experimental preparation to the one used by LUCCHESI et al. (1969), MADAN (1971) and MADAN et al. (1971) demonstrated that large doses of glucagon (30–100 μg/kg) reduced the rate of an idioventricular rhythm and in some experiments produced a complete reversal to a sinus rhythm. Figure 4 in the paper of LUCCHESI et al. (1969) indicates a trend toward a reduction in the idioventricular rhythm in the coronary ligated dog following glucagon administration. Digitalis-induced irregularities have also been reduced or reversed to a sinus rhythm by means of glucagon (COHN et al. 1970; MADAN et al. 1971). In all these studies, glucagon produced an increase in sinus rate and the overdrive effect could play an important role in the reversal of ventricular irregularities to a sinus rhythm (see GREENFIELD and ORGAIN 1967; DESANCTIS and KASTOR 1968). MOE and JELIFE (1978) have shown that entrainment of an ectopic focus by the sinoatrial node does not require a sinus rate higher than the ectopic focus since a small change in the sinus rate could cause major changes in the incidence and pattern of the ectopic beats without a direct drug action on the ectopic focus. These observations of MOE and JELIFE (1978) could explain some of the effects of glucagon on ectopic ventricular rhythms following digitalis intoxication or coronary occlusion.

Other factors besides the increase in sinus rate could play a role in the antiarrhythmic effects of glucagon. Thus, glucagon can cause a release of insulin from the pancreas (see Chap. 22) which could affect cardiac arrhythmias by its effects on potassium uptake of the heart. Glucagon enhances glycolysis in the liver and produces a short-term increase in the plasma potassium concentration. The increased insulin concentration can increase the glucose, amino acid, and potassium uptake of the heart, which could influence the excitability, conductivity, and discharge frequency from an ectopic focus in the heart (for reviews, see GREENFIELD and ORGAIN 1967; UNGER and ORCI 1979; FARAH and ALOUSI 1981).

## III. Cardiac Action Potentials

Changes in intracellular potential have been described by STEWART et al. (1969), EDMANDS et al. (1969), PRUETT et al. (1971), REYNOLD et al. (1971), KOBAYASHI et al. (1971), and PRASAD (1972, 1975 a, b). In isolated normal Purkinje fibers, ventricular and auricular muscles, glucagon in concentrations of 0.5–30 μg/ml bathing fluid shortened the total length of the action potential, reduced phases 2 and 3 of the potential and had no effect on the zero potential. However, SPILKER (1970), LÜDERITZ et al. (1971 a), and AVENHAUS et al. (1971) observed a glucagon-induced prolongation of the action potential and a reduction of the rate of rise of the zero phase of the action potential. STEWART et al. (1969) observed that glucagon can reverse the quinidine-induced changes in the refractory period and rate of rise of the sodium current. PRASAD (1977) and PRASAD and WECKWORTH (1978) reversed with glucagon some of the quinidine- and procainamide-induced arrhythmic effects on the heart. All of these data suggest that glucagon, besides its indirect heart rate effects, also has direct effect on heart membranes, which could contribute to its antiarrhythmic properties.

## IV. Coronary Blood Flow and Cardiac Oxygen Consumption

REGAN et al. (1964) and GORMAN et al. (1970) observed increases in sinus rate, cardiac contractility, and cardiac oxygen consumption, a decrease in free fatty acid extraction, an increase in lactate production, and an increase in glucose uptake following glucagon administration. The latter changes distinguished the glucagon effects from those of epinephrine. These effects of glucagon were not blocked by previous reserpine administration to the dog. Thus, these metabolic effects seem to be direct, rather than via a release of norepinephrine.

In the experiments of MARSIGLIA et al. (1970), the heart rate was kept constant, and here glucagon increased contractility, coronary blood flow, and oxygen consumption. When coronary blood flow was kept constant, a similar increase in cardiac oxygen consumption could be observed. The increases in coronary flow and oxygen consumption were qualitatively and quantitatively similar to those observed by earlier investigators. TARNOW et al. (1975) published the dose–response relationship of glucagon to heart rate contractility, coronary blood flow, and oxygen consumption of dog hearts which increased by about the same relative amount. SIMAAN and FAWAZ (1976) have shown that maximal effects of glucagon on contractility of a dog heart–lung preparation were obtained with a glucagon infusion rate of 50 μg/min. Further increases in the rate of glucagon infusion had no significant effects on contractility, but increased the sinus rate and oxygen consumption and decreased the efficiency of the heart. Pretreating the heart with veratramine decreased the heart rate effects of glucagon (FAWAZ and SIMAAN 1974), reduced the increment in the increase in oxygen consumption, and did not change the con-

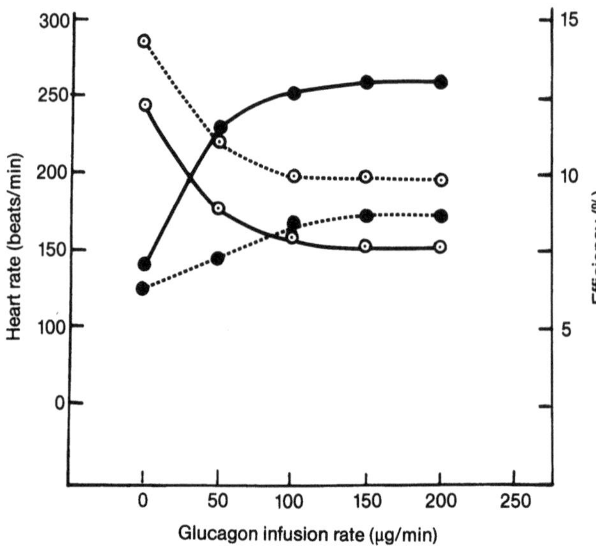


**Fig. 6.** Effect of glucagon on the heart rate and efficiency of the canine heart–lung preparation. *Full circles* heart rate; *circled dots* efficiency; *full lines* before veratramine; *broken lines* after veratramine. Veratramine (average 7 mg) was given to reduce the heart rate effects of glucagon. After SIMAN and FAWAZ (1976)

tractile effects of glucagon on the heart (Fig. 6). This suggests that a part of the increase in oxygen consumption seen with glucagon was caused by the heart rate increase and resembled the cardiac effects seen with epinephrine.

In contrast to the effects of isoproterenol, the relative increase in systolic component of coronary blood flow produced by glucagon was not observed. Coronary blood flow was increased by 75% in these unanesthetized dogs, but the ratio of systolic to diastolic flow did not change (BACHE et al. 1970). SHARMA et al. (1970) also found that glucagon increased heart rate, cardiac output, coronary blood flow, and oxygen consumption in unanesthetized dogs. However, these effects were less marked than previously observed in anesthetized dogs and were qualitatively and quantitatively similar to those observed in humans. TIBBLIN et al. (1971) have shown that the reduction in coronary flow and in total body and cardiac oxygen consumption produced by hemorrhage can be restored to near normal levels by the intravenous injection of glucagon. In canine right heart bypass preparations, glucagon increased cardiac oxygen consumption when left ventricular end-diastolic pressure (LVEDP) was below 15 mmHg and caused no significant increase in oxygen consumption if the LVEDP was above 15 mmHg, although increase in contractility and heart rate occurred in all the animals. The reduction in LVEDP was much greater in the animals with high LVEDP than in those with low pressures and this reduction in LVEDP could be correlated with a greater reduction in left ventricular circumference (BIANCO et al. 1971). These findings are in line with the observation that in the heart, cardiac oxygen consumption is related to changes in sinus rate, rate of rise of the contraction, and the change in diastolic cardiac volume. Thus, in heart failure, the reduction in diastolic volume produced by glucagon will counteract the increase in oxygen consumption caused by the increase in rate and contractile force (SONNENBLICK et al. 1968 a; see also SIMAAN and FAWAZ 1976).

The increase in coronary blood flow and cardiac oxygen consumption produced by glucagon is related to the glucagon-induced changes in rate and contractility since glucagon has no effect on the quiescent noncontracting rat heart (GMEINER and BRACHFELD 1971 a, b). BUGGE-ASPERHEIM (1972) compared adrenergic- with glucagon-induced changes and concluded that the cardiac oxygen consumption was less with glucagon than with adrenergic stimulation, probably owing to the increased cardiac work imposed by the increased aortic pressure during adrenergic inotropic stimulation of the heart and the uptake of free fatty acids. Results in human patients obtained by GOLDSCHLAGER et al. (1969) were qualitatively similar to those observed in dogs, namely, an increase in myocardial blood flow and oxygen consumption with no change in oxygen extraction. The data of MANCHESTER et al. (1970) show that glucagon effects on cardiac dynamics and oxygen consumption are less marked in humans than dogs. MOIR and NAYLER (1970) concluded that the reduction in coronary vascular resistance produced by glucagon was due to the metabolic effects induced by the increase in rate and contractility and that glucagon had no significant direct vasodilator effects on the coronary resistance vessels.

The increase in contractility and heart rate predisposed the heart to the effects of infarction (MAROKO et al. 1971; SHELL and SOBEL 1973). Thus, the administration of glucagon, which increased heart rate and contractility, significantly increased the S–T segment elevation during coronary artery occlusion, which was statistically less than that produced by isoproterenol (LEKVEN et al. 1973). However, it should be kept in mind that the increased oxygen requirements in the normal heart due to glucagon are counteracted in the failing heart by the decreased cardiac volume produced by this drug, and thus more cardiac work will be produced with the same or even less oxygen consumption (efficiency increases) (SONNENBLICK et al. 1968 a), and this will affect the degree of cardiac damage induced by the inotropic agent.

## C. Factors Which Influence Inotropic and Chronotropic Effects

### I. Species Differences

The early studies of FARAH and TUTTLE (1960) indicated that glucagon produced inotropic and chronotropic responses in the isolated hearts of dogs, cats, rats, and guinea pigs, while the rabbit heart was not responsive to glucagon. In Table 1, the literature concerning the inotropic and chronotropic effects of glucagon in various species has been summarized. It is apparent that, in contrast to other species, the rabbit heart and guinea pig ventricle are relatively insensitive to glucagon.

Quantitative differences in sensitivity of the hearts of the responsive species have been observed and Fig. 3 depicts the dose–response curve of glucagon for a number of species. However, discrepancies exist; thus, FARAH and TUTTLE (1960) and KRUTY et al. (1978) found rabbit hearts to be relatively unresponsive to glucagon, while SMITHERMAN et al. (1976) demonstrated an increase in tension de-

**Table 1.** Glucagon effects on contractility and cyclic AMP formation in cardiac muscle

| Species | Preparation | Effects on rate and contractility | Effects on cyclic AMP | Reference |
|---|---|---|---|---|
| Calf | Heart slices | – | Increased | LEE et al. (1971) |
| | Isolated heart; papillary and auricular muscle from normal heart | Increased | – | FARAH and TUTTLE (1960); GLICK et al. (1968); LUCCHESI (1968); SATTLER and VAN ZWIETEN (1972) |
| | Isolated papillary muscle from normal heart | Increased | Increased | GOLD et al. (1970) |
| Cat | Heart homogenate and particulate fractions; normal heart | – | Increased | LEVEY and EPSTEIN (1969); LEVEY et al. (1970) |
| | Heart failure; papillary muscle and particulate fraction | No change | No change | GOLD et al. (1970); LEVEY et al. (1970) |
| | Chronic heart failure | – | Increased | NOBEL-ALLEN et al. (1973) |
| Dog | Isolated heart; isolated papillary and auricular muscle | Increased | – | FARAH and TUTTLE (1960); GLICK et al. (1968); LUCCHESI (1968); KIMURA et al. (1974); CHIBA (1975); FAWAZ and SIMAAN (1974); SIMAAN and FAWAZ (1976) |
| | Intact anesthetized dog | Increased | – | GLICK et al. (1968); LUCCHESI (1968); HAMMER et al. (1973); TARNOW et al. (1975); SMITHERMAN et al. (1978); REGAN et al. (1964); WHITEHOUSE and JAMES (1966) |
| | Denervated heart | Increased | – | MATSUURA et al. (1971) |
| | Reserpinized heart | Increased | – | FARAH and TUTTLE (1960) |
| | Heart pretreated with dichloro-isoproterenol | Effects of glucagon reduced or eliminated | – | FARAH and TUTTLE (1960); WHITEHOUSE and JAMES (1966); REGAN et al. (1964); LUCCHESI (1968) |
| | Heart pretreated with propranalol | Increased | – | GLICK et al. (1968); LUCCHESI (1968); A. FARAH (1967, unpublished work) |

**Table 1.** (continued)

| Species | Preparation | Effects on rate and contractility | Effects on cyclic AMP | Reference |
|---|---|---|---|---|
| Dog | Chronic heart failure; volume overload | Reduced inotropic effect; no change in chronotropic effect | – | NEWMAN (1978) |
| Guinea pig | Isolated auricle | Increased | Increased | SPILKER (1970); MACLEOD et al. (1979, 1981); RODGERS et al. (1981); MIELKE and VAN ZWIETEN (1971); FARAH and TUTTLE (1960) |
| | Perfused heart | Increased | – | KOBAYASHI et al. (1971) |
| | Perfused heart | Increased | No change | HENRY et al. (1973, 1975) |
| | Perfused heart | Minimal change | No change | MACLEOD et al. (1981); RODGERS et al. (1981) |
| Human | Normal heart; isolated muscle | Increased | Increased | GOLDSTEIN et al. (1971) |
| | Normal heart; isolated muscle | Increased | – | LYDTIN et al. (1972); PRASAD (1972) |
| | Normal heart; homogenate | – | Increased | LEVEY and EPSTEIN (1969) |
| | Fetal heart; homogenate | – | Increased | MENON et al. (1973); DAIL and PALMER (1973) |
| | Normal intact heart | Increased | – | SMITHERMAN et al. (1978); JESSE (1972) |
| | Heart failure; isolated tissue | No change or minimal change | – | PRASAD (1975b); STRAUER (1971a, b) |
| | Chronic heart failure; isolated tissue | No change | No change | GOLDSTEIN et al. (1971) |
| Mouse | Isolated heart | Increase in rate | No change | WILDENTHAL et al. (1976) |
| | Cultured heart cells | Increase | – | BODER and JOHNSON (1972) |
| | Intact heart | Increase in rate | No change | VINICOR (1974); CLARK et al. (1976) |
| | Fetal heart | No change | No change | WILDENTHAL and WAKELAND (1973); WILDENTHAL et al. (1976) |
| Rabbit | Isolated perfused heart | No change or minimal change | – | FARAH and TUTTLE (1960); MACRI et al. (1974); KRUTY et al. (1978) |
| | Isolated papillary muscle | Increased | – | SMITHERMAN et al. (1976); KYPSON et al. (1973) |
| | Homogenate | – | Increased | BROWN et al. (1968) |

**Table 1.** (continued)

| Species | Preparation | Effects on rate and contractility | Effects on cyclic AMP | Reference |
|---|---|---|---|---|
| Rat | Isolated heart | Increased | – | FARAH and TUTTLE (1960) |
| | Isolated heart | Increased | No change | LARAIA et al. (1968) |
| | Isolated heart | Increased | Increased | MAYER et al. (1970); KREISBERG and WILLIAMSON (1964); ØYE and LANGSLET (1972); ROBISON et al. (1967); LARGIS et al. (1973); HENRY et al. (1973); BRUNT and MCNEILL (1978); MACLEOD et al. (1981); RODGERS et al. (1981); RODGERS et al. (1981) |
| | Cultured heart cells | | Increased | MOURA and SIMPKINS (1975) |
| | Heart homogenate | – | No change | LARAIA and REDDY (1969) |
| | Heart homogenate | – | Increased | MURAD and VAUGHAN (1969); CLARK et al. (1973); KEELY et al. (1975) |
| | Fetal heart | No change | No change | WILDENTHAL et al. (1976); CLARK et al. (1973) |
| Sheep | Isolated fetal heart | No change | No change | AHUMADA et al. (1976) |
| | Isolated adult heart | Increased | Increased | AHUMADA et al. (1976) |

velopment of rabbit hearts. In the guinea pig heart, auricular muscle responded to glucagon by an increase in heart rate and contractile force (SPILKER 1970; MACLEOD et al. 1979), while ventricular muscle showed only a minimal response in contractile force (MACLEOD et al. 1979, 1981; RODGERS et al. 1981), which was most likely due to the increased rate of contraction of the guinea pig ventricle (RODGERS et al. 1981). A contrary finding was that of HENRY et al. (1975), who found the increase in contractile force of guinea pig hearts induced by glucagon to be quantitatively similar to that observed in rats. No definite explanations can be given for these discrepancies and differences in the purity and source of the glucagon may have contributed to the observed differences in response. The major action of glucagon in the guinea pig heart was to increase the heart rate and any increase in contractile force of the ventricle could be ascribed to the inotropic effect of the increase in rate (SPILKER 1970; RODGERS et al. 1981).

## II. Temperature Effects

MARCUS et al. (1971) observed that, at a bath temperature of 30 °C, the response of the isolated heart tissue to glucagon was greater than that observed at 37 °C.

This temperature effect was more marked for lyophylized glucagon than for the crystalline material (MARCUS et al. 1971).

## III. Rate–Force Relationship

The effect of rate of stimulation on the inotropic effect was studied by SPILKER (1970) in isolated guinea pig left atria. Glucagon increased the contractile force at rates greater than 24 beats/min and reached a maximal effect at about 240 beats/min. The greatest relative increase in contractility was observed at a rate of 60 beats/min. The positive chronotropic effect of glucagon has been observed in all the responsive species and it is clear from the data of SPILKER (1970) and RODGERS et al. (1981) that the chronotropic effect may contribute to the inotropic effect of glucagon.

## IV. Heart Failure

The available data show that the normal heart of sensitive species in situ or in vitro responded to glucagon by an increase in the rate and contractility. However, in a number of studies, the compromised hearts did not respond or responded poorly to glucagon. In the dog heart–lung preparation, glucagon increased the heart rate and cardiac output and decreased left and right auricular and pulmonary arterial pressures and the volume of the heart (see Figs. 1 and 2). GRUHZIT and FARAH (1955) have shown that dinitrophenol-, sodium azide-, and cyanide-induced heart failures were relatively unresponsive to the cardioactive glycoside, ouabain, while spontaneous and pentobarbital-induced failures were responsive. Similar experiments have been conducted with glucagon and the data are given in Fig. 7. It can be seen that moderate or severe dinitrophenol-, azide-, and resistance-induced failures responded poorly to glucagon, while the spontaneous and pentobarbital-induced failures responded to glucagon. Heart failure produced by increasing the peripheral resistance from 80 to 170 mmHg produced lung edema, which in turn could have reduced the oxygen supply to the heart. This type of heart failure would thus probably be one induced by the increased resistance work, as well as an anoxic state. Figure 7 also shows that the severity of the failure determined the degree of recovery of the heart failure produced by a maximally effective dose of glucagon, while severe heart failure due to dinitrophenol and azide responded to the administration of epinephrine.

GOLD et al. (1970) and LEVEY et al. (1970) have shown that papillary muscle obtained from cats where the right pulmonary artery had been chronically banded did not respond to glucagon, either with a positive inotropic effect, or an increase in the concentration of right or left ventricular cyclic AMP, although the response to a catecholamine was an increase in all these parameters. Studies conducted on human cardiac tissue obtained from chronic heart failure cases did not respond like normal human heart tissue to glucagon, either by augmentation of the contractile force, or adenylate cyclase activity (PARMLEY et al. 1970; GOLDSTEIN et al. 1971; WESTLIE et al. 1971). However, NOBEL-ALLEN et al. (1973) and WINOKUR et al. (1975) have shown that in pulmonary artery banded cats with heart failure, the heart in situ and isolated papillary muscle from such hearts showed a normal in-

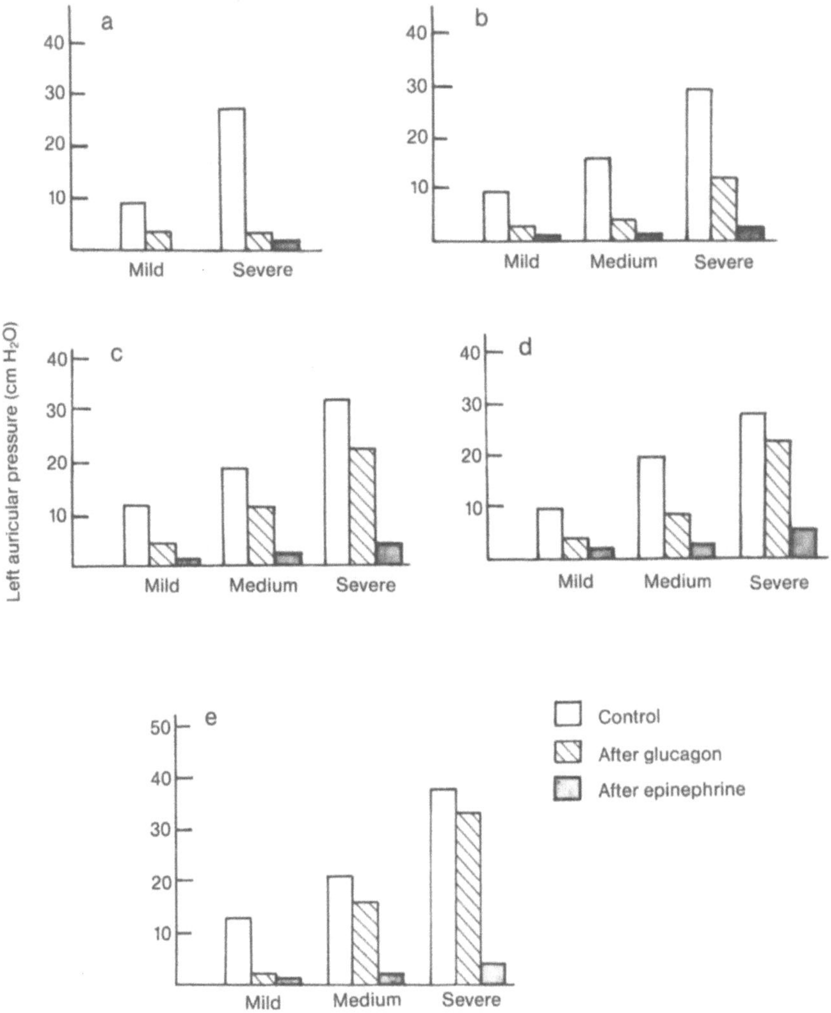


**Fig. 7 a–e.** The effect of glucagon on different types of heart failure in the canine heart–lung preparation. Temperature 37.0°–37.5 °C; blood volume 800–900 ml; resistance 90 mmHg. Diagrams illustrate: spontaneous heart failure (**a**); pentobarbital-induced failure (90–170 mg) (**b**); resistance failure (**c**); dinitrophenol-induced failure (90–210 mg) (**d**); and sodium azide-induced failure (95–320 mg) (**e**). Left auricular pressure: mild failure = 80–100 $mmH_2O$; medium failure = 180–200 $mmH_2O$; severe failure = 280–380 $mmH_2O$. Dose of glucagon was a maximally effective dose (200–250 μg); dose of epinephrine was 5 μg

crease in contractile force when glucagon was administered. STRAUER (1971 a, b) obtained papillary muscle from human failing hearts and repeated the experiments of GOLDSTEIN et al. (1971). They observed a modest increase in cardiac contractile force following the addition of a high concentration of glucagon. These human heart muscle strips were either poorly responsive to glucagon, or the concentration of glucagon used was already in the toxic range. NEWMAN (1976, 1978) studied the effects of glucagon in an acute and chronic type of volume failure by joining the aorta to the vena cava of dogs. The acute failure responded normally to glucagon, β-agonists, and calcium ion, while the chronic heart failure showed a significant reduction in the maximal contractile response to all these agents, while heart rate and blood pressure changes did not differ in this acute and chronic heart failure model.

Heart failure caused by anoxia and its reponse to glucagon was studied by SCHEUER and STEZOSKI (1971) and BUSUTTIL et al. (1974). Recovery from anoxia was enhanced by the addition of glucagon to the perfusion fluid during the anoxic period. This may be related to enhanced glycogenolysis produced by glucagon, which would maintain temporarily increased levels of utilizable substrate during the anoxic period.

NAKANO and MOORE (1971) studied the effects of glucagon in ethanol-intoxicated guinea pigs and observed that chronic administration of ethanol enlarged the heart, but reduced the inotropic response to glucagon. In the acutely ethanol-intoxicated heart, glucagon effects were similar to those observed in control hearts. In rat isolated papillary muscles and anesthetized dogs, Freon 12 ($CCl_4F_2$) directly depressed contractility of the heart and reduced the positive inotropic response of ouabain, isoproterenol, and glucagon without affecting the positive chronotropic responses of the latter two agents (WILLARD et al. 1972).

From the data presented here, it is clear that the quantitative response of the failing heart to glucagon was determined by the type and the severity, as well as the chronicity of the heart failure. The failures caused mainly by an interference with the mechanisms that couple the contraction to energy production (spontaneous failures and barbiturate-induced failure) responded well to glucagon. Heart failure induced by interfering with energy supply (sodium azide- and dinitrophenol-induced failure) responded poorly to glucagon. The data presented in Fig. 7 also indicate that the severity of the heart failure, especially failure caused by dinitrophenol and azide, determined the positive inotropic response to glucagon.

Most observations on chronic heart failure indicated a reduced inotropic response to glucagon and other inotropic agents, including β-agonists, digitalis, and calcium ion. These observations indicate that in chronic heart failure the reduced response to a variety of inotropic agents, including calcium, is probably related to a defect in the contractile proteins of the chronically failing heart.

## V. Hypertension

LARGIS et al. (1973) and STANEVA-STOYCHEVA and BOGOSLOVOVA (1976) have studied the glucagon response of hypertrophied hearts of spontaneously hypertensive rats (SHR) and have shown a reduced inotropic response compared with normal controls. Further studies by DHALLA et al. (1973) have shown a decreased glucagon and sodium fluoride sensitivity of the adenylate cyclase obtained from SHR, as compared with nonhypertensive controls. Similar findings have been reported by CHATELAIN et al. (1979b), who observed a reduced glucagon and isoproterenol sensitivity of adenylate cyclase obtained from SHR hearts. No such difference could be observed in Goldblatt hypertensive rats when compared with controls (CHATELAIN et al. 1979b). Both groups of rats showed the expected hypertrophy of the heart due to hypertension and thus the increased heart size and stress were not related to the observe change in hormonal sensitivity of the cardiac adenylate cyclase. Furthermore, the guanyl nucleotide-regulatory and the fluoride-sensitive site of the adenylate cyclase system were not impaired in the cardiac tissue of the SHR. This was interpreted by CHATELAIN et al. (1979b) to be due to a decrease in the number of receptor sites for glucagon and isoproterenol.

A number of investigators have also reported a reduced response to adrenergic stimulation of hearts obtained from SHR (FUJIWARA et al. 1972; LIMAS and LIMAS 1978) and similar findings in human hypertensive patients have been reported by JULIUS et al. (1975). LIMAS and LIMAS (1978) compared the binding of dihydroxy-prenalol $^3$H of SHR hearts with control hearts and found a reduced binding of this $\beta$-blocker in hypertensive hearts. The authors concluded that in the hypertensive heart there was a reduced number of $\beta$-adrenoceptors, which could explain the reduced $\beta$-agonist reactivity of these hearts. This decline in $\beta$-adrenoceptors occurred before the development of hypertension and reflects genetically determined differences in these hypertensive rats. SPANGLER and GROVER (1973) have observed that glucagon administration prevented the pulmonary hypertension and reduction in cardiac output in calves exposed to hypoxia-inducing conditions. The production of cyclic AMP in the smooth muscle of the lung vessels could have prevented the anoxia-induced pulmonary vasoconstriction.

## VI. Age

In the fetus, the heart gradually acquires the ability to respond to cardioactive agents and the response to these drugs very often precedes the innervation of the fetal heart. Adenylate cyclase was found in both fetal and neonatal tissue at a time preceding innervation of the heart (for a review, see GREEN et al. 1979). The response to catecholamines develops progressively and roughly parallels the increase in the basal adenylate cyclase and calcium uptake increase produced by catecholamines (COLTART and SPILKER 1972; DAIL and PALMER 1973; POLSON et al. 1977) and preceded the innervation of the heart. Thus, POLSON et al. (1977) measured changes in contractility and cyclic AMP levels in response to $\beta$-adrenergic agonists and demonstrated an increase in both these parameters in noninnervated and innervated chick embryo hearts. The increase in cyclic AMP in the noninnervated heart was greater than in the innervated heart. DOWNING et al. (1969), FRIEDMAN et al. (1969), CLARK et al. (1973), and WILDENTHAL and WAKELAND (1973) have reported that responsiveness of the heart to glucagon appears late in cardiac development and may become responsive to glucagon only some time after birth. WILDENTHAL and WAKELAND (1973) have shown that fetal mouse hearts show an increase in heart rate when glucagon was added to hearts obtained during the 17th or 18th day of gestation. However, the responses of fetal rat hearts were different from those observed in mouse hearts since rat hearts did not respond to glucagon even on the 21st day of gestation and became responsive only after birth of the rats. WILDENTHAL et al. (1976) did not observe an increase in the cyclic AMP content of hearts of fetal mice, although they responded to glucagon by an increase in heart rate. AHUMADA et al. (1976) studied the effects of glucagon on heart strips of adult sheep and newborn and fetal lambs. They observed that glucagon augmented the rate and force of contraction of adult atrial and ventricular strips, but had no effect on strips obtained from newborn or fetal lambs. Particulate fractions obtained from fetal lambs did not respond to the addition of glucagon, while fractions obtained from adult sheep showed a 43% increase in adenylate cyclase activity following the addition of glucagon. Sodium fluoride and epinephrine augmented adenylate cyclase activity in both adult sheep and fetal lambs. CLARK et al. (1973) have demonstrated adenylate cyclase activity in fetal livers before this activity appears in fetal hearts. Adenylate cyclase has been found in human fetal tissues at 12–13 weeks of gestation. A membrane preparation of human fetal cardiac tissue responded by an increase in the cyclic AMP content when sodium fluoride, epinephrine, or glucagon were added (MENON et al. 1973). DAIL and PALMER (1973) observed in human fetal hearts a basal and fluoride-sensitive adenylate cyclase as early as the 8th or 9th week of gestation; however, they were unable to detect a glucagon-stimulated cyclase until the 17th week of gestation and catecholamines either depressed or only slightly increased cyclase activity up to 17 weeks of gestation (see also PAPP et al. 1976).

These results show that fetal and newborn cardiac tissues are less sensitive to glucagon than adult heart tissues. There are species differences in the maturation rate in the responsiveness of glucagon of the developing heart and, in some species, disparities between inotropic and chronotropic effects and adenylate cyclase activation have been reported (WILDENTHAL et al. 1976). The causes of this sensitivity difference are not clear; however, a possible cause may be a low maturation of the glucagon receptors in these hearts.

## VII. Ionic Composition

A reduction in the calcium ion concentration in the perfusion fluid increased the relative response of isolated cat hearts to glucagon, although calcium influx was not significantly changed (VISSCHER and LEE 1972). In guinea pig isolated auricles, an increase in the calcium ion concentration of the perfusion fluid from 1.0 to 10 m*M* reduced the positive inotropic effect of glucagon (ANTONACCIO and LUCCHESI 1970) and even converted it to a negative inotropic effect (GREEN 1977). Since such high calcium concentration are nearly maximally positive inotropic, it is not possible to make any conclusions concerning mechanisms, especially so since high calcium concentration also reduced or even completely inhibited the effects of ouabain (CAPRIO and FARAH 1967). GREEN (1977) proposed the hypothesis that a high calcium concentration intracellularly eliminated the inotropic effects of glucagon because calcium ion inhibited the activation of the cyclic AMP system in the heart (DRUMMOND and DUNCAN 1970; SULAKHE and DHALLA 1973; TADA et al. 1975) and acted as a negative feedback mechanism on cyclic AMP production.

High potassium concentration (22 m*M*) or tetrodotoxin can block the fast sodium channels in cardiac membranes. Under these conditions, $\beta$-agonists, xanthine diuretics, histamine, sodium fluoride, and other substances that have the capacity to increase cyclic AMP, restored excitability and the contractile response of cardiac tissue, mainly via calcium/sodium-dependent slow channels in the sarcolemmal membrane (for reviews, see SCHNEIDER and SPERELAKIS 1974; SPERELAKIS et al. 1979). Glucagon did not restore the excitability of potassium-polarized

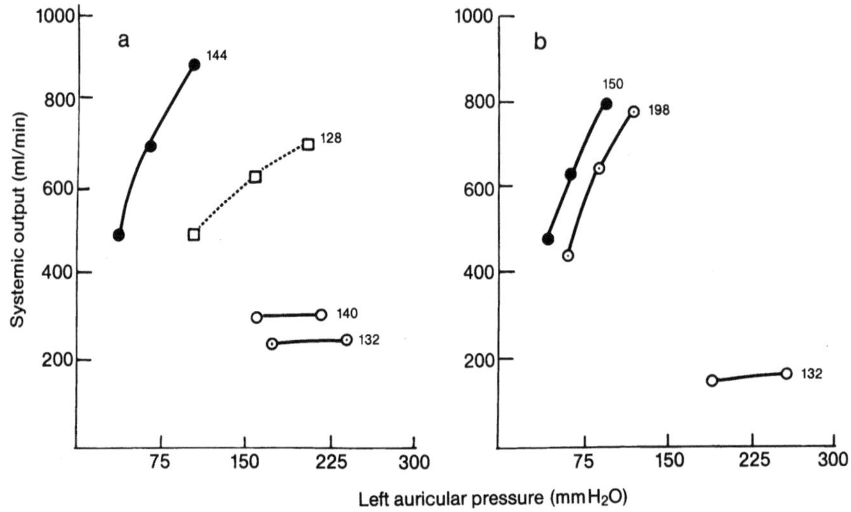


**Fig. 8 a, b.** The effect of β-adrenergic blockers on the positive inotropic and chronotropic effect of glucagon in the canine heart–lung preparation. Numbers are maximal heart rates observed (beats/min). **a** the effect of dichloroisopropylnorepinephrine (DCI). Female dog 9.8 kg; pentobarbital anesthesia; blood volume 750 ml; temperature 37.5 °C; arterial resistance = 90 mmHg. *Full circles* control; *open circles* pentobarbital-induced failure (100 mg + 10 mg DCI); *circled dots* 200 µg glucagon; *squares* 50 µg ouabain. **b** the effect of propranolol. *Full circles* control; *open circles* pentobarbital-induced failure (75 mg + 3 mg propranolol HCl); *circled dots* 200 µg glucagon

guinea pig ventricles (WATANABE and BESCH 1973, 1974). It should be kept in mind that guinea pig ventricles are relatively unresponsive to glucagon (MACLEOD et al. 1981) and thus the validity of the conclusions drawn is questionable.

## VIII. Adrenergic Blocking Agents

At the time when glucagon effects on the heart were first observed, the only β-adrenergic blocking agent available was dichloroisoproterenol (DCI), and this compound, which had both agonist and antagonist properties, was a potent inhibitor of the actions of glucagon on the dog heart (FARAH and TUTTLE 1960). These inhibitory effects of DCI have also been observed by REGAN et al. (1964), WHITEHOUSE and JAMES (1966), and by LUCCHESI (1968). However, GLICK et al. (1968), LUCCHESI (1968), SPILKER (1970), and PETERSON et al. (1978) observed that the pure β-blocker, propranolol did not significantly block the inotropic or chronotropic effects of glucagon (Fig. 8). LUCCHESI (1968) was able to show that in the intact animal, theophylline and tyramine inhibited the glucagon-induced positive inotropic and chronotropic effects. Furthermore, LUCCHESI (1968) demonstrated that the DCI inhibition of glucagon action could be reversed by propranolol. It is quite clear that glucagon is not a typical β-agonist and one must conclude that DCI, besides its β-antagonism, has antagonist properties to glucagon, which are probably related to its β-agonist actions. Another β-adrenergic antagonist, sotalol, when tested in the closed-chest anesthetized dog, blocked the positive inotropic, but not the chronotropic effects of glucagon (BERNDT et al. 1973).

## IX. The Interaction of Glucagon with Phosphodiesterase Inhibitors

This was studied by ANTONACCIO and LUCCHESI (1970) in dog and by MARCUS et al. (1971) in cat papillary muscles, and the results and conclusions are in disagreement. ANTONACCIO and LUCCHESI concluded that theophylline did not potentiate or may even have depressed the cardiac contractile response to glucagon, while MARCUS et al. (1971) claimed that the xanthine diuretic potentiated the response. BRUNT and MCNEILL (1978) restudied this problem, and, although they concluded that theophylline increased the contractile response of the isolated rat heart to glucagon, their data (Figs. 1 and 2 of their paper) are by no means convincing. Data on cyclic AMP are compatible with a potentiating effect of the xanthine diuretic on the glucagon-induced cyclic AMP concentration (BRUNT and MCNEILL 1978). WILDENTHAL and WAKELAND (1979) determined the effects of a phosphodiesterase inhibitor (RO 7-2956) on the chronotropic effects of norepinephrine and glucagon in fetal mouse hearts. A potentiation of the heart rate effects was observed with norepinephrine, but not with glucagon.

## X. Interaction of Glucagon with Cardiac Glycosides

GLICK et al. (1968) observed that the positive inotropic effect of glucagon was additive to that observed with a cardiac glycoside and did not produce arrhythmias in these arrhythmia-prone preparations. However, ANTONACCIO and LUCCHESI (1970) observed that ouabain reduced the response to glucagon and, with high doses of the glycosides, glucagon produced a negative inotropic effect in intact dogs. COHN et al. (1970) have shown that relatively large doses of glucagon could reduce, or even reverse, the ouabain-induced cardiac irregularities.

EINZIG et al. (1971) have studied the effect of glucagon on the prevention of ventricular tachycardia induced by ouabain in dogs. All dogs were pretreated with glucagon and the amount of ouabain required to produce ventricular tachycardia was determined. Pretreatment with glucagon delayed the onset of ventricular irregularities in hypokalemic, but not in normokalemic dogs. Glucagon, when given by bolus injection, converted ouabain-induced ventricular tachycardia to a sinus rhythm in 100% of the hypokalemic and in 78% of the normokalemic dogs. The overdrive phenomenon alluded to previously (BAROLD and LINHART 1970; DESANCTIS and KASTOR 1968) may have played a role in the reversal of the ventricular tachycardia, and here again there is no need to postulate a sinus rate exceeding the ventricular rate since slight changes in the sinus rate could influence the rate of discharge from an ectopic focus in the ventricle (MOE and JELIFE 1978). The increase in the dose of ouabain necessary to produce ventricular irregularity in hypokalemic dogs cannot be explained by the overdrive phenomenon since the sinus rate was higher in the normokalemic than in the hypokalemic dogs. Glucagon increased serum potassium uptake by the heart (due to insulin release from the pancreas). It may thus reduce the cardiac uptake of ouabain, which is sensitive to the concentration of potassium (MARCUS et al. 1969; COHN et al. 1967). Similar findings in normal anesthetized dogs have been described by GUPTA et al. (1978) and MADAN (1971). MADAN et al. (1971) have shown that ouabain-induced ventricular extrasystoles and tachycardia in rabbits can be delayed by the previous administration of a large dose of glucagon. Studies by GUPTA and PRASAD (1980) have shown that with an early detection method (phase/variant signature algorithm) of electrophysiologic changes, ouabain-induced changes could be reversed by glucagon. ZLOKOVIC et al. (1980) also demonstrated a protective effect of porcine glucagon against ouabain-induced irregularities seen in the rabbit heart. They attribute this protective effect of glucagon to its ability to increase the ATP concentration in the blood.

In our experience, the early glycoside-induced ventricular irregularities seen in dog heart–lung preparations could be reduced or eliminated by glucagon administration (100–250 μg/l blood). Once complete atrioventricular block had occurred, glucagon had no visible effect on the ventricular irregularities. These findings suggest that the effects of glucagon on the sinus rate must have contributed to the reduction in the idioventricular beats. As previously noted, STEWART et al. (1969) observed that glucagon had no effect on the refractory period of the normal cardiac muscle, but decreased the quinidine-induced increase in refractoriness. In line with this observation are the data which show a glucagon reversal of quinidine (PRASAD and WECKWORTH 1978) and ajmaline cardiac toxicity (HEEG and REUTER 1972). Furthermore, PRASAD (1972, 1975a) has reported that glucagon increased the membrane responsiveness, and the rate of change of the zero-phase potential ($dv/dt$), thus indicating increased intraventricular conduction in human cardiac muscle. All these data suggest that glucagon, besides its indirect effects via heart rate change probably also has direct membrane effects on cardiac muscle.

## D. Glucagon and Cyclic AMP Formation in Cardiac Muscle

The early papers of MURAD et al. (1962) and ROBISON et al. (1965) have shown that in both extracts and the isolated perfused rat heart, catecholamines raised the concentration of cyclic AMP, converted glycogen phosphorylase to its active form, and increased the contractility of the isolated perfused heart. A number of reviews have appeared which summarize the considerable literature on the effects of catecholamines on the cyclic AMP system in cardiac muscle (EPSTEIN et al. 1970, 1971; MAYER 1972, 1974; SOBEL and MAYER 1973; WOLLENBERGER 1975; TSIEN 1977; WOLLENBERGER and WILL 1978; DRUMMOND and SEVERSON 1979). Glucagon produced effects on cardiac contractility similar to those seen with catecholamines, namely, increased contractility, increased rate of development of tension, reduced time to peak tension (some reported no change or increase in time to peak tension), and relaxation time and increased heart rate. These similarities suggested that glucagon would have similar effects on the cardiac cyclic AMP system to those observed with $\beta$-adrenergic agonists.

Glucagon stimulated the formation of cyclic AMP in both intact rat myocardium and in cell preparations (ROBISON et al. 1965; BROWN et al. 1968; LARAIA et al. 1968; MURAD and VAUGHAN 1969; LEVEY and EPSTEIN 1969; LARAIA and REDDY 1969; EPSTEIN et al. 1971; LEE et al. 1971; ENTMAN 1974). In contrast to the catecholamines, the $\beta$-adrenergic blocker, propranolol did not block either the inotropic, or the cyclic AMP increase produced by glucagon. If the cyclic AMP were related to contractility, the increase in cyclic AMP should precede the increase in contractility. MAYER et al. (1970) and ØYE and LANGSLET (1972) found an apparent dissociation between the increase in cyclic AMP and contractility to glucagon; however, these authors raise the possibility that the method used may not have been sensitive enough to detect early changes. Others have found either increases or no change in cyclic AMP following glucagon administration (see Table 1). However, BRUNT and MCNEILL (1978) and MACLEOD et al. (1981) have demonstrated in isolated rat hearts an increase in cyclic AMP concomitant with the increase in cardiac contractility. BRUNT and MCNEILL (1978) have also shown that aminophylline, a phosphodiesterase inhibitor, caused a greater increase in the cyclic AMP concentration of glucagon-treated hearts, as compared with hearts treated with glucagon alone.

Glucagon administration to isolated rat or guinea pig hearts produced comparable positive inotropic effects on both of these hearts, but cyclic AMP was increased only in the rat heart (HENRY et al. 1973, 1975). When glucagon was added to broken heart cell preparations, a significant increase in cyclic AMP could be determined in the rat, but not in guinea pig hearts. However, MACLEOD et al. (1981) and RODGERS et al. (1981) have shown that, although guinea pig auricular muscle responds to glucagon by an inotropic and chronotropic effect and an increase in cyclic AMP content, the guinea pig ventricular muscle is for all practical purposes refractory to glucagon. Thus, the possible explanation of the negative data of HENRY et al. (1975) in the guinea pig heart are probably due to the fact that the whole heart was analyzed for cyclic AMP and here the nonresponding ventricular tissue would predominate. The data in the rat heart obtained by MACLEOD et al. (1981) and RODGERS et al. (1981) are similar to those reported by HENRY et al. (1975). CLARK et al. (1976) reported that glucagon increased the rate of the adult mouse heart, but did not increase the cyclic AMP content of these hearts.

ØYE and LANGSLET (1972) studied adult isolated rat hearts and determined the effects of glucagon and isoprenaline on cardiac contractile force, phosphorylase activation, and cyclic AMP levels. Both isoprenaline and glucagon increased the force of contraction and the cyclic AMP content of the rat heart, although the increase in cyclic AMP levels did not precede the increase in contractile force, but actually lagged behind, both in the isoprenaline and glucagon experiments. However, the addition of chlorpromazine, which blocks the hormone activation of adenylate cyclase (WOLFF and JONES 1970), inhibited the activation of phosphorylase without affecting the inotropic effect of glucagon. Table 1 shows that glucagon inotropic and chronotropic effects on the heart are frequently accompanied by an increase in cyclic AMP concentration. The negative correlations have been observed in hearts from premature or young animals and in mouse and guinea pig hearts. The latter discrepancy has been satisfactorily resolved by the studies of MACLEOD et al. (1981) and RODGERS et al. (1981) already discussed.

Many investigators have equated a rise in cyclic AMP concentration with a causal relation to an increase in cardiac contractility. A review by EARP and STEINER (1978) has summarized evidence that compartmentalization within the cell is possibly the cause of negative correlations. There is ample evidence indicating the compartmentalization of many of the components of the cyclic nucleotide system. Thus, TERASAKI and BROOKER (1977) have shown that, in rat atrial tissue, cyclic AMP is in free and bound forms. The bound form is found in either a particulate or soluble fraction. Addition of a β-agonist increased the free form to a greater extent than the particulate fraction. This was especially marked when the β-agonist was added in the presence of a phosphodiesterase inhibitor, indicating saturation of the binding sites in the soluble fraction. The authors suggest that these sites may represent the subunit of the cyclic AMP-dependent protein kinase. CORBIN et al. (1977) have shown that, under their experimental conditions, the cyclic AMP-dependent protein kinase in rabbit heart existed in free and bound forms. Addition of epinephrine increased the amount of the enzyme activity in the supernate, indicating that drug action will change the distribution of this cyclic AMP binding enzyme.

Immunochemical methods have been developed and the intensity of cyclic nucleotide fluorescence and its localization within the cell changed after stimulation of the cell with a hormone (see STEINER et al. 1976). These data suggest that, in heart muscle, cyclic AMP is a messenger produced by hormonal action, which binds to different proteins (protein kinases), or may penetrate deep into the cell to produce effects on nuclear receptors. The concentration changes of cyclic AMP following a stimulus may be related to the effect on receptors that are either close to or at a distance from the plasma membrane where the cyclase is located.

A negative correlation between the change in the concentration of cyclic AMP and contractile force must be interpreted with caution since the change in the cyclic AMP content of heart muscle is one of the very early steps that will lead to a change in contractile force. Cyclic AMP will be bound to protein kinases and will be inactivated by a tissue phosphodiesterase, thus the turnover rate of cyclic AMP may be increased without necessarily causing an absolute increase in the tissue concentration. Because of the compartmentalization of cyclic AMP, the protein kinases, and phosphodiesterase, as well as changes in the turnover rate of cyclic AMP, whole tissue concentrations will not have the required sensitivity to determine changes in a specific compartment of the cell. In spite of these considerations, it is rather remarkable that, in the experiments of MACLEOD et al. (1981) and RODGERS et al. (1981), correlation between the increase in contractile state and total tissue concentration of cyclic AMP can be demonstrated in rat cardiac muscle. The data on guinea pig auricular tissue were more limited, but a correlation between the increase in cyclic AMP and the contractility or sinus rate was nevertheless apparent.

Cyclic GMP (guanosine-3′,5′-monophosphate) has been implicated in the effects of negative inotropic agents on cardiac contractility (GEORGE et al. 1973; WOLLENBERGER et al. 1973; GOLDBERG et al. 1973). Working with isolated rat hearts made anoxic and then reoxygenated, BUSUTTIL et al. (1976) observed that recovery of contractility was better in the presence than in the absence of glucagon in the perfusion medium. This enhanced recovery of contractility due to glucagon was associated with a significant decrease in the cardiac levels of cyclic GMP, during both the hypoxic and reoxygenation phases. Glucagon had no effect on the cardiac concentration of cyclic AMP, either in the anoxic or reoxygenated rat heart, and these studies suggest that glucagon exerts a protective effect on the hypoxic heart, which was related to a reduction in the concentration of cyclic GMP. The importance of cyclic GMP in regulating the adenylate cyclase system has been reviewed in Chap. 13.

The evidence presented indicates that under certain circumstances inotropic or chronotropic effects of glucagon can be dissociated from an increase in cyclic AMP and compartmentalization of cyclic AMP and its binding sites could explain the negative correlations. Another possibility is that glucagon acts via cyclic AMP-dependent and cyclic AMP-independent pathways (WALSH et al. 1979). It is known that cyclic AMP plays no part in the inotropic effect of postextrasystolic potentiation (DOBSON et al. 1976), or the inotropic response of digitalis, calcium, and $\alpha$-adrenergic agonists (TSIEN 1977) and all these procedures increase intracellular calcium (ALLEN and BLINKS 1978). Agonists which do not increase the cyclic AMP content could act via a GTP-dependent regulatory protein, which could influence the gating mechanisms in the membrane and thus influence calcium entry. Furthermore, a reduction in the concentration of cyclic GMP induced by glucagon (BUSUTTIL et al. 1976) could influence the contractile state of the heart. It is known that cyclic GMP formation is calcium dependent and a reciprocal relation between the contractile state of the heart and cyclic GMP content has been demonstrated (WOLLENBERGER et al. 1973; GOLDBERG et al. 1973).

The conversion of phosphatidylethanolamine to phosphatidylcholine by catecholamines is a process which is GTP dependent, but is independent of cyclic AMP production (HIRATA et al. 1979). This methylation reaction caused a change in the membrane viscosity and thus could have influenced the membrane permeability and the contractile state of the cardiac cell. In our present state of knowledge concerning the action of glucagon on the heart, it is not possible to decide on the importance of these alternative mechanisms and dual mechanisms of the hormone could also be operative.

## E. Glucagon, Cyclic AMP, and Calcium Ion Fluxes

The relationship between ATP, the sarcoplasmic reticulum, and calcium transport was first described by HASSELBACH and MAKINOSE (1961) and EBASHI and LIPMANN (1962). They observed that an ATP-dependent pump inhibited the contractile process by lowering the calcium concentration in the medium and caused the release of bound calcium from the contractile proteins. Stimulation by cyclic AMP of calcium uptake by cardiac sarcoplasmic membranes was observed by ENTMAN et al. (1969), but was not confirmed by all investigators. Later studies by WOLLENBERGER (1972), WRAY et al. (1973), KATZ and REPKE (1973), LARAIA and MORKIN (1974), TADA et al. (1974), and HICKS et al. (1979) demonstrated that the cyclic AMP-stimulated calcium uptake was mediated by a cyclic AMP-dependent protein kinase (see WALSH et al. 1968). Further studies by KATZ and co-workers have shown that the protein kinase catalyzed the phosphorylation of a low molecular weight protein known as phospholamban (KATZ et al. 1975), which in turn stimulated calcium turnover rate by the calcium pump of the sarcoplasmic reticulum. This required ATP as a substrate and increased the amount of calcium stored in the sarcoplasmic reticulum of the heart (for reviews, see RASMUSSEN and GOODMAN 1977; WOLLENBERGER and WILL 1978; KATZ 1979; STULL and MAYER 1979; BARANY and BARANY 1981).

The increased uptake of calcium by the endoplasmic reticulum would tend to cause a relaxation of the contractile system, while the increased rate of release of calcium from this system due to phospholamban phosphorylation would increase the contractility of the cardiac muscle. Thus, drugs that increase cyclic AMP (glucagon, epinephrine) would have an effect on both calcium sequestration and calcium release from the endoplasmic reticulum of the heart. KEELY et al. (1975) have shown that glucagon increased cardiac cyclic AMP levels and protein kinase activity in rat hearts in a dose-dependent manner and ENTMAN (1974) observed that glucagon and epinephrine increased calcium accumulation in a microsomal fraction of cat myocardium. In contrast to epinephrine, this glucagon-stimulated calcium uptake was not blocked by a $\beta$-adrenergic blocking agent. NAYLER et al. (1970) determined calcium exchange in isolated dog papillary muscles and observed that the glucagon-induced increase in cardiac contractility was accompanied by an increase in calcium exchangeability. However, NAYLER et al. (1970) were unable to demonstrate an effect of glucagon on either calcium uptake or release by a dog cardiac microsomal fraction (see also MIELKE and VAN ZWIETEN 1971; VISSCHER and LEE 1972; DHALLA et al. 1973).

A new technique, utilizing the bioluminescent protein aequorin, microinjected into cardiac cells, was described by BLINKS et al. (1978). The combination of calcium ion with aequorin produced a light signal, which correlated with the intracellular concentration of calcium. With this technique, ALLEN and BLINKS (1978) have shown that a number of positive inotropic agents and procedures increased the light signal with each contraction, indicating an increased intracellular calcium concentration following each depolarization. Glucagon increased the height of the aequorin signal, decreased the time to peak light and peak tension and increased the rate of decline of the aequorin signal (BLINKS et al. 1980). It was postulated that the decline of the aequorin signal was related to the sequestration of intracellular calcium by the sarcoplasmic reticulum. FRANGAKIS and MCDANIEL (1979) have shown that glucagon and cyclic AMP increased the uptake of radiolabeled calcium in isolated rat myocytes. The rapid increase in intracellular calcium concentration in cardiac cells suggested that the release of intracellular calcium, as well as uptake from the extracellular space, plays an important role in the activation of the cardiac contractile mechanism. The decline in the aequorin signal denotes sequestration of calcium by the endoplasmic reticulum and an increase rate of loss of calcium across the sarcolemma into the extracellular space. The latter has been substantiated by BARRITT and SPIEL (1981), who have concluded that the major effect of glucagon on intracellular calcium distribution was an increase in the transfer rate of an exchangeable calcium fraction which has a short turnover time and increased the quantity of an intracellular exchangeable calcium compartment, which included the mitochondria. FRIEDMANN et al. (1980) have shown that glucagon increased calcium uptake by cardiac mitochondria and the state 3 respiration (see also BRAND and DE SELINCOURT 1980).

Another proposed site of phosphorylation by a cyclic AMP-dependent protein kinase are sarcolemmal proteins (see GLASS and KREBS 1980). Thus, SULAKHE and DHALLA (1973), SULAKHE et al. (1976), HUI et al. (1976), and SULAKHE and ST. LOUIS (1980) have demonstrated a protein kinase-dependent phosphorylation of guinea pig heart plasma membranes catalyzed by cyclic AMP, which could be reversed by a protein phosphatase. Addition of epinephrine, ATP, or cyclic AMP to these partially purified sarcolemmal membranes increased the accumulation of calcium. It is doubtful whether these membrane preparations represent pure sarcolemmal membranes and they are probably contaminated with endoplasmic reticulum and other cellular membranes. WALSH et al. (1979) purified the sarcolemma of $^{32}$P-perfused rat hearts and isolated two 36,000 and 27,000 daltons phosproteins. The 27,000 daltons protein was phosphorylated only when epinephrine was perfused into the heart and this protein was phosphorylated in vitro by a cyclic AMP-dependent protein kinase. SULAKHE and ST. LOUIS (1980) found that an exogenous kinase could phosphorylate a variety of isolated sarcolemmal proteins and it appears that during the isolation procedures denaturation of these proteins could make them far more susceptible to phosphorylation by the kinase.

Increased calcium ion movements across the membrane have been substantiated by electrophysiologic techniques. Thus, REUTER (1973, 1974a, b) has shown that cyclic AMP derivatives and norepinephrine increased the plateau height and duration of action potential of Purkinje fibers and increased the slow inward current (calcium current) and calcium conductance of the membrane; these effects on the slow inward current were related to the positive inotropic effects of these drugs (see also TSIEN 1977).

Exposing cardiac papillary muscle to either tetrodotoxin or a high potassium concentration (22 m$M$) caused an inactivation of the fast sodium current and a loss in excitability and a depolarization of the cardiac membrane. Addition of a $\beta$-adrenergic agent, cyclic AMP derivatives, sodium fluoride, or histamine restored excitability and the resultant action potential characterized by a relatively slow rate of rise and a slowly propagated impulse, which produced a contraction of the cardiac muscle. This action potential or slow potential was sensitive to the concentration of calcium and calcium inhibitors (lanthanum, manganese, verapamil; see SPERELAKIS et al. 1979). It is postulated that cyclic AMP is required for the phosphorylation of a protein acting as a gate in a sarcolemmal calcium channel. Phosphorylation of this gate protein opens up one part of this channel, while the action potential opens the second gate, thus allowing calcium entry (REUTER 1979). Studies on the excitability of potassium-depolarized guinea pig cardiac muscle have shown that glucagon did not restore excitability of the heart; however, the reports of MACLEOD et al. (1981) show that guinea pig ventricular muscle did not respond to glucagon, either by an inotropic response or an increase in cyclic AMP concentration.

Cyclic AMP-dependent phosphorylase can catalyze the phosphorylation of a number of proteins, some of which are related to the contractile state of cardiac muscle (COLE and PERRY 1975; RUBIO et a. 1975; MOIR and PERRY 1980). Thus, the inhibitory subunit of troponin (TnI) was rapidly phosphorylated in response to the addition of epinephrine to the isolated heart (ENGLAND 1976; STULL and BOSS 1977). In vitro experiments by MOIR and PERRY (1980) have shown that a serine residue of TnI was phosphorylated by cyclic AMP-dependent protein kinase. A correlation between an increase in contractile force of the perfused rat heart and the phosphorylation of TnI has been observed by ENGLAND (1976, 1977) when a $\beta$-adrenergic agent was used, but glucagon had no effect on TnI phosphorylation (ENGLAND 1976). When epinephrine effects on contractility were reversed by washing the heart, TnI phosphorylation persisted. Positive inotropic agents, such as ouabain or calcium, did not increase the phosphorylation of TnI, decreased the sensitivity of actomyosin to $Ca^{2+}$, and increased the amount of $Ca^{2+}$ needed to activate the ATPase (see EZRAILSON et al. 1977; SOLARO et al. 1981). MOPE et al. (1980) have shown that, in cardiac muscle fibers made hyperpermeable, addition of cyclic AMP increased the phosphorylation of TnI, while addition of cyclic GMP inhibited this phosphorylation by activating a phosphatase that dephosphorylates TnI. These data support the hypothesis that calcium sensitivity of cardiac muscle is controlled by a cyclic GMP-dependent phosphatase. It is thus possible that pharmacologic stimuli that induce cyclic AMP formation regulate heart contractility via a phosphorylation and dephosphorylation of TnI, and may have a relation to the effect of $\beta$-agonists on the relaxation and contraction process.

More recently, JEACOCKE and ENGLAND (1980) reported that a 150,000 daltons myofibrillar protein associated with myosin was phosphorylated to a high degree when the heart was perfused with epinephrine. The degree of phosphorylation of this protein and the increase in contractility produced by epinephrine were closely related. Incubation of cardiac myofibrils with a cyclic AMP-dependent protein kinase and ATP phosphate incorporation into this 150,000 daltons protein correlated with the phosphorylation of TnI.

Of the cardiac phospholipids, phosphatidylinositol has the highest turnover rate, which can be markedly increased by the administration of a catecholamine (GAUT and HUGGINS 1966), while glucagon increased the incorporation of $^{32}P$ into phosphatidylserine and phosphatidylethanolamine, but not into phosphatidylinositol (LO and LEVEY 1976). A relation between the high turnover rate of phospholipids in heart and a protein kinase called C kinase requires an unsaturated diacylglycerol for its activation. Arachidonic acid is a constituent of phosphatidylinositol and the hydrolysis by phospholipase C will liberate an unsaturated diacylglycerol, which could activate the C kinase. In the presence of $Ca^{2+}$, the C kinase attached itself to a membrane phosphatidylserine, phosphorylated cardiac membrane proteins, and increased the transport of $Ca^{2+}$ (LIMAS 1980). This C kinase can phosphorylate a number of muscle proteins, including phosphorylase kinase and glycogen synthetase and this kinase may be another mechanism for the control of cardiac function.

All these observations suggest that cyclic AMP-dependent kinase can phosphorylate a number of proteins which are involved in the contractility of heart muscle. Cyclic AMP-independent, but $Ca^{2+}$-dependent mechanisms could also be operative: the significance of all these various phosphorylation reactions for the increased contractility of the heart produced by glucagon is by no means clear.

## F. Effects of Glucagon on Cardiac Carbohydrate Metabolism

In the isolated heart, the metabolic effects of glucagon were similar to those observed with epinephrine (CORNBLATH et al. 1963; KREISBERG and WILLIAMSON 1964). Glucagon converted the inactive phosphorylase b to the active a form, the ratio being 1:1 at maximal glucagon activity. This conversion induced glycogenolysis and lactate production, which was observed in both normal and reserpinized hearts. Cardiac glycogen was rapidly reduced and at a steady state it was about 40% of the control value. Hexose and triose levels were elevated at a time when the glycogen content of the heart had reached a plateau.

The conversion of phosphorylase b to the active a form was affected by phosphorylase kinase and protein phosphatase reversed this process and converted phosphorylase a to its inactive b form. Phosphorylase kinase required $Ca^{2+}$ for its activity and it exists in an inactive b and an active phosphorylated a form. Phosphorylation of the b form was catalyzed by cyclic AMP-dependent kinase. The protein phosphatase dephosphorylates both phosphorylase a and phosphorylase kinase a. Activity of this phosphatase was inhibited in the presence of 5′-AMP and inorganic phosphate and this inhibition was reversed by ATP. (For details, see STULL and MAYER 1979.)

Glucagon also caused the conversion of the active glycogen synthetase to the inactive form by a phosphorylation of the active form of this enzyme. This conversion was affected directly by cyclic AMP-dependent protein kinase, which catalyzed the conversion of the a to the b form and was dephosphorylated to the active form by a protein phosphatase (see STULL and MAYER 1979).

Neither propranolol nor insulin had an effect on the glucagon-induced inhibition of glycogen synthetase. However, in spite of the lack of action of insulin on the enzyme, insulin blocked the glycogen breakdown following glucagon administration to the heart. This action of insulin is not fully understood and may be due to the inhibition of the phosphorylase by insulin. The release of insulin by glucagon administration to the intact animal limited the reduction of cardiac glycogen (BERGSTRÖM and NUTTALL 1972). Many of the effects of glucagon on glucose phosphorylation and glycolysis are likely to be secondary to its positive inotropic effect on the heart and it is possible that cyclic-AMP formed could directly activate phosphofructokinase. The increased inotropy of the heart will result in an increase in heart work, oxygen consumption, carbohydrate, and fat oxidation; however, this will activate the various enzymatic processes responsible for increased carbohydrate and lipid oxidation. The tissue concentration of ATP may decrease, that of 5′-AMP may increase and mitochondrial NADH:NAD ratio may decrease as a result of the increased work and oxygen consumption of the heart induced by glucagon (see KRUTY et al. 1978).

## G. Effects of Glucagon on Cardiac Lipid Metabolism

As reviewed in Chap. 19, glucagon is lipolytic in many species (GOODRIDGE and BALL 1965; LEFÈBVRE and LUYCKX 1969); however, in humans, it produced a modest increase in plasma free fatty acid levels, which is of short duration, and relatively large quantities have to be injected to produce this effect (SCHADE and EATON 1975). Insulin is a strong inhibitor of lipolysis, and a glucagon injection will induce a release of insulin from the pancreas and a rise in the plasma insulin concentration. This antilipolytic effect of insulin probably explains the transient course of the glucagon-induced lipolysis seen in humans and other animals. Thus, in diabetics with insulin deficiency, glucagon produced a marked increase in arterial blood glycerol concentration (an index of lipolysis), while in normal humans, a glucagon infusion produced a reduction in the arterial glycerol concentration (LILJENQUIST et al. 1974). A similar insulin–glucagon interaction has been described with rat epididymal adipose tissue in which insulin could suppress the glucagon-induced lipolysis (LEFÈBVRE and LUYCKX 1975) and these interactions of insulin and glucagon may also be operational when cardiac tissue is studied.

The data in the intact dog have shown that glucagon-induced lipolysis was minimal in contrast to that produced by catecholamines. Furthermore, catecholamines increased the free fatty acid uptake by the heart, while glucagon had no significant effects (MJÖS 1971 d; BUGGE-ASPERHEIM 1972). This uptake of free fatty acids has been related to an extra increase in cardiac oxygen consumption following catecholamine stimulation (MJÖS 1971 a, b, c, d), but was not seen following glucagon administration (MJÖS 1971 d). This increased uptake of free fatty acids may be related to the cardiac irregularities that are observed following adrenergic stimulation (OLIVER et al. 1968; KURIEN et al. 1969) and the extra increase in oxygen consumption may explain the decrease in cardiac efficiency that was frequently demonstrated following epinephrine injections into the dog heart. The minimal effects of glucagon on fatty acid uptake by the heart may explain the observation that glucagon does not increase cardiac irregularities and under certain experimental conditions even reverses experimental cardiac irregularities produced by different methods.

Fatty acids are one of the main fuels of the heart (BING 1965). Thus, JESMOK et al. (1975) have shown that glucagon, ouabain, and epinephrine increased the cardiac release of glycerol from isolated hearts about equally. Since glycerol release should be an index of triglyceride utilization, it is probable that the increase in contractility produced an increase in fatty acid oxidation, which was not related to any specific action on lipid metabolism of any of these drugs.

It is generally believed that lipoprotein lipase activity of cardiac muscle is related to triglyceride uptake by the heart or to mobilization of substrate from intracardiac triglycerides. Changes in cardiac lipoprotein lipase observed with starvation, refeeding, cold exposure, and other procedures are usually in the opposite direction to those seen in adipose tissue (BORENSZTAJN et al. 1972; 1973; RAULT et al. 1974). In the normal or starved animal, glucagon did not have a significant effect on cardiac lipoprotein lipase (RAULT et al. 1974), although exposure to a temperature of 4 °C markedly increased lipoprotein lipase activity in the rat heart. However, BORENSZTAJN et al. (1972) have shown that, in starved rats, the lipoprotein lipase activity of cardiac tissue is high, while that of adipose tissue is low. It was further shown that refeeding with glucose reduced the lipase activity of the heart, and this reduction could be prevented by an injection of glucagon. These results suggest that glucagon may participate in the regulation of myocardial lipoprotein lipase activity and may produce indirect effects via the release of insulin or epinephrine. It is of interest to note that exercise (NIKKILÄ et al. 1963), as well as adrenergic stimulation of the heart (MALLOV and CERRA 1967), increased cardiac lipoprotein lipase activity.

OSCAI (1979) administered a low (0.5 μg) and a high dose (10 μg) of glucagon to intact rats. The high dose of glucagon increased the level of heparin-nonreleasable lipoprotein lipase activity and reduced the cardiac concentration of free fatty acids and cardiac levels of triacylglycerols. The low dose glucagon resulted in a marked reduction of heparin-nonreleasable lipoprotein lipase activity and free fatty acids and increased the concentration of triacylglycerols. The authors concluded that high levels of lipoprotein lipase activity result in cardiac lipolysis, while low levels of this enzyme activity result in the synthesis of triacylglycerols. Another interpretation would be that glucagon administration increased pancreatic insulin release which counteracted and reversed the effects of the small dose, but could not completely reverse the effects of a large dose of glucagon (see also Chap. 22).

Epinephrine and glucagon can stimulate the oxidation of branched chain amino acids in hearts obtained from fasted rats perfused with substrate-free solution. Hearts of fed rats or fasted rats perfused with glucose did not show this glucagon or epinephrine effect on amino acid oxidation (BUSE et al. 1973).

Whereas insulin stimulated the uptake and incorporation of leucine and lysine into cardiac protein, glucagon inhibited this uptake and incorporation. A similar observation was made with the nonmetabolizable α-amino isobutyric acid. The major effect of glucagon was on the incorporation of the amino acids into protein, and the uptake mechanism was less sensitive to the inhibition by glucagon (HAIT et al. 1972; see also Chap. 18).

## H. Relation of the Metabolic Effects of Glucagon to Cardiac Potassium Metabolism

When glucagon was infused into intact animals, profound effects on the circulation and general metabolism were seen which may influence the function of the heart. Thus, continuous infusions of glucagon (FREY et al. 1972), or the intramuscular injection of a zinc protamine glucagon formulation into humans (KÜHN et al. 1973) produced an increase in blood glucose, immunoreactive insulin, and growth hormone. In spite of the high insulin levels, glucagon produced a reduction in glucose disappearance rate, and thus produced a carbohydrate intolerance similar to that seen in some types of diabetes. The increase in blood glucose was probably related to the glucagon-induced hepatic glycogenolysis and gluconeogenesis (CRAIG 1958; SOKAL 1966). Preceding the release of glucose from the liver, there was an increase in the hepatic vein and arterial plasma potassium concentration and a reduction in the plasma sodium concentration. This was most likely due to a release of potassium from and uptake of sodium by the liver (WOLFSON and ELLIS 1956; CRAIG 1958; BIANCO et al. 1971; LÜDERITZ et al. 1971 b; SOMLYO et al. 1971; HULSTAERT et al. 1974). This increase in plasma potassium concentration lasted only 2–5 min and was followed within 8–16 min by a hypokalemia, possibly due to the uptake of potassium by the gastrointestinal tract and heart. This uptake of potassium by the heart may play a role in the reported lack of arrhythmic potential of glucagon.

## J. Glucagon–Receptor Interactions

The second messenger hypothesis of SUTHERLAND et al. (1968) proposed that glucagon was bound to a specific receptor site on the sarcolemmal surface of the cell. This led to the activation of the intracellularly located membrane-bound adenylate cyclase which, in the presence of $Mg^{2+}$ or $Mn^{2+}$ and GTP, converted ATP to cyclic AMP. The glucagon-sensitive adenylate cyclase system contains at least three components: the specific receptor for glucagon (R), two nucleotide-regulatory components (N) containing reactive sites which are specific for GTP, and a catalytic portion (C) which, in the presence of magnesium or manganese, converts ATP to cyclic AMP (RODBELL and LONDOS 1976; LAD et al. 1979; RODBELL 1980). RODBELL has postulated that the surface receptor was linked to the internally located catalytic unit (C) by a nucleotide-regulatory component (N). This latter unit contained sites for binding GTP and mediated the effects of the receptor-bound hormone to the catalytic unit. Two types of N units have been identified: $N_s$ stimulated, while $N_i$ units inhibited the adenylate cyclase system. The unoccupied receptors inhibited the reaction of GTP with the $N_s$ units. The occupied receptor released the inhibition and $N_s$ units reacted with GTP, interacted with the catalytic unit, and activated the adenylate cyclase. The binding of GTP to $N_i$ units will inhibit the adenylate cyclase system; however, it is not known whether such an N unit exists in cardiac tissue (for details, see Chap. 13).

The intracellular cyclic AMP produced its physiologic changes by activating a cyclic AMP-dependent protein kinase. This activation was probably due to the binding of cyclic AMP to a regulatory subunit of this kinase, which enhanced the phosphorylation of a number of enzymes and proteins, utilizing ATP as a phosphate donor. The phosphorylated enzymes and proteins influenced a variety of cellular functions and these effects were reversed by a number of phosphoprotein phosphatases (Chap. 13; EXTON et al. 1977).

The binding of a peptide hormone to a receptor initiated a number of reactions, including mobility of the receptor, its aggregation, and internalization. These phenomena have been studied for a number of polypeptide hormones (insulin, glucagon, epidermal growth factor, nerve growth factor). For reviews, see PASTAN and WILLINGHAM (1981 a, b) and KING and CUATRECASAS (1981). With glucagon, the early effects on cyclic AMP, glycogenolysis, and contractility of the heart are rapidly occurring phenomena and are probably due to the surface activation of the adenylate cyclase. However, delayed phenomena, such as the effects of glucagon on cell growth, may require the internalization of the hormone–receptor complex and its attachment to an intracellular receptor, possibly located in the nucleus (see BECKER 1973; BUCHER and SWAFFIELD 1975; LEFFERT 1977; YANKNER and SHOOTER 1979; MARCHISIO et al. 1980; SAVION et al. 1981).

In all studies so far conducted, the attachment of a peptide hormone to a receptor and the aggregation of the complex, was followed by internalization of this hormone-receptor complex by a process of endocytosis. This complex was transferred to the Golgi region of the cell and at a later stage, the complex was also found in association with lysosomes and a variety of different intracellular organelles. This internalization of a peptide hormone was probably a means for hormone degradation via a receptor-mediated pathway. The loss of surface receptor following endocytosis was replenished either by a recycling of the receptor or by de novo synthesis.

Phospholipase $A_2$, when added to liver membranes, resulted in the loss of activation of the cyclase and a reduction in glucagon binding due to a change in affinity. This effect of the lipase was not observed if fluoride was the activator of the cyclic AMP system (LAD et al. 1979). The polyone antibiotic, filipin caused a marked reduction in the activation of the cyclase by glucagon as well as GTP; however, glucagon binding to the receptor was not affected. These effects of filipin and phospholipase suggest that glucagon binding and activation of the cyclase are distinct and separable systems. Furthermore, these results suggest that regulation of the receptor by nucleotides is distinct from activation of the adenylate cyclase by the nucleotide. Results obtained by DIPPLE and HOUSLAY (1979) with the polyene, amphotericin B are in agreement with this interpretation.

Further evidence that the receptor and the adenylate cyclase are independent units was obtained by hybridization techniques. The β-adrenoceptor of one cell (after inactivation of the catalytic activity) was fused to the adenylate cyclase of another cell which did not contain an intrinsic β-receptor and thus was not responsive to a β-agonist. The result of the hybridization of the cell resulted in the responsiveness of the new cell to isoproterenol and the production of cyclic AMP (ORLY and SCHRAMM 1976; SCHWARZMEIER and GILLMAN 1977). Hybridization studies with the glucagon receptor obtained from liver membranes to Friend erythroleukemia cells have been reported by SCHRAMM (1979), thus supporting the concept that the glucagon receptor and cyclase unit are separable entities (see also HOUSLAY et al. 1977). These separable units can then undergo lateral movements and these movements will probably depend on the fluidity of the membrane, which is probably controlled by the disposition of cholesterol and phospholipids in the membrane bilayer (see DIPPLE and HOUSLAY 1979; GORDON et al. 1980). Following the addition of glucagon, the two units interacted to form a two-component system which spans the bilayer membrane. With the technique of radiation inactivation in the electron beam, HOUSLAY et al. (1977) have shown that the addition of glucagon increased the target size to a value about equal to the sum of the individual sizes of the receptor and catalytic unit. This indicated that the addition of glucagon caused the receptor to interlock with the catalytic unit, thus forming a complex that was inactivated as a single target.

The rearrangement of the phospholipids in the membrane bilayer by the methyltransferases and by phospholipase probably regulate the β-adrenoceptor numbers and affinity and their coupling to the adenylate cyclase in a variety of cell types (HIRATA et al. 1979b). Phospholipase or detergent-treated solubilized enzyme systems lost activity which could be restored by the addition of phospholipids. Thus, TANAKA and STRICKLAND (1965) and MARTONOSI et al. (1968) have shown that the solubilized $Na^+$, $K^+$-ATPase lost activity and MARTONOSI et al. (1968), POHL et al. (1971), and RETHY et al. (1972) have shown that plasma membranes treated with digitonin or phospholipase A lost hormone sensitivity of the adenylate cyclase system, which could be restored by the addition of phosphatidylserine, phosphatidylcholine, and phosphatidylethanolamine. Similar phenomena have been described by LEVEY (1971a, b, 1973) and LEVEY and KLEIN (1972) for heart muscle membrane preparations and here phosphatidylserine restored glucagon and histamine responsiveness, but not norepinephrine responsiveness of the adenylate cyclase. However, the latter could be restored by the addition of phosphatidylinositol (LEVEY 1971). In a later study, LO and LEVEY (1976) reported that glucagon mediated the incorporation of labeled orthophosphate into phosphatidylserine and phosphatidylethanolamine, but not into phosphatidylcholine. Dibutyryl cyclic AMP did not increase this incorporation of phosphate into phospholipids. These findings suggest that the basic mechanism of glucagon receptor–adenylate cyclase coupling has many characteristics in common with the catecholamine receptor and that it utilizes similar biochemical pathways for the control of receptor function. In the glucagon receptor, phosphatidylserine seems to be more important than phosphatidylcholine, which is the major product in the $\beta$-adrenoceptor coupling mechanism.

LEVEY et al. (1974) and KLEIN et al. (1974) have shown that glucagon is bound to a particulate and a solubilized cardiac adenylate cyclase preparation. This binding was not dependent on the presence of a phosphatide or cyclic GMP and the labeled glucagon could be displaced by unlabeled glucagon. Glucagon binding could be suppressed by lowering the temperature or by raising the pH of the incubation medium above 8. Urea treatment did not affect the binding of glucagon to the glucagon receptor and the binding and adenylate cyclase activation were closely correlated. A considerable amount of the bound glucagon was probably not specifically related to the activation of adenylate cyclase since activation of the enzyme was already maximal after 5 min, while glucagon binding continued to increase linearly up to 40 min. At the time of maximal activation of adenylate cyclase, the amount of glucagon bound was probably less than 1% of its maximal binding to the tissue (LEVEY 1973). A highly purified glucagon-binding protein was isolated from liver membranes by GIORGIO et al. (1974). This protein, although it bound some insulin, had a high specificity for glucagon and a molecular weight of about 190,000 daltons. LEVEY et al. (1974) and LEVEY (1973) have prepared a soluble cardiac adenylate cyclase–glucagon receptor complex, which had a molecular weight of 100,000 to 200,000 daltons. Interaction of the cardiac receptor with glucagon, followed by fractionation of this material on Sephadex G-100 and Biogel P-30 produced two separate fractions: the glucagon-binding site had a molecular weight about 26,000 daltons, while the catalytic site had a molecular weight of 100,000 daltons. Based on these findings, LEVEY et al. (1974) and LEVEY (1973) proposed a cardiac model, which was modified from that of ROBISON et al. (1967) and RODBELL et al. (1969). When glucagon was bound to the receptor site, it caused a dissociation of the receptor site, which activated the catalytic site and which required the presence of phospholipids. This hypothesis is in contrast to the one proposed by HOUSLAY et al. (1977), who envisaged an interlocking of receptor with the adenylate cyclase as the final active enzyme.

At this time, it is not possible to decide which of these models is operative in the intact cell. The binding of glucagon to cells and cell membrane preparations has been utilized in the biochemical studies with glucagon and correlations between binding to a receptor and physiologic changes have been reported. That much of the binding may be nonspecific or unrelated to the functional changes in the intact cell or tissue has been pointed out by LEVEY (1973) and LEVEY et al. (1974). Early experience with the $\beta$-adrenoceptors should caution us concerning the biologic relevance of binding studies, unless these are rigidly controlled by a variety of correlations with functional changes (see BILEZIKIAN and AURBACH 1974; CUATRECASAS et al. 1974; TELL and CUATRECASAS 1974; DRUMMOND et al. 1976).

## K. Desensitization

A frequent observation has been the reduction of the sensitivity of an isolated tissue following exposure to an agonist. This desensitization, also known as tachyphylaxis, is dependent on the nature, the concentration, the time of exposure of the tissue to the agonist, and the milieu in which the isolated tissue is kept. The desensitization by catecholamines has been studied in a number of different cell types and a loss of the catecholamine-stimulated adenylate cyclase activity was observed with and without a corresponding reduction in β-adrenoceptors. In those cells where a reduction of receptors was observed, desensitization could be explained by this mechanism (see KEBABIAN et al. 1975; LEFKOWITZ and WILLIAMS 1978). However, in desensitized cells where no reduction in receptors could be observed, an uncoupling of the β-adrenoceptor from the functional component of the adenylate cyclase was probably operative since the desensitized cells still responded to the directly acting sodium fluoride, guanyl nucleotide, and cholera toxin (FISHMAN et al. 1981). Recently, TORDA et al. (1981) and YAMAGUCHI et al. (1981) have shown that repeated forced immobilization or the repeated administration of isoproterenol reduced the number of β-adrenoceptors in the heart and spleen of rats and produced a subsensitivity to the chronotropic and pressor effects of isoproterenol in pithed rats, which could be prevented by the administration of the phospholipase $A_2$ inhibitor, quinacrine. HIRATA et al. (1979a, b) and MALLORGA et al. (1980) have reported that long-term β-adrenoceptor stimulation enhanced the degradation of membrane phospholipids and decreased the number of β-adrenoceptors and these changes were prevented by quinacrine administration. These findings suggest that the turnover of membrane phospholipids may play a role in the configuration of membrane receptors. This hypothesis is strengthened by the observation of GEELEN et al. (1979) that glucagon channels diacylglycerols into phospholipids by specifically stimulating the synthesis of phosphatidylethanolamine via an increase in the rate of synthesis of phosphoethanolamine.

The conversion of phosphatidylethanolamine to phosphatidylcholine was catalyzed by two methyltransferases where *S*-adenosylmethionine serves as the methyl donor (STRITTMATTER et al. 1979; HIRATA et al. 1978). Stimulation of the synthesis of phosphatidylcholine in reticulocyte ghosts increased the number of β-adrenergic binding sites and the additon of a methyltransferase inhibitor prevented the increase in β-adrenoceptors. It was postulated by HIRATA et al. (1979a) that the methylation of phospholipids started on the cytoplasmic side and the final methylated product was located on the outer surface. A marked reduction in the viscocity of membranes has been demonstrated which correlated with the phospholipid methylation, especially the formation of the monomethyl form of the phospholipid (HIRATA and AXELROD 1978; HIRATA et al. 1979b). An increase in the methylated phospholipids occured after a short exposure of isolated astrocytoma and frog erythrocyte membranes to a β-agonist and this change correlated with an increase in the β-adrenoceptors (HIRATA et al. 1979b). However, a prolonged exposure to a β-agonist reduced the number of β-adrenoceptors, cyclic AMP formation, and the concentration of the methylated phospholipids. The reduction in these phospholipids was possibly due to the activity of phospholipase A, resulting in the accumulation of a lysophosphatidic acid. Increasing phospholipase A activity decreased the activation of cyclic AMP without an effect on the β-adrenoceptors. Lysophosphatidylcholine has been shown to inhibit the hormone-sensitive adenylate cyclase (SHIER et al. 1976; LAD et al. 1979). Furthermore, the phospholipase $A_2$ inhibitor, mepacrine blocked the desensitization of the β-agonist of both the cyclic AMP system (HIRATA et al. 1979), as well as the pharmacologic β-adrenergic effects (see YAMAGUCHI et al. 1981; TORDA et al. 1981).

These studies thus suggest that sensitization and desensitization of a tissue could be related to the synthesis and distribution of specific phospholipids in the membrane bilayer. The activity of a phospholipase could disturb this distribution by the production of lysophosphatidic acids which can inhibit cyclic AMP activity and thus reduce the sensitivity of the tissue to the agonist. No specific studies have been conducted in glucagon-desensitized heart tissue; however, the brief review of desensitization to $\beta$-adrenergic agents given in this section may be helpful in designing experiments directed toward the study of desensitization to glucagon and other agonists.

## L. Clinical Studies with Glucagon

The observations in animals that glucagon produced a positive inotropic effect with minimal arrhythmogenic potential, prompted a number of clinical studies, in both normal humans and patients suffering from a variety of cardiac conditions. The first experiments in humans were conducted by PARMLEY et al. (1968), who injected 3–5 mg glucagon intravenously into patients undergoing diagnostic cardiac catherization. PARMLEY et al. (1968) have shown an increase in the heart rate, cardiac index, rate of pressure change ($dP/dt$), but no change in the LVEDP or systemic vascular resistance. These studies were the beginning of a large number of investigations where cardiac effects of glucagon were studied in a variety of cardiac conditions. Some of these studies have been reviewed (KONES and PHILLIPS 1971a; KING et al. 1973; LAVARENNE 1974; GUERRICCHIO 1974; LUCCHESI 1977).

A few studies have been conducted in normal humans, utilizing noninvasive and invasive procedures. The studies of LINHART et al. (1968) were conducted on patients free of heart disease, and cardiodynamic measurements were conducted by standard catherization techniques. Heart rate, cardiac output, right ventricular and systolic pressure, and $dP/dt$ were increased significantly following the administration of glucagon. BYRNE et al. (1972) administered glucagon (50 µg/kg) intravenously to normal human subjects and observed an increase in heart rate and blood pressure, and a decrease in the electromechanical systole and left ventricular ejection time. The studies of BOURASSA et al. (1970) were conducted on patients with and without coronary artery disease. In subjects with normal and abnormal coronary arteries, glucagon produced a moderate increase in heart rate, mean arterial blood pressure, cardiac index, and left ventricular $dP/dt$. In individuals with significant coronary artery disease, glucagon produced a marked increase in lactate extraction, but in individuals with normal coronary arteries, glucagon administration did not change lactate extraction significantly. In both diseased and normal hearts, glucagon produced a slight reduction in the arterial potassium levels which may be related to the increase in insulin secretion produced by the injection of relatively large quantities of glucagon (see Chap. 22). The studies of TIMMIS et al. (1973) were conducted on normal humans with noninvasive procedures. The results obtained are similar to those observed by BYRNE et al. (1972) and are compatible with the interpretation that glucagon increased the contractile force and rate of the normal human heart.

Glucagon administration in acute heart failure, especially following cardiac surgery, produced equivocal improvement in the various cardiac parameters and the overall circulatory condition of the patients. In the studies conducted by SONNENBLICK et al. (1968b), PARMLEY et al. (1969), and PARMLEY and SONNENBLICK (1969), 5 mg glucagon administered intravenously increased cardiac index and the systolic ejection rate. Heart rate and mean arterial pressure were increased to a modest extent, while left ventricular and diastolic pressure were unchanged. VAUGHN et al. (1970) determined the effects of intravenous glucagon in low output failure following cardiac surgery. Significant increases in heart rate and cardiac output could be recorded, and no arrhythmias were observed following glucagon administration. On the other hand, GREGORY et al. (1969) were unable to demonstrate significant effects of glucagon on various cardiac parameters in postsurgical cases with low cardiac output. In children of age 5–12 years suffering from postsurgical low output failure, glucagon (50 μg/kg) increased the systolic and diastolic arterial pressures and produced no untoward side effects (ABBOTT 1972).

In general, the results obtained in the acute low output cardiac failure indicate that glucagon caused a dose-dependent increase in cardiac output and heart rate and was not arrhythmogenic. The dosage of glucagon was limited to a maximum of 5–10 mg since higher doses caused nausea and vomiting, and maximum cardiac effects were attained with these doses.

In chronic heart failure of various causes and intensity, the effects of glucagon on heart function were quite variable and sometimes completely absent. Thus, PARMLEY et al. (1970) studied a series of 21 patients with class I or II (criteria of New York Heart Association) heart failure and could show that cardiac index, heart rate, mean arterial pressure, and the maximum rate of rise of left ventricular pressure were increased; however, LVEDP and systemic resistance did not change significantly. The effects of glucagon could be superimposed on the effects of cardiac glycosides. In these experiments, blood glucose increased, while the serum potassium concentration declined from 4.2 to 3.7 mequiv./l. Similar results could be demonstrated when glucagon was given directly into the cardiac chambers or by continuous infusion. No cardiac irregularities due to glucagon were observed; however, nausea and vomiting were the major side effects that limited the dosage of this drug. Other investigators (KLEIN et al. 1968; LINHART et al. 1968; WILLIAMS et al. 1969; BROGAN et al. 1969; MURTAGH et al. 1970; VANDER ARK and REYNOLDS 1970; WILCKEN and LVOFF 1970; NORD et al. 1970; KONES and PHILLIPS 1971; LIKOFF 1972; LVOFF and WILCKEN 1972; LOEB et al. 1973; JESSE et al. 1975; BERTOLAZZI et al. 1976; JAESCHKE et al. 1977) have had variable results on a variety of parameters measured and, in general, the chronic types of heart failure were less responsive to glucagon than the more acute types.

The severity of the heart failure also seems to play a role. Thus, ARMSTRONG et al. (1971) administered 5 mg glucagon intravenously, and the patients with class I or II heart failure had a significantly greater increase in cardiac output than those cases classified as class III and IV heart failure. Similar results were observed by WESTLIE et al. (1971), who concluded that the cardiac response to glucagon was dependent on the cardiac output before glucagon was given (the higher the initial cardiac output, the better the response to glucagon). The animal experiments discussed previously show that a heart failure induced by interference in the metabolic machinery is much less responsive to glucagon than a heart failure caused by interference with the coupling mechanism (pentobarbital or a spontaneous failure). The degree of the experimental heart failure determined the response of the heart to glucagon, and this was especially true with the experimental heart failures produced by dinitrophenol and azide. Other studies on isolated heart tissue, obtained from human heart failure patients during surgery, also indicated that these tissues did not respond well to glucagon, although they responded well to either paired stimulation or isoproterenol (PARMLEY et al. 1970). Human clinical heart failure is poorly understood, but interference with the coupling mechanism and the metabolic machinery of the heart is likely to be involved, and it is thus not unexpected that the chronic clinical heart failure responded with great variability, not only to glucagon, but to other cardiotonic agents, such as the cardiac glycosides.

The water-soluble glucagon was usually administered intravenously. More recently, a zinc protamine glucagon formulation has been described which can be given intramuscularly and has a prolonged effect with minimal side effects. Thus, GAMBA et al. (1977) have treated refractory heart failure cases with the zinc protamine glucagon formulation, 5 mg twice daily by intramuscular injection, and have claimed good results in five of eight cases of heart failure refractory to digitalis therapy. An improvement in the arrhythmias was also observed and two cases of chronic auricular fibrillation were converted to a sinus rhythm. Similar results were obtained by KÜHN (1973) and by PICHLER et al. (1979), who observed a significant prolongation of the effects of zinc protamine glucagon on cardiac output with minimal effects on heart rate and the gastrointestinal tract (see also TRADING et al. 1969).

Glucagon has been tried in the treatment of the complications of myocardial infarction and the resultant shock state. Thus, MANCHESTER et al. (1969) and MATLOFF et al. (1970) were the first to demonstrate the beneficial effects of glucagon in canine experimental cardiac infarction. In these experiments, the administration of glucagon caused an increase in heart rate, mean arterial pressure, and aortic blood flow. The rate of left ventricular contraction increased, while the left ventricular diastolic pressure decreased, and no arrhythmogenic effects of glucagon could be observed. Similar results have been described by PURI and BING (1970) in canine cardiac infarct preparations. In the human and canine heart, glucagon increased the coronary blood flow and oxygen consumption with no change in the arteriovenous oxygen difference (MANCHESTER et al. 1970).

ASAI (1974) did an extensive study in a large number of dogs (52) and measured a number of cardiac parameters after the left coronary artery was injected with a suspension of lycopodium (40 μm). Of these animals, 78% died either of cardiac arrest or ventricular fibrillation. If glucagon (50 μg/kg) was given within 2 min after the lycopodium injection, only 30% of the animals died, and in this group coronary arterial flow and left ventricular $dP/dt$ increased, even above control values, and arrhythmic changes were infrequent. Glucagon produced favorable effects on the hemodynamics of patients in cardiogenic shock and facilitated defibrillation of an irritable myocardium.

LEDINGHAM et al. (1973) studied blood flow through the ischemic and normal regions of the canine myocardium. Glucagon or isoprenaline increased blood flow through the normal, but not through the infarcted area, probably because these vessels were already maximally dilated. On the other hand, norepinephrine and oxyfedrine increased the blood flow through both normal and infarcted areas, probably because these compounds increased the effective driving pressure by increasing the pressure in the coronary vessels (norepinephrine), or decreasing LVEDP (oxyfedrine).

In canine hearts where the circulation has been restricted, glucagon decreased lactate extraction from the blood to anaerobic levels. This was probably one of the main limitations of glucagon in spite of the fact that it increased cardiac contractility of the ischemic heart (SHAVER et al. 1974). However, if cardiac volume should decrease concomitantly with the increase in rate and contractility, this could possibly reduce the oxygen debt produced by the administration of glucagon and thus improve the overall metabolic situation for the ischemic heart (KUMAR et al. 1972). KUMAR et al. (1972) observed that the sinus tachycardia following acute infarction in dogs was reduced following the administration of glucagon, and no irregular ventricular beats were observed. Oxygen consumption and coronary blood flow were not significantly increased, although the cardiac output was significantly increased. Similar observations were made in a healing cardiac infarct preparation where glucagon decreased the left ventricular diastolic pressure and peripheral resistance and increased cardiac output and heart rate without a significant change in cardiac oxygen consumption. These data indicate that glucagon produced some beneficial effects during cardiac ischemia, which were probably related to changes in the end-diastolic pressure and the concomitant reduction in diastolic volume and oxygen consumption. As previously discussed, glucagon did not increase free fatty acid uptake by the heart. This factor may play a role in the observed reduction in ischemic damage following glucagon injection when compared with an equipotent injection of isoproterenol (LEKVEN et al. 1973).

The results of clinical studies in myocardial infarction have been summarized by KONES et al. (1971). One of the first studies to be reported was that of MANCHESTER et al. (1969), who measured coronary blood in patients by means of the $^{133}$Xe washout method and observed an increase in coronary blood flow and oxygen consumption with no change in the arteriovenous oxygen difference. Results obtained by GOLDSCHLAGER et al. (1969) indicate that, in the patient with a recent coronary artery occlusion, glucagon did not significantly change coronary blood flow, in contrast to the findings in individuals with normal coronary arteries, possibly because in these individuals the arteries were already maximally dilated. In all these cases, oxygen extraction did not change, and it was concluded that the observed coronary dilation was secondary to the increase in the cardiac dynamics produced by glucagon. BOURASSA et al. (1970) studied the effects of single does of 5 mg glucagon in patients with coronary heart disease. Contractility of the heart was increased without an increase in oxygen consumption, and lactate extraction was significantly increased. Calculated efficiency of the heart was unchanged, but a negative cardiac potassium balance was observed in these patients following the administration of glucagon. GREENBERG et al. (1972) administered glucagon to patients with angina and advanced coronary artery disease. After relatively small doses of glucagon (1–3.6 mg) were injected into the pulmonary artery, no significant changes were observed in any of the hemodynamic parameters studied. MURTAGH et al. (1970) observed improved cardiac function following glucagon administration in acute cardiac infarction. EDDY et al. (1969) administered glucagon to six patients with acute myocardial infarction; three of them had cardiogenic shock. In all cases, glucagon produced a positive inotropic effect which resulted in a temporary increase in the blood pressure with a minimal increase in heart rate, and no arrhythmias were observed. All three patients in cardiogenic shock died within 10–22 h. In the studies of DIAMOND et al. (1971), it was shown that glucagon increased the cardiac index in infarcted patients and decreased peripheral resistance, this in contrast to norepinephrine, which increased both. Studies by GOLIKOV et al. (1979) demonstrated the efficacy of glucagon in the treatment of acute myocardial infarction where it counteracted the bradycardia hypotension and disorders of cardiac rhythm and conduction. KONES et al. (1972) studied twelve patients in cardiogenic shock and administered glucagon by continuous infusion (3 mg/h); four of the patients with acute myocardial infarction recovered. The expected manifestations, namely, increased blood pressure and decreased central venous pressure, were observed following the glucagon infusion. However, some patients complained of more severe chest pains following glucagon administration. The clinical experience with glucagon in cardiogenic shock and myocardial infarction seems to indicate that this drug may be useful, especially in those cases that do not have long-lasting chronic heart disease. Its lack of arrhythmogenic properties and its effects on oxygen consumption and efficiency of the heart are distinct advantages over the more powerfully acting catecholamines.

The antiarrhythmogenic properties of glucagon as observed in experimental preparations have already been discussed. In human patients, the consensus is that glucagon is not arrhythmogenic (KONES et al. 1972), although some investigators have described an increase in the frequency of premature ventricular beats following glucagon administration (ASHLEY et al. 1970). Glucagon can be given without untoward effects to digitalized patients (BROGAN et al. 1969; NORD et al. 1970; VANDER ARK and REYNOLDS 1970; KONES and PHILLIPS 1971b). The effects of glucagon on atrioventricular conduction have been documented in a variety of animal preparations. Clinically, KONES (1971) and SANNA et al. (1975) have shown that heart block in humans can be improved by glucagon, although this is by no means a dramatic effect.

The general conclusions to be drawn from the clinical studies are that glucagon is a moderately effective inotropic agent in humans, and in contrast to the catecholamines does not have a tendency to produce arrhythmias. It does not increase free fatty acid uptake by the heart, and in the failing heart it will increase efficiency. Its peripheral dilator properties produced a modest reduction in afterload and an improvement in the general hemodynamics of the patient.

The use of glucagon in low output failure is encouraging; however, the results obtained in the more chronic types of heart failure are much less predictable. Cardiac effects of glucagon were not blocked by propranolol and related $\beta$-blockers, thus making glucagon a useful agent in treating the cardiodepressive manifestation of the $\beta$-blocking agents. Although it increased cardiac output and the rate of contraction of the left ventricle, glucagon did not consistently reduce LVEDP.

The blood flow through the ischemic area of a cardiac infarct was dependent on a driving pressure which is the resultant of the diastolic peripheral coronary pressure minus LVEDP (MARSHALL and PARRATT 1973). Thus, glucagon in contrast to norepinephrine would have a tendency to decrease the gradients under normal blood pressure conditions; however, in heart failure and shock, glucagon will frequently increase the aortic diastolic blood pressure and could conceivably increase the coronary diastolic pressure, and thus improve the blood flow through an ischemic area of the heart.

The glucagon inotropic effect was not blocked by β-adrenergic antagonists and these blockers also have antiarrhythmic properties (MADAN et al. 1971; MADAN 1977). MADAN and JAIN (1973) combined glucagon with a β-blocker for the treatment of arrhythmias which accompany myocardial infarctions. According to the authors, this was a successful combination therapy for the treatment of heart failure occurring during a myocardial infarction.

## M. Conclusions

Glucagon is a positive inotropic and chronotropic agent in hearts of a variety of species, but seems to be relatively inactive in rabbit hearts and guinea pig ventricular muscle. The contractile changes produced by glucagon resemble those observed with β-adrenergic agonists and increase the rate of rise and the maximum of the contraction, decrease the time to peak tension and relaxation time, improve auriculoventricular conduction, increase cardiac oxygen consumption, and improve the efficiency of the failing heart. Glucagon is, for all practical purposes, nonarrhythmogenic and in large doses will even suppress digitalis-induced and other types of arrhythmias. The increase in sinus rate would act as an overdrive and thus cover up extrasystoles. However, direct effects on ventricular conductive tissue have been described.

A number of factors modify the inotropic effect of glucagon. Thus, the type, severity, and chronicity of the heart failure are important factors which determine the inotropic response to glucagon. In contrast to β-agonists, glucagon's effects on the heart are not blocked by the pure β-antagonists, but they are blocked by some antagonists that also have intrinsic agonist activity. The relation between glucagon's effects on the heart and an increase in cyclic AMP have been demonstrated in a variety of species where glucagon increases cardiac contractile force. Negative correlations have also been described and have been discussed. It is likely that glucagon can increase intracellular calcium concentration via cyclic AMP-dependent and cyclic AMP-independent pathways.

The glucagon receptor and its relations to the adenylate cyclase system have been discussed and resemble in many ways the relations observed with the β-adrenergic agents, except that the glucagon receptor seems to have a separate identity from that of β-adrenoceptors. The relation of glucagon to membrane phospholipid metabolism indicates that the membrane changes produced by glucagon are probably related to the metabolism of phospholipids, especially phosphatidylserine. In the light of the desensitization studies conducted with β-agonists, it is suggested that similar mechanisms are operative to explain the desensitization of isolated cardiac muscle to repeated doses of glucagon.

Because of the cardiac activity of glucagon observed in animals, clinical studies with glucagon were conducted in human patients. In general, the human data confirm the animal experimental observations in that glucagon increased the contractility and rate of the human heart. Here again, the type, intensity, and chronicity of the heart failure seem to determine the response of the patients. Acute heart failures, in general, respond to glucagon; however, in chronic congestive heart failure, the results were rather mixed and patients frequently failed to respond. The human dose of glucagon is limited to about 5 mg since larger doses produced nausea and vomiting, requiring a reduction in dosage.

## References

Abbot TR (1972) The use of glucagon following open heart surgery in children. Br J Anaesth 44:854–857
Ahumada G, Sobel BE, Friedman WF (1976) Age-dependent mechanical and biochemical responses to glucagon. Am J Physiol 230:1590–1593
Allen DG, Blinks JR (1978) Calcium transients in aequorin-injected frog cardiac muscle. Nature 273:509–513
Antonaccio MJ, Lucchesi BR (1970) The interaction of glucagon with theophylline, $PGE_1$, isoproterenol, ouabain and $CaCl_2$ on the dog isolated papillary muscle. Life Sci 9:1081–1089
Armstrong PW, Gold HB, Daggett WM, Austen WG, Sanders CA (1971) Hemodynamic evaluation of glucagon in symptomatic heart disease. Circulation 44:67–73
Asai Y (1974) Effect of glucagon on the ischemic heart: experimental and clinical studies. Jpn Assoc Thorac Surg 22:1158–1171
Ashley WW, Kaminsky DM, Lipski JI (1970) Hemodynamic effects of glucagon in patients with fixed-rate pacemakers. Am J Cardiol 25:82–83
Avenhaus H, Lüderitz B, Strauer BE, Bolte H-D, Riecker G (1971) Cardiac effects of glucagon. Dtsch Med Wochenschr 96:702–707
Bache RJ, McHale PA, Curry CL, Alexander JA, Greenfield JC Jr (1970) Coronary and systemic hemodynamic effects of glucagon in the intact unanesthetized dog. J Appl Physiol 29:769–774
Barany M, Barany K (1981) Protein phosphorylation in cardiac and vascular smooth muscle. Am J Physiol 241:H117–H128
Barold SS, Linhart JW (1970) Recent advances in the treatment of ectopic tachycardias by electrical pacing. Am J Cardiol 25:698–706
Barritt GJ, Spiel PF (1981) Effects of glucagon on $^{45}Ca$ outflow exchange in the isolated perfused rat heart. Biochem Pharmacol 30:1–8
Becker FF (1973) Humoral aspects of liver regeneration. In: LoBue J, Gordon AS (eds) Humoral control of growth and differentiation, vol 1. Academic, London, pp 249–256
Bergström WJ, Nuttall FQ (1972) Effect of glucagon, insulin and acetylcholine on heart glycogen synthetase and phosphorylase activity. Biochim Biophys Acta 286:146–154
Berndt TB, Ansfield TJ, Alfonso S, Rowe GG (1973) Modification in dogs of the systemic and coronary hemodynamic effects of glucagon by sotalol. Am Heart J 85:671–678
Bertolazzi R, Ricciardi S, Marenco G (1976) L'impiego del glucagone nel trettamento della grave insufficienza contrattile del miocardio. Minerva Med 67:3519–3524
Bianco JA, Shanahan EA, Ostheimer GW, Guyton RA, Powell WJ Jr, Daggett WM (1971) Effects of glucagon on myocardial oxygen consumption and potassium balance. Am J Physiol 221:626–631
Bilezikian JP, Aurbach GD (1973) A $\beta$-adrenergic receptor of the turkey erythrocyte. I. Binding of catecholamine and relationship to adenylate cyclase activity. J Biol Chem 248:5575–5583
Bing RJ (1965) Cardiac metabolism. Physiol Rev 45:171–213
Blinks JR, Allen DG, Prendergast FG, Harrer GC (1978) Photoproteins as models of drug receptors. Life Sci 22:1237–1244
Blinks JR, Lee NKM, Morgan JP (1980) $Ca^{++}$ transients in mammalian heart muscle: effects of inotropic agents on aequorin signals. Fed Proc 39:854
Boder GB, Johnson IS (1972) Comparative effects of some cardioactive agents on the automaticity of cultured heart cells. J Mol Cell Cardiol 4:453–463

Borensztajn J, Samols DR, Rubinstein AH (1972) Effects of insulin on lipoprotein lipase activity in the rat heart and adipose tissue. Am J Physiol 223:1271–1275

Borensztajn J, Keig P, Rubenstein AH (1973) The role of glucagon in the regulation of myocardial lipoprotein lipase activity. Biochem Biophys Res Commun 53:603–608

Bourassa MG, Elbar J, Campeau L (1970) Effect of glucagon on myocardial metabolism in patients with an without coronary artery disease. Circulation 42:53–60

Brand MD, de Selincourt C (1980) Effects of glucagon and $Na^+$ on the control of extramitochondrial free $Ca^{2+}$ concentration by mitochondria from liver and heart. Biochem Biophys Res Commun 92:1377–1382

Brogan E, Kozonis MC, Overy DC (1969) Glucagon therapy in heart failure. Lancet 1:482–484

Brown HD, Clattopadhyay SK, Matthews WS (1968) Glucagon stimulation of adenyl cyclase activity of cardiac muscle. Naturwissenschaften 55:181–182

Brunt ME, McNeill JH (1978) The effect of glucagon on rat cardiac cyclic AMP, phosphorylase and force of contraction. Arch Int Pharmacodyn Ther 233:42–52

Bucher NLR, Swaffield MN (1975) Regulation of hepatic regeneration in rats by synergistic action of insulin and glucagon. Proc Natl Acad Sci USA 72:1157–1160

Bugge-Asperheim B (1972) Effects of increased aortic blood pressure on myocardial performance and metabolism during non-adrenergic inotropic stimulation of the heart. Scand J Clin Lab Invest 30:137–143

Buse MG, Biggers JF, Drier C, Buse JF (1973) The effect of epinephrine, glucagon, and the nutritional state on the oxidation of branched chain amino acids and pyruvate by isolated hearts and diaphragms of the rat. J Biol Chem 248:697–706

Busuttil RW, Paddock RJ, George WJ (1974) Protective effect of glucagon on the isolated perfused rat heart following severe hypoxia. Proc Soc Exp Biol Med 147:527–532

Busuttil RW, Paddock RJ, Fisher JW, George WJ (1976) Changes in cyclic nucleotide levels and contractile force in the isolated hypoxic rat heart during perfusion with glucagon. Circ Res 38:162–167

Byrne MJ, Piggot V, Spodick DH (1972) Cardiovascular responses to glucagon: physiological measurement by external recordings. Am Heart J 83:635–643

Caprio A, Farah A (1967) The effect of the ionic milieu on the response of rabbit cardiac muscle to ouabain. J Pharmacol Exp Ther 155:403–414

Chatelain P, Deschodt-Lanckman M, de Neef P, Christophe J, Robberecht P (1979a) Effect of secretin, glucagon, and vasoactive intestinal polypeptide on the hormone-sensitive rat cardiac adenylate cyclase. Arch Int Physiol Biochim 87:783–784

Chatelain P, Robberecht P, de Neef P, Claeys M, Christophe J (1979b) Low responsiveness of cardiac adenylate cyclase activity to peptide hormones in spontaneously hypertensive rats. FEBS Lett 107:86–90

Chiba SH (1975) Positive chronotropic and inotropic effects of glucagon on the canine isolated atrium. Tohoku J Exp Med 115:61–65

Chiba S (1976) Effect of secretin on pacemaker activity and contractility in the isolated blood-perfused atrium of the dog. Clin Exp Pharmacol Physiol 3:167–172

Clark CM Jr, Beatty B, Allen DO (1973) Evidence for delayed development of the glucagon receptor of adenylate cyclase in the fetal and neonatal rat heart. J Clin Invest 52:1018–1025

Clark CM Jr, Waller D, Kohalmi D, Gardner R, Clark J, Levey GS, Wildenthal K, Allen D (1976) Evidence that cyclic AMP is not involved in the chronotropic action of glucagon in the adult mouse heart. Endocrinology 99:23–29

Cohn KE, Kleiger RE, Harrison DC (1967) Influence of potassium depletion on myocardial concentration of tritiated digoxin. Circ Res 20:473–476

Cohn KE, Agmon J, Gamble OW (1970) The effect of glucagon on arrhythmias due to digitalis toxicity. Am J Cardiol 25:683–689

Cole HA, Perry SV (1975) The phosphorylation of troponin I from cardiac muscle. Biochem J 149:525–533

Collip JB (1923) Delayed manifestation of the physiological effects of insulin following administration of certain pancreatic extracts. Am J Physiol 63:391–392

Coltart DJ, Spilker BA (1972) Development of human foetal inotropic responses to catecholamines. Experientia 28:525–526
Corbin JD, Sugden PH, Lincoln TM, Keely SL (1977) Compartmentalization of adenosine 3′:5′-monophosphate and adenosine 3′:5′-monophosphate-dependent protein kinase in heart tissue. J Biol Chem 252:3854–3861
Cornblath M, Randle PJ, Parmeggiani A, Morgan HE (1963) Regulation of glycogenolysis in muscle. Effects of glucagon and anoxia on lactate production, glycogen content, and phosphorylate activity in the perfused isolated rat heart. J Biol Chem 238:1592–1597
Craig AB (1958) Observations on epinephrine and glucagon-induced glycogenolysis and potassium loss in the isolated perfused frog liver. Am J Physiol 193:425–430
Cuatrecasas P, Tell GPE, Sica V, Parikh I, Chang K-J (1974) Noradrenaline binding and the search for catecholamine receptors. Nature 247:92–97
Curry CL, Hinds JE, Hawthorne EW (1972) The effect of glucagon on the ventricular response in atrial fibrillation. A possible hazard. Am J Cardiol 29:258
Dail WG Jr, Palmer GC (1973) Localization and correlation of catecholamine-containing cells with adenyl cyclase and phosphodiesterase activities in the human fetal heart. Anat Rec 177:265–288
Daniell HB, Holl JE, Pruett JK, Bagwell EE, Woods EF (1970) Cardiovascular effects of glucagon in dogs with non-nodal pacemakers. J Electrocardiol 3:117–120
DeSanctis RW, Kastor JA (1968) Rapid intracardiac pacing for treatment of recurrent ventricular tachyarrhythmias in the absence of heart block. Am Heart J 76:168–172
Dhalla NS, Sulakhe PV, McNamara DB (1973) Studies on the relationship between adenylate cyclase activity and calcium transport by sarcotubular membranes. Biochim Biophys Acta 323:276–284
Diamond G, Forrester J, Danzig R, Parmley WW, Swan HJC (1971) Acute myocardial infarction in man. Comparative hemodynamic effects of norepinephrine and glucagon. Am J Cardiol 27:612–616
Dipple I, Houslay MD (1979) Amphotericin B has very different effects on the glucagon- and fluoride-stimulated adenylate cyclase activities of rat liver plasma membranes. FEBS Lett 106:21–24
Dobson JG, Ross J, Mayer SE (1976) The role of cyclic adenosine 3′5′-monophosphate and calcium in the regulation of contractility and glycogen phosphorylase activity in guinea pig papillary muscle. Circ Res 39:388–395
Downing SE, Talner NS, Campbell AGM, Halloran KH, Wax HB (1969) Influence of sympathetic nerve stimulation on ventricular function in the newborn lamb. Circ Res 25:417–428
Drummond GI, Duncan L (1970) Adenyl cyclase in cardiac tissue. J Biol Chem 245:976–983
Drummond GI, Severson DL (1979) Cyclic nucleotides and cardiac function. Circ Res 44:145–153
Drummond GI, Vallieres J, Drummond M (1976) Adenylate cyclase and catecholamine binding in plasma membrane enriched preparations of cardiac and skeletal muscle. Recent Adv Stud Cardiac Struct Metab 9:161–182
Earp HS, Steiner AL (1978) Compartmentalization of cyclic nucleotide-mediated hormone action. Annu Rev Pharmacol Toxicol 18:431–459
Ebashi S, Lipmann F (1962) Adenosine triphosphate-linked concentration of calcium ions in a particulate fraction of rabbit muscle. J Cell Biol 14:389–400
Eddy JD, O'Brien ET, Singh SP (1969) Glucagon and haemodynamics of acute myocardial infarction. Br Med J 4:663–665
Edmands RE, Greenspan K, Fisch C (1969) The electrophysiological aspects of epinephrine and glucagon-induced inotropy. Clin Res 17:239
Einzig S, Todd EP, Nicoloff DM, Lucas RV Jr (1971) Glucagon in prevention and abolition of ouabain-induced ventricular tachycardia in normokalemic and hypokalemic dogs. Circ Res 29:88–95
England PJ (1976) Studies on the phosphorylation of the inhibitory subunit of troponin during modification of contraction in perfused rat heart. Biochem J 160:295–304
England PJ (1977) Phosphorylation of the inhibitory subunit of troponin in perfused hearts of mice deficient in phosphorylase kinase. Biochem J 168:307–310

Entman ML (1974) The role of cyclic AMP in the modulation of cardiac contractility. Adv Cyclic Nucleotide Res 4:163–193
Entman ML, Levey GS, Epstein SE (1969) Mechanism of action of epinephrine and glucagon on the canine heart. Evidence for increase in sarcotubular calcium stores mediated by cyclic 3′,5′-AMP. Circ Res 25:429–438
Epstein SE, Skelton CL, Levey GS, Engman M (1970) Adenyl cyclase and myocardial contractility. Ann Intern Med 72:561–578
Epstein SE, Levey GS, Skelton CL (1971) Adenyl cyclase and cyclic AMP. Biochemical links in the regulation of myocardial contractility. Circulation 43:437–450
Exton JH, Cherrington AD, Hutson NJ, Assimacopoulos-Jeannet FD (1977) Reexamination of the second messenger hypothesis of glucagon and catecholamine action in liver. In: Foà PP, Bajaj JS, Foà NL (eds) Glucagon: its role in physiology and clinical medicine. Springer, Berlin Heidelberg New York, pp 321–347
Ezrailson EG, Potter JD, Michael L, Schwartz A (1977) Positive inotropy induced by ouabain, by increased frequency, by X 537 A (RO 2-2985), by calcium and by isoproterenol – the lack of correlation with phosphorylation of TnI. J Mol Cell Cardiol 9:693–698
Farah A (1938) Beitrag zur Wirkung des Insulins auf isolierte Abschnitte des Dünndarms. Arch Exp Pathol Pharmakol 188:548–553
Farah AE, Alousi AA (1981) The actions of insulin on cardial contractility. Life Sci 29:975–1000
Farah A, Tuttle R (1960) Studies on the pharmacology of glucagon. J Pharmacol Exp Ther 129:49–55
Fawaz G, Simaan J (1974) Dissociation between the chronotropic and inotropic actions of glucagon. Naunyn Schmiedebergs Arch Exp Pathol Pharmakol 283:293–301
Fishman PH, Mallorga P, Tallman JF (1981) Catecholamine-induced desensitization of adenylate cyclase in rat glioma C 6 cells. Evidence for a specific uncoupling of beta-adrenergic receptors from a functional regulatory component of adenylate cyclase. Mol Pharmacol 20:310–318
Frangakis CJ, McDaniel HG (1979) Stimulation of calcium uptake in isolated rat myocytes by C-AMP glucagon and isobutyl-methyl-xanthine. Fed Proc 38:540
Frey HM, Falch D, Forfang K, Norman N, Fremstad D (1972) Effects of longterm infusion of glucagon on carbohydrate metabolism, insulin and growth hormone secretion in patients with congestive heart failure. Diabetes 21:939–945
Fricke GR, Simon H, Esser H (1971) Die hämodynamische Wirkung von Glukagon am intakten Hundeherzen. Arch Kreislaufforsch 64:98–114
Friedman W, Sobel B, Cooper C (1969) The age dependent enhancement of the adenyl cyclase system. Proc Soc Pediatr Res 39:20
Friedmann N, Mayekar M, Wood JMcM (1980) The effects of glucagon and epinephrine on two preparations of cardiac mitochondria. Life Sci 26:2093–2098
Fujiwara M, Kuhii M, Shibeta J (1972) Differences of cardiac reactivity between spontaneously hypertensive and normotensive rats. Eur J Pharmacol 19:1–11
Gaide MS, Gelband H, Bassett AL (1981) Relaxation by glucagon of potassium contracture in cat ventricle. Experientia 37:281–282
Gamba S, Turbiglio PC, Bollati C, Bruni B (1977) Il glucagone-protamina-zinco (GPZ) nello scompenso irriducibile di cuore con aritmia. Minerva Med 68:3613–3626
Gaut ZN, Huggins CG (1966) Effect of epinephrine on the metabolism of the inositol phosphatides in rat heart in vivo. Nature 212:612–613
Gavrilescu S, Cotoi S, Pop T (1972) Effects of glucagon on atrio-ventricular conduction in man. Ital Cardiol 2:612–619
Geelen MJH, Groener JEM, De Haas CGM, Van Golde LMG (1979) Influence of glucagon on synthesis of phosphatidylcholines and phosphatidylethanolamines in monolayer cultures of rat hepatocytes. FEBS Lett 105:27–30
George WJ, Wilkerson RD, Kadowitz PJ (1973) Influence of acetylcholine on contractile force and cyclic nucleotide levels in the isolated perfused rat heart. J Pharmacol Exp Ther 184:228–235
Giorgio NA, Johnson CB, Blecher H (1974) Hormone receptors. III. Properties of glucagon binding proteins isolated from liver plasma membranes. J Biol Chem 249:428–437

Glass DB, Krebs EG (1980) Protein phosphorylation catalyzed by cyclic AMP dependent and cyclic GMP dependent protein kinases. Annu Rev Pharmacol Toxicol 20:363–368
Glick G, Parmley WW, Wechsler AS, Sonnenblick EH (1968) Glucagon: its enhancement of cardiac performance in the cat and dog and persistence of its inotropic action despite beta-receptor blockade with propranolol. Circ Res 22:789–799
Gmeiner R, Brachfeld N (1971a) Wirkungen von Glukagon auf den Kohlenhydratstoffwechsel des isolierten Rattenherzens. Arch Kreislaufforsch 65:151–161
Gmeiner R, Brachfeld N (1971b) Glukagon: Wirkungen auf den Sauerstoffverbrauch und aus die Hämodynamik des isolierten Rattenherzens. Z Kreislaufforsch 60:808–818
Gold HK, Prindle KH, Levey GS, Epstein SE (1970) Effects of experimental heart failure on the capacity of glucagon to augment myocardial contractility and activate adenyl cyclase. J Clin Invest 49:999–1006
Goldberg ND, O'Dea RF, Haddox MK (1973) Cyclic GMP. Adv Cyclic Nucleotide Res 3:155–223
Goldschlager N, Robin E, Cowan CH, Leb G, Bing RJ (1969) The effect of glucagon on the coronary circulation in man. Circulation 40:829–837
Goldstein RE, Skelton CL, Levey GS, Glancy DL, Beiser GD, Epstein SE (1971) Effect of chronic heart failure on the capacity of glucagon to enhance contractility and adenyl cyclase activity of human papillary muscles. Circulation 44:638–648
Golikov AP, Berestov AA, Maiorov WI (1979) Differential use of glucagon in the acute period of myocardial infarction. Kardiologiia 19:11–17
Goodridge AG, Ball EG (1965) Studies on the metabolism of adipose tissue. XVIII. In vitro effects of insulin, epinephrine, and glucagon on lipolysis and glycolysis in pigeon adipose tissue. Comp Biochem Physiol 16:367–381
Gordon LM, Dipple I, Sauerheber RD, Esgate JA, Houslay MD (1980) The selective effects of charged local anesthetics on the glucagon- and fluoride-stimulated adenylate cyclase activity of rat-liver plasma membranes. J Supramol Struct 14:223–234
Gorman CK, Cartier J, Goldman BS (1970) The effects of glucagon on cardiac dynamics and metabolism. Clin Res 18:713
Greeff K (1976) Einfluß von Pharmaka auf die Kontraktilität des Herzens. Verh Dtsch Ges Kreislaufforsch 42:80–92
Green RL (1977) Paradoxical effects of calcium-glucagon interaction on cardiac muscle contractility of isolated guinea pig atria. Pharmacology 15:519–528
Green TP, O'Dea RF, Mirkin BL (1979) Determinants of drug disposition and effect in the fetus. Annu Rev Pharmacol Toxicol 19:285–322
Greenberg BH, McCallister BD, Frye RL (1972) Effect of glucagon on resting and exercise haemodynamics in patients with coronary heart disease. Br Heart J 34:924–929
Greenfield JC Jr, Orgain ES (1967) The control of ventricular tachyarrhythmias by internal cardiac pacing. Ann Intern Med 66:1017–1019
Gregory J, Mueller H, Gnoj H, Ayres S, Giannelli S, Conklin E (1969) Effects of glucagon on cardiovascular dynamics and myocardial metabolism in the low output state. Clin Res 17:243
Gruhzit CC, Farah AE (1955) A comparison of the positive inotropic effects of ouabain and epinephrine in heart failure induced in the dog heartlung preparation by sodium pentobarbital, dinitrophenol, sodium cyanide and sodium azide. J Pharmacol Exp Ther 114:334–342
Guerricchio G (1974) Il glucagone. Un nuovo farmaco cardioattivo. Minerva Med 65:3799–3803
Gupta MM, Prasad K (1980) Studies on the effects of glucagon on ouabain-induced cardiac disorders using the PISA method. Adv Myocardiol 1:313–319
Gupta MM, Prasad K, Anyeung G (1978) Studies on the effects of glucagon on ouabain-induced cardiac disorders using PISA Method. J Mol Cell Cardiol 10 [Suppl 1]:30
Hait G, Kypson J, Massih R (1972) Amino acid incorporation into myocardium: effect of insulin, glucagon, and dibutyryl 3'5'-AMP. Am J Physiol 222:404–408
Hammer J, Sriusadaporn S, Freis ED (1973) Effect of glucagon on heart muscle contractility. Clin Pharmacol Ther 14:56–61
Hasselbach W, Makinose M (1961) Die Calciumpumpe der „Erschlaffungsgrana" des Muskels und ihre Abhängigkeit von der ATP-Spaltung. Biochem Z 333:518–528

Hawthorne EW, Hinds JE (1972) Experimental atrial fibrillation in conscious dogs. Am J Cardiol 29:269
Heeg E, Reuter N (1972) Wirkung von Ajmalin, N-Propyl-ajmalin and Qhinidin auf Herz und Kreislauf narkotisierter Katzen. Naunyn Schmiedebergs Arch Exp Pathol Pharmakol 272:297–306
Henry PD, Carlson EH, Sobel BE (1973) Glucagon, myocardial contractility, and cyclic AMP. Am J Cardiol 31:138
Henry PD, Dobson JG Jr, Sobel BE (1975) Dissociations between changes in myocardial cyclic adenosine monophosphate and contractility. Circ Res 36:392–400
Hicks MJ, Shigekawa M, Katz AM (1979) Mechanism by which cyclic adenosine 3′:5′-monophosphate-dependent protein kinase stimulates calcium transport in cardiac sarcoplasmic reticulum. Circ Res 44:384–391
Hirata F, Axelrod J (1978) Enzymatic methylation of phosphatidylethanolamine increases erythrocyte membrane fluidity. Nature 275:219–220
Hirata F, Viveros OH, Diliberto EJ Jr, Axelrod J (1978) Identification and properties of two methyltransferases in conversion of phosphatidylethanolamine to phosphatidylcholine. Proc Natl Acad Sci USA 75:1718–1721
Hirata F, Tallman JF Jr, Henneberry RC, Mallorga P, Strittmatter WJ, Axelrod J (1979a) Regulation of β-adrenergic receptors by phospholipid methylation. Adv Biochem Psychopharmacol 21:91–97
Hirata F, Strittmatter WJ, Axelrod J (1979b) β-Adrenergic receptor agonists increase phospholipid methylation, membrane fluidity, and β-adrenergic receptor-adenylate cyclase coupling. Proc Natl Acad Sci USA 76:368–372
Houslay MD, Ellory JC, Smith GA, Hesketh TR, Stein JM, Warren GB, Metcalf JC (1977) Exchange of partners in glucagon receptor-adenylate cyclase complexes. Physical evidence for the independent, mobile receptor model. Biochim Biophys Acta 467:208–219
Hui CW, Drummond M, Drummond GI (1976) Calcium accumulation and cyclic AMP-stimulated phosphorylation in plasma membrane enriched preparations of myocardium. Arch Biochem Biophys 173:415–427
Hulstaert PF, Beijer HJM, Brouwer FAS, Teunissen AJ, Charbon GA (1974) Glucagon: hemodynamic action related to the effect on $K^+$ and $Na^+$ metabolism. J Appl Physiol 37:556–561
Hurwitz RA (1971) Effect of glucagon on dogs with acute and chronic heart block. Am Heart J 81:644–649
Iijima T, Motomura S, Taira N, Hashimoto K (1974) Comparison of the effects of glucagon and isoprenaline on atrio-ventricular conduction and sino-atrial rate in the dog heart. Clin Exp Pharmacol Physiol 1:241–248
Jaeschke M, Kuhn H, Breithardt G (1977) Die Wirkung von Glucagon auf die systolischen Zeitintervalle und die cyclische AMP-Konzentration im Plasma bei Patienten mit chronischer Myokardinsuffizienz. Verh Dtsch Ges Kreislaufforsch 43:421
Jeacocke SA, England PJ (1980) Phosphorylation of a myofibrillar protein of $M_r$ 150,000 in perfused rat heart, and the tentative identification of this as C-protein. FEBS Lett 122:129–132
Jesmok GJ, Lech JJ, Calvert DN (1975) The effects of epinephrine, glucagon, and ouabain on glycerol release in isolated perfused rat heart. Pharmacologist 17:218
Jesse R (1972) Kontraktilität und Stoffwechsel des Myokards unter Glucagon. Verh Dtsch Ges Kreislaufforsch 78:258–261
Jesse R, Schneider KW, Deeg P (1975) Über den Einfluß intravenöser Glukagoninfusionen auf die Kontraktilität des linksventrikulären Myokards beim Menschen. Basic Res Cardiol 20:217–226
Jorpes JE (1968) The isolation and chemistry of secretin and cholecystokinin. Gastroenterology 55:157–164
Julius S, Randall OS, Esler MD, Kashima T, Ellis C, Bennett J (1975) Altered cardiac responsiveness and regulation in the normal cardiac output type of borderline hypertension. Circ Res 36:I199–I207
Katz AM (1979) Role of the contractile proteins and sarcoplasmic reticulum in the response of the heart to catecholamines: a historical review. Adv Cyclic Nucleotide Res 11:303–343

Katz AM, Repke DI (1973) Calcium-membrane interactions in the myocardium: effects of ouabain, epinephrine and 3',5'-cyclic adenosine monophosphate. Am J Cardiol 31:193–201

Katz AM, Tada M, Kirchberger MA (1975) Control of calcium transport in the myocardium by the cyclic AMP-protein kinase system. Adv Cyclic Nucleotide Res 5:453–471

Kebabian JW, Zatz M, Romero JA, Axelrod J (1975) Rapid changes in rat pineal $\beta$-adrenergic receptor: alterations in 1-[$^3$H]alprenolol binding and adenylate cyclase. Proc Natl Acad Sci USA 72:3735–3739

Keely SL, Corbin JD, Park CR (1975) Regulation of adenosine 3',4'-monophosphate-dependent protein kinase. Regulation of the heart enzyme by epinephrine, glucagon, insulin, and 1-methyl-3-isobutylxanthine. J Biol Chem 250:4832–4840

Kimura T, Kokubun M, Hashimoto K (1974) Primary effect of glucagon on positive chronotropism. Jpn J Pharmacol 24:279–283

King AC, Cuatrecasas P (1981) Peptide hormone-induced receptor mobility, aggregation, and internalization. N Engl J Med 305:77–88

King JF, Salel AF, Amsterdam EA, Massumi RA, Zelis R, Mason DT (1973) Recent advances in therapy for refractory congestive heart failure. Geriatrics 28:94–102

Klein I, Levey GS, Fletcher MA (1974) Glucagon binding and adenylate cyclase. Evidence for a dissociable receptor site. J Clin Invest 53:40A

Klein SW, Morch JE, Mahon WA (1968) Cardiovascular effects of glucagon in man. Can Med Assoc J 98:1161–1164

Kobayashi T, Nakayama R, Kimura K (1971) Effects of glucagon, prostaglandin $E_1$ and dibutyryl cyclic 3',5'-AMP upon the transmembrane action potential of guinea pig ventricular fiber and myocardial contractile force. Jpn Circ J 35:807–819

Kones RJ (1971) Glucagon in heart block. South Med J 64:459–461

Kones RJ, Phillips JH (1971a) Glucagon: present status in cardiovascular disease. Clin Pharmacol Ther 12:427–444

Kones RJ, Phillips JH (1971b) Glucagon in congestive heart failure. Chest 59:392–397

Kones RJ, Dombeck DH, Philipps JH (1972) Glucagon in cardiogenic shock. Angiology 23:525–535

Krayer O (1949) Veratramine, an antagonist to the cardiacaccelerator action of epinephrine. Proc Soc Exp Biol Med 70:631–632

Krayer O (1950) Untersuchungen über die Kreislaufwirkung der Veratrumalkaloide. Arch Exp Pathol Pharmakol 209:405–420

Krayer O, Mendez R (1942) Studies on veratrumalkaloids. I. The action of veratrine upon the isolated mammalian heart. J Pharmacol Exp Ther 74:350–364

Kreisberg RA, Williamson JR (1964) Metabolic effects of glucagon in the perfused rat heart. Am J Physiol 207:721–727

Kruty F, Gvozdjak A, Bada V, Niederland TR, Gvozdjak J, Kaplan M (1978) The effect of glucagon on the heart muscle. Relation between metabolic processes and contractility. Biochem Pharmacol 27:2153–2155

Kühn P (1973) Therapie der Herzinsuffizienz mit Zink-Protamin-Glukagon. Z Kardiol 62:728

Kühn P, Holzhey P, Niederberger M, Fritzsche H, Kroiss A, Brenner B, Kaindl F (1973) Biochemische und hämodynamische Effekte von Zink-Protamin-Glukagon am Menschen. Klin Wochenschr 51:951–957

Kumar R, Sharma GVRK, Molokhia FA, Norman JC, Inamdar AN, Messer JV, Abelmann WH, Hood WB Jr (1972) Experimental myocardial infarction. X. Efficacy of glucagon in acute and healing phase in intact concious dogs: effects on hemodynamics and myocardial oxygen consumption. Circulation 45:55–64

Kurien VA, Yates PA, Oliver MF (1969) Free fatty acids and arrhythmias during experimental myocardial infarction. Lancet 2:185

Kypson J, Hait G, Reyes C (1973) Increased glycogenolysis, lipolysis and breakdown of HTP by epinephrine /E/, glucagon /G/, histamine /H/, aminophylline /A/, and dibutyryl cyclic AMP /DCAMP/ in perfused rabbit hearts. Pharmacologist 15:232

Lad PM, Preston MS, Welton AF, Nielsen TB, Rodbell M (1979) Effects of phospholipase $A_2$ and filipin on the activation of adenylate cyclase. Biochim Biophys Acta 551:368–381

LaRaia PJ, Morkin E (1974) Adenosine 3′,5′-monophosphate dependent membrane phosphorylation, a possible mechanism for the control of microsomal calcium transport in heart muscle. Circ Res 35:298–306
LaRaia PJ, Reddy WJ (1969) Hormonal regulation of myocardial adenosine 3′,5′-monophosphate. Biochim Biophys Acta 177:189–195
LaRaia PJ, Craig RJ, Reddy WJ (1968) Glucagon: effect on adenosine 3′,5′-monophosphate in the rat heart. Am J Physiol 215:968–970
Largis EE, Allen DO, Clark J, Ashmore J (1973) Isoproterenol and glucagon effects in perfused hearts from spontaneously hypertensive and normotensive rats. Biochem Pharmacol 22:1735–1744
Lavarenne J (1974) Effects cardiovasculaires du glucagon. Therapie 29:161–184
Lavarenne J, Boucher M, Barthelemy G, Chassaing C (1974) Influence du Glucagon sur les périodes réfractaires de l'oreillette et de la zone jonctionnelle auriculo-ventriculaire. J Pharmacol (Paris) 5:409–418
Ledingham IMcA, Marshall RJ, Parratt JR (1973) Drug-induced changes in blood flow in normal and ischaemic regions of the canine myocardium. Br J Pharmacol 47:626P–627P
Lee TP, Kuo JF, Greengard P (1971) Regulation of myocardial cyclic AMP by isoproterenol, glucagon, and acetylcholine. Biochem Biophys Res Commun 45:991–997
Lefèbvre PJ, Luyckx AS (1969) Effect of insulin on glucagon enhanced lipolysis in vitro. Diabetologia 5:195–197
Lefèbvre P, Luyckx A (1975) Effect of acute kidney exclusion by ligation of renal arteries on peripheral plasma glucagon levels and pancreatic glucagon production in the anesthetized dog. Metabolism 24:169–176
Leffert HL (1977) Glucagon, insulin and their hepatic receptors: an endocrine pattern characterizing hepatoproliferative transitions in the rat. In: Foà PP, Bajaj JS, Foà NL (eds) its role in physiology and clinical medicine. Springer, Berlin Heidelberg New York, pp 305–319
Lefkowitz RJ, Williams LT (1978) Molecular mechanism of activation and desensitization of adenylate cyclase coupled β-adrenergic receptors. Adv Cyclic Nucleotide Res 9:1–17
Lekven J, Kjekshus JK, Mjös OD (1973) Effects of glucagon and isoproterenol on severity of acute myocardial ischemic injury. Scand J Clin Lab Invest 32:129–136
Levey GS (1971 a) Restoration of glucagon responsiveness of solubilized myocardial adenyl cyclase by phosphatidylserine. Biochem Biophys Res Commun 43:108–113
Levey GS (1971 b) Restoration of norepinephrine responsiveness of solubilized myocardial adenylate cyclase by phosphatidylinositol. J Biol Chem 246:7405–7410
Levey GS (1973) The role of phospholipids in hormone activation of adenylate cyclase. Recent Progr Horm Res 29:361–386
Levey GS, Epstein SE (1969) Activation of adenyl cyclase by glucagon in cat and human heart. Circ Res 24:151–156
Levey GS, Klein I (1972) Solubilized myocardial adenylate cyclase. Restoration of histamine responsiveness by phosphatidylserine. J Clin Invest 51:1578–1582
Levey GS, Prindle KH Jr, Epstein SE (1970) Effect of glucagon on adenyl cyclase activity in the left and right ventricles and liver in experimentally-produced isolated right ventricular failure. J Mol Cell Cardiol 1:403–410
Levey GS, Fletcher MA, Klein I, Ruiz E, Schenk A (1974) Characterization of $^{125}$I-glucagon binding in solubilized preparation of cat myocardial adenylate cyclase. Further evidence for a dissociable receptor site. J Biol Chem 249:2665–2673
Likoff W (1972) Treatment of the failing heart independent of digitalis or its glycosides. Semin Drug Treat 2:253–257
Liljenquist JE, Bomboy JD, Lewis SB, Sinclair-Smith BC, Felts PW, Lacy WW, Crofford OB, Liddle GW (1974) Effects of glucagon on lipolysis and ketogenesis in normal and diabetic men. J Clin Invest 53:190–197
Limas C, Limas CJ (1978) Reduced number of β-adrenergic receptors in the myocardium of spontaneously hypertensive rats. Biochem Biophys Res Commun 83:710–714
Limas CJ (1980) Phosphorylation of cardiac sarcoplasmic reticulum by a calcium-activated phospholipid-dependent protein kinase. Biochem Biophys Res Commun 96:1378–1383
Linhart JW, Barold S, Cohen LS, Hilcher FJ, Samet P (1968) Cardiovascular effects of glucagon in man. Am J Cardiol 22:706–710

Lipski JI, Kaminsky D, Donoso E, Friedberg CK (1972) Electrophysiological effects of glucagon in the normal canine heart. Am J Physiol 222:1107–1112
Lo H, Levey GS (1976) Glucagon-mediated stimulation of ($^{32}$p) orthophosphate and ($^{14}$C) serine incorporation into phosphatidylserine in cardiac muscle slices. Endocrinology 98:251–254
Loeb HS, Rahimtoola SH, Gunnar RM (1973) The failing myocardium. I. Drug management. Med Clin North Am 57:167–185
Lucchesi BR (1968) Cardiac actions of glucagon. Circ Res 22:777–787
Lucchesi BR (1977) Inotropic agents and drugs used to support the failing heart. In: Antonaccio M (ed) Cardiovascular pharmacology. Raven, New York, pp 337–375
Lucchesi BR, Stutz DR, Winfield RA (1969) Glucagon: its enhancement of atrioventricular nodal pacemaker activity and failure to increase ventricular automaticity in dogs. Circ Res 25:183–190
Lüderitz B, Bolte HD, Avenhaus H (1971 a) Einfluß von Glucagon auf das Aktionspotential an Einzelfasern des Papillarmuskels. Z Kreislaufforsch 60:130–135
Lüderitz B, Avenhaus H, Seufert CD (1971 b) Zur Wirkung von Glucagon auf die extrazelluläre Kaliumkonzentrationmessungen in Aorta, Lebervene und Sinus coronarius des Menschen. Klin Wochenschr 49:1334–1337
Lvoff R, Wilcken DEL (1972) Glucagon in heart failure and in cardiogenic shock. Experience in 50 patients. Circulation 45:534–542
Lydtin H, Leidl L, Schewe ST, Daniel W, Schierl W, Lohmöller G (1972) Kreislaufwirkungen verschiedener Applikationsformen von Glucagon. Verh Dtsch Ges Inn Med 78:1551–1554
MacLeod KM, Rodgers RL, McNeill JH (1979) Comparison of cardiac responses to glucagon in the rat and guinea pig. Proc West Pharmacol Soc 22:423–427
MacLeod KM, Rodgers RL, McNeill JH (1981) Characterization of glucagon-induced changes in rate, contractility and cyclic AMP levels in isolated cardiac preparations of the rat and guinea pig. J Pharmacol Exp Ther 217:798–804
Macri I, Rigon-Marci P, Scaliti GV (1974) Effeto di ormoni e mediatori chemici su alcuni parametri functionali del cuore di coniglio. V. Effetto del glucagone. Boll Soc Ital Biol Sper 50:1873–1877
Madan BR (1971) Effect of glucagon on ventricular arrhythmias after coronary artery occlusion and on ventricular automaticity in the dog. Br J Pharmacol 43:279–286
Madan BR (1977) Interactions of glucagon and beta-adrenoceptor antagonists in experimental ventricular arrhythmias. West Afr J Pharmacol Drug Res 4:76P–77P
Madan BR, Jain BK (1973) Combined use of glucagon and beta-adrenoceptor antagonists in ventricular arrhythmias accompanying myocardial infarction. Indian J Med Res 61:1535–1543
Madan BR, Jain BK, Gupta RS (1971) Actions and interactions of glucagon and propranolol in ouabain-induced arrhythmias in the rabbit. Arch Int Pharmacodyn Ther 194:78–82
Mallorga P, Tallman JF, Henneberry RC, Hirata F, Strittmatter WT, Axelrod J (1980) Mepacrine blocks $\beta$-adrenergic agonist-induced desensitization in astrocytoma cells. Proc Natl Acad Sci USA 77:1341–1345
Mallov S, Cerra F (1967) Effect of ethanol intoxication and catecholamines on cardiac lipoprotein lipase activity in rats. J Pharmacol Exp Ther 156:426–444
Manchester JH, Parmley WW, Matloff JM, Sonnenblick E (1969) Beneficial effects of glucagon on canine myocardial infarction and shock. Clin Res 17:252
Manchester JH, Parmley WW, Matloff JM, Leidtke J, LaRaia PJ, Herman MV, Sonnenblick EH, Gorlin R (1970) Effects of glucagon on myocardial oxygen consumption and coronary blood flow in man and dog. Circulation 41:579–588
Marchisio PC, Naldini L, Calissano P (1980) Intracellular distribution of nerve growth factor in rat pheochromocytoma PC 12 cells: evidence for a perinuclear and intranuclear location. Proc Natl Acad Sci USA 77:1656–1660
Marcus FI, Kapadia GG, Goldsmith C (1969) Alteration of the body distribution of tritiated digoxin by acute hyperkalemia in the dog. J Pharmacol Exp Ther 165:136–148
Marcus ML, Skelton CL, Prindle KH Jr, Epstein SE (1971) Potentiation of the inotropic effects of glucagon by theophylline. J Pharmacol Exp Ther 179:331–337

Maroko PR, Kjekshus JK, Sobel BE, Watanabe T, Covell JW, Ross J Jr, Braunwald E (1971) Factors influencing infarct size following experimental coronary artery occlusions. Circulation 43:67–82
Marshall RJ, Parratt JR (1973) The simultaneous measurement of blood flow and oxygen handling in normal and ischaemic areas of the myocardium in the dog. Br J Pharmacol 49:188P
Marsiglia JC, Moreyra AE, Lardani H, Cingolani HE (1970) Glucagon: its effect upon myocardial oxygen consumption. Eur J Pharmacol 12:265–270
Martonosi A, Donley J, Halpin RA (1968) Sarcoplasmic reticulum. III. The role of phospholipids in the adenosine triphosphatase activity and $Ca^{++}$ transport. J Biol Chem 243:61–70
Matloff JM, Parmley WW, Manchester JH, Berkovits B, Sonnenblick EH, Harken DE (1970) Hemodynamic effects of glucagon and intraaortic balloon counterpulsation in canine myocardial infarction. Am J Cardiol 25:675–682
Matsuura Y, Tamura M, Kato E, Uehara S, Mochizuki T (1971) Experimental studies on the effects of glucagon on the denervated transplanted heart. Hiroshima J Med Sci 20:207–213
Mayer SE (1972) Effects of adrenergic agonists and antagonists on adenylate cyclase activity of dog heart and liver. J Pharmacol Exp Ther 181:116–125
Mayer SE (1974) Effect of catecholamines on cardiac metabolism. Circ Res [Suppl 3] 34/35:129–135
Mayer SE, Namm DH, Rice L (1970) Effect of glucagon on cyclic 3′,5′-AMP, phosphorylase activity and contractility of heart muscle of the rat. Circ Res 36:225–233
Meinertz T, Nawrath H, Scholz H (1976) Possible role of cyclic AMP in the relaxation process of mammalian heart: effects of dibutyryl cyclic AMP and theophylline on potassium contractures in cat papillary muscles. Naunyn Schmiedebergs Arch Pharmacol 293:129–137
Menon KMJ, Giese S, Jaffe RB (1973) Hormone- and fluoride-sensitive adenylate cyclases in human fetal tissues. Biochim Biophys Acta 304:203–209
Mielke W, Van Zwieten PA (1971) On the positive inotropic action of glucagon in the isolated atria of the guinea-pig. J Pharm Pharmacol 23:379–381
Mjös OD (1971 a) Effect of free fatty acids on myocardial function and oxygen consumption in intact dogs. J Clin Invest 50:1386–1389
Mjös OD (1971 b) Effect of inhibition of lipolysis on myocardial oxygen consumption in the presence of isoproterenol. J Clin Invest 50:1869–1873
Mjös OD (1971 c) Free fatty acids and oxygen consumption in dogs. Scand J Clin Lab Invest 28:121–125
Mjös OD (1971 d) Effect of isoproterenol, glucagon and calcium on myocardial oxygen consumption in intact dogs. A comparative study. Scand J Clin Lab Invest 28:127–132
Moe GK, Jelife J (1978) An appraisal of 'efficacy' in the treatment of ventricular premature beats. Life Sci 22:1189–1196
Moir AJG, Perry SV (1980) Phosphorylation of rabbit cardiac-muscle troponin I by phosphorylase kinase. The effect of adrenaline. Biochem J 191:547–554
Moir TW, Nayler WG (1970) Coronary vascular effects of glucagon in the isolated dog heart. Circ Res 26:29–34
Mope L, McClellan GB, Winegrad S (1980) Calcium sensitivity of the contractile system and phosphorylation of troponin in hyperpermeable cardiac cells. J Gen Physiol 75:271–282
Moura AM, Simpkins H (1975) Cyclic AMP levels in cultured myocardial cells under the influence of chronotropic and inotropic agents. J Mol Cell Cardiol 7:71–77
Murad F, Vaughan M (1969) Effect of glucagon on rat heart adenyl cyclase. Biochem Pharmacol 18:1053–1059
Murad F, Chi Y-M, Rall TW, Sutherland EW (1962) Adenyl cyclase. III. The effect of catecholamines and choline esters on the formation of adenosine 3′,5′-phosphate by preparations from cardiac muscle and liver. J Biol Chem 237:1233–1238
Murlin JR, Clough HD, Gibbs CBF, Stokes AM (1923) Aqueous extracts of pancreas; influence on carbohydrate metabolism of depancreatized animals. J Biol Chem 56:253–296

Murtagh JG, Binnion PF, Lal S, Hutchison KJ, Fletcher E (1970) Haemodynamic effects of glucagon. Br Heart J 32:307–315
Nakano J, Moore S (1971) Effect of glucagon on the acute and chronic cardiodepressant action of ethanol. Clin Res 19:645
Nayler WG, McInnes I, Chipperfield D, Carson V, Daile P (1970) The effect of glucagon on calcium exchangeability, coronary blood flow, myocardial function, and high energy phosphate stores. J Pharmacol Exp Ther 171:265–275
Newman WH (1976) Contractility of the dog left ventricle in heart failure: length-tension curve: Response to $\beta$-agonist, $Ca^{++}$ and glucagon. Circulation 54 [Suppl 2]:155
Newman WH (1978) Volume overload heart failure: length-tension curves, and response to $\beta$-agonists, $Ca^{2+}$ and glucagon. Am J Physiol 235:H690–H700
Nikkilä EA, Torsti P, Pentilla O (1963) The effect of exercise on lipoprotein lipase activity of rat heart, adipose tissue and skeletal muscle. Metabolism 12:863–865
Nishimura A, Fortner RB, Williams JF Jr (1972) Effect of glucagon on automaticity, threshold for stimulation, and atrioventricular conduction in patients with impaired impulse formation or conduction. Am Heart J 84:359–365
Nobel-Allen N, Kirsch M, Lucchesi BR (1973) Glucagon: its enhancement of cardiac performance in the cat with chronic heart failure. J Pharmacol Exp Ther 187:475–481
Nord HJ, Fonanes HL, Williams JF (1970) Treatment of congestive heart failure with glucagon. Ann Intern Med 72:649–653
Øye I, Langslet A (1972) The role of cyclic AMP in the inotropic response to isoprenaline and glucagon. Adv Cyclic Nucleotide Res 1:291–300
Oliver MF, Kurien VA, Greenwood TW (1968) Relation between serum-free-fatty-acids and arrhythmias and death after acute myocardial infarction. Lancet 1:710–714
Orly J, Schramm M (1976) Coupling of catecholamine receptor from one cell with adenylate cyclase from another cell by cell fusion. Proc Natl Acad Sci USA 73:4410–4414
Oscai LB (1979) Role of lipoprotein lipase in regulating endogenous triacylglycerols in rat heart. Biochem Biophys Res Commun 91:227–232
Papp G, Szekeres L, Resch B, Szontagh F (1976) Action of glucagon, triiodothyromine and prostaglandins on the heart of the human fetus. Acta Physiol Acad Sci Hung 47:226–227
Parmley WW, Sonnenblick EH (1969) A role for glucagon in cardiac therapy. Am J Med Sci 258:224–229
Parmley WW, Glick G, Sonnenblick EH (1968) Cardiovascular effects of glucagon in man. N Engl J Med 279:12–17
Parmley WW, Matloff JM, Sonnenblick EH (1969) Hemodynamic effects of glucagon in patients following prosthetic valve replacement. Circulation 39:I-163–I-167
Parmley WW, Chuck L, Matloff J (1970) Diminished responsiveness of the failing human myocardium to glucagon. Cardiology 55:211–217
Pastan IH, Willingham MC (1981 a) Receptor-mediated endocytosis of hormones in cultured cells. Annu Rev Physiol 43:239–250
Pastan IH, Willingham MC (1981 b) Journey to the center of the cell: role of the receptosome. Science 214:504–509
Peterson A, Lucchesi B, Kirsh MM (1978) The effect of glucagon in animals on chronic propranolol therapy. Ann Thorac Surg 25:340–345
Pichler M, Kleinberger G, Lochs M, Magometschnigg D, Pall H (1979) Hämodynamische Wirkungen von Depot Zink-Protamin-Glukagon bei kardialer Dekompensation. Wien Klin Wochschr 91:49–51
Pohl SL, Krans HMJ, Kozyreff V, Birnbaumer L, Rodbell M (1971) The glucagon-sensitive adenyl cyclase system in plasma membranes of rat liver. VI. Evidence for a role of membrane lipids. J Biol Chem 246:4447–4454
Polson JB, Goldberg ND, Shideman FE (1977) Norepinephrine- and isoproterenol-induced changes in cardiac contractility and cyclic adenosine 3′,5′-monophosphate levels during early development of the embryonic chick. J Pharmacol Exp Ther 200:630–637
Prasad K (1972) Effect of glucagon on the transmembrane potential, contraction, and ATP-ase activity of the failing human heart. Cardiovasc Res 6:684–695
Prasad K (1975 a) Glucagon-induced changes in the action potential, contraction, and ATP-ase of cardiac muscle. Cardiovasc Res 9:355–365
Prasad K (1975 b) Electrophysiologic effects of glucagon on human cardiac muscle. Clin Pharmacol Ther 18:22–30

Prasad K (1977) Use of glucagon in the treatment of quinidine toxicity in the heart. Cardiovasc Res 11:55–63
Prasad K, Weckworth P (1978) Glucagon in procainamide-induced cardiac toxicity. Toxicol Appl Pharmacol 46:517–528
Pruett JK, Woods EF, Daniell HB (1971) Glucagon enhanced automaticity in spontaneously beating Purkinje fibres of canine false tendons. Cardiovasc Res 5:436–439
Puri PS, Bing RJ (1970) Effects of glucagon on myocardial contractility and hemodynamics in acute experimental myocardial infarction. Basis for its possible use in cardiogenic shock. Am Heart J 78:660–668
Rasmussen H, Goodman DBP (1977) Relationships between calcium and cyclic nucleotides in cell activation. Physiol Rev 57:421–509
Rault C, Fruchart JC, Dewailly P, Jaillard J, Sezille G (1974) Experimental studies on the regulation of myocardial and adipose tissue lipoprotein lipase activities in rat. Biochem Biophys Res Commun 59:160–166
Regan TJ, Lehan PH, Henneman DH, Behar A, Hellems HK (1964) Myocardial metabolic and contractile response to glucagon and epinephrine. J Lab Clin Med 63:638–647
Rethy A, Tomasi V, Trevisani A, Barnabei O (1972) The role of phosphatidylserine in the hormonal control of adenylate cyclase in rat liver plasma membranes. Biochim Biophys Acta 290:58–69
Reuter H (1973) Divalent cations as charge carriers in excitable membranes. Prog Biophys Mol Biol 26:1–43
Reuter H (1974a) Exchange of calcium ions in the mammalian myocardium: mechanisms and physiological significance. Circ Res 34:599–605
Reuter H (1974b) Localization of beta-adrenergic receptors, and effects of noradrenaline and cyclic nucleotides on action potentials, ionic currents and tension in mammalian cardiac muscle. J Physiol (Lond) 242:429–451
Reuter H (1979) Properties of two inward membrane currents in the heart. Annu Rev Physiol 41:413–424
Reynold RR, McKenney JR, O'Brien LJ (1971) Microelectrode studies of the effects of glucagon on single rat heart cells. Clin Res 19:336
Robison GA, Butcher RW, Øye I, Morgan HE, Sutherland EW (1965) The effect of epinephrine on adenosine 3′,5′-phosphate levels in the isolated perfused rat heart. Mol Pharmacol 1:168–177
Robison GA, Butcher RW, Sutherland EW (1967) Adenyl cyclase as an adrenergic receptor. Ann NY Acad Sci 139:703–723
Rodbell M (1980) The role of hormone receptors and GTP-regulatory proteins in membrane transduction. Nature 284:17–22
Rodbell M, Londos C (1976) Regulation of hepatic adenylate cyclase by glucagon, GTP, divalent cations, and adenosine. Metabolism [Suppl 1] 25:1347–1349
Rodbell M, Birnbaumer L, Pohl SL (1969) Hormones, receptors and adenyl cyclase activity in mammalian cells. In: Rall TW, Rodbell M, Condliffe P (eds) The role of adenyl cyclase and cyclic 3′,5′-AMP in biological systems. National Institutes of Health, Bethesda, pp 59–102
Rodgers RL, MacLeod KM, McNeill JH (1981) Responses of rat and guinea pig hearts to glucagon. Lack of evidence for a dissociation between changes in myocardial cyclic 3′,5′-adenosine monophosphate and contractility. Circ Res 49:216–225
Rosenblueth A, Garcia Ramos J (1947) The influence of artificial obstacles on experimental auricular flutter. Am Heart J 33:677–684
Ross G (1970) Cardiovascular effects of secretin. Am J Physiol 218:1166–1170
Rubio R, Bailey C, Villar-Palasi C (1975) Effects of cyclic AMP dependent protein kinase on cardiac actomyosin: increase in $Ca^{++}$ sensitivity and possible phosphorylation of troponin I. J Cyclic Nucleotide Res 1:143–150
Said SI, Bosher LP, Spath JA, Kontos HA (1972) Positive inotropic action of newly isolated vasoactive intestinal polypeptide (VIP). Clin Res 20:29
Sanna S, Leo P, Netter R (1975) Azione terapeutica del „glucagone“ su stati pathologici del miocardio particolarmente nei disturbi del tessuto specificao di conduzione. Boll Soc Ital Cardiol 20:1233–1249
Sattler RW, van Zwieten PA (1972) The positive inotropic action of glucagon on the cat heart in situ. Klin Wochenschr 50:531–532

Savion N, Vlodavsky I, Gospodarowicz D (1981) Nuclear accumulation of epidermal growth factor in cultured bovine corneal endothelial and granulosa cells. J Biol Chem 256:1149–1154
Schade DS, Eaton RP (1975) Modulation of fatty acid metabolism by glucagon in man. II. Effects in insulin-deficient diabetics. Diabetes 24:510–515
Scheuer J, Stezoski SW (1971) The effect of pharmacological agents upon the dynamics of anoxic perfused rat hearts. Proc Soc Exp Biol Med 137:1355–1361
Schneider JA, Sperelakis N (1974) The demonstration of energy dependence of the isoproterenol-induced transcellular $Ca^{2+}$ current in isolated perfused guinea pig hearts – an explanation for mechanical failure of the ischemic myocardium. J Surg Res 16:389–403
Schramm M (1979) Transfer of glucagon receptor from liver membranes to a foreign adenylate cyclase by a membrane fusion procedure. Proc Natl Acad Sci USA 76:1174–1178
Schwarzmeier JD, Gillman AG (1977) Reconstitution of catecholamine-sensitive adenylate cyclase activity: interaction of components following cell-cell and membrane-cell fusion. J Cyclic Nucleotide Res 3:227–238
Sharma GVRK, Kumar R, Molokhia F, Inamdar AN, Hood WB Jr, Messer JV (1970) Effect of glucagon on the myocardial metabolism and performance in the intact awake dog. Clin Res 18:328
Shaver JC Jr, Lombardo AA, Shaver VC (1974) Anaerobiosis induced by isoproterenol and glucagon in the presence of restricted coronary flow. Am Heart J 87:97–104
Shell WE, Sobel BE (1973) Deleterious effects of increased heart rate on infarct size in the conscious dog. Am J Cardiol 31:474–479
Shier WT, Baldwin JH, Nielsen-Hamilton M, Hamilton RT, Thanassi NM (1976) Regulation of guanylate and adenylate cyclase activities by lysolecithin. Proc Natl Acad Sci USA 73:1586–1590
Simaan J, Fawaz G (1976) The cardiodynamic and metabolic effects of glucagon. Naunyn Schmiedebergs Arch Exp Pathol Pharmakol 294:277–283
Smitherman T, Shapiro W, Taubert K (1976) Responsiveness of rabbit hearts to glucagon: dissociation of inotropy and adenyl cyclase activation. Clin Res 24:6A
Smitherman T, Osborn RC, Atkins JM (1978) Cardiac dose response relationship for intravenously infused glucagon in normal intact dogs and men. Am Heart J 96:363–371
Sobel BE, Mayer SE (1973) Cyclic adenosine monophosphate and cardiac contractility. Circ Res 32:407–414
Sokal JE (1966) Glucagon: an essential hormone. Am J Med 41:331–341
Solaro RJ, Robertson SP, Johnson JD, Holroyde MJ, Potter JD (1981) A Troponin-I phosphorylation: a unique regulator of the amounts of calcium required to activate cardiac myofibrils. In: Rosen OM, Krebs EG (eds) Protein phosphorylation, Cold Spring Harbor conference on cell proliferation, vol 8. Cold Spring Harbor Laboratory, New York, pp 901–911
Somlyo AV, Somlyo AP, Friedmann H (1971) Cyclic adenosine monophosphate, cyclic guanosine monophosphate, and glucagon: effect on membrane potential and ion fluxes in the liver. Ann NY Acad Sci 185:108–114
Sonnenblick EH, Ross J Jr, Braunwald E (1968a) Oxygen consumption of the heart. Newer concepts of its multifactorial determination. Am J Cardiol 22:328–336
Sonnenblick EH, Parmley WW, Matloff J (1968b) Hemodynamic effects of glucagon after prosthetic valve replacement. Circulation 37:183
Spangler RD, Grover RF (1973) Attenuation of hypoxic pulmonary hypertension in newborn calves by glucagon. Am J Cardiol 31:159
Sperelakis N, Belardinelli L, Vogel SM (1979) Electrophysiological aspects during myocardial ischemia. In: Hayase S, Murao S (eds) Proceedings of the VIIIth World Congress of Cardiology (Tokyo 1978). Myocardial metabolism during ischemia. Excerpta Medica, Amsterdam, pp 229–236
Spilker B (1970) Comparison of the inotropic response to glucagon, ouabain and noradrenaline. Br J Pharmacol 40:382–395
Staneva-Stoycheva D, Bogoslovova T (1976) Changes in some metabolic effects of isoprenaline and glucagon in normotensive and spontaneously hypertensive rats. Acta Physiol Pharmacol Bulg 2:40–45

Staub A, Sinn L, Behrens OK (1953) Purification and crystallization of hyperglycemic glycogenolytic factor (HGF). Science 117:628–629
Steiner AL, Ong S, Wedner HJ (1976) Cyclic nucleotide immunocytochemistry. Adv Cyclic Nucleotide Res 7:115–155
Steiner C, Wit AL, Damato AB (1969) Effects of glucagon on atrioventricular conduction and ventricular automaticity in dogs. Circ Res 24:167–177
Stewart JW, Myerburg RJ, Hoffman BF (1969) The effect of glucagon on quinidine-induced changes in Purkinje fibers. Circulation [Suppl 3] 40:196
Strauer BE (1971 a) Die inotropie Wirkung des Glucagons am isolierten, menschlichen Ventrikelmyokard. Klin Wochenschr 49:468–473
Strauer BE (1971 b) The influence of glucagon on myocardial mechanics of papillary muscles obtained from patients with chronic congestive heart failure. Naunyn Schmiedebergs Arch Exp Pathol Pharmakol 270:90–93
Strittmatter WJ, Hirata F, Axelrod J (1979) Phospholipid methylation unmasks cryptic $\beta$-adrenergic receptors in rat reticulocytes. Science 204:1205–1207
Stull JT, Boss JE (1977) Phosphorylation of cardiac troponin by cyclic adenosine 3′,5′-monophosphate-dependent protein kinase. J Biol Chem 252:851–857
Stull JT, Mayer SE (1979) Biochemical mechanisms of adrenergic and cholinergic regulation of myocardial contractility. In: Berne RM, Sperelakis N, Geiger SR (eds) The cardiovascular system, the heart. American Physiological Society, Bethesda, pp 741–774 (Handbook of physiology, sect 2, vol 1)
Sulakhe PV, Dhalla NS (1973) Adenyl cyclase of heart sarcotubular membrane. Biochim Biophys Acta 293:379–396
Sulakhe PV, St Louis PJ (1980) Passive and active calcium fluxes across plasma membranes. Prog Biophys Mol Biol 35:135–195
Sulakhe PV, Leung NL, St Louis PJ (1976) Stimulation of calcium accumulation in cardiac sarcolemma by protein kinase. Can J Biochem 54:438–445
Sutherland EW, Robison GA, Butcher RW (1968) Some aspects of the biological role of adenosine 3′,5′-monophosphate (cyclic AMP). Circulation 37:279–306
Tada M, Kirchberger MA, Repke DI, Katz AM (1974) The stimulation of calcium transport in cardiac sarcoplasmic reticulum by adenosine 3′,5′-monophosphate-dependent protein kinase. J Biol Chem 249:6174–6180
Tada M, Kirchberger MA, Iorio J-AM, Katz AM (1975) Control of cardiac sarcolemmal adenylate cyclase and sodium, potassium-activated adenosinetriphosphatase activities. Circ Res 36:8–17
Tada S (1929) Wirkung des Nebennieren-Pankreas- und Hypophyseninkrets auf die Bewegungen des überlebenden Darms von Kaninchen mit Funktionsstörung der Schilddrüse. Tohoku J Exp Med 14:400–414
Tanaka R, Strickland KP (1965) Role of phospholipid in the activation of $Na^+$, $K^+$-activated adenosine triphosphatase of beef brain. Arch Biochem Biophys 111:583–592
Tarnow VJ, Gethmann JW, Patschke D, Weymar A, Eberlein HJ (1975) Hämodynamik, Koronardurchblutung und Sauerstoffverbrauch des Herzens unter Glukagon. Arzneim Forsch 25:1906–1910
Tell GPE, Cuatrecasas P (1974) $\beta$-Adrenergic receptors: stereospecificity and lack of affinity for catechols. Biochem Biophys Res Commun 57:793–800
Terasaki WL, Brooker G (1977) Cardiac adenosine 3′,5′-monophosphate. Free and bound forms in the isolated rat atrium. J Biol Chem 252:1041–1050
Tibblin S, Kock NG, Schenk WG Jr (1971) Response of mesenteric blood flow to glucagon. Influence of pharmacological stimulation and blockade of adrenergic receptors. Arch Surg 102:65–70
Timmis GC, Ramos RG, Parikh R, Henke J, Gordon S (1973) The unique cardiotonic properties of glucagon. Mich Med 72:353–357
Torda T, Yamaguchi I, Hirata F, Kopin IJ, Axelrod J (1981) Quinacrine-blocked desensitization of adrenoreceptors after immobilization stress or repeated injection of isoproterenol in rats. J Pharmacol Exp Ther 216:334–338
Trading F, Nielsen P, Pingel M, Vølund A (1969) Biological and chemical properties of two glucagon preparations with prolonged action. Eur J Pharmacol 7:206–210

Trendelenberg P (1934) Schilddrüse. In: Krayer O (ed) Die Hormone, vol II. Springer, Berlin, p 49
Tsien RW (1977) Cyclic AMP and contractile activity in heart. Adv Cyclic Nucleotide Res 8:363–420
Unger RH, Orci L (1979) Glucagon: secretion, transport, metabolism, physiologic regulation of secretion, and derangements in diabetes. In: DeGroot LJ, Cahill GF Jr, Martini L, Nelson DH, Odell WD, Potts JR Jr; Steinberger E, Winegrad AI (eds) Endocrinology, vol 2. Grune and Stratton, New York, pp 959–980
Urthaler F, Isobe JH, James TN (1974) Comparative effects of glucagon on automaticity of the sinus node and atrioventricular junction. Am J Physiol 227:1415–1421
Vander Ark CR, Reynolds EW (1970) Clinical evaluation of glucagon by continuous infusion in the treatment of low output cardiac states. Am Heart J 79:481–487
Vaughn CC, Warner HR, Nelson RM (1970) Cardiovascular effects of glucagon following cardiac surgery. Surgery 67:204–211
Vinicor F (1974) Studies on the mechanism of action of glucagon in the adult mouse heart. Clin Res 22:482A
Visscher MB, Lee YCP (1972) Calcium ions and the cardiotonic action of glucagon. Proc Natl Acad Sci USA 69:463–465
Visscher MB, Müller EA (1927) Influence of insulin upon mammalian heart. J Physiol (Lond) 62:341–348
Walsh DA, Perkins JP, Krebs EG (1968) An adenosine 3′,5′-monophosphate-dependent protein kinase from rabbit skeletal muscle. J Biol Chem 243:3763–3765
Walsh DA, Clippinger MS, Sivaramakrishnan S, McCullough TE (1979) Cyclic adenosine monophosphate dependent and independent phosphorylation of sarcolemma membrane proteins in perfused rat heart. Biochemistry 18:871–877
Watanabe AM, Besch HR (1973) Effects of inotropic agents on systolic transmembrane calcium flux. Circulation [Suppl 4] 48:11
Watanabe AM, Besch HR (1974) Cyclic adenosine monophosphate modulation of slow calcium influx channels in guinea pig hearts. Circ Res 35:316–324
Westlie L, Andersen A, Jervell J, Rasmussen K, Storstein O (1971) Cardiovascular effects of glucagon. Acta Med Scand 189:179–184
Whitehouse FW, James TN (1966) Chronotropic action of glucagon on the sinus node. Proc Soc Exp Biol Med 122:823–826
Whitsitt LS, Lucchesi BR (1968) Effects of beta-receptor blockade and glucagon on the atrio-ventricular transmission system in the dog. Circ Res 23:585–595
Wilcken DEL, Lvoff R (1970) Glucagon in resistant heart failure and cardiogenic shock. Lancet 2:1315–1317
Wildenthal L, Wakeland JR (1973) Maturation of responsiveness to cardioactive drugs. Differential effects of acetylcholine, norepinephrine, theophylline, tyramine, glucagon, and dibutyryl cyclic AMP on atrial rate in hearts of fetal mice. J Clin Invest 52:2250–2258
Wildenthal K, Wakeland JR (1979) Influence of phosphodiesterase inhibitor on chronotropic effects of glucagon and norepinephrine in fetal mouse hearts. J Pharmacol Exp Ther 211:350–352
Wildenthal K, Allen DO, Karlsson J, Wakeland JR, Clark CM Jr (1976) Responsiveness to glucagon in fetal hearts. Species variability and apparent disparities between changes in beating, adenylate cyclase activation, and cyclic AMP concentration. J Clin Invest 57:551–588
Wilkerson RD, Pruett JK, Woods EF (1971) Glucagon-enhanced ventricular automaticity in dogs. Its concealment by positive chronotropism. Circ Res 29:616–625
Willard DW, Harris S, Kilen SM (1972) Inhibition by freon of inotropic actions of isoproterenol, ouabain and glucagon. Circulation 46 [Suppl 2]:238
Williams JF Jr, Childress RH, Chip JN, Border JF (1969) Hemodynamic effects of glucagon in patients with heart disease. Circulation 39:38–47
Winokur S, Nobel-Allen NL, Lucchesi BR (1975) The positive inotropic effect of glucagon in the chronically failed right ventricle as demonstrated in the isolated cat heart. Eur J Pharmacol 32:349–356

Wolff F, Jones AB (1970) Inhibition of hormone-sensitive adenyl cyclase by phenothiazines. Proc Natl Acad Sci USA 65:454–459
Wolfson SK Jr, Ellis S (1956) Effects of glucagon on plasma potassium. Proc Soc Exp Biol Med 91:226–228
Wollenberger A (1972) Cyclic nucleotides and the regulation of heart beat. In: Abstr 5th Int Congr Pharmacol, pp 231–233
Wollenberger A (1975) The role of cyclic AMP in the adrenergic control of the heart. In: Nayler WG (ed) Contraction and relaxation in the myocardium. Academic, London, pp 113–190
Wollenberger A, Will H (1978) Protein kinase-catalysed membrane phosphorylation and its possible relationship to the role of calcium in the adrenergic regulation of cardiac contraction. Life Sci 22:1159–1178
Wollenberger A, Babskii EB, Krauss E-G, Genz S, Blohm D, Bogdanova EV (1973) Cyclic changes in levels of cyclic AMP and Cyclic GMP in frog myocardium during the cardiac cycle. Biochem Biophys Res Commun 55:446–452
Woods EF, Daniell HB, Pruett JK (1970) Chronotropic responses to epinephrine (E) and glucagon (G) after acute and chronic heart block. Fed Proc 29:739
Wray HL, Gray RR, Olsen RA (1973) Cyclic 3′,5′-monophosphate stimulated protein kinase and a substrate associated with cardiac sarcoplasmic reticulum. J Biol Chem 248:1496–1498
Yamaguchi I, Torda T, Hirata F, Kopin IJ (1981) Adrenoceptor desensitization after immobilization, stress or repeated injection of isoproterenol. Am J Physiol 240:H691–H696
Yankner BA, Shooter EM (1979) Nerve growth factor in the nucleus; interaction with receptors on the nuclear membrane. Proc Natl Acad Sci USA 76:1269–1273
Zlokovic B, Andjelko I, Markovic M, Rosic N (1980) Anti-arrhythmic action of glucagon and CK-MB release from the heart in rabbits. IRCS Med Sci Biochem 8 (11):835

CHAPTER 54

# Spasmolytic Action and Clinical Use of Glucagon

B. DIAMANT and J. PICAZO

## A. Spasmolytic Effects of Glucagon in Various Species

The inhibitory effect of glucagon on the motility of the gastrointestinal tract was first described by STUNKARD et al. (1955) who noted inhibition of gastric hunger contractions in humans. The mechanism behind this effect could not be determined, but it was noted that it could be "independent of its effect on carbohydrate metabolism" since intravenous glucose injections failed to mimic the effect. Glucagon has since been found to have inhibitory action on the smooth muscle of many organs and species. WINGATE and PEARCE (1979) have compiled an extensive list of pertinent investigations focusing on the effects of glucagon on gastrointestinal motility, secretion, and absorption, and on blood flow (Table 1). Table 2 lists investigations in other areas in which glucagon has been found to have a similar smooth muscle relaxing effect.

The purity of the glucagon used in the investigations listed in Tables 1 and 2 has been mentioned in only a few instances; therefore it can only be assumed that the effects noted are specific for glucagon per se. Provided this is so, the spasmolytic and hypotonic actions of glucagon have been substantially corroborated in many in vitro and in vivo studies. By contrast, surprisingly few publications have focused on the underlying mechanism (or mechanisms) of action.

### I. Mechanisms of Action

ANURAS and COOKE (1978) studied the mechanisms of action of glucagon using strips of the isolated longitudinal and circular muscle layers of the duodenum of the opossum. They found that glucagon slightly reduced the resting tension of both layers. A more impressive effect was the inhibition of acetylcholine- (ACh)-stimulated phasic contractions of the circular muscle. This led the authors to conclude that, in the particular species studied, glucagon exerted some effect antagonistic to cholinergic excitatory receptors, besides having some direct effect on the duodenal muscles.

In contrast to the reports concerning humans (see Table 2), BEHAR (1978) found, in cats, that glucagon caused lower esophageal sphincter (LES) contractions after intravenous injections, with maximum effect after 20–40 μg/kg. The effect was not antagonized by pretreatment of the animals with atropine sulfate (30 μg/kg) nor with propanolol (1 mg/kg). The contractions were partially abolished by hexamethonium (20 mg/kg) and completely abolished by the use of hexamethonium in combination with atropine. Phentolamine (1 mg/kg) also blocked

**Table 1.** Effects of glucagon on the gastrointestinal tract (WINGATE and PEARCE 1979)

| Study | Target | Dose[a] | Effect |
|---|---|---|---|
| A. Isolated tissue | | | |
| CAMERON et al. (1970) | Human antrum | 1 μg/ml | Nil |
| GERNER and HAFFNER (1975) | Guinea pig stomach | 10 μg/ml (?) | Inhibition of pressure on distention only |
| KOWALEWSKI et al. (1976) | Perfused pig stomach | 166 μg/min i.a. | Inhibition of electrical activity |
| B. Animals: intraarterial dosage | | | |
| DANFORD (1971) | Dog | 2.5 μg/kg | Abolition of digoxin-induced vasoconstriction in superior mesenteric artery |
| FASTH and HULTÉN (1971) | (1) Cat | 10–100 μg kg$^{-1}$ min$^{-1}$ (infusion) | Splanchnic vasodilation decrease of gut motility |
| | (2) Adrenalectomized cat | | Splanchnic vasodilation normal motility |
| MACFERRAN and MAILMAN (1977) | Dog | 0.05 and 0.5 μg kg$^{-1}$ min$^{-1}$ | Increase in intestinal splanchnic flow and fluid absorption |
| C. Animals: single-dose studies | | | |
| HUBEL (1972) | Rat | 4–256 μg/kg i.v. | Increase in fluid absorption |
| JOHANSSON and SEGERSTRÖM (1972) | Rat | 1.2–9.6 mg/kg s.c. | Retardation of gastric emptying |
| KOCK et al. (1971) | Dog | 10 μg/kg i.v. | Reversion of sympathetic effect on splanchnic bed |
| LIN et al. (1973) | Dog | 50 μg/kg | Inhibition of gastric $H^+$ secretion |
| NECHELES et al. (1966) | Dog | 6–313 μg/kg i.v. | Inhibition of gastric and duodenal motility but not consistent |
| STICKNEY et al. (1958) | Rat | 7–280 μg/kg i.p. | No effect on motility |
| TIBBLIN et al. (1970) | Dog | 10 μg/kg i.v. | Increase in superior mesenteric arterial flow |
| VISNOVSKÝ (1976) | Rat | 2 mg/kg s.c. | Decrease in gastric emptying and intestinal transit |
| D. Animals: intravenous infusion studies | | | |
| BARBEZAT and GROSSMAN (1971) | Dog | 0.5 μg kg$^{-1}$ min$^{-1}$ | Increase in intestinal secretion |
| KRARUP and LARSEN (1974) | Cat | 0.1–5 μg kg$^{-1}$ min$^{-1}$ | Reversion of sympathetic effect on splanchnic bed |

**Table 1** (continued)

| Study | Target | Dose[a] | Effect |
|---|---|---|---|
| RUDO and ROSENBERG (1973) | Rat | (Chronic i.p. injection: 0.3 μg/kg every 6 h for 6 days) | Increase in intestinal sugar transport |
| SCOTT and SUMMERS (1976) | Rat | 1, 10, 100 $\mu g\ kg^{-1}\ min^{-1}$ | Inhibition of jejunal contraction and transit of higher dose levels |
| TIBBLIN et al. (1970) | Dog | 0.1 $\mu g\ kg^{-1}\ min^{-1}$ | Increase in superior mesenteric arterial flow |
| VALENZUELA (1976) | Dog | 0.1–0.4 $\mu g\ kg^{-1}\ min^{-1}$ | Decrease in intragastric pressure |
| WINGATE et al. (1977) | Dog | 0.04–0.32 $\mu g\ kg^{-1}\ min^{-1}$ | Stimulation of intestinal myoelectric activity |
| E. Humans: single-dose studies | | | |
| DOTEVALL and KOCK (1963) | Jejunum and colon | 3.6–14.3 μg/kg i.v. | Inhibition of motility for 10 min |
| KOCK et al. (1967) | Jejunum and colon | 0.7–1.4 μg/kg i.p. and i.v. | Inhibition of motility for 4–8 min |
| RATZMAN and KNOKE (1974) | Stomach | 14 μg/kg | Inhibition of basal $H^+$ secretion |
| STODDARD and DUTHIE (1976) | Stomach | 5–10 μg/kg | Biphasic effect on basic electrical rhythm |
| STUNKARD et al. (1955) | Stomach | 28 μg/kg | Inhibition of hunger contractions |
| F. Humans: infusion studies | | | |
| CHOWDHURY et al. (1976) | Recto-sigmoid in constipation | 0.17 $\mu g\ kg^{-1}\ min^{-1}$ | Similar effect to atropine on hyperactive segment |
| CHOWDHURY and LORBER (1977) | Distal colon and rectum | 0.5 $\mu g\ kg^{-1}\ min^{-1}$ | Inhibition of food- or morphine-stimulated activity |
| CORAZZIARI et al. (1973) | Duodenum and jejunum | 0.5–1.0 $\mu g\ kg^{-1}\ min^{-1}$ | Abolition of movement: duodenal but *not* jejunal stasis |
| HICKS and TURNBERG (1974) | Jejunum | 0.02 $\mu g\ kg^{-1}\ min^{-1}$ | Increase in rate of transit |
| KONTUREK et al. (1975) | Stomach | 0.09–0.7 $\mu g\ kg^{-1}\ min^{-1}$ | Inhibition of food-stimulated $H^+$ secretion |
| PAUL (1974) | Stomach and colon | 0.1 $\mu g\ kg^{-1}\ min^{-1}$ | Inhibition of motility |
| STUNKARD et al. (1955) | Stomach | 0.6 $\mu g\ kg^{-1}\ min^{-1}$ | Inhibition of hunger contractions |
| WHALEN et al. (1973) | Small intestine | Arginine infusion | Prolongation of transit with rise in plasma glucagon (81–214 pg/ml) |

[a] Abbreviations: i.a. intraarterial; i.p. intraperitoneal; i.v. intravenous; s.c. subcutaneous

**Table 2.** Effects of glucagon on other structures

| Study | Target | Dose[a] | Effect |
|---|---|---|---|
| A. Other structures within the gastrointestinal tract | | | |
| 1. Esophagus | | | |
| a. Lower esophageal sphincter (LES) pressure | | | |
| HOGAN et al. (1975) | Human | 20–100 μg bolus i.v. | Significant decrease; after 100 μg peak effect persisting 10 min or more |
| HOKE et al. (1972) | Human | 20 μg bolus i.v. | Transient dose-related decrease in LES tone |
| JAFFER et al. (1974) | Human | 10 μg/kg bolus i.v. | Maximal decrease after 10 min |
| b. Peristalsis | | | |
| HOGAN et al. (1975) | Human | 20–1,000 μg bolus i.v. | No effect |
| HOKE et al. (1972) | Human | 20 μg bolus i.v. | No effect |
| 2. Gallbladder | | | |
| CAMERON et al. (1969) | In vitro (human strip) | 10 μg/100 ml incubation | No inhibition of CCK-induced contraction |
| K.D. JØRGENSEN, J. WEIS and B. DIAMANT (1981, unpublished work) | In vitro (rabbit strip) | $10^{-7}$–$10^{-5}$ *M* incubation | Dose-dependent inhibition (75% at $10^{-5}$ *M*) of submaximal contraction induced by CCK (0.1 μg/ml) |
| JANSSON et al. (1978) | Cat | 1–20 μg $kg^{-1}$ $h^{-1}$ i.v. infusion | No changes in gallbladder motility |
| VAGNE and TROITSKAJA (1976) | Guinea pig | 1, 4, 16 μg bolus i.v. | Slight increase in CCK-induced contraction |
| 3. Sphincter of Oddi | | | |
| CARR-LOCKE and GREGG (1980) | Human | 0.016–16.0 μg $kg^{-1}$ $h^{-1}$ i.v. infusion | i. Significant pressure reduction at 0.016 μg $kg^{-1}$ $h^{-1}$. Maximal effect at 0.25 μg $kg^{-1}$ $h^{-1}$<br>ii. Slowing of sphincter wave frequencies only at ⩾1 μg $kg^{-1}$ $h^{-1}$ |
| HOGAN et al. (1977) | Human | 0.4 mg bolus i.v. | Reduction of frequency and amplitude of contractions |
| NEBEL (1975a) | Human | 1 mg bolus i.v. | Reduction of basal pressure |
| 4. Pancreatic duct sphincter | | | |
| CARR-LOCKE and GREGG (1980) | Human | 0.016–16.0 μg $kg^{-1}$ $h^{-1}$ i.v. infusion | Reduction of pressure and slowing of sphincter wave frequencies only at ⩾1 μg $kg^{-1}$ $h^{-1}$ |
| B. Structures outside the gastrointestinal tract | | | |
| 1. Ureter | | | |
| BOYARSKY and LABAY (1972) | Dog | 2.8 μg/kg bolus i.v. | Slowing or stopping of peristalsis for 1–15 min with mild diuresis |

**Table 2** (continued)

| Study | Target | Dose[a] | Effect |
|---|---|---|---|
| BOYARSKY and LABAY (1972) | In vitro (dog strip) | 1–20 μg/ml incubation | Dose-dependent decrease, eventually termination of amplitude of contractions and frequency |
| VELA NAVARRETE et al. (1979) | In vitro (horse strip) | 20–50 μg/ml incubation | Disappearance of peristalsis |
| VELA NAVARRETE et al. (1979) | Dog | 1.1 mg bolus i.v. | Disappearance of peristalsis for 90 s |
| 2. Bronchial musculature | | | |
| BLUMENTHAL and BRODY (1969) | In vitro (guinea pig) | $2.9 \times 10^{-7}$ *M* incubation | Relaxation of bronchiolar rings |
| K.D. JØRGENSEN, J. WEIS and B. DIAMANT (1981, unpublished work) | Guinea pig | 50 μg/kg bolus i.v. | Partial decrease in histamine and bradykinin constriction (Konzett-Rössler technique) |
| WARNER et al. (1971) | Dog | 1 mg bolus i.v. | Decrease in pulmonary resistance, maximal effect after 3–5 min, duration 30 min |
| 3. Fallopian tube | | | |
| CORTÉS-PRIETO et al. (1980) | Human | 0.3 mg bolus i.v. | Release of tubal spasm |
| 4. Blood vessels | | | |
| a. General peripheral resistance | | | |
| KAKU et al. (1977) | Calf | 10 μg/kg bolus i.v. | Reduction of systemic peripheral muscular resistance for 30 min, maximal reduction 31% |
| b. Hepatic artery | | | |
| LINDBERG and DARLE (1959) | Pig | 50 μg/kg bolus i.v. (femoral vein) | Increase in hepatic arterial blood flow by 80% after 2 min and 58% after 10 min; the effect persisted for 25–40 min |
| RICHARDSON and WITHRINGTON (1977) | Dog | 0.5–10 μg/min infusion in hepatic artery | Dose-related reduction (maximal 38%) of hepatic arterial vascular resistance |
| SHOEMAKER et al. (1959) | Dog | 1–3 mg/kg bolus i.v. (femoral vein) | Increase in blood flow (41%–204%) in the hepatic bed |
| c. Superior mesenteric artery | | | |
| BOND et al. (1979) | Dog | 0.25 mg/20 min infusion in mesenteric artery | Increase in blood flow to the small bowel by 418% |
| ULANO et al. (1972) | Dog | 0.5 μg $kg^{-1}$ $min^{-1}$ infusion in mesenteric artery | After hypovolemic shock: vasodilation and restoration of flow in constricted artery without altering systemic blood pressure |

**Table 2** (continued)

| Study | Target | Dose[a] | Effect |
|---|---|---|---|
| d. Renal artery | | | |
| DANFORD (1970) | Dog | 0.5 mg bolus in renal artery | Arterial dilation after 5 min and increase in blood flow (127%; maximal effect at 30 min, lasting for 60 min) |
| GAGNON et al. (1978) | In vitro (rabbit strip) | 10–100 ng/ml incubation | Dose-related depression of norepinephrine-induced (1 ng/ml) contraction |
| GLICK (1970) | Dog | 5–50 µg/kg | Increase in blood flow (12.6% ± 12%) and decrease in renal resistance (−15.6% ± 6.7%) |

[a] Abbreviation: i.v. intravenous

the effect, as did catecholamine depletion after intraperitoneal injections of reserpine (3 mg/kg, 48 and 24 h prior to the experiments).

Based on these results, it was suggested that glucagon might stimulate the preganglionic neuron of the sympathetic neural pathway to the LES, requiring an intact ganglionic cholinergic transmission and the release of norepinephrine.

BEHAR's study shows the importance of acknowledging possible species variations, and reminds one of the fact that, when studying the mechanism of the smooth muscle relaxing activity, particularly when attempting to assess the effectiveness of glucagon as a spasmolytic agent, evaluations must be carried out in humans. For the same reason, Sect. B of this chapter, which is devoted to the possible clinical potentials of glucagon, is based on studies of its effects in humans.

GAGNON et al. (1978), using isolated organs and isotonic transducers, demonstrated that glucagon relaxed aorta strips from rats and renal artery strips from rabbits. In contrast, neither contraction nor relaxation occurred in tests on stomach strips, duodenum, jejunum, ascending colon, or uterus of the rat. Similar negative results were noted in tests on aorta strips, inferior vena cava, and anterior mesenteric vein of the rabbit, rectum of the chicken, and tracheal chain and ileum of the guinea pig. The relaxant effect of glucagon in these tests was assessed on tissues contracted either with carbachol (nonvascular tissues) or with norepinephrine (vascular tissues), and the dose of glucagon tested ranged from 1 ng/ml to 1 µg/ml.

The relaxant effect of glucagon on the isolated renal artery of the rabbit was not modified by propanolol nor by cimetidine at concentrations which completely blocked the relaxing action of isoprenaline and histamine. Furthermore, no influence of diphenhydramine, phentolamine, atropine, or indomethacin was noted. These findings seem to exclude the possibility of there being an involvement of the classical autonomic nervous system transmitters or of histamine and intramural prostaglandins.

The possible involvement of cyclic adenosine-3′,5′-monophosphate (cAMP) in the relaxing action of glucagon was suggested by the finding of GAGNON et al. (1978) that phosphodiesterase inhibitors, such as papaverine and theophylline, potentiate the effect of glucagon. This correlates well with the fact that glucagon activates adenylate cyclase in various tissues. The idea that the entire glucagon molecule is needed for full biologic activity (FALOONA and UNGER 1974) was supported by the finding that [desHis$^{1}$] glucagon possesses only 11% of the activity of intact glucagon and that [desMet$^{27}$, Asn$^{28}$, Thr$^{29}$] glucagon, in concentrations nearly 40 times higher than the minimal effective dose of glucagon, was found to be completely inactive, both as an agonist and an antagonist, in the isolated renal artery of the rabbit (see also Chap. 1).

In a second paper, GAGNON et al. (1980) further substantiated the conclusion that intracellular cAMP acts as a modulator of the vascular action of glucagon on the rabbit renal artery, and focused on the known action of cAMP in inducing muscular relaxation through redistribution of $Ca^{2+}$ from contractile molecular structures (ANDERSSON 1972; RASMUSSEN 1976).

GAGNON et al. (1980) noted that cAMP added to the organ bath mimicked the action of glucagon by relaxing the rabbit renal artery when it was contracted by norepinephrine. This observation is in agreement with that of KRARUP et al. (1975), who demonstrated that the hepatosplanchnic blood flow in cats increased after the administration of glucagon as well as of cAMP. The relaxing action of both agents on the rabbit renal artery was enhanced by certain phosphodiesterase inhibitors and counteracted by inhibitors of $Ca^{2+}$ influx, such as verapamil and SKF-525A. Verapamil was found to induce a dose-dependent relaxation by itself. In contrast to that of glucagon and cAMP, this effect of verapamil could be reversed by increasing the $Ca^{2+}$ concentration in the incubation medium. This finding seemed to exclude the possibility of glucagon and cAMP having a direct effect on the influx of calcium, and implied that they and verapamil acted on different receptors. Instead, it was suggested that glucagon and cAMP might facilitate the extrusion of $Ca^{2+}$ from the cells and/or enhance the binding of intracellular $Ca^{2+}$ to structures not involved in the contractile mechanism. This was based on the assumption that verapamil and SKF-525A inhibit not only the uptake, but also the extrusion of $Ca^{2+}$ in the smooth muscle cells of the rabbit renal artery and/or the binding of $Ca^{2+}$ to cellular $Ca^{2+}$-binding sites (e.g., endoplasmatic reticulum, plasma membranes, mitochondria). Some indirect support for this concept was found in the fact that $Na_2EDTA$ relaxed the renal artery strip, an effect which could be counteracted by verapamil as well as by SKF-525A. The possible action of glucagon on $Ca^{2+}$ transport mechanisms via modulation of cAMP production as an explanation for the relaxing action of glucagon on isolated strips of the rabbit renal aorta requires further substantiation; studies of $^{45}Ca$ fluxes from preloaded strips may well prove a suitable experimental model for this purpose. The relaxing action of glucagon on renal arteries has also been shown in species other than the rabbit. DANFORD (1970) carried out renal arteriography studies in dogs before and after infusion of 0.5 mg glucagon. The administration of glucagon caused vasodilation lasting 120 min, with improved visualization of the arterial tree. Whether a similar action will be found in humans remains to be established.

Since 1978, the possible pharmacologic mechanisms behind the spasmolytic (hypotonic) action of glucagon, well known from human endoscopic studies, have been investigated in in vitro studies on guinea pig ileum at the Novo Research Institute, Bagsvaerd, Denmark. Rabbit gallbladder strips have also been used in these studies, as have, following the work of GAGNON and co-workers, strips of rabbit renal artery. In addition, results from bile flow in vivo in rabbits and rats, and from direct pressure measurements with balloons implanted in vivo in the rabbit gut, will be referred to in the following. At the time of writing, the results of these studies have not been published, but they are referred to here in order that this chapter be as complete as possible and they will appear in detail elsewhere (K.D. JØRGENSEN, J. WEISS, and B. DIAMANT 1981, unpublished work).

In our hands, glucagon did not inhibit contractions of the isolated guinea pig ileum elicited by ACh, histamine, serotonin, or $BaCl_2$. By contrast, however, it did inhibit contractions induced by transmural submaximal electrical stimulation (2 V, 0.2 Hz, 50 ms duration) using a technique frequently employed in studies on the action of endorphins and enkephalins (COWIE et al. 1978). It was found that $5 \times 10^{-8}$ $M$ (0.175 μg/ml) glucagon added to the bath inhibited the amplitude of electrically stimulated contractions by 20%, $10^{-6}$ $M$ by 50%, and $10^{-5}$ $M$ by 80%. This effect was not affected by phentolamine, propanolol, isobutylmethylxanthine (IBMX), or atropine at dose levels which by themselves had no influence on the electrically stimulated contractions. In the case of atropine, the concentration used was sufficient to block the contractions induced when ACh was added to the bath. The observations that neither $\alpha$- and $\beta$-blocking agents not IBMX influenced the inhibitory action of glucagon on the electrically stimulated guinea pig ileum do not support the concept that glucagon acts via release of catecholamines or via activation of the production of cAMP.

Prostaglandins ($PGE_1$ and $PGE_2$) may affect cholinergic neuroeffector transmission at the postganglionic postjunctional level (GUSTAFSSON et al. 1980). $PGE_1$, at concentrations which in our hands did not by themselves induce contraction of the isolated guinea pig ileum, was found to potentiate the amplitude of the contractions induced by electrical stimulation or by nicotine. Both types of stimulation were effectively inhibited by atropine and by hexamethonium, as well as by glucagon.

These data, in conjunction with the inability of glucagon to inhibit contractions induced by ACh, seem to exclude the possibility of glucagon having a direct effect on the muscle cells. Instead they suggest a neuronal action of glucagon which could, in part or totally, be an interference with the transmission of impulses through intramural ganglia. Further support for this theory comes from the finding that $PGE_1$ and $PGE_2$, at higher concentrations, induced contractions of the isolated guinea pig ileum which could also be blocked by hexamethonium, atropine, or glucagon. The spasmolytic action of glucagon could also be demonstrated in rabbits in vivo. Glucagon 100 μg/kg given as an intravenous bolus injection was found to decrease the motility and the tonus of the duodenum, jejunum, or colon, as registered by pressure measurements with balloon catheters.

Glucagon was found to be capable of relaxing contractions induced by cholecystokinin octapeptide (CCK-OP) on strips of rabbit gallbladder, but not the contractions induced by carbacholine. These studies have been followed by in vivo studies of the bile flow in rabbits in which one end of the catheter was placed in the bile duct and the other taken out at the neck. A subcutaneous glucagon injection of 100 μg/kg reduced the bile flow by about 50%, most probably owing to a decrease in the tonus of the gallbladder. This initial decrease in bile flow was followed by an increase of the flow to about 130% of the basal flow. This secondary effect may be ascribed to an additional effect of glucagon consisting of an increase of the blood flow in the mesenteric region, with a concomitant enhancement of bile production. In the rat (which lacks a gallbladder), a subcutaneous injection of 50 μg glucagon was found to give rise to a bile flow equal to 115% of the flow of placebo-treated animals over a period of 90 min. These effects correlate well with the results reported by LIN (1974) in the biliary tract of the dog. The mechanism of action of glucagon on the biliary tree has not so far been elucidated but, as will be seen from the evidence presented in Sect. A.II, this cannot be ascribed solely to an action exerted through cAMP.

## II. Structure–Activity Relationships

Studies of structure–activity relationships (SAR) for glucagon have been concerned mainly with adipokinetic and hyperglycemic effects (see Chap. 1). Very little work of this nature appears to have been done regarding the spasmolytic action of glucagon. In a recent investigation (FRANDSEN et al. 1981), the lipolytic and adenylcyclase activitating activities of des-(22-26)-glucagon were found to be below detection limits and des-(27-29)-glucagon about 500 times less potent than native porcine glucagon (for further details see Chap. 19).

In view of these results it was surprising for us to find that, with regard to the spamolytic action on the electrically stimulated guinea pig ileum, both des-(22-26)-glucagon and des-(27-29)-glucagon were equipotent with native glucagon. Furthermore we found that glucagon 1–21 obtained either through synthesis or degradation of native glucagon followed by purification, was equipotent with native glucagon as regards spamolytic effect both in vitro (electrically stimulated guinea pig ileum and CCK-OP-stimulated rabbit gallbladder strips) and in vivo (motility inhibition registered by balloon catheter in rabbit duodenum, jejunum, and colon and enhancement of bile flow in rabbits and rats). Synthetic glucagon 2–21 exerted less than 1%, and glucagon 1–20 less than 10%, of the effect of glucagon 1–21 when evaluated in the electrically stimulated guinea pig ileum model.

As noted for native glucagon, the inhibitory effect of glucagon 1–21 on the electrically stimulated guinea pig ileum was not affected by phentolamine, propanolol, atropine or IBMX. Glucagon 1–21 also inhibited $PGE_1$- and $PGE_2$-enhanced nicotine contractions of the isolated guinea pig ileum, but with a potency which was about six times less than that noted for native glucagon.

Using isolated renal artery of the rabbit, we have confirmed the spasmolytic action of glucagon after stimulation by norepinephrine, as described by GAGNON et al. (1978, 1980). With this preparation, glucagon 1–21 at concentrations up to 100 times those active for glucagon has never been found to exert any spasmolytic action. Furthermore, glucagon 1–21 does not act as an antagonist to glucagon. Glucagon ($5 \times 10^{-7}$ *M*) inhibited the contractions induced by norepinephrine by 73%, and this inhibition was not affected by the simultaneous presence of glucagon 1–21 ($2.5 \times 10^{-6}$ *M*). As noted by GAGNON et al. (1980), norepinephrine-induced contractions can be decreased by theophylline. We have tried, without success, to demonstrate a latent inhibitory action of glucagon 1–21 in the presence of theophylline.

The sensitivity of the portal vein of the rabbit to native glucagon is notably lower than that of the renal artery, since some 300 times higher concentrations of native glucagon were needed to cause a 50% inhibition of norepinephrine-induced contractions. So far we have not been able to demonstrate any inhibitory action of glucagon 1–21 in this preparation.

### III. Conclusions

Thus, it seems that in some preparations, such as certain arteries and veins, the spasmolytic action of glucagon is dependent on the full sequence of glucagon. The mechanism of action might (based on present information regarding SAR for adenylate cyclase activation) be exerted by a cAMP-dependent mechanism. In other preparations, such as the enteric smooth muscles and the biliary tree, glucagon must act via mechanisms that do not involve adenylate cyclase. Whether or not this action occurs by direct interference with ganglionic transmission remains to be established. No evidence presently available favors a direct effect on the smooth muscle. One way which may well lead to a better understanding of the spasmolytic action of glucagon 1–21 would be to perform a histochemical localization of this fragment, in the hope of finding it concentrated in the neuronal elements of those tissues which are sensitive.

## B. Clinical Use of Glucagon as a Spasmolytic or Hypotonic Drug

Only 10 years ago the action of glucagon at doses higher than physiologic on smooth muscle tone and motility was considered an untoward side effect of this metabolic hormone (LAWRENCE 1970). Since then, however, this side effect has opened up an interesting variety of possible clinical applications. Some of these have been fully confirmed in a number of experimental and clinical studies, while others remain as promising hypotheses requiring further experimentation before they can be confirmed or discarded.

When looking at the clinical potentials of glucagon, one should consider separately its diagnostic and therapeutic uses, because there are significant differences relating to the dosage patterns, diagnostic use requiring acute single administrations in most instances, whereas therapeutic use usually calls for prolonged repetitive dosing. In additon, interest in the use of glucagon in diagnostic procedures developed earlier and its specific actions in this respect are quite clearly established, while the therapeutic use of glucagon is still at a very early stage of clinical investigation.

### I. Diagnostic Applications of Glucagon

#### 1. Radiology

The use of glucagon as a premedication drug in radiologic procedures that require a diminished tone and motility of the organ under study has been reported by numerous workers. There are several general reviews describing the various indications, methods, and results (GOHEL et al. 1974; KREEL 1979 a; MILLER et al. 1979). While in most instances there is unanimity on the intravenous as the route of choice, the doses used vary considerably with different authors and techniques, ranging from 0.05 mg by bolus intravenous infusion to 2 mg by intramuscular injection.

a) Esophagus

Glucagon at doses ranging from 0.05 to 0.25 mg can facilitate a barium swallow. While glucagon does not affect esophageal motility, it does relax the LES (KREEL 1979 a) and at these low doses it has no side effects (KELVIN 1979). The relaxing action on LES could be of value in double-contrast esophagograms (CASSEL et al. 1981) or during the radiologic diagnosis of food impaction (FERRUCCI and LONG 1977; GLAUSER et al. 1979; MARKS and LOUSTEAU 1979; PILLARY et al. 1979; CRONAN and STEIN 1980; HANDAL et al. 1980; GIORDANO et al. 1981; REDDY 1981) achieving both the confirmatory diagnosis and the resolution of food impaction with a single procedure.

b) Stomach

The development of double-contrast techniques for the evaluation of the stomach (SHIRAKABE 1971) necessitated the use of motility-modifying agents in order to obtain atony. Glucagon has been used for this purpose since 1974 (for review see KREEL 1979 a) at doses ranging from 0.05 mg (KELVIN 1979) to 1 mg, most authors using intravenous doses of 0.25–0.5 mg. Doses larger than 1 mg increase the duration of action, but not the degree of hypotonia (GOHEL and LAUFER 1978). The side effects of glucagon are minimal (KETO et al. 1979), and with its use the pylorus tends to remain closed, trapping barium and gas in the stomach for a short period. In patients with a previous pyloroplasty or gastroenterostomy the simultaneous immobilization of duodenum and small bowel by glucagon facilitates better radiologic views (LAUFER 1975; OP DEN ORTH 1977). The action of glucagon on the pylorus is completely different from that of scopolamine butylbromide, which relaxes the pylorus, thereby giving rise to rapid superimpositions. In addition, as has already been said, glucagon is virtually free of side effects, and here too it is in sharp contrast to the anticholinergic agents (VIOLON and POTVLIEGE 1980). A review of pharmacologic agents used in upper gastrointestinal tract radiology in 123 institutions in the United States has shown that glucagon is the most commonly used occasional drug (75%). Of the 36 institutions routinely using premedication, 20 reported glucagon to be their drug of choice (OMINSKY and MARGULIS 1981).

c) Duodenum

The introduction of tube techniques for hypotonic duodenography (LIOTTA 1955) increased the need for improved myorelaxant drugs, as was the case with double-contrast examinations of the stomach. The tubeless duodenography performed predominantly nowadays can be carried out safely with the use of intravenous or intramuscular glucagon (CHERNISH et al. 1972 a; MILLER 1973). Intravenous doses are 0.25–0.5 mg, and intramuscular 1–2 mg. The intravenous route has the advantage of requiring a lower dose and resulting in a faster onset of action, while the intramuscular route leads to a more prolonged effect lasting up to 30 min. In duodenography, glucagon compares favorably with propantheline bromide (MILLER 1973; CARSEN and FINBY 1976; BERTRAND et al. 1977), scopolamine butylbromide (ISHII et al. 1978), and hyoscine-*N*-butylbromide (NOVAK and PROBST 1973). The lack of pyloric relaxation with glucagon, however, means that in tubeless duodenography the drug should not be given until after the duodenal loop has been filled (OP DEN ORTH 1977).

### d) Small Intestine

The radiologic evaluation of the small intestine can be achieved by follow-through examinations after esophagogastroduodenography or small bowel enema (KREEL 1979a). As glucagon paralyses the jejunum and ileum at intramuscular doses of 1–2 mg, it would interfere with transit studies. However, the lower doses more frequently used nowadays (0.1–0.2 mg intravenously) do not significantly block the motility of these organs; "if anything, the rebound 'insulin' effect increases barium transit through the small bowel" (KREEL 1979a), thus facilitating follow-through studies. As a result, in the vast majority of cases, the terminal ileum can be visualized within 60–90 min (KREEL 1975). When necessary, however, good paralyzation of the small intestine can be obtained by using a higher dose of glucagon (1 mg intravenously), which will lead to increased hypotonia. This could be particularly useful in the evaluation of melena of unknown origin, and for the detection of small occult neoplasms, early Crohn's disease, and Meckel's diverticulum, amongst other conditions (HECHT et al. 1979).

An area particularly difficult to explore is that of the terminal ileum. However, in a controlled study of 102 cases undergoing double-contrast enema, 1 mg intravenous glucagon was found to result in a significant increase in the number of cases in which a filling of the terminal ileum was achieved. Furthermore, whereas the length of the small bowel filled was only a few centimeters in the control group (50 cases), an average of 30 cm was filled in the glucagon-treated cases (VIOLON et al. 1981).

### e) Large Intestine

Single- and double-contrast barium enemas are performed with increasing frequency nowadays, and the need for spasmolytic drugs in these examinations is also growing (THOENI and MARGULIS 1978), their use resulting in less discomfort for the patient, faster procedures, and better radiologic images (KREEL 1979a; MILLER et al. 1979). Glucagon has been compared with placebo on several occasions, proving its efficacy in barium enema examinations (MILLER et al. 1974; MEEROFF et al. 1975; HARNED et al. 1976). No controlled comparative studies of glucagon and anticholinergic drugs have been reported, but several authors have expressed their preference for glucagon owing to its lack of side effects, fewer contraindications, and a more specific action on the colon (GOHEL et al. 1975; JERELE 1976; DURET and BOLLAERT 1978).

A recent review of the premedications routinely used in colon diagnostic procedures in 100 leading radiology centers showed glucagon to be used in 51%, hyoscine-*N*-butylbromide in11%, atropine in 10%, and propantheline bromide in

9% (Thoeni and Margulis 1978). As is the case in other gastrointestinal areas, the preference in this area for glucagon is based on its minimal side effects compared with anticholinergic drugs (Miller et al. 1979). Furthermore, the tendency towards a decreasing dose and, therefore, to better tolerance, also leads to better overall results.

From the earlier dose of 2 mg glucagon by intramuscular injection, the trend has been towards lower dosages, mostly using the intravenous route. Doses as low as 0.1–0.2 mg have been reported effective in double-contrast studies (Kreel 1979a). However, the dose and route of administration depend to a great extent on the technique used and to the preference of the radiologist. In children, 0.8–1.25 μg/kg body weight will produce an atonic period of 5–10 min, which is sufficient for double-contrast barium enemas (Ratcliffe 1980).

Glucagon has also been found to be effective when mixed with the barium enema. When 163 patients received 2 mg glucagon in this way the results were similar to those obtained through the traditional routes of administration, i.e., relaxation of the entire colon, disappearance of motility, passage of contrast media to terminal ileum, decreased intracolonic pressure, and less patient discomfort, (F. Heitz 1981: Utilization du glucagon par lavement, unpublished work). It remains to be clarified whether glucagon given by enema acts only locally or also, through absorption, systemically.

During barium enema fluoroscopy, glucagon also seems to help in making the differential diagnosis between functional spasm and organic lesions (Gohel et al. 1975), and to help in assessing the type and extent of the lesion (Jerele 1976). Another radiology procedure on the colon in which glucagon has been reported to be effective is in the resolution of intussusception by barium enema. Fisher and Germann (1977) reported two successful reductions when using 0.5 mg intramuscularly, while Hoy et al. (1977) satisfactorily resolved 21 of 25 cases (84%) using 0.05 mg/kg, either intravenously or intramuscularly. No controlled studies have been performed in humans, but in a recent experiment on 69 mongrel puppies, an overall reduction rate was achieved in 70%; although there was not a statistically significant difference between the animals receiving glucagon and those receiving placebo, it was reported that the administration of glucagon resulted in considerably easier reductions and in an earlier return of normal vascular supply, with, therefore, a more rapid recovery of the affected bowel (Haase and Boles 1979). Obviously, controlled human trials are needed in order to prove the value of glucagon for this indication.

#### f) Biliary System

In humans, glucagon has been found to relax both the gallbladder (Chernish et al. 1972b) and the sphincter of Oddi (Nebel 1975a). Similar results have been obtained on the sphincter of Oddi in animals (Behar 1980) and in in vitro studies (Andersson et al. 1972), but the data obtained on its effects on the gallbladder vary in different species. In vitro human gallbladder strips have been found not to respond to glucagon even at very high doses (Cameron et al. 1969). Similarly the nonstimulated gallbladder of the cat (Jansson et al. 1978) and of the guinea pig (Vagne and Troitskaja 1976) have been found not to relax after glucagon was infused. In dogs, on the other hand, glucagon has been found to decrease the intraluminal pressure and to relax the gallbladder (LIN and SPRAY 1969; LIN 1974). This divergence of results may be dose related, or due to species differences (LIN 1975). Glucagon has been used in connection with various radiologic procedures performed on the biliary tract for diagnostic purposes. As reported by CHERNISH et al. (1972 b), glucagon makes the gallbladder shadow larger, particularly when given after a fatty meal. There is, however, no indication as to what advantage this could have in connection with oral cholecystography or related procedures.

Postoperatively, the enterobiliary system can be shown by duodenal and balloon tubes and by tubeless techniques. Glucagon 0.5 mg by intravenous bolus can lead to improved results (KREEL 1979 a), particularly as it relaxes the spasm of the sphincter of Oddi produced by the high pressure often needed to opacify the entire system (CANNON and LEGGE 1979).

In intravenous cholangiography, glucagon can help differentiate functional from organic obstructions (CANNON and LEGGE 1979). Good results have also been reported in hypotonic cholangiography with 1 mg intravenous glucagon (FERRUCCI et al. 1976; EVANS and WHITEHOUSE 1979; JARRETT and BELL 1980), both T-tube and transhepatic cholangiography (CANNON and LEGGE 1979), and reflux cholangiography after sphincteroplasty and enterobiliary anastomosis (2 mg glucagon intramuscularly) (BILBAO and DOTTER 1975).

In peroperative radiology of the biliary tree, the use of intravenous glucagon (1 mg) has been found effective (BORDLEY and OLSON 1979), as it has in the functional evaluation of the sphincter of Oddi either by manodebitmetry, 0.2 mg (PIRE et al. 1976) or radiomanometry, 10 μg/kg (VINSON et al. 1977; TREFFOT et al. 1979). At higher doses, 16–17 μg/kg body weight, the results are less clear (MCCARTHY 1979) but factors such as dose, period of observation, and the concomitant use of various drugs could explain the conflicting results obtained. In relation to this last factor, glucagon has been found to reverse the choledochoduodenal sphincter spasm induced by narcotics (JONES et al. 1980).

## g) Arterial System

As glucagon acts as a potent systemic vasodilator, it was logical to presume that it could be helpful in improving the quality of arteriographic images. Most of the data collected so far in this connection have come from animal studies. Hepatic (SHOEMAKER et al. 1959; KOCK et al. 1970 a; BASHOUR et al. 1973; BRANCH 1973 a, b; LINDBERG and DARLE 1976; RICHARDSON and WITHRINGTON 1976, 1977), mesenteric (MERRILL et al. 1962; KOCK et al. 1970 b; TIBBLIN et al. 1970; MADDEN et al. 1972), and renal (DANFORD 1970; GLICK 1970; STOWE and HOOK 1970; OLSEN 1977) arterial flow have all been found to increase significantly after the administration of glucagon.

Concerning arteriography, DANFORD (1970) observed, in dogs, that the administration of 0.5 mg glucagon selectively to the renal artery resulted in a marked arterial vasodilation and in a better venous opacification of the renal parenchyma. The action of glucagon was also longer lasting (120 min) than that of other frequently used vasodilators. Similar effects have been found, also in dogs, in celiac and mesenteric angiography (DANFORD and DAVIDSON 1969). In humans, KURODA and BAUM (1970) reported that glucagon increased detail in selective angiography.

SCHMARSOW (1976), however, found in 33 patients with symptoms suggesting a pancreatic lesion, that during celiac angiography tolazoline gave better results than glucagon when the pancreatographic effect was evaluated.

h) Urinary System

In addition to increasing renal flow (see Sect. B.I.1.g), glomerular filtration (PARVING et al. 1977; UEDA et al. 1977a, b), natriuresis (STOWE and HOOK 1970; OLSEN 1977), and consequently urine output, glucagon blocks ureteral peristalsis (BOYARSKY and LABAY 1972, 1979; VELA NAVARRETE et al. 1979) and probably, at least to a certain degree, relaxes the urinary bladder (KREEL 1979a).

The possibility that this combination of effects could improve pyeloureteral visualization has been investigated in dogs (HILLMAN et al. 1980); intravenous administration of 1 mg glucagon was found to result in the abolition of ureteral peristalsis for 30 min, in better visualization of the ureters, and in a statistically significant better urogram. According to PINCKNEY and CURRARINO (1980), however, no positive results were seen in ten children who received a bolus dose of 0.5 mg by injection. Further human experiments are required in order that a conclusion may be reached on this issue, particularly in view of the many advantages that the use of glucagon could bring (HILLMAN et al. 1980).

j) Oviduct

Hysterosalpingography is a frequently performed diagnostic procedure in the evaluation of female infertility. Quite often the injection of contrast material under pressure results in spasm of the tubes and failure of the dye to spill into the peritoneal cavity. This could well lead to a false diagnosis of tubal obstruction. Several smooth muscle relaxants have been used in the past in an attempt to overcome this problem, but with little success.

Glucagon appears to be able to relax the smooth muscle layers of the oviduct, its effect here being similar to the spasmolytic effect it has on other hollow structures with muscular walls. Intramuscular glucagon 2 mg given to nine patients with tubal occlusions relaxed the spasm in five cases; it was later confirmed that three of the remaining four patients had organic obstructions (GERLOCK and HOOSER 1976). Similar good results where reported when using doses of 0.2–0.4 mg intravenous glucagon for the same purpose. Of 27 patients, 6 failed to respond after the infusion of glucagon, and all were later found to have organic obstructions (ANSARI and SHIMOURA 1978).

The reliability of this "glucagon test" has been confirmed in kymographic insufflation studies (CORTES-PRIETO et al. 1980). Reporting on 70 cases CORTES-PRIETO et al. (1981) found that the use of 0.3 mg intravenous glucagon permitted a correct diagnosis of occlusion of functional origin to be made in 26 cases in which the diagnosis would otherwise have been organic total obstruction (7 cases) or organic partial obstruction (19 cases).

### 2. Computerized Axial Tomography

Computerized axial tomography of the lower trunk region requires the use of drugs which can effectively block intestinal peristalsis, otherwise, the shifting of gas shadows creates artifacts which seriously affect the quality of the images and thus the accuracy of the diagnosis. The use of glucagon is favored by most authors for this purpose, particularly when using slow (18-s) scanners (BAERT et al. 1977; REDMAN et al. 1977; COENEN et al. 1978; KIRKPATRICK et al. 1978; MOSS et al. 1978; FOTTER and SAGER 1979; KREEL 1979b). The dosage varies, depending on technique and personal preferences; generally, intramuscular administration ranges from 1 mg (KUHNS et al. 1979) to 3 mg (CLAESSENS 1978; OSTEAUX et al. 1978; STRUYVEN et al. 1978) and intravenous administration from 0.1–0.25 mg (KREEL 1979b) to 1 mg (BAERT et al. 1977; COENEN et al. 1978; BERNARDINO et al. 1979). An alternative method consists of giving intravenous glucagon by means of a portable syringe pump. Following this technique, KREEL and BYDDER (1980) have been able to reduce the required dose by 50%. In children, the dosage may range between 9% and 91% of the adult's 2 mg intramuscular dose (BOLDT and REILLY 1977), or be 20 µg/kg up to a maximum of 0.5 mg when given intravenously (FOTTER et al. 1980).

When using glucagon in connection with abdominal computerized axial tomography it is important to bear in mind the possibility of the coexistence of unsuspected pheochromocytoma (KREEL 1979a) as the use of glucagon in this situation could act as the trigger for an acute hypertensive attack (GEELHOED 1980; see also Chap. 51).

### 3. Ultrasonography

Glucagon has been used in ultrasonographic examinations of the gastrointestinal tract for similar reasons that indicate its use in connection with gastrointestinal radiology and computerized axial tomography (BERGER et al. 1978). Besides obviating streaking artifacts of the small bowel, it produces full paralysis of the stomach and duodenum (MACMAHON et al. 1979). The fact that a fluid-distended stomach and duodenum facilitates a better visualization of the distal common bile duct (PON and COOPERBERG 1979) and the tail of the pancreas (WEIGHALL et al. 1979) has also been pointed out. The usual dose of glucagon for this indication is 1 mg given either intramuscularly or intravenously.

### 4. Endoscopy

Glucagon is used in endoscopy in order to obtain a quiet field in which to work. In an intravenous dose of 0.2 mg, it paralyzes various of the viscera and structures, relaxes some of the sphincters, and abolishes practically all gastric and intestinal secretions (PAUL et al. 1973; PAUL and FREYSCHMIDT 1974; HRADSKY et al. 1973, 1974). Interest in the use of glucagon in endoscopy has paralleled the development of this method, and reflects the search for drugs devoid of the side effects associated with the use of anticholinergic agents (MLECKO 1974). In this connection, glucagon has been found to be as effective as hyoscine butylbromide (KIIL et al. 1977) and superior to atropine alone (MELSOM et al. 1977) and to atropine plus pethidine (QVIGSTAD et al. 1979).

Finally, STERN et al. (1979) in a controlled study involving 100 patients, found hyoscine butylbromide to be superior to glucagon for antrum and pylorus relaxation, but glucagon to be more effective for duodenal relaxation. These authors concluded that glucagon should be the drug of choice for duodenal relaxation. Their suggestion that it be used in combination with hyoscine butylbromide for gastroscopy, however, does not seem justifiable.

a) Esophagogastroduodenoscopy

Almost two-thirds of all endoscopies are performed on the upper gastrointestinal tract (MYREN 1979). For the standard screening procedure very little relaxation is required and quite often a small amount of sedation is sufficient. When special procedures have to be performed, for example, control of bleeding, biopsies, or removal of small lesions, glucagon provides better conditions for a safer, more accurate operation (HRADSKY and FURUGÅRD 1976; MADSEN et al. 1976).

b) Colonoscopy

Initially, there was some interest in the potential value of glucagon in endoscopic procedures of the large bowel (KREEL 1979a). That was reasonable after the good results obtained in upper gastrointestinal endoscopy (EK 1979), and the known relaxing action of glucagon upon the colonic musculature (TAYLOR et al. 1975). However, when compared with meperidine and diazepam, glucagon (2 mg intramuscularly) seemed to offer no particular benefit as a premedication, and there was one report (NORFLEET 1978) that its use made the whole procedure more difficult for the endoscopist. FOSTER et al. (1981) found no significant difference between glucagon (1 mg intravenously) and placebo in 100 cases. By contrast, EK (1979), when comparing glucagon with placebo in a short series (13 cases), found glucagon of definite value, improving the ease of performance of total colonoscopy and decreasing patient discomfort.

With such conflicting results, it is clear that more studies are needed before the use of glucagon in colonoscopy can be either advocated or dismissed. A further point to clarify is whether, rather than using it routinely as premedication for the actual intubation procedure, glucagon should be reserved for the immobilization of the colon in special circumstances.

### 5. Endoscopic Retrograde Cholangiopancreatography

Endoscopic cannulations of the spincter of Oddi for retrograde cholangiopancreatography were first described by RABINOV and SIMON (1965) and later by MCCUNE et al. (1968). During the 1970s the frequency of endoscopic retrograde cholangiopancreatography increased dramatically and nowadays it is a standard procedure in the evaluation of biliary tree and pancreatic pathologies (KESSLER et al. 1976; RENTSCH et al. 1976; SOEHENDRA 1977; TITTOBELLO et al. 1979) and for facilitating the carrying out of related procedures such as papillotomy/sphincterotomy (SIEGEL 1979; GEENEN et al. 1981) and sphincter and ducts manometry (MECKELER and BOROW 1975; NEBEL 1975b; RÖSCH et al. 1976; HOGAN et al. 1977; RIBEIRO et al. 1977).

Glucagon is used as part of the premedication for endoscopic retrograde cholangiopancreatography, not only in order to paralyze the duodenum, but also because it relaxes the spincter of Oddi, thus facilitating cannulation of the ampulla and biliary ducts (KATON et al. 1974; KESSLER et al. 1976; LORENZ-MEYER and DOM-

BROWSKI 1976). The intravenous dose used varies according to authors, between 0.5 mg and 3 mg; 1 mg is often regarded as optimal (RENTSCH et al. 1976; MORLEY and CAMPBELL 1977; TITTOBELLO et al. 1979; ZIEGELS et al. 1979). Whether the giving of glucagon before or after the endoscope has passed the pylorus makes the procedure easier is an issue not yet resolved (KREEL 1979a). It is clear, however, that when glucagon is given in a procedure in which ampullary and ductal pressure recordings are to be performed, it should be given at least 10 min before the recordings are attempted (NEBEL 1974; CARR-LOCKE and GREGG 1981).

The possibility that the use of glucagon could also help in decreasing the incidence of pancreatitis associated with endoscopic retrograde cholangiopancreatography has also been the object of investigation. In this respect, KOCK et al. (1975) found no prophylactic effect on patients receiving 1 mg glucagon before endoscopic retrograde cholangiopancreatography followed by a glucagon infusion at the rate of 1 mg/h for 5 h. SILVIS and VENNES (1975) on their part, found in a placebo-controlled trial a significant decrease in frequency and magnitude of amylase activity after 24 h in patients receiving 1 mg intravenous glucagon. In any case, it may be assumed that, by facilitating the ease of performance of endoscopic retrograde cholangiopancreatography, glucagon decreases some of the risks inherent in this method.

## II. Therapeutic Applications of Glucagon

To date, the role of glucagon in the treatment of pathologic entities other than hypoglycemia (see Chap. 55) has been only hinted at, hypothesized, or rather briefly reported. There is a complete lack of controlled studies to prove without question that the use of glucagon can be of therapeutic value to the patient, or an improvement upon existing therapies. For this reason we shall limit ourselves here to mentioning the various indications for which different authors have felt the use of glucagon might be of clinical significance.

### 1. Gastroenterology

a) Esophagus

α) *Achalasia.* Glucagon has been found to decrease the elevated resting pressure of the LES of patients suffering from achalasia (JENNEWEIN et al. 1973; SIEWERT et al. 1973). Both groups of authors used the same dosage scheme: 60 μg/kg body weight, administered intravenously as a bolus injection. The effect was found to last about 15 min and in the ten cases treated by JENNEWEIN et al., the pressure.lowering effect of glucagon was statistically significant ($P<0.002$), ROESCH (1974) has also treated achalasia patients with glucagon, and has reported finding that 1 mg intravenous glucagon induced a marked decrease in LES pressure lasting 10–15 min. During this period, dysphagia disappeared completely. Zinc protamine glucagon, a slow-release preparation of the hormone (see Chap. 8), at a dose of 10 mg subcutaneously or intramuscularly has been found to have a more prolonged (3–4 h), but less intense effect (RÖSCH 1974). The clinical significance of this hormonal approach to the problem is perhaps rather limited (ROESCH 1974); at the present time, however, it is the only potentially successful medical treatment in sight (RISUEÑO ALVAREZ and SANCHEZ DE VEGA 1979).

β) *Esophageal Perforation.* Glucagon not only decreases LES pressure, but also acts as a potent suppressor of gastric secretion (DOTEVALL and KOCK 1963; QUIRARTE et al. 1966) and motility (STUNKARD et al. 1955). One of the major problems of esophageal perforation is mediastinitis due to the passage of the esophagogastric contents. It seems possible that glucagon, by decreasing the volume and acidity of these contents, and by blocking motility, could be of help here, particularly in elderly high risk patients.

Based on this hypothesis, PICKARD (1974) tried intravenous glucagon (1 mg, followed by 2 mg every 6 h by continous infusion) in three elderly patients with iatrogenic perforation of the esophagus. After 4 days (1 case) or 5 days (2 cases) of therapy, the perforation had sealed off completely. Of particular note is the fact that there was no need to drain the thoracic cavity as the pleural effusion was minimal in all three cases.

### b) Small Intestine

There is no reported clinical experience of the use of glucagon in diseases of the small intestine other than those referring to vascular problems, which will be dealt with in Sects. B.II.2.b and c). As stated by N.A. VOLPICELLI (see WINGATE and PEARCE 1979) there are some rather obscure points relating to small bowel motility in connection with the digestive process. How glucagon at pharmacologic doses could possibly have a positive influence on conditions such as postvagotomy diarrhea and other anomalies of the digestive function is a matter of speculation (WINGATE and PEARCE 1979).

The fact that, as reported by F. PAUL (see WINGATE and PEARCE 1979), chronic infusion of glucagon does not result in paralytic ileus in humans and that at least in one dog a postoperative ileus was relieved by the intravenous infusion of glucagon (WINGATE and PEARCE 1979), points to another area for possible therapeutic application of glucagon. In contrast, in the dog, it has been found that glucagon, infused at 0.5 to 1 mg/h, produces a continuous stimulation of the small intestine motility (EVANS et al. 1978), an action quite different from that seen at higher dosage.

### c) Large Intestine

α) *Diverticular Disease of the Colon.* One of the many features of acute diverticulitis is severe abdominal pain, due in great part to the spastic status of the hypertrophied (specially sigmoid) musculature. Because of the spasmolytic effect of glucagon on gastrointestinal musculature, DANIEL et al. (1974) tried it in the treatment of acute diverticulitis attacks in 20 consecutive cases. The patients received the glucagon in different ways: eight were given 1 mg intravenous glucagon by rapid injection, and six received a similar dose infused over 8–10 h. The injections were repeated every 4 h for a total treatment period of 36 h. The remaining six patients received 9 mg glucagon by intravenous infusion, again for a total treatment period of 36 h. All three methods were equally successful: within a mean period of 12 h all patients were pain free. The associated symptoms had also disappeared and the general condition of all patients was significantly improved.

A comparative group of 15 patients chosen retrospectively to match the study group recovered on average 4 (1.5–6) days after the beginning of the conventional treatment. Controlled clinical data are needed to confirm these results.

The quick relief of pain, particularly in cases without associated fever and leukocytosis, which DANIEL et al. (1974) reported with glucagon, points to the fact that glucagon quite likely acts simply be releasing the colonic spasm (ALMY and HOWELL 1980). It is feasible therefore that glucagon could be used as a differential diagnostic test for uncomplicated acute diverticulitis (DANIEL et al. 1974), rectal bleeding of unknown origin, and obstipation (DANIEL 1982).

*β) Colon Surgery.* Glucagon has been during end-to-end anastomosis in low anterior resections involving the use of a stapling device. Apparently the spasm, often present both on the rectal and sigmoid stumps, can be effectively relieved with glucagon (HARFORD 1979; MOSESON et al. 1980). It is interesting to note that while HARFORD gave 2 mg intravenous glucagon to relax the rectal spasm, which in his opinion facilitated the passing of the stapling device, MOSESON et al. utilized 1 mg intravenous glucagon in order to relieve the spasm of the sigmoid colon, which prevented the introduction of the distal end of the stapler. MOSESON et al. specified that after intravenous administration of 1 mg glucagon, the diameter of the sigmoid colon increased from 2–2.5 cm to 4–6 cm. Neither of these groups of authors specified the number of cases treated, but both concluded that the use of glucagon had led to an easier procedure, and to a safer and less traumatic anastomosis.

An additional benefit which could probably be derived from the use of glucagon in gastrointestinal anastomosis is its blood perfusion increasing effect; it could well be that by improving the often critical irrigation of the anastomotic ends, glucagon could protect the anastomosis and thus prevent necrosis and anastomotic leakage.

### d) Extrahepatic Biliary Tree

Pharmacologic doses of glucagon have been found to relax the sphincter of Oddi, both in animals (LIN and SPRAY 1969; LIN 1975) and in humans (NEBEL 1974; VONDRASEK and EBERHARDT 1974; HOGAN et al. 1977, 1979; BAR-MEIR et al. 1979; GEENEN et al. 1980). More recently, CARR-LOCKE and GREGG (1981) have found that glucagon also produces relaxation of the bile duct sphincter at physiologic plasma levels; it has a significant effect at an infusion rate of 0.016 $\mu g\ kg^{-1}\ h^{-1}$, and reaches a maximum effect at a dose of 0.25 $\mu g\ kg^{-1}\ h^{-1}$.

The results of experiments done with pharmacologic doses are not quite unanimous. HOGAN et al. (1977) reported a significant reduction of sphincter of Oddi pressures with intravenous glucagon pulses of 0.2 and 0.4 mg, while CSENDES et al.

(1979) found no change after an intravenous dose of 0.25 mg. The results of HOGAN et al., however, have been confirmed both by VONDRASEK and EBERHARDT (1974) and by PAUL (1979) at a dose level of 0.4 mg. Overall, it appears that the dosage of glucagon might be critical for achieving therapeutic success when treating biliary tree pathology.

α) *Biliary Colic*. PAUL (1976) was the first to report on the value of glucagon for treating patients suffering acute biliary colic. Reporting on 31 cases, PAUL found that 7 of 10 patients treated with 0.2 mg intravenous glucagon became pain free in less than 5 min, while 15 of 17 responded similarly well to a 0.5 mg dose, also by intravenous bolus injection. The remaining 4 patients, who received a 1 mg dose, became pain free very speedily. Should pain recur (as happened in just one case in his series), PAUL recommends the intravenous infusion of 5 mg glucagon for 24 h (PAUL 1979). BRANDSTÄTTER and KRATOCHVIL (1979) have also used glucagon to treat biliary colic. These authors reported on a group of 65 patients to whom they administered 1 mg glucagon as an intravenous bolus injection. Most of their patients responded to therapy within 30 s, and only eight required a second injection 20 min later. If the results of PAUL and of BRANDSTÄTTER and KRATOCHVIL could be confirmed in controlled studies, glucagon could well become a routine medication in the treatment of biliary colic (BELL 1980).

β) *Choledocholithiasis*. PAUL (1979) has also used glucagon in a continuous infusion (0.5–1 mg intravenous loading dose, followed by 3–5 mg/24 h infusion) to facilitate the passage of: (1) stones impacted in the cystic duct; (2) small gravel stones in the common bile duct; and (3) small residual stones left after endoscopic papillotomy. The use of glucagon proved successful in 3 of 5 cases with cyctic duct obstruction, in 3 of 5 with choledocholithiasis with intact papilla (the largest stone passed measured 0.5 cm), and in 56 of 61 cases following endoscopic papillotomy. The longest treatment period was 12 days.

DOMAN and GINSBERG (1981), using a glucagon infusion (1 mg intravenous loading dose, followed by a continuous maintenance drip of 5 mg/h for 8 h) treated seven patients with biliary tree stones. Six of the seven showed improvement. The only patient in which the therapy failed had a 1-cm stone in the common bile duct. The largest stone passed with an intact sphincter of Oddi measured 0.4 cm, and the largest after papillotomy 0.7 cm. Only two of the seven patients had abdominal pain; both received sustained relief after a single 8-h infusion. The high dosage scheme used on this occasion was well tolerated, with only two patients suffering mild nausea.

Finally, LATSHAW et al. (1981) reported two patients with impacted 0.6-cm calculi in which percutaneous (T-tube) removal of the stones had been unsuccessful on several attempts. After the administration of 1 mg glucagon by bolus intravenous injection, both calculi could be advanced through the sphincter of Oddi and into the duodenum with the help of a steerable French catheter. The long-term application of glucagon is probably beneficial in choledocholithiasis, both after endoscopic papillotomy (PAUL 1979) and in order to achieve a medical papillotomy

(DOMAN and GINSBERG 1981), but controlled studies are required in order to prove these assumptions (PAUL 1979).

### 2. Arterial System

#### a) Hepatic Ischemia

In several animal species (dog, pig, monkey) an average glucagon infusion dose of 10 μg glucagon/h has been shown to produce a twofold increase of hepatic blood flow (see Sect. B.I.g). This increased flow is primarily a reflection of the potent vasodilating action of glucagon on the splanchnic vascular bed. Following these reports, it was suggested that glucagon may be helpful in cases of hepatic ischemia. To date, no human clinical data are available, and the results from two animal studies are contradictory. KUPCSULIK and KOKAS (1979), using a model to study ischemic liver lesions in rats, found that high intravenous doses of glucagon (50 μg/100 g body weight) given before or after the lesion, significantly reduced ischemic lesions of the liver in both in vitro and in vivo experiments, and significantly improved the survival rate of the animals. PROCTOR et al. (1980), using dogs submitted to 120 min hemorrhagic hypotension, found that even though intravenous glucagon (1 μg/kg bolus followed by 3 μg $kg^{-1}$ $h^{-1}$ continuous infusion) significantly increased hepatic blood flow, it had a negative effect on liver energy metabolism. It is clear from this very limited amount of information that the role of glucagon therapy on hepatic ischemia must be further evaluated in animal experiments before its clinical use in humans can be justified. It should be recalled here that glucagon plays a major role in liver regeneration as discussed in detail in Chap. 21.

#### b) Gastrointestinal Ischemia

Hypotension in hypovolemic shock significantly reduces gastrointestinal mucosal blood flow and causes irreversible bowel ischemia before flow levels can return to normal following volume replacement. BOND et al. (1979) assessed the value of several vasodilating drugs (isoproterenol, nitroprusside, secretin, and glucagon) on the rate and distribution of blood flow to the gut in dogs. Intravenous glucagon 0.25 mg was found to have the greatest effect on normal dogs, increasing the flow to the stomach by 122%, to the small bowel by 418%, and to the colon by 111%. When the dogs were bled to a mean blood pressure of 40 mmHg (30% reduction in overall gut flow), 0.5 mg glucagon infused intravenously over 60 min increased flow to the same organs and especially to their mucosal layer; the increases over the prebleeding values were: stomach 105%, small bowel 78%, and colon 96%. As gastrointestinal mucosal tissues are particularly prone to irreversible necrosis in hypovolemic hypotension, these studies suggest that glucagon could prevent ischemic bowel necrosis in hypotensive patients. Glucagon given intraarterially (0.25 mg over 20 min, an 0.5 mg over 60 min) has led to similar results (BOND and LEVITT 1980). These authors emphasize that this effect is primarily the result of an intense regional vasodilation and redistribution of flow towards the gastrointestinal area (femoral and carotid flows decreased simultaneously). For this reason, glucagon should not be used in human patients in shock unless at least partial replacement of the lost volume has been carried out.

#### c) Nonocclusive Mesenteric Ischemia

Nonocclusive mesenteric ischemia is a major medical entity and a difficult problem to treat (ULANO et al. 1972). About one-third of mesenteric infarctions are nonocclusive and the mortality rate in this disease group nears 100% (SCHWAIGER et al. 1979). Treatment traditionally consists of supportive therapy and extensive intestinal resections. Most nonocclusive mesenteric ischemia cases are associated with congestive heart failure, use of digitalis, and severe shock (ULANO et al. 1972), and this probably explains the very poor prognosis for these patients.

In view of its mesenteric vasodilating effect (TIBBLIN et al. 1971), glucagon has been tried in experimental constriction of the superior mesenteric artery of the dog in order to evaluate its efficacy (DANFORD 1971; TIBBLIN et al. 1971; LEVINSKY et al. 1975). The results, when compared with various other vasodilators, have been excellent (ULANO et al. 1972; LANCIAULT et al. 1976; SCHWAIGER et al. 1979; ANDERSON et al. 1981), and glucagon is clearly the drug of choice for this indication. The recommended doses vary according to authors, as well as to the routes of administration (intramuscular, intravenous, or intraarterial).

As selective superior mesenteric arteriography is required for the diagnosis of nonocclusive mesenteric ischemia, direct intraarterial infusion is quite feasible and should be considered the route of choice. Optimal dose and length of treatment have not yet been established, but should be in order that clinical trials in humans can be started. The latest animal data indicate that doses should probably range from 0.1 to 0.2 $\mu g\ kg^{-1}\ min^{-1}$ (SCHWAIGER et al. 1979; ANDERSON et al. 1981). The length of therapy will depend to a great extent on the underlying etiology and on the degree of reversibility of the arterial spasm.

#### d) Peripheral Vascular Disease

When assessing the action of glucagon on peripheral vascular flow, most reports conclude that there is a decrease (e.g., 32% decrease on femoral artery flow), which in part is a reflection of its selective action as a potent vasodilator on the heart and on various vascular beds (LEFEBVRE and BEAUJEAN 1964; KOCK et al. 1970b). No detailed reports, either experimental or clinical, are available on the response of peripheral vascular ischemia to glucagon, but from the information which is available (KOCK et al. 1970b; BOND and LEVITT 1980) one would expect glucagon not to be of benefit in these cases.

### 3. Urinary System

#### a) Ureteral Colic

In in vitro studies, glucagon (10–15 μg/ml) has been reported to abolish ureteral peristalsis (BOYARSKY and LABAY 1972). In vivo experiments in dogs showed similar results when the animals were given 2–8 μg/kg glucagon by intravenous bolus injection (BOYARSKY and LABAY 1979). The effect started 50 s after the injection and normal peristalsis was resumed 15–20 min later. VELA NAVARRETE et al. (1979)

found similar results, also in dogs, and pointed to the possible value of glucagon for the treatment of ureteral colic, based on its spasmolytic and also on its diuretic effect which would further facilitate the passage of ureteral calculi (see also Chap. 50).

In 1977, LOWMAN et al. reported on their early clinical experience using glucagon in ten patients for the treatment of ureteral colic due to ureterolithiasis. These authors found that 1 mg intravenous glucagon provided prompt relief from pain in all patients, and that this lasted for 1–3 h. In addition, in three cases, there was a spontaneous passage of the calculi 4–8 h after the glucagon was injected.

MORISHIMA and GHAED (1979) also reported successfully treating five cases of ureteral colic due to impacted calculi with glucagon and a forced diuresis. The authors gave 1 mg glucagon by fast intravenous injection, followed by the rapid infusion of 1.5–2 l 5% dextrose in water over a period of 1 h. Pain responded to glucagon in all cases. The calculi passed in 40 min (1), 2 h (3), and 8 h (1). The fact that glucagon acts as a diuretic as well as a ureteral relaxant, makes it of potential interest for the treatment of ureterolithiasis (BOYARSKY and LABAY 1979).

## C. Closing Remarks

It is clear from the animal experiments reported here, that the spasmolytic action of glucagon in the gut and in the biliary tree is selectively inherent in glucagon 1–21, which is devoid of metabolic effects. The clinical benefits of this substance in humans have still to be more thoroughly investigated, with the aim of producing a pure spasmolytic agent with a better tolerance than those presently available.

The mechanism behind the spasmolytic action of glucagon 1–21 has not yet been established. The information available to date seems to rule out action through cAMP or through the release of catecholamines, as well as the possibility of there being a direct effect on the smooth muscle. Glucagon has been established as a most effective premedicating agent for many diagnostic procedures. Controlled clinical trials are now needed in order to establish the precise areas in which glucagon, or glucagon 1–21, will be of value as a therapeutic agent.

## References

Almy TP, Howell DA (1980) Diverticular disease of the colon. N Engl J Med 392:324–331
Anderson JH, White RI Jr, Starr FL III, Samphilipo MA Jr (1981) Superior mesenteric arterial administration of vasodilators in reversing experimentally produced nonocclusive mesenteric ischemia. Gastroenterology 80:1100
Andersson KE, Anderson R, Hedner P, Persson CGA (1972) Effect of cholecystokinin on the level of cyclic AMP and the mechanical activity in the isolated sphincter of Oddi. Life Sci 11:723–732
Andersson RGG (1972) Cyclic AMP and calcium ions in mechanical and metabolic responses of smooth muscles. Influence of some hormones and drugs. Acta Physiol Scand [Suppl] 382:1–59
Ansari AH, Shimoura H (1978) Hypotonic hysterosalpingography with glucagon. Fertil Steril 30:476–477
Anuras A, Cooke AR (1978) Effects of some gastrointestinal hormones on two muscle layers of duodenum. Am J Physiol 234:E60–E63

Baert AL, Ponette E, Pringot J, Marchal G, Dardenne A, Coenen Y (1977) C.A.T. in acute and chronic pancreatitis. In: Du Boulay GH, Mosely IF (eds) The first European seminar on computerized axial tomography in clinical practice. Springer, Berlin Heidelberg New York, p 382
Barbezat G, Grossman M (1971) Effect of glucagon on water and electrolyte movement in jejunum and ileum of dog. Gastroenterology 60:762
Bar-Meir S, Geenen JE, Hogan WJ, Dodds WJ, Stewart ET, Arndorfer RC (1979) Biliary and pancreatic duct pressures measured by ERCP manometry in patients with suspected papillary stenosis. Dig Dis Sci 24:209–213
Bashour FA, Geumei A, Nafrawi AG, Downey HF (1973) Glucagon: its effects on the hepatic arterial and venous beds in dogs. Pflügers Arch 344:83–92
Behar J (1978) Effect of glucagon on the feline lower esophageal sphincter (LES) in vivo. Gastroenterology 72:1116
Behar J (1980) Comparative pharmacological characteristics of the lower esophageal sphincter (LES) and the sphincter of Oddi (SO) in the cat in vivo. Dig Dis Sci 25:720
Bell GD (1980) Drugs used in the management of gallstones. In: Dukes MNG (ed) Side effects of drugs annual, vol 4. Excerpta Medica, Amsterdam Oxford Princeton, p 258
Berger PE, Kuhn JP, Munschauer RW (1978) Computed tomography and ultrasound in the diagnosis and management of neuroblastoma. Radiology 128:663–667
Bernardino ME, Jing BS, Wallace S (1979) Computed tomography diagnosis of mesenteric masses. AJR 132:33–36
Bertrand G, Linscheer WG, Raheja KL, Woods RE (1977) Double-blind evaluation of glucagon and propantheline bromide (pro-banthine) for hypotonic duodenography. Am J Roentgenol 128:197–200
Bilbao MK, Dotter CT (1975) Reflux cholangiography in sphincteroplasty or enterobiliary anastomosis. Radiology 115:585–588
Blumenthal MN, Brody TM (1969) Studies on the mechanism of drug-induced bronchiolar relaxation in the guinea pig. J Allergy 44:63–69
Boldt DW, Reilly BJ (1977) Computed tomography of abdominal mass lesions in children. Radiology 124:371–378
Bond JH, Levitt MD (1980) Effect of glucagon on gastrointestinal blood flow of dogs in hypovolemic shock. Am J Physiol 238:G434–G439
Bond JH, Shoenborn KE, Levitt MD (1979) Effect of glucagon on gastrointestinal blood flow in hypovolemic shock. Clin Res 27:264A
Bordley J, Olson JE (1979) The use of glucagon in operative cholangiography. Surg Gynecol Obstet 149:583–584
Boyarsky S, Labay PC (1972) Ureteral dynamics. Pathophysiology, drugs, and surgical implications. Williams and Wilkins, Baltimore
Boyarsky S, Labay PC (1979) Glucagon, ureteral colic and ureteral peristalsis. Trans Am Assoc Genitourin Surg 70:22–24
Branch RA, Nies AS, Shand DG (1973a) The influence of glucagon on regional blood flow in the rhesus monkey. Br J Pharmacol 49:149P
Branch RA, Shand DG, Nies AS (1973b) Increase in hepatic blood flow and d-propanolol clearance by glucagon in the monkey. J Pharmacol Exp Ther 187:581–587
Brandstätter G, Kratochvil P (1979) Glucagon bei Gallenkoliken. Therapiewoche 29:3362–3365
Cameron AJ, Phillips SF, Summerskill WHJ (1969) Effect of cholecystokinin, gastrin, secretin, and glucagon on human gallbladder muscle in vitro. Proc Soc Exp Biol Med 131:149–154
Cameron AJ, Phillips SF, Summerskill WHJ (1970) Comparison of effects of gastrin, cholecystokinin-pancreozymin, secretin, and glucagon on human stomach muscle in vitro. Gastroenterology 59:539–545
Cannon P, Legge D (1979) Glucagon as the hypotonic agent in cholangiography. Clin Radiol 30:49–52
Carr-Locke DL, Gregg JA (1980) Effects of exogenous secretin and glucagon on pancreatic and biliary ductal and sphincteric pressures in man. Clin Sci 59:9P

Carr-Locke DL, Gregg JA (1981) Endoscopic manometry of pancreatic and biliary sphincter zones in man. Dig Dis Sci 26:7–15
Carsen GM, Finby N (1976) Hypotonic duodenography with glucagon. Radiology 118:529–533
Cassel DG, Anderson MF, Zboralske FF (1981) Double contrast esophagrams. The prone technique. Radiology 139:737–739
Chernish SM, Miller RE, Rosenak BD, Scholz NE (1972a) Hypotonic duodenography with the use of glucagon. Gastroenterology 63:392–398
Chernish SM, Miller RE, Rosenak BD, Scholz NE (1972b) Effect of glucagon on size of visualized human gallbladder before and after a fat meal. Gastroenterology 62:1218–1226
Chowdhury AR, Lorber SH (1977) Effects of glucagon and secretin on food- and morphine-induced motor activity of the distal colon, rectum and anal sphincter. Dig Dis 22:775–780
Chowdhury AR, Dinoso VP, Lorber SH (1976) Characterization of a hyperactive segment at the rectosigmoid junction. Gastroenterology 71:584–588
Claessens J (1978) Exploration tomodensitométrique du corps entier. Brux Med 58:411–417
Coenen Y, Marchal G, Ponette E, Baert AL, Pringot J (1978) Pancreatic disease. In: Baert A, Jeanmart L, Wackenheim A (eds) Clinical computer tomography, head and trunk. Springer, Berlin Heidelberg New York, p 197
Corazziari E, Habib IF, Delle Fave G, Torsoli A (1973) Effets de certaines hormones gastro-intestinales sur la motricité duodéno-jejunale. Biol Gastroenterol 6:368–369
Cortés-Prieto J, Rey G, Rubio Martínez JM, Pesenti B, Soler A, Serrano C, Rodríguez M (1980) Glucagon administration during tubal test. 10th World congress of fertility and sterility, Madrid, 5–11 July
Cortés-Prieto J, Pesenti B, Rubio JM, Soler-Villalobos A, Rey G, Serrano C (1981) Evaluation of Rubin's test using glucagon as a spasmolytic. 3rd World congress of human reproduction, Berlin West, 22–26 March
Cowie AL, Kosterlitz HW, Waterfield AA (1978) Factors influencing the release of acetylcholine from the myenteric plexus of the ileum of the guinea pig and rabbit. Br. J Pharmacol 64:565–580
Cronan JJ, Stein S (1980) Glucagon therapy of esophageal food impaction. Conn Med 44:79–80
Csendes A, Kruse A, Funch-Jensen P, Øster MJ, Ørnsholt J, Amdrup E (1979) Pressure measurements in the biliary and pancreatic duct systems in controls and in patients with gallstones, previous cholecystectomy, or common bile duct stones. Gastroenterology 77:1203–1210
Danford RO (1970) The effect of glucagon on renal hemodynamics and renal arteriography. Am J Roentgenol 108:665–673
Danford RO (1971) The splanchnic vasoconstrictive effect of digoxin and its reversal by glucagon. In: Boley SJ (ed) Vascular disorders of the intestine. Appleton Century Crofts, New York, p 421
Danford RO, Davidson AJ (1969) The use of glucagon as a vasodilator in visceral angiography. Radiology 93:173–175
Daniel O (1982) The role of glucagon in the management of colonic disorders and colonic surgery. In: Picazo J (ed) Glucagon in gastroenterology and hepatology. MTP Press, Lancaster Boston, The Hague, p 81
Daniel O, Basu PK, Al-Samarrae HM (1974) Use of glucagon in the treatment of acute diverticulitis. Br Med J 3:720–722
Doman DB, Ginsberg AL (1981) Glucagon infusion therapy for biliary tree stones. Gastroenterology 80:1137
Dotevall G, Kock NG (1963) Effect of glucagon on gastrointestinal motility in man. Gastroenterology 45:364–367
Duret RL, Bollaert A (1978) Les modificateurs du comportement colique. In: Engelholm L, Jeanmart L, De Toeuf J, Osteaux M, Peeters JP (eds) Exploration gastro-duodénale et colique en double contraste. European Press, Gent
Ek B (1979) The use of glucagon in colonoscopy. In: Picazo J (ed) Glucagon in gastroenterology. MTP Press, Lancaster, p 53

Evans AF, Whitehouse GH (1979) The effect of glucagon on infusion cholangiography. Clin Radiol 30:499–506
Evans DF, Foster GE, Hardcastle JD, Johnson F, Wright JW (1978) The effect of glucagon on the canine duodenum and small intestine. Br J Pharmacol 64:475P
Faloona GR, Unger RH (1974) Biological and immunological activity of pancreatic glucagon and enteric glucagon-like immunoreactivity. Isr J Med Sci 10:1324–1331
Fasth S, Hultén L (1971) The effect of glucagon on intestinal motility and blood flow. Acta Physiol Scand 83:169–173
Ferrucci JT, Long JA (1977) Radiologic treatment of esophageal food impaction using intravenous glucagon. Radiology 125:25–28
Ferrucci JT, Wittenberg J, Stone LB, Dreyfuss JR (1976) Hypotonic cholangiography with glucagon. Radiology 118:466–467
Fisher JK, Germann DR (1977) Glucagon-aided reduction of intussusception. Radiology 122:197–198
Foster GE, Vellacott KD, Balfour TW, Hardcastle JD (1981) Outpatient flexible fibreoptic sigmoidoscopy, diagnostic yield and the value of glucagon. Br J Surg 68:463–464
Fotter R, Sager WD (1979) CT des Beckens und Abdomens im Kindesalter. Fortschr Roentgenstr 131:476–479
Fotter R, Sager WD, Justich E, zur Nedden D (1980) Die Bedeutung der Computertomographie in der pädiatrischen Diagnostik abdomineller und pelviner Tumoren. Roentgenblaetter 33:156–162
Frandsen EK, Grønvald FC, Heding LG, Johansen NL, Lundt BF, Moody AJ, Markussen J, Vølund Aa (1981) Glucagon: structure-function relationships investigated by sequence deletions. Hoppe Seylers Z Physiol Chem 362:665–667
Gagnon G, Regoli D, Rioux F (1978) A new bioassay for glucagon. Br J Pharmacol 64:99–108
Gagnon G, Regoli D, Rioux F (1980) Studies on the mechanism of action of glucagon in strips of rabbit renal artery. Br J Pharmacol 69:389–396
Geelhoed GW (1980) CAT scans and catecholamines. Surgery 87:719–720
Geenen JE, Hogan WJ, Dodds WJ, Stewart ET, Arndorfer RC (1980) Intraluminal pressure recording from the human sphincter of Oddi. Gastroenterology 78:317–324
Geenen JE, Vennes JA, Silvis SE (1981) Résumé of a seminar on endoscopic retrograde sphincterotomy (ERS). Gastrointest Endosc 27:31–38
Gerlock AJ, Hooser CW (1976) Oviduct response to glucagon during hysterosalpingography. Radiology 119:727–728
Gerner T, Haffner JFW (1975) X. The significance of distention for the effect of glucagon on the fundic and antral motility in isolated guinea-pig stomach. Scand J Gastroenterol [Suppl] 35:51–53
Giordano A, Adams G, Boies L Jr, Meyerhoff W (1981) Current management of esophageal foreign bodies. Arch Otolaryngol 107:249–251
Glauser J, Lilja GP, Greenfeld B, Ruiz E (1979) Intravenous glucagon in the management of esophageal food obstruction. JACEP 8:228–231
Glick G (1970) Comparison of the peripheral vascular effects of glucagon, norepinephrine, isoproterenol and dopamine. Clin Res 18:307
Gohel VK, Laufer I (1978) Double contrast examination of the postoperative stomach. Radiology 129:601–608
Gohel VK, Dalinka MK, Mandell GA, Azimi F (1974) Pharmacoradiology of the gastrointestinal tract. CRC Crit Rev Clin Radiol Nucl Med 5:69–110
Gohel VK, Dalinka MK, Coren GS (1975) Hypotonic examination of the colon with glucagon. Radiology 115:1–4
Gustafsson L, Hedqvist P, Lundgren G (1980) Pre- and postjunctional effects of prostaglandin $E_2$, prostaglandin synthetase inhibitors and atropine on cholinergic neurotransmission in guinea pig ileum and bovine iris. Acta Physiol Scand 110:401–411
Haase GM, Boles ET Jr (1979) Glucagon in experimental intussusception. J Pediatr Surg 14:664–669
Handal KA, Riordan W, Siese J (1980) The lower esophagus and glucagon. Ann Emerg Med 9:577–579

Harford FJ (1979) Use of glucagon in conjunction with the end-to-end anastomosis (EEA) stapling device for low anterior anastomoses. Dis Colon Rectum 22:452–454
Harned RK, Stelling CB, Williams S, Wolf GL (1976) Glucagon and barium enema examinations: a controlled clinical trial. Am J Roentgenol 126:981–984
Hecht HL, Hollenberg GM, Pradhan AR (1979) Glucagon-induced small intestinal hypotonia demonstrating bleeding lymphoma. Gastrointest Radiol 4:61–63
Hicks T, Turnberg LA (1974) Influence of glucagon on the human jejunum. Gastroenterology 67:1114–1118
Hillman BJ, Ovitt TW, Doering JW (1980) Improved pyeloureteral visualization using glucagon during experimental intravenous urography. Invest Radiol 15:313–317
Hogan WJ, Dodds WJ, Hoke SE, Reid DP, Kalkhoff RK, Arndorfer RC (1975) Effect of glucagon on esophageal motor function. Gastroenterology 69:160–165
Hogan WJ, Geenen JE, Schaffer RD, Dodds WJ, Stewart ET, Arndorfer RC (1977) Sphincter of Oddi pressure measurements in biliary tract disease before and after endoscopic sphincterotomy. Gastrointest Endosc 23:230
Hogan WJ, Geenen JE, Dodds WJ, Stewart ET (1979) Manometric evaluation of sphincter of Oddi motor function in humans. Invest Radiol 14:357
Hoke SE, Reid DP, Hogan WJ, Dodds WJ, Kalkhoff RK, Arndorfer RC (1972) The effect of glucagon on esophageal motor function. Clin Res 20:732
Hoy GR, Dunbar D, Boles ET Jr (1977) The use of glucagon in the diagnosis and management of ileocolic intussusception. J Pediatr Surg 12:939–944
Hradsky M, Furugård K (1976) Electrosurgical gastric polypectomy and duodenoscopy with the use of glucagon-Novo. Scand J Gastroenterol [Suppl] 38:54
Hradsky M, Stockbrügger R, Östberg H (1973) The effect of glucagon on gastric motility, the pylorus and reflux of bile into the stomach during gastroscopic examination. Scand J Gastroenterol [Suppl] 20:26
Hradsky M, Stockbrügger R, Dotevall G, Östberg H (1974) The use of glucagon during upper gastrointestinal endoscopy. Gastrointest Endosc 20:162
Hubel KA (1972) Effects of secretin and glucagon on intestinal transport of ions and water in rat. Proc Soc Exp Biol Med 139:656–658
Ishii H, Kamiya T, Oda Y, Okuno F, Takagi S, Aiso S, Yoshizawa M, Tsuchiya M (1978) The utilization of pancreatic glucagon in hypotonic duodenography (in Japanese). Jpn J Gastroenterol 75:990–996
Jaffer SS, Makhlouf GM, Schorr BA, Zeass AM (1974) Nature and kinetics of inhibition of lower esophageal sphincter pressure by glucagon. Gastroenterology 67:42–46
Jansson R, Steen G, Svanvik J (1978) A comparison of glucagon, gastric inhibitory peptide, and secretin on gallbladder function, formation of bile, and pancreatic secretion in the cat. Scand J Gastroenterol 13:919–925
Jarrett LN, Bell GD (1980) Effect of intravenous glucagon on the biliary secretion of a cholangiographic agent in man. Clin Radiol 31:657–661
Jennewein HM, Waldeck F, Siewert R, Weiser F, Thimm R (1973) The interaction of glucagon and pentagastrin on the lower esophageal sphincter in man and dog. Gut 14:861–864
Jerele JJ (1976) Use of glucagon as the hypotonic agent in barium enema examination. JAOA 76:264–271
Johansson H, Segerström A (1972) Glucagon and gastrointestinal motility in relation to thyroid-parathyroid function. Upsala J Med Sci 77:183–188
Jones RM, Fiddian-Green R, Knight PR (1980) Narcotic-induced choledochoduodenal sphincter spasm reversed by glucagon. Anesth Analg (Cleve) 59:946–947
Kaku K, Nagai I, Hongo T, Vakamudi AK, Meador JW, Akutsu T (1977) Peripheral vascular effects of glucagon, acetylcholine and dopamine with an implanted total artificial heart (TAH). Trans Am Soc Artif Intern Organs 23:611–616
Katon RM, Lee TG, Parent JA, Bilbao MK, Smith FW (1974) Endoscopic retrograde cholangiopancreatography (ERCP). Experience with 100 cases. Am J Dig Dis 19:295–306
Kelvin FM (1979) Double contrast examination of the upper gastrointestinal tract. South Med J 72:661–666

Kessler RE, Falkenstein DB, Clemett AR, Zimmon DS (1976) Indications, clinical value and complications of endoscopic retrograde cholangiopancreatography. Surg Gynecol Obstet 142:865–870
Keto P, Suoranta H, Ihamäki T, Melartin E (1979) Double contrast examination of the stomach compared with endoscopy. Acta Radiol [Diagn] (Stockh) 20:762–768
Kiil J, Andersen D, Weinreich J (1977) Relaxation of the antrum during endoscopy. A comparison between butyl scopolamine bromide (buscopan) and glucagon chloride (glucagon Novo) (in Danish). Ugeskr Laeger 139:1176–1177
Kirkpatrick RH, Wittenberg J, Schaffer DL, Black EB, Hall DA, Braitman DS, Ferrucci JT Jr (1978) Scanning techniques in computed body tomography. Am J Roentgenol 130:1069–1075
Kock NG, Darle N, Dotevall G (1967) Inhibition of intestinal motility in man by glucagon given intraportally. Gastroenterology 53:88–92
Kock NG, Roding B, Hahnloser P, Tibblin S, Schenk WG Jr (1970a) The effect of glucagon on hepatic blood flow. An experimental study in the dog. Arch Surg 100:147–149
Kock NG, Tibblin S, Schenk WG Jr (1970b) Hemodynamic responses to glucagon: an experimental study of central, visceral, and peripheral effects. Ann Surg 171:373–379
Kock NG, Tibblin S, Schenk WG (1971) Modification by glucagon of the splanchnic vascular responses to activation of the sympathicoadrenal system. J Surg Res 1:12–17
Kock H, Belohlavek D, Schaffner O, Tympner F, Rösch W, Demling L (1975) Prospective study for the prevention of pancreatitis following endoscopic retrograde cholangio-pancreatography (ERC). Endoscopy 7:221–224
Konturek SJ, Biernat J, Kwecien N, Olesky J (1975) Effect of glucagon on meal-induced gastric secretion in man. Gastroenterology 68:448–454
Kowalewsky K, O'Sullivan G, Kokodej A (1976) Effect of glucagon on myoelectrical and mechanical activity of the isolated homologous perfused porcine stomach. Pharmacology 14:115–124
Krarup N, Larsen JA (1974) The effect of glucagon on hepatosplanchnic hemodynamics, functional capacity, and metabolism of the liver in cats. Acta Physiol Scand 91:42–52
Krarup N, Larsen JA, Munck A (1975) Imitation of glucagon effects on splanchnic hemodynamics and liver function by $N^6$,2'-O-dibutyryl 3',5' cyclic AMP (DBc AMP) in cats. Acta Physiol Scand 95:110–116
Kreel L (1975) Pharmaco-radiology in barium examinations with special reference to glucagon. Br J Radiol 48:691–703
Kreel L (1979a) Glucagon in radiology. In: Picazo J (ed) Glucagon in gastroenterology. MTP Press, Lancaster, p 61
Kreel L (1979b) Patient preparation and management for computed tomography. In: Kreel L (ed) Medical imaging – CT, U/S, IS, NMR – a basic course. HM&M, Aylesbury Bucks, p 81
Kreel L, Bydder G (1980) Use of a portable syringe pump for glucagon administration in abdominal computed tomography. Radiology 136:507–508
Kuhns LR, Seigel R, Borlaza G, Rapp R (1979) Visualization of the longitudinal fold of the duodenum by computed tomography. J Comput Assist Tomogr 3:345–347
Kupcsulik P, Kokas P (1979) Ischemic damage of the liver. Part II. In vivo investigations of the prevention of the ischemic lesion of the liver. Acta Hepatogastroenterol (Stuttg) 26:284–289
Kuroda K, Baum S (1970) New advances in abdominal angiography. Surg Annu 2:113–143
Lanciault G, Fang W-G, Jacobson ED, Bowen JC (1976) Evaluation of potential agents for treatment of nonocclusive mesenteric ischemia in the dog. Circ Shock 3:239–246
Latshaw RF, Kadir S, Witt WS, Kaufman SL, White RI Jr (1981) Glucagon-induced choledochal sphincter relaxation: aid for expulsion of impacted calculi into the duodenum. AJR 137:614–616
Laufer I (1975) A simple method for routine double contrast study of the upper gastrointestinal tract. Radiology 117:513–518
Lawrence AM (1970) Glucagon in medicine: new ideas for an old hormone. Med Clin North Am 54:183–190

Lefèbvre P, Beaujean M (1964) Modifications circulatoires périphériques après administration de glucagon. Arch Int Physiol Biochim 72:9–16
Levinsky RA, Lewis RM, Bynum TE, Hanley HG (1975) Digoxin-induced intestinal vasoconstriction. The effects of proximal arterial stenosis and glucagon administration. Circulation 52:130–136
Lin TM (1974) Action of secretin, glucagon, cholecystokinin and endogenously released secretin and cholecystokinin on gallbladder and bile flow in dogs. Fed Proc 33:391
Lin TM (1975) Actions of gastrointestinal hormones and related peptides on the motor function of the biliary tract. Gastroenterology 69:1006–1022
Lin TM, Spray GF (1969) Effect of pentagastrin, cholecystokinin, caerulein, and glucagon on the choledochal resistance and bile flow of conscious dog. Gastroenterology 56:1178
Lin TM, Evans DC, Spray GF (1973) Mechanism studies of gastric inhibition by glucagon: failure of KCl and adrenergic blocking agents to prevent its action. Arch Int Pharmacodyn 202:314–324
Lindberg B, Darle N (1976) Effect of glucagon on hepatic circulation in the pig. Arch Surg 111:1379–1383
Liotta D (1955) Pour le diagnostic des tumeurs de pancreas: la duodenographie hypotonique. Lyon Chir 50:455–460
Lorenz-Meyer H, Dombrowski H (1976) Zur Wertigkeit der diagnostischen endoskopischen retrograden Cholangiopankreatikographie. Therapiewoche 26:781–784
Lowman RM, Belleza NA, Goetsch JB, Finkelstein HI, Berneike RH, Rosenfield AT (1977) Glucagon. J Urol 118:128
MacFerran SM, Mailman D (1977) Effects of glucagon on canine intestinal sodium and water fluxes and regional blood flow. J Physiol (Lond) 266:1–12
MacMahon H, Bowie JD, Beezhold C (1979) Erect scanning of pancreas using a gastric window. Am J Roentgenol 132:587–591
Madden JJ Jr, Ludewig RM, Wangensteen SL (1972) Effects of glucagon on the splanchnic and the systemic circulation. Rev Surg 29:372–373
Madsen JE, Boone WT, Livstone EM (1976) Endoscopic removal of a dental instrument from the stomach. Am J Gastroenterol 66:377–378
Marks HW, Lousteau RJ (1979) Glucagon and esophageal meat impaction. Arch Otolaryngol 105:367–368
McCarthy JD (1979) Biliary manometry as an investigative tool in biliary tract disease. In: Picazo J (ed) Glucagon in gastroenterology. MTP Press, Lancaster, p 95
McCune WS, Shorp PE, Moscowitz H (1968) Endoscopic cannulation of the ampulla of Vater. A preliminary report. Ann Surg 167:752–758
Meckeler KJH, Borow M (1975) Direct manometric study of the sphincter of Oddi in man. Gastroenterology 68:1038
Meeroff JC, Jorgens J, Isenberg JI (1975) The effect of glucagon on barium-enema examination. Radiology 115:5–7
Melsom M, Myren J, Larsen S, Moe A (1977) Comparison of glucagon and pethidine plus atropine as premedication for peroral endoscopy. Endoscopy 9:79–82
Merrill SL, Chvojka VE, Berkowitz GM, Texter EC Jr (1962) The effects of glucagon on the superior mesenteric vascular bed. Fed Proc 21:200
Miller RE (1973) Hypotonic duodenography with glucagon. Radiology 108:35–42
Miller RE, Chernish SM, Skucas J, Rosenak BD, Rodda BE (1974) Hypotonic colon examination with glucagon. Radiology 113:555–562
Miller RE, Chernish SM, Brunelle RL (1979) Gastrointestinal radiography with glucagon. Gastrointest Radiol 4:1–10
Mlecko LM (1974) Hypotonic duodenoscopy using glucagon. Gastroenterology 66:818
Morishima MS, Ghaed N (1979) Glucagon and diuresis in the treatment of ureteral calculi. Radiology 129:807–809
Morley KD, Campbell CD (1977) Initial experience with endoscopic retrograde cholangiopancreatography: the role in clinical practice. NZ Med J 86:169–173
Moseson MD, Hoexter B, Labow SB (1980) Glucagon, a useful adjunct in anastomosis with a stapling device. Dis Colon Rectum 23:25

Moss AA, Kressel HY, Korobkin M, Goldberg HI, Rohlfing BM, Brasch RC (1978) The effect of gastrografin and glucagon on CT scanning of the pancreas: a blind clinical trial. Radiology 126:711–714
Myren J (1979) The role of glucagon in different endoscopic procedures in the upper gastrointestinal tract. In: Picazo J (ed) Glucagon in gastroenterology. MTP Press, Lancaster, p 39
Nebel OT (1974) Endoscopic manometry: a new technique for the physiologic study of the human sphincter of Oddi. Gastroenterology 66:818
Nebel OT (1975a) Effect of enteric hormones on the human sphincter of Oddi. Gastroenterology 68:962
Nebel OT (1975b) Manometric evaluation of the papilla of Vater. Gastrointest Endosc 21:126–128
Necheles H, Sporn J, Walker L (1966) Effect of glucagon on gastrointestinal motility. Am J Gastroenterol 45:34–39
Norfleet NG (1978) Premedication for colonoscopy. Randomized double blind study of glucagon versus placebo. Gastrointest Endosc 24:164–165
Novak D, Probst P (1973) Glucagon versus propantheline bromide and hyoscine-N-butylbromide in tubeless hypotonic duodenography. IRCS Med Sci Libr Compend, August
Olsen UB (1977) Prostaglandin mediated natriuresis during glucagon infusion in dogs. Acta Endocrinol (Copenh) 84:429–438
Ominsky SH, Margulis AR (1981) Radiographic examination of the upper gastrointestinal tract. A survey of current techniques. Radiology 139:11–17
Op den Orth JO (1977) Tubeless hypotonic examination of the afferent loop of the Billroth II stomach. Gastrointest Radiol 2:1–5
Osteaux M, Struyven J, Huvenne R, Jeanmart L (1978) Liver. In: Baert A, Jeanmart L, Wackenheim A (eds) Clinical computer tomography, head and trunk. Springer, Berlin Heidelberg New York, p 169
Parving HH, Noer J, Kehlet H, Mogensen CE, Svendsen PAA, Heding LG (1977) The effect of short-term glucagon infusion on kidney function in normal man. Diabetologia 13:323–325
Paul F (1974) Quantitative Untersuchungen der Wirkung von Pankreas Glukagon und Secretin auf die Magen-Darm-Motorik mittels elektromanometrischer Simultanregistrierungen beim Menschen. Klin Wochenschr 52:983–989
Paul F (1976) Intravenöse Langzeitinfusion von Glukagon zur Abtreibung von Gallenwegskonkrementen. Fortschr Endoskopie 161–163
Paul F (1979) The role of glucagon in the treatment of biliary tract pathology. In: Picazo J (ed) Glucagon in gastroenterology. MTP Press, Lancaster, p 107
Paul F, Freyschmidt J (1974) Zur Anwendung von kristallinem Pankreas-Glukagon bei endoskopischen und röntgenologischen Untersuchungen des Intestinaltrakts. Z. Gastroenterol 12:614–615
Paul F, Misaki F, Seifert E (1973) Crystalline pancreatic glucagon – a new spasmolytic agent: results of comparative endoscopic and electromanometric investigations in the proximal gastrointestinal tract. Endoscopy 5:199–204
Pickard R (1974) Glucagon in management of perforated esophagus. Br Med J 4:232
Pillary G, Bank S, Katzka I, Fulco JD (1979) Meat bolus impaction of the lower esophagus associated with a paraesophageal hernia. Successful noninvasive treatment with intravenous glucagon. Am J Gastroenterol 71:287–289
Pinckney LE, Currarino G (1980) Excretory urography with glucagon. Radiology 135:513–514
Pire JC, Burde A, Flament JB, Rives J (1976) Utilisation des morphinomimétiques en chirurgie biliaire. Nouv Presse Med 5:1838
Pon MS, Cooperberg PL (1979) Oral water and intravenous glucagon – to aid ultrasonic visualization of the common bile duct. J Can Assoc Radiol 30:173–174
Proctor HJ, Wood JJ, Palladino WG (1980) The effect of glucagon on hepatic cellular energetics during a low flow state. Surgery 87:369–374
Quirarte C, Woodward ER, Dragstedt LR (1966) Glucagon and inhibition of gastric secretion. Arch Surg 93:475–479

Qvigstad T, Larsen S, Myren J (1979) Comparison of glucagon, atropine, and placebo as premedication for endoscopy of the upper gastrointestinal tract. Scand J Gastroenterol 14:231–235
Rabinov KR, Simon M (1965) Peroral cannulation of ampulla of Vater for direct cholangiography and pancreaticography. Preliminary report of a new method. Radiology 85:693–697
Rasmussen H (1976) Ions and second messengers. In: Weissman G, Clayborne R (eds) Cell membranes, biochemistry, cell biology and pathology. HP Books, Tucson, p 203
Ratzmann KP, Knoke M (1974) Glukagon und seine Wirkung auf den Gastrointestinaltrakt. Z Inn Med 29:94–97
Reddy AN (1981) Noninvasive management of esophageal meat impaction. Gastrointest Endosc 27:202–203
Redman HC, Federal WA, Castellino RA, Flatstein E (1977) Evaluation of normal and abnormal lymph nodes at computerised tomographic scanning of the abdomen and pelvis. In: Du Boulay GH, Mosely IF (eds) The first European seminar on computerized axial tomography in clinical practice. Springer, Berlin Heidelberg New York, p 406
Rentsch I, Gärtner U, Müller P, Kerk L, Klatte E, Lange G, Pintarelli F (1976) Die ERCP – ein neues Routineverfahren in der Oberbauchdiagnostik? Bremer Aerzteblatt 29/6
Ribeiro BF, Cotton PB, Dicanari JB, Roberts M, Laurence B (1977) Duodenoscopic manometry of the bile duct and sphincter of Oddi. Gut 18:A407
Richardson PDI, Withrington PG (1976) The vasodilator actions of isoprenaline, histamine, prostaglandin $E_2$, glucagon and secretin on the hepatic arterial vascular bed of the dog. Br J Pharmacol 57:581–588
Richardson PDI, Withrington PG (1977) The effects of glucagon, secretin, pancreozymin and pentagastrin on the hepatic arterial vascular bed of the dog. Br J Pharmacol 59:147–156
Risueño Alvarez JC, Sánchez de Vega D (1979) Acalasia. Rev Clin Esp 6:15–21
Rösch W (1974) LES response to glucagon and zinc protamine glucagon in achalasia. 5th World congress of gastroenterology, Mexico City, 13–19 October
Rösch W, Kock H, Demling L (1976) Manometric studies during ERCP and endoscopic papillotomy. Endoscopy 8:30–33
Rudo ND, Rosenberg IH (1973) Chronic glucagon administration enhances intestinal transport in the rat. Proc Soc Exp Biol Med 142:521–525
Schmarsow R (1976) Der pankreatographische Effekt bei der Pankreasangiographie nach Verabreichung von Glukagon. Fortschr Roentgenstr 124:310–314
Schwaiger M, Fondacaro JD, Jacobson ED (1979) Effects of glucagon, histamine, and perhexiline on the ischemic canine mesenteric circulation. Gastroenterology 77:730–735
Scott LD, Summers RW (1976) Correlation of contractions and transit in rat small intestine. Am J Physiol 230:132–137
Shirakabe H (1971) Double contrast studies of the stomach. Bunkodo, Tokyo
Shoemaker WC, Van Itallie TB, Walker WF (1959) Measurement of hepatic glucose output and hepatic blood flow in response to glucagon. Am J Physiol 196:315–318
Siegel JH (1979) Endoscopic management of choledocholithiasis and papillary stenosis. Surg Gynecol Obstet 148:747–752
Siewert R, Früh E, Waldeck F (1973) Senkung des Druckes im unteren Ösophagussphinkter bei der Achalasie durch Glucagon. Dtsch Med Wochenschr 98:2045–2046
Silvis SE, Vennes JA (1975) The role of glucagon in endoscopic cholangiopancreatography. Gastrointest Endosc 21:162–163
Soehendra N (1977) Technik, Schwierigkeiten und Ergebnisse der endoskopisch-retrograden Cholangio-Pankreatikographie (ERCP). Chirurg 48:98–104
Stern AI, Korman MG, Hansky J, Littlejohn G, Schmidt GT (1979) Comparison of buscopan or glucagon in upper gastrointestinal endoscopy. Annual scientific meeting of the gastroenterological society of Australia, Brisbane, 14–15 May
Stickney JC, Northrup DW, Van Liere EJ (1958) Effect of glucagon on propulsive motility of rat small intestine. Fed Proc 17:157
Stoddard CJ, Duthie HL (1976) Effect of vagotomy on the response of gastric myoelectrical activity to glucagon and food. Scand J Gastroenterol [Suppl] 42:77–83

Stowe NT, Hook JB (1970) Role of alterations in renal hemodynamics in the natriuretic action of glucagon. Arch Int Pharmacodyn 183:429–438
Struyven J, Osteaux M, Huvenne R, Jeanmart L (1978) Kidneys. In: Baert A, Jeanmart L, Wackenheim A (eds) Clinical computer tomography, head and trunk. Springer, Berlin Heidelberg New York, p 213
Stunkard AJ, Van Itallie TB, Reis BB (1955) The mechanism of satiety: effect of glucagon on gastric hunger contractions in man. Proc Soc Exp Biol Med 89:258–261
Taylor I, Duthie HL, Cumberland DC, Smallwood R (1975) Glucagon and the colon. Gut 16:973–978
Thoeni RF, Margulis AR (1978) The state of radiographic technique in the examination of the colon: a survey. Radiology 127:317–323
Tibblin S, Kock NG, Schenk WG Jr (1970) Splanchnic hemodynamic responses to glucagon. Arch Surg 100:84–89
Tibblin S, Kock NG, Schenk WG (1971) Response of mesenteric blood flow to glucagon. Influence of pharmacological stimulation and blockade of adrenergic receptors. Arch Surg 102:65–70
Tittobello A, Testoni PA, Evangelista A, Ballarin E (1979) Diagnostic value of endoscopic retrograde cholangiography and pancreatography in comparison with percutaneous transhepatic cholangiography. In: Gentilini P, Popper H, Sherlock S, Teodori U (eds) Problems in intrahepatic cholestasis. Karger, Basel, p 191
Treffot M-J, Quilichini F, Vinson M-F (1979) Biliary surgery, radiomanometry and glucagon. In: Picazo J (ed) Glucagon in gastroenterology. MTP Press, Lancaster, p 87
Ueda J, Hitoshi N, Mizuo M, Youichi A (1977a) Effects of glucagon on the renal hemodynamics of dogs. Eur J Pharmacol 41:209–212
Ueda J, Hitoshi N, Mizuo M, Youichi A (1977b) Effects of glucagon on the renal hemodynamics in the dog. Jpn Circ J 4:991–998
Ulano HB, Treat E, Shanbour LL, Jacobson ED (1972) Selective dilation of the constricted superior mesenteric artery. Gastroenterology 62:39–47
Vagne M, Troitskaja V (1976) Effect of secretin, glucagon and VIP on gallbladder contraction. Digestion 14:62–67
Valenzuela JE (1976) Effect of intestinal hormones and peptides on intragastric pressure in dogs. Gastroenterology 71:766–769
Vela Navarrete R, García Sacristán A, Illera M, González Azpeitia JA, Jiménez J (1979) Efecto del glucagon sobre la motilidad ureteral: estudios in vitro e in vivo. Rev Clin Esp 155:351–354
Vinson M-F, Treffot M-J, Quilichini F (1977) Radiomanométrie biliaire. Intérêt du glucagon. Nouv Presse Med 6:2897
Violon D, Potvliege R (1980) Double contrast study of the upper digestive tract with the use of glucagon. J Belge Radiol 63:581–588
Violon D, Steppe R, Potvliege R (1981) Improved retrograde ileography with glucagon. Am J Roentgenol 136:833–834
Visnovský P (1976) Effect of corticosterone, glucagon and growth hormone on gastrointestinal propulsive motility in rats (in Czech). Bratisl Lek Listy 65:394–399
Vondrasek P, Eberhardt G (1974) Endoskopische Druckmessungen mittels Halbleitertechnik. Z Gastroenterol 12:453–458
Warner WA, Begley L, Penman RW (1971) Effect of glucagon on pulmonary airway resistance in dogs. Fed Proc 30:2010
Weighall SL, Wolfman NT, Watson N (1979) The fluid-filled stomach: a new sonic window. J Clin Ultrasound 7:353–356
Whalen GE, Wu WC, Ganeshappa KP, Wall MJ, Kalkhoff RK, Soergel KH (1973) The effect of endogenous glucagon on human small bowel function. Gastroenterology 64:822
Wingate DL, Pearce E (1979) The physiological role of glucagon in the gastrointestinal tract. In: Picazo J (ed) Glucagon in gastroenterology. MTP Press, Lancaster, p 19
Wingate DL, Morris D, Thomas PA (1977) Glucagon stimulates intestinal motor activity. Gut 18:A966–A967
Ziegels Ch, Jacquet N, Whalen C, Delforge M (1979) Opacification par voie rétrograde des canaux biliaires et pancréatiques. Rev Med Liege 34:928–936

CHAPTER 55

# Glucagon in the Diagnosis and Treatment of Hypoglycaemia

V. MARKS

## A. Introduction

Two of glucagon's many pharmacological properties are utilised clinically for the diagnosis and treatment of hypoglycaemia. They are its ability: (1) to raise the blood glucose concentration by increasing hepatic glycogenolysis; and (2) to stimulate insulin secretion by direct action on pancreatic B-cells. Though still widely used for the treatment of acute episodes of iatrogenic hypoglycaemia in insulin-dependent subjects, the advent of more specific and more effective treatments has reduced almost to zero the usefulness of glucagon as a therapy for chronic hypoglycaemia. Similarly, the availability of improved diagnostic procedures for hyperinsulinism, glycogen storage disease and the various endocrinopathies associated with, and causative of, spontaneous hypoglycaemia has reduced the usefulness of glucagon in the differential diagnosis of these disorders to that of an adjunct rather than a key.

## B. Normal Response to Glucagon

### I. Blood Glucose

The intravenous or intramuscular administration of glucagon after an overnight fast to healthy children and adults is normally followed, within a few minutes, by a rise in blood glucose concentration. This is substantially larger in arterial (and capillary) than in venous blood owing to increased peripheral assimilation of glucose under the influence of insulin liberated from the pancreas as a result of the dual stimulatory effects of hyperglucagonaemia and hyperglycaemia. Maximum blood glucose levels are generally achieved within 15–30 min and occur somewhat earlier after intravenous than after intramuscular injection. Few normal subjects have a smaller than 2 mmol/l rise in venous blood glucose concentration under these circumstances. More prolonged fasting results in a progressively smaller glycaemic response which approximates zero (HARO et al. 1965) after 48–96 h without food (shorter in children) but which returns, almost to prestarvation levels, if fasting is continued (ARKY et al. 1970).

### II. Plasma Insulin

Plasma insulin levels rise immediately following the rapid intravenous injection of glucagon in healthy children and adults and reach their maximum value within 3–

10 min of commencing the injection (SAMOLS and MARKS 1966; MARRI et al. 1966; MARKS 1971). The rise in plasma insulin commences before any change in blood glucose concentration becomes apparent and is not determined by it. A rise in plasma insulin concentration of less than 30 or more than 100 mIU/l, following the rapid intravenous injection of glucagon 30 µg/kg body weight (to a maximum of 1 mg) occurs in less than 10% of healthy overnight fasted subjects. Like that of blood glucose, the insulinaemic response to glucagon is reduced by starvation but, unlike the former, is exaggerated by obesity and certain pharmacological agents, e.g. sulphonylureas and glucocorticoids.

## C. Glucagon in the Treatment of Hypoglycaemia

Glucagon owes its place in the treatment of hypoglycaemia exclusively to its ability to liberate glucose from the liver by initiating glycogenolysis through activation of liver phosphorylase. Its ability to accelerate gluconeogenesis, which is probably more important in glucose homeostasis, plays no part in this action, nor does glucagon have any effect upon peripheral glucose utilisation except possibly to accelerate it secondarily to glucagon-stimulated insulin secretion (SAMOLS et al. 1965).

Though once considered to be largely, if not entirely, mediated by cyclic AMP (cAMP), promotion of glycogenolysis by glucagon is now thought to be, at least in part, independent of cAMP (GRILL et al. 1979) though the exact mechanism is not understood (see also Chap. 14). The hyperglycaemic effect of glucagon is abolished or diminished when, for any reason, the quantity of glycogen in the liver is reduced or is otherwise unavailable for conversion into glucose. Conversely, it is exaggerated when the quantity of glycogen in the liver is increased, either by previous ingestion of a high carbohydrate diet or by long-continued and unopposed insulin action (FINEGOLD et al. 1980).

Glucagon, unlike adrenaline (epinephrine), to which it was once likened, has no effect upon muscle phosphorylase and cannot, therefore, assist in the transference of carbohydrate from the much larger stores of glycogen that are present in skeletal muscle – from which they cannot, because of the lack of glucose-6-phosphatase, be liberated as glucose – to the liver, where they can. It is probably for this reason, as well as through its ability to suppress insulin secretion, that adrenaline enhances the hyperglycaemic effect of glucagon when the two hormones are given together (VAN ITALLIE and BENTLEY 1955).

The hyperglycaemic effect of exogenous glucagon is generally poorly maintained during continuous intravenous infusions (SAMOLS et al. 1966) and this is, probably only in part, due to increased insulin secretion from the subject's own B-cells suppressing hepatic glucose release and increasing its utilisation peripherally. Experiments by RIZZA and GERICH (1979) on normal human volunteers suggest that it is, in part, due to a decrease in hepatic responsiveness to glucagon whose persistence is, however, shown by the fall in hepatic glucose output that promptly ensues when the glucagon infusion is stopped.

## I. Insulin-Induced Hypoglycaemia

### 1. Insulin Coma Therapy

Glucagon was used to terminate insulin-induced hypoglycaemia in patients undergoing SAKEL's therapy for schizophrenia within a few years of its becoming available for clinical trial. SCHULMAN and GREBEN (1957) were the first to describe their experiences. They gave intravenous, subcutaneous or intramuscular injections of crystalline glucagon in doses ranging from 1 to 15 mg to patients rendered unconscious by large doses of insulin. They concluded that glucagon was more readily regulated and less cumbersome to use than either intravenous or gavage glucose for restoring consciousness and was most effective when given intravenously at a dose of 0.2 mg/kg body weight. Subsequent authors described equally good or better results with considerably smaller doses of glucagon. ESQUIBEL et al. (1958), for example, used 5 mg crystalline glucagon in 5 ml saline given intramuscularly to interrupt no less than 2,475 insulin-induced comas in 63 psychiatric patients without mishap. Recovery from coma was slower with glucagon than with intravenous glucose and considered to be more acceptable by the patients. Further reports of the efficacy of glucagon in terminating therapeutically induced hypoglycaemia continued to appear (BRAUN and PARKER 1959; LAQUEUR and LABURT 1960; WAIFE 1960; ARIEFF et al. 1960) right up to the time of its abandonment in favour of more easily administered and effective treatments for schizophrenia and other psychiatric disorders. All authors agreed that following restoration of consciousness by glucagon it was essential to ensure that the patient ate in order to prevent a relapse into hypoglycaemic coma which was, probably correctly, attributed to continued absorption of insulin from the injection site and its persistant action in the tissues.

### 2. Accidental Hypoglycaemia

In contrast to the seemingly uniformly successful application of glucagon to restoration of consciousness in patients undergoing insulin coma therapy, its use in diabetic patients is not always attended by such prompt recovery from hypoglycaemia. ELRICK et al. (1958) treated 41 patients, including several diabetics, with either 1 or 2 mg of glucagon by intramuscular or intravenous injection and observed complete recovery from hypoglycaemic coma within 5–20 min in all but one of them. Though similarly favourable results were published by SHIPP et al. (1964) and by HAUNZ (1962), all of whom studied children holidaying in a summer camp for diabetic children, this was not the experience of MACCUISH et al. (1970). They gave glucagon, 1–2 mg, either by intravenous or intramuscular injection, to 100 consecutive diabetic patients who had arrived at the Accident and Emergency Department of the Royal Infirmary in Edinburgh, and who were both hypoglycaemic and unable to take glucose by mouth. Of the 100 patients, 41 regained consciousness sufficiently to allow them to swallow a glucose solution within 30 min of the first glucagon injection; the remaining 59 were given 25 g glucose intravenously (as a 50% solution), 36 were awake within 15 min and another 4 responded to a second injection of glucose. Despite these measures, 19 patients remained unconscious, though all but 2 eventually recovered after further treatment with glucose, steroids and intravenous mannitol (HOFFBRAND and SEVITT 1966; MACCUISH et al. 1970). Patients who failed to benefit from glucagon showed a mean rise in blood glucose concentration of less than 0.5 mmol/l and in none of them did the blood glucose level exceed 2.2 mmol/l before intravenous glucose therapy was administered. These patients differed, as a group, from those who recovered in having been unconscious for slightly longer (an average of 60 min compared with 49 min) before treatment was begun, but the overlap was enormous. There was no discernible difference in either the type or quantity of insulin used by the glucagon-responsive and non-glucagon-responsive patients, nor, in those who did recover, was there any perceptible difference in the rate of response to intravenous or intramuscular glucagon administration. In patients in whom glucagon produced benefit, 1 mg was as effective as 2 mg in achieving hyperglycaemia and restoring consciousness and produced fewer undesirable side effects such as nausea, vomiting and abdominal pain.

Reasons for the differences in behaviour between the patients studied by MACCUISH et al. (1970) and those investigated by others, include the longer period of unconsciousness suffered before treatment with glucagon was begun and their seemingly greatly reduced hepatic glycogen stores. This may possibly have been due to their patients' too rigid adherence to a low carbohydrate diet and which may, itself, have contributed to their precipitation into coma under the influence of their usual dose of insulin.

In view of the continued and widespread use of glucagon as a therapeutic agent for interrupting accidental hypoglycaemic coma (GIBBS et al. 1958; HAUNZ 1962; SHIPP et al. 1964) it is somewhat surprising that further controlled clinical trials of its effectiveness have not been carried out. Nevertheless, it is possible, on the basis of what evidence is available, to recommend the use of glucagon, 1 mg by intramuscular injection, as the drug of choice for treating iatrogenic hypoglycaemic coma. It is safe, easily administered and, at worst, causes only a slight delay in restoring normoglycaemia with intravenous glucose. Only when glucagon fails to produce sufficient restoration of consciousness in 20–30 min to permit glucose to be taken by mouth are intravenous glucose and other resuscitative measures necessary.

## II. Sulphonylurea-Induced Hypoglycaemia

Sulphonylurea-induced hypoglycaemia tends to occur mainly in elderly diabetic subjects with impaired renal function and receiving treatment with long-acting compounds such as chlorpropamide. In younger subjects, it occurs almost exclusively in those who have taken a sulphonylurea with suicidal intent or have been given chlorpropamide for the treatment of diabetes insipidus. In either event, the drug accumulates in the body and hypoglycaemia, once it has appeared, is notoriously difficult to treat, requiring prolonged hyperglycaemic therapy (BERGMAN 1965).

Studies on patients who had attempted suicide with sulphonylureas led CREUTZFELDT et al. (1969) to postulate that impaired hepatic glucose release, rather than increased peripheral utilisation of glucose, might be responsible for the hypoglycaemia and that glucagon could have a unique role to play in the treatment of sulphonylurea-induced hypoglycaemia. DAVIES et al. (1967), on the basis of their experience with a 17-year-old girl who had deliberately taken between 5.0 and 7.5 g chlorpropamide, concluded that "glucagon should always be given in addition to glucose in the treatment of severe hypoglycaemia induced by sulphonylurea drugs" since without it they had been unable to maintain normoglycaemia despite giving glucose constantly by intravenous infusion. The contrary view was expressed by MARRI et al. (1968) who described collapse due to hypoglycaemia in a 65-year-old woman within 30 min of receiving 1 mg glucagon intravenously. Until then, she had been receiving treatment, with limited success, by intermittent intravenous glucose infusions for sulphonylurea- (chloropropamide)-induced coma. The sudden fall in blood glucose concentration that followed glucagon injection in their patient was accompanied by loss of consciousness and collapse and was attributed to a massive release of endogenous insulin in response to glucagon. A peak plasma insulin concentration of 1,080 mIU/l was reached within 10 min of the glucagon injection and hyperinsulinaemia persisted for at least a further 30 min. No such dramatic increase in plasma insulin was observed following glucagon administration to a 16-year-old boy with chlorpropamide-induced hypoglycaemia (JOHNSON et al. 1977) who responded favourably to treatment with diazoxide.

Though other investigators (FREY and ROSELUND 1970; CODACCIONI et al. 1971; DOWELL and IMRIE 1972; ALARIC 1973; FORREST 1974; FORMAN et al. 1974) have reported their experience with glucagon for the treatment of sulphonylurea-induced hypoglycaemia, no clear consensus has emerged as to the benefit, or otherwise, of including it in the therapeutic regimen which consists essentially of providing glucose intravenously in sufficient quantity to maintain the blood glucose concentration within the range 5–10 mmol/l. The amount of glucose required to achieve this is often much larger than uninformed opinion might suppose, but certainly no greater than is sometimes required to maintain normoglycaemia in patients harbouring non-insulin-secreting, hypoglycaemia-producing neoplasms, some of whom have been known to need up to 1,500 g/day intravenous glucose.

The availability of diazoxide – a specific antagonist to sulphonylurea-induced insulin secretion (MARIANI and LOUBATIERES 1972) – as an antidote to sulphonylurea-induced hypoglycaemia (PFEIFFER et al. 1976; JOHNSON et al. 1977; JACOBS et al. 1978; PFEIFFER et al. 1978) – has reduced further discussion of the role of glucagon in this condition to one of acedemic interest only.

## III. Spontaneous Hypoglycaemia

### 1. Recovery from Coma

Glucagon can sometimes be used to restore consciousness sufficiently to permit the ingestion of a glucose-containing drink in patients with spontaneous hypoglycaemia. It is, however, effective only in those varieties of hypoglycaemia in which liver glycogen stores are both preserved and available for conversion into glucose. In practice, this restricts its clinical usefulness mainly to cases of hyperinsulinism, hypoglycaemia due to extrapancreatic neoplasms and factitious hypoglycaemia. The information obtained by observing the blood glucose and plasma insulin responses to glucagon injection may, however, throw light on the pathogenic mechanism involved, and thereby prove diagnostically useful at the cost of little more than a transient delay in effecting recovery by injecting glucose intravenously.

## 2. Treatment and Prevention of Recurrent Attacks

There have been several attempts to use continuous intravenous infusions or intermittent, intramuscular injections of either regular or specifically prepared glucagon formulations for the treatment of intractable hypoglycaemia. These are discussed in the following sections.

### a) Hyperinsulinism

α) *Insulinoma.* LANDAU et al. (1958) treated a 28-year-old obese woman with hypoglycaemia, caused by a disseminated islet cell carcinoma, with glucagon continuously over a 55-day period. During this time she received a total of 900 mg glucagon by intravenous infusion in addition to glucose, the daily requirement for which rose from 336 to 720 g during the course of her terminal illness. Whenever glucagon was omitted from the infusion, the requirement for glucose increased by about 15%–30%. Despite the large dose of glucagon employed, the patient experienced no episodes of nausea, vomiting, decreased intestinal motility, electrolyte imbalance or other harmful reactions attributable to glucagon administration. BECK-CHRISTIANSEN (1964) used subcutaneous injections of crystalline glucagon five times daily, increasing to eight times daily, to maintain tolerable normoglycaemia in a 50-year-old woman with metastatic insulinoma over an 18-month period. LJUNGSTRÖM (1964) used 3–4 daily injections, each containing 0.5 mg crystalline glucagon, successfully to treat hypoglycaemia of unknown aetiology in a 71-year-old woman. On this regime, which was coupled with a low carbohydrate and high protein diet, the patient lost 24 kg in weight and improved clinically. Although there was a modest rise in blood glucose concentration on this regime, the possibility that cerebral adaptation to the utilisation of ketones as an alternative fuel might have contributed to her clinical improvement cannot be dismissed.

A zinc glucagon preparation, which was never marketed commercially, was used, with limited success, by ROTH et al. (1966) to treat two patients with refractory hypoglycaemia. In one, a benign occult insulinoma was eventually removed from the pancreatic remnant left behind after three previously unsuccessful operations; in the second patient, no cause was found. Batches of zinc glucagon, given as a 5 mg dose by intramuscular injection, differed markedly in their hyperglycaemic potential. Some were quite ineffective whilst others retained their hyperglycaemic properties throughout the trial. WEINGES (1959) used a similar zinc glucagon preparation to treat a patient with metastatic insulinoma and achieved some alleviation of hypoglycaemia, but was unable to produce persistent normoglycaemia.

β) *Nesidioblastosis.* Various infants suffering from what is now recognised as nesidioblastosis have been treated with zinc glucagon preparation in an attempt to alleviate their more or less constant hypoglycaemia, usually with limited success. In the child described by FRASIER et al. (1965) severe rebound hypoglycaemia regularly occurred 5–8 h after the last injection of glucagon whether it was given as crystalline glucagon in oil or as zinc glucagon. A more successful outcome was described following the use of zinc glucagon by KUSHNER et al. (1963) in a preliminary report of a 15-month-old child who was unresponsive to adrenocorticotropic hormone. Follow-up in this case was limited to 3 months and was not much longer in a similar case reported by ROSENBLOOM et al. (1966). It is questionable, therefore, in the absence of further reports, whether the initial beneficial effects were maintained. There is, however, one report (NABBEN 1968) of zinc glucagon having been used successfully for 4 years in a child with leucine-sensitive nesidioblastosis in whom it maintained its clinical effectiveness until spontaneous remission occurred.

### b) Glucagon Deficiency

Although glucagon deficiency (see also Chap. 42) has been postulated (MCQUARRIE 1954; MCQUARRIE et al. 1950; GROLLMAN et al. 1964; GOTLIN and SILVER 1970; WAGNER et al. 1969), on the basis of unreliable histological methods for the investigation of pancreatic islet morphology, as the probable basis of many cases of hypoglycaemia in childhood, this is now known not to be so. The majority of children suffering from what was once called "idiopathic hypoglycaemia of childhood" (MCQUARRIE 1954) have either nesidioblastosis (YAKOVAC et al. 1971) or some variety of ketotic hypoglycaemia (COLLE and ULSTRÖM 1964; CORNBLATH and SCHWARTZ 1976).

More recently, VIDNES and OYASAETER (1977) and KOLLEE et al. (1978) have described cases of hypoglycaemia which they attributed to glucagon deficiency, but only in the former case was the evidence at all persuasive; in the latter, nesidioblastosis seemed more likely. Both patients responded well to twice daily injections of glucagon – given mainly as a long-acting zinc glucagon preparation – in doses of 0.4 mg (VIDNES and OYASAETER 1977) or 3.5 mg (KOLLEE et al. 1978) at a time, for as long as they were followed up, which was 2 and 16 months, respectively. No further cases of hypoglycaemia attributed to glucagon deficiency have been reported in detail.

### c) Extrapancreatic Neoplasm

Glucagon is generally quite effective in raising the blood glucose concentration and improving cerebration in patients unconscious from hypoglycaemia caused by extrapancreatic neoplasms (MARKS and SAMOLS 1966) providing that they are not also malnourished (MARKS et al. 1974) and/or suffering from severe liver disease. The rise in blood glucose concentration produced in response to glucagon is, however, only rarely accompanied by a corresponding rise in plasma insulin concentration which, when it does occur, is invariably smaller than normal (FRIEND and HALES 1965; MARKS and SAMOLS 1966; PAULLADA et al. 1968; FRERICHS et al. 1970; COLWELL and WILBER 1971; CHOWDHURY and BLEICHER 1973). The use of repeated glucagon injections to reduce the frequency and severity of hypoglycaemic attacks in a 68-year-old patient with primary carcinoma of the liver was described by MUHE et al. (1969), but there have been no further reports of its use for this purpose.

d) Glycogen Storage Disease

Early expectations (ROSSI 1959; SOKAL et al. 1962; SPELLBERG 1969) that glucagon might prove useful in the long-term treatment of hypoglycaemia due to glycogen storage disease type I (von Gierke's disease) have not been fulfilled and, in view of the associated hyperlactacidaemia, its use for this purpose has been abandoned.

### IV. Current Status of Glucagon for Treatment of Hypoglycaemia

Glucagon deservedly enjoys a special position in the treatment of acute neuroglycopenia caused by inadvertent insulin overdosage. It can be administered in a supramaximum dose of 1 mg by simple intramuscular injection and can, for this reason, be given safely, by relatively unskilled personnel after only minimal instruction. It is, therefore, ideally suited for use by parents of diabetic children or the relatives and friends of more mature diabetic patients who, for one reason or another, are liable to suffer from unheralded hypoglycaemic attacks. Recovery of consciousness sufficiently to permit glucose to be taken safely by mouth is usually achieved within 15–30 min of the injection, in all but the most refractory cases, which generally require more extensive resuscitative measures. There is no advantage in giving glucagon intravenously over giving it intramuscularly (TAYLOR et al. 1978) though the subcutaneous route should be avoided if possible because of the unpredictability of absorption.

Recovery of consciousness produced by glucagon should ordinarily be followed by the ingestion of a meal. Glucagon generally fails to produce a hyperglycaemic response in patients whose hypoglycaemia is caused by conditions other than hyperinsulinism or an extrapancreatic neoplasm and the presence of ketonuria or ketonaemia is a contraindication to glucagon therapy. In particular, glucagon is contraindicated for the treatment of alcohol-induced hypoglycaemia in children in whom the delay in restoring blood glucose concentrations to normal by intravenous glucose may have serious consequences (MADISON 1968). There appears to be no place for glucagon in the long-term treatment of hypoglycaemia from any of the known causes of spontaneous or iatrogenic hypoglycaemia except possibly that due to genuine glucagon deficiency (VIDNES and OYASAETER 1977) if and when that condition can be shown to exist.

## D. Glucagon in the Diagnosis of Hypoglycaemia

### I. Differential Diagnosis of Hypoglycaemia

The ability of glucagon to produce a rise in blood glucose concentration when given to normal fasting human subjects, including children, and its inability to do so in patients with severe liver disease (VAN ITALLIE and BENTLEY 1955; ELRICK et al. 1957) led to its early introduction as a diagnostic test for liver disease in general (VAN ITALLIE 1956) and of glycogen storage disease in particular (PINCUS and RUTMAN 1953; HUBBLE 1954; CARSON and KOCH 1955; BENEDETTI 1971; BENEDETTI and KOLB 1966). The more recent application of glucagon to the diagnosis of hypoglycaemia is based on its insulinotropic properties (SAMOLS et al. 1965, 1966; MARKS and SAMOLS 1968a) but its use for this purpose has lessened as tests of islet cell function based upon suppression rather than stimulation of B-cell activity have become available (HORWITZ and RUBENSTEIN 1974; TURNER 1976; TURNER and HEDING 1977; SERVICE et al. 1977; SCARLETT et al. 1977; BRENNAN et al. 1980; MARKS and ROSE 1981).

## 1. In Adults

MARKS (1960) observed that patients with hypoglycaemia due to insulinoma behaved in a characteristic manner in response to exogenous glucagon and suggested that this might be due to direct stimulation of insulin secretion by glucagon (MARKS 1962). This suggestion was subsequently confirmed by SAMOLS et al. (1965) and others (LANGS and FRIEDBERG 1965; TURNER and MACINTYRE 1966; LEFEBVRE and LUYCKX 1966; PORTE et al. 1966; DEVRIM and RECANT 1966; BENEDETTI et al. 1967) using a variety of techniques both in vivo and in vitro.

The original intramuscular test for insulinoma, which was based solely upon the blood glucose response and depended upon the appearance of a hypoglycaemic rebound 90–150 min after the injection of glucagon (ALIVISATOS and MCCULLAGH 1955; MARKS 1960; MARRACK et al. 1961), was displaced by the introduction of an intravenous test (MARKS and SAMOLS 1968a) some years later. This proved equally, or more, reliable (SAMOLS and MARKS 1966; MCKIDDIE et al. 1969; MARKS and SAMOLS 1969; KHURANA et al. 1971; KUMAR et al. 1974; OHNEDA et al. 1975) than most other provocative tests, i.e. tolbutamide, L-leucine and glucose tolerance tests, for the diagnosis of insulinoma apart from the prolonged fast test and the measurement of plasma C-peptide levels during spontaneous or insulin-induced hypoglycaemia (MARKS and ROSE 1981).

Typically, insulinoma patients are hypoglycaemic after an overnight fast and their blood glucose level rises normally or supranormally – depending on whether venous or capillary (arterial) blood glucose levels are measured – in response to the rapid intravenous or intramuscular injection of 1 mg glucagon. Peak blood glucose levels occur 20–40 min after the injection, fall gradually thereafter, usually to well below the fasting value, and are associated with the appearance of acute neuroglycopenic symptoms. Plasma insulin and C-peptide levels are usually inappropriately high at commencement of the test and show an excessive and prolonged rise in response to glucagon in roughly 70% of insulinoma patients, the majority of the non-responding tumours belonging to histological categories III and IV, i.e. the mainly malignant and poorly differentiated ones (CREUTZFELDT et al. 1973, 1976).

In contrast to normal subjects who have been rendered artificially hypoglycaemic by insulin and show either a greatly reduced or absent insulinaemic response to glucagon (GOLDFINE et al. 1972), insulinoma patients experience a hyperinsulinaemic response to glucagon even in the presence of hypoglycaemia. The reason for this difference in insulinotropic response to glucagon between normal and tumorous B-cells is unknown.

A small proportion of seemingly ordinary, benign insulinomas fail to respond in the characteristic manner to glucagon (MARKS and SAMOLS 1968a; MCKIDDIE et al. 1969; PI-SUNYER et al. 1969; LINS and EFENDIC 1979), exhibiting a normal, or even subnormal, rise in plasma insulin following glucagon injection whilst still retaining their capacity to respond excessively to some, or none, of the other traditional insulinotropic stimuli. Attempts to distinguish benign from malignant insulinomas, and single from multiple tumours, on the basis of the intravenous glucagon test are misplaced and may be seriously misleading.

An exaggerated insulinaemic response to glucagon occurs in many conditions other than insulinoma, the most important being obesity, acromegaly, Cushing's syndrome and treatment with exogenous glucocorticoids or sulphonylureas (MARKS and SAMOLS 1968 b; KUMAR et al. 1974; WIDSTRÖM and CERASI 1973). The insulinaemic response to glucagon cannot be used, therefore, as a primary diagnostic procedure for insulinoma, but rather as an aid to the differential diagnosis of fasting hypoglycaemia since insulinoma alone among the causes of this condition is associated with an exaggerated insulinaemic response to glucagon.

Patients with hypoglycaemia caused by extrapancreatic neoplasms generally experience a normal or supranormal rise in blood glucose concentration, but an impaired or absent rise in plasma insulin concentration (FRIEND and HALES 1965; MARKS and SAMOLS 1966; PAULLADA et al. 1968; FRERICHS et al. 1970; KREISBERG et al. 1970; COLWELL and WILBER 1971; CHOWDHURY and BLEICHER 1973). Following successful removal of the tumour, the insulinaemic response to glucagon usually returns to normal, leading to the suggestion by MARKS (1976) that a possible cause of hypoglycaemia in this condition is the elaboration by the tumour of a somatostatin-like substance.

Patients with hypoglycaemia of endocrine origin generally have reduced glycaemic and insulinaemic responses to glucagon (MARKS and SAMOLS 1968 a; KUMAR et al. 1974) whilst those with the various types of exclusively reactive hypoglycaemia behave normally. Factitious hypoglycaemia due to sulphonylurea abuse may be associated with an exaggerated insulinaemic response to glucagon in some cases (MARRI et al. 1968; KUMAR et al. 1974) and be extremely difficult to distinguish, with certainty, from hypoglycaemia due to insulinoma without prolonged observation and intense investigation of the patient under controlled conditions. Alcohol-induced (fasting) hypoglycaemia is associated with a failure of blood glucose to rise normally after glucagon administration and a reduced, or absent, insulinaemic response (MARKS and SAMOLS 1968 b). Both responses return to normal following recovery from the acute episode.

## 2. In Children

The glucagon test is usually performed in children as an aid to the differential diagnosis of fasting hypoglycaemia, especially that due to inborn errors of metabolism. It differs from that carried out in adults mainly in that less importance attaches to the insulinaemic than to the glycaemic and lactataemic responses, especially under differing dietary conditions. The dose of glucagon usually employed, namely 30 µg/kg body weight, is supramaximal for both glycaemic and insulinaemic responses and is not critical; nor is its route of administration, which may be either intramuscular or intravenous.

Amongst the many causes of spontaneous hypoglycaemia in infants and young children (CORNBLATH and SCHWARTZ 1976; MARKS and ROSE 1981), only that due to endogenous hyperinsulinism (or maliciously administered exogenous insulin) responds to glucagon with a normal or supranormal rise in blood glucose concentration (FINEGOLD et al. 1980). The diagnostic importance of this observation has been stressed by FINEGOLD et al. (1980) who emphasised the value of performing glucagon tests on children during symptomatic hypoglycaemic episodes rather than during periods of normoglycaemia when they often fail to reveal anything useful.

Insulinomas are rare in children under the age of 4 years and the commonest cause of hyperinsulinism in infants is a functional derangement of insulin secretion, now commonly referred to as nesidioblastosis (YAKOVAC et al. 1971) but which, contrary to former belief, probably does not have a well-defined morphological basis (JAFFE et al. 1980; MARKS and ROSE 1981). Nesidioblastosis resembles insulinoma in three important respects: namely, it has the capacity to cause severe *fasting* hypoglycaemia by *continued* secretion of insulin in the presence of a low blood concentration, and it is associated with a high incidence of *leucine* sensitivity (MARKS and ROSE 1981). Neither tolbutamide nor glucagon provokes excessive insulin secretion in the majority of cases of nesidioblastosis – as they do in most patients with insulinomas – even when, as often happens, the abnormality of islet tissue is confined to a circumscribed area of the pancreas and presents a gross anatomical appearance reminiscent of, and readily confused with, an insulinoma. Detailed histopathological and immunohistological examination usually permits differentiation of focal or localised nesidioblastosis from insulinoma though many of the case reports currently in the literature purporting to describe insulinomata in the newborn are really examples of focal nesidioblastosis.

Infants and children with nesidioblastosis only rarely exhibit an exaggerated hyperinsulinaemic response to glucagon, despite often being grossly obese. They do, however, show a normal or greater than normal rise in blood glucose concentration in response to glucagon given during a fast-induced hypoglycaemic episode, unlike those with almost all other diseases causing spontaneous hypoglycaemia in this age group (FINEGOLD et al. 1980). Children with any and all of the various kinds of ketotic hypoglycaemia, for example, have a subnormal or absent rise in blood glucose concentration under these circumstances. Indeed, this is so characteristic a feature that "ketotic hypoglycaemia" (COLLE and ULSTRÖM 1964; STEPHENSON and HAINSWORTH 1966; GRUNT et al. 1970) has often been dubbed "glucagon-unresponsive hypoglycaemia" (ROSENBLOOM 1972; FALORNI et al. 1979).

Glucagon unresponsiveness invariably reflects either a temporary or permanent impediment to the mobilisation of hepatic glycogen and its conversion into glucose. Although it has many causes the commonest is acute or chronic dietary carbohydrate deprivation. Others include interference with gluconeogenesis, glycogenesis or glycogenolysis by an enzymic deletion secondary to an inborn error of metabolism or to inhibition by exogenous toxins. Whatever the cause, liver glycogen depletion is an important cause of hypoglycaemia and ketosis in children although in only a tiny minority do neuroglycopenic symptoms intervene and register as disease. Replenishment of hepatic glycogen stores by feeding, or the administration of intravenous glucose to those patients already in coma, is followed by restoration of the hyperglycaemic response to glucagon except in those cases in which there is a permanent enzymatic blockade to glycogenolysis. Fast-induced ketosis is not necessarily associated with a reduced glycaemic response to glucagon (NITZAN and KOWADLO-SILBERGELD 1966) which appears only to occur when liver glycogen stores have been depleted and the blood glucose concentration has fallen to hypoglycaemic levels (HABBICK et al. 1971; CORNBLATH and SCHWARTZ 1976). There is no reason to believe, despite frequent assertions to the contrary (BROBERGER et al. 1959; TIETZE et al. 1972; SIZONENKO et al. 1972; ZUPPINGER 1975; KERR 1980) that adrenomedullary hyporesponsiveness to hypoglycaemia is anything more than an epiphenomenon observed in some, but not all, children with glucagon-unresponsive and other types of fasting hypoglycaemia (MARKS and ROSE 1981).

The insulinaemic response to glucagon is less helpful in the differential diagnosis of hypoglycaemia in children than in adults. The insulinotropic effect of glucagon is either grossly reduced or absent during hypoglycaemic episodes in children with glucagon-unresponsive (i.e. ketotic) hypoglycaemia, but is restored to normal by refeeding. Those whose hypoglycaemia is due to glycogen storage disease may, like those with nesidioblastosis, show a normal insulinaemic response to glucagon (CROCKFORD et al. 1966), even when it is administered during hypoglycaemia, but whereas children with nesidioblastosis have a normal or supranormal rise in blood glucose under these circumstances, those with glycogenosis do not.

## II. Glycogen Storage Disease

The failure of the blood glucose concentration to rise in response to the injection of glucagon in patients with glycogen storage disease was amongst the first clinical observations to be made following its purification and release for clinical pharmacological research (PINCUS and RUTMAN 1953; HUBBLE 1954; SCHULMAN and SATUREN 1954; CARSON and KOCH 1955; LOWE et al. 1962; PERKOFF et al. 1962). It was, for a long time, used as one of the mainstays in the diagnosis and differential diagnosis of the glycogen storage diseases, but has latterly been displaced from its preeminent position by biochemical analysis of liver biopsies. The distinction between the various forms of glycogenosis is important not only because their prognoses are so different, but also because treatments that are suitable for one variety are not necessarily so for another (SPELLBERG 1969; MAHLER 1976; HOWELL 1978; FERNANDEZ 1980). FERNANDEZ et al. (1974) proposed a simple diagnostic scheme for the differential diagnosis of glycogen storage disease based upon the sequential use of glucose, galactose and glucagon which employs not only blood glucose, but also blood lactate measurements. Though useful, it is now generally believed that the results of such indirect tests should be supplemented, in the majority of cases, by liver biopsy.

### 1. Type I (Glucose-6-phosphatase Deficiency)

Children with glycogen storage disease, type I (von Gierke's disease, glucose-6-phosphatase deficiency), typically show an absent or only subnormal rise in blood glucose concentration following the administration of glucagon, which is scarcely modified, if at all, by prior feeding (SOKAL et al. 1961; STEINITZ 1967). The extent of the rise in blood glucose concentration reflects the severity or completeness of the reduction in glucose-6-phosphatase activity and is inversely related to the rise in plasma lactate, which is almost pathognomonic for this condition (SOKAL et al. 1961; ZUPPINGER and ROSSI 1969). Alcohol has been reported (LOWE and MOSOVICH 1965), paradoxically, to restore the hyperglycaemic response to glucagon in patients with glycogen storage disease type I, though this is a far from universal finding (ZUPPINGER 1975).

Glucagon has been used in attempts to prevent the occurrence of hypoglycaemia in children with von Gierke's disease and to reduce liver size. It has not proved beneficial, however, for this purpose and may, by aggravating the lactacidaemia, be harmful or even dangerous (SOKAL et al. 1962; ZUPPINGER and ROSSI 1969). This is scarcely surprising since there is no evidence of glucagon deficiency in glycogen storage disease, rather the reverse, hyperglucagonaemia being an almost invariable accompaniment of hypoglycaemia in this condition.

### 2. Type III (Debrancher Enzyme Deficiency)

Children and adults with glycogen storage disease type III (limit dextrinosis, debrancher enzyme deficiency) typically show an attenuated or absent hyperglycaemic response to glucagon during hypoglycaemic episodes, or after an overnight fast, with restoration to normal for a few hours following the ingestion of a carbohydrate-containing meal (HUG 1962; HUG et al. 1963; ZUPPINGER 1975). A "double-barrel" type of glucagon test for glycogenosis type III was suggested on theoretical grounds, by HUG (1962) and later confirmed as useful by him and his coworkers (HUG et al. 1963). It was, however, subsequently shown to be unreliable by LIMBECK and KELLEY (1965) who warned that glucagon tests should not be relied upon to differentiate the various types of liver glycogen disease. Glucagon does not normally produce as large a rise in plasma lactate levels in limit dextrinosis as in glycogen storage disease type I, but there is some overlap between them and so it cannot be used in differential diagnosis.

### 3. Types VI (Liver Phosphorylase Deficiency) and IX (Phosphorylase Kinase Deficiency)

The main disability suffered by patients with either glycogen storage disease type VI (liver phosphorylase deficiency) or type IX (phosphorylase kinase deficiency) is usually simple, asymptomatic hepatomegaly; only rarely do they develop signs and symptoms of hypoglycaemia and, even then, usually only in response to dietary or other types of stress. The glycaemic response to glucagon is invariably reduced or absent but, unlike that in von Gierke's disease, is unaccompanied by a rise in plasma lactate (MAHLER 1976; HOWELL 1978; FERNANDEZ 1980).

### 4. Glycogen Synthase Deficiency

Subnormal or absent glycaemic responses to glucagon (LEWIS et al. 1963; AYNSLEY-GREEN et al. 1977) are observed in patients with glycogen synthase deficiency whilst they are fasting but, for some unexplained reason, not after they have had a meal.

### 5. Other Types of Glycogen Storage Disease

The blood glucose response to glucagon is normal in all other varieties of glycogen storage disease, i.e. types II, IV, VII, and VIII, except pseudoglycogenosis type I (CHALMERS et al. 1978) which resembles von Gierke's disease in all clinical and biochemical aspects except for the presence of normal hepatic glucose-6-phosphatase activity.

## III. Disorders of Gluconeogenesis

Four distinct inborn errors of metabolism which result in defective gluconeogenesis and are associated, either regularly or infrequently, with the appearance of hypoglycaemia during fasting, have been described. Glucose-6-phosphatase deficiency (von Gierke's disease) is the commonest, and has already been considered. Fructose-1,6-diphosphatase deficiency, is the next commonest (BAKER and WINEGRAD 1970; FROESCH 1978; BAERLOCHER et al. 1978) and is characterised biochemically by the fact that hypoglycaemia is provoked not only by fasting, but also by fructose, glycerol, alanine or dihydroxyacetone administration, whether given orally or intravenously. Glucagon does not usually provoke a rise in blood glucose concentration when given during a hypoglycaemic episode, but does so when administered 8–12 h, or less, after a meal. In some cases, however, glucagon does cause a rise in blood glucose concentration even when administered during a hypoglycaemic attack (TAUNTON et al. 1978).

Pyruvate carboxylase deficiency (HOMMES et al. 1980), which seldom causes hypoglycaemia, and phosphoenolpyruvate carboxykinase deficiency (FISER et al. 1974; HOMMES et al. 1976; VIDNES and SOVIK 1976), which invariably does, are both extremely rare and too little information is available, at the present time, to permit valid conclusions to be drawn as to the diagnostic value, if any, of glucagon in these conditions.

## IV. Sugar-Induced Hypoglycaemia

Reactive hypoglycaemia is an extremely common response (MARKS and ROSE 1981) to the administration of an oral glucose load, providing it is large enough, but a rare one to a mixed meal (CHARLES et al. 1981). It is, nevertheless, an extremely common diagnosis in some parts of the world, notably the United States, and differentiation from insulinoma and other causes of hypoglycaemia may present problems. Patients in whom reactive hypoglycaemia occurs only in response to the ingestion of large oral glucose or sucrose loads, especially when they are combined with alcohol, as in gin and tonic (O'KEEFE and MARKS 1977; JOFFE et al. 1981), but never during fasting, have been said (FOÀ et al. 1980) to have an increased incidence of impaired glucagon secretion, but this in contrary to most other workers' experience (LEFEBVRE et al. 1976). Such patients invariably exhibit both normal glucose and insulinaemic responses to exogenous glucagon and all other tests of carbohydrate metabolism currently available also yield normal results (MARKS and ROSE 1981).

Hypoglycaemia develops regularly in response to the administration of fructose or sucrose in patients with hereditary fructose intolerance and somewhat less regularly after galactose or lactose administration (GREENMAN 1950; MORTENSEN and SONDERGAARD 1954) in those with galactosaemia. During episodes of fructose-induced hypoglycaemia, patients with hereditary fructose intolerance (PERHEENTUPA et al. 1962; CORNBLATH and SCHWARTZ 1976; FROESCH 1978) do not respond to glucagon with a rise in blood glucose, but they do so at all other times.

In a so far unique case of glycerol intolerance which was not associated with fructose-1,6-diphosphatase deficiency, the usual cause of this condition, glycerol, caused a rapid and profound fall in blood glucose concentration (MACLAREN et al. 1975) when administered orally or intravenously. Overnight fasting was also associated with the appearance of hypoglycaemia which was responsive to glucagon on some occasions, but not on others. When given shortly after a meal, glucagon invariably produced a normal rise in blood glucose concentration.

### V. Liver Disease

Liver disease, though often cited as one of the more common causes of hypoglycaemia, is in fact rarely so – and even then it usually occurs only as a terminal event in patients with acute hepatic failure (MARKS and ROSE 1981). In those rare individuals with chronic liver disease in whom hypoglycaemia does develop, glucagon produces a subnormal or zero rise in blood glucose concentration (MARRACK et al. 1961) and a normal, or rarely supranormal, rise in plasma insulin (SAMOLS and HOLDSWORTH 1968).

Hypoglycaemia occurs in patients with primary hepatoma, sometimes as the presenting symptom. Patients with this type of hypoglycaemia usually, though not invariably, show a grossly diminished hyperglycaemic response to glucagon (MCFADZEAN and YOUNG 1969), unlike those with hypoglycaemia produced by other extrapancreatic neoplasms who, except during the terminal cachectic stage of their illness (MARKS et al. 1974), usually respond normally. This difference is, however, of little diagnostic significance.

## E. Conclusions

Glucagon still has an important role to play in the emergency treatment of insulin-induced hypoglycaemic coma in diabetic subjects, especially in children, in whom its capacity for administration by intramuscular instead of exclusively by intravenous injection – as glucose must be – is especially valuable. It has, however, failed to find a significant place for itself in the long-term treatment of either iatrogenic or spontaneous hypoglycaemia. Its use as a diagnostic agent in hypoglycaemia, which was once of considerable importance, especially for the diagnosis of insulinoma and of glycogen storage disease, has diminished with the greater availability of more reliable and specific procedures (MARKS and ALBERTI 1976). It does, however, still provide a useful tool for investigating pathogenic mechanisms in cases in which these are otherwise obscure.

## References


Alaric R (1973) Hypoglycémie prolongée au chlorpropamide. Union Med Can 102:1290–1291

Alivisatos JG, McCullagh EP (1955) Studies with glucagon in patients with insulin sensitivity. JAMA 159:1098–1105

Arieff A, Crawford J, Adams J, Smith D (1960) Glucagon in insulin coma therapy: its use in a small psychiatric unit of a general hospital. Q Bull Northwest Univ Med Sch 34:7–10

Arky RA, Finger M, Veverbrants E, Braun AP (1970) Glucose and insulin response to intravenous glucagon during starvation. Am J Clin Nutr 23:69–95

Aynsley-Green A, Williamson DH, Gitzelmann R (1977) Hepatic glycogen synthetase deficiency: definition of syndrome and enyzme studies on a 9-year-old girl. Arch Dis Child 52:573–579

Baerlocher K, Gitzelmann R, Steinman B, Gitzelmann-Cumarasamy N (1978) Hereditary fructose intolerance in early childhood: a major challenge. Survey of 20 symptomatic cases. Helv Paediatr Acta 33:465–487

Baker L, Winegrad AI (1970) Fasting hypoglycaemia and metabolic acidosis associated with deficiency of hepatic fructose-1,6-diphosphatase activity. Lancet 2:13–16

Beck-Christiansen O (1964) Metastasizing insulinoma treated with glucagon. Dan Med Bull 11:70–72

Benedetti A (1971) The glucagon plus galactose tolerance. In: Austoni M, Scandellari C, Federspil G, Trisotto A (eds) Current topics on glucagon. Cedam, Padova, pp 177–191

Benedetti A, Kolb FO (1966) Metabolic effects and epinephrine in four adults with type I glycogen storage disease. Diabetes 15:529

Benedetti A, Simpson RG, Grodsky GM, Forsham PH (1967) Exaggerated insulin response to glucagon in simple obesity. Diabetes 16:666–669

Bergman H (1965) Hypoglycaemic coma during sulphonylurea therapy. Acta Med Scand 177:287–298

Braun M, Parker M (1959) The use of glucagon in the termination of therapeutic insulin coma. Am J Psychiatry 115:814–815

Brennan MO, Service FJ, Carpenter AM, Rubenstein AH, Edis AJ (1980) A complex case of hypoglycemia diagnosed by C-peptide suppression test. In: Andreani D, Lefèbvre PJ, Marks V (eds) Current views on hypoglycaemia and glucagon. Academic, London, pp 321–330

Broberger O, Jungner I, Zetterström R (1959) Studies in spontaneous hypoglycemia in childhood: failure to increase the epinephrine secretion in insulin-induced hypoglycemia. J Pediatr 55:713–719

Carson MJ, Koch R (1955) Clinical studies with glucagon in children. J Pediatr 47:161–170

Chalmers RA, Ryman BE, Watts RWE (1978) Studies on a patient with in vivo evidence of type I glycogenosis and normal enzyme activities in vitro. Acta Paediatr Scand 67:201–207

Charles MA, Hofeldt F, Shackleford A, Waldeck N, Dobson LE, Bunker D, Coggins JT, Eichner H (1981) Comparison of oral glucose tolerance tests and mixed meals in patients with apparent idiopathic postabsorptive hypoglycemia: absence of hypoglycemia after meals. Diabetes 30:465–470

Chowdhury F, Bleicher SJ (1973) Studies of tumor hypoglycemia. Metabolism 22:663–674

Codaccioni JL, Rubin P, Mattei A, Vague P (1971) Quatre hypoglycémies graves donc une mortelle au cours du traitement par glibenclamide. Diab Metab 19:37–41

Colle E, Ulström RA (1964) Ketotic hypoglycemia. J Pediatr 64:632–651

Colwell JA, Wilber JF (1971) Studies of insulin and growth hormone secretion in a subject with hepatoma and hypoglycemia. Diabetes 20:607–614

Cornblath M, Schwartz R (1976) Disorders of carbohydrate metabolism in infancy, 2nd edn. Saunders, Philadelphia

Creutzfeldt W, Frerichs H, Perings F (1969) Serum insulin levels in hypoglycaemic shock due to attempted suicide with tolbutamide and insulin. Ger Med Monthly 14:14–19

Creutzfeldt W, Arnold R, Creutzfeldt C, Deuticke U, Frerichs H, Track NS (1973) Biochemical and morphological investigations of 30 human insulinomas. Correlation between the tumour content of insulin and proinsulin-like components and the histological and ultrastructural appearance. Diabetologia 9:217–231
Creutzfeldt W, Creutzfeldt C, Frerichs H, Track NS, Arnold R (1976) Histochemistry, ultrastructure and hormone content of human insulinomas. In: Andreani D, Lefèbvre P, Marks V (eds) Hypoglycemia. Proceedings of the European Symposium, Rome. Thieme, Stuttgart, pp 7–18
Crockford PM, Porte D, Wood FC, Williams RH (1966) Effect of glucagon on serum insulin, plasma glucose and free fatty acids in man. Metabolism 15:114–122
Davies DM, MacIntyre A, Millar E, Bell SM, Mehra SK (1967) Need for glucagon in severe hypoglycaemia induced by sulphonylurea drugs. Lancet 1:363–364
Devrim S, Recant L (1966) Effect of glucagon on insulin release in vitro. Lancet 2:1227–1228
Dowell RC, Imrie AH (1972) Chlorpropamide poisoning in non-diabetics. Scott Med J 17:305–309
Elrick H, Arai Y, Yearwood-Drayton V (1957) Observation on actions of combined glucagon-insulin infusion in diabetic patients. J Clin Invest 36:887–888
Elrick H, Witten TA, Arai Y (1958) Glucagon treatment of insulin reactions. N Engl J Med 258:476–480
Esquibel AJ, Kurland AA, Mendelsohn D (1958) The use of glucagon in terminating insulin coma. Dis Nerv Syst 19:485–486
Falorni A, Massi-Benedetti F, Sposito M, Barboni G, Latom M (1979) Insulin and glucagon secretion in the ketotic (idiopathic glucagon unresponsive) hypoglycemia of childhood. J Endocrinol Invest 2:51–57
Fernandez J (1980) Hepatic glycogenosis: diagnosis and management. In: Burman D, Holton JB, Pennock CA (eds) Inherited disorders of carbohydrate metabolism. MTP Press, Lancaster, pp 297–312
Fernandez J, Koster JF, Grose WFA, Sorgedrager N (1974) Hepatic phophorylase deficiency: its differentiation from other hepatic glycogenoses. Arch Dis Child 49:186–191
Finegold DN, Stanley CA, Baker L (1980) Glycemic response to glucagon during fasting hypoglycaemia: an aid to the diagnosis of hyperinsulinism. J Pediatr 96:257–260
Fiser RH, Melsher HL, Fischer DA (1974) Hepatic phosphoenolypyruvate carboxykinase (PEPCK) deficiency: a new cause of hypoglycemia in childhood (Abstr). Pediatr Res 8:432
Foà PP, Dunbar JC, Klein SP, Levy SH, Malik MA, Campbell BB, Foà NL (1980) Reactive hypoglycemia and A-cell (pancreatic) glucagon deficiency in the adult. JAMA 244:2281–2285
Forman BH, Feeney E, Boas L (1974) Drug induced hypoglycemia. JAMA 229:522
Forrest JAH (1974) Chlorpropamide overdosage. Delayed and prolonged hypoglycemia. Clin Toxicol 7:19–24
Frasier SD, Smith FG, Nash A (1965) The use of glucagon-gel in idiopathic spontaneous hypoglycemia of infancy. Pediatrics 35:120–123
Frerichs H, Willms B, Kasper H, Creutzfeldt C, Creutzfeldt W (1970) Contribution to the pathogenesis of tumour hypoglycaemia. Eur J Clin Invest 1:2–11
Frey HMM, Roselund B (1970) Studies in patients with chlorpropamide-induced hypoglycemia. Diabetes 19:930–937
Friend JAR, Hales CN (1965) Spontaneous hypoglycaemia and sarcoma. Acta Endocrinol (Copenh) 50:233–238
Froesch ER (1978) Essential fructosuria: hereditary fructose intolerance and fructose 1,6-diphosphatase deficiency. In: Stanbury JB, Wyngaarden JB, Fredrickson DS (eds) The metabolic basis of inherited disease, 4th edn. McGraw-Hill, New York, pp 121–136
Gibbs GE, Ebers DW, Meckel BR (1958) Use of glucagon to terminate insulin reactions in diabetic children. Nebr State Med J 43:56–57
Goldfine ID, Cerasi E, Luft R (1972) Glucagon stimulation of insulin release in man: inhibition during hypoglycemia. J Clin Endocrinol Metab 35:312–315
Gotlin RW, Silver HK (1970) Neonatal hypoglycaemia, hyperinsulinism, and absence of pancreatic alpha-cells. Lancet 1:1346

Greenman L (1950) Alterations in blood glucose following intravenous galactose. J Biol Chem 183:577–585
Grill V, Cerasi E, Wahren J (1979) Role of cyclic AMP in glucagon-induced stimulation of hepatic glucose output in man. Scand J Clin Invest 39:689–696
Grollman A, McCaleb WE, White FN (1964) Glucagon deficiency as a cause of hypoglycemia. Metabolism 13:686–690
Grunt JA, McGarry ME, McCollum AT, Gould JB (1970) Studies of children with ketotic hypoglycemia. Yale J Biol Med 42:420–438
Habbick BF, McNeish AS, Stephenson JBP (1971) Diagnosis of ketotic hypoglycaemia of childhood. Arch Dis Child 46:295–300
Haro EN, Blum SF, Faloon WW (1965) The glucagon response of fasting obese subjects. Metabolism 14:976–984
Haunz EA (1962) Feigned insulin reactions in diabetic children prevented by glucagon. Lancet 82:263–268
Hoffbrand BI, Sevitt LH (1966) Use of mannitol in prolonged coma due to insulin overdose. Lancet 1:802
Hommes FA, Bendien K, Elema JD, Bremer HJ, Lombeck I (1976) Two cases of phosphoenolpyruvate carboxykinase deficiency. Acta Paediatr Scand 65:233–240
Hommes FA, Schrijver J, Dias T (1980) Pyruvate carboxylase deficiency; studies on patients and on an animal model system. In: Burman D, Holton JB, Pennock CA (eds) Inherited disorders of carbohydrate metabolism. MTP Press, Lancaster, pp 269–286
Horwitz DL, Rubenstein AH (1974) Insulin suppression (letter). Lancet 2:1021
Howell RR (1978) The glycogen storage diseases. In: Stanbury JB, Wyngaarden JB, Fredrickson DS (eds) The metabolic basis of inherited disease. Blakiston, New York, pp 137–159
Hubble D (1954) Glucagon and glycogen storage disease of the liver. Lancet 1:235–237
Hug G (1962) Glucagon tolerance test in glycogen storage disease. J Pediatr 60:545–549
Hug G, Krill CE, Perrin EV, Guest GM (1963) Cori's disease (amylo 1,6 glucosidase deficiency): report of a case in a negro child. N Engl J Med 268:113–120
Jacobs RF, Nix RA, Paulus JE, Kiel EA, Fiser RH (1978) Intravenous infusion of diazoxide in the treatment of chlorpropamide-induced hypoglycemia. J Pediatr 93:801–803
Jaffe R, Hashida Y, Yunis EJ (1980) Pancreatic pathology in hyperinsulinemic hypoglycemia of infancy. Lab Invest 42:356–365
Joffe BI, Roach L, Baker S, Shires R, Sandler M, Seftel HC (1981) Failure to induce reactive hypoglycaemia by drinking a starch-based alcohol beverage (sorghum beer). Ann Clin Biochem 18:22–24
Johnson SF, Schade DS, Peake GT (1977) Chlorpropamide-induced hypoglycemia. Successful treatment with diazoxide. Am J Med 63:799–804
Kerr DS (1980) Epinephrine deficiency in children: fasting metabolism and response to 2-deoxyglucose. In: Andreani D, Lefèbvre PJ, Marks V (eds) Current views on hypoglycaemia and glucagon. Academic Press, London New York, pp 409–412
Khurana RC, Klayton R, Jung Y, Gonzalez AR, Dhawer VPS, Corredor DG, Sieracki JC, Danowski TS (1971) Insulin and glucose patterns in control subjects and in proved insulinoma. Am J Med Sci 262:115–118
Kollee LA, Monnens LA, Cejka V, Wilms RH (1978) Persistent neonatal hypoglycaemia due to glucagon deficiency. Arch Dis Child 53:422–424
Kreisberg RA, Hershman JM, Spenney JG, Boshell BR, Pennington LF (1970) Biochemistry of extrapancreatic tumor hypoglycemia. Diabetes 19:248–258
Kumar D, Mehtalia SD, Miller LV (1974) Diagnostic use of glucagon-induced insulin response. Studies in patients with insulinoma or other hypoglycemic conditions. Ann Intern Med 80:697–701
Kushner RS, Lemli L, Smith DW (1963) Zinc glucagon in the management of idiopathic hypoglycemia. J Pediatr 63:1111–1115
Landau BR, Levie HJ, Hertz R (1958) Prolonged glucagon administration in a case of hyperinsulinism due to disseminated isletcell carcinoma. N Engl J Med 259:296–288
Langs HM, Friedberg D (1965) Stimulation of insulin secretion by glucagon. Clin Res 13:548

Laqueur HP, LaBurt HA (1960) Experiences with low-zinc insulin, with semilente insulin, with glucagon and adrenalin-thiamin in insulin coma treatment. J Neuropsychiatry 2:86–92
Lefèbvre P, Luyckx A (1966) Glucagon-stimulated insulin release. Lancet 1:1040
Lefèbvre PJ, Luyckx AS, Lecomte MJ (1976) Studies on the pathogenesis of reactive hypoglycemia: role of insulin and glucagon. In: Andreani D, Lefèbvre PJ, Marks V (eds) Hypoglycemia. Proceedings of the European Symposium. Thieme, Stuttgart, pp 91–98
Lewis GM, Spencer-Peet J, Stewart KM (1963) Infantile hypoglycaemia due to inherited deficiency of glycogen synthetase in liver. Arch Dis Child 38:40–48
Limbeck GA, Kelley VC (1965) "Double-barrel" glucagon test: correlation with enzyme assays in limit dextrinosis. Am J Dis Child 109:162–164
Lins PE, Efendic S (1979) Responses of patients with insulinomas to stimulators and inhibitors of insulin release that have been linked with cyclic adenosine monophosphate. Diabetes 28:190–195
Ljungström B (1964) Glucagon treatment of spontaneous hyperglycemia (in Norwegian). Nord Med 71:177–179
Lowe CU, Mosovich LL (1965) The paradoxical effect of alcohol on carbohydrate metabolism in four patients with liver glycogen disease. Pediatrics 35:1005–1008
Lowe CU, Sokal JE, Mosovich LL, Sarcione EJ, Dobray BH (1962) Studies in liver glycogen disease. Effects of glucagon and other agents on metabolic pattern and clinical status. Am J Med 33:4–19
MacCuish AC, Munro JF, Duncan LJP (1970) Treatment of hypoglycaemic coma with glucagon, intravenous dextrose, and mannitol infusion in a hundred diabetics. Lancet 2:946–949
MacLaren NK, Cowles C, Ozand PT, Shuttee R, Cornblath M (1975) Glycerol intolerance in a child with intermittent hypoglycemia. J Pediatr 86:43–49
Madison LL (1968) Ethanol-induced hypoglycemia. Adv Metab Disord 3:85–109
Mahler RF (1976) Disorders of glycogen metabolism. Clin Endocrinol Metab 5:579–598
Mariani MM, Loubatières A (1972) Récentes expèriences concernant l'antagonisme entre certains sulfamides hypoglycemiants et le diazoxide. In: Austoni M, Scandellari C, Trissotto A, Federspil G (eds) Hypoglycaemia and diazoxide. Cedam, Padova, pp 33–49
Marks LJ, Steinke J, Podolsky S, Egdahl RH (1974) Hypoglycemia associated with neoplasia. Ann NY Acad Sci 230:147–160
Marks V (1960) Response to glucagon by subjects with hyperinsulinism from islet cell tumours. Br Med J 1:1539–1540
Marks V (1962) The investigation of hypoglycaemia. In: Pyke DA (ed) Disorders of carbohydrate metabolism. Pitman, London, pp 229–239
Marks V (1971) The biological significance of the insulinotropic effect of glucagon in man. In: Austoni M, Scandellari C, Federspil G, Trissotto A (eds) Current topics on glucagon. Cedam, Padova, pp 63–71
Marks V (1976) Hypoglycaemia. 2. Other causes. Clin Endocrinol Metab 5:769–782
Marks V, Alberti KGMM (1976) Selected tests of carbohydrate metabolism. Clin Endocrinol Metab 5:805–820
Marks V, Rose FC (1981) Hypoglycaemia, 2nd edn. Blackwell, Oxford
Marks V, Samols E (1966) Hypoglycaemia of non-endocrine origin (non-islet cell tumours). Proc R Soc Med 59:338–340
Marks V, Samols E (1968 a) Glucagon test for insulinoma: a chemical study in 25 cases. J Clin Pathol 21:346–352
Marks V, Samols E (1968 b) Glucagon mediated insulin release in man. In: Levine R, Pfeiffer EF (eds) Mechanism and regulation of insulin secretion. Il Ponte, Milan, pp 285–308
Marks V, Samols E (1969) Diagnostic tests for evaluating hypoglycaemia. Excerpta Medica Int Congr Ser 172:864–872
Marrack D, Rose FC, Marks V (1961) Glucagon and tolbutamide tests in the recognition of insulinomas. Proc R Soc Med 54:749–751
Marri G, Tyler J, Marks V, Samols E (1966) Stimolazione della secrezione insulinica nell'umo mediante glucagone. Minerva Med 57:2733–2737

Marri G, Cozzolino G, Palumbo R (1968) Glucagon in sulphonylurea hypoglycaemia. Lancet 1:303
McFadzean AJS, Young RTT (1969) Further observations on hypoglycaemia in hepatocellular carcinoma. Am J Med 47:220–235
McKiddie MT, Buchanan KD, Abernethy RJ (1969) Plasma insulin studies in the diagnosis of insulinoma. Scott Med J 14:200–208
McQuarrie I (1954) Idiopathic spontaneously occurring hypoglycemia in infants. Clinical significance of problems and treatment. Am J Dis Child 87:399–428
McQuarrie I, Bell ET, Zimmerman B, Wright WS (1950) Deficiency of alpha cells of pancreas as possible etiological factor in familial hypoglycemosis. Fed Proc 9:337
Mortensen O, Søndergaard G (1954) Galactosemia (galactose disease). Acta Paediatr Scand 43:467–477
Muhe E, Schricker KT, Raithel D (1969) Die Behandlung spontaner Hypoglycämien bei einem inoperablen Leberzellkarzinom durch Glucagon. Dtsch Med Wochenschr 94:1781–1785
Nabben FAE (1968) Transient hypoglycemia with leucine hypersensitivity not based on increased insulin production. Treatment with zinc glucagon (in Dutch). Maandschr Kindergeneeskd 36:2–17
Nitzan M, Kowadlo-Silbergeld A (1966) Responses to glucagon and epinephrine, and glycogen reserves in children with nondiabetic ketosis. Isr J Med Sci 2:683–689
Ohneda A, Maruhama Y, Itabashi H, Horigome K, Yanbe A, Ishii S, Chiba M, Kai Y, Abe R, Yamagata S (1975) Diagnostic value of intravenous glucagon test in insulinoma. Tohoku J exp Med 116:205–211
O'Keefe SJD, Marks V (1977) Lunchtime gin and tonic: a cause of reactive hypoglycaemia. Lancet 1:1286–1288
Paullada JJ, Lisci-Garmilla A, Gonzales-Angulo A, Jurado-Mendoza J, Quijano-Narezo M, Gomez-Peralta L, Doria-Medina M (1968) Hemangiopericytoma associated with hypoglycemia. Am J Med 44:990–999
Perheentupa J, Pitkanen E, Nikkila EA, Somersalo O, Hakosalo J (1962) Hereditary fructose intolerance: a clinical study of four cases. Ann Paediatr 8:221–235
Perkoff GT, Parker VJ, Hann RF (1962) The effects of glucagon in three forms of glycogen storage disease. J Clin Invest 41:1099–1105
Pfeifer MA, Wolter CF, Samols E (1978) Management of chlorpropamide-induced hypoglycemia with diazoxide. South Med J 71:606–608
Pfeiffer EM, Thum C, Raptis S, Beischer W, Ziegler R (1976) Hypoglycemia in diabetes. In: Andreani D, Lefèbvre P, Marks V (eds) Hypoglycemia. Procedings of the European Symposium. Thieme, Stuttgart, pp 112–126
Pincus IJ, Rutman JZ (1953) Glucagon, the hyperglycemic agent in pancreatic extracts. Arch Intern Med 92:666–667
Pi-Sunyer FX, van Itallie TB, Zintel HA (1969) Insulin stimulatory tests in a patient with islet cell adenoma. Am J Surg 118:95–99
Porte D, Graber AL, Kuzuya T, Williams RH (1966) The effect of epinephrine on immunoreactive insulin levels in man. J Clin Invest 45:228–236
Rizza RA, Gerich JE (1979) Persistent effect of sustained hyperglucagonaemia on glucose production in man. J Clin Endocrinol Metab 48:352–355
Rosenbloom AL (1972) Ketotic (idiopathic glucagon unresponsive) hypoglycaemia: diazoxide effects. Arch Dis Child 47:544–549
Rosenbloom AL, Smith DW, Cohan RC (1966) Zinc glucagon in idiopathic hypoglycemia of infancy. Am J Dis Child 112:107–111
Rossi E (1959) Glucagon. Triangle 4:13
Roth H, Thier S, Segal S (1966) Zinc-glucagon in the management of refractory hypoglycemia due to insulin-producing tumors. N Engl J Med 274:493–497
Samols E, Holdsworth D (1968) Disturbances of carbohydrate metabolism: liver disease. In: Dickens F, Randle PJ, Whelan WJ (eds) Carbohydrate metabolism and its disorders. Academic, London, pp 289–336
Samols E, Marks V (1963) Insulin assay in insulinomas. Br Med J 1:507–510

Samols E, Marks V (1966) Application of insulin radioimmunoassay in diagnosis and clinical investigation. In: Proceedings of the conference on problems connected with the preparation and use of labelled proteins in tracer studies, Euratom, Pisa, pp 285–300
Samols E, Marri G, Marks V (1965) Promotion of insulin secretion by glucagon. Lancet 1:415–516
Samols E, Marri G, Marks V (1966) Interrelationship of glucagon, insulin, and glucose. The insulinogenic effect of glucagon. Diabetes 15:855–866
Scarlett JA, Mako ME, Rubenstein AH, Blix PM, Goldman J, Horwitz DL, Tager H, Jaspan JB, Stjernholm MR, Olefsky JM (1977) Factitious hypoglycemia: diagnosis by measurement of serum C-peptide immunoreactivity and insulin-binding antibodies. N Engl J Med 297:1029–1032
Schulman JL, Greben SE (1957) The effect of glucagon on the blood glucose level and the clinical state in the presence of marked insulin hypoglycemia. J Clin Invest 36:74–80
Schulman JL, Saturen P (1954) Glycogen storage disease of the liver. I. Clinical studies during the early neonatal period. Pediatrics 14:632–645
Service FJ, Horwitz DL, Rubenstein AH, Kuzaya H, Mako ME, Reynolds C, Molnar GD (1977) C-peptide suppression test for insulinoma. J Lab Clin Med 90:180–186
Shipp JC, Delcher HK, Munroe JF (1964) Treatment of insulin hypoglycemia in a diabetic camp. A comparison of glucagon (1 and 2 mg) and glucose. Diabetes 13:645–648
Sizonenko PC, Paunier L, Vallotton MB, Terraz M, Scholer-Markovic D (1972) Childhood hypoglycaemia: plasma glucose and renin response to deoxyglucose for assessment of adrenal medulla responsiveness. Helv Paediatr Acta 27:565–573
Sokal JE, Lowe CU, Sarcione EJ, Mosovich LL, Doray BH (1961) Studies of glycogen metabolism in liver glycogen disease (von Gierke's disease). Six cases with similar metabolic abnormalities and responses to glucagon. J Clin Invest 40:364–374
Sokal JE, Lowe CJ, Sarcione EJ (1962) Liver glycogen disease (von Gierke's disease). Arch Intern Med 109:612–624
Spellberg MA (1969) Treatment of glycogen storage disease. Am J Gastroenterol 52:45–47
Steinitz K (1967) Laboratory diagnosis of glycogen disease. Adv Clin Chem 9:227–354
Stephenson JBP, Hainsworth IR (1966) Ketotic hypoglycaemia in childhood. Proc Assoc Clin Biochem 5:80–81
Taunton OD, Greene HL, Stifel FB, Hofeldt FD, Lufkin EG, Hagler L, Herman Y, Herman RH (1978) Fructose-1,6-diphosphatase deficiency, hypoglycemia and response to folate therapy in a mother and her daughter. Biochem Med 19:260–276
Taylor JR, Sherratt HSA, Davies DM (1978) Intramuscular or intravenous glucagon for sulphonylurea hypoglycaemia. Eur J Clin Pharmacol 14:125–127
Tietze HU, Zurbrügg RP, Zuppinger JA, Joss EE, Käser H (1972) Occurrence of impaired cortisol regulation in children with hypoglycemia associated with adrenal medullary hyporesponsiveness. J Clin Endocrinol Metab 34:948–958
Turner DS, McIntyre N (1966) Stimulation by glucagon of insulin release from rabbit pancreas in vitro. Lancet 1:351–352
Turner RC (1976) The diagnosis of insulinomas and other causes of fasting hypoglycaemia. In: Andreani D, Lefèbvre P, Marks V (eds) Hypoglycemia: proceedings of the european symposium. (Hormone and Metabolic Research Supplement Series) Thieme, Stuttgart, pp 40–45
Turner RC, Heding LG (1977) Plasma pro-insulin, C-peptide and insulin in diagnostic suppression tests for insulinomas. Diabetologia 13:571–577
van Itallie TB (1956) Glucagon: physiological and clinical considerations. N Engl J Med 254:794–803
van Itallie T, Bentley WBA (1955) Glucagon-induced hyperglycemia as an index of liver function. J Clin Invest 34:1730–1737
Vidnes J, Oyasaeter S (1977) Glucagon deficiency causing severe neonatal hypoglycemia in a patient with normal insulin secretion. Pediatr Res 11:943–947
Vidnes J, Søvik O (1976) Gluconeogenesis in infancy and childhood. III. Deficiency of the extramitochondrial form of hepatic phosphoenolpyruvate carboxykinase in a case of persistent neonatal hypoglycaemia. Acta Paediatr Scand 65:307–312

Wagner T, Spranger J, Brunck HJ (1969) Kongenitaler-α-Zellmangel als Ursache einer chronischen infantilen Hypoglykämie? Monatsschr Kinderheilkd 117:236–238
Waife SO (1960) An integrated view of glucagon. J Mich Med Soc 59:1519–1523
Weinges KF (1959) Der Einfluß eines protrahiert wirkenden Glucagons auf den Blutzucker, das anorganische Serumphosphat und die Gesamtaminosäuren im Serum. Naunyn Schmiedebergs Arch J Exp Pathol 237:22–26
Widström A, Cerasi E (1973) On the action of tolbutamide in normal man. III. Interaction of tolbutamide with glucagon, aminophylline and arginine in stimulating insulin response. Acta Endocrinol (Copenh) 72:532–544
Yakovac WC, Baker L, Hummeler K (1971) Beta cell nesidioblastosis in idiopathic hypoglycemia of infancy. Pediatrics 79:225–231
Zuppinger KA (1975) Hypoglycemia in childhood. Evaluation of diagnostic procedures. Karger, Munich
Zuppinger KA, Rossi E (1969) Metabolic studies in liver glycogen disease with special reference to lactate metabolism. Helv Med Acta 35:406–422

CHAPTER 56

# Miscellaneous Pharmacologic Effects of Glucagon

P. J. LEFEBVRE

## A. Introduction

As detailed in Chaps. 34–38, the physiologic role of glucagon as a key hormone in the regulation of glucose metabolism is now largely accepted. In addition, and as reviewed in Chaps. 42–49, glucagon is also involved in the pathophysiology of numerous disorders, including diabetes (Chap. 44), the glucagonoma syndrome (Chap. 43), and some hypoglycemic states (Chap. 42). From the pharmacologic point of view and at doses which are usually a thousand times greater than the levels measured in plasma, glucagon appears as a useful agent for the diagnosis of various disorders, such as pheochromocytoma (Chap. 51) or some forms of hypoglycemia (Chap. 55). As reviewed in Chap. 54, glucagon, by its inhibitory effect on smooth muscle motility is frequently used as an aid in radiodiagnosis. On the therapeutic side, glucagon remains most useful in the treatment of certain forms of hypoglycemia (Chap. 55) or cardiac failure (Chap. 53). In this last chapter, we will briefly review some topics where clear-cut actions of glucagon have been described but which, until now, have not led to large scale clinical use. The actions of glucagon on inflammatory reactions, on food intake and body weight, on bronchial motility, on erythropoiesis, and on tumor growth fall into this category.

## B. The Antiinflammatory Action of Glucagon

### I. In Animal Experiments

A systematic investigation of the effects of glucagon on experimental inflammation was performed between 1960 and 1963 by LEFEBVRE and his co-workers. Injected intraperitoneally at a dose of 100 μg/100 g body weight in rats, glucagon significantly inhibited the edema induced by the injection in the paw of histamine, formalin, and dextran, but not serotonin (LEFEBVRE 1960; LEFEBVRE and VAN CAUWENBERGE 1962). The antiedematous effect of glucagon was totally abolished by previous adrenalectomy (LEFEBVRE 1961 a) or by administration of an adrenolytic drug (R-3248, 4′-fluoro-4-[1-(4-acetylaminomethyl-4-phenyl)piperidine]butyrophenone, Janssens Pharmaceutical Products, Beerse, Belgium) at a dose of 500 μg/100 g body weight (LEFEBVRE and VAN CAUWENBERGE 1962). Since: (1) at the doses used, glucagon markedly stimulates the release of catecholamines from the adrenal medulla (LEFEBVRE and DRESSE 1961; DRESSE and LEFEBVRE 1961); (2) the antiphlogistic action of glucagon is abolished by adrenalectomy and reproduced by intraperitoneal injection of epinephrine (LEFEBVRE 1961 a, 1962); and (3) the antiphlogistic action of epinephrine was also abolished by the same adrenolytic drug (LEFEBVRE 1962), the inhibitory effect exerted by glucagon on the edema induced by local injection of histamine, formalin, or dextran has been attributed to a glucagon-induced adrenal catecholamine release (LEFEBVRE 1962; LEFEBVRE and VAN CAUWENBERGE 1962). It should be noted that the inhibitory effect of glucagon on histamine- or formalin-induced local edema is also abolished by previous hypophysectomy (LEFEBVRE 1961 b). GARCIA LEME et al. (1975) confirmed that glucagon, administered subcutaneously at doses of 50 and 100 μg/100 g body weight, reduced the edema resulting from injection in the rat paw of carrageenan or dextran. They also found that the effect was inhibited by adrenalectomy, but reported that it was not modified by adrenodemedullation. They concluded that glucagon may exert its antiinflammatory effect through the release of adrenal corticosteroids, a mechanism considered unlikely by LEFEBVRE et al. (1961). In various models of "chronic inflammation", glucagon exerted no significant effect; the models investigated comprised the cotton pellet-induced granuloma, the granuloma pouch of Selye, and the neoformation of connective tissue into Ivalon sponges implanted in the rat (LEFEBVRE and LAPIERE 1963) as well as the lathyrism induced in rats by the injection of aminoacetonitrile (FRANCHIMONT et al. 1961).

### II. In Humans

Attempts have been made to use glucagon for the relief of acute inflammation in patients with rheumatoid arthritis and related disorders. HELMER et al. (1957) administered high doses of glucagon (12.5 mg intravenously over a 10-h period daily for 3 days) in three patients with rheumatoid arthritis. A decrease in joint pain with increased mobility and a decrease in joint fluid have been observed, but the improvement lasted only 3–7 days after discontinuation of treatment. A similar slight, but transient improvement induced by glucagon (10 mg daily; intravenous route; 8–10-h infusion) has been reported in another group of three patients with rheumatoid arthritis by LEFEBVRE (1964); however, no objective clinical, biologic, or histologic (synovial) changes have been observed in these patients. We are not aware of any more recent or more prolonged study of the effects of glucagon in this type of patient.

## C. The Effect of Glucagon on Food Intake and Body Weight

As reviewed by GALLOWAY (1972), glucagon, in early studies, was reported to inhibit food intake and to reduce body weight. At doses of 0.125–4.0 mg given intravenously to seven normal subjects in 16 separate experiments, glucagon promptly inhibited gastric contractions and reduced hunger (STUNKARD et al. 1955). SCHULMAN et al. (1957) and PENICK and HINKLE (1961) reported a significant reduction in food intake and a simultaneous weight loss in two small groups of patients treated with glucagon, contrasting with a weight gain when a placebo was given. The marked depressing effect of glucagon on appetite of healthy humans was confirmed in another study by PENICK and HINKLE (1963). As reviewed in detail in

Chap. 54, there are now good reasons to believe that the inhibitory effect exerted by glucagon on appetite and food intake may be secondary to the marked inhibitory action of glucagon on gastrointestinal motility. However, experiments performed in animals suggest that other mechanisms, maybe involving a "satiety signal", may also participate in the glucagon-induced suppression of feeding (BALAGURA et al. 1975; MARTIN and NOVIN 1977; MARTIN et al. 1978; DE CASTRO et al. 1979; VANDERWEELE et al. 1979, GEARY et al. 1981; GEARY 1982).

A recent study of LANGHANS et al. (1982) has shown that, in rats, intraperitoneal injections of antibodies to pancreatic glucagon at the onset of the first meal after food deprivation increased meal size 63% and meal duration 74%. The antibodies also reduced the increase in hepatic vein blood glucose that occurred during meals in control rats, but did not affect the prandial increase in portal vein blood glucose. The results suggest that, under these conditions, pancreatic glucagon is necessary for the normal termination of meals.

## D. Glucagon as a Bronchodilator

A possible bronchodilating effect of glucagon was investigated by LOCKEY et al. (1969) in ten patients with reversible airway obstruction during mild to moderate attacks of asthma. Prompt improvement was observed in nine of ten patients in less than 30 min after intravenous glucagon injection. Additional improvement resulted from the use of isoproterenol. Endogenous catecholamine release (see Chap. 51) may be involved in this effect of glucagon which was confirmed by OPPOLZER and KUMMER (1973), EL NAGGAR and COLLINS (1974), IMBRUCE et al. (1975), and DIEZ-JARILLA et al. (1981). Experimentally, BLUMENTHAL and BRODY (1969) clearly showed glucagon-induced bronchiolar relaxation in the guinea pig. The intimate mechanism of glucagon on bronchial relaxation is still obscure and a possible mediatory role of cyclic nucleotides remains to be established (MURAD 1974; MURAD and KIMURA 1974).

## E. The Effect of Glucagon on Erythropoiesis

NAETS and GUNS (1980) reported that after administration of 2 × 200 μg long-acting zinc protamine glucagon in rats, erythropoiesis was markedly inhibited. Total normoblast counts per femur, reticulocytes, and $^{59}$Fe uptake into red cells were respectively to 35%, 50%, and 17% of control values. Similar results were observed with male and female mice injected twice daily with 50 μg glucagon. The erythropoietic response of mice to hypoxia was also inhibited. Response of polycytemic mice to exogenous erythropoietin was reduced after glucagon injection, an effect proportional to the logarithm of the glucagon dose. Analysis of the mechanisms involved suggested that glucagon acts mainly at the level of erythroid cell differentiation. These findings led NAETS and GUNS (1980) to suggest that hyperglucagonemia was responsible for the anemia frequently observed in glucagonomas (see Chap. 43). The effect of glucagon in primary or secondary human erythrocytemia has not been investigated until now.

# F. Glucagon as an Antitumoral Agent

Early studies have demonstrated that glucagon has inhibitory properties on the growth of tumors in laboratory animals. SALTER et al. (1958) reported that glucagon inhibited by 20%–40% the growth of WALKER's carcinoma in the rat, an effect which was potentiated by insulin. This antitumoral effect of glucagon was confirmed on WALKER's carcinoma by GOLD (1978) and on other animal tumors by JOHNSON and WRIGHT (1959). KLEIN et al. (1974) have reported that glucagon added to culture of MORRIS hepatoma cells interfered with the activity of the adenylate cyclase. In contrast, MURAKAMI and MASUI (1980) observed that glucagon stimulated by 57% the growth in a serum-free, synthetic medium of a human colon carcinoma cell line. PAVELIĆ and PAVELIĆ (1980) reported that glucagon injected into mice with mammary aplastic carcinoma retarded the growth of the tumor and prolonged the mean survival time of the animals. Glucagon stimulated the plaque-forming capacity and phagocytosis in tumor-bearing animals. Cyclophosphamide treatment abolished the antitumor effect of glucagon, while the effect of the hormone was enhanced in animals pretreated with *Corynebacterium parvum*. The authors concluded that the tumor-retarding effects of glucagon were mediated mainly by maintaining high B-type reactivity and phagocytosis. PAVELIĆ and VUK-PAVLOVIĆ (1981) confirmed that glucagon-stimulated immunity and phagocytosis retarded the growth of murine tumors. To the best of our knowledge, the inhibitory properties of glucagon on the growth of tumors have not been investigated clinically in humans.

# G. Other Effects of Glucagon: Paget's Disease of Bone and Muscular Dystrophy

Some observations suggest that glucagon may be useful in the tratment of Paget's disease of bone (CONDON 1971; CHRISTIANSEN and TØNNESEN 1974; HADJIPAVLOU et al. 1977, 1978; RYAN 1977; CHAKRAVORTY 1979; CONDON et al. 1981). Experiments performed in rats showed that glucagon had no effect on immobilization osteoporosis in rats but, nevertheless, significantly increased the femoral weight in both immobilized and intact hindlimbs (WOODWARD and JOWSEY 1972). Glucagon administered on a daily basis at doses ranging from 2 to 20 μg/day to mature muscular dystrophic mice was found to cause beneficial effects in terms of mean survival and mean maximum weight; glucagon also improved clinical appearance in dystrophic mice, permitting weight bearing on hindlimbs and improved locomotion (POPE 1973). We are not aware of any clinical study of the effect of glucagon on muscular dystrophy in humans.

*Acknowledgments.* We thank Dr. B. DIAMANT, Novo Research Institute, Bagsvaerd, Denmark for his help in surveying the literature covered in this chapter.

# References

Balagura S, Kanner M, Harrell LE (1975) Modifications of feeding patterns by glucodynamic hormones. Behav Biol 13:457–465

Blumenthal MN, Brody TM (1969) Studies on the mechanism of drug-induced bronchiolar relaxation in the guinea-pig. J Allergy 44:63–69

Chakravorty NK (1979) Treatment of Paget's disease of bone. Gerontology 25:151–158
Condon JR (1971) Glucagon in the treatment of Paget's disease of bone. Br Med J 4:719–721
Condon JR, Surtees J, Robinson V (1981) Control of osteitis deformans using glucagon, calcitonin and mithramycin. Postgrad Med J 57:84–88
Christiansen C, Tønnesen KH (1974) Zinc-protamine-glucagon in the treatment of Paget's disease of bone: preliminary report. Acta med Scand 196:495–496
De Castro JM, Paullin SK, Delugas GM (1979) Insulin and glucagon as determinants of body weight set point and microregulation in rats. J Comp Physiol Psychol 92:571–579
Diez-Jarilla JL, Gonzales-Macias J, Lazo-Guzman FJ, De Castro del Pozo S (1981) Beta-blockade in asthma (letter). Br Med J 283:309
Dresse A. Lefèbvre P (1961) Nouvelle mise en évidence de la libération par le glucagon de l'adrénaline surrénalienne. CR Soc Biol 155:1168–1169
El Naggar M, Collins VJ (1974) Spirometry following glucagon and isoproterenol in chronic obstructive lung disease. Crit Care Med 2:82–85
Franchimont P, Lefèbvre P, Van Cauwenberge H (1961) Effets de la sérotonine, d'un de ses inhibiteurs l'UML 491 et du glucagon sur l'ostéolathyrisme expérimental du rat. CR Soc Biol 155:427–431
Galloway J (1972) The pharmacology and clinical use of glucagon. In: Lefèbvre P, Unger RH (eds) Glucagon. Molecular physiology, clinical and therapeutic implications. Pergamon, Oxford, pp 299–318
Garcia Leme J, Morato M, Souza MZA (1975) Anti-inflammatory action of glucagon in rats. Br J Pharmacol 55:65–68
Geary M (1982) Pancreatic glucagon and postprandial satiety in the rat. Physiol Behav 28:313–322
Geary N, Langhans W, Scharrer E (1981) Metabolic concomitants of glucagon-induced suppression of feeding in the rat. Am J Physiol 241:R330–R335
Gold J (1978) Effect of high-dose glucagon on tumor growth and survival time in cancer bearing animals (abstract). Proc Am Assoc Cancer Res 19:8
Hadjipavlou AG, Danais S, Greenwood F, Siller TN, Tsoukas GM (1978) Les effets cliniques et métaboliques de la mithramycine et du glucagon sur le traitement de la maladie de Paget. Union Med Can 107:849–856
Hadjipavlou AG, Tsoukas GM, Siller TN, Danais S, Greenwood F (1977) Combination drug therapy in treatment of Paget's disease of bone: clinical and metabolic response. J Bone Joint Surg 59:1045–1051
Harries AD (1981) Beta-blockade in asthma (letter). Br Med J 282:1321
Helmer OM, Kirtley WR, Ridolfo AS (1957) Clinical and metabolic changes induced by glucagon in patients with rheumatoid arthritis (abstract). J Lab Clin Med 50:824
Imbruce R, Goldfedder A, Maguire W, Briscoe W, Nair S (1975) The effect of glucagon on airway resistance. J Clin Pharmacol 15:680–684
Johnson SI, Wright HF (1959) Antitumor activity of glucagon. Cancer Res 19:557–560
Klein I, Levey GS, Bricker LA, Morris HP (1974) Glucagon and epinephrine activation of adenylate cyclase and glucagon binding in Morris hepatomas. Endocrinology 94:279–282
Langhans W, Zieger U, Scharrer E, Geary N (1982) Stimulation of feeding in rats by intraperitoneal injection of antibodies to glucagon. Science 218:894–896
Lefèbvre P (1960) Influence du glucagon sur l'oedème local provoqué chez le rat par certaines substances phlogistiques. CR Soc Biol 154:2154–2156
Lefèbvre P (1961 a) Influence de la surrénalectomie sur l'inhibition par le glucagon des réactions oedémateuses du rat. CR Soc Biol 155:410–412
Lefèbvre P (1961 b) Effets de l'hypophysectomie sur les réactions oedémateuses du rat et sur leur inhibition par le glucagon. CR Soc Biol 155:1149–1151
Lefèbvre P (1962) Glucagon et inflammation expérimentale. Ann Endocrinol 23:275–279
Lefèbvre P (1964) Usage du glucagon dans le domaine du diagnostic et de la thérapeutique. Med Hyg 22:439–441
Lefèbvre P, Dresse A (1961) Influence du glucagon sur le taux des catécholamines surrénaliennes chez le rat. CR Soc Biol 155:412–414

Lefèbvre P, Lapière ChM (1963) Influence du glucagon sur le développement du granulome à l'ouate et de la poche granulomateuse de Selye. Son action sur la colonisation d'éponges en Ivalon inplantées chez le rat. Arch Int Pharmacodyn Ther 141:145–152
Lefèbvre P, Palem-Vliers M, Van Cauwenberge H (1961) Le glucagon stimule-t-il le cortex surrénalien du rat? CR Soc Biol 155:1726–1728
Lefèbvre P, Van Cauwenberge H (1962) Glucagon et réactions oedémateuses du rat. Arch Int Pharmacodyn Ther 138:222–229
Lockey SD Jr, Reed CE, Ouellette JJ (1969) Bronchodilating effect of glucagon in asthma (abstract). J Allerg 43:177–178
Martin JR, Novin D (1977) Decreased feeding in rats following hepatic portal infusion of glucagon. Physiol Behav 19:461–464
Martin JR, Novin D, Vanderweele DA (1978) Loss of glucagon suppression of feeding after vagotomy in rats. Am J Physiol 234:E314–E318
Murad F (1974) Mechanism of action of some bronchodilators. Cyclic nucleotide metabolism in tracheal preparations. Am Rev Respir Dis 110:111–118
Murad F, Kimura H (1974) Cyclic nucleotide levels in incubations of guinea-pig trachea. Biochim Biophys Acta 343:275–280
Murakami H, Masui H (1980) Hormonal control of human colon carcinoma cell growth in serum-free medium. Proc Natl Acad Sci USA 77:3464–3468
Naets JP, Guns M (1980) Inhibitory effect of glucagon on erythropoiesis. Blood 55:997–1002
Oppolzer R, Kummer I (1973) Glukagon und seine Wirkung auf die Atemwegswiderstände bei Asthmatikern. Wien Z Inn Med 54:309–311
Pavelić K, Pavelić J (1980) Glucagon suppressed proliferation rate of mammary aplastic carcinoma in mice. Horm Metab Res 12:243–246
Pavelić K, Vuk-Pavlović S (1981) Retarded growth of murine tumors in vivo by insulin- and glucagon-stimulated immunity and phagocytosis. J Natl Cancer Inst 66:889–892
Penick SB, Hinkle LE Jr (1961) Depression of food intake induced in healthy subjects by glucagon. N Engl J Med 264:893–897
Penick SB, Hinkle LE Jr (1963) The effect of glucagon, phenmetrazine and epinephrine on hunger, food intake and plasma nonesterified fatty acids. Am J Clin Nutr 13:110–114
Pope RS (1973) Glucagon treatment for muscular dystrophy in the mouse. Am J Physiol 225:518–520
Ryan WG (1977) Paget's disease of bone. Annu Rev Med 28:143–152
Salter JM, De Meyer R, Best CH (1958) Effect of insulin and glucagon on tumor growth. Br Med J 2:5–7
Schulman JL, Carleton JL, Whitney G, Whitehorn JC (1957) Effect of glucagon on food intake and body weight in man. J Appl Physiol 11:419–421
Stunkard AJ, Van Itallie TB, Reis BB (1955) The mechanism of satiety: effect of glucagon on gastric hunger contractions in man. Proc Soc Exptl Biol Med 89:258–261
Vanderweele DA, Geiselman PJ, Novin D (1979) Pancreatic glucagon, food deprivation and feeding in intact and vagotomized rabbits. Physiol Behav 23:155–158
Woodward AH, Jowsey J (1972) The effects of glucagon on immobilization osteoporosis in rats. Endocrinology 90:1399–1401

# Subject Index

# Handbook of Experimental Pharmacology

Continuation of "Handbuch der experimentellen Pharmakologie"



Volume 19
**5-Hydroxytryptamie and Related Indolealkylamines**

Volume 20: Part 1
**Pharmacology of Fluorides I**

Part 2
**Pharmacology of Fluorides II**

Volume 21
**Beryllium**

Volume 22: Part 1
**Die Gestagene I**

Part 2
**Die Gestagene II**

Volume 23
**Neurohypophysial Hormones and Similar Polypeptides**

Volume 24
**Diuretica**

Volume 25
**Bradykinin, Kallidin and Kallikrein**

Volume 26
**Vergleichende Pharmakologie von Überträgersubstanzen in tiersystematischer Darstellung**

Volume 27
**Anticoagulantien**

Volume 28: Part 1
**Concepts in Biochemical Pharmacology I**

Part 3
**Concepts in Biochemical Pharmacology III**

Volume 29
**Oral wirksame Antidiabetika**

Volume 30
**Modern Inhalation Anesthetics**

Volume 32: Part 2
**Insulin II**

Volume 34
**Secretin, Cholecystokinin, Pancreozymin and Gastrin**

Volume 35: Part 1
**Androgene I**

Part 2
**Androgens II and Antiandrogens/Androgene II und Antiandrogene**

Volume 36
**Uranium - Plutonium - Transplutonic Elements**

Volume 37
**Angiotensin**

Volume 38: Part 1
**Antineoplastic and Immunosuppressive Agents I**

Part 2
**Antineoplastic and Immunosuppressive Agents II**

Volume 39
**Antihypertensive Agents**

Volume 40
**Organic Nitrates**

Volume 41
**Hypolipidemic Agents**

Volume 42
**Neuromuscular Junction**

Volume 43
**Anabolic-Androgenic Steroids**

Springer-Verlag
Berlin
Heidelberg
New York

MIX
Papier aus verantwortungsvollen Quellen
Paper from responsible sources
FSC® C105338

If you have any concerns about our products,
you can contact us on
**ProductSafety@springernature.com**

In case Publisher is established outside the EU,
the EU authorized representative is:
**Springer Nature Customer Service Center GmbH**
**Europaplatz 3, 69115 Heidelberg, Germany**

Printed by Libri Plureos GmbH
in Hamburg, Germany